ENVIRONMENTAL CHEMISTRY

ENVIRONMENTAL CHEMISTRY

Fifth Edition

Colin Baird
University of Western Ontario

Michael Cann
University of Scranton

W. H. Freeman and Company • New York

Executive Editor: Jessica Fiorillo
Development Editor: Brittany Murphy
Marketing Manager: Alicia Brady
Media and Supplements Editor: Dave Quinn
Senior Media Producer: Keri Fowler
Editorial Assistant: Nicholas Ciani
Senior Project Editor: Vivien Weiss
Photo Editor: Ted Szczepanski
Photo Researcher: Cecilia Varas
Art Director: Diana Blume
Illustrations: Macmillan Publishing Solutions
Senior Illustration Coordinator: Bill Page
Production Coordinator: Susan Wein
Composition: MPS Ltd.
Printing and Binding: RR Donnelley

Library of Congress Control Number: 2011945363

ISBN-13: 978-1-4292-7704-4
ISBN-10: 1-4292-7704-1

Printed in the United States of America
First printing

W. H. Freeman and Company
41 Madison Avenue
New York, NY 10010
Houndmills, Basingstoke RG21 6XS, England
www.whfreeman.com

Contents

Chapter 14   Dioxins, Furans, and PCBs   623

Chapter 15   Other Toxic Organic Compounds of Environmental Concern   663

Preface

To the Student

There are many definitions of environmental chemistry. To some, it is solely the chemistry of Earth's natural processes in air, water, and soil. More commonly, as in this book, it is concerned principally with the chemical aspects of problems that humankind have created in the natural environment. Part of this infringement on the natural chemistry of our planet has been a result of the activities of our everyday lives. In addition, chemists, through the products that they create and the processes by which they make these products, have also had a significant impact on the chemistry of the environment.

Chemistry has played a major role in the advancement of society and in making our lives longer, healthier, more comfortable, and more enjoyable. The effects of human-made chemicals are ubiquitous and in many instances quite positive. Without chemistry there would be no pharmaceutical drugs, no computers, no automobiles, no TVs, no DVDs, no lights, no synthetic fibers. However, along with all the positive advances that result from chemistry, copious amounts of toxic and corrosive chemicals have been produced and dispersed into the environment. Historically, chemists as a whole have not always paid enough attention to the environmental consequences of their activities.

But it is not just the chemical industry, or even industry as a whole, that has emitted substances into the air, water, and soil that are troublesome. The fantastic increase in population and affluence since the Industrial Revolution has overloaded our atmosphere with carbon dioxide and toxic air pollutants, our waters with sewage, and our soil with garbage. We are exceeding the planet's natural capacity to cope with waste, and in many cases, we do not know the consequences of these actions. As a character in Margaret Atwood's novel *Oryx and Crake* (McClelland and Stewart, 2003) stated, "The whole world is now one vast uncontrolled experiment."

During your journey through the chapters in this text, you will see that scientists do have a good handle on many environmental chemistry problems and have suggested ways—although sometimes very expensive ones—to keep us from inheriting the whirlwind of uncontrolled experiments on the planet. Chemists have also become more aware of the contributions of their own profession and industry in creating pollution and have created the concept of *green chemistry* to help minimize their environmental footprint in the future.

To illustrate these efforts, case studies of their initiatives have been included in the text. However, as a prelude to these studies, the Introduction discusses something of the history of environmental regulations—especially in the United States—and the principles, as well as an illustrative application, of the green chemistry movement that has developed. As concerns over such issues as food, water, energy, climate change, and waste production escalate, the concept of *sustainability* is rapidly moving from the wings to center stage on the world agenda. Sustainability is introduced in the following Introduction section and issues related to sustainability are blended throughout the text.

Although the science underlying environmental problems is often maddeningly complex, the central aspects of it can usually be understood and appreciated with only introductory chemistry as background preparation. However, students who have not had some introduction to organic chemistry are encouraged to work through the Background Organic Chemistry section in the online Appendix, particularly before tackling Chapters 13 to 15. Furthermore, the listing of general chemistry concepts that will be used in each chapter should assist in identifying topics from the earlier course material that would be worth reviewing.

To the Instructor

Environmental Chemistry, Fifth Edition, has been revised, updated, and expanded in line with comments and suggestions made by a variety of users and reviewers of the fourth edition. Since some instructors prefer to cover chapters in an order different from ours, each chapter's opening outline lists previously introduced concepts that will be used again, which should facilitate reordering. Furthermore, we have divided the material into smaller subsections and numbered them. The *Detailed Chemistry of the Atmosphere* chapter has been repositioned to the end of the book since many instructors do not teach from it, although in a course, it can readily follow Chapter 3. In addition, following discussions with our reviewers, in Chapter 13 we have deleted some of the descriptive information about pesticides that are no longer in use.

We have expanded the coverage of topics related to climate change, especially the generation of sustainable, renewable energy—which is now covered in two chapters, the first on biofuels and other alternative fuels, and the second on solar energy. As a consequence, this edition could be used as the text for a number of types of courses in addition to Environmental Chemistry. For example, a one-semester *Energy and the Environment* course might use Chapters 3 through 9. Instructors who do not cover policy implications of energy and climate change topics could skip the first and last parts of Chapter 6.

As in previous editions, the background required to solve both in-text and end-of-chapter problems is either developed in the text or would have been covered previously in a general chemistry course—as listed for each chapter at its beginning. Where appropriate, hints are given to start students on the solution. The *Solutions Manual* to the text includes worked solutions to most problems (other than Review Questions, which are designed to direct students back to descriptive material within each chapter).

New to This Edition

Our philosophy in revising the textbook this time has been to make it more user-friendly (both for instructors and for students) as well as to bring it up-to-date. Furthermore, we have expanded the coverage of energy production (especially for biofuels), the generation and disposal of CO_2, and innovative ways to combat climate change.

New Features

• **Green text**—to emphasize the most important statements, definitions, and conclusions.

• **Greater use of bullets and tables**—to cover points most readily covered in a list or sequence.

• **Subsection numbering**—to allow instructors to assign material to be covered or skipped more easily and students to find particular topics more easily.

• **Breaking the text into smaller subsections and shorter paragraphs**—to promote student understanding and allow maximum instructor flexibility.

• **More schematic diagrams**—to promote student comprehension of the more complicated chemistry and appeal to a variety of learning styles.

• **An *Activity* has been inserted into many chapters**—these Web- or library-based miniprojects could be assigned to individual students or to a group to report on.

• **Marginal notes**—to supplement the main text with additional interesting material and to indicate which Review Questions are relevant to the material at hand.

• **More hints and background**—added to the more difficult in-text Problems and Additional Problems.

• **Parts III and IV have been interchanged**—so that water chemistry appears earlier in the book, as preferred by many instructors.

• **Detailed mathematical material has been repositioned**—toward the end of the chapter in many cases, so instructors have flexibility in coverage.

• **Increased international coverage**—to give all students a better perspective on environmental problems and solutions around the world. For example, there is increased coverage of gaseous and particulate air pollution and CO_2 emissions and air quality standards in developed as well as developing countries.

• **An Appendix has been added**—to review the balancing of redox equations and assignment of oxidation numbers (states).

• **Organic Chemistry Appendix has been moved**—to the textbook's Web site at www.whfreeman.com/envchem5e.

New Green Chemistry Cases

- A Nonvolatile, Reactive Coalescent for the Reduction of VOCs in Latex Paints

- Development of Bio-Based Toners

- Recycling Carbon Dioxide: A Feedstock for the Production of Chemicals and Liquid Fuels

- Bio-Based Liquid Fuels and Chemicals

- *Spinetoram*, an Improvement on a Green Pesticide

New Material on Climate Change and CO_2

Substantial sections on the following topics have been added:

- Geoengineering the Climate (by chemical and physical means)

- Energy and CO_2 Intensity Parameters and Predicted Global Trends

- Carbon Capture and Storage (CCS)—The Sequestration of CO_2

- Shale-Gas Production and the Alberta Oil Sands

- The *Deepwater Horizon* Disaster

- Biodiesel Production from Algae and Other Sources

- Renewable Energy (Solar, Wind) Storage by Chemical Means

- Dye-Sensitized Solar Cells

- The Fukushima Nuclear Accident, and the Storage of Spent Nuclear Fuel

Significant additions have also been made on many other topics, including:

- A new box reviewing the calculation of reaction rates

- Smoke from wood stoves and new technology for developing countries

- Removing CO_2 from ambient air

- Biodegradable plastics

- The alternative theory to LRTAP

- E-waste and its disposal and recycling

Updates have been made throughout the book, especially concerning:

- Melanoma rates and UV-A protection in sunscreens

- The polar ozone holes and ODS concentration declines

- Smog, SO_2 emission rates, and air-quality standards around the world

- Catalytic converters for diesel-powered vehicles
- Particulate pollution and the atmospheric brown cloud
- Sea-level rises and the melting of glaciers
- Point-of-use water disinfection
- Desalination box—expanded to incorporate recent advances and news
- Increased and updated coverage on by-products of chlorination, including in swimming pools
- Material on arsenic in drinking water updated and expanded in geographic scope
- Origin of lead in drinking water from transit pipes
- New information concerning the effect of lead on children's health
- New fire retardants

Supplements

The book companion Web site at www.whfreeman.com/envchem5e offers Case Studies that let students explore current environmental controversies and a Background Organic Chemistry section that provides a necessary introduction for those students who have not taken organic chemistry. Here, instructors can also access PowerPoint slides of all art, tables, and graphs from the text.

The *Solutions Manual* (1-4641-0646-0) includes worked solutions to almost all problems (other than Review Questions, which are designed to direct students back to the appropriate material within each chapter).

To All Readers of the Text

The authors are happy to receive comments and suggestions about the content of this book from instructors and students. Please contact Colin Baird at ncolinbaird@gmail.com and Michael Cann at cannm1@scranton.edu.

Acknowledgments

The authors wish to express their gratitude and appreciation to a number of people who in various ways have contributed to this fifth edition:

To the students and instructors who have used previous editions of the text, and via their reviews and e-mails have pointed out subsections and problems that needed clarifying or extending.

To W. H. Freeman Executive Editor for the third, fourth, and fifth editions, Jessica Fiorillo; Senior Project Editor Vivien Weiss; and Development Editor Brittany Murphy—for their encouragement, ideas, insightful suggestions, patience, and organizational abilities. To Margaret Comaskey for her careful copyediting and suggestions again in this edition, to Cecilia Varas for finding the photographs and for obtaining permissions for figures and photographs, to Diana Blume for design, and to Susan Wein for coordinating production.

Colin Baird wishes to express his thanks . . .

To his colleagues at the University of Western Ontario and elsewhere who made valuable suggestions and supplied information and answered queries on various subjects: Myra Gordon, Ron Martin, Martin Stillman, Garth Kidd, Duncan Hunter, Roland Haines, Edgar Warnhoff, Marguerite Kane, Currie Palmer, Rob Lipson, Dave Shoesmith, Felix Lee, Peter Guthrie, Geoff Rayner-Canham, and Chris Willis.

To his daughter, Jenny, and his granddaughters, Olivia and Sophie, for whom and for others of their generations this subject really matters.

Mike Cann wishes to express his thanks . . .

To his students (especially Marc Connelly and Tom Umile) and fellow faculty at the University of Scranton, who have made valuable suggestions and contributions to his understanding of green chemistry and environmental chemistry.

To Joe Breen, who was one of the pioneers of green chemistry and one of the founders of the Green Chemistry Institute.

To Paul Anastas and Tracy Williamson (both of the U.S. Environmental Protection Agency), whose boundless energy and enthusiasm for green chemistry are contagious.

To his loving wife, Cynthia, who has graciously and enthusiastically endured countless discussions of green chemistry and environmental chemistry.

To his children, Holly and Geoffrey, and his grandchildren, McKenna, Alexia, Alan Joshua, Samantha, and Arik, who, along with future generations, will reap the rewards of sustainable chemistry.

Both authors wish to express thanks to the reviewers of the fourth edition, as well as draft versions of sections of the fifth edition of the text, for their helpful comments and suggestions:

Samuel Melaku Abegaz, *Columbus State University*
John J. Bang, *North Carolina Central University*
James Boulter, *University of Wisconsin–Eau Claire*

George P. Cobb, *Texas Tech University*
David B. Ford, *University of Tampa*
Chaoyang Jiang, *University of South Dakota*

Joseph P. Kakareka, *Florida Gulf Coast University*

Michael E. Ketterer, *Northern Arizona University*

Cielito DeRamos King, *Bridgewater State University*

Rachael A. Kipp, *Suffolk University*

Min Li, *California University of Pennsylvania*

Kerry MacFarland, *Averett University*

Matthew G. Marmorino, *Indiana University–South Bend*

Robert Milofsky, *Fort Lewis College*

Jim Phillips, *University of Wisconsin–Eau Claire*

Ramin Radfar, *Wofford College*

A. Lynn Roberts, *Johns Hopkins University*

Kathryn Rowberg, *Purdue University–Hammond*

John Shapley, *University of Illinois*

Joshua Wang, *Delaware State University*

Darcey Wayment, *Nicholls State University*

Chunlong ("Carl") Zhang, *University of Houston–Clear Lake*

Introduction to Environmental Problems, Sustainability, and Green Chemistry

In this book you will study the chemistry of the air, water, and soil, as well as the effects of anthropogenic activities on the chemistry of the Earth. In addition, you will learn about sustainability and green chemistry, which aims to design technologies that lessen the ecological footprint of our activities.

If mankind is to survive, we shall require a substantially new manner of thinking.

Albert Einstein

Environmental chemistry deals with the reactions, fates, movements, and sources of chemicals in the air, water, and soil. In the absence of humans, the discussion would be limited to naturally occurring chemicals and processes. Today, with the burgeoning population of the Earth, coupled with continually advancing technology, human activities have an ever-increasing influence on the chemistry of the environment. The earliest humans, and even those living little more than a century ago, must have thought of the Earth as so vast that human activity could scarcely have any more than local effects on the soil, water, and air. Today we realize that our activities can have not only local and regional, but also global, consequences.

The quotation from Einstein that begins this section was in reference to the dawn of the nuclear age and the concomitant threat of nuclear war. Today, Einstein's words are just as appropriate from the perspective that the effects upon the Earth of our current consumption of resources and accompanying production of waste cannot be sustained. The environmental impact (I) of humans may be thought of as a function of population (P), affluence (A), and technology (T).

$$I = P \times A \times T$$

The last 100 years have been witness to rapid growth in all of these areas, leading to the "perfect environmental storm." It took until 1800 for the human population of the Earth to reach 1 billion. Since that time there has been a seven-fold increase in population, with projections of 9 billion people by 2050. By the end of today, there will be an additional 200,000 people on this planet to feed, clothe, and shelter. Although many people still live in abject poverty, in terms of sheer numbers, never have so many lived so well.

China and India, the world's two most populous countries with one-third of the world's population, have recently had unprecedented economic growth, as evidenced by their GDP growth rate of about 10% for several years. This has lifted many of their people out of poverty and elevated their lifestyles. Unfortunately, their model for rising affluence is the same consumption/waste paradigm common in the West. The accompanying consumption of both renewable and nonrenewable resources and the production of pollution are simply not sustainable for so many across the globe.

Fueled by human ingenuity and innovation, the last 100 years have also witnessed more technological advances than all of preceding human history. Remarkable discoveries include humans walking on the moon over 40 years ago, drugs and medical advances that have helped to increase our life expectancy in the United States from 47 years in 1900 to 79 years today, electronic devices that were not even imaginable a century ago, agricultural advances that allow us to feed 7 billion people, transportation that allows us to eat dinner in New York and breakfast the following morning in London, and the discovery of DNA and the human genome project that have unlocked many of the secrets of life. However, most of these technological advances have been made with little attention to their local, regional, and even global environmental consequences. This combination of exponential population growth, dramatic rise in affluence, and unprecedented technological advancement has left a legacy of toxic waste dumps, denuded landscapes, daunting climate change, spent natural resources, and accelerated extinction of species. Never has a group of living organisms had such a far-reaching and significant impact on the environment of the Earth.

There are now many indications that we have exceeded the carrying capacity of the Earth—that is, the ability of the planet to convert our wastes back into resources (often called nature's interest) as fast as we consume its natural resources and produce waste. Some say that we are living beyond the "interest" that nature provides us and dipping into nature's capital. In short, many of our activities are not sustainable.

As we write these introductory remarks, we are reminded of the environmental consequences of human activities that impact the areas where we live and beyond. Colin spends his summers on a small island just off the north Atlantic coast in Nova Scotia, while Mike spends a few weeks each winter on the west coast of southern Florida, a few kilometers from the Gulf of Mexico. Although these locations are a great distance apart, if predictions are correct, both may be permanently submerged by the end of this century as a result of rising sea levels brought about by enhanced global warming (see Chapters 6 and 7). The public footbridge that links Colin's island to the mainland is treated with creosote, and the local residents no longer harvest mussels from the beds below for fear they may be contaminated with PAHs (Chapter 15). Colin's well on this island was tested for arsenic, a common pollutant in that area of abandoned gold mines (Chapter 12). To the north, the once robust cod fishing industry of Newfoundland has collapsed due to overfishing.

Mike lives in northeastern Pennsylvania on a lake where the wood in his dock is preserved with the heavy metals arsenic, chromium, and copper

(Chapter 12). Within a short distance are two landfills (Chapter 16), which take in an excess of 8,000 tonnes of garbage per day (from municipalities as far as 150 kilometers away), as well as two *Superfund* Sites (Chapter 16) and a nuclear power plant that generates plutonium and other radioactive wastes for which there is no working disposal plan in the United States (Chapter 9). Furthermore, within the last couple of years, natural gas wells have sprung up like weeds as drillers use a hydraulic fracturing process (fracking) (Chapter 6) that may leave a legacy of contaminated groundwater (Chapter 11) in many states in the United States.

Colin's home in London, Ontario, is within an hour's drive of Lake Erie, famous for nearly having "died" of phosphate pollution (Chapter 11), and nuclear power plants on Lake Huron. Nearby farmers grow corn to supply to a new factory that produces ethanol for use as an alternative fuel (Chapter 7), and in Ottawa, a Canadian company has built the first demonstration plant to convert the cellulose from agricultural residue into ethanol (Chapter 7).

On sunny days we both apply extra sunscreen because of the thinning of the ozone layer (Chapters 1 and 2) and suffer the effects on our eyes and lungs of ozone-polluted ground-level air each summer (Chapters 3 and 4). Three of the best salmon rivers in North America in Nova Scotia must be stocked each season because the salmon no longer migrate up the acidified waters. Many of the lakes and streams of the beautiful Adirondack region of upstate New York are a deceptively beautifully crystal clear, only because they are virtually devoid of plant and animal life, again because of acidified waters (Chapter 4).

Environmental issues like these probably have parallels that exist where you live, and learning more about them may convince you that environmental chemistry is not just a topic of academic interest, but one that touches your life every day in very practical ways. Many of these environmental threats are a consequence of anthropogenic activities over the last 50 to 100 years.

In 1983 the United Nations charged a special commission with developing a plan for long-term sustainable development. In 1987 the report titled "Our Common Future" was issued. In this report (more commonly known as the *The Brundtland Report*), the following definition of **sustainable development** is found:

> Sustainable development is development that meets the needs of the present without compromising the ability of future generations to meet their own needs.

Although there are many definitions of sustainable development (or sustainability), this is the most widely used. The three intersecting areas of sustainability are focused on society, the economy, and the environment. Together they are known as the *triple bottom line*. In all three areas, consumption (particularly of natural resources) and the concomitant production of waste are central issues.

The concept of an "ecological footprint" is an attempt to measure the amount of biologically productive space that is needed to support a particular human lifestyle. Currently there are about 4.5 acres of biologically productive space for each person on the Earth. This land provides us with the resources that we need to support our lifestyles and to receive the waste that we generate and convert it back into resources. If the entire population of 7 billion people lived like Colin and Mike, rather typical North Americans, the total ecological footprint would require more than four planet Earths. Obviously, everyone on the planet can't live in as large and as inefficient a house, drive as many kilometers in such an inefficient vehicle, consume as much food (in particular, meat) and energy, create as much waste, etc., as those living in developed countries.

As developing countries such as China and India (with a combined total of over 2 billion people and two of the fastest growing economies in the world) expand economically, they look to the lifestyles of the 1 billion people on the planet that live in developed countries. Factor in the expected increase in global population to 9 billion by 2050 and clearly this is not sustainable development. The people of the world (including and in particular those in developed countries) must strive to develop a lifestyle that is sustainable. This does not necessarily mean a lower standard of living for those in the developed world, but it does mean finding ways (more efficient technologies along with conservation) to reduce our consumption of natural resources and the concomitant production of waste.

There is now a widespread movement toward the growth and implementation of sustainable, or green, technologies. These technologies seek to reduce energy and resource consumption, use and expand renewable resources, and reduce the production of waste. In chemistry, these developments are known as *green chemistry*, which we will describe later in this introduction and will see as a theme throughout this text.

Our ecological footprint in many cases is not limited to our backyard. As mentioned above, the consequences of our activities may be regional and even global. As we will see in Chapter 4, the burning of coal to produce electricity in the midwestern United States produces acid rain that falls in Ontario; in turn, emissions from Ontario are responsible for producing much of the acid rain in northern New York State. Rising global temperatures (Chapters 5 and 6), due in part to the burning of fossil fuels, have significant adverse impacts on those who use little, if any, fossil fuels.

One of these groups is the Inuit, who inhabit the northern reaches of Canada, Russia, Greenland, and Alaska. These people depend on hunting and fishing for sustenance. Ironically, the northern latitudes of the planet have experienced some of the most significant temperature rises due to global warming—warming that has resulted in major changes in the surrounding flora and fauna and that has significantly altered the Inuits' way of life. The atmosphere of our planet is a *commons*, or perhaps more appropriately described as an *open resource*. We all use and benefit from this commons, but no one is directly responsible for it. Its use as a dumping ground for pollutants often affects more than those who are doing the dumping, a concept known as the **tragedy of the commons.**

What we perceive as normal is primarily what we encounter in our everyday lives. But of course, things change, sometimes in seconds or over millennia.

To the untrained "eye," most environmental changes are not that noticeable. But what we now think of as normal may not have been so 100 years ago or even 50 years ago. In the 1600s, English fishermen were quoted as saying the cod off Newfoundland were "so thick by the shore that we hardly have been able to row a boat through them." In 1951 factory fishing began, and in a mere 50 years the cod industry off Newfoundland, the area's main economic activity, was dead, leading not only to environmental but to economic disaster. To today's Newfoundland teenagers this is the norm, although to their parents and grandparents this is far from what they grew up with. This is an example of **shifting baselines,** as well as another example of the tragedy of the commons. The melting ice sheets and loss of habitat for caribou that the Inuit are experiencing is also an example of shifting baselines.

The triple bottom line, ecological footprint, the tragedy of the commons, and shifting baselines are all examples of concepts that are commonly used in discussing sustainability. We will encounter these and other sustainability concepts throughout this book. We suggest that you make a list of these concepts (Table 0-1) and as you read the text keep a record of where and in what context these are encountered.

TABLE 0-1	Sustainability Concepts

Triple Bottom Line (TBL): Although corporations have traditionally been solely focused on the economic (prosperity) bottom line, many (in this age of a greater corporate social responsibility) are adopting a wider corporate strategy that also includes the social (equality) and environmental (quality) bottom lines. This is also called **people, profits, and planet.**

Tragedy of the Commons: In 1968, biologist Garrett Hardin put forth the argument that a common (open) resource (e.g., water, air, land) used by rational individuals for their own good will result in decimation of that resource.

Systems Thinking: Requires one to understand an entire system and how aspects of the system are interconnected. This understanding will allow one to realize that introducing change may have unintended consequences far beyond the original intent of the change. This is particularly true of environmental systems and is a major theme of this book.

Life-Cycle Assessment (LCA): Provides an inventory of materials and energy (inputs) that are consumed and the waste and emissions produced during the entire life cycle of a product, from acquiring the materials (e.g., mining) needed to produce the product to disposing of the product; i.e., from cradle to grave or better yet, *cradle to cradle*. After identification of the inputs and releases at each step of the LC, an analysis of the impact on the environment (in some cases, both social and economic impacts) can determine the steps that can be taken to minimize inputs and releases, and thus the impact on the environment.

Cradle-to-Cradle: At the end of a product's life cycle, rather than being disposed of (as in cradle-to-grave), the spent product becomes the material to produce another product, thus mimicking the regenerative approach of nature.

(continued on p. xxiv)

TABLE 0-1	Sustainability Concepts *(continued)*

Ecological Footprint: A measure of the biologically productive space (both land and water) that is required to support a lifestyle. You can test for your ecological footprint at http://myfootprint.org.

Carbon Footprint: A measure of the amount of greenhouse gases (in carbon dioxide equivalents) that are produced from various activities such as transportation, manufacturing, food production, and heating and cooling.

Water Footprint: Also known as *virtual water*, an indication of the amount of water required (both direct and indirect) to produce a particular product (e.g., a cup of coffee, an automobile, a computer chip). For more information on how water footprint is assessed, visit http://www.waterfootprint.org.

Precautionary Principle: Even in the absence of scientific consensus, if an action or policy is likely to cause harm to people or the environment, then the burden of proof that this action causes no harm falls to the individuals taking the action.

External Costs: Also known as *externalities*, these are costs (or benefits) that are not reflected in the price of a good or service. An example might be the environmental cost of emitting a pollutant into the environment during the manufacture of a product. This environmental cost is paid for, not by the person using the product, but by all of the people who live in the commons where the pollutant was released.

A Brief History of Environmental Regulation

In the United States, many environmental disasters came to a head in the 1960s and 1970s. In 1962, the deleterious effects of the insecticide DDT were brought to the forefront by Rachel Carson in her seminal book, *Silent Spring* (Houghton Mifflin, 1962). In 1969, the Cuyahoga River, which runs through Cleveland, Ohio, was so polluted with industrial waste that it caught fire. The Love Canal neighborhood in Niagara Falls, New York, was built on the site of a chemical dump, and in the mid-1970s, during an especially rainy season, toxic waste began to ooze into the basements of area homes and drums of waste surfaced. The U.S. government purchased the land and cordoned off the entire Love Canal neighborhood. These distressing events were brought into the homes of Americans on the nightly news, and along with other environmental disasters they became rallying points for environmental reform.

This era saw the creation of the U.S. Environmental Protection Agency (EPA) in 1970, the celebration of the first Earth Day, also in 1970, and a mushrooming number of environmental laws. Before 1960, there were approximately 20 environmental laws in the United States; now there are over 120. Most of the earliest of these were focused on conservation or setting aside land from development. The focus of environmental laws changed dramatically starting in the 1960s. Some of the most familiar U.S. environmental legislation include the *Clean Air Act* (1970) and the *Clean Water Act* (originally known as the Federal Water Pollution Control Act Amendments of 1972). One of the major provisions of these acts was to set up pollution-control programs. In effect, these programs attempted to control the release of toxic and other harmful chemicals into the environment. The Comprehensive Environmental Response, Compensation and Liability Act (also known as the *Superfund Act*) set up a procedure and provided funds for cleaning up toxic waste sites. These acts thus focused on dealing with pollutants after they were produced and are known as "end-of-the-pipe solutions" and "command and control laws."

The risk due to a hazardous substance is a function of the exposure to and the hazard of the substance:

$$\text{Risk} = f \, (\text{exposure} \times \text{hazard})$$

The end-of-the-pipe laws attempt to control risk by preventing exposure to these substances. However, exposure controls inevitably fail, which points out the weakness of these laws. The Pollution Prevention Act of 1990 is the only U.S. environmental act that focuses on the paradigm of *prevention* of pollution at the source: if hazardous substances are not used or produced, then their risk is eliminated. There is also no need to worry about controlling exposure, controlling dispersion into the environment, or cleaning up hazardous chemicals.

 ## Green Chemistry

The U.S. Pollution Prevention Act of 1990 set the stage for **green chemistry.** Green chemistry became a formal focus of the U.S. EPA in 1991, playing an integral part in the EPA's setting a new direction by which the agency worked with and encouraged companies to voluntarily find ways to reduce the environmental consequences of their activities. Paul Anastas and John Warner defined *green chemistry* as the design of chemical products and processes that reduce or eliminate the use and generation of hazardous substances. Moreover, green chemistry seeks to

• reduce waste (especially toxic waste),

• reduce the consumption of resources and ideally use renewable resources, and

• reduce energy consumption.

Anastas and Warner also formulated the *Twelve Principles of Green Chemistry*. These principles provide guidelines for chemists in assessing the environmental impact of their work.

The 12 Principles of Green Chemistry

1. It is better to **prevent waste** than to treat or clean up waste after it is formed.

2. Synthetic methods should be designed to **maximize the incorporation of all materials** used in the process into the final product.

3. Wherever practicable, synthetic methodologies should be designed to use and generate **substances that possess little or no toxicity** to human health and the environment.

4. Chemical products should be designed to **preserve efficacy of function while reducing toxicity.**

5. The use of **auxiliary substances** (e.g., solvents, separation agents, etc.) **should be made unnecessary** whenever possible and innocuous when used.

6. **Energy requirements** should be recognized for their **environmental and economic impacts and should be minimized.** Synthetic methods should be conducted at ambient temperature and pressure.

7. A raw material **feedstock should be renewable** rather than depleting whenever technically and economically practical.

8. Unnecessary **derivatization** (blocking group, protection/deprotection, temporary modification of physical/chemical processes) should be **avoided** whenever possible.

9. **Catalytic reagents** (as selective as possible) are superior to stoichiometric reagents.

10. **Chemical products** should be designed so that at the end of their function they **do not persist in the environment** and instead break down into innocuous degradation products.

11. Analytical methodologies need to be further developed to allow for **real-time, in-process** monitoring and control prior to the formation of hazardous substances.

12. Substances and the form of a substance used in a chemical process should be chosen so as to **minimize the potential for chemical accidents,** including releases, explosions, and fires.

  In most of the chapters in the text, real-world examples of green chemistry are discussed. During these discussions, you should keep in mind the Twelve Principles of Green Chemistry and determine which of them are met by the particular example. Although we won't consider all of the principles at this point, a brief discussion of some of them is beneficial.

• Principle 1 is the heart of green chemistry and places the emphasis on the prevention of pollution at the source rather than cleaning up waste after it has been produced.

- Principles 2–5, 7–10, and 12 focus on the materials that are used in the production of chemicals and the products that are formed.

  ○ In a chemical synthesis, in addition to the desired product(s), unwanted by-products are often formed and then usually discarded as waste. Principle 2 encourages chemists to look for synthetic routes that maximize the production of the desired product(s) while at the same time minimizing the production of unwanted by-products (see the synthesis of ibuprofen discussed later).

  ○ Principles 3 and 4 stress that the toxicity of materials and products should be kept to a minimum. As we will see in later discussions of green chemistry, Principle 4 is often met when new pesticides are designed with reduced toxicity to nontarget organisms.

  ○ During the course of a synthesis, chemists employ not only compounds that are actually involved in the reaction (reactants) but also auxiliary substances such as solvents (to dissolve the reactants and to purify the products) and agents that are used to separate and dry the products. These materials are usually used in much larger quantities than the reactants, and they contribute a great deal to the waste produced during a chemical synthesis. When they are designing a synthesis, Principle 5 reminds chemists to consider ways to minimize the use of these auxiliary substances.

  ○ Many organic chemicals are produced from petroleum, which is a nonrenewable resource. Principle 7 urges chemists to consider ways to produce chemicals from renewable resources such as plant material (biomass).

  ○ As we will see in Chapter 13, DDT is an effective pesticide. However, a major environmental problem is its stability in the natural environment. DDT degrades only slowly. Although it has been banned in most developed countries since the 1970s (in the United States since 1972), it can still be found in the environment, particularly in the fatty tissues of animals. Principle 10 stresses the need to consider the lifetime of chemicals in the environment and the need to focus on materials (such as pesticides) that degrade rapidly in the environment to harmless substances.

- Many chemical reactions require heating or cooling and/or a pressure higher or lower than atmospheric pressure. Performing reactions at other than ambient temperature and pressure requires energy; Principle 6 reminds chemists of these considerations when designing a synthesis.

Presidential Green Chemistry Challenge Awards

To recognize outstanding examples of green chemistry, the **Presidential Green Chemistry Challenge Awards** were established in 1996 by the U.S. EPA. Generally, five awards are given each year at a ceremony held at the

National Academy of Sciences in Washington, D.C. The awards are given in the following three Focus Areas.

1. The use of alternative synthetic pathways for green chemistry, such as

    - catalysis/biocatalysis,
    - natural processes, such as photochemistry and biomimetic synthesis, and
    - alternative feedstocks that are more innocuous and renewable (e.g., biomass).

2. The use of alternative reaction conditions for green chemistry, such as

    - solvents that have a reduced impact on human health and the environment, and
    - increased selectivity and reduced wastes and emissions.

3. The design of safer chemicals that are, for example

    - less toxic than current alternatives, and
    - inherently safer with regard to accident potential.

Real-World Examples of Green Chemistry

To introduce you to the important and exciting world of green chemistry, real-world cases of green chemistry are incorporated throughout this book. These examples are winners of Presidential Green Chemistry Challenge Awards. As you explore these examples, it will become apparent that green chemistry is very important in lowering the ecological footprint of chemical products and processes in the air, water, and soil.

We begin our journey into this important topic by briefly exploring how green chemistry can be applied to the synthesis of *ibuprofen*, an important everyday drug. In this discussion, we will see how the redesign of a chemical synthesis can eliminate a great deal of waste and pollution and reduce the amount of resources required.

Before discussing the synthesis of ibuprofen, we must first take a brief look at the concept of **atom economy,** developed by Barry Trost of Stanford University, who won a Presidential Green Chemistry Challenge Award for it in 1998. Atom economy focuses our attention on Green Chemistry Principle 2 by asking the question: *How many of the atoms of the reactants are incorporated into the final desired product and how many are wasted?* As we will see in our discussion of the synthesis of ibuprofen, when chemists synthesize a compound, not all the atoms of the reactants are utilized in the desired product. Many of these atoms may end up in unwanted products (by-products), which are in many instances considered waste. These waste by-products may be toxic and can cause considerable environmental damage if not disposed of properly. In the past, waste products from chemical and other processes have been discarded with little thought, resulting in environmental disasters such as the Love Canal.

Before we take on the synthesis of ibuprofen, let us look at a simple illustration of the concept of atom economy using the production of the desired compound, 1-bromobutane (compound 4) from 1-butanol (compound 1).

$$H_3C-CH_2-CH_2-CH_2-OH + Na-Br + H_2SO_4 \longrightarrow$$
$$H_3C-CH_2-CH_2-CH_2-Br + NaHSO_4 + H_2O$$

If we inspect this reaction, we find that not only is the desired product formed, but so are the unwanted by-products sodium hydrogen sulfate and water (compounds 5 and 6). On the left side of this reaction, we have printed in green all the atoms of the reactants that are utilized in the desired product; the remaining atoms (which become part of our waste by-products) are printed in black. Adding up all of the green atoms on the left side of the reaction, we get 4 C, 9 H, and 1 Br (reflecting the molecular formula of the desired product, 1-bromobutane).

The molar mass of these atoms collectively is 137 g mol^{-1}, the molar mass of 1-bromobutane. Adding up *all* the atoms of the reactants gives 4 C, 12 H, 5 O, 1 Br, 1 Na, and 1 S, and the total molar mass of all these atoms is 275 g mol^{-1}. If we take the molar mass of the atoms that are utilized, divide by the molar mass of all the atoms, and multiply by 100, we obtain the % **atom economy,** here 50%. Thus we see that half of the molar mass of all the atoms of the reactants is wasted and only half is actually incorporated into the desired product.

> % **atom economy** = (molar mass of atoms utilized/ molar mass of all reactants) × 100 = (137/275) × 100 = 50%

This is one method of accessing the efficiency of a reaction. Armed with this information, a chemist may want to explore other methods of producing 1-bromobutane that have a greater % atom economy. We will now see how the concept of atom economy can be applied to the preparation of ibuprofen.

Ibuprofen is a common analgesic and anti-inflammatory drug found in such brand name products as *Advil*, *Motrin*, and *Medipren*. The first commercial synthesis of ibuprofen was by the Boots Company PLC of Nottingham, England. This synthesis, which has been used since the 1960s, is shown in Figure 0-1. Although a detailed discussion of the chemistry of this synthesis is beyond the scope of this book, we can calculate the atom economy of this synthesis and obtain some idea of the waste produced. In Figure 0-1, the atoms printed in green are those that are incorporated into the final desired product, ibuprofen, whereas those in black type end up in waste by-products.

We can inspect the structures of each of the reactants and determine that the total of all the atoms in the reactants is 20 C, 42 H, 1 N, 10 O, 1 Cl, and 1 Na. The molar mass of all these atoms totals 514.5 g mol^{-1}. We can also determine that the number of atoms of the reactants utilized in the ibuprofen (the atoms printed in green) is 13 C, 18 H, and 2 O (the molecular formula of ibuprofen). These atoms have molar mass of 206.0 g mol^{-1} (the molar mass of the ibuprofen). The ratio of the molar mass of the utilized

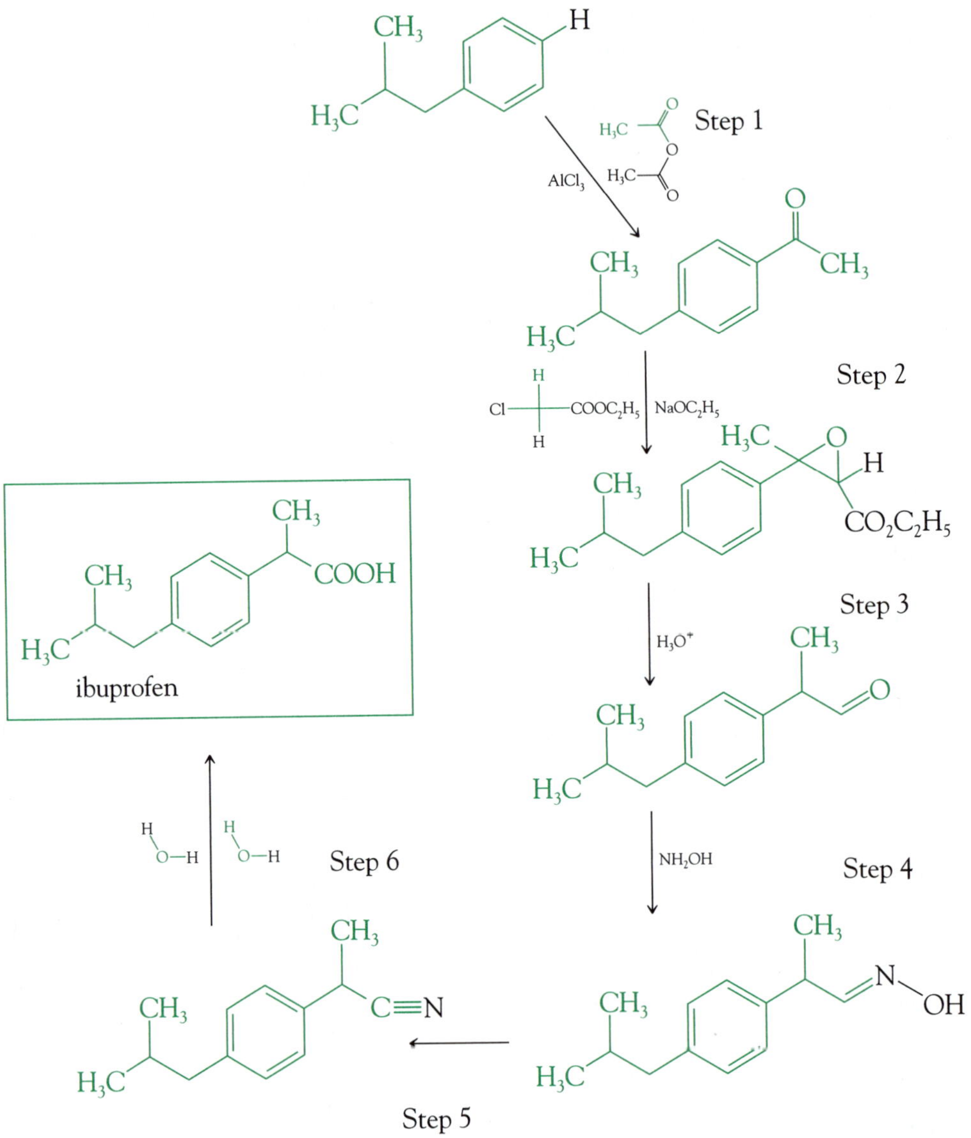

FIGURE 0-1 The Boots Company synthesis of ibuprofen. [Source: M. C. Cann and M. E. Connolly, *Real-World Cases in Green Chemistry* (Washington, D.C.: American Chemical Society, 2000).]

atoms to the molar mass of all the reactant atoms, multiplied by 100, gives an atom economy of 40%:

$$\% \textbf{ atom economy} = (\text{molar mass of atoms utilized}/$$
$$\text{molar mass of all reactants}) \times 100$$
$$= (206.0/514.5) \times 100 = 40\%$$

Only 40% of the molar mass of all the atoms of the reactants in this synthesis ends up in the ibuprofen; 60% is wasted. Because more than 30 million pounds of ibuprofen are produced each year, if we produced all the ibuprofen by this synthesis, there would be over 35 million pounds of unwanted waste produced just from the poor atom economy of this synthesis.

A new synthesis (Figure 0-2) of ibuprofen was developed by the BHC Company (a joint venture of the Boots Company PLC and Hoechst Celanese Corporation), which won a Presidential Green Chemistry Challenge Award in 1997. This synthesis has only three steps as opposed to the six-step Boots synthesis and is less wasteful in many ways. One of the most obvious improvements is the increased atom economy. The molar mass of all the atoms of the reactants in this synthesis is 266.0 g mol^{-1} (13 C, 22 H, 4 O; note that the HF, Raney nickel, and the Pd in this synthesis are used in only catalytic amounts and thus do not contribute to the atom economy), whereas the utilized atoms (printed in green) again weigh 206.0 g mol^{-1}. This yields a % atom economy of 77%.

$$\textbf{\% atom economy} = (\text{molar mass of atoms utilized}/\text{molar mass of all reactants}) \times 100$$
$$= (206.0/266.0) \times 100 = 77\%$$

A by-product from the acetic anhydride (reactant 2) used in step 1 is acetic acid. It is isolated and utilized, which increases the atom economy of this synthesis to more than 99%. Additional environmental advantages of the BHC synthesis include the elimination of auxiliary materials (Principle 5), such as solvents and the aluminum chloride promoter (replaced with the catalyst HF, Principle 9), and higher yields. Thus the green chemistry of the BHC Company synthesis lowers the environmental impact for the synthesis of ibuprofen by lowering the consumption of reactants and auxiliary substances while simultaneously reducing the waste. Other improved syntheses that are winners of Presidential Green Chemistry Challenge Awards include the pesticide *Roundup*, the antiviral agent *Cytovene*, and the active ingredient in the antidepressant *Zoloft*.

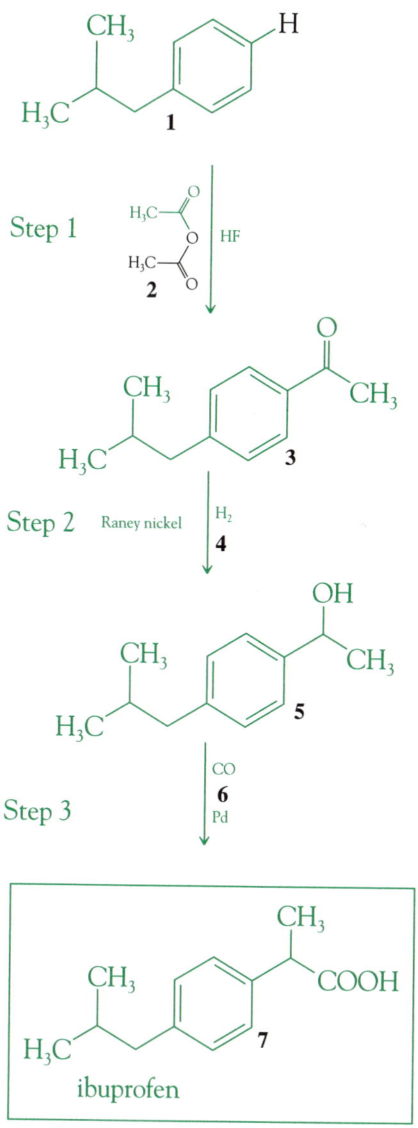

Green chemistry provides a paradigm for reducing both the consumption of resources and the production of waste, thus moving toward sustainability. One of the primary considerations in the manufacture of chemicals must be the environmental impact of the chemical and the process by which it is produced. Sustainable chemistry must become part of the psyche

FIGURE 0-2 The BHC Company synthesis of ibuprofen. [Source: M. C. Cann and M. E. Connolly, *Real-World Cases in Green Chemistry* (Washington, D.C.: American Chemical Society, 2000).]

of not only chemists and scientists, but also business leaders and policy-makers. With this in mind, real-world examples of green chemistry have been incorporated throughout this text to expose you (our future scientists, business leaders, and policymakers) to sustainable chemistry.

Further Readings

1. Anastas, P. T., and J. C. Warner, *Green Chemistry Theory and Practice* (New York: Oxford University Press, 1998).

2. Cann, M. C., and M. E. Connelly, *Real-World Cases in Green Chemistry* (Washington, D.C.: American Chemical Society, 2000).

3. Cann, M. C., and T. P. Umile, *Real-World Cases in Green Chemistry, vol. 2* (Washington, D.C.: American Chemical Society, 2007).

4. Cann, M. C., "Bringing State of the Art, Applied, Novel, Green Chemistry to the Classroom, by Employing the Presidential Green Chemistry Challenge Awards," *Journal of Chemical Education* 76 (1999): 1639–1641.

5. Cann, M. C., "Greening the Chemistry Curriculum at the University of Scranton," *Green Chemistry* 3 (2001): G23–G25.

6. Ryan, M. A., and M. Tinnesand, eds., *Introduction to Green Chemistry* (Washington, D.C.: American Chemical Society, 2002).

7. Kirchhoff, M., and M. A. Ryan, eds., *Greener Approaches to Undergraduate Chemistry Experiments* (Washington, D.C.: American Chemical Society, 2002).

8. World Commission on Environment and Development, *Our Common Future* [The "Bruntland Report"] (New York: Oxford University Press, 1987).

9. Wackernagel, M., and W. Rees, *Our Ecological Footprint: Reducing Human Impact on the Earth* (Gabriola Island, BC: New Society Publishers, 1996).

PART I

ATMOSPHERIC CHEMISTRY AND AIR POLLUTION

Contents of Part I

Introduction

We begin this book by considering stratospheric ozone depletion, chronologically the first truly global environmental problem—one that threatened life around the world and that required international agreements to solve. We then turn to ground-level air pollution, which by contrast is primarily a local or regional environmental problem. In Part II we return to global problems—namely, the climate change brought about by rising concentrations of greenhouse gases, and the role of energy production and use—problems with which our global society currently is wrestling and which will require a myriad of approaches to resolve. In all cases, we will concentrate our attention on sustainable solutions, and how a knowledge of environmental chemistry is necessary to devise them.

As you study the first two chapters, you will be struck by the difficulty in living up to the *precautionary principle*: that the burden of proof in introducing a new product or process falls on the manufacturer to ensure that it causes no harm to the public or the environment. Indeed, the replacement of the highly toxic gas sulfur dioxide in refrigerators by CFCs (chlorofluorocarbons) in the 1930s was considered to be a boon to public safety. Only decades later was the serious flaw of their role in the depletion of stratospheric ozone discovered. As you shall see, knowledge of atmospheric chemistry has been uppermost in the development of CFC replacements.

The production of smog and acid rain was an *unintended consequence* of the combustion of fossil fuels in power plants and vehicles. In Chapters 3 and 4 we shall see how scientists have determined the reactions that produce air pollution and invented devices such as catalytic converters that, at least in developed countries, have drastically reduced its magnitude. ●

Stratospheric Chemistry
The Ozone Layer

In this chapter, the following introductory chemistry topics are used:

- ⮑ Moles; concentration units including mole fraction
- ⮑ Ideal gas law; partial pressures
- ⮑ Thermochemistry: ΔH, ΔH_f; Hess' law
- ⮑ Kinetics: Rate laws; reaction mechanisms, activation energy, catalysis

Introduction

The **ozone layer** is a region of the atmosphere that is called "Earth's natural sunscreen" because it filters out harmful ultraviolet (UV) rays from sunlight before they can reach the surface of our planet and cause damage to humans and other life forms. Any substantial reduction in the amount of this ozone would threaten life as we know it. Consequently, the appearance in the mid-1980s of a large "hole" in the ozone layer over Antarctica represented a major environmental crisis. Although steps have been taken to prevent its expansion, the

A young girl applies sunscreen to protect her skin against UV rays from the Sun. [Source: Lowell George/CORBIS.]

hole will continue to appear each spring over the South Pole; indeed, one of the largest holes in history occurred in 2006. Thus it is important that we understand the natural chemistry of the ozone layer, the subject of this chapter. The specific processes at work in the ozone hole, and the history of the evolution of the hole, are elaborated upon in Chapter 2.

We begin by considering how the concentrations of atmospheric gases are reported and the region of the atmosphere where the ozone is concentrated.

1.1 Regions of the Atmosphere

The names of chemicals important to a chapter are printed in **bold,** along with their formulas, when they are introduced. The names of chemicals less important in the present context are printed in *italics*.

The main components (ignoring the normally ever-present but variable water vapor) of an unpolluted version of the Earth's atmosphere are **diatomic nitrogen**, N_2 (about 78% of the molecules); **diatomic oxygen**, O_2 (about 21%); *argon*, Ar (about 1%); and **carbon dioxide**, CO_2 (presently about 0.04%).

This mixture of chemicals seems unreactive in the lower atmosphere even at temperatures or sunlight intensities well beyond those naturally encountered at the Earth's surface.

The lack of noticeable reactivity in the atmosphere is deceptive. In fact, many environmentally important chemical processes occur in air, whether clean or polluted. In this chapter and the next, these reactions will be explored in detail. In Chapters 3 and 4, reactions that occur in the **troposphere, the region of the sky that extends from ground level to about 15 kilometers altitude, and contains 85% of the atmosphere's mass, are discussed. In this chapter we will consider processes in the stratosphere, the portion of the atmosphere from approximately 15 to 50 kilometers altitude (i.e., 9–30 miles) that lies just above the tropo-sphere.** The chemical reactions to be considered are vitally important to the continuing health of the ozone layer, which is found in the bottom half of the stratosphere. The ozone concentrations and the average temperatures at altitudes up to 50 kilometers in the Earth's atmosphere are shown in Figure 1-1.

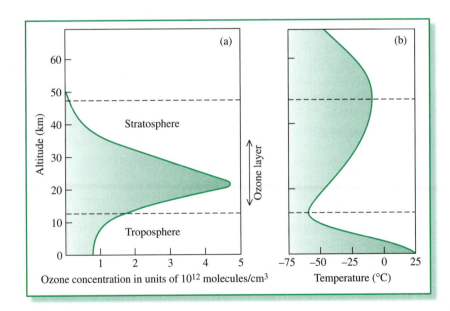

FIGURE 1-1 Variation with altitude of (a) ozone concentration (for mid-latitude regions) and (b) air temperature for various regions of the lower atmosphere.

The stratosphere is defined as the region that lies between the altitudes where the temperature trends display reversals: the bottom of the stratosphere occurs where the temperature first stops decreasing with height and begins to increase, and the top of the stratosphere is the altitude where the temperature stops increasing with height and begins to decrease. The exact altitude at which the troposphere ends and the stratosphere begins varies with season and with latitude.

1.2 Environmental Concentration Units for Atmospheric Gases

Two types of concentration scales are commonly used for gases present in air. For *absolute* concentrations, the most common scale is the number of molecules per cubic centimeter of air. The variation in the concentration of ozone with altitude on the molecules per cubic centimeter scale is illustrated in Figure 1-1a. Absolute concentrations are also sometimes expressed in terms of the partial pressure of the gas, which is stated in units of atmospheres or kiloPascals or bars. According to the ideal gas law ($PV = nRT$), partial pressure is directly proportional to the molar concentration n/V, and hence to the molecular concentration per unit volume, when different gases or components of a mixture are compared at the same Kelvin temperature T. The absolute concentration scale of moles per liter, familiar to all chemists from its use for liquid solutions, is rarely used for gases because they are so dilute.

Relative concentrations are usually based on the chemists' familiar mole fraction scale (called *mixing ratios* by physicists), which is also the molecule fraction scale. Because the concentrations for many constituents are so small, atmospheric and environmental scientists often re-express the mole fraction or molecule fraction as a *parts per* _____ value. Thus, a concentration of 100 molecules of a gas such as carbon dioxide dispersed in one million (10^6) molecules of air would be expressed as 100 parts per million, i.e., 100 ppm, rather than as a molecule or mole fraction of 0.0001. Similarly, ppb and ppt stand for parts per billion (one in 10^9) and parts per trillion (one in 10^{12}), respectively.

It is important to emphasize that for *gases*, these relative concentration units express the number of *molecules* of a pollutant (i.e., the "solute" in chemists' language) that are present in one million or billion or trillion *molecules* of air. Since, according to the ideal gas law, the volume of a gas is proportional to the number of molecules it contains, the "parts per" scales also represent the *volume* a pollutant gas would occupy, compared to that of the stated *volume of air*, if the pollutant were to be isolated and compressed until its pressure equaled that of the air. In order to emphasize that the concentration scale is based upon molecules or volumes rather than upon mass, a v (for volume) is sometimes shown as part of the unit, e.g., 100 ppm$_v$ or 100 ppmv.

The Physics, Chemistry, and Biology of UV

To understand the importance of atmospheric ozone, we must consider the various types of light energy that emanate from the Sun and consider how UV light in particular is selectively filtered from sunlight by gases in air. This leads us to consider the effects on human health of UV, and quantitatively how energy from light can break apart molecules. With that background, we then can investigate the natural processes by which ozone is formed and destroyed in air.

1.3 Absorption of Light by Molecules

The chemistry of ozone depletion, and of many other processes in the stratosphere, is driven by energy associated with light from the Sun. For this reason, we begin by investigating the relationship between light absorption by molecules and the resulting activation, or energizing, of the molecules that enables them to react chemically.

An object that we perceive as black in color absorbs light at all wavelengths of the visible spectrum, which runs from about 400 nm (violet light) to about

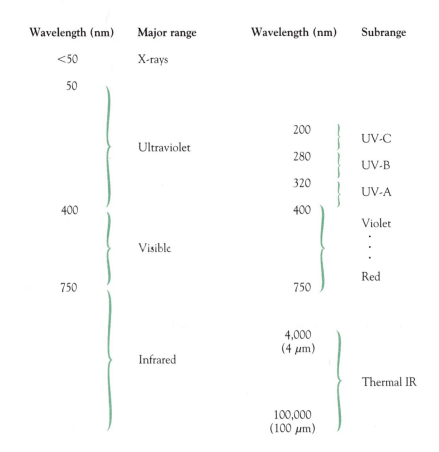

Wavelength (nm)	Major range	Wavelength (nm)	Subrange
<50	X-rays		
50		200	UV-C
	Ultraviolet	280	UV-B
		320	UV-A
400		400	Violet
	Visible		⋮
750		750	Red
	Infrared	4,000 (4 μm)	Thermal IR
		100,000 (100 μm)	

FIGURE 1-2 The electromagnetic spectrum. The ranges of greatest environmental interest in this book are shown.

750 nm (red light); note that one nanometer (nm) equals 10^{-9} meter. Substances differ enormously in their propensity to absorb light of a given wavelength because of differences in the energy levels of their electrons. Diatomic molecular oxygen, O_2, does not absorb visible light very readily, but it does absorb some types of **ultraviolet (UV) light, which is light having wavelengths between about 50 and 400 nm.**

The most environmentally relevant portion of the electromagnetic spectrum is illustrated in Figure 1-2. Notice that the UV region begins at the violet edge of the visible region, hence the name *ultraviolet*. The division of the UV region into components will be discussed later in this chapter. At the other end of the spectrum, beyond the red portion of the visible region, lies **infrared** light, which will become important to us when we discuss the greenhouse effect in Chapter 5.

An absorption spectrum such as that illustrated in Figure 1-3 is a graphical representation that shows the relative fraction of light that is absorbed by a given type of molecule as a function of wavelength. Here, the efficient light-absorbing behavior of O_2 molecules for the UV region between 70 and 250 nm is shown; some minuscule amount of absorption continues beyond 250 nm, but in an ever-decreasing fashion (not shown). Notice that the fraction of light absorbed by O_2 (given on a logarithmic scale in Figure 1-3) varies quite dramatically with wavelength. This sort of selective absorption behavior is observed for all atoms and molecules, although the specific regions of strong absorption and of zero absorption vary widely, depending upon the structure of the species and the energy levels of their electrons.

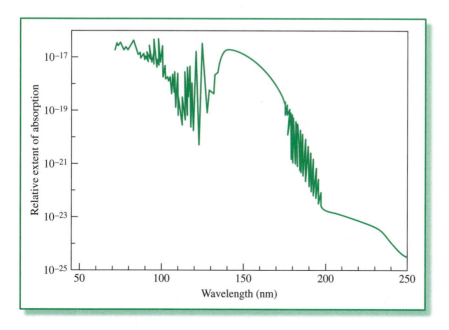

FIGURE 1-3 Absorption spectrum of O_2. [Source: T. E. Graedel and P. J. Crutzen, *Atmospheric Change: An Earth System Perspective* (New York: W. H. Freeman, 1993).]

1.4 Filtering of Sunlight's UV Component by Atmospheric O_2 and O_3

As a result of its light-absorption characteristics, the O_2 gas that lies *above* the stratosphere filters from sunlight most of the UV light from 120 to 220 nm; the remainder of the light in this range is filtered by the O_2 in the stratosphere. Also fortunately for life on the surface, ultraviolet light that has wavelengths *shorter* than 120 nm is filtered in and above the stratosphere by O_2 and other constituents of air such as N_2. **Thus, no UV light having wavelengths shorter than 220 nm reaches the Earth's surface.** This screening protects our skin and eyes, and, in fact, protects all biological life from extensive damage by this part of the Sun's output.

Diatomic oxygen also filters some, but not all, of sunlight's UV in the 220−240-nm range. Rather, ultraviolet light in the whole 220−320-nm range is filtered from sunlight mainly by molecules of **ozone,** O_3, that are spread through the middle and lower stratosphere. The absorption spectrum of ozone in this wavelength region is shown in Figure 1-4. Since its molecular constitution, and thus its set of energy levels, is different from that of diatomic oxygen, its light absorption characteristics also are quite different.

Ozone, aided to some extent by O_2 at the shorter wavelengths, filters out all of the Sun's ultraviolet light in the 220−290-nm range, which overlaps the 200−280-nm region known as UV-C (see Figure 1-2). However, ozone can only absorb a fraction of the Sun's UV light in the 290−320-nm range, since, as you can infer from Figure 1-4b, its inherent ability to absorb light of these wavelengths is quite limited. The remaining amount of the sunlight of such wavelengths, 10−30% depending upon latitude, penetrates the atmosphere to the Earth's surface. **Thus ozone is not completely effective in shielding us from light in the UV-B region, defined as that which lies from 280 to 320 nm.** Since the absorption by ozone falls off in an almost exponential manner with wavelength in this region (see Figure 1-4b),

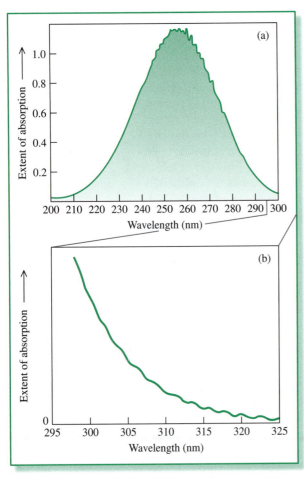

FIGURE 1-4 Absorption spectrum of O_3: (a) from 200 to 300 nm and (b) from 295 to 325 nm. Note that different scales are used for the extent of absorption in the two cases. [Sources: (a) Redrawn from M. J. McEwan and L. F. Phillips, *Chemistry of the Atmosphere* (London: Edward Arnold, 1975). (b) Redrawn from J. B. Kerr and C. T. McElroy, *Science* 262: 1032–1034. Copyright 1993 by the AAAS.]

the fraction of solar UV-B that reaches the troposphere increases with increasing wavelength.

Because neither ozone nor any other constituent of the clean atmosphere absorbs significantly in the **UV-A** range, i.e., 320–400 nm, most of this, the least biologically harmful type of ultraviolet light, does penetrate to the Earth's surface.

Nitrogen dioxide gas does absorb UV-A light but it is present in such small concentration in clean air that its net absorption of such light is quite small.

1.5 The Deleterious Effects of UV Light on Human Skin

A reduction in stratospheric ozone concentration allows more UV-B light to penetrate to the Earth's surface. A 1% decrease in overhead ozone results in a 2% increase in UV-B intensity at ground level. This increase in UV-B is the principal environmental concern about ozone depletion, since it leads to detrimental consequences to many life forms, including humans. Exposure to UV-B causes human skin to sunburn and suntan; overexposure can lead to skin cancer, the most prevalent form of cancer. Increasing amounts of UV-B may also adversely affect the human immune system and the growth of some plants and animals.

Most biological effects of sunlight arise because UV-B can be absorbed by DNA molecules, which then may undergo damaging reactions. By comparing the variation in wavelength of UV-B light of differing intensity arriving at the Earth's surface with the absorption characteristics of DNA as shown in Figure 1-5, it can be concluded that the major detrimental effects of sunlight absorption (the product of the two curves) will occur at about 300 nm. Indeed, in light-skinned people, the skin shows maximum UV absorption from sunlight at about 300 nm.

Most skin cancers in humans are due to overexposure to UV-B in sunlight, so any decrease in ozone is expected to yield eventually an increase in the incidence of this disease. Fortunately, the great majority of skin cancer cases are not the often-fatal (25% mortality rate) **malignant melanoma,** but rather one of the slowly spreading types that can be treated, and that collectively affect about one in four Americans and three in four Australians at some point in their lives. The incidence rate of nonmelanoma skin cancer is exponentially related to exposure to UV.

The incidence of the *malignant* melanoma form of skin cancer, which over their lifetime affects about one in one hundred Americans, is thought to be related to short periods of very high UV exposure, particularly early in life, which could overwhelm the skin cells' ability to deal with high concentrations of very reactive molecules formed by the sunlight. Especially susceptible are fair-skinned, fair-haired, freckled people who burn easily and who

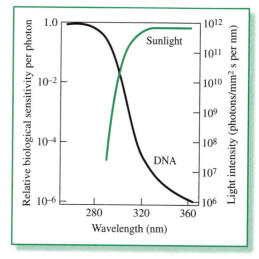

FIGURE 1-5 The absorption spectrum for DNA and the intensity of sunlight at ground level versus wavelength. The degree of absorption of light energy by DNA reflects its biological sensitivity to a given wavelength. [Source: Adapted from R. B. Setlow, *Proceedings of the National Academy of Science USA 71* (1974): 3363–3366.]

have moles with irregular shapes or colors. Consequently, the highest incidence rates for malignant melanoma are in the United States, Canada, South Africa, Australia and New Zealand, the United Kingdom, and western European countries, including Scandinavia.

The incidence of malignant melanoma is also related to latitude. White males living in sunny climates such as Florida or Texas are twice as likely to die from this disease as those in the more northerly states. As Figure 1-6 shows, Australian women have about twice the mortality rate from malignant melanoma as their counterparts in the United Kingdom and Canada, though the dramatic increases from the 1950s to about 1990 seem to have stopped and partially reversed, at least in Australia. (Rates for men are even higher and also showed the temporal increase.) Curiously, indoor workers—who have intermittent exposure to the Sun—are more susceptible than are tanned, outdoor workers! The lag period between first exposure and melanoma is 15–25 years. If malignant melanoma is not

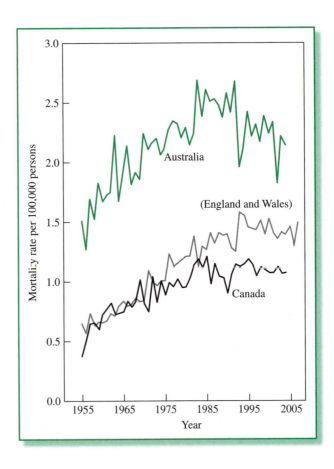

FIGURE 1-6 Mortality rates due to skin melanoma for females (all ages). [Source: International Agency for Research on Cancer.]

treated early, it can spread via the bloodstream to body organs such as the brain and the liver.

By creating graphs using software from the World Health Organization's database (http://www-dep.iarc.fr/WHOdb/WHOdb.htm), compare the mortality rate trends over time for malignant melanoma for your country and several others of interest to you. Are the rates for males higher than for females in all the cases you investigate? Are the rates increasing or decreasing over time? Summarize your findings in a one-page report or a few PowerPoint slides.

1.6 Sunscreens

The term *full spectrum* is sometimes used to denote sunscreens that block UV-A as well as UV-B light. The use of sunscreens that block UV-B, but not UV-A, may actually lead to an *increase* in melanoma skin cancer, since sunscreen usage allows people to expose their skin to sunlight for prolonged periods without burning. The substances used in sunscreen lotions are either particles that reflect or scatter sunlight (e.g., zinc oxide, titanium dioxide) or complex organic compounds that absorb its UV component before it can reach the skin. Sunscreens were one of the first consumer products to use *nanoparticles*, which are tiny particles only a few dozen or a few hundred nanometers (10^{-9} m) in size. Since such particles are so tiny and do not absorb or reflect visible light, the sunscreens appear transparent. A potential drawback to some of the sunscreen ingredients is that they can produce reactive oxygen species, such as OH, if they absorb some of the sunlight rather than reflecting it all. Indeed, titanium dioxide particles used in sunscreens are coated or doped with materials that prevent such processes.

Products proposed as potential sunscreens are eliminated if they undergo a fast irreversible chemical reaction when they absorb sunlight, because this would quickly reduce the effectiveness of the application and because the reaction products could be toxic to the skin. Also, the commonly used sunscreen component *PABA* (*p*-aminobenzoic acid) is no longer generally used because of evidence that it can itself cause cancer.

The **SPF** (Sun Protection Factor) of a sunscreen measures the multiplying factor by which a person can stay exposed to the Sun without burning. Thus an SPF of 15 means that he or she can stay in the Sun fifteen times longer than without the sunscreen. To receive that protection, however, the sunscreen must be reapplied at least every few hours. The SPF system measures the effectiveness of the sunscreen against UV-B. A newer scale, not yet fully adopted, rates the sunscreen's effectiveness against UV-A using a star system: 4 stars corresponds to the highest protection (providing at least 80%

of the sunscreen's corresponding UV-B protection), 3 stars ("high"), 2 stars ("medium"), 1 star ("low"), and 0 stars (no protection).

Because of the long time lag (30–40 years) between exposure to UV and the subsequent manifestation of nonmalignant skin cancers, it is unlikely that effects from ozone depletion are observable as yet. The rise in skin cancer that has occurred in many areas of the world—and that is still occurring, especially among young adults—is probably due instead to greater amounts of time spent by people outdoors in the Sun over the past few decades. For example, the incidence of skin cancer among residents of Queensland, Australia, most of whom are light-skinned, rose to about 75% of the population as lifestyle changes increased their exposure to sunlight years before ozone depletion began. As a consequence of its experience with skin cancer, Australia has led the world in public health awareness of the need for protection from ultraviolet exposure.

1.7 Other Environmental Effects of UV Light

In addition to skin cancer, UV exposure has been linked to several other human conditions. The front of the eye is the one part of the human anatomy where ultraviolet light can penetrate the human body. However, the cornea and lens filter out about 99% of UV from light before it reaches the retina. Over time, the UV-B absorbed by the cornea and lens produces highly reactive molecules called free radicals that attack the structural molecules and can produce a cataract, which is a clouding in the crystalline lens of the eye, and which leads to loss of color vision and eventually to blindness. Indeed, there is some evidence that increased UV-B levels give rise to an increased incidence of eye cataracts, particularly among the non-elderly (see Figure 1-7). UV exposure has also been linked to increase in the rate of macular degeneration, the gradual death of cells in the central part of the retina. Increased UV-B exposure also leads to a suppression of the human immune system, probably with a resulting increase in the incidence of infectious diseases, although this has not yet been extensively researched.

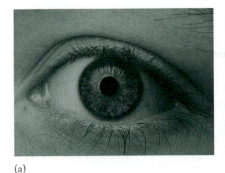
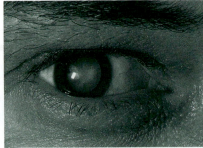

FIGURE 1-7 (a) A normal human eye and (b) a human eye with cataract. [Sources: (a) Martin Dohrn/ Photo Researchers; (b) Sue Ford/ Photo Researchers.]

(a)   (b)

However, sunlight does have some positive effects on human health. *Vitamin D*, which is synthesized from precursor chemicals by the absorption of UV by the skin, is an anticancer agent. Insufficient vitamin D can reduce the rate of bone growth and regeneration—since the vitamin is required for calcium utilization by the body—and thereby lead to rickets in children and to increased fragility among middle-aged and elderly adults. Low levels of vitamin D have been found to lead to an increased risk of colorectal and pancreatic cancers. Sunlight intensity is so weak during winter months in mid- and high-latitudes (which includes most of Canada, the United States, and northern Europe) that sufficient vitamin D cannot be synthesized by the body during that period.

Indeed, a 2007 survey of Canadians found that a very high percentage of nonwhites (whose higher levels of skin melanin protect them more against UV absorption) had insufficient levels of vitamin D in their blood, as did about a third of those of European extraction. It is speculated that some of the generally higher cancer incidence in northern countries compared to that in more southern countries may arise from vitamin D deficiency rather than from pollution.

Humans are not the only organisms affected by ultraviolet light. It is speculated that increases in UV-B exposure can interfere with the efficiency of photosynthesis, and plants may respond by producing less leaf, seed, and fruit. All organisms that live in the first five meters or so below the surface in bodies of clear water would also experience increased UV-B exposure arising from ozone depletion and may be at risk. It is feared that production of the microscopic plants called phytoplankton near the surface of seawater may be at significant risk from increased UV-B; this would affect the marine food chain for which it forms the base. Experiments indicate that there is a complex interrelationship between plant production and UV-B intensity, since the latter also affects the survival of insects that feed off the plants.

Review Questions 1–4 at the end of this chapter refer to the material covered above.

Stratospheric Chemistry: The Ozone Layer

1.8 Variation in Light's Energy with Wavelength

As Albert Einstein realized, **light can be considered not only a wave phenomenon but also to have particle-like properties in that it is absorbed (or emitted) by matter only in finite packets, now called photons.** The quantity of energy, E, associated with each photon is related to the frequency, ν, and the wavelength, λ, of the light by the formulas

$$E = h\nu \quad \text{or} \quad E = hc/\lambda \quad \text{since} \quad \lambda\nu = c$$

Here, h is Planck's constant (6.626218×10^{-34} J s) and c is the speed of light (2.997925×10^8 m s^{-1}). From the equation, it follows that **the shorter the wavelength of the light, the greater the energy it transfers to matter when**

In terms of photon energy, UV-C > UV-B > UV-A > visible > infrared.

absorbed. Ultraviolet light is high in energy content, visible light is of intermediate energy, and infrared light is low in energy. Furthermore, UV-C is higher in energy than UV-B, which in turn is more energetic than is UV-A.

For convenience, the product hc in the equation above can be evaluated on a molar basis to yield a simple formula relating the energy absorbed by 1 mole of matter when each of its molecules absorbs one photon of a particular wavelength of light. If the wavelength is expressed in nanometers, the value of hc is 119,627 kJ mol^{-1} nm, so the equation becomes

$$E = 119,627/\lambda$$

where E is in kJ mol^{-1} if λ is expressed in nm.

The photon energies for light in the UV and visible regions are of the same order of magnitude as the enthalpy (heat) changes, $\Delta H°$, of chemical reactions, including those in which atoms dissociate from molecules. For example, it is known that the dissociation of molecular oxygen into its monatomic form requires an enthalpy change of 498.4 kJ mol^{-1}:

$$O_2 \longrightarrow 2\,O \qquad \Delta H° = 498.4 \text{ kJ mol}^{-1}$$

In general, we can calculate enthalpy changes for any reaction by recalling from introductory chemistry that for any reaction, $\Delta H°$ equals the sum of the enthalpies of formation, $\Delta H_f°$, of the products minus those of the reactants:

$$\Delta H° = \Sigma \Delta H_f° \text{ (products)} - \Sigma \Delta H_f° \text{ (reactants)}$$

In the case of the reaction above,

$$\Delta H° = 2\,\Delta H_f° \text{ (O, g)} - \Delta H_f° \text{ (O}_2\text{, g)}$$

From data tables, we find that $\Delta H_f°$ (O, g) $= +249.2$ kJ mol^{-1}, and we know that $\Delta H_f°$ (O$_2$, g) $= 0$ since O$_2$ gas is the stablest form of the element. By substitution,

$$\Delta H° = 2 \times 249.2 - 0 = 498.4$$

To a good approximation, for a dissociation reaction, $\Delta H°$ is equal to the energy required to drive the reaction. Since all the energy has to be supplied by one photon per molecule (see below), the corresponding wavelength for the light is

$$\lambda = 119,627 \text{ kJ mol}^{-1} \text{ nm}/498.4 \text{ kJ mol}^{-1} = 240 \text{ nm}$$

Thus any O$_2$ molecule that absorbs a photon from light of wavelength 240 nm or shorter has sufficient excess energy to dissociate.

$$O_2 + \text{UV photon } (\lambda < 240 \text{ nm}) \longrightarrow 2\,O$$

Reactions that are initiated by energy in the form of light are called **photochemical reactions.** The oxygen molecule in the above reaction is variously said to be *photochemically dissociated* or *photochemically decomposed* or to have undergone *photolysis.*

Atoms and molecules that absorb light (in the ultraviolet or visible region) immediately undergo a change in the organization of their electrons. They are said to exist temporarily in an electronically **excited state,** and to denote this, their formulas are followed by a superscript asterisk (*). However, **atoms and molecules generally do not remain in the excited state, and therefore do not retain the excess energy provided by the photon, for very long.** Within a tiny fraction of a second, they must either use the energy to react photochemically or return to their **ground state**—the lowest energy (most stable) arrangement of the electrons. They quickly return to the ground state either by themselves emitting a photon or by converting the excess energy into heat that becomes shared among several neighboring free atoms or molecules as a result of collisions (i.e., molecules must "use it or lose it").

$$M + photon \longrightarrow M^* \begin{cases} \nearrow reaction \\ \longrightarrow M + photon \\ \searrow M + heat \end{cases}$$

Consequently, **molecules normally cannot accumulate energy from several photons until they receive sufficient energy to react; all the excess energy required to drive a reaction usually must come from a single photon.** Therefore light of 240 nm or less in wavelength can result in the dissociation of O_2 molecules, but light of longer wavelength does not contain enough energy to promote the reaction at all, even though certain wavelengths of such light can be absorbed by the molecule (see Figure 1-3). In the case of an O_2 molecule, the energy from a photon of wavelength greater than 240 nm can, if absorbed, temporarily raise the molecules to an excited state, but the energy is rapidly converted to an increase in the energy of motion of it and of the molecules that surround it.

$$O_2 + photon \ (\lambda > 240 \text{ nm}) \longrightarrow O_2^* \longrightarrow O_2 + heat$$

$$O_2 + photon \ (\lambda < 240 \text{ nm}) \longrightarrow O_2^* \longrightarrow 2 \text{ O} \quad \text{or} \quad O_2 + heat$$

PROBLEM 1-1

What is the energy, in kilojoules per mole, associated with photons having the following wavelengths? What is the significance of each of these wavelengths? [*Hint:* See Figure 1-2.]

(a) 280 nm   (b) 400 nm   (c) 750 nm   (d) 4000 nm   ●

PROBLEM 1-2

The $\Delta H°$ for the decomposition of ozone into O_2 and atomic oxygen is $+105 \text{ kJ mol}^{-1}$:

$$O_3 \longrightarrow O_2 + O$$

What is the longest wavelength of light that could dissociate ozone in this manner? By reference to Figure 1-2, decide the region of sunlight (UV, visible, or infrared) in which this wavelength falls.   ●

PROBLEM 1-3

Using the enthalpy of formation information given below, calculate the maximum wavelength that can dissociate NO_2 to NO and atomic oxygen. Recalculate the wavelength if the reaction is to result in the complete dissociation into free atoms (i.e., N + 2 O). Is light of these wavelengths available in sunlight?

ΔH_f° values (kJ mol^{-1}): NO_2: +33.2; NO: +90.2; N: +472.7; O: +249.2 ●

Of course, in order for a sufficiently energetic photon to supply the energy to drive a reaction, it first must be absorbed by the molecule. As you can infer from the examples of the absorption spectra of O_2 and O_3 (Figures 1-3 and 1-4), there are many wavelength regions in which molecules simply do not absorb significant amounts of light. Thus, for example, because ozone molecules do not absorb visible light near 400 nm, shining light of this wavelength on them does not cause them to decompose, even though 400-nm photons carry sufficient energy to dissociate them to atomic and molecular oxygen (see Problem 1-2). Furthermore, as discussed above, just because molecules of a substance absorb photons of a certain wavelength and such photons are sufficiently energetic to drive a reaction does not mean that the reaction necessarily will occur; the photon energy can be diverted by a molecule into other processes undergone by the excited state. Thus **the availability of light with sufficient photon energy is a necessary but not a sufficient condition for reaction to occur with any given molecule.**

1.9 Creation of Ozone in the Stratosphere

In this and the next section, the formation of ozone in the stratosphere and its destruction by noncatalytic processes are analyzed. As we shall see, the formation reaction generates sufficient heat to determine the temperature in this region of the atmosphere.

Far above the stratosphere, the air is very thin and the concentration of molecules is so low that most oxygen exists in atomic form, having been dissociated from O_2 molecules by UV-C photons from sunlight. The eventual collision of oxygen atoms with each other leads to the re-formation of O_2 molecules, which subsequently dissociate photochemically again when more sunlight is absorbed.

$$O_2 + UV\text{-}C \longrightarrow 2\,O$$

$$O + O \longrightarrow O_2$$

In the stratosphere itself, the intensity of the UV-C light is much less since much of it is filtered by the diatomic oxygen that lies above. In addition, since the air is denser than it is higher up, the molecular oxygen concentration is much higher in the stratosphere. For this combination of reasons, **most stratospheric oxygen exists as O_2 rather than as atomic oxygen.** Because the concentration of O_2 molecules is relatively large and the concentration of atomic oxygen is so small, the most likely fate of the stratospheric oxygen atoms that are created by the photochemical decomposition of O_2 is *not* their mutual collision to re-form O_2 molecules. Rather, **the oxygen atoms are more likely at such altitudes to collide and react with undissociated, intact diatomic oxygen molecules, an event that results in the production of ozone:**

$$O + O_2 \longrightarrow O_3 + \text{heat}$$

Indeed, **this reaction is the source of all the ozone in the stratosphere.** During daylight hours, ozone is constantly being formed by this process, the rate of which depends upon the amount of UV light and consequently the concentration of oxygen atoms and molecules at a given altitude.

At the bottom of the stratosphere, the abundance of O_2 is much greater than that at the top because air density increases progressively as one approaches the surface. However, relatively little of the oxygen at this level is dissociated and thus little ozone is formed because almost all the high-energy UV has been filtered from sunlight before it descends to this altitude. For this reason, **the ozone layer does not extend much below the stratosphere.** Indeed, the ozone present in the lower stratosphere is largely formed at higher altitudes and over equatorial regions, and transported there.

In contrast, at the top of the stratosphere, the UV-C intensity is greater but the air is thin and therefore relatively little ozone is produced, since the oxygen atoms collide and react with each other rather than with the small number of intact O_2 molecules. Consequently, the production of ozone reaches a maximum where the product of UV-C intensity and O_2 concentration is greatest. The maximum density of ozone occurs lower—at about 25 km over tropical areas, 21 km over mid-latitudes, and 18 km over subarctic regions—since much of it transported downward after its production. Collectively, most of the ozone is located in the region between 15 and 35 km, i.e., the lower and middle stratosphere, known informally as the **ozone layer** (see Figure 1-1a).

A third molecule, which we will designate as M, such as N_2 or H_2O or even another O_2 molecule, is required to carry away the heat energy generated in the collision between atomic oxygen and O_2 that produces ozone. Thus the reaction above is written more realistically as

$$O + O_2 + M \longrightarrow O_3 + M + \text{heat}$$

The release of heat by this reaction results in the temperature of the stratosphere as a whole being higher than the air that lies below or above it, as indicated in Figure 1-1b. Notice from Figure 1-1b that within the stratosphere, the air at a given altitude is cooler than that which lies above it. The general name for this phenomenon is a **temperature inversion.** Because cool air is denser than hot air (ideal gas law), it does not rise spontaneously due to the force of gravity; consequently, vertical mixing of air in the stratosphere is a very slow process compared to mixing in the troposphere. The air in this region therefore is *strati*fied—hence the name *strato*sphere.

In contrast to the stratosphere, there is extensive vertical mixing of air within the troposphere. The Sun heats the ground, and hence the air in contact with it, much more than it does the air a few kilometers higher. It is for this reason that the air temperature falls with increasing altitude in the troposphere; the rate of decline of temperature with height is called the *lapse rate*. The less dense, hotter air rises from the surface and results in extensive vertical exchange of air within the troposphere.

PROBLEM 1-4

Given that the total concentration of molecules in air decreases with increasing altitude, would you expect the *relative* concentration of ozone, on the ppb scale, to peak at a higher or a lower altitude or the same altitude compared to the peak for the absolute concentration of the gas? ●

1.10 Destruction of Stratospheric Ozone

The results for Problem 1-2 show that photons of light in the visible range and even in portions of the infrared range of sunlight possess sufficient energy to split an oxygen atom from a molecule of O_3. However, such photons are *not* efficiently absorbed by ozone molecules and consequently their dissociation by such light is not important, except in the lower stratosphere where little UV penetrates.

As we have seen previously, ozone *does* efficiently absorb UV light with wavelengths shorter than 320 nm, and the excited state thereby produced does undergo a dissociation reaction. Thus absorption of a UV-C or UV-B photon by an ozone molecule in the stratosphere results in the decomposition of that molecule. **This photochemical reaction accounts for much of the ozone destruction in the middle and upper stratosphere:**

This is one destruction reaction of ozone.

$$O_3 + \text{UV photon } (\lambda < 320 \text{ nm}) \longrightarrow O_2{}^* + O^*$$

The oxygen atoms produced in the reaction of ozone with UV light have an electron configuration that differs from the lowest energy configuration, and therefore exist in an electronically excited state; the oxygen molecules from the reaction also are produced in an excited state.

PROBLEM 1-5

By reference to the information in Problem 1-2, calculate the longest wavelength of light that decomposes ozone to O^* and $O_2{}^*$, given the following thermochemical data:

$$O \longrightarrow O^* \quad \Delta H^\circ = 190 \text{ kJ mol}^{-1}$$

$$O_2 \longrightarrow O_2^* \quad \Delta H^\circ = 95 \text{ kJ mol}^{-1}$$

[*Hints:* Express the overall reaction of O_3 decomposition as a sum of simpler reactions for which ΔH° values are available, and combine their ΔH° values according to Hess' law, which states that ΔH° for an overall reaction is the sum of the ΔH° values for the simpler reactions that are added together.] ●

Most oxygen atoms produced in the stratosphere by photochemical decomposition of ozone or of O_2 subsequently react with intact O_2 molecules to re-form ozone. However, some of the oxygen atoms react instead with intact ozone molecules and in the process destroy them, since they are converted to O_2:

$$O_3 + O \longrightarrow 2\,O_2$$

This is the second destruction reaction of ozone.

In effect, the unbonded oxygen atom extracts one oxygen atom from the ozone molecule. This reaction is inherently inefficient since, although it is exothermic, its activation energy is 17 kJ mol^{-1}, a sizable one for atmospheric reactions to overcome. Consequently, **few collisions between O_3 and O occur with sufficient energy to result in reaction.**

The ozone production and destruction processes discussed above constitute the so-called **Chapman mechanism** (or *cycle*), shown in Figure 1-8. Recall that the series of simple reaction steps that document how an overall chemical process, such as ozone production and destruction, occur at the molecular level is called a **reaction mechanism.**

To summarize the processes, ozone in the stratosphere is constantly being formed, decomposed, and re-formed during daylight hours by a series of reactions that proceed simultaneously, though at very different rates depending upon altitude. Ozone is produced in the stratosphere because there is adequate UV-C from sunlight to dissociate some O_2 molecules and thereby produce oxygen atoms, most of which collide with other O_2 molecules and form ozone. The ozone gas filters UV-B and UV-C from sunlight but is destroyed temporarily by this process or by reaction with oxygen atoms. The average lifetime of an ozone molecule at an altitude of 30 km is about half an hour, whereas it is months in the lower stratosphere.

Ozone is not formed below the stratosphere due to a lack of the UV-C required to produce the O atoms necessary to form O_3, because this fraction of sunlight has been absorbed by O_2 and O_3 in the stratosphere. Above the stratosphere, oxygen atoms predominate and usually collide with other O atoms to eventually reform O_2 molecules.

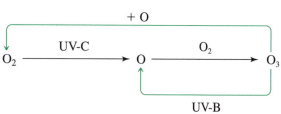

FIGURE 1-8 The Chapman mechanism.

Perhaps the alternative
name *ozone screen* is more
appropriate than ozone
layer.

Even in the ozone-layer portion of the stratosphere, O_3 is *not* the gas of greatest abundance or even the dominant oxygen-containing species; its relative concentration never exceeds 10 ppm. Thus, the term *ozone layer* is something of a misnomer. Nevertheless, this tiny concentration of ozone is sufficient to filter all the remaining UV-C and much of the UV-B from sunlight before it reaches the lower atmosphere.

As in the case of stratospheric ozone, it is not uncommon to find that the concentration of a substance, natural or synthetic, in some compartment of the environment or in an organism does not change much with time. This does not necessarily mean that there are no inputs or outputs of the substance. More often, the concentration does not vary much with time because the input rate and the rate at which the substance decays or is eliminated from some compartment in the environment have become equal: we say that the substance has achieved a **steady state.** Equilibrium is a special case of the steady state; it arises when the decay process is the exact opposite of the input. The mathematical implications of the steady state in common situations involving reactive substances are explored in Box 1-3, located at the end of this chapter.

Review Questions 5–11 are
based upon material in the
preceding section.

Catalytic Processes of Ozone Destruction

In the early 1960s, it was realized that there are mechanisms for ozone destruction in the stratosphere in addition to the processes described in the Chapman mechanism. These additional processes all involve catalysts that are present in air. In the material that follows, we investigate two general reaction mechanisms by which stratospheric ozone is catalytically destroyed, paying particular attention to the role of chlorine and bromine.

1.11 Mechanism I of Ozone Destruction

There exist a number of atomic and molecular species, designated in general as X, that react efficiently with ozone by abstracting (removing) an oxygen atom from it:

$$X + O_3 \longrightarrow XO + O_2$$

This is the first catalytic
destruction mechanism for
ozone.

In those regions of the stratosphere where the atomic oxygen concentration is appreciable, the XO molecules react subsequently with oxygen atoms to produce O_2 and to re-form X:

$$XO + O \longrightarrow X + O_2$$

The **overall reaction** corresponding to this reaction mechanism is obtained by algebraically summing the successive steps that occur in air over and over again an equal number of times. In the case of the additional steps of the mechanism, the reactants in the two steps are added together and become the reactants of the overall reaction, and similarly for the products:

$$X + O_3 + XO + O \longrightarrow XO + O_2 + X + O_2$$

Molecules that are common to both sides of the reaction equation, in this case X and XO, are then cancelled, and common terms collected, yielding the balanced overall reaction

$$O_3 + O \longrightarrow 2\,O_2 \quad \text{overall reaction}$$

Thus the species X are catalysts for ozone destruction in the stratosphere, since they speed up a reaction (here, between O_3 and O), but are eventually re-formed intact and are able to begin the cycle again —with, in this case, the destruction of further ozone molecules.

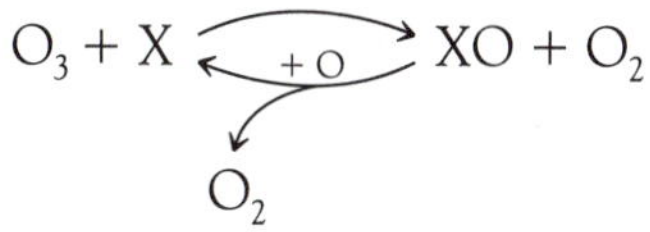

$$O_3 + X \underset{+\,O}{\longleftrightarrow} XO + O_2$$
$$O_2$$

As previously discussed (Chapman cycle), the above overall reaction can occur as a simple collision between an ozone molecule and an oxygen atom even in the absence of a catalyst, but almost all such direct collisions are ineffective in producing reaction. The X catalysts greatly increase the efficiency of this reaction and thereby decrease the steady-state concentration of ozone. **All the environmental concerns about ozone depletion arise from the fact that we have inadvertently increased the stratospheric concentrations of several X catalysts by the release at ground levels of certain gases, especially those containing chlorine and bromine.** Such an increase in the catalyst concentration leads to a reduction in the concentration of ozone in the stratosphere by the mechanism shown above and by one discussed later.

Most ozone destruction by the catalytic mechanism (i.e., the combination of sequential steps) described above, hereafter designated **Mechanism I,** occurs in the middle and upper stratosphere, where the ozone concentration is low to start with. Chemically, all the X catalysts are **free radicals, which are atoms or molecules containing an odd number of electrons.** As a consequence of the odd number, one electron is not paired with one of opposite spin character (as occurs for all the electrons in almost all stable molecules). **Free radicals are usually very reactive,** since there is a driving force for their unpaired electron to pair with one of the opposite spin even if it is located in a different molecule.

An analysis of which free-radical reactions are feasible in air and which are not is given in Box 1-1.

BOX 1-1   The Rates of Free-Radical Reactions

The rate of a given chemical reaction is affected by a number of parameters, most notably the magnitude of the activation energy required before the reaction can occur. Thus reactions with appreciable activation energies are inherently very slow processes and can often be ignored compared to alternative, faster processes for the chemicals involved. In gas-phase reactions involving simple free radicals as reactants, the activation energy exceeds that imposed by their endothermicity by only a small amount. Thus we can assume, conversely, that all exothermic free-radical reactions will have only a small activation energy (Figure 1a). Therefore, exothermic free-radical reactions usually are fast (providing, of course, the reactants exist in reasonable concentrations in the atmosphere). An example of an exothermic free-radical reaction with a small energy barrier is

$$Cl + O_3 \longrightarrow ClO + O_2$$

The activation energy here is only 2 kJ mol^{-1}.

Reactions involving the combining of two free radicals generally are exothermic, since a new bond is formed, so they too proceed quickly with little activation energy, provided that the radical concentrations are high enough that the reactants do in fact collide with each other at a fast rate.

In contrast, endothermic reactions in the atmosphere will be much slower since the activation barrier must of necessity be much larger (see Figure 1b). At atmospheric temperatures, few if any collisions between the molecules would have sufficient energy to overcome this large barrier and allow reaction to occur. An example is the endothermic reaction

$$OH + HF \longrightarrow H_2O + F$$

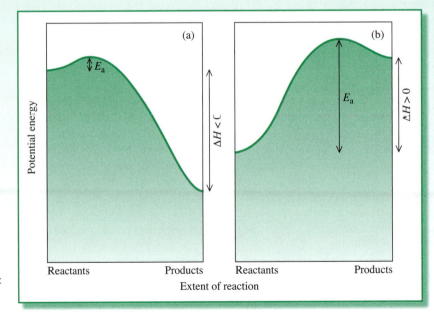

FIGURE 1 Potential energy profiles for typical atmospheric free-radical reactions, showing (a) exothermic and (b) endothermic patterns.

Its activation energy must be at least equal to its $\Delta H° = 69$ kJ mol^{-1}, and consequently the reaction would be so very slow at stratospheric temperatures that we can ignore it completely.

Draw an energy profile diagram, i.e., one similar to Figure 1b, for the abstraction from water of a hydrogen atom by ground-state atomic oxygen, given that the reaction is endothermic by about 69 kJ mol^{-1}. On the same diagram, show the energy profile for the reaction of O* with H_2O to give the same products, given that O* lies above ground-state atomic oxygen (O) by 190 kJ mol^{-1}. From these curves, predict why abstraction by O* occurs quickly but that by O is extremely slow in the atmosphere.

1.12 Catalytic Destruction of Ozone by Nitric Oxide and Hydroxyl

The catalytic destruction of ozone occurs even in a "clean" atmosphere (one unpolluted by artificial contaminants) since small amounts of the X catalysts have always been present in the stratosphere. **One important "natural" version of X—i.e., one of the species responsible for catalytic ozone destruction in a nonpolluted stratosphere—is the free-radical molecule nitric oxide, NO.** It is produced when molecules of **nitrous oxide,** N_2O, rise from the troposphere to the stratosphere, where they may eventually collide with an excited oxygen atom produced by photochemical decomposition of ozone. Most of these collisions will yield $N_2 + O_2$ as products, but a few of them result in the production of nitric oxide:

$$N_2O + O^* \longrightarrow 2\,NO$$

This reaction is the origin of stratospheric NO.

We can ignore the possibility that NO produced in the troposphere will migrate to the stratosphere; as explained in Chapter 3, the gas is efficiently oxidized to nitric acid, which is then readily washed out of the tropospheric air, before this process can occur.

The NO molecules that are the products of the above reaction catalytically destroy ozone by extracting an oxygen atom from ozone and forming **nitrogen dioxide,** NO_2, i.e., they act as X in Mechanism I:

$$NO + O_3 \longrightarrow NO_2 + O_2$$
$$\underline{NO_2 + O \longrightarrow NO + O_2}$$
$$\text{overall} \quad O_3 + O \longrightarrow 2\,O_2$$

The calculation of the rates of reaction steps, such as those in this mechanism, is discussed in Box 1-2.

BOX 1-2   Calculating the Rates of Reaction Steps

The quantitative rate at which reactions occur in generating products and consuming reactants can be calculated from experimental numerical constants previously determined for the process.

As an example, consider the gas-phase reaction between nitric oxide and ozone to produce nitrogen dioxide and molecular oxygen:

$$NO + O_3 \longrightarrow NO_2 + O_2$$

Since this is a simple one-step reaction, we know from general principles that its rate is proportional to the product of the concentrations of the reactants, each raised to the power of its coefficient (here both are 1). Thus the **rate law** for this process is

$$\text{rate} = k\,[NO]\,[O_3]$$

The parameter k is the **rate constant** for the process.

From experiment, we know that at an atmospheric temperature of about $-50°C$, the value of $k = 6.5 \times 10^{-15}$ molecules^{-1} cm^3 sec^{-1}. Typical stratospheric concentrations, in the same units as those for which k is given, are

$$[NO] = 1.0 \times 10^9 \text{ molecules cm}^{-3}$$

and

$$[O_3] = 3.0 \times 10^{12} \text{ molecules cm}^{-3}$$

By substitution of these numerical values into the rate law for the reaction, we obtain

$$\begin{aligned}\text{rate} = &(6.5 \times 10^{-15} \text{ molecules}^{-1} \text{ cm}^3 \text{ sec}^{-1}) \\ &\times (1.0 \times 10^9 \text{ molecules cm}^{-3}) \\ &\times (3.0 \times 10^{12} \text{ molecules cm}^{-3})\end{aligned}$$

and so

$$\text{rate} = 2.0 \times 10^7 \text{ molecules cm}^{-3} \text{ sec}^{-1}$$

Thus about 20 million molecules of ozone react with the same number of nitric oxide molecules in every cubic centimeter per second.

The rate constants for reactions can be calculated at any given temperature if values for the **Arrhenius equation** are known for it. In particular, the variation of k with Kelvin temperature T is given by

$$k = A\,e^{-E/RT} \qquad \text{Arrhenius equation}$$

The **pre-exponential term** A has the same units as does k, since the exponential term overall has no units. The term E is the reaction's **activation energy,** given in units of *joules* (not kilojoules) *per mole* when R is expressed in units of joules per mole per Kelvin:

$$R = 8.3 \text{ J K}^{-1} \text{ mol}^{-1}$$

For the reaction between NO and O_3, from experiment we have the values

$$A = 1.8 \times 10^{-12} \text{ molecules}^{-1} \text{ cm}^3 \text{ sec}^{-1}$$

and

$$E = 10.4 \text{ kJ mol}^{-1} = 10{,}400 \text{ J mol}^{-1}$$

Thus we could, for example, recalculate the value of the rate constant when the temperature rises to $-30°C$.

$$T = 273 + t = 273 - 30 = 243 \text{ K}$$

so at this temperature, the value of the exponent is

$$\begin{aligned}-E/RT = &-10400 \text{ J mol}^{-1}/(8.3 \text{ J K}^{-1} \text{ mol}^{-1} \\ &\times 243 \text{ K}) \\ = &-5.16\end{aligned}$$

and so the value of the rate constant is

$$\begin{aligned}k = &(1.8 \times 10^{-12} \text{ molecules}^{-1} \text{ cm}^3 \text{ sec}^{-1}) \\ &\times \exp(-5.16) \\ = &1.0 \times 10^{-14} \text{ molecules}^{-1} \text{ cm}^3 \text{ sec}^{-1}\end{aligned}$$

As expected, the value of k—and therefore of the reaction rate—has increased with increasing temperature.

PROBLEM 1

Calculate the rate of the reaction between NO and O_3 at $-30°C$ using the value determined above for k and assuming NO and O_3 concentrations of 5.0×10^9 and 5.0×10^{12} molecules cm^{-3}, respectively.

PROBLEM 2

Using the same concentrations as in Problem 1, recalculate the rate of the reaction at $-60°C$.

PROBLEM 1-6

Not all XO molecules such as NO_2 survive long enough to react with oxygen atoms; some are photochemically decomposed to X and atomic oxygen, which then reacts with O_2 to re-form ozone. Write out the three steps (including one for ozone destruction) for this process and add them together to deduce the net reaction. Does this sequence destroy ozone overall, or is it a *null cycle*, which is defined as one that involves a sequence of steps with no chemical change overall?   ●

Another important X catalyst in the stratosphere is the hydroxyl free radical, OH. It originates from the reaction of excited oxygen atoms, O*, with water or *methane*, CH_4, molecules:

$$O* + CH_4 \longrightarrow OH + CH_3$$

The methane originates from emissions from the surface, a small fraction of which survive sufficiently long to migrate up to the stratosphere.

PROBLEM 1-7

Write out the two-step mechanism by which the hydroxyl free radical catalytically destroys ozone by Mechanism I. By adding together the steps, deduce the overall reaction.   ●

PROBLEM 1-8

By analogy with its reaction with methane, write a balanced equation for the reaction by which O* produces OH from water vapor.   ●

1.13 Destruction of Ozone Without Atomic Oxygen: Mechanism II

A factor that minimizes the catalyzed gas-phase destruction of ozone by Mechanism I is the requirement for atomic oxygen to complete the cycle by reacting with XO in order to permit the regeneration of the X catalyst in a usable form.

$$XO + O \longrightarrow X + O_2$$

As discussed above, the concentration of oxygen atoms is very low in the lower stratosphere (15−25-km altitude), so the gas-phase destruction of ozone by reactions that require atomic oxygen is sluggish there.

There is another general catalytic sequence, henceforth designated **Mechanism II,** that depletes ozone in the lower stratosphere, particularly when the concentrations of the X catalysts are relatively high. **It accounts for the majority of ozone depletion by man-made chemicals, especially in ozone holes.** First, two ozone molecules are destroyed by the same catalysts as discussed previously and by the same initial reaction:

$$X + O_3 \longrightarrow XO + O_2$$

$$X' + O_3 \longrightarrow X'O + O_2$$

We have used X′ to symbolize the catalyst in the second equation to indicate that it need not be chemically identical to X, the one in the first equation. Either X or X′ must be a chlorine atom, whereas the other one can be atomic chlorine or bromine. Mechanism II is not known to operate if either X or X′ is NO.

In the steps that follow the first, the two molecules XO and X′O that have an added oxygen atom react with each other. As a consequence, the catalysts X and X′ are ultimately regenerated, usually after the combined, but unstable, molecule XOOX′ has formed and been decomposed by either heat or light:

$$XO + X'O \longrightarrow [XOOX'] \longrightarrow X + X' + O_2$$

By convention in chemistry, a species shown in square brackets is one with a transient existence.

When we sum these steps, the *overall reaction* is seen to be

$$2\,O_3 \longrightarrow 3\,O_2$$

We shall see several examples of catalytic Mechanism II in operation in the ozone holes (Chapter 2) and in the mid-latitude lower stratosphere. Indeed, most ozone loss in the lower stratosphere occurs according to this net reaction. Mechanisms I and II are summarized in Figure 1-9.

Finally, we note that while the rate of production of ozone from oxygen depends only upon the concentrations of O_2 and O_3 and of UV light at a given altitude, what determines the rate of ozone destruction is somewhat more complex. The rate of ozone decomposition by UV-B c. by catalysts

FIGURE 1-9 Summary of catalytic ozone destruction by Mechanisms I and II.

Mechanism I

$$X + O_3 \rightarrow XO + O_2$$
$$\underline{XO + O \rightarrow X + O_2}$$
$$O_3 + O \rightarrow 2\,O_2 \quad \text{overall}$$

Mechanism II

$$X + O_3 \rightarrow XO + O_2$$
$$X' + O_3 \rightarrow X'O + O_2$$
$$\underline{XO + X'O \rightarrow \rightarrow X + X' + O_2}$$
$$2\,O_3 \rightarrow 3\,O_2 \quad \text{overall}$$

depends upon ozone's concentration multiplied by the sunlight intensity or the catalyst concentration, respectively. In general, the concentration of ozone will rise until the net rate of destruction just meets the rate of production, and then will remain constant at this steady-state level as long as the intensity of sunlight remains the same. If, however, the rate of destruction is temporarily increased by the introduction of additional molecules of a catalyst, the steady-state concentration of ozone must then decrease to a new, lower value at which the rates of formation and destruction are again equal.

However, it should be clear from the discussion above that due to its constant re-formation reactions, atmospheric ozone *cannot* be permanently and totally destroyed, no matter how great the level of catalyst. It should also be realized that any decrease in the concentration of ozone at higher altitudes allows more UV penetration to lower altitudes, which produces more ozone there; thus there is some "self-healing" of total ozone loss.

1.14 Atomic Chlorine and Bromine as X Catalysts

Atomic chlorine, Cl, is a free radical and an efficient X catalyst. As we shall see in detail in the next chapter, its concentration in the stratosphere was greatly increased over the twentieth century by the release at the Earth's surface of synthetic chlorine-containing gases. These gases were commercially produced in great volume because they are efficient, nonflammable refrigerants and propellants. However, they are so stable that they eventually rise from ground level to the stratosphere where they decompose, yielding atoms of chlorine.

The unintended consequence of this catalytic destruction process is the ozone hole—the massive destruction of ozone that now occurs annually above the South Pole. Thus the stratosphere, even though it lies far above the Earth's surface, has not escaped our ecological footprint. The process that produces the ozone hole, detailed here and in Chapter 2, is highly complex, and requires a *systems* approach for its analysis. In recent times, scientists and engineers have attempted to better anticipate the long-range environmental consequences of new products and processes, having learned their lesson from the ozone hole and other environmental disasters.

However, synthetic gases are not the only suppliers of chlorine to the ozone layer. There always has been some chlorine in the stratosphere as a result of the slow upward migration of the **methyl chloride** gas, CH_3Cl (also called *chloromethane*), produced at the Earth's surface, mainly in the oceans by the interaction of chloride ion with decaying vegetation. Recently another large source of methyl chloride, from tropical plants, has been discovered; this may be the missing source of the compound for which scientists had been searching.

Only a portion of the methyl chloride molecules are destroyed in the troposphere. When intact molecules of it reach the stratosphere, they are

As discussed in Table 0-1 on pages xxiii–xxiv, this systems-approach emphasis on environmental, economic, and social consequences is known as the *triple bottom line*.

photochemically decomposed by UV-C or attacked by OH radicals. In either case, atomic chlorine, Cl, is eventually produced:

$$CH_3Cl \xrightarrow{UV\text{-}C} Cl + CH_3$$

or

$$OH + CH_3Cl \longrightarrow Cl + \text{other products}$$

Chlorine atoms are efficient X catalysts for ozone destruction by Mechanism I:

$$Cl + O_3 \longrightarrow ClO + O_2$$
$$ClO + O \longrightarrow Cl + O_2$$
$$\overline{\text{overall}\quad O_3 + O \longrightarrow 2\,O_2}$$

Each chlorine atom can catalytically destroy many tens of thousands of ozone molecules in this manner. **At any given time, however, the great majority of stratospheric chlorine normally exists not as Cl or as the free radical chlorine monoxide, ClO, but as a form that is not a free radical and that is inactive as a catalyst for ozone destruction. The two main catalytically inactive (or reservoir) molecules containing chlorine in the stratosphere are hydrogen chloride gas, HCl, and chlorine nitrate gas, ClONO$_2$.**

The chlorine nitrate is formed by the combination of chlorine monoxide and nitrogen dioxide; after a few days or hours, a given $ClONO_2$ molecule is photochemically decomposed back to its components, and thus the catalytically active ClO is re-formed.

$$ClO + NO_2 \underset{\text{sunlight}}{\rightleftharpoons} ClONO_2$$

However, under normal circumstances, more chlorine exists at steady state as $ClONO_2$ than as ClO.

The other catalytically inactive form of chlorine, HCl, is formed when atomic chlorine abstracts a hydrogen atom from a molecule of stratospheric methane:

$$Cl + CH_4 \longrightarrow HCl + CH_3$$

This reaction is slightly endothermic, so its activation energy is nonzero, and it therefore proceeds at a slow but significant rate (see Box 1-1). (The *methyl free radical*, CH$_3$, does not operate like the X catalysts since it combines with an oxygen molecule and is finally degraded to carbon dioxide by reactions discussed in Chapter 3.) Eventually, each HCl molecule is reconverted to the active form, i.e., atomic chlorine, by reaction with the hydroxyl radical:

$$OH + HCl \longrightarrow H_2O + Cl$$

Again, usually much more chlorine exists as HCl than as atomic chlorine at any given time under normal steady-state conditions.

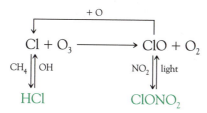

Reaction summary

When the first predictions concerning stratospheric ozone depletion were made in the 1970s, it was not realized that about 99% of stratospheric chlorine usually is tied up in the inactive forms. When the existence of inactive chlorine was discovered in the early 1980s, the predicted amounts of stratospheric ozone loss in the future were lowered appreciably. As we shall see, however, there are conditions under which inactive chlorine can become temporarily activated and massively destroy ozone, a discovery which was not made until the late 1980s.

Although there has always been some chlorine in the stratosphere due to the natural release of CH_3Cl from the surface, in recent decades the amount has been completely overshadowed by much larger quantities of chlorine released into air during the production or use of synthetic chlorine-containing gaseous compounds. Most of these substances are *chlorofluorocarbons* (CFCs); their nature, usage, and replacements for them will be discussed in detail in Chapter 2.

As with methyl chloride, large quantities of **methyl bromide,** CH_3Br, are also produced naturally and some of it eventually reaches the stratosphere, where it is decomposed photochemically to yield atomic bromine. Like chlorine, bromine atoms can catalytically destroy ozone by Mechanism I:

$$Br + O_3 \longrightarrow BrO + O_2$$

$$BrO + O \longrightarrow Br + O_2$$

In contrast to chlorine, almost all the bromine in the stratosphere remains in the active free-radical forms Br and BrO, since the inactive forms, **hydrogen bromide,** HBr, and **bromine nitrate,** $BrONO_2$, are efficiently decomposed photochemically by sunlight. In addition, the formation of HBr from the attack of atomic bromine on methane is a slower reaction than the analogous process involving atomic chlorine, since it is much more endothermic and therefore has a higher activation energy:

$$Br + CH_4 \longrightarrow HBr + CH_3$$

A lower percentage of stratospheric bromine exists in inactive form than does chlorine because of the slower speed of this reaction and because of the efficiency of the photochemical decomposition reactions. For that reason, stratospheric bromine is more efficient at destroying ozone than is chlorine (by a factor of 40 to 50), but there is much less of it in the stratosphere, so overall it is less important.

When molecules such as HCl and HBr eventually diffuse from the stratosphere back into the upper troposphere, they dissolve in water droplets and are subsequently carried to lower altitudes and are transported to the ground by rain. Thus, although the lifetime of chlorine and bromine in the stratosphere is long, it is not infinite and the catalysts are eventually removed. However, the average chlorine atom destroys about 10,000 molecules of ozone before it is removed!

Review Questions 12–16 are based upon material in the preceding section.

BOX 1-3   The Steady-State Analysis of Atmospheric Reactions

The Steady-State Approximation

If we know the nature of the creation and destruction reaction steps for a reactive substance, we can sometimes algebraically derive a useful equation for its steady-state concentration.

As a simple example, consider the formation and destruction of oxygen atoms *above* the stratosphere. As mentioned before, the atoms are formed by the photochemical dissociation of molecules of diatomic oxygen:

$$O_2 \longrightarrow 2\,O \qquad \text{(i)}$$

The atoms re-form diatomic oxygen when two of them collide simultaneously with a third molecule, M, which can carry away most of the energy released by the newly formed O_2 molecule:

$$O + O + M \longrightarrow O_2 + M \qquad \text{(ii)}$$

Recall from introductory chemistry that the rates of the individual steps in reaction mechanisms can be calculated from the concentrations of the reactants and the rate constant, k, for the step. Thus the rate of reaction (i) equals $k_i\,[O_2]$. The rate constant k_i here incorporates the intensity of the light impinging upon the molecular oxygen. Thus since two O atoms are formed for each O_2 molecule that dissociates,

rate of formation of O atoms = $2\,k_i[O_2]$

The rate of destruction of oxygen atoms by reaction (ii) is

rate of destruction of O atoms = $2\,k_{ii}[O]^2[M]$

where we square the oxygen atom concentration because two of them are involved as reactants in the step.

The net rate of change of O atom concentration with time equals the rate of its formation minus the rate of its destruction:

rate of change of [O] = $2\,k_i[O_2] - 2\,k_{ii}[O]^2[M]$

When atomic oxygen is at a steady state, this net rate must be zero, and thus the right-hand side of the equation above must also be zero. As a consequence, it follows that

$$k_{ii}[O]^2[M] = k_i[O_2]$$

By rearrangement of this equation, we obtain a relationship between the steady-state concentrations of O and of O_2:

$$[O]_{ss}^2/[O_2]_{ss} = k_i/(k_{ii}[M])$$

We see now why the ratio of oxygen atoms to diatomic molecules increases as we go higher and higher above the stratosphere: it is because the air pressure drops, and therefore so does [M], so the O_2 re-formation rate decreases.

Steady-State Analysis of the Chapman Mechanism

After this introduction, we now are ready to apply the steady-state analysis to the Chapman mechanism described by Figure 1-8. The four reactions of concern are shown again below. Notice that the recombination of O atoms, i.e., reaction (ii) above, is not included because its rate in the mid- and low-stratosphere is not competitive with other reactions, since the oxygen atom concentration is small there.

$$O_2 \longrightarrow 2\,O \qquad (1)$$
$$O + O_2 + M \longrightarrow O_3 + M \qquad (2)$$
$$O_3 \longrightarrow O_2 + O \qquad (3)$$
$$O_3 + O \longrightarrow 2\,O_2 \qquad (4)$$

Noting that O is produced or consumed in all four reactions, we obtain four terms in its overall rate expression and assume it is in a steady state:

$$\text{rate of change of [O]} = 2\,\text{rate}_1 - \text{rate}_2 + \text{rate}_3 - \text{rate}_4$$
$$= 0 \qquad (A)$$

Other useful information about concentrations can be obtained by considering the steady-state expression for the ozone concentration:

$$\text{rate of change of } [O_3] = \text{rate}_2 - \text{rate}_3 - \text{rate}_4$$
$$= 0 \qquad (B)$$

If we add together the expressions for the rates of change in [O] and in $[O_3]$, i.e., equations (A) and (B) above, we find that the rates for reactions 2 and 3 cancel, and we obtain

$$2\,\text{rate}_1 - 2\,\text{rate}_4 = 0$$

Using the expressions for these two rates in terms of reactant concentrations, we find

$$2\,k_1[O_2] - 2\,k_4[O_3][O] = 0$$

or

$$[O_3][O] = k_1[O_2]/k_4 \qquad (C)$$

Another useful expression can be obtained by subtracting equation (B) from (A). We obtain

$$2\,\text{rate}_1 - 2\,\text{rate}_2 + 2\,\text{rate}_3 = 0$$

which by rearrangement and cancellation becomes

$$\text{rate}_3 = \text{rate}_2 - \text{rate}_1$$

It is known from experiment that rate_2 (and rate_3) are much larger than rate_1, so the latter can be neglected here, giving simply

$$\text{rate}_3 = \text{rate}_2$$

Using the expressions for these two reaction rates in terms of the concentrations of their reactants,

$$k_3[O_3] = k_2[O]\,[O_2][M]$$

Rearranging this equation, we can solve for the ratio of ozone to atomic oxygen:

$$[O_3]/[O] = k_2[O_2][M]/k_3 \qquad (D)$$

Equations (C) and (D) give us two equations in the two unknowns, [O] and $[O_3]$. Multiplying their left sides together and equating the result to the product of their right sides eliminates [O] and leaves us with an equation for the ozone concentration:

$$[O_3]^2 = [O_2]^2[M]k_1k_2/k_3k_4$$

or, taking the square root of both sides, we obtain an expression for the steady-state concentration of ozone in terms of the diatomic oxygen concentration:

$$[O_3]_{ss}/[O_2]_{ss} = [M]^{0.5}(k_1k_2/k_3k_4)^{0.5} \qquad (E)$$

(continued on p. 32)

Thus the steady-state ratio of ozone to diatomic oxygen depends on the square root of the air density through [M]. The ratio is also proportional to the square root of the product of the rate constants for the reactions, 1 and 2, in which atomic oxygen and then ozone are produced, and inversely proportional to the square root of the product of the ozone destruction reaction rate constants. Substitution of numerical values for the rate constants k and for [M] into equation (E) predicts the correct order of magnitude for the ozone/diatomic oxygen ratio, i.e., about 10^{-4} in the mid-stratosphere. Ozone never is the main oxygen-containing species in the atmosphere, not even in "the ozone layer."

Equation (E) predicts that the concentration of ozone relative to that of diatomic oxygen should fall slowly as we climb in the atmosphere, given that it is proportional to the square root of the air density, through the [M] dependence. This occurs because the formation reaction of ozone, through step 2, will slow down as [M] declines. This decline with increasing altitude is observed in the upper stratosphere and above. Below about 35 km, however, the more important change in the terms of equation (E) involves k_1, and consequently the $[O_3]/[O_2]$ ratio is not simply proportional to $[M]^{0.5}$.

The rate constant k_1 incorporates the intensity of sunlight capable of dissociating diatomic oxygen into its atoms. Since the UV-C sunlight required ($\lambda < 242$ nm) is successively filtered by absorption as the light beam descends toward the Earth's surface, the value of k_1 declines especially rapidly in the low stratosphere and below. Thus the concentration of ozone predicted by applying the steady-state analysis to the Chapman mechanism successfully predicts that the ozone concentration will peak in the stratosphere. However, as discussed above, the actual peak of ozone concentration (~ 25 km, above the equator) occurs rather lower in the stratosphere than the altitude of maximum production (~ 40 km) because horizontal air movement transports ozone downward.

Substitution of equation (E) into (C) allows us to deduce an expression for the steady-state concentration of free oxygen atoms:

$$[O]_{ss} = (k_1 k_3 / k_2 k_4)^{0.5} / [M]^{0.5}$$

Thus the concentration of atomic oxygen is predicted to increase with altitude, as [M] declines—as in our previous analysis for the upper atmosphere—and as k_1 and k_3 increase, since UV light intensity increases with increasing altitude. Indeed, atomic oxygen dominates over ozone at high altitudes, whereas below about 50 km, ozone is always dominant.

The production of ozone through reaction (2) is critically dependent upon the supply of free oxygen atoms in reaction (1). The rate of oxygen atom production, in turn, is highly dependent upon the intensity of UV-C sunlight. As we have noted, this intensity falls sharply as we descend through the stratosphere. The UV-C light intensity also depends strongly upon latitude, being strongest over the equator and declining continuously toward the poles. Thus ozone production is greatest over the equator.

The qualitative behavior of the variation of ozone concentration with altitude predicted by equation (E) is correct, but the predicted amounts of ozone exceed the observed—by about a factor of two near the peak concentration. Scientists eventually found that they had underestimated the rate of the ozone destruction reaction (4) by about a factor of four, since there are catalysts in the stratosphere that greatly speed up the overall reaction.

PROBLEM 1

Consider the following 3-step mechanism for the production and destruction of excited oxygen atoms, O*, in the atmosphere:

$$O_2 \xrightarrow{\text{light}} O + O*$$

$$O* + M \longrightarrow O + M$$

$$O* + H_2O \longrightarrow 2\ OH$$

Develop an expression for the steady-state concentration of O* in terms of the concentrations of the other chemicals involved.

PROBLEM 2

Perform a steady-state analysis for d(Cl)/dt and for d[ClO]/dt in the following mechanism:

$$Cl_2 \longrightarrow 2\ Cl \qquad (1)$$

$$Cl + O_3 \longrightarrow ClO + O_2 \qquad (2)$$

$$2\ ClO \longrightarrow 2\ Cl + O_2 \qquad (3)$$

$$ClO + NO_2 \longrightarrow ClONO_2 \qquad (4)$$

Obtain expressions for the steady-state concentrations of Cl and ClO, and hence for the rate of destruction of ozone.

PROBLEM 3

Perform a steady-state analysis on the 3-step reaction mechanism below. Assume that both ozone and atomic oxygen are in a steady state, and derive an expression for the ratio $[NO_2]/[NO]$.

$$NO_2 \longrightarrow NO + O$$

$$O + O_2 \longrightarrow O_3$$

$$NO + O_3 \longrightarrow NO_2 + O_2$$

Review Questions

Test your knowledge of some of the factual information in this chapter. If the answer to a question is not obvious to you, use the Index to find the subtopic involved and review that material.

1. Which three gases constitute most of the Earth's atmosphere?

2. What range of altitudes constitutes the troposphere? the stratosphere?

3. What is the wavelength range for visible light? Does ultraviolet light have shorter or longer wavelengths than visible light?

4. Which atmospheric gas is primarily responsible for filtering sunlight in the 120–220-nm region? Which, if any, gas absorbs most of the Sun's rays in the 220–320-nm region? Which absorbs primarily in the 320–400-nm region?

5. What is the name given to the finite packets of light absorbed by matter?

6. What are the equations relating photon energy E to light's frequency ν and wavelength λ?

7. What is meant by the expression *photochemically dissociated* as applied to stratospheric O_2?

8. Write the equation for the chemical reaction by which ozone is formed in the stratosphere. What are the sources for the different forms of oxygen used here as reactants?

9. Write the two reactions that, aside from the catalyzed reactions, contribute most significantly to ozone destruction in the stratosphere.

10. What is meant by the phrase *excited state* as applied to an atom or molecule? Symbolically, how is an excited state signified?

11. Explain why the phrase *ozone layer* is a misnomer.

12. Define the term *free radical*, and give two examples relevant to stratospheric chemistry.

13. What are the two steps, and the overall reaction, by which X species such as ClO catalytically destroy ozone in the middle and upper stratosphere via Mechanism I?

14. What is meant by the term *steady state* as applied to the concentration of ozone in the stratosphere?

15. Explain why, atom for atom, stratospheric bromine destroys more ozone than does chlorine.

16. Explain why ozone destruction via the reaction of O_3 with atomic oxygen does not occur to a significant effect in the lower stratosphere.

Additional Problems

The problems given within the chapter, and the more elaborate ones given here, are designed to test your problem-solving abilities.

1. A possible additional mechanism that could exist for the creation of ozone in the high stratosphere begins with the creation of (vibrationally) excited O_2 and ground-state atomic oxygen from the absorption of photons with wavelengths less than 243 nm. The $O_2{}^*$ reacts with a ground-state O_2 molecule to produce ozone and another atom of oxygen. What is the net reaction from these two steps? What do you predict is the fate of the two oxygen atoms, and what would be the overall reaction once this fate is included?

2. In the nonpolluted atmosphere, an important mechanism for ozone destruction in the lower stratosphere is

$$OH + O_3 \longrightarrow HOO + O_2$$
$$HOO + O_3 \longrightarrow OH + 2\,O_2$$

Does this pair of steps correspond to Mechanism I? If not, what is the overall reaction?

3. A proposed mechanism for ozone destruction in the late spring over northern latitudes in the lower stratosphere begins with the photochemical decomposition of $ClONO_2$ to Cl and NO_3, followed by photochemical decomposition of the latter to NO and O_2. Deduce a catalytic ozone destruction cycle, requiring no atomic oxygen,

that incorporates these reactions. What is the overall reaction?

4. Deduce possible reaction step(s), none of which involve photolysis, for Mechanism II following the $X + O_3 \longrightarrow XO + O_2$ step such that the sum of all the mechanism's steps does not destroy or create any ozone.

5. As will be discussed in Chapter 2, atomic chlorine is produced under ozone-hole conditions by the dissociation of diatomic chlorine, Cl_2. Given that diatomic chlorine gas is the stablest form of the element, and that the ΔH_f° value for atomic chlorine is $+121.7$ kJ mol^{-1}, calculate the maximum wavelength of light that can dissociate diatomic chlorine into the monatomic form. Does such a wavelength correspond to light in the visible or the UV-A or the UV-B region?

6. Under conditions of low oxygen atom concentration, the radical HOO can react reversibly with NO_2 to produce a molecule of $HOONO_2$:

$$HOO + NO_2 \longrightarrow HOONO_2$$

(a) Deduce why the addition of nitrogen oxides to the lower stratosphere could lead to an *increase* in the steady-state ozone concentration as a consequence of this reaction.

(b) Deduce how the addition of nitrogen oxides to the middle and upper stratosphere could *decrease* the ozone concentration there as a consequence of other reactions.

(c) Given the information stated in parts (a) and (b), in what regions of the stratosphere should supersonic transport airplanes fly if they emit substantial amounts of nitrogen oxides in their exhaust?

7. At an altitude of about 35 km, the average concentrations of O* and of CH_4 are approximately 100 and 1×10^{11} molecules cm^{-3}, respectively, and the rate constant k for the reaction between them is approximately 3×10^{-10} cm^3 molecules^{-1}s^{-1}. Calculate the rate of destruction of methane in molecules per second per cubic centimeter and in grams per year per cubic centimeter under these conditions. [*Hint:* Recall that the rate law for a simple process is its rate constant k times the product of the concentrations of its reactant concentrations.]

8. The rate constants for the reactions of atomic chlorine and of hydroxyl radical with ozone are given by 3×10^{-11} $e^{-250/T}$ and 2×10^{-12} $e^{-940/T}$,

where T is the Kelvin temperature. Calculate the ratio of the rates of ozone destruction by these catalysts at 20 km, given that at this altitude the average concentration of OH is about 100 times that of Cl and that the temperature is about $-50°C$. Calculate the rate constant for ozone destruction by chlorine under conditions in the Antarctic ozone hole, when the temperature is about $-80°C$ and the concentration of atomic chlorine increases by a factor of one hundred to about 4×10^5 molecules cm^{-3}.

9. The Arrhenius equation (see Box 1-2, and note that in energy terms, $R = 8.3$ J K^{-1} mol^{-1}) relates reaction rates to temperature via the activation energy. Calculate the ratio of the rates at $-30°C$ (a typical stratospheric temperature) for two reactions having the same Arrhenius A factor and initial concentrations, one of which is endothermic and has an activation energy of 30 kJ mol^{-1} and the other which is exothermic with an activation energy of 3 kJ mol^{-1}.

2

The Ozone Holes

In this chapter, the following introductory chemistry topics are used:

➲ Kinetics: mechanisms; catalysis; reaction order

➲ Structural organic chemistry (see online Appendix)

Background from Chapter 1 used in this chapter:

➲ Photochemical decomposition

➲ Ozone destruction mechanism II

➲ Free radicals

Introduction

In Chapter 1, the gas-phase chemistry of the unpolluted stratosphere was explored. Since the late 1970s, however, the normal functioning of the stratosphere's ozone screen—and the protection it provides us—has been periodically upset by anthropogenic chlorine-containing chemicals in the atmosphere. Most famously, these substances now cause an "ozone hole" to open each spring season above the South Pole. In addition, ozone levels in the stratosphere over the North Pole, and to some extent even over our heads, have also been depleted. In this chapter the extent of these stratospheric ozone losses is documented, and the special chemical processes that produce such destruction are described. We also document how knowledge of this chemistry led to action by humankind to prevent even more drastic loss of ozone, which should eventually heal the stratosphere.

The Ozone Hole and Mid-Latitude Ozone Depletion

We begin our discussion of stratospheric ozone depletion by describing how the amount of overhead ozone is reported, and the history of how the ozone hole over the Antarctic was first discovered.

2.1 Dobson Units for Overhead Ozone

Ozone, O_3, is a gas that is present in small concentrations throughout the atmosphere. The total amount of atmospheric ozone that lies over a given

point on Earth is measured in terms of Dobson units (DU). One Dobson unit is equivalent to a 0.01-mm (0.001-cm) thickness of pure ozone at the density it would possess if it were brought to ground-level (1 atm) pressure and 0°C temperature. On average, this total overhead ozone at temperate latitudes amounts to about 350 DU; thus, if all the ozone were to be brought down to ground level, the layer of pure ozone would be only 3.5 mm thick. Because of stratospheric winds, ozone is transported from tropical regions, where most of it is produced, toward polar regions. Thus, ironically, the closer to the Equator you live, the less the total amount of ozone that protects you from ultraviolet light. Ozone concentrations in the tropics usually average 250 DU, whereas those in subpolar regions average 450 DU, except of course when holes appear in the ozone layer over such areas. There is natural seasonal variation of ozone concentration, with the highest levels in the early spring and the lowest in the fall.

2.2 History of the Annual Ozone Hole Above Antarctica

The **Antarctic ozone hole** was discovered by Dr. Joe C. Farman and his colleagues in the British Antarctic Survey. They had been recording ozone levels over this region since 1957. Their data indicated that the total amounts of ozone each October had been gradually falling each year, especially during the mid-September to mid-October period, with precipitous declines beginning in the late 1970s. This is illustrated in Figure 2-1b, where the average minimum daily amount of overhead ozone for this annual period is plotted against the year. The months from September to November correspond to the *spring* season at the South Pole, and follow a period of very cold 24-hour nights common to polar winters.

By the mid-1980s, the springtime loss in ozone at some altitudes over Antarctica was complete, and resulted in a loss of more than 50% of the total overhead amount. It is therefore appropriate to speak of a "hole" in the ozone layer that now appears each spring over the Antarctic and that lasts for several months. The average geographic area covered by the ozone hole has increased substantially since it began (see Figure 2-1a), and now is comparable in size to that of the North American continent.

Initially, it was not clear whether the hole was due to a natural phenomenon involving meteorological forces or to a chemical mechanism involving air pollutants. In the latter possibility, the suspect chemical was chlorine, produced mainly from gases released into the air in large quantities as a consequence of their use, for example, in air conditioners. Scientists had predicted that the chlorine would destroy ozone, but only to a small extent and only after several decades had elapsed.

The discovery of the Antarctic ozone hole came as a complete surprise to everyone. As a result of subsequent research, however, it was confirmed that the hole indeed does occur as a result of chlorine pollution.

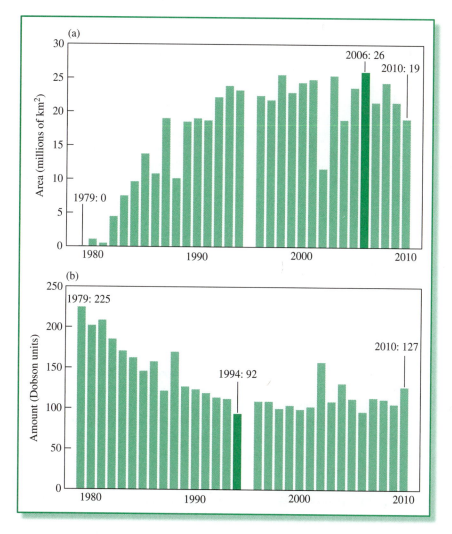

FIGURE 2-1 Historic evolution of the Antarctic ozone hole. (a) Area covered by the hole (average for September 7 to October 13) and (b) minimum overhead ozone amounts (average for September 21 to October 16). No data were acquired during the 1995 season. [Source: NASA, at http://ozonewatch.gsfc.nasa.gov/]

The complicated chemical processes that cause ozone depletion are now understood, and are discussed in this chapter. Based upon this knowledge, we can predict that the hole will continue to reappear each spring until about the middle of this century, and that a corresponding hole may appear above the Arctic region.

As a consequence of these discoveries, governments worldwide moved quickly to legislate a phase-out in production of the responsible chemicals so that the situation did not become much worse by the development of even more severe ozone depletion over populated areas, with the corresponding threat to the health of humans and other organisms that this increase would bring.

FIGURE 2-2 Changes in
average overhead ozone
amounts at different
latitudes. (a) Increases
1996–2005; (b) decreases
1979–1995. [Source:
E. C. Weatherland and
S. B. Anderson, *Nature* 441
(2006): 39.]

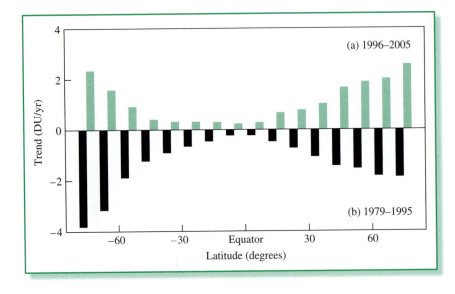

2.3 Ozone Depletion in Temperate Areas

Ozone was being depleted not just in the air above the Poles but to some extent worldwide. The average overhead ozone loss at mid-latitudes amounted to about 3% in the 1980s. As indicated by the lengths of the vertical bars in Figure 2-2b, the losses during the 1980s and early 1990s were greater the higher the latitude both in the northern and the southern hemispheres. However, this trend to ozone loss was reversed in the period from 1996 to 2005, the gains in the northern hemisphere in this period approximately cancelling the earlier losses (Figure 2-2a).

The higher the latitude, the nearer to the closer Pole.

Research reported in 2009 found that over the 1997–2008 decade, ozone recovery (of ~1%) had begun but only in the *upper* stratosphere, compared to losses of 14% there over the two preceding decades. The reversal coincides with the beginning of a gradual decrease in chlorine concentration at those altitudes, where the first recovery had indeed been expected to occur. Australian researchers reported in 2011 that once natural variations in atmospheric circulation are taken into account, the extent of ozone depletion in the ozone hole itself due to chemical change appears to have been slowly declining since the late 1990s.

Review Questions 1 and 2 are based on the material in the above section.

Ozone depletion and the possibility of an ozone hole above the North Pole is discussed in Section 2.7.

The Chemistry of Ozone Depletion

As discussed previously, scientists discovered in 1985 that stratospheric ozone over Antarctica is reduced by about 50% for several months each year, due mainly to the action of chlorine. An episode of this sort, during which there is said to be a hole in the ozone layer, occurs from September to early November, corresponding to spring at the South Pole. The hole has been appearing since about 1979, as was shown in Figure 2-1b, which illustrates the variation in the minimum September–October ozone concentrations above the Antarctic as a function of year. Extensive research in the late 1980s led to an understanding of the chemistry of this phenomenon.

In the following sections, we discuss the peculiar process by which chlorine in the stratosphere becomes activated to destroy ozone and look at the detailed mechanism by which destruction occurs. We then consider the various measures of ozone-hole size, which allow us to investigate whether the hole above the Antarctic has been declining over time, whether a hole exists above the North Pole, and the effects of the holes on the amount of UV light to which we are exposed at ground level.

2.4 The Activation of Catalytically Inactive Chlorine

The ozone hole occurs as a result of special polar winter weather conditions in the lower stratosphere, where ozone concentrations usually are highest, that temporarily convert all the chlorine that is stored in the catalytically inactive forms HCl and $ClONO_2$ into the active forms Cl and ClO (Chapter 1). Consequently, the high concentration of active chlorine causes a massive, though temporary, annual depletion of ozone.

The conversion of inactive to active chlorine occurs at the surface of particles formed by a solution of water; sulfuric acid, H_2SO_4; and nitric acid, HNO_3, the latter formed by combination of hydroxyl radical, OH, with nitrogen dioxide, NO_2, gas. The same conversion reactions could potentially occur in the gas phase but are so slow there as to be of negligible importance; they become rapid only when they occur on the surfaces of cold particles.

In most parts of the world, even in winter, the stratosphere is cloudless. Condensation of water vapor into liquid droplets or solid crystals that would constitute clouds doesn't normally occur in the stratosphere since the concentration of water in that region is exceedingly small, although there are always small liquid droplets, consisting largely of sulfuric acid, present, as well as some solid sulfate particles. However, the temperature in the lower stratosphere drops so low ($-80°C$) over the South Pole in the sunless winter months that condensation, forming particles, does occur. The usual stratospheric warming mechanism—the release of heat by the $O_2 + O$ reaction—is absent because of the lack of production of atomic oxygen from O_2 and O_3 when there is total darkness.

In turn, because the polar stratosphere becomes so cold during the total darkness at midwinter, the air pressure drops since it is proportional to the Kelvin temperature, according to the ideal gas law $PV = nRT$. This pressure phenomenon, in combination with the Earth's rotation, produces a vortex, a whirling mass of air in which wind speeds can exceed 300 km (180 miles) per hour. Since matter cannot penetrate the vortex, the air inside it is isolated and remains very cold for many months. At the South Pole, the vortex is sustained well into the springtime (October). (The vortex around the North Pole usually breaks down in February or early March, before much sunlight returns to the area, but recently there have been exceptions to this generalization, as discussed later.)

The particles produced by condensation of the gases within the vortex form **polar stratospheric clouds,** or PSCs. As the temperature drops, the first crystals to form are small ones containing water and sulfuric and nitric acids. When the air temperature drops a few degrees further, below $-80°C$ (193 K), a larger type of crystal—consisting mainly of frozen water ice and perhaps also nitric acid—also forms.

Chemical reactions that lead ultimately to ozone destruction occur in a thin aqueous layer present at the surface of the PSC ice crystals:

- Upon contact, gaseous **chorine nitrate,** $ClONO_2$, reacts at the surface with water molecules to produce **hypochlorous acid,** HOCl:

$$ClONO_2(g) + H_2O(aq) \longrightarrow HOCl(aq) + HNO_3(aq)$$

- Also in the aqueous layer, gaseous **hydrogen chloride,** HCl, dissolves and forms ions:

$$HCl(g) \xrightarrow{\substack{\text{aqueous}\\\text{layer}}} H^+(aq) + Cl^-(aq)$$

- Reaction of the two new forms of dissolved chlorine, one oxidized (Cl^-) and the other reduced (HOCl), produces **molecular chlorine,** Cl_2, which escapes to the surrounding air:

$$Cl^-(aq) + HOCl(aq) \longrightarrow Cl_2(g) + OH^-(aq)$$

This process is illustrated schematically in Figure 2-3. Overall, when the steps are added together, the process corresponds to the *net reaction*

$$HCl(g) + ClONO_2(g) \longrightarrow Cl_2(g) + HNO_3(aq)$$

since the ions H^+ and OH re-form water. Similar reactions probably also occur on the surface of solid particles.

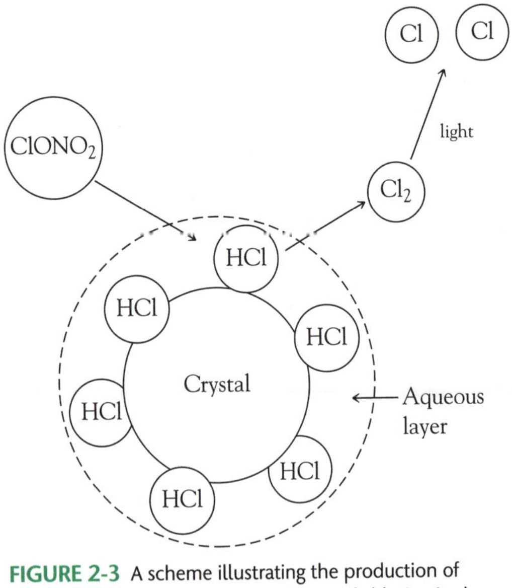

During the dark winter months, molecular chlorine accumulates within the vortex in the lower stratosphere, and eventually becomes the predominant chlorine-containing gas. **Once a little sunlight reappears** in the very early Antarctic spring, or the air mass moves to the edge of the vortex where there is some sunlight, **the chlorine molecules are decomposed by the light into atomic chlorine,** Cl:

$$Cl_2 + sunlight \longrightarrow 2 Cl$$

FIGURE 2-3 A scheme illustrating the production of molecular chlorine from inactive forms of chlorine in the winter and spring in the stratosphere in polar regions.

Similarly, any gaseous HOCl molecules released from the surface of the crystals undergo photochemical decomposition to produce hydroxyl radicals and atomic chlorine:

$$HOCl + sunlight \longrightarrow OH + Cl$$

Massive destruction of ozone by the atomic chlorine produced in these reactions then ensues via catalytic reactions.

Since stratospheric temperatures above the Antarctic remain below $-80°C$ even in the early spring, **the crystals persist for months.** Any of the Cl that is converted back to HCl by the reaction with methane is subsequently reconverted to Cl_2 on the crystals and then back to Cl by sunlight. Inactivation of **chlorine monoxide,** ClO, by conversion to $ClONO_2$ does *not* occur, since all the NO_2 necessary for this reaction is temporarily bound as nitric acid in the crystals. The larger crystals move downward under the influence of gravity into the upper troposphere, thereby removing NO_2 from the lower stratosphere over the South Pole, and further preventing the deactivation of chlorine. This *denitrification* of the lower stratosphere extends the life of the Antarctic ozone hole and increases the ozone depletion.

Only when the PSCs and the vortex have vanished does chlorine return predominantly to the inactive forms. The liberation of HNO_3 from the remaining crystals into the gas phase results in its conversion to NO_2 by the action of sunlight:

$$HNO_3 + UV \longrightarrow NO_2 + OH$$

More importantly, air containing normal amounts of NO_2 mixes with polar air once the vortex breaks down in late spring. The nitrogen dioxide then quickly combines with chlorine monoxide to form the catalytically inactive chlorine nitrate. Consequently, **the catalytic destruction cycles largely cease operation and the ozone concentration builds back up toward its normal level a few weeks after the PSCs have disappeared and the vortex has ceased.** Thus, the ozone hole closes for another year, though the ozone levels nowadays never quite return to their natural levels, even in the fall.

However, before the ozone levels build back up in the spring, some of the ozone-poor air mass can move away from the Antarctic and mix with surrounding air, temporarily lowering the stratospheric ozone concentrations in adjoining geographic regions, such as Australia, New Zealand, and the southern portions of South America. Indeed, this occurred over Tierra del Fuego, located at the southern tip of South America, for several weeks in November 2009.

2.5 Reactions That Create the Ozone Hole

In the lower stratosphere—the region where the PSCs form and chlorine is activated—the concentration of free oxygen atoms is small; few atoms are produced there on account of the scarcity of the UV-C light that is required to dissociate O_2. Furthermore, any atomic oxygen atoms produced in this way immediately collide with the abundant O_2 molecules to form ozone, O_3. Thus, ozone-destruction mechanisms based upon the $O_3 + O \longrightarrow 2\ O_2$ reaction, even when catalyzed, are not important here.

Rather, **most of the ozone destruction in the ozone hole occurs via the process called Mechanism II** in Chapter 1, with both X and X′ being atomic chlorine and with the overall reaction being $2\ O_3 \longrightarrow 3\ O_2$. Thus the sequence starts with the reaction of chlorine with ozone:

$$\textbf{\textit{Step 1:}}\quad Cl + O_3 \longrightarrow ClO + O_2$$

Confirmation that ozone destruction occurs by this reaction is evident in Figure 2-4, in which the experimental ClO and O_3 concentrations are plotted as a function of latitude for part of the Southern Hemisphere during the spring of 1987. As anticipated if step 1 is the process by which ozone destruction occurs, the two species display opposing trends, i.e., they anticorrelate very closely.

• At sufficient distances away from the South Pole (which is at 90°S), the concentration of ozone is relatively high and that of ClO is low, since chlorine is mainly tied up in inactive forms.

• However, as one travels closer to the Pole and enters the vortex region, the concentration of ClO suddenly becomes high and simultaneously that of O_3 falls off sharply (Figure 2-4): most of the chlorine has been activated and most of the ozone has consequently been destroyed. The latitude at which the concentrations both change sharply marks the beginning of the ozone hole, which continues through to the region above the South Pole.

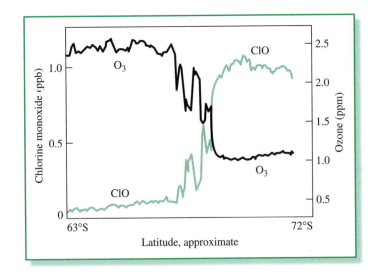

FIGURE 2-4 Stratospheric ozone and chlorine monoxide concentrations versus latitude near the South Pole (90°S) on September 16, 1987. [Source: Reprinted with permission from P. S. Zurer, *Chemical and Engineering News* (30 May 1988): 16. Copyright 1988 by the American Chemical Society.]

The anticorrelation of ozone and ClO concentrations shown in Figure 2-4 was considered by researchers to be the "smoking gun," proving that anthropogenic chlorine compounds such as CFCs emitted into the atmosphere indeed caused the formation of the ozone hole.

In the next reaction in the Mechanism II sequence, **two ClO free radicals, produced in two separate step 1 events, combine temporarily to form a nonradical dimer, dichloroperoxide, ClOOCl (or Cl_2O_2):**

$$\textit{Step 2a:} \quad 2\ ClO \longrightarrow Cl—O—O—Cl$$

The rate of this reaction becomes high, which is important to ozone loss by this mechanism, because the chlorine monoxide concentration has risen steeply due to the activation of the chlorine.

Once the intensity of sunlight has risen appreciably in the Antarctic spring, the dichloroperoxide molecule, ClOOCl, absorbs UV light and splits off one chlorine atom. The resulting ClOO free radical is unstable, and so it subsequently decomposes (in about a day), releasing the other chlorine atom:

$$\textit{Step 2b:} \quad ClOOCl + UV\ light \longrightarrow ClOO + Cl$$

$$\textit{Step 2c:} \quad ClOO \longrightarrow O_2 + Cl$$

Adding steps 2a, 2b, and 2c we see that **the net result is the conversion of two ClO molecules to atomic chlorine** via the intermediacy of the transient dimer ClOOCl, which corresponds to the second stage of Mechanism II:

$$\textit{Step 2 overall:} \quad 2\ ClO \longrightarrow [ClOOCl] \xrightarrow{\text{light}} 2\ Cl + O_2$$

Thus, **by these processes, ClO returns to the ozone-destroying form of chlorine, Cl.**

If we add the overall reaction step 2 to two times step 1 (the factor of 2 being required to produce the two intermediate ClO species needed in reaction 2a so that none remains in the overall equation), we obtain the *overall reaction*

$$2\ O_3 \longrightarrow 3\ O_2$$

Thus a complete catalytic ozone destruction cycle exists in the lower stratosphere under these special weather conditions, i.e., when a vortex is present. The cycle also requires very cold temperatures, since under warmer conditions ClOOCl is unstable and reverts back to two ClO molecules before it can undergo photolysis, thereby short-circuiting any ozone destruction. Before appreciable sunlight becomes available in the early spring, most of the chlorine exists as ClO and Cl_2O_2 since step 2b requires fairly intense light levels; such an atmosphere is said to be primed for ozone destruction.

Simplified Mechanism II for ozone hole:

$$Cl + O_3 \longrightarrow ClO + O_2$$
$$2\ ClO \xrightarrow{UV} 2\ Cl + O_2$$

About three-quarters of the ozone destruction in the Antarctic ozone hole occurs by the mechanism set forth above, in which chlorine is the only catalyst. This ozone-destruction cycle contributes greatly to the creation of the ozone hole. Each chlorine atom destroys about 50 ozone molecules per day during the spring.

The slow step in the mechanism is number 2a, which is the combination of 2 ClO molecules. Since the rate law for step 2a is second order in ClO concentration (i.e., its rate is proportional to the square of the ClO concentration), it proceeds at a substantial rate, and the destruction of ozone is significant, only when the ClO concentration is high. The abrupt appearance of the ozone hole is consistent with the quadratic rather than linear dependence of ozone destruction upon chlorine concentration by the Cl_2O_2 mechanism. Let us hope that there are not many more environmental problems whose effects will display such nonlinear behavior and which would similarly surprise us!

PROBLEM 2-1

A minor route for ozone destruction in the ozone hole involves Mechanism II with bromine as X′ and chlorine as X (or vice-versa). The ClO and BrO free radical molecules produced in these processes then collide with each other and rearrange their atoms to eventually yield O_2 and atomic chlorine and bromine. Write out the mechanism for this process, and add up the steps to determine the overall reaction.

PROBLEM 2-2

Suppose that the concentration of chlorine continues to rise in the stratosphere, but that the relative increase in bromine does not increase proportionately. Will the dominant mechanism involving dichloroperoxide or the "chlorine plus bromine" mechanism of Problem 2-1 become relatively more important or less important as the destroyer of ozone in the Antarctic spring?

PROBLEM 2-3

Why is the mechanism involving dichloroperoxide of negligible importance in the destruction of ozone, compared to the one that proceeds by ClO + O, in the upper levels of the stratosphere?

In the lower stratosphere above Antarctica, an ozone destruction rate of about 2% per day occurs each September due to the combined effects of the various catalytic reaction sequences. As a result, by early October almost all

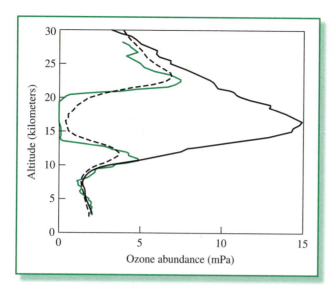

FIGURE 2-5 The typical vertical distribution of ozone over Antarctica in mid-spring (October) in 1962–1971 (black curve, before the ozone hole started), in the 1991–2001 period (dashed curve), and in 2001 (green curve). Ozone partial pressure is in millipascals. [Source: WMO/UNEP Scientific Assessment of Ozone Depletion 2006.]

the ozone is wiped out between the altitudes of 15 and 20 km, just the region in which its concentration normally is highest over the Pole. This result is illustrated in Figure 2-5, which shows the measured partial pressure of ozone in October as a function of altitude over the Antarctic in the years before ozone depletion occurred (black curve), in the years when depletion was only partial (dashed curve) and in 2001 (green curve), by which time depletion at these levels was total.

In summary, the special vortex weather conditions in the lower stratosphere above the Antarctic in winter cause denitrification and led to the conversion of inactive chlorine into Cl_2 and HOCl. These two compounds produce atomic chlorine when sunlight appears. The chlorine atoms efficiently destroy ozone via Mechanism II. Once the vortex disappears in the late spring, the ice particles on which the activation of chlorine compounds occurs disappear, the chlorine return to inactive forms, and the hole heals.

The seasonal evolution and decline of the Antarctic ozone hole in 2010 is illustrated in Figure 2-6. Even though stratospheric temperatures were low enough (below 193 K) to produce crystals before July (Figure 2-6c), significant ozone depletion did not start to occur until August (Figures 2-6a, b), presumably when sunshine first hit the primed chlorine. Maximum depletion and hole size occurred in late September/early October. Stratospheric temperatures start to rise appreciably thereafter, and were sufficient to start melting the crystals by mid-October (Figure 2-6c). The hole starts to collapse about a month later.

Review Questions 2–5 are based on the material in the above sections.

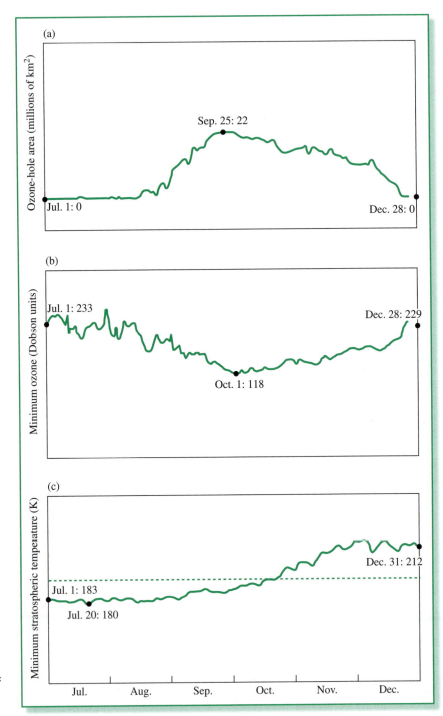

FIGURE 2-6 Evolution of the 2010 Antarctic ozone hole. (a) Area covered by the hole, (b) minimum daily amount of overhead ozone, and (c) minimum daily temperature in the lower stratosphere. The dashed line is the temperature at which ice particles form/melt. [Source: NASA, at http://ozonewatch.gsfc.nasa.gov/]

Polar Ozone Holes

2.6 The Size of the Antarctic Ozone Hole

Because (as explained later) the stratospheric concentration of chlorine continued to increase until the end of the twentieth century, the extent of Antarctic ozone depletion increased from the early 1980s at least until the 1990s. There are several relevant measures of the extent of ozone depletion:

• One measure is the *surface area* covered by low ozone; Figure 2-1a shows the area that lies within the 220-DU contour line for the mid-September to mid-October period as a function of year. This area grew rapidly and approximately linearly during the 1980s; the size of the hole in maximum depletion years (1998, 2006) has been somewhat larger than in that period, though overall there has been neither an overall increase nor a decrease since the early 1990s.

• Similarly, the sharp decrease in the *minimum amount of overhead ozone* in the spring that occurred from 1978 to the late 1980s was replaced by a slower decline. The minimum ozone has been remarkably constant since the mid-1990s (see Figure 2-1b), though 2002 was an exception in both amount of depletion and hole size.

• The average *length of time* that ozone depletion occurs has also increased in recent years. Some reduction in ozone levels is now usually seen both in mid-winter (at least in the outer portions of the continent where there is some sunshine at that time) and in the summer as well as the spring, and, indeed, there is now some persistence of the depletion from one year to the next.

• The *vertical region* over which almost total ozone depletion occurs, 12–22 km, has not increased since the mid-1990s.

Natural variations in conditions, such as the solar cycle and polar temperatures, may well mask any signs of recovery expected in the ozone hole for the next few decades. The various reactions that lead to catalytic ozone destruction by atomic chlorine by various mechanisms are summarized in Figure 2-7.

ACTIVITY

Using the information to be found at www.ozonewatch.gsfc.nasa.gov and other websites you may find useful, compare the history of the most recent Antarctic ozone hole to the time evolution of the 2010 hole in Figure 2-6. Did the maximum depletion, maximum area, and minimum temperature exceed 2010 values and did they occur at about the same time as they did in 2010? Photocopy or download Figure 2-1 and manually add data for more recent years to the two bar graphs. Are there definitive signs yet from your data that the hole is becoming smaller in area or that depletion is lessening?

FIGURE 2-7 A summary of the main ozone destruction reaction cycles operating in the Antarctic ozone hole.

Ozone destruction step

$$O_3 + Cl \longrightarrow O_2 + ClO$$

Atomic chlorine reconstitution

Mid-stratosphere

$$ClO + O \longrightarrow Cl + O_2$$

Ozone hole/low stratosphere

$$2\ ClO \longrightarrow ClOOCl$$

$$ClOOCl + UV \longrightarrow ClOO + Cl$$

$$ClOO \longrightarrow Cl + O_2$$

Inactivation of chlorine

$$Cl + CH_4 \longrightarrow HCl + CH_3$$

$$ClO + NO_2 \longrightarrow ClONO_2$$

Activation of chlorine on particle surfaces

$$HCl(g) \xrightarrow{H_2O} H^+(aq) + Cl^-(aq)$$

$$H_2O(aq) + ClONO_2(g) \longrightarrow HOCl(aq) + HNO_3(aq)$$

$$Cl^-(aq) + HOCl(aq) \longrightarrow Cl_2(g) + OH^-(aq)$$

$$Cl_2(g) + sunlight \longrightarrow 2\ Cl(g)$$

$$H^+(aq) + OH^-(aq) \longrightarrow H_2O(aq)$$

2.7 Stratospheric Ozone Destruction over the Arctic Region

Given the similarity in climate, it may seem surprising that an ozone hole above the Arctic did *not* start to form at the same time as in the Antarctic. Episodes of partial springtime ozone depletion over the Arctic region have occurred several times since the mid-1990s. The phenomenon is less severe than in Antarctica: the reasons for this are that **the stratospheric temperature over the Arctic does not fall as low nor for as long and that air circulation to surrounding areas is not as limited.** The flow of tropospheric air over mid-latitude mountain ranges (Himalayas, Rockies) in the northern hemisphere creates waves of air that can mix with polar air, warming the Arctic stratosphere.

Because the air is generally not as cold, polar stratospheric clouds form less frequently over the Arctic, and do not last as long, as over the Antarctic. In the past, only small crystals were formed; these are not large enough to fall out of the stratosphere and thereby denitrify it. However, during the extended polar night, the chlorine nitrate and hydrogen chloride do react on the surface of the small particles to produce molecular chlorine, which then dissociates to atomic chlorine, and which by reaction with an ozone molecule becomes chlorine monoxide.

Before the mid-1990s, the vortex containing the cold air mass above the Arctic broke up by late winter; therefore NO_2-containing air mixed with vortex air before much sunlight returned to the polar region in the spring. Since the stratospheric air temperature usually rose above $-80°C$ by early March, the nitric acid in the particles was converted back to gaseous nitrogen dioxide before the intense spring sunlight could drive the Cl_2O_2 mechanism. Due to increases in NO_2 from both these sources, the activated chlorine was mostly transformed back to $ClONO_2$ before it could destroy much ozone. Thus the total extent of ozone destruction over the Arctic area was much less than that over the Antarctic in the past.

The extent of winter-spring ozone loss over Arctic regions has been very inconsistent, with almost no depletion in some winters, but significant depletion in others, as indicated in Figure 2-8. Significant losses occurred in March of several successive years in the mid-1990s, but the total ozone-column loss did not exceed 25% again until 2010. A record 40% loss was then observed in 2011, a winter which, although relatively warm at ground level, was colder than usual at stratospheric levels. Interestingly, the amount of ozone loss has been found to correlate linearly with the area associated with polar stratospheric clouds: the greater the area, the more the loss of ozone in a given year.

For reasons that will be explained in Chapter 5, both the depletion of ozone and the increase in carbon dioxide levels themselves cool the stratosphere, and this will lead to even *more* depletion if cooling occurs in the springtime and thereby extends the period in which PSCs remain. Some

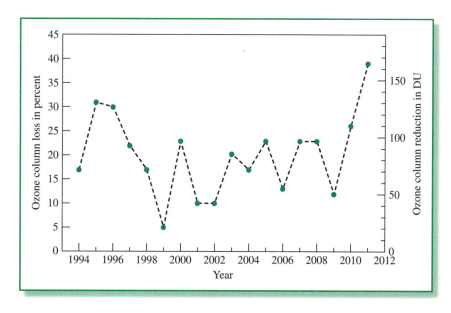

FIGURE 2-8 Column loss over the Arctic of overhead ozone in springtime. [Source: Global Observing Systems Information Center.]

scientists predict that the recovery with time from ozone depletion will be slower in the Arctic than the Antarctic because of the cooling effects of CO_2 and O_3. Scientists do not yet know whether the abrupt cooling in the winters that produced record ozone depletion was due largely to the effects of increased CO_2 or not.

Because the magnitude of ozone depletion above the Arctic in recent winters was about the same as that observed over the South Pole in the early 1980s, some atmospheric scientists have stated that an Arctic ozone hole now forms in some years. Since depletion of overhead ozone is never 100% complete, the definition of what conditions constitute a "hole" is somewhat arbitrary.

The chemistry underlying *mid-latitude losses* in stratospheric ozone is discussed in Box 2-1.

BOX 2-1   The Chemistry Behind Mid-Latitude Decreases in Stratospheric Ozone

Scientists have had a harder time tracking down the source of this mid-latitude ozone depletion than for that over polar regions. As in Antarctica, almost all the ozone loss in nonpolar regions occurs in the lower stratosphere. Some scientists have speculated that reactions leading to ozone destruction could occur not only on ice crystals but also on the surfaces of other particles present in the lower stratosphere. They suggested that the reactions could occur on cold liquid droplets consisting mainly of sulfuric acid that occur naturally in the lower stratosphere at all latitudes. The liquid droplets would have to be cold enough for significant uptake by them of gaseous HCl to occur, or no net reaction would take place. There always exists a small background amount of the acid, due to the oxidation of the naturally occurring gas *carbonyl sulfide*, COS, some of which survives long enough to reach the stratosphere.

However, the dominant though erratic source of the H_2SO_4 at these altitudes is direct injection into the stratosphere of sulfur dioxide gas emitted from volcanoes, followed by its oxidation to the acid. Indeed, a steep decline in ozone in 1992–1993 followed the June 1991 massive eruption of Mt. Pinatubo in the Philippines, and measurable ozone depletion was noted for several years after the eruption of El Chichon in Mexico in 1982. Both these volcanic eruptions temporarily increased the concentration of sulfuric acid droplets in the lower stratosphere.

The other relevant reaction that takes place on the surface of the sulfuric acid droplets results in some denitrification of stratospheric air. In the gas-phase steps of the sequence, ozone itself converts some nitrogen dioxide, NO_2, into *nitrogen trioxide*, NO_3, which then combines with other NO_2 molecules to form *dinitrogen pentoxide*, N_2O_5:

$$NO_2 + O_3 \longrightarrow NO_3 + O_2$$

$$NO_2 + NO_3 \longrightarrow N_2O_5$$

These gas-phase processes normally are reversible and do not remove much NO_2 from the air, but in the presence of high levels of aqueous liquid droplets, a conversion of N_2O_5

to nitric acid occurs instead:

$$N_2O_5 + H_2O(\text{droplets}) \xrightarrow[\text{droplets}]{H_2SO_4} 2\ HNO_3$$

By this mechanism, much of the NO_2 that normally would be available to tie up chlorine monoxide as $ClONO_2$ becomes unavailable for this purpose; hence a greater proportion of the chlorine atoms occur in the catalytically active form and destroy ozone.

In the mid-latitude lower stratosphere, the most important catalytic ozone destruction reactions involving halogens employ Mechanism II, with X being atomic chlorine or bromine and X' being hydroxyl radical,

$$Cl + O_3 \longrightarrow ClO + O_2$$

$$OH + O_3 \longrightarrow HOO + O_2$$

$$ClO + HOO \longrightarrow HOCl + O_2$$

$$HOCl \xrightarrow{\text{sunlight}} OH + Cl$$

and similarly for the case where bromine replaces chlorine. The reaction sequence involving collision of ClO with BrO discussed for the Antarctic ozone hole is also operative here.

PROBLEM 1

Deduce the overall reaction equation for the reaction sequence given above.

This mechanism explains why, in the current high-chlorine lower stratosphere, large volcanic eruptions can deplete mid-latitude stratospheric ozone for a few years, but it does not account for the overall trend of decreasing ozone in the 1980s. Some of the decrease was probably due to the above mechanism operating on the background concentration of sulfuric acid particles in the lower stratosphere; its magnitude would have increased continuously in this time period since the chlorine levels were continuously increasing. Chlorine and bromine increases combined resulted in about a 4% decline in mid-latitude ozone levels in the 1979–1995 period. However, much of the gradual decline over mid-latitudes is believed to be due to other factors, such as springtime dilution of ozone-depleted polar air and its transport out of the polar regions, changes in the solar cycle, and both natural and anthropogenic changes in the pattern of atmospheric transport and temperatures.

2.8 Increases in UV at Ground Level

Experimentally, the amount of UV-B from sunlight (see Chapter 1) reaching ground level increases by a factor of three to six in the Antarctic during the early part of the spring because of the appearance of the ozone hole. Biologically, the most dangerous UV doses under hole conditions occur in the late spring (November and December), when the Sun is higher in the sky than in earlier months and low overhead ozone values still prevail. Abnormally high UV levels have also been detected in southern Argentina when ozone-depleted stratospheric air from the Antarctic travelled over the area. Indeed, the stalling of the hole's edge over Tierra del Fuego in 2009 exposed its residents to double their normal levels of ultraviolet light for three weeks. Higher-than-normal UV levels would have been experienced during sunny days in southern Finland in the spring of 2011 due to the significant Arctic ozone depletion at the time.

Increases in ground level UV-B intensity have also been measured in the spring months in mid-latitude regions in North America, Europe, and New Zealand. Calculations indicate that the extent of UV increases since the 1980s over mid- and high-latitude regions at times have amounted to 6 to 14%. The most definitive experimental evidence comes from New Zealand, where long-term summertime increases in UV-B, but, as expected, not UV-A, had amounted to 12% by 1998–1999.

The situation over mid-latitudes is complicated by the facts that some UV-B is absorbed by the ground-level ozone produced by pollution reactions (as explained in Chapter 3), thereby masking any changes in UV-B that are due to small amounts of stratospheric ozone depletion, and that records of UV received at the Earth's surface were started only in the 1990s.

Review Questions 6 and 7 are based on the material in the above sections.

The Chemicals That Cause Ozone Destruction

The increase in levels of stratospheric chlorine and bromine that occurred in the last half of the twentieth century was due primarily to the release into the atmosphere of organic compounds containing chlorine and bromine that are **anthropogenic**, that is to say they are man-made. These anthropogenic contributions to stratospheric halogen levels completely overshadowed the natural input. In this section, we investigate

- why the levels of chlorine and bromine increased due to the release into the air of compounds having certain characteristics,

- how international agreements were put in place to control such substances,

- what is the strategy underlying the formulations of compounds that are the new replacements for the original halogen compounds, and the practical difficulties and controversy about phasing-out methyl bromide, and

- how two practical replacements developed by Green Chemistry for the now-banned chemicals can be employed.

The chlorine- and bromine-containing compounds that give rise to increased levels of the halogens in the stratosphere are those that do not have a **sink—i.e., a natural removal process** such as dissolution in rain or oxidation by atmospheric gases—in the troposphere. After a few years of travelling in the troposphere, they begin to diffuse into the stratosphere, where eventually they undergo photochemical decomposition by UV-C from sunlight, and thereupon release their halogen atoms.

The variation in the total concentration of stratospheric chlorine and bromine atoms, expressed as the equivalent of chlorine in terms of ozone-destruction power, measured over the course of the last quarter century and projected to the middle of the twenty-first century, is illustrated by the

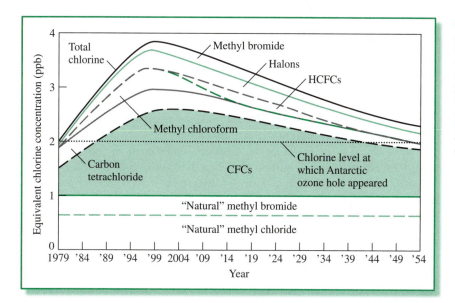

FIGURE 2-9 Actual and projected concentration of stratospheric chlorine versus time, showing the contributions of various gases. Note that the ozone-depleting effects of bromine atoms in halons and methyl bromide have been converted to their chlorine equivalents. [Source: DuPont.]

topmost curve in Figure 2-9. The peak chlorine equivalent concentration of about 3.8 ppb occurred in the late 1990s, and was almost four times as great as was the "natural" level due to methyl chloride and methyl bromide releases from the sea. The Antarctic ozone hole first appeared when the chlorine concentration reached about 2 ppb (dotted horizontal line).

2.9 CFC Decomposition Increases Stratospheric Chlorine

As is clear from inspection of Figure 2-9, the recent increase in stratospheric chlorine was due primarily to the use and release of **chlorofluorocarbons,** compounds containing chlorine, fluorine, and carbon (only), which are commonly called **CFCs.** In the 1980s, about 1 million tonnes (i.e., metric tons, 1000 kg each) of CFCs were released annually into the atmosphere. These compounds are nontoxic, nonflammable, nonreactive, and have useful condensation properties (suiting them for use as coolants, for example). Because of these favorable characteristics they found a multitude of uses. Large volumes of several CFCs were manufactured commercially and employed worldwide throughout the mid- to late-1900s. Most of the compounds produced eventually leaked, or were released upon disposal, from the devices in which they were originally placed, and entered the atmosphere as gases.

CFCs have no tropospheric sink, so all molecules of them eventually rise to the stratosphere. In contrast to intuitive expectation, this vertical transport in the atmosphere is *not* affected by the fact that the mass of these molecules is greater than the average molecular mass of nitrogen and oxygen in air, because the differential force of gravity is much less than that due to the constant collisions of other molecules, which randomize the directions of even heavy molecules.

The CFC molecules eventually migrate to the middle and upper parts of the stratosphere where there is sufficient unfiltered UV-C from sunlight to photochemically decompose them, thereby releasing chlorine atoms. CFCs do not absorb sunlight with wavelengths greater than 290 nm, and generally require that of 220 nm or less for photolysis. The CFCs must rise to the mid-stratosphere before decomposing, since UV-C does not penetrate to lower altitudes. Because vertical motion in the stratosphere is slow, their atmospheric lifetimes are long. It is because of their long stratospheric lifetimes that the chlorine concentration in Figure 2-9 falls so slowly with time.

PROBLEM 2-4

Reactions of the type

$$OH + CF_2Cl_2 \longrightarrow HOF + CFCl_2$$

are conceivable tropospheric sinks for CFCs. Can you deduce why they don't occur, given that C—F bonds are much stronger than O—F bonds?   ●

2.10 Other Chlorine-Containing, Ozone-Depleting Substances

Another widely used carbon-chlorine compound that lacks a tropospheric sink—although some of it ends up dissolving in ocean waters—was **carbon tetrachloride,** CCl_4, which also is photochemically decomposed in the stratosphere, thereby producing chlorine atoms. Like CFCs, then, it is classified as an **ozone-depleting substance** (ODS). Commercially, carbon tetrachloride was used as a solvent and as an intermediate in the manufacture of several important CFCs, and during production some was lost to the atmosphere. Its use as a dry-cleaning solvent was discontinued in most developed countries some decades ago, but until recently has continued in many other countries. Because of its relatively long atmospheric lifetime (26 years), it will continue to make a significant contribution to stratospheric chlorine for several more decades (Figure 2-9).

Methyl chloroform, CH_3—CCl_3, or *1,1,1-trichloroethane*, was produced in large quantities and used in metal cleaning in such a way that much of it was released into the atmosphere. Although about half of it is removed from the troposphere by reaction with the hydroxyl radical, the remainder survives long enough to migrate to the stratosphere. Because its average lifetime is only five years and its production has been largely phased out, its concentration in the atmosphere has declined rapidly since the 1990s. According to Figure 2-9, the contribution of methyl chloroform to stratospheric chlorine was substantial in the 1990s, but by 2010 became negligible.

2.11 Green Chemistry: The Replacement of CFC and Hydrocarbon Blowing Agents with Carbon Dioxide in Producing Foam Polystyrene

Polystyrene is a common polymer that is used to make many everyday items. This polymer varies in appearance from a rigid solid plastic to foam polystyrene. Rigid plastic polystyrene is used in disposable silverware; audiocassette, CD, and DVD cases; and appliance casings. Foam polystyrene is utilized as insulation in coolers and houses, foam cups, meat and poultry trays, egg cartons, and in some countries is still used in fast-food containers. Globally about 10 million tonnes of polystyrene are produced on an annual basis, with approximately half used to produce the foam form.

In order to produce foam polystyrene, the melted polymer is combined with a gas under pressure. This mixture is then extruded into an environment of lower pressure where the gas expands, leaving a foam that is about 95% gas and 5% polymer. In the past, CFCs were employed as blowing agents for rigid plastic foams, and foam polystyrene is no exception. When these foams are crushed or they degrade, the CFCs are released into the atmosphere where they can migrate to the stratosphere and act to destroy ozone. Low-molecular-weight hydrocarbons, such as pentane, have also been used as blowing agents; although these compounds do not deplete the ozone layer, they do contribute to ground-level smog when they are emitted into the atmosphere, as we will see in Chapter 3. Low-molecular-weight hydrocarbons are also very flammable and reduce worker safety.

The search for the replacement of CFC and hydrocarbon blowing agents led the Dow Chemical Company of Midland, Michigan, to develop a process employing 100% carbon dioxide as a blowing agent for polystyrene foam sheets. For this discovery, Dow was the recipient of a Presidential Green Chemistry Challenge Award in 1996. **Carbon dioxide,** CO_2, is not flammable nor does it deplete the ozone layer. Nonetheless, we will see in Chapter 5 that it is a greenhouse gas and thus contributes to the environmental problem of global warming, so one might wonder whether we are trading one environmental problem for another. However, waste carbon dioxide from other processes (natural gas production and the preparation of ammonia) that would otherwise be emitted into the atmosphere can be captured and used as a blowing agent. In addition, we will see in Chapter 5 that CFCs not only dramatically affect the ozone layer but also are greenhouse gases significantly more potent than is carbon dioxide.

Dow Chemical found an added advantage in the polystyrene foam sheets made with carbon dioxide in that they remained flexible for a much longer time than those made with CFCs. This results in less breakage during use and a longer shelf life. In addition, foam sheets made with CFCs had to be degassed of the CFCs prior to recycling them, while carbon dioxide rapidly escapes from the polystyrene leaving a sheet composed of 95% air and 5% polystyrene within a few days.

2.12 CFC Replacements

Compounds such as CFCs and CCl_4 have no tropospheric sinks because they do not undergo any of the normal removal processes. They are not soluble in water and thus they are not rained out of the air; they are not attacked by the hydroxyl radical or any other atmospheric gases and so do not decompose; and they are not photochemically decomposed by either visible or UV-A light.

The compounds being implemented as the direct replacements for CFCs all contain hydrogen atoms bonded to carbon. Consequently a majority (though not necessarily 100%) of the molecules will be removed from the troposphere by a sequence of reactions which begins with hydrogen abstraction by OH:

$$OH + H\overset{|}{\underset{|}{C}}\!\!\longrightarrow H_2O + \text{C-centered free radical} \longrightarrow$$
$$CO_2 \text{ and other products eventually}$$

Reactions of this type are discussed in more detail in Chapters 3 and 17. Because methyl chloride, methyl bromide, and methyl chloroform each contain hydrogen atoms, a fraction of such molecules are removed in the troposphere before they have a chance to rise to the stratosphere.

The *temporary* replacements for CFCs employed in the 1990s and the early years of the twenty-first century contained hydrogen, chlorine, fluorine, and carbon; they were called **HCFCs, hydrochlorofluorocarbons.** The most important example was CHF_2Cl, the gas called *HCFC-22* (or just *CFC-22*). It was employed in modern domestic air conditioners and in some refrigerators and freezers, and found some use in blowing foams such as those used in food containers. Since it contains a hydrogen atom and thus is mainly removed from air before it can rise to the stratosphere, its long-term ozone-reducing potential is small—only 5% of that of the CFC that it replaced. This advantage is offset, however, by its property of decomposing to release chlorine more quickly than does the CFC, so its *short-term* potential for ozone destruction is greater than that implied by this percentage. But because most HCFC-22 is destroyed within a few decades after its release, it is responsible for almost no *long-term* ozone destruction. However, most concerns about stratospheric ozone destruction are centered on the next few decades, before substantial reduction of stratospheric chlorine occurs from the phase-out of CFCs.

Notice the contribution of HCFCs to the curves in Figure 2-9. They should be significant only from the late 1990s until about 2030. The tropospheric concentration of HCFC-22 rose in a linear fashion with time in the late 1900s and early 2000s.

Reliance exclusively on HCFCs as CFC replacements would eventually lead to a renewed buildup of stratospheric chlorine, because the volume of HCFC consumption would presumably rise with increasing world population and affluence. **Products that are entirely free of chlorine, and that therefore pose no hazard to stratospheric ozone, are the ultimate replacements for CFCs and HCFCs.**

Hydrofluorocarbons, HFCs, substances that contain hydrogen, fluorine, and carbon, are the main long-term replacements for CFCs and HCFCs. The compound $CH_2F—CF_3$, called *HFC-134a*, has an atmospheric lifetime of several decades before finally succumbing to OH attack. HFC-134a is now used as the working fluid in new refrigerators and automobile air conditioners produced in developed countries. The product called *R-410a* is a 50–50 mixture of the two hydrofluorocarbons CH_2F_2 and CHF_2CF_3, and is used in new air-conditioning units for homes and other buildings in these countries.

Unfortunately, one atmospheric degradation pathway for some HFCs, and for several HCFCs as well, produces **trifluoroacetic acid,** TFA, CF_3COOH, as an intermediate, which is then removed from the air by rainfall. Some scientists worry that TFA represents an environmental hazard to wetlands since it will accumulate in aquatic plants and could inhibit their growth. However, some of the TFA in the environment arises from the degradation under heating of polymers such as Teflon, not from CFC replacements. Polyfluorocarboxylic acids, of which the acid form of TFA is an example, have been used in certain commercial products, but are now being phased out, as discussed in Chapter 15.

Another environmental concern with HFCs involves their accumulation in air after inadvertent release during use. While present in the troposphere, before they are destroyed, HFCs contribute to global warming by enhancing the greenhouse effect, a topic discussed in detail in Chapter 5. Outside of North America, industry usually uses cyclopentane or isobutane rather than an HFC as a refrigerant. These hydrocarbons have a much shorter lifetime in air than HFCs. Some environmentalists hope that developing countries follow the hydrocarbon rather than the HFC route when they start to manufacture goods requiring coolants. Fully fluorinated compounds are unsuitable replacements for CFCs because they have no tropospheric or stratospheric sinks, and if released into the air, they would contribute to global warming for very long periods of time.

The decisions discussed above regarding replacements for CFCs represent an attempt by governments and industry to employ *systems thinking*. In particular, the choices for the replacements considered detailed knowledge of the stratospheric ozone system and the influence of chlorine, bromine, and fluorine upon it. Attempts were made to minimize further harm to the ozone layer while still allowing developing countries to implement refrigeration systems. The unintended consequence of increasing atmospheric greenhouse gas concentrations when CFC replacements were sought was minimized by choosing hydrofluorcarbons or hydrocarbons with limited atmospheric lifetimes, and by requiring that refrigeration systems be much better sealed than in the past to prevent chronic loss of refrigerant to the atmosphere.

The concept of systems thinking was explained in Table 0-1 on pages xxiii–xxiv.

2.13 Halons

Halon chemicals are bromine-containing, hydrogen-free substances such as CF$_3$Br and CF$_2$BrCl. Because they have no tropospheric sinks, they eventually rise to the stratosphere. There they are photochemically decomposed, with the release of atomic bromine (and chlorine, if present), which, as we have already discussed, is an efficient X catalyst for ozone destruction. Thus, halons also are ozone-depleting substances. Bromine from halons will continue to account for a significant fraction of the ozone-destroying potential of stratospheric halogen catalysts for decades to come (Figure 2-9).

Other brominated fire retardants are discussed in Chapter 15.

Halons are used in fire extinguishers. They operate to quell fires by releasing atomic bromine, which combines with the free radicals in the combustion to form inert products and less-reactive free radicals. The halons release their bromine atoms even at moderately high temperatures, since their C—Br bonds are relatively weak. Since they are nontoxic and leave no residues upon evaporation, halons are very useful for fighting fires, particularly in inhabited, enclosed spaces, such as military aircraft and those housing electronic equipment, such as computer centers.

The substitution of other chemicals for halons in the testing of the extinguishers drastically reduces halon emissions to the atmosphere, since only a minority of the releases will be from the fighting of actual fires. Fine sprays of water can be substituted for halons in fighting many fires.

2.14 Can Stratospheric Fluorine Destroy Ozone?

Fluorine atoms and **hydrogen fluoride,** HF, are liberated in the stratosphere as a result of the decomposition of CFCs, HCFCs, HFCs, and halons. In principle, the fluorine atoms could catalytically destroy ozone (see Problem 2-5). However, the reaction of atomic fluorine with methane and other hydrogen-containing molecules in the stratosphere is rapid, and produces HF, a very stable molecule. Because the H—F bond is much stronger than is O—H, the reactivation of fluorine by the attack of the hydroxyl radical on hydrogen fluoride molecules is very endothermic. Consequently its activation energy is high and the reaction is extremely slow at atmospheric temperatures (see Box 1-1). Thus atomic fluorine is quickly and permanently deactivated before it can destroy any significant amount of ozone.

PROBLEM 2-5

(a) Write the set of reactions by which atomic fluorine could operate as X catalysts by Mechanisms I and II in the destruction of ozone. (b) An alternative to the second step of Mechanism I in the case of X = F is the reaction of FO with ozone to give atomic fluorine and two molecules of oxygen. Write out this mechanism, and deduce its overall reaction.

PROBLEM 2-6

The free radical CF_3O is produced during the decomposition of HCF-134a. Show the sequence of reactions by which it could destroy ozone acting as an X catalyst in a manner reminiscent of OH. (Note that it is too short-lived to actually destroy much ozone.)

2.15 International Agreements That Restrict ODSs

In contrast to almost all other environmental problems, such as global warming/climate change (Chapters 5 and 6), international agreement on remedies to stratospheric ozone depletion was obtained and successfully implemented in a fairly short period of time. Invoking the *precautionary principle* to minimize possible harm to humans and the environment, the use of CFCs in most aerosol products was banned in the late 1970s in North America and some Scandinavian countries.

The precautionary principle was discussed in Table 0-1.

This decision was taken on the basis of predictions, made by Sherwood Rowland and Mario Molina, chemists at the University of California, Irvine, concerning the effect of chlorine on the thickness of the ozone layer. There was no experimental indication of any depletion at the time of their prediction. Rowland and Molina, together with the German chemist Paul Crutzen, were jointly awarded the Nobel Prize in Chemistry in 1995 to honor their work in researching the science underlying ozone depletion.

The growing awareness of the seriousness of chlorine buildup in the atmosphere led to international agreements to phase out CFC production in the world. The breakthrough came at a conference in Montreal, Canada, in 1987 that gave rise to the **Montreal Protocol;** this agreement has been strengthened at several follow-up conferences. **As a result of this international agreement, all ozone-depleting chemicals are now destined for phase-out in all nations.**

All legal CFC *production* in developed countries ended in 1995. Developing countries had been allowed until 2010 to reach the same goal. Figure 2-10 shows how the tropospheric concentrations of the most widely used CFCs have changed in recent decades. The level of CFC-11 ($CFCl_3$), the average atmospheric lifetime of which is about 50 years, peaked about 1993, six years after its production started a precipitous decline. Its concentration has dropped slowly since then; the level of CFC-12 (CF_2Cl_2), which has a lifetime of more than 100 years, did not peak until about 2002.

The production of carbon tetrachloride and methyl chloroform has been phased out. The atmospheric level of CH_3CCl_3 has dropped already to a small fraction of its peak in the 1990s (Figure 2-10, light green curve) but that of CCl_4 has declined very slowly due to a lack of a tropospheric sink (Figure 2-10, dark green curve). Developed countries have agreed to end production of HCFCs by 2030, and developing countries by 2040, with no increases allowed after 2015. The atmospheric concentration of the widely used HCFC-22 rose during the early 2000s, but may have levelled off (Figure 2-10, black curve).

FIGURE 2-10 Tropospheric concentrations of CFCs and other chlorine-containing ODSs. [Source: NOAA, at http://www.esrl.noaa.gov/gmd/hats/]

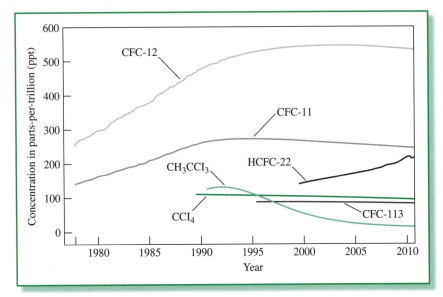

Halon production was halted in developed countries in 1994 by the terms of the Montreal Protocol. However, use of existing stocks continues, as do releases from fire-fighting equipment. In addition, China and Korea—which, as developing countries, had until 2010 to terminate production—increased their production of these chemicals in the 1990s. For these reasons, the atmospheric concentration of halons continued to rise, but seems now to have levelled off.

The other bromine-containing ODS is the pesticide gas **methyl bromide,** CH_3Br. Scientifically, we do not yet have a good handle on atmospheric methyl bromide. In particular:

• Significant new natural emission sources of the gas to the atmosphere continue to be discovered. Consequently, even the approximate ratio of synthetic/natural emissions is uncertain, as is the lifetime of about one year.

• The tropospheric concentration of the gas has changed much more since 1999 than had been anticipated by production levels and controls. Its concentration is currently declining, albeit slowly.

Methyl bromide was added to the Montreal Protocol during the 1992 revision of the international treaty. It was agreed that developed countries would phase out methyl bromide production and importation completely in 2005. Its consumption in all developing countries combined, which amounted to less than half the U.S. usage, was to have been frozen at 1995–1998 levels in 2002, to have been reduced by 20% in 2005, and is to be completely eliminated by 2015. However, its phase-out has been strongly resisted by some U.S. farmers, and planned reductions have been deferred. The pros and cons of implementing the Montreal Protocol controls on this controversial chemical are discussed in the Case Study *Strawberry Fields— The Banning of Methyl Bromide* on the website associated with this chapter.

Recently the soil fumigant **methyl iodide,** CH_3I, was approved in the United States as a replacement for methyl bromide. Although not a threat to the ozone layer, the use of methyl iodide is very controversial because it is highly toxic and difficult to control.

As a direct result of the implementation of the gradual phase-out of ozone-depleting substances, the total tropospheric concentration of chlorine peaked in 1994, and had declined by about 10% by 2007. Much of the initial drop was due to the phase-out of methyl chloroform, which has a short atmospheric lifetime (Figure 2-10, light green curve); since it has now been almost eliminated, the overall rate of decline of tropospheric chlorine has slowed. The concentrations of CFCs are slow to decrease because they were used in many applications such as foams and cooling devices that have only slowly emitted them to the atmosphere, a process that continues even today.

The stratospheric chlorine equivalent level was predicted to have peaked, at less than 4 ppb, at the turn of the century, with a gradual decline predicted thereafter (see Figure 2-9). Observations in 2000 indicated that the actual chlorine content in the stratosphere had peaked, but the bromine abundance was still increasing. The slowness in the decline of the stratospheric chlorine level is due to

- the long time it takes molecules to rise to the middle or upper stratosphere and to then absorb a photon and dissociate to atomic chlorine,

- the slowness of the removal of chlorine and bromine from the stratosphere, and

- the continued input of some chlorine and bromine into the atmosphere.

Because ozone is formed (and destroyed) in rapid natural processes, its level responds very quickly to a change in stratospheric chlorine concentration. Thus the Antarctic ozone hole probably will not continue to appear after the middle of the twenty-first century, that is, once the chlorine equivalent concentration is reduced back to the 2 ppb level it had in the years before the hole began to form (Figure 2-9). More recent projections predict the Antarctic hole area will start to decrease in about 2023, but the full recovery will not happen until about 2070.

Without the Montreal Protocol agreements, catastrophic increases in chlorine, to many times the present level, would have occurred, particularly since CFC usage and atmospheric release in developing countries would have increased dramatically. A further doubling of stratospheric chlorine levels would probably have led to the formation of a substantial ozone hole each spring over the Arctic region. By 2030, the stratospheric

Review Questions 9–13 are based on material in the above sections.

chlorine level probably would have reached 9 ppb, resulting in mid-latitude losses of about 10–15%. And with the increase in ozone depletion would have come a catastrophic increase in skin cancers.

PROBLEM 2-7

Given that their C—H bonds are not quite as strong as those in CH_4, can you rationalize why ethane, C_2H_6, or propane, C_3H_8, is a better choice than methane to inactivate atomic chlorine in the stratosphere?   ●

PROBLEM 2-8

No controls on the release of CH_3Cl, CH_2Cl_2, or $CHCl_3$ have been proposed. What does that imply about their atmospheric lifetimes, compared to those for CFCs, CCl_4, and methyl chloroform?   ●

2.16 Green Chemistry: Harpin Technology— Eliciting Nature's Own Defenses Against Diseases

Earlier in the chapter, we learned that methyl bromide is used as a pesticide (more specifically, as a soil fumigant), and that some of it finds its way into the stratosphere where it becomes involved in the destruction of the ozone layer. An interesting development, which offers an alternative to methyl bromide, is known as *Harpin Technology*. This technology was developed by EDEN Bioscience Corporation in Bothell, Washington, for which it was awarded a Presidential Green Chemistry Challenge Award in 2001.

Harpin is a naturally occurring bacterial protein that was isolated from the bacteria *Erwinia Amylovora* at Cornell University. When applied to the stems and leaves of plants, harpin elicits the plant's natural defense mechanisms to diseases caused by bacteria, viruses, nematodes, and fungi. Hypersensitive response (HR), which is induced by harpin, is an initial defense by plants to invading pathogens that results in cell death at the point of infection. The dead cells surrounding the infection act as a physical barrier to the spread of the pathogen. In addition, the dead cells may release compounds that are lethal to the pathogen.

Pests often build up immunity to pesticides. However, since harpin does not directly affect the pest, it is unlikely that immunity to it will occur. In addition to using traditional pesticides to control the infestation of plants, more recently a second approach to this problem has been to develop genetically altered plants. The DNA in such plants has been altered to provide the plant with a means to ward off various pests. Although this approach is often quite successful, it is not without its critics, especially in Europe, where genetically altered plants face serious restrictions. In contrast, harpin has no effect on the plant's DNA: it simply activates defenses that are innate to the plant.

Traditional pesticides are generally made by chemists employing lengthy chemical syntheses, which invariably create large quantities of waste, which are often toxic. In addition, the compounds (chemical feedstocks) from which the pesticides are produced are derived from petroleum. Approximately 2.7% of all petroleum is used to produce chemical feedstocks, and thus the production of these compounds is in part responsible for the depletion of this nonrenewable resource. In contrast, harpin is made from a genetically altered benign laboratory strain of the *Escherichia coli* bacteria through a fermentation process. After the fermentation is complete, the bacteria are destroyed and the harpin protein is extracted. Most of the wastes are biodegradable. Thus the production of harpin produces only nontoxic biodegradable wastes and does not require petroleum.

Harpin has very low toxicity. In addition, it is applied at 0.002−0.06 kg/acre, which represents an approximately 70% reduction in quantity when compared to conventional pesticides. Harpin is rapidly decomposed by UV light and microorganisms, which is in part responsible for its lack of contamination and buildup in soil, water, and organisms and the fact that it leaves no residue in foods.

An added benefit of harpin is that it also acts as a plant growth stimulant. Harpin is thought to aid in photosynthesis and nutrient uptake, resulting in increased biomass, early flowering, and enhanced fruit yields. Harpin is sold as a 3% solution in a product called *Messenger*.

Review Questions

1. What is a *Dobson Unit*? How is it used in relation to atmospheric ozone levels?

2. If the overhead ozone concentration at a point above the Earth's surface is 250 DU, what is the equivalent thickness in millimeters of pure ozone at 1.0 atm pressure?

3. Describe the process by which chlorine becomes activated in the Antarctic ozone-hole phenomenon.

4. What are the steps in Mechanism II by which atomic chlorine destroys ozone in the spring over Antarctica?

5. Describe the reasons why the Antarctic ozone hole closes in late spring/early summer.

6. Explain why full-scale ozone holes have not yet been observed over the Arctic.

7. What are two effects to human health that scientists believe will result from ozone depletion?

8. Define what is meant by a tropospheric *sink*.

9. Explain what *CFCs* were and some of their uses. Did they have a tropospheric sink? Why did their emissions in air lead to an increase in stratospheric chlorine?

10. Explain what *HCFCs* are and state what sort of reaction provides a tropospheric sink for them. Is their destruction in the troposphere 100% complete? Why are HCFCs not considered to be suitable long-term replacements for CFCs?

11. What types of chemicals are proposed as long-term replacements for CFCs?

12. Chemically, what are *halons*? What was their main use?

13. What gases are being phased out according to the Montreal Protocol agreements?

Green Chemistry Questions

1. The development of carbon dioxide as a blowing agent for foam polystyrene won a Presidential Green Chemistry Challenge Award.

(a) Which of the three focus areas (see page xxviii) for these awards does this award best fit into?

(b) List two of the twelve principles of green chemistry (see pages xxiii–xxiv) that are addressed by the green chemistry of the carbon dioxide process.

2. What environmental advantages does the use of carbon dioxide as a blowing agent have over the use of CFCs and hydrocarbons?

3. Does the carbon dioxide that is used as a blowing agent contribute to global warming?

4. The development of harpin won a Presidential Green Chemistry Challenge Award.

(a) Which of the three focus areas (see page xxviii) for these awards does this award best fit into?

(b) List four of the twelve principles of green chemistry (see pages xxiii–xxiv) that are addressed by the green chemistry of the use of harpin.

5. Why is there little concern that pests will develop immunity to harpin?

6. Why is harpin not expected to accumulate in the environment?

Additional Problems

1. (a) Some authors use milliatmospheres centimeter (matm cm) rather than the equivalent Dobson Unit to express the unit for the amount of overhead ozone; 1 matm cm = 1 DU. Prove that the number of moles of overhead ozone over a unit area on the Earth's surface is proportional to the height of the layer, as specified in the definition of Dobson Units, and that 1 DU is equal to 1 matm cm.

(b) Calculate the total mass of ozone that is present in the atmosphere if the average overhead amount is 350 Dobson Units, and given that the radius of the Earth is about 6400 km. [Hints: The volume of a sphere, which you can approximate the Earth to be, is $4\pi r^3/3$. You may assume that ozone behaves as an ideal gas.]

2. The chemical formula for any CFC, HCFC, or HFC can be obtained by adding 90 to its code number. The three numerals in the result represent the number of C, H, and F atoms, respectively. The number of Cl atoms can then be determined using the condition that the number of H, F, and Cl atoms must add up to $2n + 2$, where n is the number of C atoms. From this information, deduce the formulas for compounds with the following codes:

(a) 12   **(b)** 113   **(c)** 123   **(d)** 124

3. Using the information discussed in Problem 2 above, deduce the code numbers for each of the following compounds:

(a) CH_3CCl_3 **(b)** CCl_4 **(c)** CH_3CFCl_2

4. Using the information in Problem 2, show that 134 is the appropriate label for CH_2FCF_3. Why is an a or b designation also required to uniquely characterize the latter compound? What would be the code numbers for the HCFs in R-410a, namely CH_2F_2 and CHF_2CF_3? Does the number 410 correspond to the code number for either of these compounds?

5. The chlorine dimer mechanism is not implicated in significant ozone destruction in the lower stratosphere at mid-latitudes even when the particle concentration becomes enhanced by volcanoes. Deduce two reasons why this mechanism is not important under these conditions.

6. When Mechanism II for ozone destruction operates with X = Cl and X′ = Br, the radicals ClO and BrO react together to reform atomic chlorine and bromine (see Problem 2-1). A fraction of the latter process proceeds by the intermediate formation of BrCl, which undergoes photolysis in daylight. At night, however, all the bromine eventually ends up as BrCl, which does not decompose and restart the mechanism until dawn. Deduce why all the bromine exists as BrCl at night, even though only a fraction of the ClO with BrO collisions yields this product.

7. Explain what changes are observed in the UV-B intensity at ground level during ozone hole episodes.

8. What would be the advantages of using hydrocarbons rather than HFCs or HCFCs as aerosol propellant to replace CFCs? What is their major disadvantage? What type of agent should be added to aerosol cans containing hydrocarbon propellants to overcome this disadvantage and make them safer?

9. Consider the following set of compounds: $CFCl_3$, $CHFCl_2$, CF_3Cl, and CHF_3. Assuming that equal numbers of moles of each were released into the air at ground level, rank these four compounds in terms of their potential to catalytically destroy ozone in the stratosphere. Explain your ranking.

The Chemistry of Ground-Level Air Pollution

In this chapter, the following introductory chemistry topics are used:

- ➲ Ideal gas law
- ➲ Equilibrium concept, including redox reactions and their balancing
- ➲ Acid–base theory, including pH and weak acid calculations

Background from previous chapters used in this chapter:

- ➲ Excited states
- ➲ Photon energies, UV types (A, B, C)
- ➲ Gas-phase catalysis
- ➲ Sink concept
- ➲ Temperature inversions

this most excellent canopy, the air,
look you, this excellent o'erhanging
firmament, this magestic roof fretted
with golden fire, why, it appears
no other thing to me than a foul
and pestilent congregation of vapours

Wm. Shakespeare, Hamlet, Act II, Scene 2

Introduction

As one travels from city to city in various parts of the world, the most obvious environmental difference among them is often the extent of their air pollution. Some cities seem pristine, while others are blanketed by a haze that restricts visibility and induces coughing and tearing. As we shall see in this chapter and the next, the chemical nature of the air pollution, the origin of its reactants and the processes they undergo, and its effect on human health all vary considerably from place to place.

Although people often think industry must be the source of most air pollution—and the generation of electric power by coal can produce significant amounts of emissions—in modern times it is often exhaust from vehicles that is the main culprit. The most manifest sign of vehicular air pollution is the black smoke emanating from the tailpipes of diesel trucks and buses. This sight is more common now in developing countries, since such pollution has largely been controlled in developed nations. Indeed, over the past decades, as urban populations and vehicle densities have grown rapidly in developing countries, air pollution there has dramatically worsened. In general, serious regulation of air pollution is not attempted until a country has achieved a reasonably high degree of affluence.

One of the most important features of the Earth's atmosphere is that it is an oxidizing environment, a phenomenon due to the large concentration of diatomic oxygen, O_2, which it contains. Almost all the gases that are released into the air, whether they are "natural" substances or "pollutants," are eventually completely oxidized in the atmosphere, and the end products subsequently deposited on the Earth's surface. The oxidation reactions are vital to the cleansing of the air.

In this chapter, the chemistry underlying the pollution of tropospheric air is examined. As background, we begin the chapter by discussing the concentration units by which gases in the lower atmosphere are reported, and the constitution and chemical reactivity of clean air. The effects of polluted air upon the environment and upon human health are discussed in Chapter 4.

3.1 Concentration Units for Atmospheric Pollutants

Air contains tiny, invisible suspended particles as well as gases. **The particles found suspended in air are usually heterogeneous mixtures,** and consequently no molar mass can be associated with them. Concentration scales for such solids do *not* report the number of atoms or molecules, but rather the *mass* of such particles found in a particular volume of air. **The usual concentration unit for particles in air is micrograms (of particles) per cubic meter (of air),** μg m^{-3}. This absolute concentration scale can also be used for gases.

There is no consensus regarding the appropriate units by which to express concentrations of gases in air. In Chapter 1, ratios involving numbers of molecules—the "parts per" system—were emphasized as a measure. Other measures are often also encountered and will be used in this chapter:

Molecules of a gas per cubic centimeter of air, molecules cm^{-3}

Micrograms of a substance per cubic meter of air, μg m^{-3}

Moles of a gas per liter of air, moles L^{-1}

The absolute concentration scale *moles per liter*, familiar to all chemists from its use for liquid solutions, is itself rarely used for gases because they are so dilute.

Given the lack of a consensus on a single appropriate scale, it is important to be able to convert gas concentrations from one scale to another. This form of manipulation is discussed in Box 3-1. Note that gas pressures cited in units of atmospheres are synonymous with concentrations on the "parts per"

BOX 3-1   The Interconversion of Gas Concentrations

The number of moles of a substance is proportional to the number of the molecules of it (Avogadro's number, 6.02×10^{23}, is the proportionality constant), and the partial pressure of a gas is proportional to the number of moles of it. Thus a concentration, for example, of 2 ppm for any pollutant gas present in 1 atm air means

2 molecules of the pollutant in 1 million molecules of air

2 moles of the pollutant in 1 million moles of air

2×10^{-6} atm partial pressure of pollutants per 1 atm total air pressure

2 L of pollutant in 1 million liters of air (when the partial pressures and temperatures of pollutant and air have been adjusted to be equal)

Let us convert a concentration of 2 ppm to its value in molecules (of pollutant) per cubic centimeter (cm³) of air for conditions of 1 atm total air pressure and 25°C. Since the value of the numerator, 2 molecules, in the new concentration scale is the same as in the original, all we need to do is establish the volume, in cubic centimeters, that 1 million molecules of air occupy. This volume is easy to evaluate using the ideal gas law ($PV = nRT$), since we know that

$$P = 1.0 \text{ atm}$$

$$T = 25 + 273 = 298 \text{ K}$$

$$n = (10^6 \text{ molecules})/$$

$$(6.02 \times 10^{23} \text{ molecules mol}^{-1})$$

$$= 1.66 \times 10^{-18} \text{ mol}$$

and the gas constant $R = 0.082$ L atm mol⁻¹ K⁻¹.

Now $PV = nRT$, so

$$V = nRT/P$$

$$= 1.66 \times 10^{-18} \text{ mol}$$

$$\times 0.082 \text{ L atm mol}^{-1} \text{ K}^{-1}$$

$$\times 298 \text{ K atm}^{-1}$$

$$= 4.06 \times 10^{-17} \text{ L}$$

Since 1 L = 1000 cm³, then $V = 4.06 \times 10^{-14}$ cm³, so it follows that the concentration in the new units is 2.0 molecules/(4.06×10^{-14} cm³), or 4.9×10^{13} molecules cm⁻³.

In general, the most straightforward strategy to use to change the value of a concentration a/b from one scale to its value p/q on another is to independently convert the units of the numerator a to the units of the numerator p (both of which involve only the pollutant) and then convert the denominator b to its new value q (both of which involve the total air sample).

To convert a value in molecules cm⁻³ or ppm to mol L⁻¹, we must change the molecules of pollutant to the number of moles of pollutant; for a pollutant concentration, again of 2 ppm, we can write

moles of pollutant

$$= (2 \text{ molecules} \times 1 \text{ mol})/$$

$$(6.02 \times 10^{23} \text{ molecules})$$

$$= 3.3 \times 10^{-24} \text{ mol}$$

Thus the molarity is (3.3×10^{-24} mol)/(4.06×10^{-17} L), or 8.2×10^{-8} M.

An alternative way to approach these conversions is to use the definition that 2 ppm means 2 L of pollutant per 1 million liters of air and to find the number of moles and

(continued on p. 72)

molecules of pollutant contained in a volume of 2 L at the stated pressure and temperature.

A unit often used to express concentrations in polluted air is micrograms per cubic meter, i.e., $\mu g\ m^{-3}$. If the pollutant is a pure substance, we can interconvert such values into the molarity and the "parts per" scales, provided that the pollutant's molar mass is known.

Consider as an example the conversion of $320\ \mu g\ m^{-3}$ to the ppb scale if the pollutant is SO_2, the total air pressure is 1.0 atm, and the temperature is 27°C. Initially the concentration is

$$\frac{320\ \mu g\ of\ SO_2}{1\ m^3\ of\ air}$$

First we convert the numerator from grams of SO_2 to moles, since from there we can obtain the number of molecules of SO_2:

$$320 \times 10^{-6}\ g\ SO_2 \times \frac{1\ mol\ SO_2}{64.1\ g\ SO_2}$$

$$\times \frac{6.02 \times 10^{23}\ molecules\ of\ SO_2}{1\ mol\ SO_2}$$

$$= 3.01 \times 10^{18}\ molecules\ of\ SO_2$$

Then, using the ideal gas law, we can change the volume of air to moles and then molecules, using $1\ L = 1\ dm^3 = (0.1\ m)^3$:

$$n = PV/RT = 1.0\ atm \times 1.0\ m^3$$

$$\times \frac{\dfrac{1\ L}{(0.1\ m)^3}}{\dfrac{0.082\ L\ atm}{mol\ K \times 300\ K}}$$

$$= 40.7\ mol$$

Now $40.7\ mol \times 6.02 \times 10^{23}\ molecules/mol = 2.45 \times 10^{25}$ molecules, or 2.45×10^{16} billion molecules of air.

Thus the SO_2 concentrations is

$$\frac{3.01 \times 10^{18}\ molecules\ of\ SO_2}{2.45 \times 10^{16}\ billion\ molecules\ of\ air} = 123\ ppb$$

Note that the conversion of moles to molecules was not strictly necessary, as Avogadro's number cancels from numerator and denominator. As stated previously, ppb refers to the ratio of the number of moles as well as to the ratio of the number of molecules.

It is vital in all interconversions to distinguish between quantities associated with the pollutant and those of air.

PROBLEM 1

Convert a concentration of 32 ppb for any pollutant to its value on

(a) the ppm scale,

(b) the molecules cm^{-3} scale, and

(c) the molarity scale.

Assume 25°C and a total pressure of 1.0 atm.

PROBLEM 2

Convert a concentration of 6.0×10^{14} molecules cm^{-3} to the ppm scale and to the moles per liter (molarity) scale. Assume 25°C and 1.0 atm total air pressure.

PROBLEM 3

Convert a concentration of 40 ppb of ozone, O_3, into

(a) the number of molecules cm^{-3}, and

(b) micrograms m^{-3}.

Assume the air mass temperature is 27°C and its total pressure is 0.95 atm.

PROBLEM 4

The average outdoor concentration of carbon monoxide, CO, is about 1000 $\mu g\ m^{-3}$. What is this concentration expressed on the ppm scale? On the molecules cm^{-3} scale? Assume that the outdoor temperature is 17°C and that the total air pressure is 1.04 atm.

scales after correction for the magnitude of the denominator (of 1.0 atm usually). So, for example, a partial pressure of 0.002 atm in air is equivalent to 2000 ppm, since $0.002\ atm \times 10^6 = 2000$.

3.2 The Chemical Fate of Trace Gases in Air

From various natural sources—including fires, lightning, anaerobic biological decay, and emissions from volcanoes—our atmosphere regularly receives inputs of many gases including the partially oxidized compounds **carbon monoxide, CO, nitric oxide, NO,** and **sulfur dioxide, SO_2,** and several simple compounds of hydrogen combined with another element in a highly reduced form, such as **ammonia, NH_3, hydrogen sulfide, H_2S,** and **methane, CH_4.**

The concentrations of these gases do not build up in clean air because there are not only sources for them but also sinks, which result in their continual destruction. For the gases mentioned above, the destruction processes are oxidation reactions that occur in air. However, none of the gases reacts *directly* with diatomic oxygen molecules because the activation energy for such processes is too high. Rather, **their reactions begin when they are attacked by the hydroxyl free radical, OH,** even though the concentration of this species in air is exceedingly small, a few million molecules per cubic centimeter on average (see Problem 3-1). Its concentration has remained constant in air over the last few decades at least.

Recall that a free radical has one electron in the outermost shell of one of its atoms that neither participates in a bond to another atom nor is part of a nonbonding electron pair.

The presence of an unpaired electron makes most free radicals, including OH, very reactive. The Lewis structure for the hydroxyl free radical is

$$: \overset{\displaystyle .}{\underset{\displaystyle ..}{O}} — H$$

In clean tropospheric air, as in the stratosphere, the **hydroxyl radical is produced when a small fraction of the excited oxygen atoms resulting from the photochemical decomposition of trace amounts of atmospheric ozone, O_3, react with gaseous water to abstract one hydrogen atom from each H_2O molecule:**

$$O_3 \xrightarrow{\text{UV-B}} O_2 + O^*$$

$$O^* + H_2O \longrightarrow 2\ OH$$

See Additional Problem 2 for the lifetime calculation.

The average tropospheric lifetime of a hydroxyl radical is only about one second, since it reacts quickly with one or other of many atmospheric gases.

Because the lifetime of hydroxyl radicals is short and sunlight is required to form more of them, the OH concentration drops quickly at nightfall.

PROBLEM 3-1

In one study, the concentration of OH in air at the time was found to be 8.7×10^6 molecules per cubic centimeter. Calculate its molar concentration, and its concentration in parts per trillion, assuming that the total air pressure is 1.0 atm and the temperature is 15°C. ●

In its reaction with otherwise-stable gases whose molecules contain multiple bonds, OH adds itself to them, thereby forming a larger free radical. For example, hydroxyl adds to carbon monoxide molecules, forming the transient free radical HOCO:

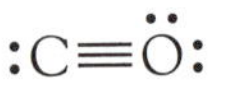

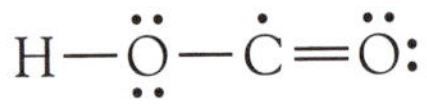

$$OH + CO \longrightarrow HOCO$$

Most collisions of OH and CO molecules are ineffective in promoting a reaction. Consequently, the average lifetime of a carbon monoxide molecule in air is a month or two.

Molecular oxygen reacts quickly with transient free radicals such as HOCO once they are formed, thereby involving itself in the oxidation process. In the present case, O_2 *abstracts* a hydrogen atom from the free radical, thereby forming the **hydroperoxy free radical,** HOO, and the fully oxidized product CO_2, **carbon dioxide:**

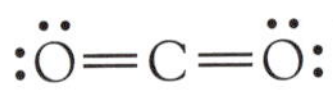

$$O_2 + HOCO \longrightarrow HOO + CO_2$$

The hydroperoxy radical produced in the atmospheric oxidation of carbon monoxide, and indeed of most molecules, is in turn converted back to the hydroxyl radical by its oxidation of nitric oxide, NO, which is present in adequate concentration for this purpose in all but the very cleanest air:

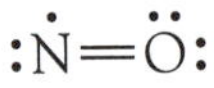

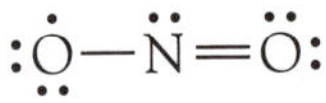

$$HOO + NO \longrightarrow OH + NO_2$$

The general cycle of OH/HOO formation and consumption in the atmospheric oxidation of various molecules is summarized by the diagram below; in the case of some organic molecules, sunlight is required for intermediate steps in the mechanism:

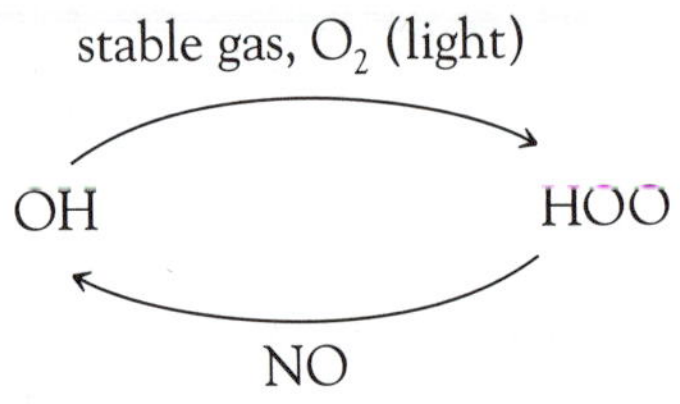

stable gas, O_2 (light)

OH   HOO

NO

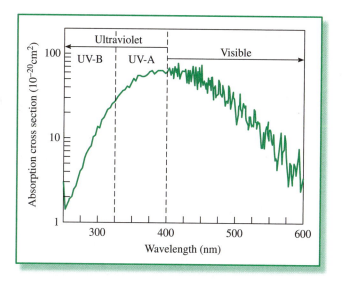

FIGURE 3-1 Absorption spectrum for gaseous NO_2. [Source: J. H. Seinfeld and S. N. Pandis, "*Atmospheric Chemistry and Physics*," John Wiley & Sons, New York, 1998.]

Although suspected for decades of playing a pivotal role in air chemistry, the presence of OH in the troposphere was confirmed only relatively recently since its concentration is so very small. The great importance of the hydroxyl radical to tropospheric chemistry arises because it, not O_2, initiates the oxidation of almost *all* reduced gases. Without OH and its related reactive species HOO, most naturally occurring gases, and pollutant gases such as the unburned hydrocarbons emitted from vehicles would not be efficiently removed from the troposphere.

Indeed, OH has been called the "tropospheric vacuum cleaner" or "detergent."

The reactions that OH initiates correspond to a flameless, ambient-temperature "burning" of the reduced gases of the lower atmosphere. If these gases were to accumulate, the atmospheric composition would be quite different, as would the forms of life that would be viable on Earth. Interestingly, hydroxyl is unreactive toward molecular oxygen—in contrast to the behavior of O_2 with many free radicals—and to molecular nitrogen, thus it survives long enough to react with so many other species.

Within a few minutes, most of the **nitrogen dioxide,** NO_2, produced in the OH/HOO cycle during the daytime absorbs UV-A from sunlight (see its spectrum in Figure 3-1) and photochemically decomposes to nitric oxide and atomic oxygen (this also occurs in the stratosphere, as mentioned in Chapter 1):

Light with wavelength shorter than 394 nm has sufficient energy to decompose NO_2 by this reaction.

$$NO_2 + UV\text{-}A \longrightarrow NO + O$$

From the viewpoint of the nitrogen oxides, the cycle of NO oxidation by HOO and the reduction of NO_2 by sunlight are summarized on the next page:

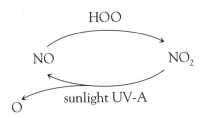

Some NO is oxidized instead by organic peroxy free radicals, ROO, as discussed in Chapter 17.

The oxygen atoms produced in this cycle quickly react with molecular oxygen to form ozone. As is the case in the stratosphere, **this reaction is the only source of ozone in the troposphere:**

$$O + O_2 \longrightarrow O_3$$

NO_2^* itself may react with water molecules to produce OH radicals directly, rather than exclusively by prior ozone production.

In summary, stable gases in the air that are not already fully oxidized react directly with OH, rather than O_2, even though it is present in tiny concentration. The OH is originally produced from reaction of O* from ozone photodecomposition, the ozone having been created from the oxygen atom produced by photochemical decomposition of NO_2. After it is used for reaction initiation, the OH is transformed into HOO, which is recycled back to OH by reaction with NO.

Review Questions 1–4 are based on the material above.

Urban Ozone: The Photochemical Smog Process

3.3 The Origin and Occurrence of Smog

Many urban centers in the world undergo episodes of air pollution during which relatively high levels of ground-level ozone—an undesirable constituent of air if present in appreciable concentrations at low altitudes in the air that we breathe—are produced as a result of the light-induced chemical reaction of pollutants. This phenomenon is called **photochemical smog,** and is sometimes characterized as "an ozone layer in the wrong place," to contrast it with the beneficial stratospheric ozone discussed in Chapter 1. The word *smog* is a combination of *smoke* and *fog*. The process of smog formation involves hundreds of different reactions, involving dozens of chemicals, occurring simultaneously. Indeed, urban atmospheres have been referred to as giant chemical reactors. The most important reactions that occur in such air masses will be discussed in greater detail in Chapter 17. In the following material, we investigate the nature and origin of the pollutants—especially nitrogen oxides—and see how they combine to produce photochemical smog.

The chief original reactants in an episode of photochemical smog are molecules of nitric oxide, NO, and of unburned hydrocarbons and partially oxidized hydrocarbons that are emitted into the air as pollutants from internal combustion engines; nitric oxide is also released from electric power

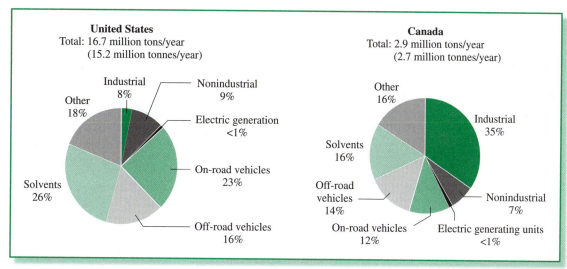

United States
Total: 16.7 million tons/year
(15.2 million tonnes/year)

Canada
Total: 2.9 million tons/year
(2.7 million tonnes/year)

FIGURE 3-2 VOC emission sources in North America in 2006. [Source: International Joint Commission, *Canada–United States Air Quality Agreement: 2008 Progress Report*, Washington, D.C. and Ottawa, Ontario, 2008.]

plants. The concentrations of these chemicals are orders of magnitude greater than are found in clean air.

Collectively, the **substances, including hydrocarbons and their derivatives, that readily vaporize into the air are called volatile organic compounds,** or VOCs, many of which react in photochemical smog.

The emissions, by sector, of VOCs in the United States and Canada are illustrated by the pie charts in Figure 3-2. Emissions from on-road transportation especially have fallen drastically over the last two decades.

Gaseous hydrocarbons and partially oxidized hydrocarbons are VOCs that are present in urban air as a result of the evaporation of solvents, liquid fuels, and other organic compounds. For example, vapor is released into the air when a gasoline tank is filled unless the hose's nozzle is specially designed to minimize this loss. Evaporated, unburned gasoline is also emitted from the tailpipe of a vehicle before its catalytic converter has been warmed sufficiently to operate. Two-cycle engines such as those in outboard motor boats are particularly notorious for emitting significant proportions of their gasoline unburned into the air. Personal watercraft manufactured in the 1990s, before pollution controls came into effect, emitted more smog-producing emissions in a day's operation than an automobile of the same era driven for several years! In many regions, new lawn mowers are required to be outfitted with a catalytic converter, though this issue is controversial since some mower manufacturers claim that a hot converter could pose a fire hazard to the engine. Some outboard motors and domestic firepit fireplaces are also now equipped with catalytic converters.

Formally, VOCs are defined as organic compounds having boiling points that lie between 50°C and 260°C.

Another vital ingredient in producing photochemical smog is **sunshine, which serves indirectly to increase the concentration of free radicals that participate in the chemical processes of smog formation. Substances that are emitted directly into air are called primary pollutants;** in the smog reaction, these are NO and VOCs, most of which are relatively innocuous with regard to human health. **Substances into which the primary ones are transformed are called secondary pollutants;** in smog they include ozone, **nitric acid,** HNO_3, and partially oxidized (and in some cases nitrated) organic compounds, which are much more toxic than the reactants. An approximate overall chemical equation for the smog reaction is

Overall reaction:

$$VOCs + NO + O_2 + sunlight \longrightarrow \longrightarrow \text{mixture of } O_3, HNO_3, \text{organics}$$
Primary pollutants   Secondary pollutants

Other than those that absorb sunlight and subsequently decompose, most atmospheric molecules that are transformed in air begin by reacting with the hydroxyl free radical. The most reactive VOCs in urban air are *alkene* hydrocarbons, since they contain a carbon–carbon double bond, C=C; and *aldehydes*, which contain a C=O bond. These compounds are particularly reactive since their reactions with OH, in which the hydroxyl radical adds to a carbon atom participating in the double bond, analogous with the case of carbon monoxide discussed previously, are very fast.

Hydroxyl initiates the reaction in the atmosphere of a hydrogen-containing molecule that does not contain a multiple bond by abstracting a hydrogen atom from it, thereby forming a water molecule and leaving a free radical fragment of the hydride. For example, the first reaction of methane molecules in air is their loss of hydrogen to hydroxyl:

$$CH_4 + OH \longrightarrow CH_3 + H_2O$$

The fragment CH_3 contains an odd number of electrons, and is the **methyl free radical.** Such free radicals *do* react with diatomic oxygen in a sequence of reactions that involves their oxidation to carbon monoxide (and ultimately to carbon dioxide):

$$CH_4 \xrightarrow{OH} CH_3 \xrightarrow{O_2} \longrightarrow \text{other intermediates} \xrightarrow{O_2} CO$$

As a result of reactions in the sequence in which peroxy free radicals oxidize NO to NO_2, the hydroxyl free radical is eventually regenerated in the same manner as it is in the CO oxidation. Thus OH acts as a *catalyst* in the atmospheric oxidation of most species and is effective in air even in tiny concentration.

In smog episodes, aldehydes are among the intermediates, and their photochemical decomposition by UV-A produces additional free radicals. Since on average more than 1.0 molecule of HOO is produced in each cycle as a result, more than 1.0 OH molecule is created by their reaction with nitric oxide. **Over time, the total concentration of OH and HOO free radicals**

Recall that a catalyst is defined as a substance that speeds up a chemical reaction, but is regenerated during the process.

builds up during a smog reaction, thereby accelerating it. Notice that the operation of the OH/HOO cycle and the smog it produces depends upon the *simultaneous* presence of NO and of reactive VOCs; without one or the other of these key reactants the cycle could not proceed nearly as quickly. **The NO and VOCs here are said to act in synergism; their overall effect is much greater than would be the sum of either acting in isolation.**

When the reaction sequence of oxidizing CH_4 to CO is combined with that of CO oxidation to CO_2, the overall reaction is seen to be the complete oxidation of methane to carbon dioxide and water, the same reaction as occurs when natural gas is burned in air:

$$CH_4 + 2\,O_2 \longrightarrow CO_2 + 2\,H_2O$$

Details of the complete sequences by which such free-radical reactions oxidize the gases emitted into clean and polluted air are explained in Chapter 17.

3.4 Ground-Level Ozone in Smog

Photochemical smog is a widespread phenomenon in the modern world. In order for a city to generate photochemical smog, several conditions must be fulfilled.

- First, there must be substantial vehicular traffic in order to emit sufficient NO, reactive hydrocarbons, and other VOCs into the air.

- Second, there must be warmth and ample sunlight in order for the crucial reactions, some of them photochemical, to proceed at a rapid rate.

- Finally, there must be relatively little movement of the air mass so that the reactants are not quickly diluted or swept away.

For reasons of geography (e.g., the presence of mountains) and dense population, cities such as Los Angeles, Denver, Mexico City, Tokyo, Athens, Sao Paulo, and Rome all fit the bill splendidly and consequently are subject to frequent smog episodes. Indeed, the photochemical smog phenomenon was first observed in Los Angeles in the 1940s and has generally been associated with that city ever since, although pollution controls have partially alleviated the city's smog problem in recent decades.

3.5 An Episode of Photochemical Smog

It is instructive to follow a heavy episode of photochemical smog, as occurred in Los Angeles in the 1960s, through a day in order to understand the sequence of processes that are important at different stages. Refer to Figure 3-3, in which the concentrations of various chemicals are plotted against time of day, while reading the following.

- As discussed above, it is predominantly NO rather than NO_2 that is released from vehicles and power plants into the air. One can see the

FIGURE 3-3 Time-of-day (diurnal) variation in the concentration of gases during days of marked eye irritation in Los Angeles in the 1960s. [Source: Redrawn from D. J. Speeding, *Air Pollution* (Oxford: Oxford University Press, 1974).]

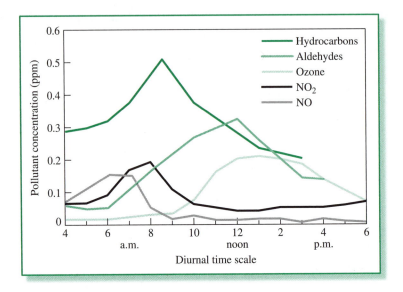

concentration of NO rising from emissions from early-morning vehicle traffic in Figure 3-3. Over a period of hours, the NO is gradually oxidized to NO_2 by the buildup of HOO, as discussed above. One might have expected the ozone concentration to have increased quickly as NO_2 underwent photochemical decomposition, but in the morning the ozone is quickly destroyed by its reaction with NO to form NO_2 and oxygen:

$$O_3 + NO \longrightarrow O_2 + NO_2$$

The reactive hydrocarbons in emissions are beginning their oxidation, as evidenced from the rapid buildup in the concentration of one of their intermediate oxidation products, the aldehydes (medium-green curve).

PROBLEM 3-2

By adding up the three reactions, show that the net result of the photochemical decomposition of NO_2, the formation of ozone from atomic oxygen, and the above reaction constitute no overall reaction, i.e., a null process.

• By mid-morning, lesser amounts of NO enter the air since rush-hour traffic is over and most of what has been emitted has been converted to NO_2. At this point, the level of OH free radicals has increased sufficiently that the concentration of hydrocarbons (dark green curve) begins to decline. At the same time, the concentration of ozone begins to build rapidly since it is no longer quickly destroyed by reaction with NO since the latter is quickly re-oxidized by radicals to NO_2 instead. At about noon, the production of aldehydes is overtaken by the photochemical and free-radical decomposition of existing stock, leading to its subsequent decline.

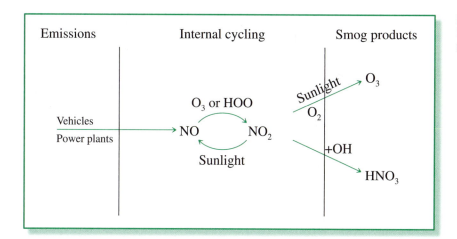

FIGURE 3-4 Nitrogen-based components of photochemical smog.

- The concentration of NO_2 peaks in the morning but remains at a lower level throughout the rest of the day due to continuing production from NO emissions. There exists a sink for NO_2 when the concentration of free radicals is high: it combines with hydroxyl radical to produce nitric acid, one of the end-products of smog:

$$NO_2 + OH \longrightarrow HNO_3$$

Some NO_2 also reacts with hydrocarbon chains to nitrate them. Overall, the smog process from the viewpoint of the nitrogen emissions as NO_2 chains is represented in Figure 3-4, showing NO continuously flowing into the atmospheric reactor and ozone and nitric acid being produced following a period of cycling of the nitrogen oxides.

ACTIVITY

Search the web for information about ground-level ozone levels in whichever large city you inhabit most frequently. If historical records dating back a few years are available, create a file of days with the highest levels each year. Is the number of high-ozone days about the same each year? Do several high-ozone days occur in sequence or are they isolated incidents? Beside each high-ozone day, find and list the maximum afternoon temperature from weather records. Is there a correlation of maximum ozone and maximum temperature? Why might you expect such a correlation to exist? If ozone and/or nitrogen dioxide levels by the hour are available for your city, plot them against time for a typical high-ozone day. Do they follow the trends in Figure 3-3?

3.6 Nitrogen Oxide Production During Fuel Combustion

Given the roles of NO and NO_2 in photochemical smog where their concentrations are many times those of background, it is important to understand their origins as atmospheric pollutants. Nitrogen oxide gases are produced by two different reactions whenever a fuel is burned in air with a hot flame.

- Some nitric oxide is produced from the oxidation of nitrogen atoms contained in molecules of the fuel itself; it is called **fuel NO.** About 30–60% of a fuel's nitrogen is converted to NO during combustion. However, most fuels do not contain much nitrogen, so this process accounts for only a small fraction of NO emissions.

- Nitric oxide is produced by the oxidation at high combustion temperatures of atmospheric nitrogen, and it is called **thermal NO.** At high flame temperatures, some of the nitrogen and oxygen gases in the air passing through the flame combine to form NO:

$$N_2 + O_2 \xrightarrow{\text{hot flame}} 2\,NO$$

LeChatelier's principle; see also Additional Problem 4.

The higher the flame temperature, the more NO is produced. Since this reaction is very endothermic, its equilibrium constant is very small at normal temperatures but increases rapidly as the temperature rises.

One might expect that the relatively high concentrations of NO that are produced under combustion conditions would revert back to molecular nitrogen and oxygen as the exhaust gases cool, since the equilibrium constant for the above reaction is much smaller at lower temperatures. However, the activation energy for the reverse reaction is also quite high, so the process cannot occur to an appreciable extent except at high temperatures. The flow of gases through the combustion zone is so rapid that the NO does not have sufficient time to react at the high temperatures its reaction requires. **Thus the relatively high concentrations of nitric oxide produced during combustion are maintained in the cooled exhaust gases. Equilibrium cannot be quickly re-established and the nitrogen is "frozen" as NO.**

Because the reaction between N_2 and O_2 has a high activation energy, it is negligibly slow *except* at the very high temperatures such as occur in the modern combustion engines of vehicles—particularly when they are traveling at high speeds—and in power plants. Very little NO is produced by the burning of wood and other natural materials since the flame temperatures involved in such combustion processes are relatively low.

Two distinct mechanisms are involved in the initiation of the reaction of molecular nitrogen and oxygen to produce thermal nitric oxide; in one it is atomic oxygen that attacks intact N_2 molecules, whereas in the other it is free radicals, such as CH, that are derived from the decomposition of the fuel. The initial reaction steps of the first mechanism are

$$O_2 \rightleftharpoons 2\,O$$

$$O + N_2 \longrightarrow NO + N$$

Recall from introductory chemistry that the rate of a reaction step is proportional to the product of the concentrations of its reactants.

The rate of the second, slower step is proportional to [O] [N_2]. However, since, from the equilibrium in the first step, [O] is proportional to the square root of $[O_2]$, it follows that the rate of NO formation will be proportional to $[N_2] [O_2]^{1/2}$. Consequently, this process is relatively slow under oxygen-poor conditions.

The nitric oxide released into air is gradually oxidized to nitrogen dioxide over a period of minutes to hours, the rate depending upon the concentration of the pollutant gases present. **Collectively, NO and NO$_2$ in air is referred to as NO$_X$,** pronounced "nox." The yellow-brown color in the atmosphere of a smog-ridden city is due in part to the nitrogen dioxide present, since this gas absorbs visible light, especially near 400 nm (see its spectrum in Figure 3-1), removing sunlight's purple component while allowing most yellow light to be transmitted. The small levels of NO$_X$ in clean air result in part from the operation of the above reactions in the very energetic environment of lightning flashes and in part from the release of NO$_X$ and of ammonia, NH$_3$, from biological sources. NO$_X$ is also emitted from coniferous trees when sunlight shines on them and when the ambient concentrations of these gases are low.

The sources of anthropogenic NO$_X$ emissions in North America are shown by sector in Figure 3-5. The quantities from on-road transportation, and from electric power generation in the United States, fell substantially over the last dozen years.

Although our analysis above has identified ozone as the main product of smog, the situation is actually more complicated, as a detailed study in Chapter 16 indicates. For example, nitrogen dioxide reacts with organic free radicals to produce organic nitrates. Many of the products form particles or are incorporated into them, as discussed in Section 3.24.

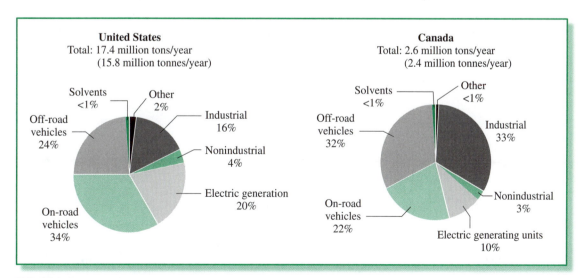

FIGURE 3-5 Anthropogenic NO$_X$ sources in North America, by sector in 2006.
[Source: International Joint Commission, *Canada–United States Air Quality Agreement: 2008 Progress Report,* Washington, D.C. and Ottawa, Ontario, 2008.]

3.7 Governmental Goals for Reducing Ozone Concentrations

The ozone level in clean air amounts to only about 30 ppb. Many countries individually, as well as the *World Health Organization*, WHO, have established goals for **maximum allowable ozone concentrations in air averaged over an 8-hr period.** The United States adopted an 8-hr ozone standard of 80 ppb in 1997, but this was revised downward to 75 ppb in 2008, and may be dropped to the 60–70-ppb level eventually. The standard in Canada is 65 ppb. The European Union standard of 120 μg m^{-3}, when converted to ppb units, is close to the U.S. limit (Problem 3-3).

PROBLEM 3-3

Convert into ppb units the EU ozone standard of 120 μg m^{-3} and the WHO standard of 100 μg m^{-3}, assuming summertime air temperatures of 27°C. ●

PROBLEM 3-4

Calculate the air temperature at which the EU ozone level of 120 μg m^{-3} is equivalent to the current U.S. standard of 75 ppb. Assume a total air pressure of 1.0 atm. ●

3.8 Photochemical Smog Around the World

Many major cities in North America, Europe, and Japan exceed ozone levels of 120 ppb typically for 5 to 10 days each summer. The levels of ozone in Los Angeles air used to reach 680 ppb. Figure 3-6 illustrates the decline in

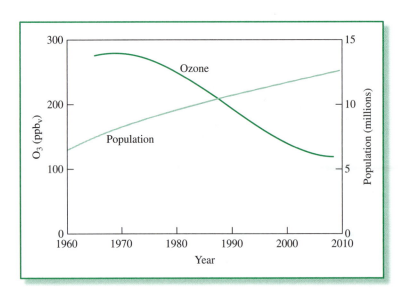

FIGURE 3-6 Maximum atmospheric ozone concentration (left axis) and human population (right axis) in Los Angeles over a half-century. [Source: D. D. Parish and T. Zhu, *Science* 326 (2009): 674–675.]

maximum ozone levels (dark green curve) in Los Angeles over time—by more than a factor of two over the last half-century—notwithstanding the continual increase in population of the area over that time period (light green curve). Presumably the Los Angeles pollution level would have increased greatly over the half-century, given the increase in the number of vehicles, if pollution controls had not been instituted.

The air in Mexico City is so polluted by ozone, particulate matter, and other components of smog, and by airborne fecal matter, that it is responsible for thousands of premature deaths annually; indeed, in the center of the city residents can purchase pure oxygen from booths to help them breathe more easily! In 1990 Mexico City exceeded the WHO air guidelines on 310 days, although peak smog levels have steadily declined since the 1990s and never reached the maximums attained in Los Angeles. In contrast to temperate areas where photochemical smog attacks occur almost exclusively in the summer—when the air is sufficiently warm to sustain the chemical reactions—Mexico City suffers its worst pollution in the winter months, when temperature inversions prevent pollutants from escaping. Some of the smog in Mexico City originates from *butenes* that are a minor component of the liquefied gas that is used for cooking and heating in homes, some of which apparently leaks into the air.

Butenes contain a C=C bond and consequently are very reactive.

Athens and Rome, as well as Mexico City, attempt to limit vehicular traffic during smog episodes. One strategy used by Athens and Rome is to allow only half the vehicles to be driven on alternate days, the allocation being based upon the license plate numbering (odd or even numbers). France plans to ban access in cities including Paris to vehicles made before 1997, when strict emission standards were instituted in Europe. Low-emission zones of various types are already in existence in parts of Sweden, Germany, Italy, and London, England and have succeeded in reducing urban air pollution.

Owing to the rapid increase in the number of vehicles on their roads and their generally warm climates, ground-level ozone levels and episodes of photochemical smog occur more and more frequently in developing countries. Peak ozone levels have been increasing in Beijing since the 1990s, especially from May to October. Other Chinese megacities such as Shanghai and Guangzhou also now suffer from high ozone episodes.

When hot summertime weather conditions produce large amounts of ozone in urban areas but do not allow much vertical mixing of air masses as they travel to rural sites, elevated ozone levels are often observed in eastern North America and western Europe in zones that extend for 1000 km (600 miles) or more. Thus **ozone control is a regional rather than a local air-quality problem.** Indeed, on occasion, polluted air from North America moves across the Atlantic to Europe, northern Africa, and the Middle East; that from Europe can move into Asia and the Arctic; and especially in springtime, that from Asia can reach the west coast of North America and contribute to increasing ozone levels there. Some analysts believe that by 2100, even the background level of ozone throughout the Northern Hemisphere will exceed current ozone standards.

Due to the long-range transport of primary and secondary pollutants in air currents, many areas which themselves generate few emissions are subject to regular episodes of high ground-level ozone and other smog oxidants. Indeed, some rural areas, and even small cities, that lie in the path of such polluted air masses experience *higher* levels of ozone than do nearby larger urban areas! This occurs because **in the larger cities some of the ozone transported from elsewhere is eliminated by reaction with nitric oxide released locally by cars into the air,** as illustrated previously in the reaction of NO with O_3. Ozone concentrations of 90 ppb are common in polluted rural areas. This phenomenon occurs in northern European countries such as Denmark, where ozone transported from southern countries reacts with NO emitted from vehicles in the Danish cities, but remains unreacted over rural areas, which consequently are more polluted by ozone than are urban areas.

Considerable amounts of ozone are transported from its origin in the U.S. Midwest to surrounding states and Canadian provinces, especially around the Great Lakes, such as the farmland in southwest Ontario, which receives ozone-laden air from industrial regions in the United States that lie across Lake Erie.

A plot of ozone concentration contours for summer afternoon smog conditions in North America in about 1990 is shown in Figure 3-7. At each point along any solid line, the concentration of ozone has the same value; hence contours connect regions having equal levels of ozone.

The highest levels (100 ppb) occurred in the Los Angeles and New York–Boston areas, but note the 80-ppb contour over a wide area south of the Great Lakes and into the Southeast, as well as one surrounding Houston.

Imported O_3 + local NO $\longrightarrow$ NO_2 + O_2.

Contour diagrams like that in Figure 3-7 are similar to geographic maps in which adjacent locations of equal altitude are connected by lines.

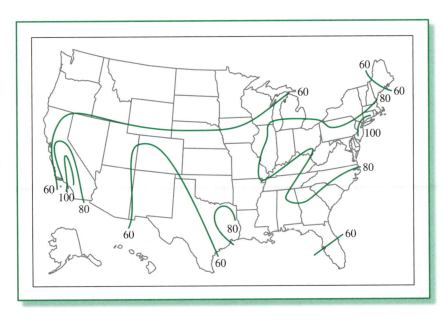

FIGURE 3-7 Ninetieth percentile contours of summer afternoon ozone concentrations (ppb) measured in surface air over the United States. Ninetieth percentile means that concentrations are higher than this 10% of the time. [Adapted from A. M. Fiore, D. J. Jacob, J. A. Logan, and J. H. Yin, "Long-Term Trends in Ground-Level Ozone over the Contiguous United States, 1980–1995," *J. Geo-phys. Res.* 103 (1998): 1471–1480.]

Ozone levels were particularly high over Houston—reaching 250 ppb on occasion—because of emissions of highly reactive VOCs containing C=C bonds from the region encompassing the petrochemical industry. By the late 2000s, the most ozone-polluted cities in the United States were

- Los Angeles, and seven other California cities ranging from Bakersfield north of it to San Diego to its south

- Houston, Texas

- Charlotte, North Carolina

The photochemical production of ozone also occurs during dry seasons in rural tropical areas where the burning of biomass for the clearing of forests or brush is widespread. Although most of the carbon is transformed immediately to CO_2, some methane and other hydrocarbons are released, as is some NO_X. Ozone is produced when these hydrocarbons react with the nitrogen oxides under the influence of sunlight.

Review Questions 5–7 are based on material in the sections above.

Improving Air Quality: Photochemical Smog

The history of attempts to improve the quality of the air that we breathe stretches back many centuries. In the developed world, episodes of the type of smog (mainly soot and sulfur dioxide) produced by unregulated coal burning began to be reduced in the mid-twentieth century and have largely disappeared there, as discussed further in Chapter 4. As we shall see below, the conquest of photochemical smog is proving to be a harder objective.

3.9 Limiting VOC and NO Emissions to Reduce Ground-Level Ozone

In order to improve the air quality in urban environments that are subject to photochemical smog, the quantity of reactants, principally NO_X and hydrocarbons containing C=C bonds, plus other reactive VOCs, emitted into the air must be reduced. The control strategies that have been put in place in the United States have resulted in some reduction in ozone levels in the past few decades, notwithstanding the huge increase in total vehicle-miles driven—up to 100% more in the last 25 years—that has occurred.

For economic and technical reasons, the most common control strategy has been to reduce hydrocarbon emissions. However, except in downtown Los Angeles, the percentage reduction in ozone and other oxidants that is thereby achieved usually has been much less than the percentage reduction in hydrocarbons. This happens because usually there is **initially an over-abundance of hydrocarbons relative to the amount of nitrogen oxides,** and cutting back hydrocarbon emissions simply reduces the excess without slowing down the reactions significantly. In other words, it is usually the nitrogen oxides, rather than reactive hydrocarbons, that are the species that

Nitrogen oxides are the *limiting reactants* in these areas.

determine the overall rate of the reaction. This is especially true for rural areas that lie downwind of polluted urban centers.

Due to the large number of reactions that occur in polluted air, the functional dependence of smog production upon reactant concentration is complicated, and the net consequence of making moderate decreases in primary pollutants is difficult to deduce. Computer modeling indicates that NO_X reduction, rather than VOC reduction, would be much more effective in reducing ozone in most of the eastern United States. However, Mexico City's ozone is limited by VOCs, even though there are numerous sources of them.

An example of the predictions that arise from the modeling studies is shown in Figure 3-8. The relationships between the NO_X and the VOC concentrations that produce contours for three different values for the concentration of ozone are shown. Notice that the same concentration of ozone results from many different combinations of VOC and NO_X. Point A represents a typical set of conditions in which the ozone production is NO_X *limited*. For example, reducing the concentration of VOCs from 1.2 ppm to 0.8 ppm has virtually no effect on the ozone concentration, which remains at about 160 ppb since the contour in this region is almost linear and runs parallel to the horizontal axis. However, a reduction of the NO_X level, from about 0.03 ppm at point A to a little less than half this amount, which corresponds to dropping down to the curve directly below it in the figure, cuts the predicted ozone level in half, from 160 ppb to 80 ppb. Chemically, NO_X-limited conditions occur when, due to the high concentration of VOC reactants, an abundance of peroxy free radicals HOO and ROO are produced, which quickly oxidize NO emissions to NO_2:

$$HOO + NO \longrightarrow OH + NO_2$$

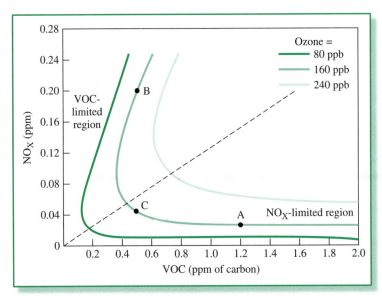

FIGURE 3-8 The relationship between NO_X and VOC concentrations in air and the resulting levels of ozone produced by their reaction. Points A, B, and C denote conditions discussed in the text. [Source: Redrawn from National Research Council, *Rethinking the Ozone Problem in Urban and Regional Air Pollution* (Washington, DC: National Academy Press, 1991).]

The nitrogen dioxide then photochemically decomposes to produce the free oxygen atoms that react with O_2 to produce ozone, as previously discussed (see Figure 3-4).

In the portion of the *VOC-limited* region that lies to the left of the diagonal dashed line of Figure 3-8, there is a large excess of NO_X; under such conditions, the OH radical tends to react with NO_2, and so less of it is available to initiate the reaction of more VOCs:

$$OH + NO_2 \longrightarrow HNO_3$$

Consequently, lowering the NO_X concentration actually produces *more* ozone, not less, since more OH is thereby available to react with the VOCs, although production of other smog reaction products such as nitric acid is thereby reduced. Thus, for example, when the VOC concentration is about 0.5 ppm, lowering the NO_X concentration from 0.21 ppm—corresponding to point B on Figure 3-8—even by two-thirds of this amount is predicted to increase slightly the ozone level beyond 160 ppb; further reductions do not begin to decrease ozone until NO_X reaches about 0.05 ppm. Indeed, ozone concentrations on weekends, when there is much less truck traffic, are higher in Los Angeles and many other cities, since the very high weekday levels of NO_2 are lower then and so less of it is available to combine with and provide a sink for OH.

In situations where the NO_X is less abundant but VOCs are relatively plentiful, i.e., to the right side of the dashed line in Figure 3-8, reducing NO_X does reduce ozone. Thus, when the VOC level is 0.5 ppm, the ozone concentration falls back to 160 ppb when the NO_X is reduced to 0.04 ppm (point C), and declines more with further decreases of NO_X.

PROBLEM 3-5

Using Figure 3-8, and assuming a NO_X concentration of 0.20 ppm, estimate the effect on ozone levels of reducing the VOC concentration from 0.5 to 0.4 ppm. Do your results support the characterization of that zone of the graph as VOC limited?

PROBLEM 3-6

Using Figure 3-8, again with an initial VOC concentration of 0.50 ppm, estimate the effect on ozone levels of lowering the NO_X concentration from 0.20 to 0.08 ppm. Explain your results in terms of the chemistry discussed above.

Some urban areas such as Atlanta, Georgia, and others located in the southern United States, incorporate or border upon heavily wooded areas whose trees emit enough reactive hydrocarbons to sustain smog and ozone production, even when the concentration of **anthropogenic** hydrocarbons, i.e., those that result from human activities, is low. Deciduous trees and shrubs emit the gas *isoprene*, whereas conifers emit *pinene* and *limonene*; all three hydrocarbons contain C=C bonds.

In urban atmospheres, the concentration of these natural reactive hydrocarbons normally is much less than that of the anthropogenic hydrocarbons, and it is not until the latter are reduced substantially that the influence of these natural substances becomes noticeable. In areas affected by the presence of vegetation, then, only the reduction of emissions of nitrogen oxides will reduce photochemical smog production substantially. As an air mass moves from an urban area to a rural one downwind, it often changes from being *VOC-limited* to being *NO_X limited*, since there are few sources of nitrogen oxides, but often substantial ones of reactive VOCs, outside of cities, and since the reactions that consume nitrogen oxides occur more quickly than do those that consume VOCs.

Although hydrocarbons with C=C bonds and aldehydes are the most reactive types in photochemical smog processes, other VOCs play a significant role after the first few hours of a smog episode have passed and the concentration of free radicals has risen. For this reason, control of emissions of *all* VOCs is required in areas with serious photochemical smog problems. Gasoline, which is a complex mixture of hydrocarbons, is now formulated in order to reduce its evaporation, since gasoline vapor has been found to contribute significantly to atmospheric concentrations of hydrocarbons. The control of VOCs in air is discussed in more detail in Chapter 11.

Regulations in California (with Los Angeles especially in mind) limit the use of hydrocarbon-containing products such as barbeque-grill starter fluid, household aerosol sprays, and oil-based paints that consist partially of a hydrocarbon solvent that evaporates into the air as the paint dries. The air quality in this region has improved because of current emission controls, but the increase in vehicle-miles driven and the hydrocarbon emissions from non-transportation sources such as solvents have thus far prevented a more complete solution. Research has also indicated that any substantial increase in the emissions of methane to the atmosphere could prolong and intensify the periods of high ozone in the United States, even though CH_4 is usually considered to be a rather unreactive VOC.

The blue hazes that are observed over forested areas such as the Great Smoky Mountains in North Carolina and the Blue Mountains in Australia result from the reaction of natural hydrocarbons in sunlight—in the *absence* of much NO_X and hence largely without its involvement—to produce *carboxylic acids* that condense to form suspended particles of the size that scatter sunlight and thereby produce a haze. Some of the ozone molecules present above the forests react with the C=C bond in the natural hydrocarbons to first produce aldehydes, which then are further oxidized in air to the corresponding carboxylic acids. Eventually, the acids in the aerosol are attacked by hydroxyl radicals, which initiate their decomposition, if the haze is not rained out of the air beforehand.

Recall that carboxylic acids have the general formula RCOOH.

3.10 Catalytic Converters for Gasoline Engines

Over the last decades, automobile manufacturers have employed several strategies to decrease VOC and NO_X emissions from their vehicles and thereby meet governmental standards. One early technique for NO_X control was to lower the temperature of the combustion flame and thereby decrease the rate of creation of thermal nitric oxide. The temperature lowering was achieved by recirculating a fraction of the engine emissions back through the flame.

In recent decades, more complete control of NO_X emissions from gasoline-powered cars and trucks has been achieved using **catalytic convert-ers** placed just ahead of the muffler in the vehicle's exhaust system. The original **two-way converters** controlled only carbon-containing gases, including carbon monoxide, CO, by completing their combustion to carbon dioxide. However, by use of a surface impregnated with a platinum-rhodium catalyst, the modern **three-way converter** changes nitrogen oxides back to elemental nitrogen and oxygen using unburned hydrocarbons and the com-bustion intermediates CO and H_2 as reducing agents:

$$2\,NO \longrightarrow N_2 + O_2 \quad \text{overall}$$

via, for example,

$$2\,NO + 2\,H_2 \longrightarrow N_2 + 2\,H_2O$$

> Recall that the reaction of N_2 with O_2 has a high activation energy, so its rate is very dependent upon temperature.

> Several hundred dollars' worth of precious metals are present in each catalytic converter.

> Reduction

PROBLEM 3-7

Write and balance reactions in which NO is converted to N_2 (a) by CO, and (b) by C_6H_{14}. [*Hint:* The other reaction product is CO_2, plus H_2O in the lat-ter case.] ●

The carbon-containing gases in the exhaust are catalytically oxidized almost completely to CO_2 and water by the oxygen that is present:

$$2\,CO + O_2 \longrightarrow 2\,CO_2$$

$$C_nH_m + (n + m/4)\,O_2 \longrightarrow n\,CO_2 + m/2\,H_2O$$

$$CH_2O + O_2 \longrightarrow CO_2 + H_2O$$

> Oxidation

The catalyst is dispersed as very tiny crystallites, initially less than 10 nm in size. Oxygen sensors in the vehicle's exhaust system are monitored by a computer chip that controls the intake air/fuel ratio of the engine to the stoi-chiometric amount required by the fuel in order to ensure a high level of con-version of the pollutants. The whole process is illustrated in Figure 3-9a. If the air/fuel mix is not very close to the stoichiometric ratio, the warmed catalyst will not be effective for reduction (if there is too much air), causing nitrogen oxides will be emitted into the air, or for oxidation (if there is too little air), causing CO and hydrocarbons to be emitted, as illustrated in Figure 3-9b.

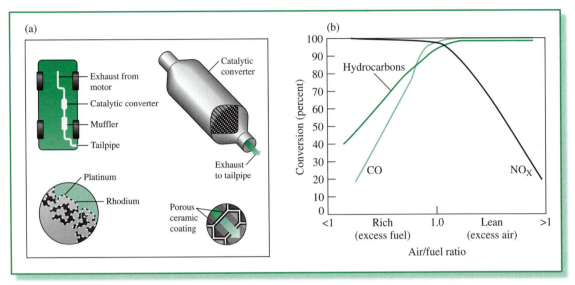

FIGURE 3-9 (a) Modern catalytic converter for automobiles, with its position in the exhaust system indicated. [Source: L. A. Bloomfield, "Catalytic Converter," *Scientific American* (February 2000): 108.] (b) Efficiency in conversion of catalytic converter versus air/fuel ratio. [Source: B. Harrison, "Emission Control," *Education in Chemistry* 37 (2000): 127.]

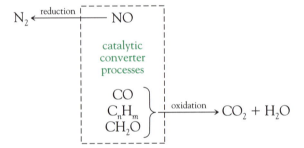

Some progress has been reported in the use of less-valuable metals, such as copper and chromium, instead of the expensive platinum-group metals as catalysts in catalytic converters. Although the metals are recycled from old converters, a portion is inevitably lost in the process. Scientists have expressed concern about the environmental problem of widely broadcasting the tiny particles of platinum, palladium, and rhodium that are lost from the converters themselves during their operation.

The catalyst that reduces nitric oxide to nitrogen also reduces **sulfur dioxide,** SO_2, to **hydrogen sulfide,** H_2S. The emitted gases include H_2S and other reduced sulfur compounds, which collectively often give vehicle emissions their characteristic odor of rotten eggs. In addition, the small amounts of sulfur-containing molecules in gasoline—and diesel fuel—can partially deactivate catalytic converters if sulfate particles produced from them during the gasoline's combustion become attached to and thereby cover the sites of the catalyst metal, deactivating them. The maximum annual average sulfur levels in gasoline, amounting to several hundred parts per million in the past, have been reduced to 30 ppm in both the United States and Canada, and to 10 ppm in the European Union. By contrast, the maximum sulfur in gasoline for India's cities is 150 ppm, though there are plans to lower it to 50 ppm in the future.

The sulfur is removed from gasoline during refining, usually by **hydrodesulfurization,** itself a catalytic process, which reacts organic sulfur-containing molecules in the gasoline with **hydrogen** gas, H_2, to produce hydrogen sulfide, which then is removed. Alternatively, the sulfur-containing molecules may be removed from the fuel by absorbing them during the refining process.

In the first few minutes after a vehicle's engine has been started up, the catalysts are cold, so the converters cannot operate effectively and there are bursts of emissions from the tailpipe. Indeed, approximately 80% of all the emissions from converter-equipped cars are produced in the first few minutes after starting. Once an engine has warmed up and the catalysts have been heated to about 300°C by engine exhaust, three-way catalysts convert 80–90% of the hydrocarbons, CO, and NO_X to innocuous substances before the exhaust gases are released into the atmosphere. However, fuel-rich mixtures are fed to the engine in the first minute or so after the vehicle engine is started, and also when high acceleration occurs, so carbon monoxide and unburned hydrocarbons are emitted directly into the air under these oxygen-starved conditions.

Research and development is underway to develop catalytic converters that would convert start-up emissions so that these are not released into the air. Various approaches being investigated include

• devising a converter that will operate at lower temperatures or that can be preheated so it begins to operate immediately,

• storing pollutants until the engine and converter are heated, and

• recirculating engine exhaust through the engine until the reactions are more complete.

Older cars (with no converters or just two-way converters) still on the road continue to pollute the atmosphere with nitrogen oxides even during their normal operation.

The maximum amounts of emissions that can legally be released from light-duty motor vehicles such as cars have gradually been decreased in order to improve air quality. The U.S. EPA sets regulations for the maximum emissions per mile of CO, NO_X, total and non-methane hydrocarbons for gasoline-powered vehicles. Some governments have recently instituted mandatory inspections of exhaust systems to ensure that they continue to operate properly.

Methane is much slower to react than are other hydrocarbons, so its emissions are often excluded from regulations.

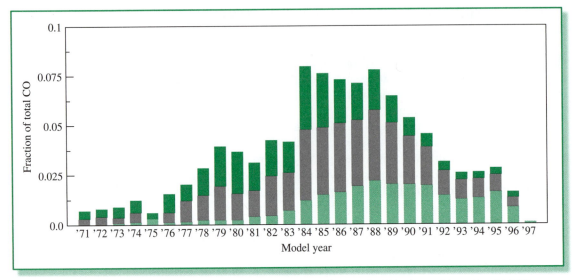

FIGURE 3-10 Fleet-weighted CO contribution of vehicles by model year in mid-1990s Denver, Colorado, traffic analysis. Light green represents vehicles rated "good" in terms of their CO emission levels, gray represents vehicles rated "fair," and dark green those rated "poor." [Source: G. A. Bishop et al., "Drive-By Motor Vehicle Emissions: Immediate Feedback in Reducing Air Pollution," *Environ. Sci. Technol.* 34 (2000): 1110.]

Vehicles whose catalytic converters have been damaged or tampered with produce most of the emissions: typically, 50% of the hydrocarbons and carbon monoxide are released from 10% of the cars on the road. For example, a study of soot emissions from cars in The Netherlands established that 5% of vehicles accounted for 43% of pollution. A mid-1990s survey of carbon monoxide emitted by passing traffic in Denver, Colorado produced the results shown in Figure 3-10. The great majority of Denver's vehicles of ages up to about 12 years were rated "good" (light green in the figure) in terms of their control of tailpipe emissions. Most of the carbon monoxide came from cars 6–12 years old (Figure 3-10), because they were so numerous and because of the presence in that fleet of cars with "poor" or "fair" emission levels (dark green and gray in Figure 3-10, respectively). In recent decades, the gradual deterioration of catalytic converters is less pronounced, so the occurrence of such high emitters is now far less common.

3.11 Air Quality Standards

The legislated Air Quality Standards for the common, important air pollutants are listed in Table 3-1. Different jurisdictions differ slightly in the maximum concentrations they permit; usually an area is considered to be in compliance if these maxima are only rarely exceeded. Notice that for several pollutants different values are listed depending upon the time period over which the concentration occurs, the assumption being that human health can tolerate a higher concentration of the pollutant for a short period.

TABLE 3-1	Air Quality Standards, in ppm, for Gases						
Gas	WHO	USA	Canada	EU	Australia	China	India
Ozone, O_3 8 hr	0.051	0.075	—	0.061	(0.08 over 4 hr)	—	0.051
1 hr	—	—	0.082	—	0.10	0.061–0.10	0.092
Carbon monoxide, CO 8 hr	—	9	13	9	9	(3.5–5.3 over 1 day)	5.3
1 hr		35	31				3.5
Nitrogen dioxide, NO_2 1 hr	0.11	0.10	0.21	0.11	0.12	—	—
1 yr	0.021	0.053	0.053	0.021	0.03	0.021–0.042	0.021
Sulfur dioxide, SO_2, 1 hr	(0.19 for 10 minutes)	0.075	0.33	0.13	0.20	—	—
24 hr	0.008	0.14	0.115	0.05	0.08	0.019–0.096	0.031
1 yr	—	0.03	0.023	—	0.02	0.008–0.039	0.019

Note: Since all listed species are gases, the ppm scale has the moles/moles basis, equivalent to volume/volume.

The U.S. EPA has traced the history of air quality improvement in the United States over the last three decades (www.epa.gov/airtrends). For example, the concentration of carbon monoxide over many areas exceeded the 9-ppm maximum in 1980, but by 1990 there were few exceedences, and nowadays the national average is 2 ppm. The decrease in NO_2 and SO_2 concentrations have been also been continuous but in a less dramatic fashion, although almost all sites were always in compliance. However, improvement in average ozone levels has been much harder to achieve, with many of the monitoring sites still not in compliance and the national average barely meeting the quality standard even in the late 2000s.

The control over the past half-century of carbon monoxide levels in urban areas of developed countries has been one of the real success stories in environmental management. Most of the reduction has resulted from the use of catalytic converters on vehicles, from the ever-tightening regulations on vehicular emissions, and from the natural continuing increase in the fraction of vehicles built after standards had been tightened. The introduction of **oxygenated** substances, which are hydrocarbons in which some of the atoms have been replaced by oxygen, into American gasoline has also reduced CO emissions from vehicles (as will be discussed further in Chapter 7).

Average new-vehicle emissions before any emission controls were introduced in the United States were about 38 g CO km^{-1}, compared to the present standard of 1.5 g km^{-1}. Generally speaking, carbon monoxide emissions from cars are greatest from cold engines (and cold catalytic converters) and when the vehicle has an increased load due to rapid acceleration or climbing a hill, since under these conditions, a rich fuel mix supplies insufficient oxygen to completely oxidize the gasoline.

There are some cities in developed countries that still are vulnerable to CO concentration exceedences in winter due to meteorological and topological conditions. The temperature inversions in Fairbanks, Alaska, for example, produce several days each winter with high carbon monoxide levels. People such as traffic police who work outdoors in areas of high vehicular traffic can be exposed to elevated CO levels for long periods.

In the past, emission standards were applied only to passenger vehicles. However, starting in 2004, new U.S. regulations required for the first time that gasoline-powered sport-utility vehicles (SUVs)—which now account for about half of new-vehicle sales—and light trucks also meet emission standards.

Motorcycles, and the three-wheeled *tuk-tuk* taxis common in Asia, are gasoline powered and their emissions, especially of CO and unburned hydrocarbons, make substantial contributions to air pollution. In a recent study in Switzerland, scooters and motorcycles (some fitted with catalytic converters) were found to emit more CO and hydrocarbons in urban driving, and significantly more NO_X on the highway, per kilometer travelled, than cars.

In general, most areas of most *developed* countries now meet their current air quality standards most of the time. This result represents a major achievement in an important environmental problem; without emission regulations and devices put into place over the last few decades, air quality, especially in urban areas of these countries, would have worsened considerably with the increase in the number of vehicles and the use of electricity generated by fossil-fuel combustion. In this respect at least, the continuing use of fossil fuels has become sustainable by regulations and technology formulated as a result of the systems analysis of the chemistry underlying air pollution discovered by chemical research.

Unfortunately, air quality in the urban areas of many *developing* countries has generally deteriorated in recent times; it will be a major challenge for these areas to institute controls to clean up their air. India in 2009 updated its national ambient Air Quality Standards, which in most cases are more stringent than EU values. China uses a three-grade system for air quality standards—hence the ranges quoted in Table 3-1. About half its cities met its intermediate standards in 2005.

3.12 Catalytic Converters for Diesel Engines

Most trucks and buses are powered by diesel engines, and similarly for half the cars in Europe and many regions elsewhere outside North America. **The catalytic converters used on vehicles with diesel engines in the past were much less effective than those on gasoline-powered vehicles.** They typically removed only about half the gaseous hydrocarbon emissions, compared to 80–90% removal for gasoline engine emissions. This difference was due to the less-active catalyst formulations that had to be used with diesels because of the high sulfur content of diesel fuel; more-active catalysts would have oxidized the sulfur dioxide gas to sulfate particles, which would have covered the catalyst surface and rendered it ineffective. Much more effective catalytic converters are now being installed on diesel equipment for use with low-sulfur fuel. In addition, the loss of engine exhaust gases and particles to the air—including that inside the vehicle—through crankcase emissions is being eliminated by rerouting the emissions back into the engine.

Catalytic converters used in the past for diesel engines did not convert NO_X since diesel engines are operated "fuel lean," i.e., with excess oxygen present, and hence the required chemically reducing conditions for NO_X control did not exist. However, lowering the engine operating temperature by recirculating some of the exhaust gases through the engine reduces the amount of NO_X produced, as was also done in the past with automobiles, as mentioned above.

Fuel economy is maximized by running the engine under lean conditions.

In one modern scheme for NO_X removal from diesel exhaust, the gas is first completely oxidized to NO_2—as CO and hydrocarbons are oxidized with the engine running under "lean" conditions, i.e., with excess air, most of the time. The NO_2 is subsequently passed over an absorber such as *barium oxide*, BaO, on which it is temporarily stored as *barium nitrate*, $Ba(NO_3)_2$. Periodically, when the absorber becomes saturated, the vehicle's computer system temporarily switches conditions to fuel-rich, at which time unburned diesel fuel passes over the stored NO_X, creating reducing conditions and thereby promoting the catalytic reduction of NO_X to N_2.

A completely different scheme has been introduced by Mercedes-Benz and BMW for some of their diesel cars and trucks. Each vehicle carries a supply of the reducing agent **urea,** $CO(NH_2)_2$. This substance (heated, if necessary, in cold weather) is injected slowly into a catalyst to react with NO_X, converting the mixture to nitrogen, water, and carbon dioxide.

Urea is resupplied to the vehicle's storage when it is serviced. The structural formula for urea is $O{=}C(NH_2)_2$.

PROBLEM 3-8

Deduce the balanced redox equation that converts urea and nitric oxide into N_2, CO_2, and water. [*Hint:* Assume for simplicity that the carbon is +4 in both reactant and product. Note that the techniques from introductory chemistry for assigning oxidation numbers and for balancing redox equations are summarized in the Appendix.]   ●

New emission regulations in North America demand substantial reductions in NO_X emissions from diesel-powered vehicles. The U.S. EPA has also proposed that train locomotives and ships powered by diesel should also be forced to reduce substantially their emission of nitrogen oxides.

In addition to CO, hydrocarbon, NO_X, and SO_2 gases, diesel-engine exhaust also includes significant quantities of solid and liquid particles, the maximum emissions of which per mile is regulated in the United States. The liquid particles consist of unburned fuel and lubricating oil, plus some sulfuric acid produced from sulfur in the fuel. The catalytic converters formerly used on diesel-engine vehicles were designed to oxidize the carbon-containing gases and liquids without oxidizing the SO_2 further to sulfuric acid and sulfates.

Diesel fuel intended for new on-road vehicles in the United States and Canada had its maximum allowed sulfur level drop from 500 ppm to a maximum of 30 ppm sulfur *for low-sulfur diesel fuel* and in 2010 to 15 ppm S for *ultra-low-sulfur diesel fuel*. The maximum has been lowered to 50 ppm in the European Union, which is the target also for India, although its current limit is set at 150 ppm. Lowering the sulfur level in diesel fuel will also reduce particle emissions to some extent. Ironically, some of the soot carbon reacts with nitrogen dioxide in the exhaust gas to oxidize it to gases:

$$C(s) + NO_2 \longrightarrow CO + NO$$

A filter—sometimes known as a **particle trap**—is required to achieve a suitable reduction in diesel exhaust particles. The traps can physically remove up to 90% of small-particle emissions, thus preventing them from escaping into the air from the exhaust system of light-duty diesel vehicles. In order to prevent a buildup of solids, which would restrict engine exhaust flow or melt the trap, the system is designed so that the soot will ignite and burn away once a temperature of at least 500°C is attained. Alternatively, tiny amounts of a metal catalyst compound containing iron or copper are added to lower the temperature at which ignition will occur and assure more continuous regeneration of the filter. Eventually the filter requires cleaning since nonorganic components of the exhaust build up over time.

Scientists and engineers currently are developing a new type of internal combustion engine that combines the best aspects of gasoline and diesel technologies. In the *homogeneous charge compression ignition* (HCCI) engine, the fuel and the air are well mixed before ignition, thereby preventing formation of soot particles, as occurs in diesel engines. High compression of the air/fuel mix allows combustion to begin at many locations in the cylinder, and produces a much greater efficiency than in gasoline engines, thereby reducing CO_2 emissions. Since the engine is run with an even larger excess of air than in a diesel, and no "hot spots" occur where combustion temperatures are high, much smaller amounts of NO_X than normal are produced.

Eventually, vehicles using internal combustion may well be replaced by emission-free ones powered by fuel cells. Such a prospect is discussed in Chapter 7 when we consider various alternative fuels.

3.13 Control of Nitric Oxide Emissions from Power Plants

Vehicles emit most of the NO_X in Mexico City, with rather little coming from power plants.

In the United States, large quantities of NO_X are emitted currently from vehicles (on-road and off-road) and from electric power plants, and collectively they constitute the majority of the anthropogenic sources of these gases (see Figure 3-5). In this section, we investigate the processes that are in use to decrease the quantity of nitrogen oxides emitted into the air by power plants.

• To reduce their NO_X production, some power plants use special burners designed to **lower the temperature of the flame.** Alternatively, the recirculation of a small fraction of the exhaust gases through the combustion zone has the same effect, as previously discussed for vehicles.

• Nitric oxide formation in power plants can also be greatly reduced by **having the combustion of the fuel occur in stages.** In the first, high-temperature stage, no excess oxygen is allowed to be present, thus limiting its ability to react with N_2. In the second stage, additional oxygen is supplied to complete the fuel's combustion but under lower-temperature conditions, so that again little NO is produced. Up to half the NO that would be produced in one-stage combustion is eliminated by this method.

• Other power plants, especially those in Japan and Europe, have been fitted with large-scale versions of **catalytic converters that change NO_X back to N_2** before the release of stack gases into the air. The reduction of NO_X to N_2 in these **selective catalytic reduction** systems is accomplished to 80–95% completion by adding ammonia, NH_3, to the cooled gas stream. This highly reduced compound of nitrogen combines in a redox reaction with the partially oxidized compound NO to produce N_2 gas in the presence of oxygen:

$$4\,NH_3 + 4\,NO + O_2 \longrightarrow 4\,N_2 + 6\,H_2O$$

However, tight control is needed to regulate the addition of ammonia in order to prevent its inadvertent oxidation to NO_X. The catalytic process occurs at 250°C to 500°C, depending upon the catalyst used, and is >80% effective. A small amount of unreacted ammonia is emitted into the air. Urea can also be used as the reducing agent, as in the catalytic converters on diesel cars (Section 3-12, and Problem 3-8). Indeed, reduction can be accomplished *without* expensive catalysts, though with much less efficient nitric oxide removal, by reacting urea with the uncooled gases at about 900°C. A combination of staged combustion and selective noncatalytic reduction can be used to eliminate up to 90% of NO_X emissions.

• The **wet scrubbing of exhaust gases** by an aqueous solution can also be used to prevent NO_X from being emitted into outside air. Since NO itself is rather insoluble in water and in typical aqueous solutions, half or more of the NO_X must be in the form of the much more soluble NO_2 for such techniques to be effective. Solutions of **sodium hydroxide,** NaOH, react with equimolar amounts of NO and NO_2 to produce an aqueous solution of **sodium nitrite,** $NaNO_2$:

$$NO + NO_2 + 2\,NaOH \longrightarrow 2\,NaNO_2 + H_2O$$

Deduce the balanced reaction in which ammonia reacts with nitrogen dioxide to produce molecular nitrogen and water. Using the balanced equation, calculate the mass of ammonia that is required to react with 1000 L of air at 27°C and 1 atm pressure containing 10 ppm of NO_2. [*Hint:* Recall from introductory chemistry that $PV = nRT$, and that a balanced equation indicates the number of moles of the substances that react with each other.] ●

3.14 Future Reductions in Smog-Producing Emissions

Although direct emissions of five (CO, VOCs, SO_2, particulate matter, and lead) of the six major air pollutants in the United States fell significantly between 1970 and 2000, emissions of NO_X grew by 20%, with half that increase occurring during the 1990s. Since energy consumption grew 45% and vehicle distances traveled grew 143% in that period, restrictions on nitrogen oxide emissions have achieved some success in controlling some of the growth in this pollutant, but not enough to prevent an overall increase. Emissions of NO from electric power plants, their other major source, have recently been falling somewhat.

As a consequence of the increase in overall NO emissions, ground-level ozone concentrations increased in the southern (especially around Houston) and north-central regions of the United States in the 1990s. The latter effect can be seen in Figure 3-7, where high ozone levels center around the New York–Boston area and a few midwestern sites, and somewhat lesser levels cover most of the East Coast and Midwest, extending into southern Ontario. High ozone levels have been a problem in Southern California for many decades.

To help reduce the incidence of summertime smog in south-central Canada and the northeastern United States, the two nations have signed an Annex to their *Air Quality Agreement*. The United States was committed to reduce NO_X emissions originating in northern and northeastern states by 35% by 2007, and also to reduce VOC emissions during summer months when most smog forms in this region. Canada agreed to reduce its NO_X emissions from power plants in southern Ontario by 50% by the same date. The drastic lowering of the allowed sulfur levels in gasoline should assist in the reduction of NO_X from vehicles. Emission standards for SUVs, trucks, and buses are also being tightened to make them more in line with those for regular automobiles.

The long blackout of electrical power that occurred in August, 2003 in eastern North America yielded some interesting information concerning the contribution of power plants to air pollution in that region. Measurements over Pennsylvania taken 24 hours after the blackout began found that SO_2 levels were down 90%, and ozone levels down about 50%, compared to a similar hot, sunny day a year earlier, and that visibility increased by about 40 km because haze from particulates had decreased by 70%.

The *Gothenburg Protocol*, which controls the release of many pollutants in Europe, was expected to have reduced NO_X emissions there by more than 40% by 2010, compared to 1990 levels. Great Britain—which saw its emissions decline by the late 1990s by almost 40% compared to their peak in the late 1980s—had to reduce their emissions by another third from 1998 levels by 2010 in order to meet these regulations. European VOC emissions were due to drop by 40% according to the protocol.

Review Questions 8–12 are based on material in the above sections.

Green Chemistry: Strategies to Reduce VOCs Emanating from Organic Solvents

In addition to their role in paints, organic solvents are used in many different products and processes, in both commercial and household applications. More than 15 billion kilograms of organic solvents are used worldwide each year in such areas as the electronics, cleaning, automotive, chemical, mining, food, and paper industries. These liquids include not only hydrocarbon solvents but also halogenated solvents. Both types of solvents contribute not only to air pollution as VOCs but also to water pollution (Chapter 11). Some halogenated solvents contribute to the depletion of the ozone layer, as seen in Chapter 2.

The following three sections give examples of green chemistry illustrating different strategies for reducing VOCs and their emissions. The first illustrates a method for reducing the amount of VOCs in paints, while the second and third are examples of replacing organic solvents with solvents that do not produce VOCs.

3.15 Green Chemistry: A Nonvolatile, Reactive Coalescent for the Reduction of VOCs in Latex Paints

Paints generally consist of three major components: pigment, binder, and solvent. The pigment gives the paint color and may be natural or synthetic, and may be organic or inorganic. The binder, also known as the vehicle or resin, provides adhesion, binds the pigment, and provides such properties as toughness, durability, flexibility, and gloss. The primary function of the solvent is to act as the carrier for the nonvolatile components. The solvent may be water (in latex or water-based paints) or organic (in oil-based or alkyd paints). During the course of drying, the solvent evaporates into the surrounding air and is the major source of VOCs from oil-based paints.

Although oil-based paints have been known for centuries, latex paints became commercially available only in the 1950s. The introduction of latex

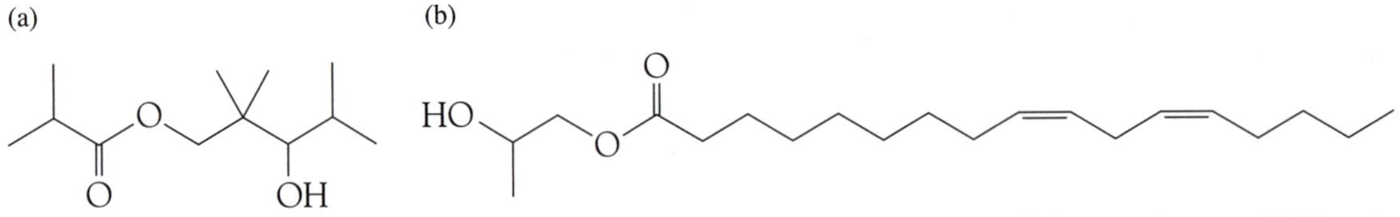

(a)

(b)

FIGURE 3-11 Structures of (a) 2,2,4-trimethyl-1, 3-pentanediol monoisobutyrate (TMB) and (b) propylene glycol monoester of linoleic acid (Archer RC™).

paints greatly reduced the amount of VOCs compared to oil-based paints. Recent, more stringent environmental regulations at the federal and state levels have led to significant efforts to reduce VOCs even from latex paints in the United States.

One significant source of VOCs in latex paints is an additive that acts as a *coalescent*. The resin in latex paints is composed of very small particles of an organic polymer suspended in water. Coalescents are organic compounds that are absorbed by the resin particles and soften or plasticize these particles. When the paint is spread, the water evaporates and these particles flow together to form a thin, uniform film that adheres to the surface of the material being painted. The paint hardens as the coalescent is slowly emitted into the atmosphere. Paint contains 2–3% coalescent by mass. Since more than 2 billion liters (600 million gallons) of latex paint are used in the United States each year, more than 50 million kilograms (120 million pounds) of coalescents are emitted into the atmosphere annually. Worldwide emissions of these VOCs are estimated to be more than three times this value.

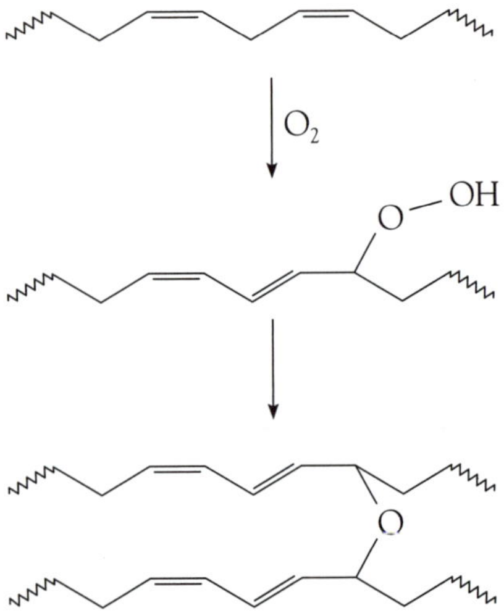

In 2005, the Archer Daniels Midland Company (ADM) won a Presidential Green Chemistry Challenge Award for the development of coalescents (Archer RC™) that actually bind through covalent bonds to each other and to the resin, and thereby become part of the paint film; they are therefore not emitted as VOCs. Figure 3-11a shows the structure of TMB, a common coalescent, while Figure 3-11b is the structure of the coalescent developed by ADM. Both of these ester compounds have low polarities and are thus absorbed by the resin particles. However, (b) is an ester of propylene glycol and linoleic acid. Linoleic acid is a major component of linseed oil and, like linseed oil, it can undergo auto-oxidative cross-linking in the presence of oxygen due to the carbon–carbon double bonds present, as illustrated by the prototype reaction shown in Figure 3-12.

The oxidized, cross-linked polymer chains have low vapor pressure and furthermore are absorbed into and bonded to the resin, further decreasing their volatility and ability to be emitted.

FIGURE 3-12 Auto-oxidation cross-linking of an unsaturated system.

Traditional coalescents such as TMB are made from petroleum feed-stocks. Another green feature of the coalescent developed by ADM is that it can be produced from the renewable bio-feedstocks linoleic acid and propylene glycol. Linoleic acid is readily available from corn and sunflower oils, and although propylene glycol has traditionally been pro-duced from petroleum, it can be formed from glycerin, a by-product of the synthesis of biodiesel (as we will see in Chapter 7 in another example of green chemistry).

3.16 Green Chemistry: The Replacement of Organic Solvents with Supercritical and Liquid Carbon Dioxide; Development of Surfactants for This Compound

Discovering solvents with less environmental impact, and even designing processes that use no solvents at all, are the subjects of many green chemistry initiatives. Carbon dioxide, CO_2, is one solvent that is receiving consider-able attention as a replacement for traditional organic solvents. Although carbon dioxide is a gas at room temperature and pressure, it can be liquefied easily by the application of pressure. In addition to liquid carbon dioxide, there is considerable interest in supercritical carbon dioxide (a discussion of supercritical fluids can be found in Box 3-2) as a solvent in the electronics industry. The decaffeination of coffee and tea with carbon dioxide is a well-known application of this solvent.

Liquid carbon dioxide is attractive as a solvent due to its low viscosity and polarity and its wetting ability. Because of its low polarity, carbon diox-ide is able to dissolve many small organic molecules. However, larger mole-cules including oils, polymers, waxes, greases, and proteins are generally insoluble in it. To increase the solubility of compounds in water, surfactants such as soaps and detergents have been developed which allow this very polar solvent to dissolve less polar materials such as oils and grease. In an analogous fashion, surfactants for carbon dioxide have been developed which increase the range of materials that will dissolve in it.

Joseph DeSimone, of the University of North Carolina and North Caro-lina State University, earned a Presidential Green Chemistry Challenge Award in 1997 for his preparation and development of polymeric surfactants for carbon dioxide. DeSimone is currently the director of the National Sci-ence Foundation Science and Technology Center for Environmentally Responsible Solvents and Processes. This center focuses on discovering ways to replace conventional organic solvents and water with carbon dioxide in a multitude of processes. An example of a surfactant developed by DeSimone

BOX 3-2   Supercritical Carbon Dioxide

The supercritical fluid state of matter is produced when gases or liquids are subjected to very high pressures and, in some cases, to elevated temperatures. At pressures and temperatures at or beyond the *critical point*, separate gaseous and liquid phases of a substance no longer exist. Under these conditions, only the supercritical state, with properties that lie between those of a gas and those of a liquid, exists. For carbon dioxide, the *critical pressure* is 72.9 atm and the *critical temperature* is only 31.3°C, as illustrated in the phase diagram in Figure 1. Depending upon exactly how much pressure is applied, the physical properties of the supercritical fluid vary between those of a gas (relatively lower pressures) and those of a liquid (higher pressures); the variation of properties with P or T is particularly acute near the critical point. Thus the density of supercritical carbon dioxide varies over a considerable range, depending upon how much pressure (beyond 73 atm) is applied to it

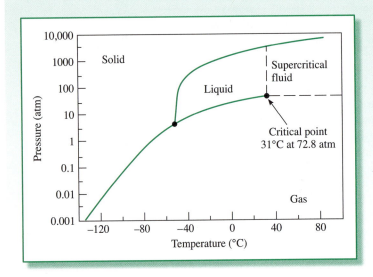

FIGURE 1 Phase diagram for carbon dioxide.

is the block copolymer shown in Figure 3-13a. This molecule has nonpolar regions, which are CO_2-philic, and polar regions, which are CO_2-phobic. When dissolved in carbon dioxide, the CO_2-philic regions orient themselves to interact with the surrounding carbon dioxide solvent, while the CO_2-phobic regions aggregate with one another. The overall result is the formation of a structure know as a *micelle* (Figure 3-13b). Polar substances that normally do not dissolve in carbon dioxide will dissolve in the center polar region of the micelle.

DeSimone was one of founders of a dry-cleaning chain that uses liquid carbon dioxide, along with surfactants that he developed, to clean clothes. The spent liquid carbon dioxide is drained from the clothes after the wash cycle (in much the same as the wash water in our washing machines at home is drained off after the wash cycle) and the carbon dioxide is allowed to evaporate by simply reducing the pressure. The carbon dioxide vapors are then captured, liquefied by increasing the pressure, and reused for another wash. Carbon dioxide is plentiful and inexpensive, since it can be recovered as a by-product from natural gas wells or ammonia production. Capture of carbon dioxide from these processes puts to good use this compound which would normally be released to the atmosphere and contribute to global warming (see Chapter 6). By way of contrast, most dry cleaners in North America presently use *perchloroethylene*, $Cl_2C{=}CCl_2$, known as PERC, as the solvent. PERC is a VOC, since it has a high vapor pressure and readily escapes into the troposphere if not carefully controlled. PERC is also is a groundwater contaminant (see Chapter 11) and is a suspected human carcinogen.

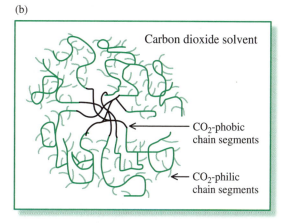

(a)

CO₂-phobic chain segment

CO₂-philic chain segment

(b)

FIGURE 3-13 A copolymer surfactant for carbon dioxide. (b) A micelle in liquid carbon dioxide. [Source: M. C. Cann and M. E. Connelly, *Real-World Cases in Green Chemistry* (Washington, DC: American Chemical Society, 2000).]

3.17 Green Chemistry: Using Ionic Liquids to Replace Organic Solvents: Cellulose, a Naturally Occurring Polymer Replacement for Petroleum-Derived Polymers

Cellulose (Figure 3-14) is a polymer of glucose that makes up about 40% of all organic matter on Earth. About 700 billion tonnes of cellulose exist on Earth, with another 40 billion tonnes produced each year by plants as the major component of biomass from atmospheric carbon dioxide and water via photosynthesis. This removal of carbon dioxide from the atmosphere helps to mitigate some of the global warming caused by anthropogenic emissions of the gas.

Many polymers produced from crude oil are ubiquitous in our everyday lives, including **polyethylene terephthalate** (PET), which is found in beverage bottles and polyester clothing, **polyethylene,** which is employed in making plastic bags and milk jugs, **polyvinyl chloride,** which is found as plastic

FIGURE 3-14 Structure of cellulose.

FIGURE 3-14 Structure of cellulose.

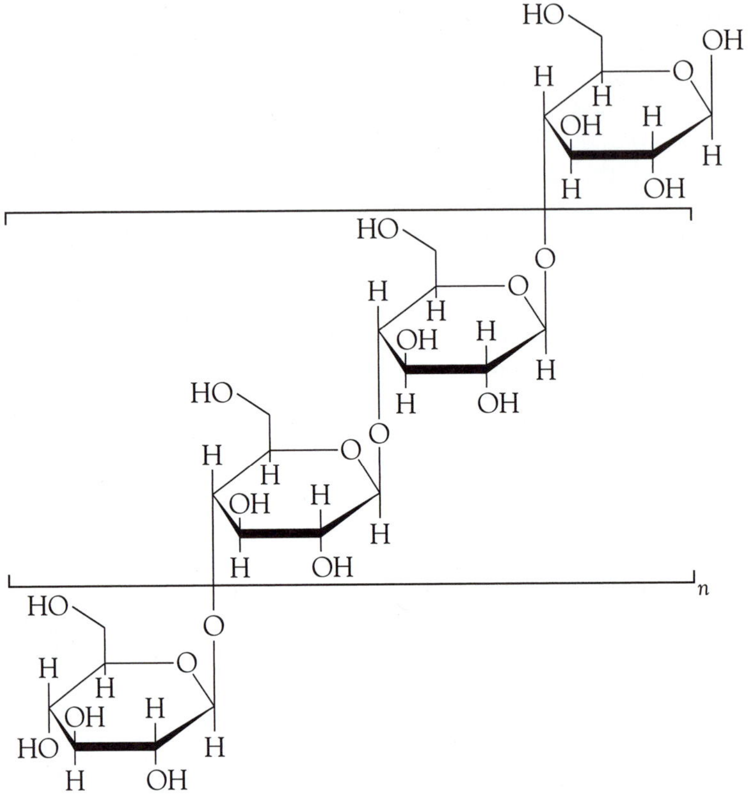

pipes and shower curtains, and **polystyrene,** which we discussed in the green chemistry section in Chapter 2. Hundreds of millions of kilograms of these petrochemical-based polymers are produced each year, requiring as raw material approximately 700 million barrels of crude oil. As the price of conventional crude oil increases and the supply declines, a major focus of green chemistry is the production of organic chemicals, including polymers, from biomass (see the green chemistry section in Chapter 7). An even more intriguing opportunity is to use naturally occurring polymers such as cellulose to replace crude oil in these syntheses.

The use of cellulose is severely limited by its insolubility in water and in traditional organic solvents. The strong intra- and inter-chain hydrogen bonding between the numerous hydroxyl groups on the cellulose polymer are likely the reason for this insolubility, which results in very poor processability for cellulose. Consequently, only about 0.1 billion tonnes of cellulose has been used annually as a feedstock for further processing.

In the previous green chemistry section, the replacement of traditional organic solvents with supercritical and liquid carbon dioxide was discussed. A very interesting and relatively unknown group of compounds that is of growing interest as replacements for traditional organic solvents are called *room temperature ionic liquids*, or just **ionic liquids** (ILs). Most ionic compounds have characteristically high melting points due to their strong network of ionic bonding. For example, sodium chloride (table salt) has a melting point of 801°C. In contrast, a few ionic compounds have melting points below or moderately above ($\leq$100°C) room temperature; such compounds are known as (room temperature) ionic liquids. ILs are generally composed of bulky ions that have dispersed rather than localized charges and large nonpolar groups (Figure 3-15). As a consequence, their oppositely charged ions have only weak attractive interactions with one another, which results in the low melting points of these compounds.

One very attractive characteristic of ILs is their very low vapor pressure, in contrast to most organic solvents, which because of their significant vapor pressures are VOCs and contribute to tropospheric pollution. Because they are ionic, many ILs are nonvolatile and thus their potential to replace VOCs is of significant interest. Ionic liquids may also be purified and recycled, thereby adding to the green characteristics of these solvents. In addition, they are nonflammable, and many are stable up to 300°C, making them attractive for reactions and processes which require high temperatures.

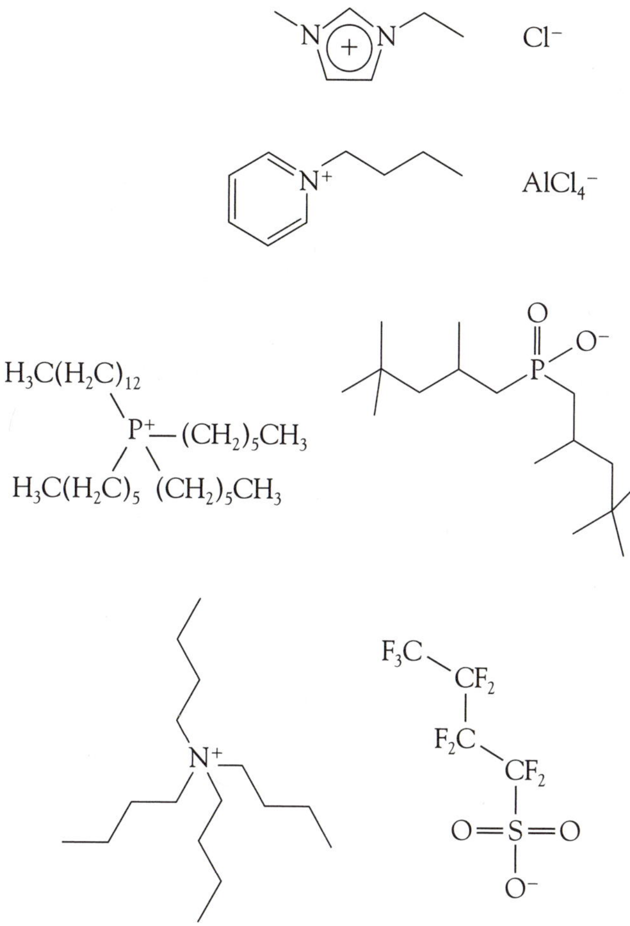

Another strong interest of the green chemistry community is the use of microwave ovens to facilitate chemical processes and reactions. Conventional heat sources—such as heating mantles, Bunsen burners, and oil baths—heat materials from the outside in, transferring energy (in turn) from the heat source to the bottom of the reaction vessel, to the solvent inside the beaker, and finally to the dissolved reactants. In each step, heat energy is lost to the surroundings as it is transferred. A microwave-absorbing reactant or solvent, however, can be targeted by microwaves and therefore can be directly heated by irradiation in a microwave oven or reactor. Thus with microwave heating the contents may be heated directly without heating the vessel. Most people have experienced this phenomenon when heating a cup

FIGURE 3-15 Ion pairs in four typical ionic liquids.

of water in a household microwave oven. The water heats quite quickly while the cup remains relatively cool. Chemists have found that many reactions and processes can be accelerated in a microwave oven, whose efficiency of heating has the potential to reduce energy requirements.

In order to heat effectively via a microwave source, a substance must be polar and/or ionic. ILs heat up very quickly in a microwave since by nature they are ionic. They can reach temperatures as high as 300°C in 15 sec of microwave heating.

Robin Rogers and his group at The University of Alabama won a Presidential Green Chemistry Challenge Award in 2005 for their discovery that certain ILs readily dissolve cellulose when heated with a microwave oven. The process they developed involves the use of gentle, pulsed microwave heating in a domestic microwave oven to expedite the dissolution of cellulose in ILs. Their studies indicate that with the IL 1-butyl-3-methylimidazolium chloride, they can produce solutions with up to 25% (by mass) cellulose.

There is evidence that the chloride ion in this compound disrupts the internal hydrogen bonding in the cellulose, thereby leading to dissolution. The addition of small amounts of water solvates the chloride ions, allowing the hydrogen bonding of the cellulose to resume. The cellulose then precipitates from the solution, and can then be deposited as films, membranes, and fibers.

By dispersing additives in the IL either before or after the dissolution of cellulose, composite or encapsulated cellulose-based materials can be formed when the polymer is regenerated. For example, *laccase*, an enzyme found in fungi that degrades polyphenolic compounds, has been encapsulated in a cellulose support without loss of its activity. The enzyme, when supported on a cellulose film, can be immersed in an aqueous reaction environment and easily removed at the end of the reaction by removing the film, which can then be reused.

The Rogers group has also successfully suspended many other materials in cellulose. These include dyes that can be used to detect metals such as mercury, and magnetite (Fe_3O_4), which produces a composite with uniform magnetic properties. Using this method, cellulose can be combined with other polymers to produce blends. When mixed with polypropylene, a composite that has excellent tear properties is formed. The use of this material for packaging offers significant promise. Encapsulation of medically active compounds along with magnetic materials has the potential to produce microcapsules that can be directed to specific parts of the body. In addition, titanium dioxide infusion into cellulose fibers can produce clothing and bedding with antibacterial properties. The development of cellulose-based materials has the potential to produce new materials with novel characteristics that require less use of petroleum than conventional polymers.

Green Chemistry
Questions 1–7 at the end
of the chapter are based on
the three cases above.

Improving Air Quality: Sulfur-Based Emissions

On a global scale, most SO_2 is produced by volcanoes and by the oxidation of various sulfur-containing gases produced by the decomposition of plants. Because this natural sulfur dioxide is mainly emitted high into the atmosphere or far from populated centers, the background concentration of the gas in clean air is quite small, about 1 ppb. However, a sizable additional amount of sulfur dioxide is emitted into ground-level air, particularly over land masses in the Northern Hemisphere, due to industrial activities.

3.18 Sulfur Dioxide and Hydrogen Sulfide Sources and Abatement

Globally, **the main anthropogenic source of SO_2 is the combustion of coal,** a solid which, depending upon the geographic area from which it is mined, contains 1 to 6% sulfur. **All the sulfur present in coal when it is combusted is oxidized to sulfur dioxide** and mixed with the carbon dioxide and other emission gases. In most countries, including the United States, the major use of coal is to generate electricity. Indeed, electric generation accounts for the majority of SO_2 emissions in the United States (Figure 3-16).

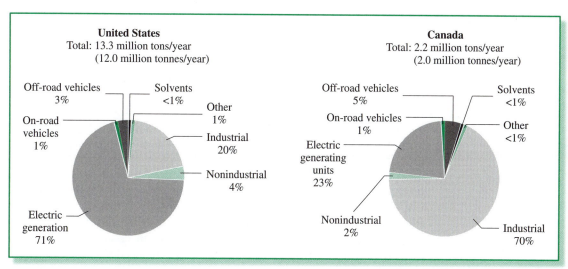

FIGURE 3-16 Anthropogenic SO_2 emissions by sources in North America in 2006.
[Source: International Joint Commission, *Canada–United States Air Quality Agreement: 2008 Progress Report,* Washington, D.C. and Ottawa, Ontario, 2008.]

Approximately half of the sulfur is trapped as inclusions in the incombustible mineral content of the coal. If the coal is pulverized before combustion, much of this type of sulfur can be mechanically removed, as discussed in the next section. The remaining sulfur, which usually amounts to about 1% of the coal's mass, is bonded in the complex organic structure of the solid and *cannot* be removed without expensive processing that breaks covalent bonds.

Large **point sources—individual sites that emit large amounts of a pollutant—of SO_2 are also associated with the nonferrous smelting industry,** where ores are converted into free metals. Many valuable and useful metals, such as copper and nickel, occur in nature as ores containing the **sulfide ion, S^{2-}.** In the first stage of their conversion to the free metals, they were usually "roasted" in air to remove the sulfur; this was converted to SO_2 and traditionally was released into the air. For example,

$$2 \, NiS(s) + 3 \, O_2(g) \longrightarrow 2 \, NiO(s) + 2 \, SO_2(g)$$

Ores such as copper sulfide can be smelted in a process that uses pure oxygen forced into the smelting chamber, and the very concentrated sulfur dioxide that is obtained from the reaction can be readily extracted, liquefied, and sold as a by-product rather than being released into the air. On the other hand, the SO_2 concentration in the waste gases from conventional roasting processes (such as that used for nickel) is high. Consequently, it is feasible to pass the gas over an oxidation catalyst that converts much of the SO_2 to *sulfur trioxide*, to which water can be added to produce commercial concentrated **sulfuric acid, H_2SO_4:**

$$2 \, SO_2(g) + O_2(g) \longrightarrow 2 \, SO_3(g)$$

$$SO_3(g) + H_2O(aq) \longrightarrow H_2SO_4(aq)$$

The latter reaction (which represents only initial reactants and end-product) is in fact accomplished in two steps (not shown) in order to ensure that none of the substances escape into the environment: first the trioxide is combined with sulfuric acid, and then water is added to the resulting solution.

The alternative to sulfur dioxide control—to simply allow the pollutant gas from smelting to be emitted into the air—can cause devastation from SO_2 to the plant life in the surrounding area unless extremely high smokestacks are used. The tallest such stacks in the world are located at Sudbury, Ontario, and reach 400 m high. However, using tall stacks simply solves a local SO_2 problem at the expense of creating a problem downwind. For example, emissions from mainland North America can sometimes be detected in Greenland.

Sulfur occurs to the extent of a few percent in crude oil (with higher concentrations in tar sands and shale oil), but is reduced to the level of a few tens or hundreds of ppm in gasoline and diesel fuel. Some sulfur dioxide is emitted into air directly as SO_2 or indirectly as H_2S by the petroleum industry when oil is refined and natural gas is cleaned before delivery. The huge increase in usage of gasoline and diesel fuel over the past few decades has not produced the corresponding increase in SO_2 into the air since a greater and greater fraction of the sulfur has been removed by refineries; indeed, the amount of sulfur dioxide equivalent removed by refineries now exceeds that emitted into the air.

The predominant component in natural gas wells is sometimes H_2S rather than CH_4! The substantial amounts of hydrogen sulfide obtained from its removal from oil and natural gas are often converted to solid, *elemental sulfur,* an environmentally benign substance, using the gas-phase process called the **Claus reaction:**

$$2\ H_2S + SO_2 \longrightarrow 3\ S + 2\ H_2O$$

One-third of the molar amount of hydrogen sulfide extracted from the fossil fuel is first combusted to sulfur dioxide to provide the other reactant for this process. Huge amounts of elemental sulfur are produced by sulfur removal, especially from natural gas. Notice the analogy between the Claus reaction and the selective catalytic reduction process for nitric oxide control (Section 3-13): they both involve the reaction together of the oxidized and reduced forms of an element (N or S) to form the innocuous elemental form.

It is very important to remove hydrogen sulfide from gases before their dispersal in air because it is a highly poisonous substance, more so than sulfur dioxide. The concentration of H_2S sometimes becomes elevated in the area surrounding natural gas wells during *flaring*—the burning off of gas that cannot be immediately captured. Flaring burns only about 60% of the hydrogen sulfide content of the gas, so the remainder becomes dispersed into the surrounding air. Hydrogen sulfide is also a common pollutant in the air emissions from pulp and paper mills.

In addition to H_2S, several other smelly gases containing sulfur in a highly reduced state are emitted as air pollutants in petrochemical processes; these include CH_3SH, $(CH_3)_2S$, and CH_3SSCH_3. The term **total reduced sulfur is used to refer to the total concentration of sulfur from H_2S and these three compounds.**

The word *reduced* is not used here to denote a decrease in sulfur emissions, but rather to the oxidation number of the sulfur.

3.19 Clean Coal: Reducing Sulfur Dioxide Emissions from Power Plants

When the sulfur dioxide present in exhaust gases is dilute, as in the case of power-plant emissions, its extraction by oxidation is not feasible. Instead, the SO_2 gas is removed by an acid–base reaction between it and **calcium carbonate** (limestone), $CaCO_3$, or **calcium oxide** (lime), CaO, in the form of wet, crushed solid.

Reaction of water with the CaO produces the basic hydroxide $Ca(OH)_2$.

The emitted gases are either passed through a slurry of the wet solid, or bombarded by jets of the slurry. In some applications called *dry scrubbing*, a very fine spray of aqueous calcium oxide or calcium carbonate is used to trap the sulfur dioxide from the emission gases; its water component evaporates upon contact with the hot emission gases.

Almost all the sulfur dioxide gas can be removed by such *scrubber* processes, more formally known as **flue-gas desulfurization.** In some operations, notably in Japan and Germany, the product is fully oxidized by reaction with air, and the resulting **calcium sulfate,** $CaSO_4$, is dewatered and sold as *gypsum*. Usually, the product is a mixture of **calcium sulfite,** $CaSO_3$, and calcium sulfate as a slurry which is then partially dewatered or as a dry solid if granular calcium oxide was used. It is then usually buried in a landfill. The reactions with calcium carbonate are

$$CaCO_3 + SO_2 \longrightarrow CaSO_3 + CO_2$$

$$2\ CaSO_3 + O_2 \longrightarrow 2\ CaSO_4$$

Alternatively, the sulfur dioxide can be captured by the use of slurries of a base such as *sodium sulfite* or *magnesium oxide* or amine salts, and these reactant compounds and concentrated SO_2 gas later regenerated by thermally decomposing the product. Wet scrubbing removes more of the SO_2 (>90%) than does dry scrubbing (>70%), so the latter technique is useful mainly with low-sulfur coals.

Recently, **clean coal technologies** have been developed to use coal in ways that are cleaner and often more energy efficient than those employed in the past. In the various technologies, the cleaning can occur precombustion, during combustion, postcombustion, or by conversion of the coal to another fuel.

• **In precombustion cleaning, the coal has some of the sulfur associated with its mineral content—which exists usually in the form of pyritic sulfur, FeS_2—removed** so it cannot subsequently produce sulfur dioxide. The coal first is ground to a very small particle size, effectively into separate mineral particles and carbon particles. Since they have different densities, the two types of particles can be separated by mixing the pulverized solid in a liquid of intermediate density—namely water—and allowing the fuel portion to rise to the top, where it can be skimmed off, dried, and then combusted. Up to half the pyritic sulfur is removed from the coal in this way.

As an alternative to such physical cleaning, biological or chemical methods can be employed. For example, bacteria cultured to eat the organic sulfur in coal can be utilized. Chemically, the sulfur can be leached from coal with a hot sodium or potassium caustic solution.

• **In combustion cleaning, the combustion conditions can be modified so as to reduce the formation of pollutants, and/or pollutant-absorbing substances are injected into the fuel to capture pollutants as they form.** In **fluidized-bed combustion,** pulverized coal and limestone ($CaCO_3$) are mixed and then suspended on jets of air ("fluidized") in the combustion chamber. About half the sulfur dioxide is thereby captured in solid form as calcium sulfite and sulfate before the gas can escape. This procedure allows for much-reduced combustion temperatures and therefore also greatly reduces the amount of nitrogen oxides that are formed and released. Fluidized-bed combustion finds limited use since its capture efficiency is not high enough to meet emission standards from the combustion of high-sulfur coal.

• Some of the advanced techniques used in postcombustion cleaning—such as the use of granular calcium oxide or calcium carbonate or sodium sulfite solutions—have already been described. In the *SNOX* process developed in Europe, cooled flue gases are mixed with ammonia gas to remove the nitric oxide by its catalytic reduction to molecular nitrogen (by the reaction discussed in Section 3.13). The resulting gas is reheated and the sulfur dioxide oxidized catalytically to sulfur trioxide, which subsequently is hydrated by water to sulfuric acid, condensed, and removed.

In **coal conversion,** the fuel is first gasified by reaction with steam, as will be described in Chapter 7. The gas mixture is cleaned of pollutants, and the purified gas is then burned in a gas turbine that generates electricity. The waste heat of the combustion gases is used to produce steam for a conventional turbine and thus generate more electricity. Alternatively, the gasified coal can be converted into liquid fuels suitable for vehicular use.

Sulfur dioxide emissions from power plants can also be minimized by burning oil, natural gas, or low-sulfur coal for the fuel, though they usually are more expensive than is high-sulfur coal.

PROBLEM 3-10

What mass of calcium carbonate is required to react with the sulfur dioxide that is produced by burning one tonne (1000 kg) of coal that contains 5.0% sulfur by mass?

PROBLEM 3-11

Write the balanced reaction whereby sodium hydroxide can be used to scrub sulfur dioxide from exhaust gases by a reaction that produces water and sodium sulfite, Na_2SO_3. What substance would you have to react with this solution to produce calcium sulfite and regenerate the sodium hydroxide? Write the balanced equation for the latter process, and deduce the net reaction for the cycle.

3.20 Governmental Goals for Reducing Sulfur Dioxide Emissions

The history, by region, of anthropogenic SO_2 emissions into the air over the past century and a half is shown in Figure 3-17. Europe—including the eastern

FIGURE 3-17 Global anthropogenic SO₂ emissions by region. [Source: S. J. Smith et al., *Atmospheric Chemistry and Physics* 10 (2010): 16111–16151.]

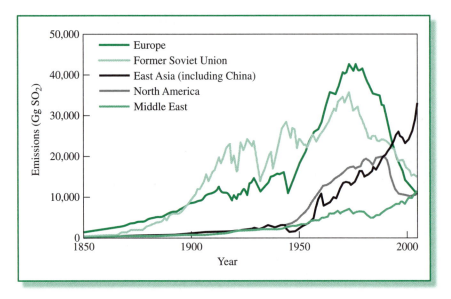

countries—was the largest source of SO_2 emissions through the second half of the twentieth century, but their contribution—along with that from North America, which had been the second-largest emitter—have greatly diminished. East Asia became the leading contributor of the emissions in the 1990s, and continues to be so today.

The global total of SO_2 emissions (not shown; equal to the sum of all the curves in Figure 3-17 plus smaller contributions from other regions) reached its maximum in the 1970s and had been steadily declining from then until the turn of the millennium, but now is again increasing. This recent increase has occurred due to the rise in emissions from coal burning, which has always been the dominant source. Petroleum production and combustion, including emissions from fueling international shipping, is still globally the second largest contributor of SO_2 emissions, with metal smelting the third largest.

The distribution of anthropogenic SO_2 emissions by sector for the United States and Canada is shown in the pie charts in Figure 3-16; emissions for the United States are dominated by electricity generation (coal burning) and those for Canada by smelting (classed as "Industrial" on the graphs). Because of federal regulations, the amount of sulfur dioxide emitted into the air in North America has fallen substantially from its peak level in 1973.

The 1991 *Air Quality Accord* between the United States and Canada required both countries to reduce substantially their sulfur dioxide emissions beyond those of previous laws and agreements. Such emissions in the United States are restricted in accordance with the *Clean Air Act* and its amendments.

The average concentration of sulfur dioxide in air over the United States fell from 13.2 ppb to 7.5 ppb from 1975 to 1991, and further to 3.3 ppm by 2008. Whereas Phase I (1995 deadline) of the Clean Air Act imposed controls on only the largest coal-fired plants, those of Phase II, which began in 2000, imposed more stringent requirements, and applies to almost all plants. There is an SO_2 tonnage limit for each power plant that emits this gas, based upon the power it produces, and a cap on overall national emissions as well. By 2009, the SO_2 emissions in the United States had dropped by 67% compared to 1980 levels.

The reductions in sulfur dioxide emissions by power plants in the U.S. Midwest has been achieved at lower-than-expected cost, due in part to the availability of cheap, low-sulfur coal (1% S, versus more than 3% S in the high-sulfur coal used previously) and inexpensive scrubbers, and in part to the implementation of a system of tradable emission permits. This permit system, in operation since the early 1980s, allows industries to buy emission allowances if they need to exceed their allowed levels, or to sell excess allowances on the open market (through the Chicago Board of Trade) if they do not need their whole allowance. A similar program has been initiated for nitrogen oxide emissions.

In Europe, the EU issued a directive in 1988 specifying reductions from large power plants of 50–70% of their SO_2 emissions from 1980 levels by 2003. This cutback was achieved mainly by switching from coal to natural gas in power stations, and by the use of low-sulfur coal and the scrubbing of emissions in facilities where coal was still burned. According to the 1999 Gothenburg Protocol, Europe's sulfur dioxide emissions are to be cut beyond 1990 levels by another 63% by 2020.

Much of the increase in global SO_2 emissions in the early 2000s was due largely to those from China (included in East Asia in Figure 3-17), which increased by 53% from 2000 to 2006. This was due in large part to the expansion of coal burning, which supplies about two-thirds of China's energy. Emissions from this source began to decline around 2006 when SO_2 began to be removed by flue-gas desulfurization—more than half its power plants now have scrubbers—and by closure of many small coal-fired power plants. The decline was reflected in atmospheric SO_2 concentrations above eastern China, which began to decrease after 2007. Nevertheless, China is now the world's leading source of the gas.

Japan initiated tight controls on SO_2 emissions in the 1970s, and by 1980 their power plants had almost eliminated such emissions by the widespread installation of scrubbers. The rate of emissions from Japan and from South Korea both now continue in gradual decline.

Ironically, volcanic releases of the gas from Japan outrank its anthropogenic sources.

By the turn of the millennium, the high rate of SO_2 emissions from the former Soviet Union had declined by half (Figure 3-17), presumably due more to economic problems than to intentional controls, but have begun to increase again.

3.21 The Oxidation of Sulfur Dioxide in Suspended Water Droplets

Although some sulfur dioxide in air is oxidized by gas-phase reactions, most of it is converted to sulfuric acid after it has dissolved in tiny suspended water droplets present in clouds, mists, etc. The uncatalyzed oxidation of dissolved aqueous SO_2 by dissolved O_2 proceeds at very slow rate unless a catalyst such as Fe^{3+} or the ions of other transition metals are also present in the droplets.

The most important oxidizing agents in the airborne droplets, though present only in tiny concentrations, are not molecular oxygen but the dissolved atmospheric gases ozone and hydrogen peroxide, H_2O_2. The concentration of these two pollutants is much greater in air masses undergoing photochemical smog than in clean air.

In general, the concentration of a dissolved gas in the liquid phase can be determined by considering the equilibrium between its two forms. Thus for hydrogen peroxide, we have

$$H_2O_2\,(g) \rightleftharpoons H_2O_2(aq)$$

The useful form of the equilibrium constant for such processes is the **Henry's law constant, K_H, which is equal to the concentration of the dissolved species divided by the partial pressure of the gas.** For the above reaction, we have

$$K_H = \frac{[H_2O_2]}{P_{H_2O_2}}$$

If the concentration is expressed as a molarity, and the unit of pressure is atmospheres, then from experimental data

$$K_H = 7.4 \times 10^4 \text{ M atm}^{-1}$$

Using this information, we can determine the molarity of H_2O_2 in a raindrop for typical clean-air conditions of 0.1 ppb, i.e., equivalent to 0.1×10^{-9} atm.

$$[H_2O_2] = K_H\,P_{H_2O_2}$$
$$= 7.4 \times 10^4 \text{ M atm}^{-1} \times 0.1 \times 10^{-9} \text{ atm}$$
$$= 7.4 \times 10^{-6} \text{ M}$$

For unknown reasons, hydrogen peroxide levels in suspended aerosol droplets in the air at several California locations are orders of magnitude larger even than predicted by Henry's law.

Although a concentration of 7.4 μM seems tiny by comparison with values routinely encountered in laboratories, it is sufficient to oxidize dissolved SO_2 at an appreciable rate. The hydrogen peroxide concentration in smoggy air is an order of magnitude or more larger than in clean air, so its concentration in water droplets rises accordingly, as does the rate of oxidation of sulfur dioxide.

The calculation of the solubility of SO_2 in raindrops is more complicated, since in the aqueous phase it exists as **sulfurous acid, H_2SO_3:**

$$SO_2(g) + H_2O(aq) \rightleftharpoons H_2SO_3(aq)$$

The Henry's law expression does *not* include the concentration of the solvent, water:

$$K_H = \frac{[H_2SO_3]}{P_{SO_2}}$$

Since $K_H = 1.0$ M atm^{-1} for SO_2, and since its concentration in a typical sample of air is about 0.1 ppm, i.e., equivalent to 0.1×10^{-6} atm, the equilibrium concentration of sulfurous acid is

$$[H_2SO_3] = K_H P_{SO_2}$$

$$= 1.0 \text{ M atm}^{-1} \times 0.1 \times 10^{-6} \text{ atm.}$$

$$= 1.0 \times 10^{-7} \text{ M}$$

This value of about 10^{-7} M for the equilibrium concentration of H_2SO_3 is deceptive since it by no means represents *all* the sulfur dioxide that dissolves in a water droplet (see Figure 3-18). Sulfurous acid is a weak acid whose ionization to the **bisulfite ion,** HSO_3^-, must also be considered in calculating the solubility of sulfur dioxide:

$$H_2SO_3 \rightleftharpoons H^+ + HSO_3^-$$

The **acid dissociation** (or ionization) **constant** K_a for H_2SO_3 is equal to 1.7×10^{-2}, where K_a is related to concentrations by the expression

$$K_a = \frac{[H^+][HSO_3^-]}{[H_2SO_3]}$$

The concentrations in such expressions are equilibrium values. Since the equilibrium molarity of H_2SO_3 is determined in the raindrop by its interchange with SO_2 in air, we can substitute that known value into the K_a expression:

$$K_a = \frac{[H^+][HSO_3^-]}{1.0 \times 10^{-7}}$$

Rearranging the equation to solve for the ion concentrations, which from stoichiometry are equal in value, we obtain

$$[HSO_3^-]^2 = 1.7 \times 10^{-2} \text{ M} \times 1.0 \times 10^{-7} \text{ M}$$

and hence

$$[HSO_3^-] = 4 \times 10^{-5} \text{ M}$$

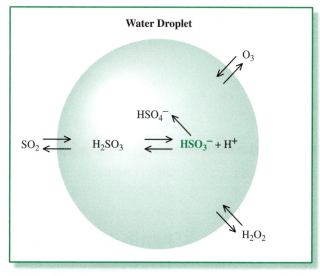

FIGURE 3-18 Dissolution of atmospheric gases SO_2, O_3, and H_2O_2 into a water droplet and their subsequent reactions.

Most oxidation of dissolved
sulfur dioxide occurs via
bisulfite ion, as indicated in
Figure 3-18.

Thus, the equilibrium ratio of bisulfite ion to sulfurous acid in water is about $400:1$. Consequently, the total dissolved sulfur dioxide is about 4×10^{-5} M, rather than just the 1×10^{-7} M that represents the contribution from the un-ionized acid.

Since the concentration of hydrogen ion produced by the reaction is also 4×10^{-5} M, the pH of such raindrops is 4.4. Rain does not become much more acidic than this if no strong acids are dissolved in the droplets.

PROBLEM 3-12

Bisulfite ion can act as a weak acid and ionize further:

$$HSO_3^- \rightleftharpoons H^+ + SO_3^{2-}$$

Given that K_a for HSO_3^- is 1.2×10^{-7}, calculate the concentration of SO_3^{2-} that is present in the raindrops of pH 4.4 discussed above. [*Hint:* The concentrations of bisulfite and hydrogen ion will be very close to their values established previously.]

PROBLEM 3-13

Calculate the pH of rainwater in equilibrium with SO_2 in a polluted air mass for which the sulfur dioxide concentration is 1.0 ppm. [*Hint:* Recall the relationship between partial pressure and ppm concentration discussed earlier in the chapter.]

PROBLEM 3-14

Calculate the concentration of SO_2 that must be reached in polluted air if the dissolved gas is to produce a pH of 4.0 in raindrops without any oxidation of the sulfur.

PROBLEM 3-15

Review Questions 13–17 are
based on the material in the
preceding sections.

(a) Confirm by calculation that the pH of CO_2-saturated water at 25°C is 5.6, given that the CO_2 concentration in air is 390 ppm. For carbon dioxide, the Henry's law constant $K_H = 3.4 \times 10^{-2}$ M atm^{-1} at 25°C. The K_a for carbonic acid, H_2CO_3, is 4.5×10^{-7} at that temperature. (b) Recalculate the pH for a carbon dioxide concentration of 560 ppm, i.e., double that of the preindustrial age.

Particulates in Air Pollution

The black smoke released into the air by a diesel truck is often the most obvious form of pollution that we routinely encounter. The smoke is composed largely of particulate matter. **Particulates are tiny solid or liquid** particles—other than those of pure water—that are temporarily suspended in air and that are usually individually invisible to the naked eye. Collectively, however, such particles often form a haze that restricts visibility. Indeed, on many summer days the sky over North American and European cities is milky white rather than blue. More importantly, **breathing air that contains particulates is known to be hazardous to human health**. Ozone and particulate matter have the greatest negative effects on human health of all air pollutants in most parts of the world, as discussed in Chapter 4. In the material that follows, we investigate the wide range of sizes of the suspended particles and their origins.

3.22 Particulate Size

The particles that are suspended in a given mass of air are neither all of the same size or shape nor do they all have the same chemical composition. The smallest suspended particles are about 0.002 μm (2 nm) in their dimensions; by contrast, the length of typical gaseous molecules is 0.0001 to 0.001 μm (0.1 to 1 nm). The upper limit for suspended particles corresponds to dimensions of about 100 μm (0.1 mm). When atmospheric water droplets coalesce to particles bigger than this, they are raindrops and fall out of the air so quickly they are not considered to be "suspended." The range of particle sizes for common types of suspended particulates is illustrated in Figure 3-19.

Although few of the particles suspended in air are exactly spherical in shape, it is convenient and conventional to speak of all particles as if they were so. Indeed, the **diameter** of particulates is their most important property. **Qualitatively, individual particles are classified as coarse or as fine depending upon whether their diameters are greater or less than 2.5 μm, respectively.** (About 100 million particles of diameter 2.5 μm would be required to cover the surface of a small coin.)

There are many common names for atmospheric particles: "dust" and "soot" refer to solids, whereas "mist" and "fog" refer to liquids, the latter denoting a high concentration of water droplets. **An aerosol is a collection of particulates, whether solid particles or liquid droplets, dispersed in air.** A true aerosol (as opposed, say, to the fairly large droplets from a hair-spray dispenser) consists of very small particles: their diameters are less than 100 μm.

Intuitively, one might think that all particles should settle out under the influence of gravity and be deposited onto the Earth's surface rapidly, but this is not true for the smaller ones due to air turbulence. According to *Stokes' law*, the rate, in distance per second, at which particles settle increases with the square of their diameter. In other words, a particle half the diameter of another falls four times more slowly. The small ones fall so slowly they are suspended almost indefinitely in air (unless they stick to some object they encounter). As we shall see later, the very small ones aggregate to form larger ones, usually still in the fine-size category. Fine particulates usually remain

FIGURE 3-19 Sizes of common airborne fine and coarse particulates. [Source: Adapted from J. G. Henry and G. W. Heinke, *Environmental Science and Engineering* (Upper Saddle River, NJ: Prentice Hall, 1989).]

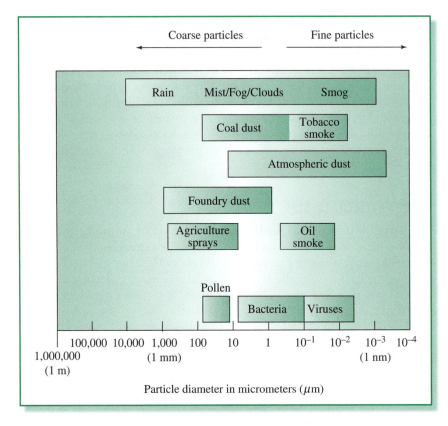

airborne for days or weeks, whereas coarse particulates settle out fairly rapidly. In addition to this sedimentation process, **particles also are commonly removed naturally from tropospheric air by their incorporation into falling raindrops, usually within a week or two.**

3.23 Sources and Composition of Coarse Particles

The primary-versus-secondary distinction made between atmospheric gaseous pollutants (Section 3.3) is also applied to suspended particles. **Most coarse particles are primary,** although they often begin their existence as even coarser matter, since they originate chiefly from the disintegration of larger pieces of matter. Minerals constitute one important type of the coarse particulates in air. Because many of the large particles in atmospheric dust, particularly in rural areas, originate as soil or rock, their elemental composition is similar to that of the Earth's crust, namely high concentrations of Al, Ca, Si, and O in the form of aluminum silicates, some of which also contain the calcium ion (Chapter 16).

- Wind storms in deserts sweep large amounts of fine sand into the air. Dust storms in Asia, whose effects reach as far away as North America, are increasing due to the continuing transformation of fertile land into desert as a consequence of global warming, deforestation, and overgrazing.

- The wind generates coarse particles by the mechanical disintegration of leaf litter.

- Pollen released from plants also consists of coarse, primary particles.

- Wildfires and volcanic eruptions generate both fine and coarse particulate matter.

- Near and above oceans, the concentration of solid NaCl is very high, since sea spray leaves sodium chloride particles airborne when the water evaporates. Indeed, sea-salt aerosols are by far the largest mass of primary particles in air, followed by soil dusts and debris from natural fires.

Although most coarse particulates originate with natural sources, human activities such as stone crushing in quarries and land cultivation result in particles of rock and topsoil being picked up by the wind. Coarse particles in many areas are basic, reflecting the calcium carbonate and other such salts in soils.

3.24 Sources and Composition of Fine Particles

Primary fine particles of anthropogenic origin include ones generated by the wearing of tires and vehicles brakes, and the dust from metal smelting. The incomplete combustion of carbon-based fuels such as coal, oil, gasoline, and diesel fuel produces many fine soot particles, which are mainly crystallites (miniature crystals) of carbon. Consequently, one of main sources of carbon-based primary atmospheric particulates, both fine and coarse, is the exhaust from vehicles, especially those having diesel engines. About half the organic content from heavy-duty diesel vehicles is **elemental carbon** (EC); this soot is commonly seen as the black smoke that emanates from such equipment. Most carbon-containing emissions from gasoline-powered engines are composed of **organic compounds** (OC) rather than elemental carbon.

Whereas coarse particles result mainly from the breakup of larger ones, **fine particles are formed mainly by chemical reactions between gases and by the coagulation of even smaller species, including molecules in the vapor state;** they are mainly secondary particles in nature. Although most of the mass of atmospheric fine particles arises from natural sources, that over urban areas often has mainly an anthropogenic origin.

The average organic content of fine particles is generally greater than that of coarse ones. In areas subject to photochemical smog, **a substantial fraction of the organic compounds in the particulate phase are formed from the reaction of VOCs and nitrogen oxides in the photochemical smog reaction,** and correspond to partially oxidized hydrocarbons that have

The main final products from photochemical smog are gaseous ozone and fine suspended particulates containing OC.

incorporated oxygen to form carboxylic acids, etc. and nitrogen to form nitro groups, etc.

Aromatic hydrocarbons with at least seven carbon atoms (e.g., toluene) that enter the air of cities as VOCs from the evaporation of gasoline also form aerosols. Hydrocarbons having fewer than seven carbons give oxidation products with substantial vapor pressures that remain in the gas phase.

Research performed following the 2010 Deepwater Horizon oil spill discovered that, although VOC hydrocarbons in the surface oil formed organic aerosols relatively quickly, longer-chain semivolatile and intermediate volatility organic compounds reacted to form the bulk of the organic aerosol, even though they were slower to vaporize.

The other important fine particles suspended in the atmosphere consist predominantly of inorganic compounds of sulfur and of nitrogen. Much of the natural sulfur in air originates as *dimethyl sulfide*, $(CH_3)_2S$, emitted from the oceans. A by-product of its oxidation in air is *carbonyl sulfide*, COS, a long-lived trace atmospheric component that also results from the atmospheric oxidation of *carbon disulfide*, CS_2, and from direct emissions from oceans and biomass. Some of the COS makes its way into the stratosphere, where it is oxidized and produces the natural sulfate aerosol found at those altitudes. Both dimethyl sulfide and hydrogen sulfide are oxidized in air mainly to SO_2.

Sulfur dioxide gas is also emitted directly in large quantities both by natural sources such as volcanoes and as pollution from power plants and smelters. It becomes oxidized over a period of hours or days to sulfuric acid and sulfates in the air. Sulfuric acid itself travels in air not as a gas but as an aerosol of fine droplets, since it has such a great affinity for water molecules. A huge volcanic eruption in Iceland in 1783 produced enough sulfuric acid particles to blanket Europe in a "great dry fog" for the entire summer, killing many people.

In Iceland itself, fluoride emitted from the volcano was a worse problem, since it proved fatal to crops, livestock, and people.

Another natural source of atmospheric particles has been discovered. Alkyl iodine compounds such as CH_2I_2 are emitted by seaweed into the air above coastal regions. Absorption of the ultraviolet component of light is sufficient to detach iodine atoms from such gaseous molecules. In subsequent reactions analogous to those of chlorine in the stratosphere, the iodine atoms react with ozone to form *iodine monoxide*, IO, which in turn dimerizes to form I_2O_2. The dimer and other iodine-oxygen compounds condense to form fine particles.

PROBLEM 3-16

By analogy with the reactions of atomic chlorine discussed in Chapter 2, write balanced equations for the reaction of atomic iodine with ozone and for the dimerization of IO.   ●

3.25 The Neutralization of Acids in Air

Fine particles in many areas are acidic, due to their content of sulfuric and nitric acids. The nitric acid is the end-product of the oxidation of nitrogen-containing atmospheric gases such as NH_3, NO, and NO_2, and the sulfuric acid comes from the oxidation of SO_2. Because HNO_3 has a much higher vapor pressure than does H_2SO_4, there is less condensation of nitric acid onto preexisting particles than occurs with H_2SO_4.

Both sulfuric and nitric acids in tropospheric air often eventually encounter ammonia gas that is released as a result of biological decay processes occurring at ground level. The **atmospheric acids undergo an acid–base reaction with the ammonia,** which transforms them into the soluble salts **ammonium sulfate,** $(NH_4)_2SO_4$, and **ammonium nitrate,** NH_4NO_3. Since sulfuric acid contains two hydrogen ions, the neutralization reaction occurs in two-stages, the first producing **ammonium bisulfate,** NH_4HSO_4:

$$H_2SO_4(aq) + NH_3(g) \longrightarrow NH_4HSO_4(aq)$$

$$NH_4HSO_4(aq) + NH_3(g) \longrightarrow (NH_4)_2SO_4(aq)$$

The ammonia that results from animal urine originates in the liquid as urea, $CO(NH_2)_2$, which subsequently hydrolyzes:

$$CO(NH_2)_2 + H_2O \longrightarrow 2\ NH_3 + CO_2$$

The neutralization of acidity by ammonia gas released into the air from livestock, from the use of fertilizers, and by carbonate ion suspended in air from the dust raised by farming activities are the main reason that precipitation over the central United States is not particularly acidic, and similarly for that over regions of China. However, some acidification results from the ionization of the **ammonium ion,** NH_4^+, a weak acid that is produced by ammonia neutralization:

$$NH_4^+ \rightleftharpoons NH_3 + H^+$$

Although the nitrate and sulfate salts initially are formed from acids in aqueous particles, evaporation of the water can result in the production of solid particles. The predominant ions in fine particles are the anions **sulfate,** SO_4^{2-}, **bisulfate,** HSO_4^-, and **nitrate,** NO_3^-, and the cations ammonium, NH_4^+, and hydrogen ion, H^+. Aerosols dominated by oxidized sulfur compounds are called **sulfate aerosols.**

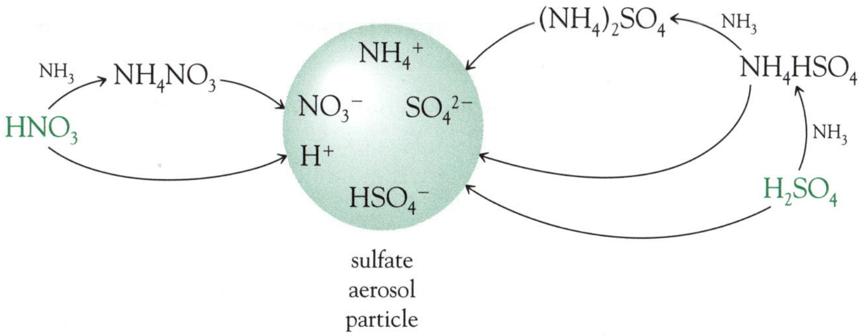

sulfate
aerosol
particle

On the west coast of North America, nitrate rather than sulfate is the predominant anion since more pollution results initially from nitrogen oxides than from sulfur dioxide, since coal mined in the western United States tends be low in sulfur. In Great Britain, most of the fine particles in the winter months originate as soot from car exhaust and pollution from industry, whereas in the summer they arise from the oxidation of sulfur and nitrogen oxides.

If there is substantial ammonia gas in the air, nitric acid will react with it to form ammonium nitrate solid in the particulate phase. Recent simulations of smog formation in Southern California indicate that, although reductions in VOC concentrations without any change in NO_X would reduce ozone formation, the production of nitrate-based particulates would actually *increase* because more nitrogen dioxide would then react to produce nitric acid and then nitrate ion. The simultaneous control of ozone and particulates presents regulators with a formidable challenge!

In summary, coarse particles are usually either soot or inorganic (soil-like) in nature, whereas fine ones are mainly either soot, sulfate, or nitrate aerosols. Fine particles are usually acidic due to the presence of unneutralized acids and are eventually neutralized by ammonia, whereas coarse ones are usually basic also because of their soil content.

3.26 Smoke from Wood Stoves

The burning of wood in domestic fireplaces and wood stoves produces large quantities of particulates, which are emitted from the chimneys into outdoor air. Indeed, in residential neighborhoods where wood is the predominant fuel used for heating, wood stoves contribute up to 80% of the fine particles in the air during the winter months. Some municipalities have controversially limited the number of wood stoves permitted in a given geographical subregion in order to deal with the problem. Outdoor wood-fired boilers, used to heat water for saunas and swimming pools, have grown so much in popularity that the particulates they emit have become a significant problem.

Newer wood stoves and boilers have emission controls incorporated into their design, which greatly reduce the quantity of particulates and gases they release into the air. For example, U.S. EPA-certified wood stoves reduce the amount of smoke from 15–30 (uncontrolled) down to 2–7 grams per hour of operation. The control is accomplished in one of two ways:

• A catalytic converter, consisting of a coated ceramic honeycomb structure is located near the top of the firebox. Smoke and gases from the wood fire below are passed through it and more fully oxidized before continuing upward to the chimney.

• More commonly, the firebox is designed so that near its top, secondary combustion of the smoke occurs before it is emitted into the outside air. This is accomplished by having some of the air entering the stove diverted through tubes under and around the fire, thereby heating it (see Figure 3-20). The hot air is released near the top of the firebox, where the smoke combusts. A baffle is used above this area to divert the air and maximize the amount of time it spends in the secondary combustion area before being released up the chimney.

Because these new stove designs combust much more of the fuel value of the wood, rather than emitting much of it as smoke, they are about 50% more energy-efficient than those of older design.

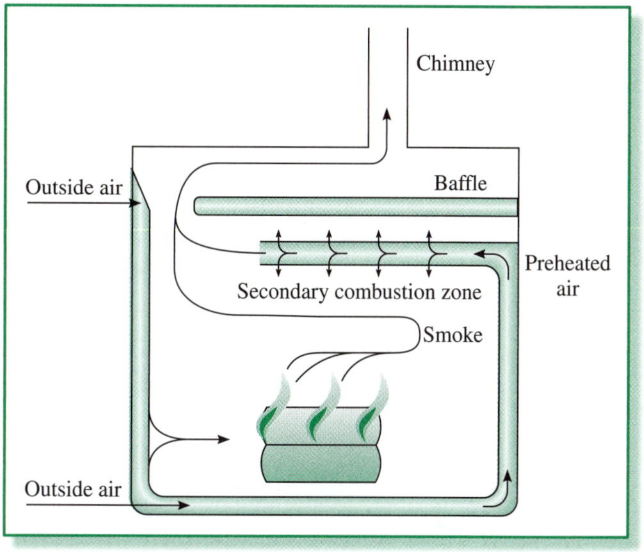

3.27 Smoke over Large Areas of Land

FIGURE 3-20 Noncatalytic wood stove (schematic) with preheated air and baffle for more complete combustion.

Serious episodes of smoky haze pollution over large areas of land have occurred in recent years in Southeast Asia, especially in Malaysia, Singapore, and Indonesia. The smoke originates mainly from forest fires—many of them in Indonesia—that are intentionally started in order to clear land that can be subsequently used for agriculture and to grow trees for their rubber, palm oil, or pulp content. A secondary source of the smoke is the smoldering underground fires that slowly burn in underground coal and peat deposits.

Indeed, there are estimated to be a quarter million individual coal fires currently burning in Indonesia, as well as many in China and India, and there are also many peat fires in Malaysia. The fires are initiated when an outcropping of coal or a peat deposit that has dried after draining is ignited, typically during one of the fires set to clear the land. Fires can also be ignited by lightning strikes in coal exposed to the air, and even by spontaneous combustion when the surface pyrite is oxidized and the heat released by this reaction sets the carbon ablaze. These underground fires can continue to burn for decades after the original forest fires have stopped.

A so-called *atmospheric brown cloud* of particles and gases from forest fires, vehicle exhausts, and domestic cookers—especially in rural areas—that burn wood, dung, and agricultural waste overhangs most of East and Southeast Asia annually from December to May, the main season for home heating. The brown cloud over the Indian Ocean consists mainly of smoke from the burning of dried manure in cooking fires. This haze lowers sunlight levels at the surface by up to 15%, with a corresponding decline in the yield of crops such as rice and an alteration to rainfall and monsoon patterns. In contrast to the pollution aerosol over North America and Europe, to which it is comparable in magnitude, the "black carbon" content of the brown cloud is significant. The absorption of sunlight by this elemental carbon alters the local hydrological cycle and hence the weather over the northern Indian Ocean. The lack of nitric oxide produced in the low-temperature flames of burning biomass currently limits ozone production over the area, but that will likely be reversed in the future with increased use of fossil fuels for vehicles.

Large forest fires in northern Canada produce huge quantities of carbon monoxide and VOCs, which have been found to travel as far as the U.S. Southeast and which may well increase ozone and particulate concentrations in the air of this region.

Air Quality Indices and Size Characteristics for Particulate Matter

As we shall see in Chapter 4, the effect of particles suspended in air upon human health depends significantly upon the size of the particles involved. In the material that follows, we investigate the pollution indexes used by governmental agencies to characterize the level of particulate air pollution present in an air sample as well as the effect of particle size on visibility through air masses.

3.28 The PM Indices

When air quality is monitored, the most common measure of the concentration of suspended particles is the **PM index, which is the amount of particulate matter that is present in a given volume**. Since the matter involved usually is not homogeneous, no molar mass for it can be quoted and thus concentrations are given in terms of the mass, rather than the number of moles, of particles present. The usual units are *micrograms* of particulate matter *per cubic meter* of air, i.e., $\mu g\ m^{-3}$. Because smaller particles have a greater detrimental effect on human health than do larger ones, as we shall see later in this chapter, usually **only those having a specified diameter or less are collected and reported; this cut-off diameter, in μm, is listed as the subscript to PM.**

Government agencies in many countries, including the United States and Canada, monitor PM_{10}, **which is the total concentration of all particles having diameters less than 10 μm,** which corresponds to all of the fine-particle range plus the smallest members of the coarse range. These are called **inhalable** particles since they can be breathed into the lungs. A typical value for PM_{10} in an urban setting is 20–30 $\mu g\ m^{-3}$.

More commonly, regulators in developed countries now use the $PM_{2.5}$ **index, i.e., that which includes all and only fine particles, which are also called respirable particles.** The respirable range includes only particles that can penetrate deep into the lungs, where there are no natural mechanisms such as the cilia that line the walls of bronchial tubes to catch particles and move them up and out. Urban $PM_{2.5}$ values are usually in the 10–20 $\mu g\ m^{-3}$ range in North America, although background concentrations are only 1–5 $\mu g\ m^{-3}$.

Some researchers have introduced the term $PM_{10-2.5}$, which, although termed "coarse" corresponds only to the smallest of the particles covered by the conventional definition of coarse particles. **The new term ultrafine is applied to particles with very small diameters, usually taken to be less than 0.1 μm.** **Nanoparticles** are still smaller, less than about 0.02 μm. (Figure 1 in Box 3-3 shows the distribution of particles in an air mass and visually summarizes the size definitions.) Most ultrafine particles are anthropogenic in origin. In the past, the **total suspended particulates,** abbreviated TSP, which is the concentration of all particulates suspended in air, was often reported instead of a PM index.

As the importance of suspended particulate matter to human health (discussed in Chapter 4) became more and more apparent in the late twentieth century, some governments set air quality standards for PM_{10}. Now that $PM_{2.5}$ is recognized to be more relevant than PM_{10} and the instrumentation for its measurement is widely available, standards for it have been implemented in some countries such as the United States and have been proposed for others, and in some cases have replaced PM_{10} regulation. The status of PM regulations, as of 2010, is summarized in Table 3-2.

Governments generally do not attempt to regulate overall emissions of particulate matter since so much of it is secondary in origin, not primary. An exception is the emission of particulates from vehicles.

Particulate air pollution globally is a serious problem, especially in cities in developing countries; as indicated in Figure 3-21, few of the world's megacities currently come close to meeting the World Health Organization annual standard for PM_{10} (horizontal green line) of 20 μg m^{-3}. Several other Chinese and Indian cities and Cairo are also among the most particulate-polluted cities in the world. Given the continuing increase in vehicles, most lacking emission controls, in developing countries, and the fact that many of

TABLE 3-2	Particulate Matter Air Standards, in units of μg m^{-3}			
Country or Organization	$PM_{2.5}$ 24-hour	$PM_{2.5}$ Annual	PM_{10} 24-hour	PM_{10} Annual
WHO	25	10	50	20
USA	35	15	150	
Canada	(30)			
EU	—	(25)	50	40
Australia	(25)	(8)	50	
China	—	—	50–250	40–150
India	60	40	100	60

Note: Figures in round brackets are tentative standards.

FIGURE 3-21 Particulate
air quality in major world
cities *Note*: Values are
annual PM_{10} means. The
green horizontal line is the
WHO air quality guideline
for this parameter. [Source:
D. D. Parish and T. Zhu, *Science*
326 (2009): 674–675.]

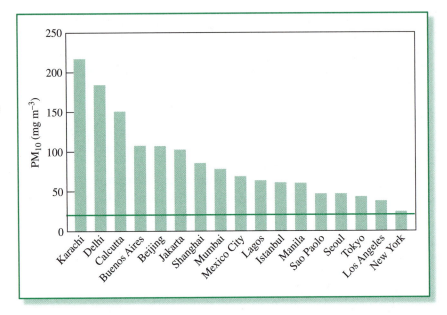

FIGURE 3-21 Particulate air quality in major world cities *Note*: Values are annual PM_{10} means. The green horizontal line is the WHO air quality guideline for this parameter. [Source: D. D. Parish and T. Zhu, *Science* 326 (2009): 674–675.]

them are diesel-powered, it seems likely that PM levels will increase over time. Most Chinese cities failed to meet the daily and annual WHO guidelines for $PM_{2.5}$, Beijing and Shanghai by factors of ten and six, respectively, in the early 2000s.

Ironically, although levels of primary particulate matter and that formed from SO_2 decreased over China in the late 2000s due to government controls and an economic downturn, by 2009 particulate levels were again increasing due to their formation as secondary pollutants from reaction of NO_X emissions from the increasing number of vehicles on the road. Almost all sites in the United States meet its annual PM_{10} standard, although not the lower limit more generally adopted, and there are still some areas that miss the $PM_{2.5}$ limit, notwithstanding gradual improvement over the last decade.

Many large cities in sub-Saharan Africa suffer from high levels of particulate air pollution from vehicle exhaust, although little regular monitoring is yet in place there. In Nairobi, Kenya, for example, the air sampling during the mid-2000s revealed very high levels of $PM_{2.5}$, with enhanced black carbon. Most of the vehicles in Nairobi at the time had been imported from Asia as used cars and had no pollution control devices.

PROBLEM 3-17

What would be the correct PM symbol for an index that included only ultrafine particles? What would be the PM symbol for the TSP index? Numerically, would the value for the ultrafine component of a given air mass be larger or smaller than its TSP?

As discussed in detail in Box 3-3, the distribution of particles suspended in air peaks in the micrometer region because **smaller particles coagulate** to form ones of this size and further growth is slow, and because much larger ones rapidly settle out. The large increase in surface area that occurs when a large particle is split into smaller ones is explored in Problem 3-18.

Review Questions 18–21 are based on material in the preceding sections.

PROBLEM 3-18

Let k be a given measure of length; then suppose a cubic particle of dimension $3k \times 3k \times 3k$ is split up into 27 cubes of size $k \times k \times k$. Calculate the relative increase in surface area when this occurs by comparing the surface area (length times width) of the six faces of the larger cube to the sum of all those of the smaller ones. From your answer, deduce whether the total surface area of a given mass of atmospheric particle is larger or smaller when it occurs as a large number of small particles rather than a small number of large ones.

BOX 3-3   The Distribution of Particle Sizes in an Urban Air Sample

Because particles suspended in the atmosphere are of different origins and compositions, and were often formed and interacted over a period of time in haphazard ways, there is a wide distribution of particle sizes present in any air mass.

One way of looking at the distribution of sizes is to plot the *number* of particles having a given diameter against the diameter; this is done in the dark green curve of Figure 1 for a typical urban air sample.

The peak in the distribution occurs at just under 0.01 μm. The oddly shaped net distribution is presumably the sum of several symmetrical ("bell shaped") distributions having peaks at different diameters. The particles of the distribution at the smallest diameter are primary in nature, formed by the condensation of vapors of pollutants formed by chemical reactions, such as the sulfuric acid formed by the oxidation of gaseous sulfur dioxide and the soot particles formed by combustion. The coagulation of such particles into larger ones (which can occur in minutes) and the deposition of gas molecules onto them result in the broad distribution of secondary particles. Particles of this size also are created when water in aqueous droplets containing dissolved solids evaporate. Growth beyond this size is slow because the larger the particle, the slower it moves and thus the less likely it is to encounter and coagulate with particles of comparable size. Growth by condensation of gases is also slow for larger particles since their surface-to-mass ratio is smaller than for small particles.

The particles associated with the right-side tail of the distribution are mainly soot or consist of material produced by mechanical disintegration of soil particles, etc. Numerically, there are few particles of mass *larger* than a few microns in diameter because each has such a large mass and because the larger ones quickly settle out of the air (Stokes' law), although particles that then settle on

(continued on p. 130)

roadways often become resuspended by the action of vehicular traffic.

The plots of particle *numbers* can be misleading for some purposes because tiny particles of very small mass and surface area dominate the samples and thus the distributions. One alternative way to represent the data in a more meaningful way is to plot the total *mass* of all particles of a given size in an air sample against the diameter to see how mass is distributed among the different sizes. This type of plot is shown by the light green curve in Figure 1.

The distribution function for mass is displaced to larger diameters compared to that for particle numbers for the following reason: the mass (or volume) of a particle is proportional to the cube of its diameter d (since for a sphere volume is proportional to the cube of the radius), so the height of the curve at any diameter of the distribution in the light green curve in Figure 1 corresponds to the value for the number distribution for this air mass times d^3. Very small particles, even though numerous, do not collectively have a large mass. Consequently, in the mass distributions, the peak heights for larger particles are higher than are those for

smaller ones; the whole distribution appears to have shifted to higher diameters. Two symmetrical distribution curves, one centered in the fine region at about 0.4 μm and the other in the coarse region at about 9 μm, appear to be superimposed to produce the final bimodal "double-humped" distribution. Notice that the total mass of the coarse-particle range (i.e., the sum of the area under the dark green curve for $d > 2.5$ μm) in Figure 1 is greater than that for the fine region; this ratio is even larger for clean, rural air masses.

PROBLEM 1

The distribution of particle surface area against size is proportional to the *square* of the diameter times the number distribution, since the surface area of a sphere is proportional to the square of its radius or diameter. Given the distribution for the particle volumes and masses (Figure 1), which is d^3 times the number distribution rather than d^2 times it, qualitatively predict the shape of the distribution curve and approximate location of peak(s) for total particle surface area.

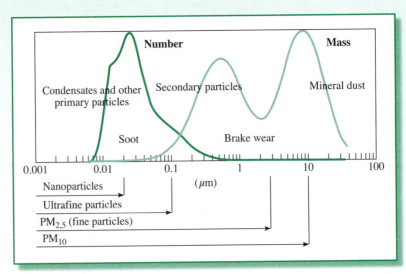

FIGURE 1 The distribution of particle numbers and of particle masses versus particle size for airborne particles. Particles in the condensates, soot, and brake-wear categories originate mainly from vehicular traffic. The mineral dust comes from road abrasion and from agricultural and natural sources. [Source: Modified from J. Fenger, *Atmospheric Environment* 43 (2009): 13–22.]

Review Questions

1. In the "micrograms per cubic meter" concentration scale, to what substances do micrograms and cubic meters refer?

2. What chemical substance initiates the air oxidation of stable molecules? How is it initially formed, and how is it reformed?

3. In general terms, what is meant by *photochemical smog*? What are the initial reactants in the process? Why is sunlight required?

4. What is meant by a *primary pollutant* and by a *secondary pollutant*? Give examples.

5. How does OH react with a stable molecule containing a C=C bond? With an alkane?

6. What is meant by the term *synergism*? Give an example.

7. What is the chemical reaction by which *thermal NO* is produced? From which two sources does most urban NO arise? What is meant by the term NO_X? What is meant by *fuel NO*?

8. Describe the strategies by which reduction of urban ozone levels has been attempted. What difficulties have been encountered in these efforts? Is photochemical smog strictly a localized urban problem?

9. What is meant by geographic regions that are *VOC-limited*? NO_X-*limited*?

10. Describe the operation of the *three-way catalyst* in transforming emissions released by an automobile engine. Does the catalyst operate when the engine is cold? Why is it important for converters that the level of sulfur in gasoline be minimized?

11. Describe the manner in which emissions from diesel-powered vehicles can be controlled, including the use of catalytic converters.

12. Describe the reaction used in the *selective catalytic reduction* of nitrogen oxides. What other techniques are used for NO_X emission control from power plants?

13. What are the main anthropogenic sources of sulfur dioxide? Describe the strategies by which these emissions can be reduced. What is the *Claus reaction*?

14. What species are included in the air pollution index called *total reduced sulfur*?

15. Describe the three strategies used in *clean coal*.

16. What two species, other than O_2, are active oxidizing agents of sulfur dioxide in atmospheric water droplets?

17. State *Henry's law*.

18. Define the term *aerosol*, and differentiate between *coarse* and *fine particulates*. What are the usual origins of these two types of atmospheric particles?

19. What are the usual chemical components of a *sulfate aerosol*?

20. Write a balanced equation illustrating the reactions that occur between one molecule of ammonia and (a) one molecule of nitric acid and (b) one molecule of sulfuric acid.

21. What are the usual concentration units for suspended particulates? What would the designation PM_{40} mean? What do the terms *respirable* and *ultrafine* mean?

 ## Green Chemistry Questions

1. The development of a low-VOC coalescent for paints by ADM won a Presidential Green Chemistry Challenge Award.

(a) Into which of the three focus areas (see page xxviii) for these awards does this award best fit?

(b) List two of the twelve principles of green chemistry that are addressed by this discovery. Justify each of your answers.

2. What are the environmental advantages of the coalescent developed by ADM?

3. Consider the structure (Figure 3-11b) of the coalescent developed by ADM. Using your knowledge of organic chemistry, explain why this molecule undergoes reaction with oxygen so readily. Would you expect linoleic acid to react in a similar manner? Stearic acid?

4. *PERC* replaced gasoline and kerosene in the dry-cleaning process.

(a) Describe any environmental problems or worker hazards that would be associated with these solvents.

(b) Would these same environmental problems or workers hazards be eliminated by the use of PERC?

(c) By the use of carbon dioxide?

5. The development of surfactants for carbon dioxide by Joseph DeSimone won a Presidential Green Chemistry Challenge Award.

(a) Which of the three focus areas (see page xxviii) for these awards does this award best fit into?

(b) List two of the twelve principles of green chemistry (see pages xxiii–xxiv) that are addressed by the green chemistry developed by DeSimone.

6. The ions in ionic liquids (ILs) have weak ionic attractions for one another. This weak interaction is due to one or more factors including

• the presence of bulky nonpolar groups which prevent the close interaction of the charged regions of the ions, and

• delocalized and/or dispersed charges resulting in low charge density.

Inspect the (ILs) in Figure 3-15 and discuss the structural features of these compounds that result in weak interactions between the oppositely charged ions.

7. The discovery of the dissolution of cellulose with ionic liquids and the formation of various cellulose composites by Robin Rogers won a Presidential Green Chemistry Challenge Award.

(a) Which of the three focus areas (see page xxviii) for these awards does this award best fit into?

(b) The use of an abundant, naturally occurring polymer, a microwave heat source, and ionic liquids are three important green chemistry aspects of this study. For each of these aspects list at least two of the twelve principles of green chemistry (see pages xxiii–xxiv) that are addressed in this study.

Additional Problems

1. The rate constant for the oxidation of nitric oxide by ozone is 2×10^{-14} molecule^{-1} cm^3 sec^{-1}, whereas that for the competing reaction in which it is oxidized by oxygen, i.e.,

$$2\,NO + O_2 \longrightarrow 2\,NO_2$$

is 2×10^{-38} molecule^{-2} cm^6 sec^{-1}. For typical concentrations encountered in morning smog episodes, namely 40 ppb for ozone and 80 ppb for nitric oxide, deduce the rates of these two reactions and decide which one is the dominant process. [*Hint:* The concentrations of the reactants must be expressed in units appropriate to the rate constant.]

2. In a particular air mass, the concentration of OH was found to be 8.7×10^6 molecules cm^{-3}, and that of carbon monoxide was 20 ppm.

(a) Calculate the rate of the reaction of OH with atmospheric CO at 30°C, given that the rate constant for the process is $5 \times 10^{-13} e^{-300/T}$ molecule^{-1} cm^3 sec^{-1}.

(b) Estimate the half-life of an OH molecule in air, assuming that its lifetime is determined by its reaction with CO. [*Hint:* Re-express the rate law as a pseudo-first-order process with the level of CO fixed at 20 ppm. Consult your introductory chemistry textbook to find the relationship between the half-life of a substance and the rate constant for its first-order decay.]

3. In the overall reaction that produces nitric oxide from N_2 and O_2, the slow step in the mechanism is the reaction between atomic oxygen and molecular nitrogen to produce nitric oxide and atomic nitrogen.

(a) Write out the chemical equation for the slow step and the rate law equation for it.

(b) Given that its rate constant at 800°C is 9.7×10^{10} L mol^{-1} sec^{-1}, and that its activation energy is 315 kJ mol^{-1}, calculate the amount by which the rate constant increases if the temperature is raised to 1100°C.

4. At combustion temperatures, the equilibrium constant for the reaction of N_2 with O_2 is about 10^{-14}. Calculate the concentration of nitric oxide that is in equilibrium with atmospheric levels of nitrogen and oxygen. Repeat the calculation for normal atmospheric temperatures, at which the equilibrium constant is about 10^{-30}. Given that the concentration of NO that exits from the combustion zone in a vehicle is much higher than this latter equilibrium value, what does that imply about equilibrium in the reaction mixture? [*Hint:* Use the stoichiometry of the reaction to reduce the number of unknowns in the expression for K.]

5. The concentration of ozone in ground-level air can be determined by allowing the gas to react with an aqueous solution of potassium iodide, KI, in a redox reaction that produces molecular iodine, molecular oxygen, and potassium hydroxide.

(a) Deduce the balanced reaction for the overall process.

(b) Determine the ozone concentration, in ppb, in a 10.0-L sample of outdoor air if it required 17.0 μg of KI to react with it.

6. The percentage of sulfur in coal can be determined by burning a sample of the solid and passing the resulting sulfur dioxide gas into a solution of hydrogen peroxide, which oxidizes it to sulfuric acid, and then titrating the acid. Calculate the mass percent of sulfur in a sample if the gas from an 8.05-g sample required 44.1 mL of 0.114 M NaOH in the titration of the diprotic acid.

7. Calculate the volume, at 20°C and 1.00 atm, of SO_2 produced by the conventional roasting of 1.00 tonnes (1,000 kg) of nickel sulfide ore, NiS. What mass of pure sulfuric acid could be produced from this amount of SO_2?

8. Ironically, SO_2 could be extracted from gas emissions by passing it through a solution of sulfite ion, SO_3^{2-}.

(a) Assuming sulfite ion acts as a base and the sulfur dioxide is present in water initially as sulfurous acid, write an acid–base reaction between the species.

(b) Devise a scheme by which dilute sulfur dioxide in an emission gas could be captured by an aqueous solution of sulfite ion, and later released as a concentrated stream of SO_2.

9. Assuming its concentration in air is 2.0 ppb, calculate the molar solubility of SO_2 in raindrops whose pH is fixed (by the presence of strong acids) to be 4.0, 5.0, and 6.0. The data required for the calculations is present in Section 3.21 of the text.

10. The sulfur species that undergoes oxidation in water droplets is the bisulfite ion, HSO_3^-, so the rate of oxidation is proportional to its concentration multiplied by that of the oxidizing agent. Predict how changes in pH in the droplet will affect the rate of oxidation (a) if O_3 reacts with bisulfite ion, and (b) if hydrogen peroxide in the protonated form $H_3O_2^+$, formed in the equilibrium

$$H_2O_2 + H^+ \rightleftharpoons H_3O_2^+$$

is the species that reacts with bisulfite.

11. The settling rate of particulates in air is directly proportional to the square of their diameters (Stokes' law), provided that their densities are equal. If emitted particulates with a specific diameter are found to settle out after two days, how long would it take particulates of the same material with half the diameter to settle out if they are emitted from the same tall chimney?

4

The Environmental and Health Consequences of Polluted Air—Outdoors and Indoors

In this chapter, the following introductory chemistry topics are used:

- ➲ pH and acid–base concepts
- ➲ Balancing of redox equations

Background from previous chapters used in this chapter:

- ➲ Photochemical smog
- ➲ Coarse and fine particulates, aerosols; PM_x indexes
- ➲ Thermal NO
- ➲ Temperature inversions; UV types
- ➲ ppm, ppb, and $\mu g\ m^{-3}$ concentration scales for gases

Introduction

Smog, whether sulfur-based or photochemical, often has unpleasant odors due to some of its gaseous components. More seriously, the initial pollutants, intermediates, and final products of the reactions in smog affect human health and can cause damage to plants, animals, and some materials. In this chapter, the detrimental effects on humans and other animals, plants, and materials of the gases and particles in polluted air—including the air we encounter indoors—and methods by which air pollution can be combatted, are described. Included in the discussions are the environmental effects of acid rain, a phenomenon that results from polluted air.

4.1 Haze

The most obvious manifestation of photochemical smog is a yellowish-brownish-gray haze that is due to the presence in air of small water droplets containing products of chemical reactions that occur among pollutants in air. This haze, familiar to most of us who live in urban areas, now extends periodically to once-pristine areas such as the Grand Canyon in Arizona.

Particles whose diameter is about that of the wavelength of visible light, i.e., 0.4–0.8 μm, can scatter light—reflect it in random directions—and consequently interfere with its transmission, thereby reducing visual clarity, long-distance visibility, and the amount of sunlight reaching the ground. **A high concentration in air of particles of diameters between 0.1 μm and 1 μm produces a haze.** Indeed, one conventional technique of measuring the extent of particulate pollution in an air mass is to determine its haziness. The existence of smog in the air can often be determined by simply looking at buildings or hills in the distance, and seeing if their appearance is partially masked by haze.

The widespread haze in the Arctic atmosphere in winter is due to sulfate aerosols that originate from the burning of coal, especially in Russia and Europe. The enhanced haziness in summertime over much of North America is due mainly to sulfate aerosols arising from industrialized areas in the United States and Canada. Haze over China, produced by air pollution, so reduces sunlight intensity that it may be cutting food production by as much as 30% across a third of the country.

Fine particles are largely responsible for the haze in Los Angeles and other locations subject to episodes of photochemical smog. The smog aerosols contain nitric and sulfuric acids that have been neutralized to salts. Also present in these aerosols are organic carbon products that are formed in the photochemical smog reactions; however, intermediates formed from fuel molecules having short carbon chains usually have high enough vapor pressures that they exist as gases rather than condense onto particles. The typical composition of the fine component of an aerosol suspended over continental areas is illustrated in Figure 4-1.

Since **most fine particles in urban air are secondary,** their number can be controlled only by reducing emissions of the primary pollutant gases from which they are created. Thus governments have successively required more and more stringent emission controls on vehicles, power plants, etc., as discussed in

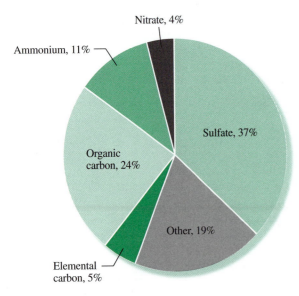

FIGURE 4-1 Typical composition of fine continental aerosol. [Adapted from J. Heintzenberg, *Tellus* 41B (1989): 149–160.]

Chapter 3. The switch to low-sulfur gasoline and diesel fuels have made catalytic converters on vehicles more efficient in reducing emissions.

Acid Rain

One of the most serious environmental problems facing many regions of the world is acid rain. This generic term covers a variety of phenomena, including *acid fog* and *acid snow*, all of which correspond to atmospheric precipitation of substantial acidity. In this section, the nature of the acids present in precipitation is discussed.

4.2 Natural and Anthropogenic Acid Rain

The phenomenon of acid rain was discovered by Angus Smith in Great Britain in the mid-1800s, but then it was essentially forgotten until the 1950s. It refers to **precipitation that is significantly more acidic than "natural" (i.e., unpolluted) rain,** which itself is often mildly acidic due to the presence in it of dissolved atmospheric carbon dioxide, which forms **carbonic acid,** H_2CO_3.

$$CO_2(g) + H_2O(aq) \rightleftharpoons H_2CO_3(aq)$$

The weak acid H_2CO_3 then partially ionizes to release a **hydrogen ion,** H^+, with a resultant reduction in the pH of the system:

$$H_2CO_3(aq) \rightleftharpoons H^+ + HCO_3^-$$

Because of this source of acidity, the pH of unpolluted, "natural" rain is about 5.6 (see Problem 3-15a). Only rain that is appreciably more acidic than this—i.e., with a pH of less than 5—is considered to be truly "acid" rain since, because of natural trace amounts of strong acids, the acidity level of rain in clean air can be a little greater than that owing to carbon dioxide alone.

Strong acids such as **hydrochloric acid,** HCl, produced by emissions of hydrogen chloride gas by volcanic eruptions can produce "natural" acid rain temporarily in regions such as Alaska and New Zealand. On the other hand, the pH of unpolluted rain can be somewhat *greater* than 5.6 due to the presence of weakly basic substances such as carbonates originating with airborne soil particles that have partially dissolved in the droplets.

The two predominant acids in acid rain are sulfuric acid, H_2SO_4, and nitric acid, HNO_3, both of which are strong acids. Generally speaking, acid rain is precipitated far downwind from the source of the corresponding primary pollutants, namely **sulfur dioxide,** SO_2, and **nitric oxide,** NO. **The strong acids are created during the transport of the air mass that contains the primary pollutants.**

$$SO_2 \xrightarrow[H_2O]{O_2} H_2SO_4$$

$$NO_X \xrightarrow[H_2O]{O_2} HNO_3$$

Overall reactions

The late economist-philosopher John Kenneth Galbraith once noted, "Acid rain falls on the just and the unjust and also equally on the rich and poor." Economists recognize this effect as an externality or an **external cost.** The economic, environmental, and social costs of acid rain are not reflected in the price of fossil fuels, but this cost is borne by all who reside in "the commons."

Consequently, acid rain is a pollution problem that does not respect state or national boundaries because of the long-range transport that the atmospheric pollutants often undergo. For example, most acid rain that falls in Norway, Sweden, and the Netherlands originates as sulfur and nitrogen oxides emitted in other countries in Europe. Indeed, the modern recognition of acid rain as a problem stems from observations made in Sweden in the 1950s and 1960s, which were due to emissions from outside its borders.

4.3 The Acids and Acidity of Acid Rain

Acid rain has a variety of ecologically damaging consequences, and the presence of acid particles in air may also have direct effects on human health. However, **the effects of acid rain on soil vary dramatically from region to region.** In this section, we investigate the chemical processes underlying the ecological effects of acid rain.

Nitric oxide is not especially soluble in water, and the acid (sulfurous) that sulfur dioxide produces upon dissolving in water is a weak one. Consequently, **the primary pollutants NO and SO_2 themselves do not make rainwater particularly acidic.** However, some of the mass of these primary pollutants is converted over a period of hours or days into the secondary pollutants nitric acid and sulfuric acid, both of which are very soluble in water and are strong acids. Indeed, virtually all the acidity in acid rain is due to the presence of these two acids. In eastern North America, sulfuric acid greatly predominates because some electrical power is generated from power plants that use high-sulfur coal. In western North America, nitric acid attributable to vehicle emissions is predominant, since the coal mined and burned there is low in sulfur.

Figure 4-2 shows a contour map of the average pH in precipitation in different regions of the world. The lowest pH ever recorded, 2.4, occurred for a rainfall in April 1974 in Scotland. Indeed, central-west Europe, including the United Kingdom, has a serious acid rain problem, as can be seen from the pH = 4.0 and 4.5 contours surrounding the area in Figure 4-2. In North America, the greatest acidity occurs in the eastern United States and in southern Ontario, since both regions lie in the path of air originally polluted by emissions from power plants in the Ohio valley. On the other hand, much of the acidity that falls in upper New York State stems from emissions in southern Ontario.

In addition to the acids delivered to ground level during precipitation, a comparable amount is deposited on the Earth's surface by means of **dry deposition, the process by which nonaqueous chemicals are deposited onto solid and liquid surfaces at ground level when air containing them passes over the surfaces.** Much of the original SO_2 gas is *not* oxidized in the air but rather is removed by dry deposition from air before reaction can occur: oxidation and conversion to sulfuric acid occurs after deposition. **Wet deposition** processes encompass the transfer of pollutants to the Earth's surface by rain, snow, or fog—i.e., by aqueous solutions.

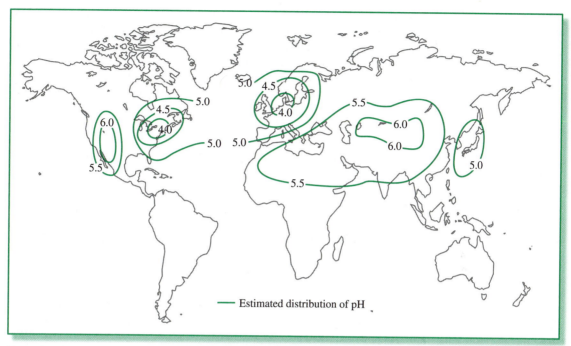

— Estimated distribution of pH

FIGURE 4-2 Global pattern of acidity of precipitation. [Source: Redrawn from J. H. Seinfeld and S. N. Pandis, *Atmospheric Chemistry and Physics* (Chichester: John Wiley, 1998).]

In recent years, a new source of sulfuric acid in lakes has appeared—the oxidation of sulfur in shallow wetlands dried up by global warming and thereby exposed to the air.

As Problem 4-1 shows, the oxidation of **ammonium ion,** NH_4^+, to nitrate ion produces hydrogen ions. Indeed, the large emissions of ammonia into the air from manure in areas of livestock and poultry farming result in the atmospheric deposition of ammonium ion, which then is oxidized by soil microbes. The resulting H^+ contributes to the acidification of soil. Large-scale agriculture in China produces extensive emissions of ammonia, so acidity from ammonium ion presents a problem there.

PROBLEM 4-1

Deduce the balanced redox half-reaction of conversion of ammonium ion, NH_4^+, to nitrate ion, NO_3^-, and thereby show that H^+ is also produced in this process.

Note that the techniques for balancing redox equations are reviewed in the Appendix.

4.4 Neutralization of Acid Rain by Soil

The extent to which acid precipitation affects biological life in a given area depends strongly upon the composition of the soil and bedrock in that

area. If the bedrock is limestone or chalk, the acid can be efficiently neutralized ("buffered"), since these rocks are composed of **calcium carbonate,** $CaCO_3$, which acts as a base and reacts with acid, producing **bicarbonate ion,** HCO_3^-, as an intermediate:

$$CaCO_3(s) + H^+(aq) \longrightarrow Ca^{2+}(aq) + HCO_3^-(aq)$$

$$HCO_3^-(aq) + H^+(aq) \longrightarrow H_2CO_3(aq) \longrightarrow CO_2(g) + H_2O(aq)$$

The reactions here proceed almost to completion, owing to the excess of H^+ that is present. Thus the rock dissolves, producing carbon dioxide and replacing the hydrogen ion by calcium ions. Neutralization by calcium carbonate and similar compounds that are commonly present as suspended particles in atmospheric dust is the mechanism by which carbonic acid in normal rainfall, and acid rain over some areas, has a greater than expected pH. These same reactions are responsible for the deterioration of limestone and marble statues; fine detail, such as ears, noses, and other facial features, are gradually lost as a result of reaction with acid and with sulfur dioxide itself.

> The milky-white color of India's famed Taj Mahal is losing its luster and turning pale due to acid rain.

In contrast, **areas strongly affected by acid rain are those having granite or quartz bedrock,** since the soil there has little capacity to neutralize the acid. Figure 4-3 shows areas of North America having low soil alkalinity, i.e., low amounts of basic compounds with which acids can react. Large areas susceptible to acidity are the Precambrian Shield regions of Canada and Scandinavia. Acid rain resulting from the massive development of the tar sands to produce synthetic crude oil in northern Alberta, and the SO_2 and NO_X emissions that result, are now affecting areas in Manitoba and northern Saskatchewan that lie upwind from them, since the soils in these two areas have very little neutralizing capacity (Figure 4-3).

Acidity from precipitation leads to the deterioration of soil. When the pH of soil is lowered, plant nutrients such as the cations of potassium, calcium, and magnesium are exchanged with H^+ and thereupon leached from it. Lakes rely on calcium ion produced by weathering of land-based soil and rocks that later is washed into them. The initial effect of acid rain therefore is to *increase* the calcium ion concentration in nearby lakes, but eventually the Ca^{2+} level *declines* once the acid has leached most calcium from the soil. Waterborne life that

FIGURE 4-3 Regions of North America with low soil alkalinity for neutralizing acid rain. [Source: D. J. Jacob, *Introduction to Atmospheric Chemistry* (Princeton, NJ: Princeton University Press, 1999), p. 233.]

depends on calcium can be threatened by the subsequent low levels of the element.

China now has serious acid rain problems due to its high emissions of SO_2 from power plants. Acidification is more serious in southern and southwestern areas than in northern China, where airborne alkaline dust swept in from desert areas neutralizes the acidity. Some of the acid rain that originates in China is carried by the wind to Japan and on occasion all the way to North America.

> The neutralizing dust does not reach southern China.

4.5 Neutralization of Acid Rain by Anthropogenic Actions

Although as a consequence of governmental regulations sulfur dioxide emission levels fell significantly in recent decades in both Europe and North America, there has *not* been as large a corresponding change in the pH of the precipitation, especially in northeastern North America. Ironically, **the lack of corresponding reduction in acidity is attributed to a decline over the same period of fly ash emissions from smokestacks and of other solid particles, all of which are alkaline and in the past neutralized a fraction of the sulfur dioxide and sulfuric acid** in the same way that calcium carbonate operates in soil. The reduction in sulfate deposition in the northeastern United States and south-central Canada from the early 1980s to the late 1990s is shown in Figures 4-4a and 4-4b, respectively. Overall, sulfate deposition rates had fallen by 43% across the eastern United States by 2009.

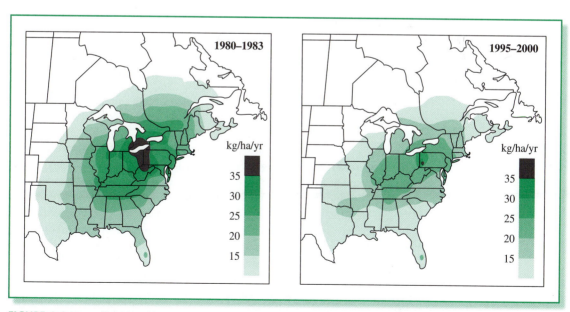

FIGURE 4-4 Wet sulfate deposition in eastern North America as a four-year mean (kilograms per hectare per year). [Source: Canadian National Atmospheric Chemistry Database, Meteorological Service of Canada, Environment Canada.]

Because of acid rainwater falling and draining into them, tens of thousands of lakes in the Precambrian Shield regions of both Canada and Sweden have become strongly acidified, as have lesser numbers in the United States, Great Britain, and Finland. Lakes in Ontario are particularly hard hit, since they lie directly in the path of polluted air and since the soil there contains little limestone. In a few cases, attempts have been made to neutralize the acidity by adding limestone or **calcium hydroxide,** $Ca(OH)_2$, to the lakes; however, this process must be repeated every few years to sustain an acceptable pH. Adding **phosphate ion,** PO_4^{3-}, to lakes can also control acidity, since it stimulates plant growth during the natural denitrification process by which **nitrate ion,** NO_3^-, is converted to reduced nitrogen with the consumption of large quantities of hydrogen ions, as shown in the reduction half-reaction

$$2\,NO_3^- + 12\,H^+ + 10\,e^- \longrightarrow N_2 + 6\,H_2O$$

Nitrate neutralizes H^+; ammonium ion produces it.

In Australia, soil acidity has a completely different origin. Acidification is associated there with the removal, by harvest of plant and animal crops and by soil leaching, of nitrate ion. Presumably, the loss of nitrate prevents its natural buffering of acidity by the reaction shown above. As in Canadian lakes, the effects of the acidification have been partially reversed in Australia by the addition of lime to the soil.

The regulatory scheme used in the United States of requiring reductions in sulfur dioxide emissions in certain designated geographical regions has been extended by European scientists and regulators into **the concept of critical load, which recognizes that different levels of risk from acid rain are faced in different regions.** Geographical areas having buffering capacity can withstand a much greater input of acid rain before damage occurs than those without the capacity. Thus, higher sulfur dioxide emissions from a particular region can be allowed if the area in which the resulting sulfuric acid is usually deposited has a high critical load. To determine the critical loads for different regions, scientists use computer models that incorporate soil chemistry, rainfall, topography, etc. The use of the concept has had great success in Sweden, for example.

In using critical loads, pollution control becomes effects-based rather than source-based. Although the critical loads concept has been implemented in regulations in Europe and Canada, and is favored by many scientists and politicians in the United States, it has not been implemented there. Although significant progress has been made in reducing SO_2 emissions, and more reductions are scheduled both for it and for NO_X, scientists predict that these efforts will be insufficient to allow a full recovery to lakes and forests in northeastern United States and south-central Canada.

Acid rain is still a serious environmental problem in many areas of the world, especially where unregulated coal-fueled power plants are numerous and there is no source of alkalinity to neutralize it. Although acid rain is less of a problem now in developed countries than it was in the late twentieth century, it is becoming a serious concern in many developing countries.

4.6 Environmental Effects of Acid Rain

Acid rain in the United States was considered initially to be a problem for its northeastern region. Indeed, one of the hardest-hit regions is the Catskill Mountains in New York State, whose surface rocks consist of calcium-poor sandstone and from which most of the nutrients have now been leached. At the Hubbard Brook Experimental Forest in New Hampshire, half the calcium and magnesium in soil was leached by 1996 and, as a result, vegetative growth almost stopped. However, as a consequence of the reduction in SO_2 emissions, by 2003 over half the lakes in the Adirondack Mountains of New York State showed some significant recovery from acid rain. Unfortunately, full recovery for these lakes is predicted to take another 25 to 100 years to achieve.

Acid rain now is also a concern in the southeastern United States. Here soils are generally thicker and were able to neutralize more acid. However, much of that leaching ability now has been exhausted and acid levels in many waterways have increased substantially. It has been discovered that the recovery of such soils, and of those in Germany, is slowed once acid precipitation has declined, because previously stored sulfate ion is then released, causing more cation leaching and penetration of acidity deeper into the ground.

Acidification reduces the ability of some plants to grow, including those in fresh-water systems. Because of the decrease in this productivity in lakes and streams that feed them, the amount of **dissolved organic carbon** (DOC) in the surface water has declined. The DOC contains molecules that absorb some of the ultraviolet from sunlight; thus, **a decline in DOC levels has allowed more penetration of UV light into the lower layers of the lake.** In addition, global warming has resulted in the drying up of some streams that supplied DOC to lakes. Furthermore, stratospheric ozone depletion has also allowed more UV to reach the Earth's surface, including lakes, in the first place. Thus fresh-water lakes have suffered a "triple whammy" from global environmental problems.

ACTIVITY

Using the internet and/or library resources, discover the region geographically closest to you that has or has had a problem with acid rain. What is or were the main sources of emissions that caused the acidity? What steps have been taken to reduce the emissions, and to remediate the soil and water of excess acidity?

4.7 Release of Aluminum into Water Bodies by Acid Rain

Acidified lakes characteristically have elevated concentrations of dissolved aluminum ion, Al^{3+}, and it is now known that **many of the biological effects of acid rain are due to increased levels of aluminum ion dissolved in water rather than to the hydrogen ion itself.**

Aluminum is called *aluminium* in many parts of the world.

Aluminum ions are leached from rocks in contact with acidified water by reaction with the hydrogen ions; under normal, near-neutral pH conditions, the aluminum is immobilized in the rocks by their insolubility.

$$\text{Al compounds}(s) \xrightarrow{\text{H}^+} \text{Al}^{3+}$$

Plots of dissolved aluminum concentration versus water acidity for lakes in the Adirondack Mountains of New York State and for lakes in Sweden are illustrated in Figure 4-5 and confirm this origin. (The chemistry underlying these processes is further discussed in Chapter 10, as are the reasons why natural waters have pH values of 7 or 8, rather than the 5.3 of rain, and why aluminum mobilization leads to dieback of trees.)

Scientists believe that both the acidity itself and the high concentrations of aluminum *together* are responsible for the devastating decreases in fish populations that have been observed in many acidified water systems. Different types of fish and aquatic plants vary in their tolerances for aluminum and acid, so the biological composition of a lake varies as it gradually becomes increasingly acidic. Generally speaking, fish reproduction is severely diminished even at low levels of acidity that can, however, be tolerated by adult fish. Very young fish, hatched in early spring, are also subject to the shock of very acidic water that occurs when the acidic winter snow all melts in a short time and enters the water systems.

Healthy lakes have a pH of about 7 or a little higher; few fish species survive and reproduce when the pH drops much below 5. As a result, many lakes and rivers in affected areas are now devoid of their valuable fish; e.g., 30% of the salmon rivers in Nova Scotia are too acidic for Atlantic salmon to survive. The water in many acidified lakes is crystal clear due to the death of most of the flora and fauna.

However, aluminum levels in drainage from soils at medium-to-high elevations in the United States declined significantly over the period from 1984 to 1998, and if the trend continues, will not be a threat to fish by about 2012.

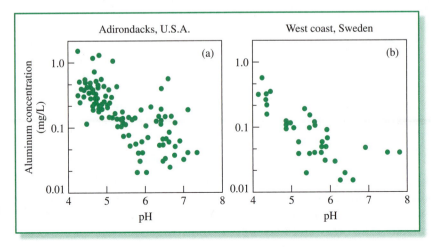

FIGURE 4-5 Aluminum concentrations versus pH of the water in different fresh-water lakes in (a) the Adirondacks and (b) western Sweden. Notice the logarithmic vertical axis. [Source: M. Havas and J. F. Jaworski, *Aluminum in the Canadian Environment* (Ottawa: National Research Council of Canada Report 24759, 1986).]

4.8 Effects of Air Pollution on Trees and Crops

Air pollution can have a severe effect on trees. The phenomenon of forest decline was first observed on a large scale in western Germany and occurs mainly at high altitudes. However, the cause-and-effect relationship behind this forest decline has been very difficult for scientists to untangle. As discussed above, acidification of the soil can leach nutrients from it and, as occurs in lakes, solubilize aluminum. This element may interfere with the uptake of nutrients by trees and other plants. Apparently, **both the acidity of the rain falling on affected forests and the tropospheric ozone and other oxidants in the air to which they are exposed pose significant stresses to the trees.** These two stresses alone will not kill them, but when combined with drought, temperature extremes, disease, or insect attack, the trees become much more vulnerable.

Forests at high altitudes are most affected by acid precipitation, possibly because they are exposed to the base of low-level clouds, where the acidity is most concentrated. Fogs and mists are more acidic even than precipitation, since there is much less total water to dilute the acid. For example, white birch trees along the shores of Lake Superior experience dieback in regions where acid fog occurs, as it frequently does there. Deciduous trees (i.e., those that lose their leaves annually) affected by acid rain gradually die from their tops downward; the outermost leaves dry and fall prematurely and are not replenished the following spring. The trees become weakened as a result of these changes, and become more susceptible to other stressors. In some regions of Europe and North America, forest soils are limed in order to combat the effects of acidity on trees.

Ground-level ozone itself has an effect on some agricultural crops due to its ability to attack plants. Apparently the ozone reacts with the gas (ethene) that the plants emit, generating free radicals that then damage plant tissue. The rate of photosynthesis is slowed, and hence the total amount of plant material is reduced, by the action of ozone. As in the case of trees, air pollution acts as a stressor to plants.

The collective damage from air pollution to North American crops, e.g., alfalfa in the United States and white beans in Canada, is estimated to be $3 billion a year—another example of *external costs* arising from environmental problems. Other crops whose yields are adversely affected by current levels of ozone include wheat, corn, barley, soybeans, cotton, and tomatoes. Ozone has also caused crop losses in many regions of China. The fraction of the world's cereal crops that are grown in regions of high ozone, and therefore subject to damage, is predicted to more than triple by 2025.

Review Questions 1–6 are based on the material covered above.

The Human Health Effects of Outdoor Air Pollutants

Breathing polluted air can have a dramatic, negative influence on human health. In this section, the most important of these effects of outdoor air pollutants are described, and the variation in concentration of the dominant air pollutants between different countries is discussed.

The effect that pollutants have on human health cannot be deduced from general principles of biology or physiology, but must be established by experimentation. One can imagine experiments involving animals or human volunteers in which the health effects of exposure to brief periods of artificially produced high-level pollution are studied. However, the extrapolation of information gained from short-term studies of high-level pollution to long-term exposures at low levels is difficult. In particular, **for some pollutants, there may exist a threshold pollutant concentration, or exposure below which a particular health effect does not occur.** In these cases predictions obtained by assuming simple direct proportionality between exposure and effect would be unwarranted. In addition, there could be deleterious effects of chronic exposure that do not come into play when exposure, even intense exposure, to pollutants occurs only for brief periods of time.

For these reasons, the best information regarding the effects of pollutants on health comes from the large scale "experiment" in which we are all enrolled as "test animals"—namely, living in a society in which we are routinely exposed to these pollutants for our whole lives. Because the level of exposure to any given pollutant varies considerably from place to place, scientists can collect information on health and on pollution levels in different locations, and correlate them using statistics to establish the effect of one upon the other.

For example, several North American studies have statistically linked the rate of hospitalization for congestive heart failure among elderly people to the daily carbon monoxide concentration in outside air. Mexico City currently has the highest levels of carbon monoxide among the world's most polluted cities. However, both CO and NO_2 are usually more of a problem in indoor air, and will be discussed in detail in a later section.

It is a historical characteristic that **once an undeveloped country starts industrial development, its outdoor air quality worsens significantly.** The situation continues to deteriorate **until a significant degree of affluence is attained, at which point emission controls are enacted and enforced, and the air begins to clear.** Thus, although the quality of air is now improving with time in most developed countries, it is worsening in the larger cities of developing countries.

Mexico City and several urban areas in China, especially Beijing, are generally considered to have the worst urban air pollution in the world at present. In 2002, thirteen of the twenty world cities having the highest averages for airborne particulate matter were located in China, five in India, plus Cairo, Egypt, and Jakarta, Indonesia. Half the respiratory disease in China is caused by air pollution. A report by the United Nations Environmental Program (UNEP) estimates that deaths worldwide from all forms of air pollution amounted to 2.7–3.0 million in 2001, a figure that may rise to 8 million by 2020.

As would be expected, **the major effects on human health from air pollution occur in and through the lungs.** For example, asthmatics suffer worse episodes of their disease when the sulfur dioxide or the ozone or the particulate concentration rises in the air that they breathe. A study in California found that **asthma can be caused by air pollution,** specifically by ozone, and especially among highly active children, who naturally inhale more air into their lungs. A study in British Columbia found that exposure to air pollution—especially NO_X and CO—from traffic during gestation and the first year of life increases the risk of developing asthma in childhood.

4.9 The Health Effects of Soot-and-Sulfur-Dioxide Smog

In the middle decades of the twentieth century, several Western industrialized cities experienced wintertime episodes of smog from soot-and-sulfur pollution that were so serious that the death rate increased noticeably. For example, in London, England in December 1952 about 4,000 people died within a few days—plus 8,000 more in the next few months—as a result of the high concentrations of these pollutants that had built up in a stagnant, foggy air mass trapped by a temperature inversion close to the ground. Those at most risk were elderly persons already suffering from bronchial problems, and young children. A ban of household coal burning, from which most of the pollutants originated, has now largely eliminated such problems. Scientists are still unsure whether the main sulfur-containing agent that caused such serious problems in London was the SO_2, the sulfuric acid droplets, or the sulfate particulates.

Today, due to pollution controls, *soot-and-sulfur smogs* are no longer a major problem in Western countries. For example, deaths from bronchitis have fallen by more than half in the United Kingdom, the result of changes in air quality (and smoking habits). However, the quality of winter air in some areas of what was the Eastern bloc of countries, such as southern Poland, the Czech Republic, and eastern Germany, until recently was very poor on account of the burning of large amounts of high-sulfur "brown coal" for both industrial and home-heating purposes.

Brown coal can contain up to 15% sulfur.

For example, although the acceptable limit for the concentration of SO_2 in air is $80 \ \mu g \ m^{-3}$ in many countries, the level of this gas in Prague surpassed $3,000 \ \mu g \ m^{-3}$ on occasion in the late twentieth century. Indeed, four out of five children admitted to the hospital in some areas of Czechoslovakia in the early 1990s were there for treatment of respiratory problems. However, the average SO_2 level in Prague decreased by about 50% from the early 1980s to the early 1990s, and overall in the Czech Republic SO_2 emissions are now only a few percent of 1990 levels. The tremendous improvement of air quality in eastern Germany since 1990, where mean SO_2 levels have dropped from 113 to $6 \ \mu g \ m^{-3}$, has seen a decrease in childhood respiratory infections and an increase in lung function.

The effects of sulfur dioxide are also evident in cities such as Athens, where the death rate increases by 12% when the concentrations of the gas exceeds 100 μg m^{-3}. Carved detail on the ancient statues and monuments of Athens is also being seriously eroded by sulfur dioxide and its secondary pollutants. And the air in London is not so improved that it does not affect human health. A recent study concluded that one in every 50 heart attacks was triggered by outdoor air pollution, from a combination of smoke, CO, SO_2, and NO_2.

European cities are not the only ones affected by air pollution. Both sulfur dioxide and particulate matter levels regularly exceed *World Health Organization* guidelines (see Tables 3-1 and 3-2) in Beijing, Seoul, and Mexico City. In many large cities in the developing world, coal is still the predominant fuel and in many of them diesel-powered vehicles substantially worsen the problem. In Beijing, high SO_2 emissions from coal-burning furnaces in buildings, plus smoke from smelters on the edges of the city, plus windblown dust and sand from the Gobi Desert, combine to produce poor air quality. Indeed, there are a number of cities in China in which the air quality is among the poorest in the world. According to recent projections, if no attempts are made to reduce SO_2 emissions as industrialization increases, then by 2020 the concentration of the gas in Mumbai and the Chinese cities of Shanghai and Chongqing will be about four times the WHO maximum safe limit.

Although acute smog episodes from soot and sulfur-based chemicals have been eliminated in the West, many residents in these countries still are chronically exposed to measurable levels of suspended particles containing sulfuric acid and sulfates due to the long-range transport of these substances from industrialized regions that still emit SO_2 into the air. For example, research has shown a positive correlation between atmospheric concentrations of oxidized sulfur and ozone and hospital admissions for respiratory problems in southern Ontario. There is some evidence that the acidity of the pollution is the main active agent in causing lung dysfunction, including wheezing and bronchitis in children. Asthmatic individuals are adversely affected by acidic sulfate aerosols, even at very low concentrations.

AP Photo/Mikhail Metzel

It has been speculated that pollution due to SO_2 and sulfates causes a decrease in resistance to colon and breast cancer in people living in northern latitudes. The suggested mechanism of this action is a reduction in the amount of available UV-B that is necessary to form vitamin D, which is a protective agent for both types of cancers. Since sulfur dioxide absorbs UV-B and sulfate particles scatter it, significant concentrations of either substance in air will reduce the amount of UV-B reaching ground level. Thus too little UV-B can have detrimental health effects, just as too much of it does—as was outlined in Chapter 1.

4.10 The Health Effects of Photochemical Smog

Photochemical smog, which arises from nitrogen oxides, is now more important than is soot-and-sulfur smog in most cities, particularly those with high population and vehicle density. As discussed in Chapter 3, it consists of gases such as ozone, and an aqueous phase containing water-soluble organic and inorganic compounds in the form of suspended particles. In contrast to "London smogs," which chemically were reducing in nature due to the presence of sulfur dioxide, **photochemical smogs are oxidizing.**

PROBLEM 4-2

Deduce the balanced half-reaction in which SO_2 acts as a reducing agent in an acidic environment, given that it is oxidized to sulfate ion, SO_4^{2-}, in the process.   ●

Ozone itself is a harmful air pollutant. In contrast to sulfur-based chemicals, its effect on the robust and healthy is as serious as on those with pre-existing respiratory problems. Experiments with human volunteers have shown that **ozone produces transient irritation in the respiratory system,** giving rise to coughing, nose and throat irritation, shortness of breath, and chest pains upon deep breathing. People with respiratory problems can often tell from their symptoms—such as the tightening of the chest or the beginning of a cough—when the air quality is poor. Even healthy young people often experience such symptoms while exercising outdoors by cycling or jogging during smog episodes. Indeed, there is evidence that the daily race times of cross-country runners increase with increasing ozone concentration in the air that they inhale.

A few percent of the day-to-day fluctuations in mortality rate in Los Angeles is explained by variations in the concentrations of air pollutants. An analysis of 95 urban centers in the United States discovered that a period of high ozone concentrations increased daily cardiovascular and respiratory mortality by about 0.5% per 10-ppb increase following a few days of continuous exposure. It is not yet clear what, if any, long-term lung dysfunction results from exposure to ozone. Exposure to ozone produces a number of indirect health effects as well—including a decrease in sperm count.

One anticipated effect of ozone is a decreased resistance to disease from infection because of the destruction of lung tissue; chronic exposure to high levels of urban ozone leads to the premature aging of lung tissue. At the molecular level, **ozone readily attacks substances containing components with C=C bonds,** such as occur in biological tissues of the lung. As discussed later, the fine particulates produced in the photochemical smog process can have a deleterious health effect on humans.

Ozone is not the only gas in photochemical smog that can affect health. A recent study of patients of a respiratory disease clinic in Toronto discovered that the subsequent rate of mortality from circulatory disease suffered by the patients increased the higher their exposure to traffic-related air pollution, as measured by the concentration of NO_2 in the air near their homes. This study reinforced similar results found from studies in Holland, Norway, and California regarding the toxicity and health effects of vehicle emissions.

Finally, we note that there are some positive effects of photochemical air pollution upon human health! For example, the rate of skin cancers in areas heavily polluted by ozone is probably reduced because of the ability of the gas to filter UV-B from sunlight.

4.11 Particulates as Health Risks

Particulate matter in the form of smoke from coal burning has been an air pollution problem for hundreds of years, especially in the United Kingdom. John Evelyn wrote in his January, 1684 diary that "London by reason of the excessive coldness of the air, hindering the ascent of the smoke, was so filled with the fuliginous [sooty] steam of sea-coal, that hardly could one see across the street, and this filling the lungs with its gross particles exceedingly obstructed the breast, so as one would scarce breathe." Indeed, unsuccessful attempts to control coal burning and punish offenders had begun in the thirteenth century in Britain.

One important reason that inhaled particles cause so much damage to human health is their ability to carry organic molecules and metal ions on their surfaces into the lungs and to release them there. **Substances that dissolve into the body of a particle are said to be absorbed by it; those that simply stick loosely to the surface of the particle are said to be adsorbed** (see Figure 4-6). An important example of the latter is represented by the adsorption of large organic molecules onto the surfaces of carbon ("soot") particles, as discussed later in Chapter 15. Many insoluble airborne particles are surrounded by a film of water, which can itself dissolve other substances.

Large particles—coarse ones, according to the definition in Chapter 3—**are of less concern to human health than are small ("fine") ones** for several general reasons:

- Since coarse particles settle out quickly, human exposure to them via inhalation is reduced.

- When inhaled, coarse particles are efficiently *filtered* by the nose (including its hairs) and throat and generally do not travel as far as the lungs. In contrast, inhaled fine particles usually travel through to the lungs (which is why they are called *respirable*), can be adsorbed on cell surfaces there, and can consequently affect our health.

- The ratio of surface area to mass of large particles is smaller than that of small ones, and thus, gram for gram, their ability to transport adsorbed gas molecules to any parts of the respiratory system and there to catalyze chemical and biochemical reactions is correspondingly smaller.

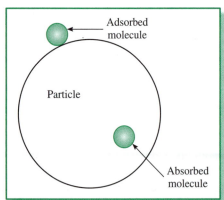

FIGURE 4-6 Contrast between adsorption and absorption of molecules on and in an airborne particle (schematic).

Adsorbed molecule

Particle

Absorbed molecule

Perhaps Shakespeare was referring to this type of air pollution in the quotation from *Hamlet* that opened Chapter 3.

• Devices such as electrostatic precipitators, spray towers, and cyclone collectors that are used to remove particulates from air are efficient only for coarse particles. Thus, although a device may remove 95% of the total particulate mass, the reduction of surface area and of respirable particles is a much lower fraction; see Problem 4-3. *Baghouse filters*, which are finely woven fiberglass fabric bags through which air is forced (as in a vacuum cleaner), are highly efficient in removing fine particles in the 1-μm size range, and all the larger ones as well.

PROBLEM 4-3

An air-filtering device is tested and is found to remove all particles larger than 1 μm in diameter, but almost none of the smaller ones. Calculate the percentage of surface area removed by the device for a sample of particulates, 95% of the mass of which is particles of diameter 10 μm and 5% of which is particles of diameter 0.1 μm. Assume all particles are spherical and of equal density. [*Hint:* Recall that the surface area of a sphere is $4\pi r^2$. Calculate the surface areas of particles of each size from this formula.]

The air pollution parameter that correlates most strongly with increases in the rate of disease or mortality in most regions is the concentration of respirable (fine) particulates, $PM_{2.5}$. Epidemiological studies have shown consistent and significant strong associations between mortality from both acute and long-term exposure to airborne particulate concentrations in the air, especially to the fine component ($PM_{2.5}$), whether produced by vehicle exhaust or power-plant emissions. Indeed, **particulate-based air pollution is found to have a greater influence on degrading human health than that produced directly by any gaseous pollutant, even ozone.**

Acute means short-term exposure at a high level.

The strongest links between human health and exposure to airborne particulate matter are based on studies involving cities in the United States, and are the subject of the Case Study *The Effect of Urban Air Particulates on Human Mortality* at the website associated with this chapter.

An even more recent study of teachers in California found a correlation between the level of long-term exposure to $PM_{2.5}$ and the likelihood of death from cardiopulmonary diseases; the most dangerous fine particulates were discovered to be sulfates and organic carbon. A new study of data from across the United States confirms that, although $PM_{2.5}$ particulates play the strongest role among air pollutants in increasing acute mortality from stroke, heart attacks, cardiovascular disease, and respiratory problems, the smallest of the coarse ones (i.e., those included in $PM_{10-2.5}$) also cause some increase in the death rate, although there are regional variations in the toxicity of these larger particles.

A number of studies have correlated day-to-day urban *morbidity* (sickness) rates, as measured by hospital admission rates, for respiratory problems, against the pollution levels during the same short time period. For example, there have

Erik Simonsen/Getty Images

The California wildfires of 2008.

been several reports concerning the immediate effects on the population of southern Ontario of the pollutants—ozone gas and sulfate particulates—to which they are most exposed. In one study, the average number of hospital admissions for respiratory problems correlated best with the ozone level of the previous day, and to a slightly lesser degree with the sulfate level from the previous day, for the summers of 1983–1988. Air pollution was found to account for about 6% of summertime hospital respiratory admissions, a magnitude close to that found in previous investigations in Ontario and New York State. Respiratory admissions correlated significantly with both PM_{10} and ozone concentrations in Spokane, Washington—an area where atmospheric sulfur dioxide is essentially nonexistent and therefore can be ruled as the true culprit in causing the illnesses.

Airborne particulates in both the $PM_{2.5}$ and $PM_{10-2.5}$ ranges collected from the air during wildfires in California and Alaska have been found to be much more toxic than comparable quantities from ambient air in studies performed on mice. The toxicity apparently results from carbon-centered free radicals, which may have been produced from reactive metals on the particles.

Although most of the studies relating human health to air pollution have not dealt with cancer, it is nevertheless of some concern. The exhaust from diesel engines has been classified as "likely to be carcinogenic to humans" by the U.S. EPA. Studies in California and in Seattle conclude that 70% or more of the risk to human health arises from hazardous organic pollutants in the diesel exhaust.

Review Questions 7–9 are based on the material in the preceding section.

Indoor Air Pollution

The levels of some common air pollutants often are greater indoors than outdoors, although pollutant concentrations do vary significantly from one building to another. Since most people spend more time indoors than outdoors, exposure to indoor air pollutants is an important environmental concern, and may cause more problems to human health than does poor outdoor air.

In the material that follows, we investigate the various indoor air pollutants that are thought to have the most serious effects on human health. In order that the necessary background material will have been covered, a discussion of several other indoor air pollutants of interest to human health is deferred until later chapters: radon is discussed in Chapter 9, pesticides in Chapter 13, and PAHs in Chapter 15. Chloroform in indoor air is considered when water purification is discussed, along with indoor air contamination by chlorinated organic solvents, in Chapter 11.

4.12 Formaldehyde

The most controversial indoor organic air pollutant gas is **formaldehyde,** $H_2C{=}O$. It is a widespread trace constituent of the atmosphere since it occurs as a stable intermediate in the oxidation of methane and of other VOCs. While its concentration in clean outdoor air is too small to be important—about 10 ppb in urban areas, except during episodes of photochemical smog—the level of formaldehyde gas *indoors* is often orders of magnitude greater, in certain cases exceeding 1,000 ppb (1 ppm). The indoor formaldehyde concentration is usually in the 5–20-ppb range in American homes.

The chief sources of indoor exposure to this gas are emissions from cigarette smoke and from synthetic materials that contain formaldehyde resins (a type of plastic) used in urea formaldehyde foam insulation and in the adhesive employed in manufacturing plywood and particleboard (chipboard). Many useful resins (which are rigid polymeric materials) are prepared by combining formaldehyde with another organic substance. Formaldehyde itself is used in the dyeing and gluing of carpets, carpet pads, and fabrics. **In the first few months and years after their manufacture, such materials release small amounts of free formaldehyde gas into the surrounding air.** Consequently, new prefabricated structures such as mobile homes that contain chipboard generally have much higher levels of formaldehyde in their air than do older, conventional homes. Many manufacturers of pressed-wood products have now modified their production processes in order to reduce the rate at which formaldehyde is released.

The rate of formaldehyde emission from synthetic materials increases with temperature and relative humidity, and declines as the materials age. Initially, formaldehyde temporarily trapped as a gas or simply adsorbed onto the materials is released into the surrounding air. There is also release of formaldehyde due to the rearrangement and dissociation of amide end-groups on resin polymers, from R—NH—CH_2OH to R—NH_2 + H_2CO. Later, slow but continuous reactions of water vapor in humid air with the methylene bridges joining amide groups within the polymer backbone provide a constant emission of formaldehyde:

$$R{-}NH{-}CH_2{-}NH{-}R + H_2O \longrightarrow 2\ R{-}NH_2 + H_2CO$$

Formaldehyde has a pungent odor, with a detection threshold in humans of about 100 parts per billion, i.e., 0.1 ppm. At somewhat higher levels, many people report problems of irritation to their eyes, especially if they wear contact lenses, and to their noses, throats, and skin. The formaldehyde in cigarette smoke can cause eye irritation. Common symptoms of *acute* formaldehyde exposure include coughing, wheezing, bronchitis, and chest pains. Chronic exposure to low levels of formaldehyde produces similar effects and respiratory symptoms.

Formaldehyde's odor is often noticeable in stores that sell carpets and synthetic fabrics.

Formaldehyde in air may cause children to develop asthma, and to have more respiratory infections and allergies and asthma, although evidence for these effects is controversial. Dampness in homes, allowing the proliferation of dust mites, fungi, and bacteria, also plays a large role in increasing lower respiratory tract illnesses, especially in children.

Formaldehyde is thought to be the most important VOC in producing what is known as sick building syndrome. This term is used to describe situations in which the occupants of a building experience acute health and comfort effects that seemed to be linked to the time they spend in a particular building, though no specific illness or cause is apparent. Complaints commonly include

- headaches;
- irritation of eyes, nose, or throat; dry cough;
- dizziness and nausea; fatigue;
- difficulty in concentrating; and
- dry or itchy skin.

Indoor air can also be contaminated with some of the newer organic compounds in fire retardants, etc., as will be discussed in Chapter 15.

In addition to VOCs emitted from indoor sources, other factors that contribute to the syndrome include inadequate ventilation, pollutants entering from outside the building, and biological contamination of the air from bacteria, molds, pollen, and viruses that have bred in stagnant water that has accumulated in air vents, etc.

Compounds chemically related to formaldehyde, namely the ketones **acetone,** $(CH_3)_2C{=}O$, and **2-butanone** (also called *methyl ethyl ketone*, MEK) are also commonly found in indoor air in U.S. homes, owing to their use as solvents in nail polish, paint removers, etc.

Formaldehyde is established as a **carcinogen** (a cancer-causing agent) in test animals and may also be carcinogenic to humans; it was classified as a *probable human carcinogen* by the U.S. EPA in 1987. In 2010, the EPA released a draft report concerning the health effects, including cancer, of formaldehyde. The expected cancer sites are in the respiratory system, including the nose, and cancers at these sites have been found in some people who are exposed to the gas in occupational settings.

However, studies of human populations exposed to formaldehyde have led to no clear-cut conclusions concerning an increase in cancer frequency arising from nonoccupational exposure to it. From animal studies, an upper limit to the possible effect in humans can be estimated: it corresponds to an increase in the cancer rate of one or two cases per 10,000 people after 10 years of living in a high-formaldehyde house or trailer. However, the lower limit to the effect could well be a zero increase in cancer rate. **In summary, no scientific consensus has yet been reached on the dangers to human health of low-level exposure to formaldehyde.**

4.13 Benzene and Other Gasoline-Related Hydrocarbons

Benzene

Like formaldehyde, benzene is classified as a **hazardous air pollutant,** HAP, a group of compounds also called *air toxics.* **Benzene,** C_6H_6, is a stable, volatile liquid hydrocarbon that through the modern age has found a variety of uses.

It is a minor constituent of gasoline, and was commonly used as a solvent for many organic products, including paints and inks. The public is exposed to benzene vapor indoors from the use of solvents and gasoline, through smoking (mainly for the smoker but to a lesser extent for those inhaling second-hand smoke), and from the importing of benzene from outdoor air into the house. The levels of benzene generally are smaller outdoors and in large buildings than in individual homes, especially those with smokers living in them. A significant fraction of benzene vapor exposure occurs while riding in motor vehicles and refueling them at gas stations.

Benzene is classified by the U.S. EPA as a *known human carcinogen.* Chronic exposure at high occupational levels increases the rate of leukemia to individuals. Indeed, there were many deaths among workers in the first half of the twentieth century from exposure to benzene from petroleum-based solvents, such as those used in the rubber and glue industries; in the making of paints, adhesives, and coatings; and in dry cleaning. It also causes *aplastic anemia,* a condition in which an individual is chronically tired and is especially susceptible to infections, because the bone marrow produces insufficient red blood cells. There continues to be some uncertainty, however, about whether occupational or domestic exposure to low levels of benzene vapor does indeed increase the risk for leukemia and multiple myeloma. It has been estimated that benzene accounted for one-quarter of cancer deaths caused by air toxics in the United States.

Because of the serious health problems it causes, the use of benzene as a solvent has largely been phased out. In addition, its maximum allowable level in gasoline has been reduced. Benzene can be replaced in many applications by **toluene,** $C_6H_5CH_3$, which consists of molecules of benzene with one hydrogen atom replaced by a methyl group. The —CH_3 group in toluene provides liver enzymes with a site that is much easier to attack and thereby initiate metabolism than any of the very strong bonds in benzene itself. Toluene, the corresponding dimethylated benzenes called *xylenes,* the *1,2,4-trimethylated benzene,* and *ethylbenzene* are all present in modern unleaded gasoline, and are very commonly detected in indoor air, as are the nonaromatic hydrocarbons *cyclohexane* and *decane.* The concentration of toluene usually greatly exceeds that of benzene itself. However, there is evidence that methylated benzenes are demethylated in automotive catalytic converters, and that as a consequence, additional benzene is emitted into the air under some operating conditions.

Another gaseous indoor and outdoor pollutant of concern is **1,3-butadiene,** which has the structure $CH_2=CH—CH=CH_2$. This hydrocarbon

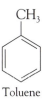

1,3-Butadiene

CH₃ (Toluene structure)

Toluene

is classified as an air toxic since there is evidence that it causes cancer—leukemia and *non-Hodgkin's lymphoma* especially—and may also negatively affect human reproduction. It is produced as a by-product of the incomplete combustion of fuels, in forest fires and wildfires, and is a component of cigarette smoke.

4.14 Nitrogen Dioxide

Indoor concentrations of NO_2 often exceed outdoor levels in homes that contain stoves, space heaters, and water heaters that are fueled by gas. The flame temperature in these appliances is sufficiently high that some nitrogen and oxygen in the air combine to form (thermal) NO, which eventually is oxidized to nitrogen dioxide. In one study it was established that NO_2 levels in homes that use gas for cooking or that have a kerosene stove average 24 ppb, compared to 9 ppb of NO_2 for homes that have neither. Peak concentrations near gas cooking stoves can exceed 300 ppb.

$$N_2 + O_2 \longrightarrow NO \longrightarrow NO_2$$

Some nitric oxide is also released from the burning of wood and other biomass fuels since these natural materials contain nitrogen. However, the flame temperature in burning such fuels is much lower than in burning gas, so relatively little thermal NO is produced from nitrogen in the air.

Nitrogen dioxide is soluble in biological tissue and is an oxidant, so its effects on health, if any, are expected to occur in the respiratory system. There have been many studies of the effects on respiratory illness in children owing to exposure to low levels of NO_2 emitted by gas appliances, but the results of different studies are not mutually consistent and are inadequate for establishing a cause-and-effect relationship. One study found that a 15-ppb increase in the mean NO_2 concentration in a home leads to about a 40% increase in lower respiratory system symptoms among children aged 7 to 11 years. Nitrogen dioxide is the only oxide of nitrogen that is detrimental to health at concentrations likely to be encountered in residences.

Nitrogen dioxide is presumably responsible for the finding that indoor concentrations of **nitrous acid,** HNO_2, exceed those found outdoors, since the gas reacts with water to form nitrous and nitric acids:

$$2\,NO_2 + H_2O \longrightarrow HNO_2 + HNO_3$$

$$HNO_2 + O_3 \longrightarrow HNO_3 + O_2$$

Indoor nitrous acid concentrations were found to correlate inversely with ozone gas concentrations, presumably because the acid is oxidized to nitric acid by the gas.

4.15 Carbon Monoxide

Carbon monoxide, CO, is a colorless, odorless gas whose concentration indoors can be greatly increased by the incomplete combustion of carbon-containing fuels such as wood, gasoline, kerosene, or gas. High indoor concentrations usually are the result of a malfunctioning combustion appliance, such as a kerosene heater. Even properly functioning kerosene or gas heaters in poorly ventilated rooms can result in CO levels in the 50–90-ppm range. Average indoor and outdoor CO concentrations usually amount to a few parts per million, though elevated values in the 10–20-ppm range are common in parking garages due to the carbon monoxide emitted by motor vehicles. Exhaust fumes containing high levels of CO and other pollutants can enter homes having attached garages. In developing countries, carbon monoxide poisoning is a serious hazard when biomass fuels are used to heat poorly ventilated rooms in which people sleep.

The major danger from carbon monoxide arises from its ability, when inhaled, to complex strongly with the hemoglobin in blood and thus to impair its ability to transport oxygen to cells. Hemoglobin's affinity for CO is 234 times that for oxygen, and once one CO is bound to a given hemoglobin molecule, the rate of release of its remaining oxygen molecules to cells is reduced. Recent research established that mental functioning is reduced during short-term exposure to high levels of CO, and perhaps also as a result of long-term exposure to low concentrations, because the brain, like the heart, is a body organ with a high requirement for oxygen.

One important feature of the reduction in vehicular pollutant emissions over the last few decades has been the substantial decline in accidental death from acute carbon monoxide poisoning. In the United States alone, it has been estimated that tens of thousands of deaths have been avoided as a result.

Smoking tobacco is a significant source of carbon monoxide indoors. Although nonsmokers usually have less than 1% of their hemoglobin tied up as the complex with CO, the figure for smokers is many times this value because of the carbon monoxide that they inhale during smoking. Studies have shown that increased mortality from heart disease can result even if only several percent of hemoglobin is chronically tied up as the CO complex. Exposure to very high concentrations of CO results in headache, fatigue, unconsciousness, and eventually death (if such exposure is sustained for long periods).

Low-priced, easily installed carbon monoxide detectors suitable for warning residents in homes and offices when high CO levels occur are now widely available. However, scientists have begun to worry about the poorly known health effects of chronic exposure to low levels of CO and the fact that such exposure may be quite common.

4.16 Smoke from Cooking Stoves

The use of highly polluting indoor stoves—used for cooking and in some regions also for heating and lighting—is widespread in rural areas of many developing countries, involving an estimated three billion people. The lack of a chimney over the stove, or even of a hole in the ceiling for smoke to escape, means that the occupants—especially the women and small children, who spend more time indoors—are exposed to carbon monoxide and to high levels of particulate pollution for extended periods every day.

Smoke production is especially pronounced as a result of inadequate ventilation when the fuels used for the stove or for an open fire pit are wood, coal, animal dung, crop residues, and other forms of unprocessed biomass, presumably because such fuels are fairly high in moisture content and consequently their combustion is rather incomplete. In some countries, the women use blowpipes to give the fire more oxygen, but they inhale smoke in the process.

As a result of being exposed to and inhaling the smoke, the young children are particularly susceptible to eye problems and both acute and chronic respiratory infections such as pneumonia. Indeed, the smoke is a major killer of children under 5 years of age. Children whose families use fire pits often suffer severe burns when they accidently fall into them. The women also suffer from respiratory ailments such as lung cancer, emphysema and *chronic obstructive pulmonary disease* (COPD), and chronic bronchitis. Cooking smoke from biomass fuels increases asthma rates among elderly men and women. Overall, **more than 1.5 million people die prematurely each year from indoor air pollution from cooking stoves,** including almost half a million in China alone.

The World Health Organization has designated the indoor cooking stove problem to be one of the four most critical global environmental problems. In response, a number of public and private organizations are sponsoring an initiative called *The Global Alliance for Clean Cookstoves*, whose aim is have 100 million homes in the developing world adopt clean, fuel-efficient stoves and fuels by 2020. The particulate emissions from traditional cookstoves can be reduced by 90% by switching from wood to charcoal, although this is not an efficient use of the wood supply where it is scarce.

China has set an indoor air quality PM_{10} standard of 150 μg m^{-3}, still three times that recommended by the WHO.

4.17 Environmental Tobacco Smoke

It is well established that **smoking tobacco is the leading cause of lung cancer and is one of the main contributors to heart disease.** Nonsmokers are often exposed to cigarette smoke, although in lower concentrations than smokers since it is diluted by air. This **environmental tobacco smoke,** ETS ("second-hand smoke"), has been the subject of many investigations in order to determine whether or not it is harmful to people who are exposed to it.

ETS consists of both gases and particles. The concentration of some toxic products of partial combustion is actually *higher* in sidestream smoke than in mainstream, since combustion occurs at a lower temperature—and so is less complete—in the smoldering cigarette compared to one through which air is being inhaled. Since the sidestream smoke is usually diluted by air before being inhaled, however, the concentrations of pollutants reaching the lungs of nonsmokers are much lower than those reaching the lungs of smokers themselves.

The chemical constitution of tobacco smoke is complex: it contains thousands of components, several dozen of which are carcinogens. The gases present in cigarette smoke include

- carbon monoxide, carbon dioxide;

- formaldehyde and several other aldehydes, ketones, and carboxylic acids;

- nitrogen oxides, hydrogen cyanide, ammonia, and a number of organic nitrogen compounds;

- methyl chloride;

- 1,3-butadiene;

- toluene, benzene, and several hundred different polycyclic aromatic hydrocarbons (PAHs), to be discussed in Chapters 7 and 15; and

- cadmium and radioactive elements such as polonium (see Chapter 9).

Included in the nitrogen compounds are several **nitrosamines,** organic nitrogen compounds of formula R_2N—N=O, which together with the PAHs, are probably the most important respiratory carcinogens in the smoke.

The particulate phase of cigarette smoke is called the tar, and much of it is respirable in size. The zone in a cigarette that actively burns, as occurs when a smoker inhales a puff, is quite hot (700–950°C) and produces CO and H_2 as well as the expected CO_2 and water vapor. Immediately downstream of this area is a cooler zone (200–600°C) where smoke constituents such as nicotine distill out of the tobacco. When this vapor cools further along the cigarette path toward the smoker, much of it condenses to aerosol particles that constitute the particulate phase of the smoke.

Many people experience irritation of their eyes and airways from exposure to ETS. The gaseous components of ETS, especially formaldehyde, hydrogen cyanide, acetone, toluene, and ammonia, cause most of the odor and irritation. Exposure to ETS aggravates the symptoms of many people who suffer from asthma or from *angina pectoris*, chest pains brought on by exertion. ETS, particularly when it originates with maternal smoking, is known to induce new cases of asthma in children, especially those of preschool age. Indeed, the rate of children being diagnosed with asthma has declined over the last decade, due at least in part to a decline in the number of homes in which adults smoke.

Some studies have established correlations between the rate of acute respiratory illness and the level of *indoor* $PM_{2.5}$ (which would include the total amount of respirable particulates from all sources, including tobacco smoke). **Passive smoking**—which involves inhalation of sidestream as well as already exhaled smoke—is believed by scientists to cause bronchitis, pneumonia, and other infections such as those of the ear in up to 300,000 infants, and several thousand instances of *sudden infant death*, in the United States each year. Second-hand smoke may even reduce the cognitive abilities of children, whether they are exposed prenatally or when young. Being exposed to second-hand smoke, whether on the job or by living with a smoker, approximately doubles a nonsmoker's chance of developing asthma.

In 1993, the U.S. EPA classified ETS as a *known human carcinogen*, and estimated that it causes about 3,000 lung cancer deaths annually. ETS is also considered to be responsible for killing as many as 60,000 Americans annually from heart disease. In a study of American nurses, it was found that nonsmoking women regularly exposed to ETS had a 91% greater rate of heart attacks than women who had no exposure. Apparently, the smoke leads to hardening of the arteries, a main cause of heart attacks.

An analysis of all recent studies on passive smoking led to the conclusion that the risks of developing lung cancer and heart disease are each increased by about one-quarter for nonsmoking spouses of smokers. Longtime workers in bars and restaurants in which smoking is permitted also have an increased rate of lung cancer, even if they themselves do not smoke. A British study estimated that ETS kills 140,000 Europeans annually through cancer and heart disease. In 2010, the World Health Organization estimated that ETS globally kills 600,000 people each year.

4.18 Asbestos

The term **asbestos refers to a family of six naturally occurring silicate minerals that are fibrous.** Structurally, they are composed of long double-stranded networks of silicon atoms connected through intervening oxygen atoms. The net negative charge of this silicate structure is neutralized by the presence of cations such as magnesium.

The most commonly used form of asbestos, **chrysotile,** has the formula $Mg_3Si_2O_5(OH)_4$. It is a white solid whose individual fibers are curly. Chrysotile, mined mainly in Russia, China, Brazil, Canada (Quebec), and Kazakhstan, is the principal type of asbestos used in the world. It has been employed in huge quantities because of its resistance to heat, its strength, and its relatively low cost. Common applications of asbestos include its use as insulation and spray-on fireproofing material in public buildings, in automobile brake-pad lining, as an additive to strengthen cement used for roofing and pipes, and as a woven fiber in fireproof cloth.

The use of asbestos has been sharply reduced since the 1970s in developed countries because it is now recognized from studies on the health of asbestos miners and other asbestos workers to be a human carcinogen. It causes *mesothelioma*, a normally rare, incurable cancer of the lining of the chest or abdomen. In addition, airborne asbestos fibers and cigarette smoke act **synergistically:** their combined effect is greater than the sum of their individual effects (in this case equal to the product of the two) in causing lung cancer.

There is much controversy concerning whether chrysotile should be banned outright from further use and whether or not existing asbestos insulation in buildings should be removed. Many experts feel that existing asbestos should be left in place unless it becomes damaged enough that there is a chance that its fibers will become airborne. Indeed, its removal can increase dramatically the levels of airborne asbestos in a building unless extraordinary precautions are taken.

Some environmentalists, however, feel that existing asbestos is a ticking time bomb—that it should be removed as soon as possible, as one can never predict when building insulation will be damaged.

Most of the initial concern about asbestos was related to **crocidolite,** *blue asbestos*, and **amosite,** *brown asbestos*. Evidence implicating crocidolite in causing cancer in humans was already well established several decades ago. It is a material with thin, straight, and relatively short fibers that more readily penetrate lung passages, and it is a more potent carcinogen than the white form. Crocidolite and amosite are mined in South Africa and Australia and were not used much in North America, but were used in many areas of Europe, including the United Kingdom.

More than 50 countries, including the European Union and Australia, have now banned *all* forms of asbestos. Some environmentalists and physicians worry that, although workers in developed countries wear masks and overalls and handle white asbestos properly to greatly minimize their exposure to it, these practices are not common yet in developing countries, in some of which it is still used in construction. Canada, among other countries, has resisted efforts by agencies of the United Nations to place chrysotile on the list of most hazardous substances, even though there is strong evidence that it causes cancer.

One scientist has stated: "Removing asbestos is like waking up a pit bull terrier by poking a stick in its ear. We should let sleeping dogs lie."

Review Questions 10–15 are based on the material in the preceding section.

Review Questions

1. Discuss the relationship between atmospheric particulates and haze.

2. What is *acid rain*? What two acids predominate in it? Explain why the predominant acid in acid rain differs in eastern and western North America.

3. What is the difference between *dry* and *wet deposition*?

4. Using chemical equations, explain how acid rain is neutralized by limestone that is present in soil. Describe ways in which humans have tried to neutralize acidified lakes.

5. Explain what is meant by the expression *critical load*.

6. Describe the effects of acid precipitation upon (a) dissolved levels of aluminum, (b) fish populations, and (c) trees.

7. Explain the difference between soot-and-sulfur-dioxide smog and photochemical smog in terms of the chemicals involved and the health effects they cause.

8. What is the difference in meaning between *absorbed* and *adsorbed* when they refer to particulates?

9. List four important reasons why coarse particles usually are of less danger to human health than are fine particles.

10. What are the main sources of formaldehyde in indoor air? What are its effects? What are the characteristics of *sick building syndrome*?

11. What is meant by the phrase *air toxic*? What are the main exposure routes to, and effects on health of, benzene? What chemicals can be used to replace benzene?

12. What are the main sources of nitrogen dioxide and of carbon monoxide in indoor air?

13. Why is smoke from cooking stoves an important health problem in many developing countries?

14. What are some of the constituents of tobacco smoke? What is meant by *tar* in the smoke?

15. What are the three forms of asbestos called? Why is asbestos of environmental concern?

Additional Problems

1. A sample of acidic precipitation is found to have a pH of 4.2. Upon analysis, it is found to have a total sulfur concentration of 0.000010 M. Calculate the concentration of nitric acid in the sample, and from the ratio of nitric to total acid, decide whether the air sample probably originated in eastern or in western North America.

2. If the pH of rainfall in upstate New York is found to be 4.0, and if the acidity is half due to nitric acid and half to the two hydrogen ions released by sulfuric acid, calculate the masses of the primary pollutants nitric oxide and sulfur dioxide that are required to acidify 1 liter of such rain.

3. The pH in a lake of size 3.0 km × 8.0 km and an average depth of 100 m is found to be 4.5. Calculate the mass of calcium carbonate that must be added to the lake water in order to raise its pH to 6.0.

4. The pH of a sample of rain is found to be 4.0. Calculate the percentage of HSO_4^- that is ionized in this sample, given that the acid dissociation constant for the second stage of ionization of H_2SO_4 is 1.2×10^{-2} mol L^{-1}. Repeat the calculation for a pH of 3.0. Is the trend shown by these calculations consistent with qualitative predictions made according to Le Châtelier's

principle (which states that the position of equilibrium shifts so as to minimize the effect of any stress)? [*Hint:* Write the expression for the acid dissociation constant for the weak acid in terms of the concentrations of the reactants and products, and use the stoichiometry of the balanced equation to reduce the number of unknowns to one.]

5. Calculate the mass of fine particles inhaled by an adult each year, assuming that he/she inhales about 350 L of air per hour and that the average $PM_{2.5}$ index of this air is 10 μg m^{-3}. Assuming that each particle has a diameter of about 1 μm, and that the density of the particles is about 0.5 g mL^{-1}, calculate the total surface area of this annual load of particles. [*Hint:* The surface area of a spherical particle is equal to $4\pi r^2$, where r is its radius.]

6. The detection threshold of formaldehyde by humans is about 100 ppb. Would a typical human be able to detect formaldehyde at a concentration of 250 μg m^{-3} if the air temperature was 23°C and the pressure 1.00 atm?

7. What mass of formaldehyde gas must be released from building materials, carpets, etc. in order to produce a concentration of 0.50 ppm of the gas in a room having dimensions of 4 m × 5 m × 2 m?

PART II

ENERGY AND CLIMATE CHANGE

Contents of Part II

Introduction

By general consensus, the most serious environmental problem facing humanity today is global climate change—the continuous increases in average global air temperatures and changing weather patterns, resulting in drought for some areas and flooding for others. Most atmospheric scientists are now convinced that this climate change results mainly from the anthropogenic enhancement of the greenhouse effect caused by ever-increasing levels of carbon dioxide and other "greenhouse gases."

In this part of the book, we investigate the mechanism of the greenhouse effect, the atmospheric gases that cause it, and the role of suspended atmospheric particles in either reinforcing or mitigating global warming. The role of fossil fuels in energy production in modern societies, and with it the cumulative increase that has occurred in atmospheric CO_2 levels, is then analyzed. We then concentrate our attention on the search for sustainable solutions to the carbon dioxide problem—such as burying the gas emissions underground or in the sea or avoiding their generation by using energy sources such as biofuels or hydrogen, wind, photovoltaics, solar power towers, or even nuclear energy. As we shall see, chemistry and chemists play an important role in all these topics. ●

5

The Greenhouse Effect

In this chapter, the following introductory chemistry topics are used:

⮞ Combustion

Background from previous chapters used in this chapter:

⮞ Sunlight wavelength regions (UV, visible, IR)

⮞ Absorption spectra

⮞ ppm, ppb, ppt concentration scale for gases

⮞ CFCs and their replacements

⮞ Tropospheric ozone

⮞ Aerosols

Introduction

Everyone has heard the prediction that the greenhouse effect will significantly affect climates around the world in the future. The terms *greenhouse warming* and *global warming* in ordinary usage simply mean that average global air temperatures are expected to increase by several degrees as a result of the buildup of carbon dioxide and other greenhouse gases in the atmosphere. Indeed, most atmospheric scientists believe that such **global warming** has been under way already for some time and is largely responsible for the air temperature increase of about two-thirds of a Celsius degree that has occurred since 1860.

The phenomenon of rapid global warming—with its demands for large-scale adjustments—is generally considered to be our most crucial worldwide environmental problem, although both positive and negative effects would be associated with any significant increase in the average global temperature. Unlike stratospheric ozone depletion, which has manifested itself in spectacular fashion in the form of the ozone hole, the phenomenon of global warming due to the greenhouse effect has yet to be observed in a fashion that convinces everyone of its existence. No one currently is sure of the extent or timing of future temperature increases, nor is it likely that reliable predictions for individual regions will ever be available much in advance of the events in question. If current models of the atmosphere are correct, however, significant warming will occur in coming decades. It is important that we understand the factors that are driving this increase so that we can, if we wish, take steps to avoid potential catastrophes caused by rapid climate change in the future.

In this chapter, the mechanism by which global warming could arise is explained, and the nature and sources of the chemicals that are responsible for the effect are analyzed. The extent of the atmospheric warming to date, and other indications that change is underway, are also discussed. The predictions concerning global warming in the future, and an analysis of steps that could be taken to minimize it, will be presented here and in Chapter 6.

The Mechanism of the Greenhouse Effect

5.1 The Earth's Energy Source

The Earth's surface and atmosphere are kept warm almost exclusively by energy from the Sun, which radiates energy as light of many types. In its radiating characteristics, the Sun behaves much like a **blackbody,** i.e., an object that is 100% efficient in emitting and in absorbing light. **The wavelength, λ_{peak}, in micrometers, at which the maximum emission of energy occurs by a radiating blackbody decreases inversely with increasing Kelvin temperature T** according to the relationship

$$\lambda_{peak} = 2{,}897/T$$

Since for the surface of the Sun, from which the star emits light, the temperature $T \sim 5800$ K, from the equation it follows that λ_{peak} is about 0.50 μm, a wavelength that lies in the visible region of the spectrum (and corresponds to green light). Indeed, the maximum observed solar output (see the dashed portion of the curve in Figure 5-1) occurs in the range of visible light, i.e., that of wavelengths between 0.40 and 0.75 μm.

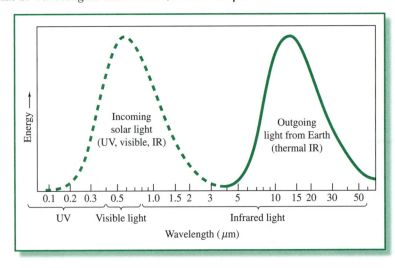

FIGURE 5-1 Wavelength distributions (using different scales) for light emitted by the Sun (dashed curve) and by the Earth's surface and troposphere (solid curve). [Source: Redrawn from J. Gribbin, "Inside Science: The Greenhouse Effect," *New Scientist,* supplement (22 October 1988).]

Beyond the "red limit," the maximum wavelength for visible light, the Earth receives **infrared light** (IR) in the 0.75–4-μm region from the Sun. Of the energy received at the top of the Earth's atmosphere from the Sun, slightly over half the total is IR and most of the remainder is visible light. At the opposite end of the visible wavelength spectrum from IR, beyond the "violet" limit, lies **ultraviolet light** (UV), which has wavelengths less than 0.4 μm, and is a minor component of sunlight, as discussed in Chapter 1.

Of the total incoming sunlight of all wavelengths that impinges upon the Earth, about 50% is absorbed at its surface by water bodies, soil, vegetation, buildings, etc. A further 20% of the incoming light is absorbed by water droplets in air (mainly in the form of clouds) and by molecular gases—the UV component by stratospheric **ozone,** O_3, and **diatomic oxygen,** O_2, and the IR by **carbon dioxide,** CO_2, and especially by water vapor. A small amount of sunlight is absorbed by suspended particulates of black soot.

> Much of the sunlight absorbed at the surface becomes converted to heat that evaporates water from oceans, lakes, and rivers and from vegetation.

The remaining 30% of incoming sunlight is reflected back into space by clouds, suspended particles, ice, snow, sand, and other reflecting bodies, without being absorbed. **The fraction of sunlight reflected back into space by an object is called its albedo,** which therefore is about 0.30 for the Earth overall. Clouds are good reflectors, with albedos ranging from 0.4 to 0.8. Snow and ice are also highly reflecting surfaces for visible light (high albedos), whereas bare soil and bodies of water are poor reflectors (low albedos). Thus the melting of sea ice in polar regions to produce open water greatly increases the fraction of sunlight absorbed there and decreases the Earth's overall albedo. Planting trees in snow-covered forests reduces the albedo of the surface and may actually contribute to global warming.

5.2 Historical Temperature Trends

The trends in average surface temperature for the past 2000 years, as reconstructed for most of that period from indirect evidence such as tree ring growth, is shown in Figure 5-2a. (The Medieval Warm Period early in the millennium was apparently restricted to the north Atlantic region, so it is not very evident on the global plot.) Notice the consistently downward trend in temperature until the beginnings of the Industrial Revolution.

The warming of the climate during the twentieth century stands in stark contrast to the gradual cooling trend in the previous 900 years of the millennium, producing a "hockey stick" shape to the temperature plot in Figure 5-2a. The trends in global average surface temperature over the last century and a half are illustrated in detail in Figure 5-2b. Air temperature did not increase *continuously* throughout the twentieth century. A significant warming trend occurred in the 1910–1940 period, due mainly to a lack of volcanic activity and a slight increase in the intensity of sunlight. This period was followed by some cooling over the next three decades, due mainly to aerosols resulting from increased volcanic activity. These decades were succeeded in turn by a warming period that has been sustained from about

FIGURE 5-2 (a) Reconstruction of changes in yearly average global surface temperatures, relative to the 1961–1990 average. [Source: M. E. Mann and P. D. Jones, *Geophysical Research Letters* (2003):30, 1820.] (b) Global average land–ocean temperatures at the Earth' surface (5-year running average). [Source: adapted from Makiko Sato, *Updating the Climate Science*] (c) Linear temperature trends covering the 25, 50, 100, and 150 years to 2005. [Source: Adapted from S. Solomon et al., eds., *Climate Change 2007: The Physical Science Basis, Intergovernmental Panel on Climate Change,* Cambridge University Press.]

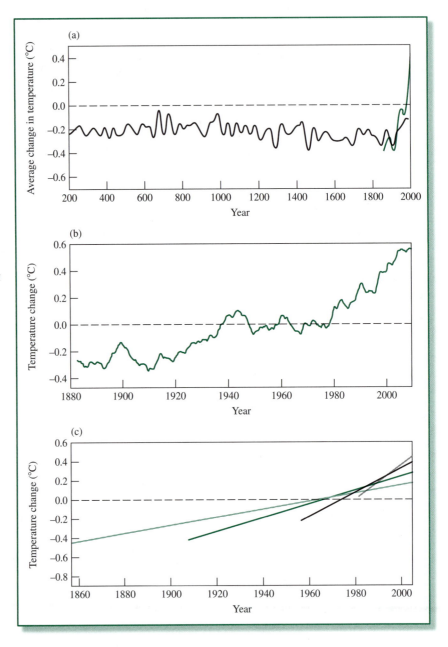

The years 1998, 2002, 2003, 2006, and 2007 were next warmest, all in a tie for third-place.

1970 to the present, and which has so far amounted to about 0.6°; this is attributed almost entirely to anthropogenic influences, as discussed in detail in this chapter. The 2000–2010 period was the hottest at least since 1850, when instrumental records began. The warmest years on record (to 2010) were 2005 and 2010.

That the *rate* of warming has increased over the past century and a half is illustrated by the four lines in Figure 5-2c, which correspond to the optimum straight lines drawn through the temperature variations of Figure 5-2b for various periods ending in 2005. The four periods and their rates of warming are:

Period	Period Length (years)	Rate of Warming, per Decade
1856–2005	150	0.045°C
1906–2005	100	0.074°C
1956–2005	50	0.128°C
1981–2005	25	0.177°C

ACTIVITY

Using whatever local records are available online, in the library, or from the weather office, tabulate average monthly temperatures (and precipitation amounts, if also available) over the last few decades for the town, city, or state/province in which you live. Plot the averages against year for a selected month each season—say, January, April, July, and October—and write a short note discussing any trends you observe. Are there signs of global warming in your region?

5.3 Earth's Energy Emissions and the Greenhouse Effect

Like any warm body, the Earth emits energy; indeed, the amount of energy that the planet absorbs and the amount that it releases into space must be equal over the long term if its temperature is to remain level. (Currently the planet is absorbing slightly more than it is emitting, thereby warming the air and the oceans.) The emitted energy (see the solid portion of the curve in Figure 5-1) is neither visible nor UV light, because the Earth is not hot enough to emit light in these regions. Since the temperature of the Earth's surface is approximately 300 K, then according to the equation above for λ_{peak}, if the Earth acted like a blackbody, its wavelength of maximum emission would be about 10 μm. Indeed, the Earth's emission does peak in that general region, actually at about 13 μm, and consists of infrared light having wavelengths starting at about 5 μm and extending, albeit weakly, beyond 50 μm (Figure 5-1, solid curve). The 5–100-μm range is called the **thermal infrared** region since such energy is a form of heat, the same kind of heat energy a heated iron pot would radiate.

Infrared light is emitted both at the Earth's surface and by its atmosphere, though in different amounts at different altitudes since the emission rate is very temperature sensitive: in general, **the warmer a body, the more energy it emits per second. The rate of release of energy as light by a** blackbody increases in proportion to the fourth power of its Kelvin temperature:

$$\text{rate of energy release} = kT^4$$

where k is a proportionality constant. Thus, doubling its absolute temperature increases sixteenfold (2^4) the rate at which a body releases energy. More realistically, for contemporary surface conditions of planet Earth, a one-degree rise in temperature would increase the rate of energy release by 1.3%.

Recall that
$T(K) = t(°C) + 273.$

PROBLEM 5-1

Calculate the ratio of the rates of energy release by two otherwise-identical blackbodies, one of which is at 0°C and the other at 17°C. At what temperature is the rate of energy release twice that at 0°C?   ●

Some gases in air absorb thermal infrared light—though only at characteristic wavelengths—and therefore the IR emitted from the Earth's surface and atmosphere does not all escape directly to space. Very shortly after its absorption by atmospheric gases such as CO_2, the IR photon may be re-emitted. Alternatively, the absorbed energy may quickly be redistributed as heat among molecules that collide with the absorber molecule, and it may be eventually re-emitted as IR by them. Whether re-emitted immediately by the initial absorber molecule or later by other ones in the area, the direction of the photon is completely random (Figure 5-3). Consequently, some of this thermal IR is redirected back toward the Earth's surface, and is reabsorbed there or in the air above it. Indeed, the atmosphere is fairly opaque to infrared, in contrast to its near transparency to sunlight.

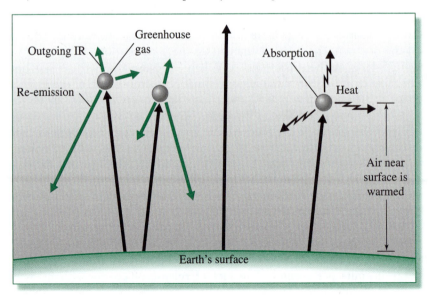

FIGURE 5-3 The greenhouse effect: outgoing IR absorbed by greenhouse gases is either re-emitted (left side of diagram) or converted to heat (right side).

By absorbing IR photons and redistributing the energy as heat to surrounding molecules, the air temperature in the region of the absorbing molecule increases. However, this air mass does not heat up without limit as its molecules trap more and more of the outgoing infrared light, because there is an opposing phenomenon that prevents such a catastrophe. As explained above, the rate of energy emission increases with temperature, so the molecules that have shared the excess energy themselves emit more and more energy as infrared light as they warm up (Figure 5-3). The water droplets and vapor in clouds are also very effective in absorbing infrared light emitted from beneath them. The temperature at the tops of clouds is quite cool relative to the air beneath them, so clouds do not radiate as much energy as they absorb. Overall, air temperatures increase only enough to re-establish the planetary equality between incoming and outgoing energy.

The phenomenon of interception of outgoing IR by atmospheric constituents and its dissipation as heat to increase the temperature of the atmosphere (as illustrated in Figure 5-3) **is called the greenhouse effect.** It is responsible for the average temperature at the Earth's surface and the air close to it being about $+15°C$ rather than about $-18°C$, the temperature it would be if there were no IR-absorbing gases in the atmosphere. The surface is as much warmed by this indirect mechanism as it is by the solar energy it absorbs directly! **The very fact that our planet is not entirely covered by a thick sheet of ice is due to the natural operation of the greenhouse effect,** which has been going on for billions of years.

The atmosphere operates in the same way as a blanket, retaining within the immediate region some of the heat released by a body and thereby increasing the local temperature. The phenomenon that worries environmental scientists is that increasing the concentration of the trace gases in air that absorb thermal infrared light (piling on more blankets, so to speak) would result in the absorption and conversion to heat of an even greater fraction of the outgoing thermal infrared energy than occurs at present, and would thereby increase the average surface temperature well beyond 15°C. This phenomenon is sometimes referred to as the **enhanced greenhouse effect** (or *artificial global warming*) to distinguish its effects from the greenhouse effects that have been operating naturally for millennia.

The principal constituents of the atmosphere—N_2, O_2, and Ar—are incapable of absorbing infrared light; the reasons for this will be discussed in the following section. **The atmospheric gases that in the past have produced most of the greenhouse warming are water vapor (responsible for about two-thirds of the effect) and carbon dioxide (responsible for about one-quarter).** Indeed, the absence of water vapor and of clouds in the dry air of desert areas leads to low nighttime temperatures there since so little of the outgoing IR is redirected back to the surface, even though the daytime temperatures are quite high on account of direct absorption of solar energy by the surface. More familiar to people living in temperate climates is the crisp chill in winter air on cloudless days and nights. Cloudy nights are usually warmer than clear ones because clouds reradiate IR that they have absorbed from emissions from the surface.

5.4 A Very Simple Model of the Greenhouse Effect

The greenhouse effect may be better understood by considering the following approximate model. Using physics, the temperature of an Earth that had no greenhouse gases in its air but was balanced with respect to incoming and out-going energy is calculated to be $-18°C$, or 255 K. Since, according to the equation above, the rate of energy emission from such a planet would be $k (255)^4$, it follows that the rate of energy *input* from the Sun, whether or not the Earth's atmosphere contains greenhouse gases, is *also* $k (255)^4$. Overall, the real Earth acts as if about 60% of the energy it emits as infrared light is eventually transmitted into space, the remainder being the fraction that is not only absorbed by greenhouse gases, but also reradiated downward and further heats the surface and atmosphere. Thus the rate at which the Earth loses energy to space as IR is not simply kT^4, but rather $0.6 kT^4$. Since we know that

rate of loss of energy from Earth = rate of energy input from Sun

it follows for the real Earth that

$$0.6 kT^4 = k (255)^4$$

Cancelling T and taking the fourth root of both sides, we obtain an expression for the temperature:

$$T = (255)/0.6^{0.25}$$

so

$$T = 290 \text{ K}$$

From this model, the Earth's calculated surface temperature is 290 K, i.e., $+17°C$, an increase of 35 degrees by the operation of the natural greenhouse effect.

In reality, however, **very little of the IR emitted at or near the Earth's surface escapes into space.** Rather it is absorbed by the air close to the ground, and then re-emitted. A less approximate, but still simple model of the atmosphere that incorporates this effect is discussed in Box 5-1. The IR from the air close to the ground that is emitted upward is mainly absorbed by the next layer of air, which is heated by it, though to a lesser extent than is the layer underneath, and is partially re-emitted. With increasing altitude, the fraction of the IR received from the air lying below a given level is less and less likely to be absorbed, since the atmosphere becomes thinner and thinner. Thus more and more of the IR is likely to pass upward through into space. Indeed, very little of the IR emitted upward into the upper troposphere is absorbed since the air is thin at such altitudes.

Because less and less IR is absorbed with increasing altitude, less and less is degraded into heat; there is therefore a natural tendency for the air to cool

BOX 5-1   A Simple Model of the Greenhouse Effect

The calculation in the main text of the Earth's surface temperature assumed a specific value for the fraction of IR emitted from the surface that was transmitted through the atmosphere to outer space. However, this fraction—and the temperature—can be calculated from the physics of the situation. Consider a model Earth containing an atmosphere that consists of a single, uniform layer of air that is completely nontransparent to outgoing IR emitted from the surface, i.e., an atmosphere that absorbs all the IR and that converts it temporarily to heat. The layer of air itself is assumed to act as a blackbody that emits IR equally upward into space and downward back to the surface (Figure 1).

If a balance is to be achieved on Earth between incoming and outgoing energy, the air mass in the model must emit twice as much energy ($2X$) per second as is absorbed (X) from sunlight by the surface, since only half the air's energy escapes upward and is released into space. Define $rate_b$ as the total energy release rate from the air mass, and $rate_a$ as the absorption rate from sunlight. Since we know that the rate of energy emission rises with the fourth power of the temperature, it follows that for any two Kelvin temperatures T_a and T_b involving the same type of blackbody, the ratio of rates of energy emission is given by

$$(rate_b/rate_a) = (T_b/T_a)^4$$

In our case, the rate ratio must be 2/1, and we know that T_a is 255 K. Thus

$$(T_b/255)^4 = 2.0$$

Taking square roots of both sides twice, we obtain

$$(T_b/255) = 2.0^{0.25} = 1.189$$

so

$$T_b = 303 \text{ K}$$

This simple model predicts the Earth's surface (and air) temperature to be 303 K, or 30°C, compared to the actual value of 15°C. The model is unrealistic, and leads to an overestimation of the greenhouse effect, because it assumes that *all* the IR escaping from the surface is absorbed, and that the atmosphere is completely uniform and acts exactly as a blackbody. Earth's actual surface temperature of 15°C is more consistent with a slightly more complicated model, in which about one-third of the IR emitted from the surface passes through the atmosphere unabsorbed, and about two-thirds is absorbed by the air and subsequently re-emitted. The most accurate model consists of a sequence of several layers of air, with temperatures decreasing with altitude, each layer acting as a blackbody.

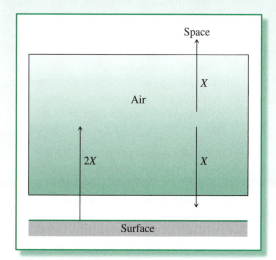

FIGURE 1 Energy released by the Earth's surface and absorbed and released by the atmosphere according to the model.

the higher the altitude. In reality, other factors such as convection currents in air also play an important role in determining the decline in temperature with altitude. The temperature at the top of the troposphere, from which the emitted IR reaches outer space, is only $-18°C$, so overall the real Earth does radiate into space at the same temperature as was computed for the planet if it had no greenhouse gases. Thus the Earth emits the same amount of energy—equal to the amount absorbed from sunlight—into space with or without the operation of the greenhouse effect.

5.5 Earth's Energy Balance

The current energy inputs and outputs from the Earth—in watts (i.e., joules per second) per square meter of its surface and averaged over day and night, over all latitudes and longitudes, and over all seasons—are summarized in Figure 5-4. A total of 342 watts per square meter ($W\ m^{-2}$) are present in sunlight outside the Earth's atmosphere. Of this, 235 $W\ m^{-2}$ are absorbed by the atmosphere and the surface; this much energy must be re-emitted into space if the planet is to maintain a steady temperature. **Because of the presence of greenhouse gases however, emission of only 235 $W\ m^{-2}$ from the surface would not be sufficient to ensure this balance.** Because absorption of IR by greenhouse gases heats the surface and lower atmosphere however, the amount of IR released by them is *increased*. Given the current concentration of greenhouse gases in air, the balance is achieved and 235 $W\ m^{-2}$ escapes from the top of the atmosphere into space if 390 $W\ m^{-2}$ are emitted from the surface, i.e., when 155 $W\ m^{-2}$ of IR does not escape into space.

Ironically, an increase in CO_2 concentration is predicted to cause a *cooling* of the stratosphere. This phenomenon occurs for two reasons.

- First, more outgoing thermal IR is absorbed at low altitudes (the troposphere), so less is left over to be absorbed by and warm the gases in the stratosphere.

- Second, at stratospheric temperatures CO_2 emits more thermal IR upward to space and downward to the troposphere than it absorbs as photons—most of the absorption at these altitudes is due to water vapor and ozone—so increasing its concentration cools the stratosphere.

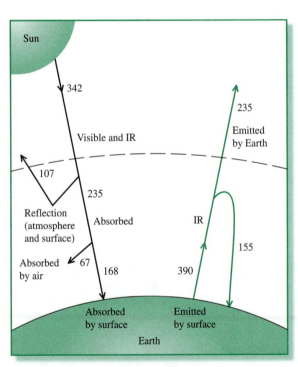

FIGURE 5-4 Globally and seasonally averaged energy fluxes to and from the Earth, in watts per square meter of surface. [Source: Data from Chapter 1 of J. T. Houghton et al., *Climate Change 1995—The Science of Climate Change (Intergovernmental Panel on Climate Change)* (Cambridge: Cambridge University Press, 1996).]

The observed cooling of the stratosphere has been taken to be a signal that the greenhouse effect is indeed undergoing enhancement.

Molecular Vibrations: Energy Absorption by Greenhouse Gases

Light is most likely to be absorbed by a molecule when its frequency almost exactly matches the frequency of an internal motion within the molecule. For frequencies in the infrared region, the relevant internal motions are the **vibrations** of the molecule's atoms relative to each other.

5.6 Types of Molecular Vibrations

The simplest vibrational motion in a molecule is **the oscillatory motion of two bonded atoms X and Y relative to each other. In this motion, called a bond-stretching vibration, the X to Y distance increases beyond its average value R, then returns to R, then contracts to a lesser value, and finally returns to R,** as illustrated in Figure 5-5a. Such oscillatory motion occurs in all bonds of all molecules under all temperature conditions, even at absolute zero. A huge number (about 10^{13}) of such vibrational cycles occur each second. The exact frequency of the oscillatory motion depends primarily upon the type of bond—i.e., whether it is single or double or triple—and upon the identity of the two atoms involved. For many bond types, e.g., for the C—H bond in methane and the O—H bond in water, the stretching frequency does not fall within the thermal infrared region. The stretching frequency of carbon–fluorine bonds does, however, occur within the thermal infrared range (4 to 50 μm) and thus any molecules in the atmosphere with C—F bonds will absorb outgoing thermal IR light and enhance the greenhouse effect.

The other relevant type of vibration is an oscillation in the distance between two atoms X and Z bonded to a common atom Y but not bonded to each other. Such motion alters the XYZ bond angle from its average

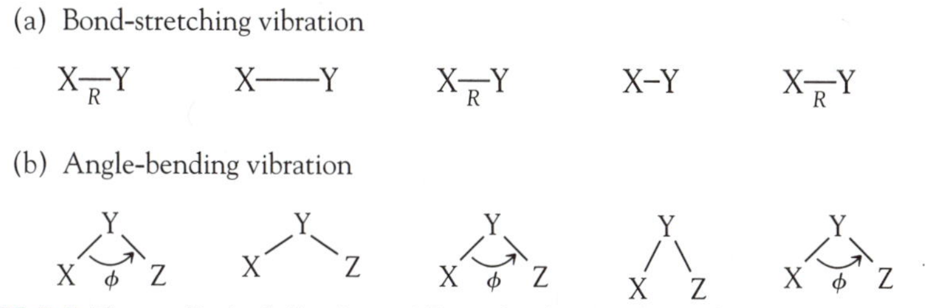

(a) Bond-stretching vibration

(b) Angle-bending vibration

FIGURE 5-5 The two kinds of vibrations within molecules. Bond stretching (a) is illustrated for a diatomic molecule XY. The variable R represents the average value of the X–Y distance. In (b), the angle-bending vibration is shown for a triatomic molecule XYZ. The average XYZ angle is indicated by ϕ.

value ϕ, and is called a **bending vibration.** All molecules containing three or more atoms possess bending vibrations. The oscillatory cycle of bond angle increase followed by a decrease and then another increase, etc., is illustrated in Figure 5-5b. The frequencies of many types of bending vibrations in most organic molecules occur within the thermal infrared region.

If infrared light is to be absorbed by a molecule during a vibration, there must be a difference in the position in the molecule between its center of positive charge—that of its nuclei—and the center of negative charge—that of its electron "cloud,"—at some point during the motion. More compactly stated, in order to absorb IR light, the molecule must have a dipole moment during some stage of the vibration. Technically, there must be a *change* in the magnitude of the dipole moment during the vibration, but this is more or less guaranteed to be the case if there is a nonzero dipole moment at any point in the vibration.

The centers of charge coincide in free atoms and (by definition) in homonuclear diatomic molecules like O_2 and N_2, so these molecules have dipole moments of zero at all times in their stretching vibration. Thus *argon* gas, Ar, *diatomic nitrogen* gas, N_2, and diatomic oxygen, O_2, do not absorb IR light.

For carbon dioxide, during the vibratory motion in which both CO lengths lengthen and shorten simultaneously, i.e., synchronously, there is at no time any difference in position between the centers of positive and negative charges, since both lie precisely at the central nucleus of this linear molecule. Consequently, during this vibration, called the **symmetric stretch,** the molecule cannot absorb IR light. However, in the **antisymmetric stretch** vibration in CO_2, the contraction of one CO bond occurs when the other is expanding, or vice versa, so that during the motion the centers of charge no longer necessarily coincide. Consequently, IR light at this vibration's frequency *can* be absorbed since, at some points in the vibration, the molecule does have a dipole moment.

symmetric
stretch

antisymmetric
stretch

Similarly, the bending vibration in a CO_2 molecule, in which the three atoms depart from a colinear geometry, is a vibration that can absorb IR light at that frequency, since the centers of positive and negative charge do not coincide when the molecule is nonlinear.

Molecules with three or more atoms generally have some vibrations that absorb IR, since even if their average shape is highly symmetric with a zero dipole moment, they undergo some vibrations that reduce this symmetry and produce a nonzero dipole moment. For example, CH_4 molecules have an average structure that is exactly tetrahedral, and hence a zero average dipole moment, because the polarities of the C—H bonds exactly cancel each other in this geometry. The zero dipole is maintained during the vibration in which all four bonds simultaneously stretch or contract. However, during the vibrational motions in which some of the bonds stretch while others contract, and those in which some H—C—H bond angles become greater than tetrahedral while others become less, the molecule has a nonzero dipole moment. Molecules of CH_4 undergoing such unsymmetrical vibrations can absorb infrared light.

PROBLEM 5-2

Deduce whether the following molecules will absorb infrared light due to internal vibrational motions:

(a) H_2 (b) CO   (c) Cl_2 (d) O_3 (e) CCl_4 (f) NO

PROBLEM 5-3

None of the four diatomic molecules listed in Problem 5-2 actually absorbs much, if any, of the Earth's outbound light in the *thermal* infrared region. What does this imply about the frequencies of the bond-stretching vibrational motion of those molecules that can in principle absorb IR light?

Review Questions 1–6 are based on the preceding material.

The Major Greenhouse Gases

5.7 Carbon Dioxide: Absorption of Infrared Light

As stated previously, the absorption of light by a molecule occurs most efficiently when the frequencies of the light and of one of its vibrations match almost exactly. However, light of somewhat lower or somewhat higher frequency than that of the vibration is absorbed by a collection of molecules. This ability of molecules to absorb infrared light over a short range of frequencies rather than just at a single frequency occurs because it is not only the energy associated with vibration which changes when an infrared photon is absorbed; there is also a change in the energy associated with the rotation (tumbling) of the molecule about its internal axes. This **rotational energy of a molecule can be either slightly increased or slightly decreased when IR light is absorbed to increase its vibrational energy. Consequently, photon absorption occurs at slightly higher or lower frequencies than that corresponding to the vibration.** Generally, the absorption tendency of a gas falls off as the light's frequency lies farther and farther in either direction from the vibrational frequency.

    The absorption spectrum for carbon dioxide in a portion of the infrared range is shown in Figure 5-6. For CO_2, the maximum absorption of light in the thermal infrared range occurs at a wavelength of 15.0 μm, which corresponds to a frequency of 2×10^{13} cycles per second (Hertz). The absorption occurs at this particular frequency since it matches that of one of the vibrations in a CO_2 molecule, namely the OCO angle-bending vibration. Carbon

FIGURE 5-6 The infrared absorption spectrum for carbon dioxide. The scale for wavelength is linear when expressed in wavenumbers, which have units of cm^{-1}; wavenumber = 10,000/ wavelength in nm. [Source: Redrawn from A.T. Schwartz et al., *Chemistry in Context: Applying Chemistry to Society*, American Chemical Society (Dubuque, IA: Wm. C. Brown, 1994).]

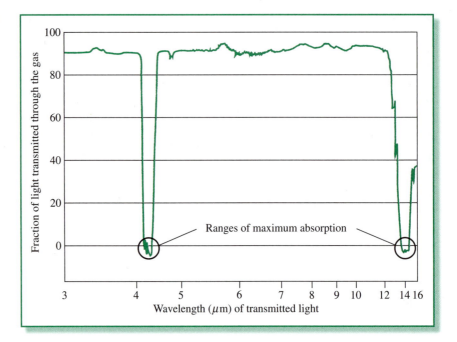

FIGURE 5-6 The infrared absorption spectrum for carbon dioxide. The scale for wavelength is linear when expressed in wavenumbers, which have units of cm^{-1}; wavenumber = 10,000/ wavelength in nm. [Source: Redrawn from A.T. Schwartz et al., *Chemistry in Context: Applying Chemistry to Society*, American Chemical Society (Dubuque, IA: Wm. C. Brown, 1994).]

dioxide also strongly absorbs IR light having a wavelength of 4.26 μm, which corresponds to the 7×10^{13} cycles per second (Hertz) frequency of the anti-symmetric OCO stretch vibration.

PROBLEM 5-4

Calculate the energy absorbed per mole and per molecule of carbon dioxide when it absorbs infrared light at (a) 15.0 μm, and at (b) 4.26 μm. Express the per mole energies as fractions of that required to dissociate CO_2 into CO and atomic oxygen, given that the enthalpies of formation of the three gaseous species are -393.5, -110.5, and $+249.2$ kJ mol^{-1}, respectively. [*Hint*: Recall the relationship between wavelength and energy in Section 1.8. Avogadro's constant = 6.02×10^{23}.]   ●

The carbon dioxide molecules that are now present in air collectively absorb about half of the outgoing thermal infrared light having wavelengths in the 14–16-μm region, together with a sizable portion of that in the 12–14- and 16–18-μm regions. It is because of CO_2's absorption that in Figure 5-7 the black curve, which represents the amount of IR light that actually escapes from our atmosphere, falls so steeply around 15 μm; the vertical separation between the green and black curves is proportional to the amount of IR of a given wavelength that is being absorbed rather than escaping. Increasing further the CO_2 concentration in the atmosphere will prevent more of the remaining IR from escaping, especially in the "shoulder"

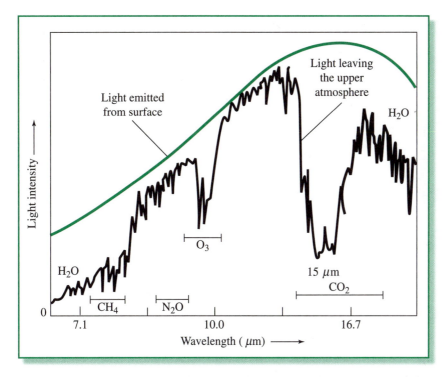

FIGURE 5-7 Experimentally measured intensity (black curve) of thermal IR light leaving the Earth's surface and lower atmosphere (above the Sahara desert) compared with the theoretical intensity (green curve) that would be expected without absorption by atmospheric greenhouse gases. The regions in which the various gases have their greatest absorption are indicated. [Source: E. S. Nesbit, *Leaving Eden* (Cambridge: Cambridge University Press, 1991).]

regions around 15 μm, and hence will further warm the air. (Although carbon dioxide also absorbs IR light at 4.3 μm due to the antisymmetric stretching vibration, there is little energy emitted from the Earth at this wavelength—see Figure 5-1—and thus this potential absorption is not very important.)

5.8 Carbon Dioxide: Past Concentration and Emission Trends

Measurements of air trapped in ice-core samples from Antarctica indicate that the atmospheric concentration of carbon dioxide in preindustrial times (i.e., before about 1750) was about 280 ppm. The concentration had increased by almost 40%, to 390 ppm, by 2010. A plot of the increase in the annual atmospheric CO_2 concentration over time is shown in Figure 5-8. The insert to the figure shows details of the increase in recent times. In the period from 1975 to 2000, the concentration grew at an average annual rate of about 0.4%, or 1.6 ppm—almost double the rate of the 1960s. The rate of increase in the first decade of the twenty-first century rose to about 2.0 ppm annually.

Seasonal fluctuations in CO_2 concentrations (see insert to Figure 5-8) are due to the spurt in the growth of vegetation in the spring and summer, which removes CO_2 from air, and the vegetation decay cycle in fall and winter, which

The insert to Figure 5-8 is called the *Keeling curve*, named after the American chemist Charles Keeling, who began collecting accurate CO_2 concentrations in remote locations in 1958.

FIGURE 5-8 The historic variation in the atmospheric concentration of carbon dioxide. The insert shows the trend, with seasonal fluctuations, in recent times. [Source: Main graph: Adapted from J. L. Sarmiento and N. Gruber, "Sinks for Anthropogenic Carbon," *Physics Today* 55 (August 2002): 30; Inset: NOAA.]

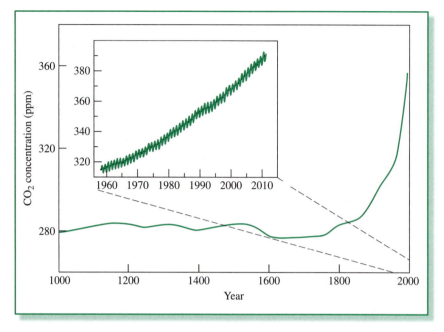

increases it. In particular, huge quantities of CO_2 are extracted from the air each spring and summer by the process of plant photosynthesis:

$$CO_2 + H_2O \xrightarrow{\text{sunlight}} O_2 + \text{polymeric } CH_2O$$

The term *polymeric CH$_2$O* used for the product in this equation is an umbrella word for plant fiber, typically the cellulose that gives wood its mass and bulk. The CO_2 "captured" by the photosynthetic process is no longer free to function as a greenhouse gas—or as any gas—while it is packed away in this polymeric form. The carbon that is trapped in this way is called **fixed carbon.** However, the biological decay of this plant material, the reverse of the reaction, which occurs mainly in the fall and winter, frees the withdrawn carbon dioxide. The global carbon dioxide fluctuations follow the seasons of the Northern Hemisphere, since there is so much more land mass—and hence much more vegetation—there compared to the Southern Hemisphere.

Much of the considerable increase in anthropogenic contributions to the increase in carbon dioxide concentration in air is due to the combustion of fossil fuels—chiefly coal, oil, and natural gas—which were formed eons ago when plant and animal matter was covered by geological deposits before it could be broken down by air oxidation. The topic of fossil fuels and their combustion is discussed in greater depth in Chapter 6.

A significant amount of carbon dioxide is added to the atmosphere when forests are cleared and the wood burned in order to provide land for agricultural use. This sort of activity occurred on a massive scale in temperate climate zones in past centuries (consider the immense deforestation that accompanied the settlement of the United States and southern Canada) but has now shifted largely to the tropics. The greatest single amount of current deforestation occurs in Brazil and involves both rain forest and moist deciduous forest, but the annual rate of deforestation on a percentage basis is actually greater in Southeast Asia and in Central America than in South America.

Overall, deforestation accounts for about one-quarter of the annual anthropogenic release of CO_2, the other three-quarters originating mainly with the combustion of fossil fuels. Notwithstanding forestry harvesting operations, the total amount of carbon contained in the forests of the Northern Hemisphere (including their soils) is increasing, and in the 1980s the annual increment approximately equaled the decreases in stored carbon discussed above in Asia and South and Central America.

PROBLEM 5-5

Carbon dioxide is also released into the atmosphere when calcium carbonate rock (limestone) is heated to produce the quicklime, i.e. calcium oxide, used in the manufacture of cement:

$$CaCO_3(s) \longrightarrow CaO(s) + CO_2(g)$$

Calculate the mass, in metric tons, of CO_2 released per metric ton of limestone used in this process. What is the mass of carbon that the air gains for each gram of carbon dioxide that enters the atmosphere? (Notice that at least as much again carbon dioxide is released from combustion of the fossil fuel needed to heat the limestone as is released from the limestone itself.) ●

5.9 Carbon Dioxide: Atmospheric Lifetime

The lifetime for a carbon dioxide molecule emitted into the atmosphere is a complex quantity since, in contrast to most gases, it is not decomposed chemically or photochemically. On average, within a few years of its release into the air, a CO_2 molecule will likely dissolve in surface seawater or become absorbed by a growing plant. However, many such carbon dioxide molecules are rereleased back into the air a few years later on average, so this disposal is only a *temporary* sink for the gas.

The only permanent sink for CO_2 is dissolution in the deep waters of the ocean and/or precipitation there as insoluble calcium carbonate. However, the top few hundred meters of sea water mixes slowly with deeper waters; thus carbon dioxide that is freshly dissolved in surface water requires hundreds of years to penetrate to the ocean depths. Consequently, although the oceans will ultimately dissolve much of the increased CO_2 now in the air, the time scale associated with this permanent sink is very long, hundreds of years.

Because the processes involving the interchange of carbon dioxide between the air and the biomass and shallow ocean waters, and between shallow and deep seawater, are complicated, it is not possible to quote a meaningful average lifetime for the gas in air alone. Rather, we should think of new CO_2 fossil-fuel emissions as being rather quickly allocated among air, the shallow ocean waters, and biomass, with interchange among these three compartments occurring continuously. Then slowly, over a period of many decades and even centuries, almost all this new carbon dioxide will eventually enter its final sink, the deep ocean.

In effect, the atmosphere rids itself of almost half of any new carbon dioxide within a decade or two, but requires a much longer period of time to dispose of the rest. It is commonly quoted as taking 50–200 years for the carbon dioxide level to adjust to its new equilibrium concentration if a source of it increases. In summary, the effective lifetime of additional CO_2 in the atmosphere should be considered to be long, of the order of many decades or centuries, rather than the few years required for its initial dissolution in seawater or absorption by biomass.

5.10 Carbon Dioxide: Inputs and Outputs

The annual fossil fuel/cement release had increased to 8.7 Gt, and that from deforestation had dropped to slightly less than 1 Gt, by 2008.

The annual inputs and outputs of anthropogenic carbon dioxide to and from our atmosphere, *averaged* over the decade 2000–2009, are summarized in Figure 5-9. (Releases and absorption by natural processes are overall in balance, and are not included on the diagram even though they constitute about 94% of the flow.) Fossil-fuel combustion and cement production released 7.7 gigatonnes (Gt—i.e., billions of tonnes, equivalent to Petagrams, 10^{15} g) of the carbon component (only) of CO_2 into the air.

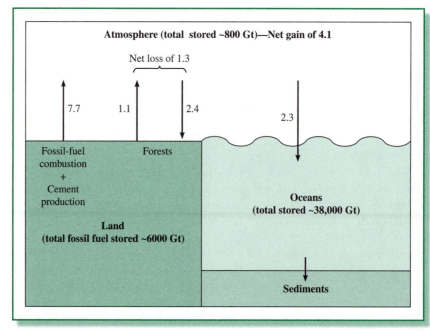

FIGURE 5-9 Annual fluxes of CO_2 to and from the atmosphere, in units of Gigatonnes (Gt) of carbon, averaged over the 2000–2009 period. [Data source: Adapted from *Carbon Budget*, Global Carbon Project, www.globalcarbonproject.org.]

The upper layers of the ocean absorbed tens of gigatonnes of carbon and released about 10% less than this quantity, giving a net average absorption by this principal sink of 2.3 Gt. The carbon released by slash-and-burn tropical deforestation and other land use changes amounted to 1.3 Gt less than that absorbed by forest growth and soil storage. **Because overall less than half the anthropogenic CO_2 emissions are quickly removed, over the short and medium terms the gas continues to accumulate in the atmosphere,** in this period by 4.1 Gt annually.

The variations over the last century and a half in the various annual sources and sinks for CO_2 are summarized in Figure 5-10. The growth of the total annual *emissions*, in terms of the mass of carbon, of carbon dioxide from fossil-fuel combustion, cement production, and deforestation since the start of the Industrial Revolution is given by the top of the uppermost curve in Figure 5-10. Historically, the emission rate from fossil fuels grew rapidly in the second half of the twentieth century, the rate of increase being about five times greater for that in the first half. Another sharp increase occurred early in the twenty-first century.

Notice in Figure 5-10 the year-to-year fluctuations in the amount of the gas absorbed by the oceans and especially in that absorbed by land sinks (biomass). Although the fraction of the new emissions that remains in the atmosphere undergoes substantial variations from year to year, its average increment is increasing with time. In particular, **the uptake by oceans and land is not keeping up with the increase in emissions.**

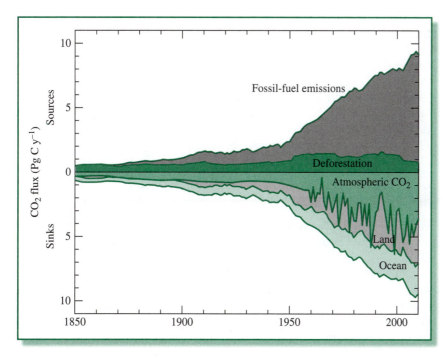

FIGURE 5-10 Annual fluxes of anthropogenic CO_2 from various sources and to various sinks from 1850 to 2010. The unit of Petagram (Pg), 10^{15} g, is equivalent to 1 Gigatonne (Gt). [Data source: Adapted from *Carbon Budget 2009*, Global Carbon Project, www.globalcarbonproject.org.]

The increase in growth rate of certain types of trees due to the increased concentration of carbon dioxide in the air is called **CO₂ fertilization.** Some scientists suspect that the rate of photosynthesis is speeding up as the level of CO_2 and the air temperature increase and that the formation of greater amounts of fixed carbon represents an important sink for the gas. Indeed, an increase in the biomass of northern temperate forests is the most likely sink to account for the annual atmospheric CO_2 loss for which scientists had previously been unable to assign a cause. This increased activity in photosynthesis has been confirmed by satellite data for the region between 45°N and 70°N. Boreal (evergreen) forests of the Northern Hemisphere currently store almost 1 Gt of carbon into standing biomass alone. Much of the increase in the biomass of temperate forests at high latitudes has occurred in the soil, especially as peat. However, it has been found recently that climate warming accelerates the decomposition to CO_2 of existing peat deposits in northern zones.

The anthropogenic releases of CO_2 amount only to about 4% of the enormous amounts produced by nature, so a very small variation in the rate at which carbon is absorbed into biomass or released from it could have a large effect on the residual amount of CO_2 that accumulates in the atmosphere. Unfortunately, scientists still do not completely understand the global carbon cycle. As Figure 5-8 indicates, however, there is no doubt that the atmospheric CO_2 concentration is increasing.

PROBLEM 5-6

(a) Given that the atmospheric burden of carbon (as CO_2) increases by about 4.1 Gt annually, calculate the increase in the ppm concentration of carbon dioxide that this brings about. (b) Given that its total concentration was 390 ppm in 2010, calculate the total mass of CO_2 that was present in the air. After converting to mass of carbon, does your answer agree with the value listed in Figure 5-9? Note that the mass of the atmosphere $= 5.1 \times 10^{21}$ g, and that air's average molar mass $= 29.0$ g mol^{-1}. [*Hint:* Express the amounts of CO_2 and air as moles, and recall the definition of ppm units in moles.] ●

5.11 Water Vapor: Its Infrared Absorption and Role in Feedback

Water molecules, always abundant in air, absorb thermal IR light through their H—O—H bending vibration; the peak in the spectrum for this absorption occurs at about 6.3 μm. As a consequence, almost all the relatively small amount of outgoing IR in the 5.5–7.5-μm region is intercepted by water vapor (see Figure 5-7). (The symmetric and antisymmetric stretching vibrations for water occur close together, at 2.8 and 2.7 μm, both outside the thermal IR region.) Absorption of light leading to increases in the rotational energy of water molecules, without any change in vibrational energy, removes thermal IR of wavelength 18 μm and longer. In fact, **water is the most important greenhouse gas in the Earth's atmosphere,** in the sense that it produces more greenhouse warming than does any other gas, although on a per molecule basis it is a less efficient IR absorber than is CO_2.

Although human activities, such as the burning of fossil fuels, produce water as product, the concentration of water vapor in air is determined primarily by temperature and by other aspects of the weather. Virtually all the H_2O in the troposphere arises from the evaporation of liquid and solid water on the Earth's surface and in clouds. The rate at which water evaporates and the maximum amount of water vapor that an air mass can hold both increase sharply with increasing temperature. In fact, the equilibrium vapor pressure of liquid water increases exponentially with temperature. **The rise in air temperature that is caused by increases in the concentration of the other greenhouse gases, and by other global warming factors, heats the surface water and ice, and thereby causes more evaporation to occur.** Indeed, the average atmospheric content of water vapor has increased since at least the 1980s.

The consequent increase in the water vapor concentration from global warming owing to increased CO_2 etc. produces an additional amount of global warming due to $H_2O(g)$ that is comparable in magnitude to the combined original amount due to the other greenhouse gases, because water vapor is a greenhouse gas! This behavior of water is an example of the general phenomenon called **positive feedback: the operation of a phenomenon produces a result that itself further amplifies the result.** Feedback is a reaction to change; with positive feedback, the reaction accelerates the pace of future change. Positive feedback sometime called a self-reinforcing cycle.

On the other hand, **a system whose output reduces the subsequent level of output displays negative feedback.** An example of negative feedback from daily life is the attempt by a business to raise its profits by increasing its prices. However, the rise in price often results in a reduction in demand for the item of concern, and the rise in profits is less than anticipated. (No value judgment as to the desirability of the effect is implied by the terms *positive* and *negative*; only the increase or decrease in the pace of change is meant.)

Since it comes about as an indirect effect of increasing the levels of other gases, and since it is not within our control or an initiating factor in the process, the warming increment due to water is usually incorporated without further comment into the direct warming effects of the other gases. Consequently, water is not usually listed explicitly among gases whose increasing concentrations are enhancing the greenhouse effect.

Water in the form of liquid droplets in clouds also absorbs thermal IR. However, clouds also reflect some incoming sunlight, both UV and visible, back into space. It is not yet clear whether the additional cloud cover produced by increasing the atmospheric water content will have a net positive or a net negative contribution to global warming. Clouds over tropical regions are known to have a zero net effect on temperature, but those in northern latitudes produce a net cooling effect since their ability to reflect sunlight outweighs their ability to absorb IR. Thus, if increased air temperatures produce more of this latter type of cloud, the enhancement by the greenhouse effect global warming would be damped. However, no one is sure that additional northern cloud cover would occur at the same altitudes and act in the same manner as do the current clouds. Overall, the net effect of clouds on global warming is still subject to some uncertainty.

5.12 The Atmospheric Window

As a result of absorption of IR light of other wavelengths, mainly by carbon dioxide, methane, and water, **it is essentially only infrared light from 8 to 13 μm that escapes the atmosphere efficiently** (see Figure 5-7). Since light of these wavelengths passes unimpeded, **this portion of the spectrum is called the atmospheric window.**

The injection into the atmosphere, even in trace amounts, of gases that can absorb thermal IR light will lead to additional global warming, i.e., to enhancing the greenhouse effect. Of special concern are pollutant gases that absorb thermal IR in the atmospheric window region, since the absorption by H_2O and CO_2 in other regions is already so great that there remains little such light for trace gases to absorb. In particular, **the fraction of light absorbed by a gas as it passes through the atmosphere is logarithmically related to its concentration, C,** by the Beer–Lambert law. Thus the additional global warming produced by carbon dioxide is logarithmically related to the increases in its concentration. However, because logarithmic functions are almost linear near $C = 0$, the warming produced by *trace gases* is linearly proportional to their concentration increases (see Additional Problem 5).

In considering which potential pollutants might contribute to global warming, recall that we can dismiss free atoms and homonuclear diatomic molecules, as they cannot absorb IR light. Heteronuclear diatomic molecules such as CO and NO are also not of direct concern since their only vibration—bond stretching—has a frequency that lies outside the thermal IR region. In general, however, most long-lived gases consisting of molecules with three or more atoms are of concern since they possess many vibrations that absorb IR, one or more of which usually fall in the thermal IR region. The important trace greenhouse gases, i.e., those whose concentration is small in absolute terms but whose ability even at these levels to warm the air is substantial, are detailed below, following a discussion of their average lifetimes in the atmosphere. Those to be discussed are substances whose natural concentrations are zero or low and can therefore be substantially affected by anthropogenic emissions.

Review Questions 7–10 are based on the material in the preceding sections.

Other Greenhouse Gases

5.13 Methane: Absorption and Sinks

After carbon dioxide and water, methane, CH_4, is the next most important greenhouse gas. A methane molecule contains four C—H bonds and is tetrahedral. Although C—H bond-stretching vibrations occur well outside the thermal IR region, **H—C—H bond-angle-bending vibrations absorb at 7.7 μm, near the edge of the thermal IR window; consequently atmospheric methane absorbs IR in this region.**

In contrast to the century-long lifetime of carbon dioxide emissions, **molecules of methane in air have an average lifetime of slightly less than a decade. As discussed in Chapters 3 and 17, the dominant sink for atmospheric methane, accounting for almost 90% of its loss from air, is its reaction with molecules of hydroxyl, OH,** the very reactive free radical gas present in air in tiny concentration:

Details of this transformation are covered in Chapter 17.

$$CH_4 + OH \longrightarrow CH_3 + H_2O$$

This reaction is the first step of a sequence that transforms methane ultimately to CO and then to CO_2.

$$CH_4 \longrightarrow \longrightarrow CH_2O \longrightarrow \longrightarrow CO \longrightarrow \longrightarrow CO_2$$

The other two sinks for methane gas are its reaction with soil and its loss to the stratosphere, to which a few percent of emissions eventually rise. Methane reacts there with OH, or atomic chlorine or bromine, or excited atomic oxygen; reaction with the latter produces hydroxyl radicals and eventually water molecules:

$$O^* + CH_4 \longrightarrow OH + CH_3$$

$$OH + CH_4 \longrightarrow H_2O + CH_3$$

Stratospheric water vapor acts as a significant greenhouse gas. About one-quarter of the total global warming caused by increased methane emissions is not brought about directly, but rather is due to this effect in the stratosphere by which the region's water content is increased. Owing to decreased levels of ozone and increased levels of carbon dioxide, the stratosphere has undergone cooling in recent decades; the increase in water vapor has reduced the amount of this cooling, and so has contributed to the warming of the atmosphere overall.

Per molecule, increasing the amount of methane in air causes a much larger warming effect than does adding more carbon dioxide, since each CH_4 molecule is much more likely to absorb a thermal IR photon that passes through it than does a CO_2 molecule. The effect of the methane is restricted to the first decade or two after its release however, because it is highly likely to be oxidized during this period. When considered over the century after its emission, a kilogram of methane is still about 23 times more effective in raising air temperature than the same mass of carbon dioxide; the ratio is about three times that value over the first 20 years. However, because CO_2 is so long-lived in air and its concentration has increased 80 times more, methane has been less important in heating the atmosphere. To date, methane is estimated to have produced about one-third as much global warming as has carbon dioxide.

5.14 Methane: Emission Sources

About 70% of current methane emissions are anthropogenic in origin and originate from several different source types, including energy production, agriculture, and waste disposal. Most of the natural sources involve plant decay, as do some of the anthropogenic emissions.

Most of the methane produced from plant decay results from the process of anaerobic decomposition, which is decomposition of formerly living matter in the absence of air, i.e., under oxygen-starved conditions. This process converts cellulose (approximate empirical formula CH_2O) into methane and carbon dioxide:

$$2\,CH_2O \longrightarrow \longrightarrow CH_4 + CO_2$$

The original names for methane were *swamp gas* and *marsh gas.*

Anaerobic decomposition occurs on a huge scale where plant decay operates under water-logged conditions, e.g., in natural wetlands such as swamps and bogs, and in rice paddies.

Wetlands are the largest *natural* source of methane emissions, although emissions from this source have decreased sharply over the past centuries as wetlands have been drained. The huge increase in rice production over the same period has presumably led to correspondingly large increases in methane emissions from this source.

• **The expansion of wetlands that occurs by the deliberate flooding of land to produce more hydroelectric power adds to the total natural emissions of the gas.** Deep, small reservoirs produce and emit much less methane than do shallow ones that contain large volumes of flooded biomass, such as those in the Brazilian Amazon, especially if the trees are not first removed. Indeed, the combined global warming effect of the methane and carbon dioxide produced by a large, shallow reservoir created to generate hydroelectric power can, for many years, exceed the carbon dioxide that would have been emitted if a coal-fired power plant used to generate the same amount of electrical power! Hydroelectric power is not a zero-emission form of energy production if land is flooded to create it.

• **The anaerobic decomposition of the organic matter in garbage in landfills (garbage dumps) is another important source of methane emissions to air.** Food waste in the landfill produces the greatest amount of methane. In some communities, methane from landfills is collected and burned to generate heat, rather than being allowed to escape into the air. Although the combustion of methane produces an equal number of molecules of carbon dioxide, because the *per molecule* effect of CO_2 molecules is so much smaller than of CH_4 molecules, the net greenhouse effect of the emission is thereby greatly reduced to that corresponding to the amount of CO_2 absorbed from the air when the plant matter was growing.

- **The burning of biomass,** such as forests and grasslands in tropical and semitropical areas, releases methane to the extent of about 1% of the carbon consumed, along with larger amounts of carbon monoxide (both compounds are products of incomplete—poorly ventilated—combustion).

- **Ruminant animals—including cattle, sheep, and certain wild animals—produce huge amounts of methane as a by-product** in their stomachs when they digest the cellulose in their food. **The animals subsequently emit the methane into the air** by belching or flatulence. The decrease in the population of some methane-emitting wild animals (e.g., buffalo) in recent centuries has been far exceeded by the huge increase in the population of cattle and sheep. The net result has been a large increase in emissions of methane from animal sources.

Septic tanks are also a source of methane emissions.

- **Methane is released into air when natural gas pipelines leak, when coal is mined and the CH_4 trapped within it is released into the air, and when the gases dissolved in crude oil are released—or incompletely flared—into the air when the oil is collected or refined.** Emissions from these sources have likely leveled off in the last decade. The technique by which scientists determine the component to atmospheric methane arising from fossil-fuel sources is discussed in Box 5-2.

In summary, there are six different significant sources of atmospheric methane, of which natural wetlands make the greatest contribution (~25%). The current relative importance of the five important anthropogenic sources of atmospheric methane is thought to be

energy production/distribution ~ ruminant animal livestock
> rice production ~ biomass burning ~ landfills

5.15 Methane: Concentration Trends and Possible Future Increases

Historically (i.e., before 1750), the methane concentration in air was approximately constant at about 0.75 ppm, i.e., 750 ppb. It has more than doubled since preindustrial times, to about 1.8 ppm. Almost all of this increase occurred in the twentieth century because the emissions grew quickly, by a factor of four, during the twentieth century. **The rise in the atmospheric CH_4**

BOX 5-2   Determining the Emissions of "Old Carbon" Sources of Methane

The relative abundances of carbon isotopes in atmospheric carbon dioxide can be used to help deduce its origin by the following logic. The carbon in all living matter contains a small, constant fraction of a radioactive isotope, carbon-14 (^{14}C), taken in via the carbon cycle when photosynthesis captures atmospheric CO_2 and when animals in turn feed off plant matter. This fact underlies the radiocarbon dating methods used by archaeologists and anthropologists: When an organism dies, its ^{14}C decays at a known first-order rate that makes the date of its death calculable. (The assumptions justifying these methods are that biotic carbon and atmospheric carbon in CO_2 are balanced—in equilibrium with one another—and that the level of atmospheric ^{14}C is constant. The principles underlying radioactive decay are discussed in Chapter 9.)

However, in the case of atmospheric methane, the average fraction of ^{14}C is less than the value found in living tissue. This indicates that a significant fraction of the CH_4 escaping into air must be "old carbon" that has been trapped in the ground for so long that its ^{14}C content has diminished to almost zero as a result of radioactive decay through the ages. Most methane containing old carbon is released into the air as a by-product of the mining, processing, and distribution of fossil fuels. Methane trapped in coal is released into the atmosphere when this material is mined, as is methane in oil when it is pumped from the ground. The transmission of natural gas, which is almost entirely methane, involves losses into the air due to leakage from pipelines and is the largest of the atmospheric sources of old carbon. Measurements of the methane levels in the air of various cities have indicated that much of the loss from pipelines in the past occurred in eastern Europe. Finally, there is probably a small contribution to the old carbon source from methane trapped in permafrost in far northern latitudes; the methane was formed by the decay of plant matter that lived there many thousands of years ago when the polar climate was much warmer than it is today.

level that has occurred since preindustrial times is presumed to be the consequence of emissions from such human activities as increased food production and fossil-fuel use, as discussed above. As in the case of carbon dioxide, post-World War II emission rates increased annually much more quickly than had been the case previously. By the early 1990s, however, the rate of growth in concentration had declined rapidly, and for a decade was almost zero; the methane concentration in air was almost constant in that period (Figure 5-11). However, the concentration began to rise again starting in 2007.

Since the rate of change in concentration of a substance is proportional to the difference between the rate of change in the emission rate and the rate of change in the destruction rate, change in either one or both of the two rates could be responsible for minor variations in the methane concentration; neither rate can be measured directly very accurately. Natural gas pipelines carry ~90% methane, about 1.5% of which is lost to the atmosphere.

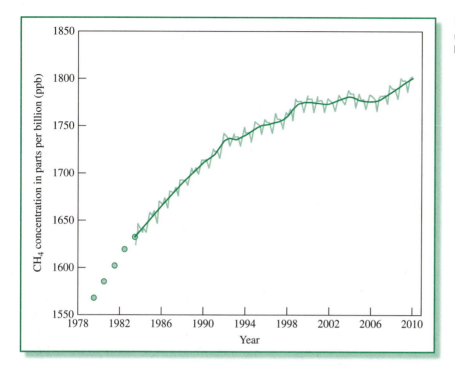

FIGURE 5-11 Atmospheric methane concentration [Source: NOAA.]

Part of the decline in the emission rate of methane into the air in the early 1990s was likely due to much-decreased emissions from pipelines in the former Soviet Union, which a few decades ago leaked much more gas than at present. However, increasing use of fossil fuels in northern Asia may now have replaced some of these emissions. A variety of reasons have been advanced to account for the increases in atmospheric methane concentration in the 2007–2010 period, but scientists have not yet agreed as to the dominant cause.

PROBLEM 5-7

The concentration of atmospheric methane is 1.8 ppm, and the rate constant for the reaction between CH_4 and OH is 3.6×10^{-15} cm^3 molecule^{-1} s^{-1}. Calculate the rate, in teragrams per year, of methane destruction by reaction with hydroxyl radical, the concentration of which is 8.7×10^5 molecules cm^{-3}. See Problem 5-6 for additional data. ●

Some scientists have speculated that **the rate of release of methane into air could greatly increase in the future as a consequence of temperature rises from the enhanced greenhouse effect.** For instance, higher temperatures would accelerate the anaerobic biomass decay of plant-based matter, as occurs in a common landfill. In turn, the additional release of methane would itself cause a further rise in temperature. This is another example of positive feedback.

Methane release from biomass decay among the extensive bogs and tundra in Canada, Russia, and Scandinavia could also increase with increasing air temperature and would also constitute positive feedback. However, the rate of biomass decay and of plant growth, and thus of CH_4 production, also depends on soil moisture and therefore on rainfall, which probably would be affected by climate change in an as yet uncertain direction in these regions, so the net feedback from these sources could be positive or negative.

There is much methane currently immobilized in the permafrost of far northern regions; it was produced from the decay of plant materials during warm spells in the region but became trapped due to glaciation as temperatures became lower and lower at the start of the last ice age. Melting of the permafrost due to global warming could release large amounts of this methane. Melting would also allow the decomposition of organic matter currently present in the permafrost, with the consequent release of more methane.

In addition, **there are monumental amounts of methane trapped at the bottom of the oceans, on continental shelves, in the form of methane hydrate.** This substance has the approximate formula $CH_4 \cdot 6H_2O$ and is an example of a **clathrate compound, i.e., a rather remarkable structure that forms when small molecules occupy vacant spaces (holes) in a cage-like polyhedral structure formed by other molecules.** In the present case, methane is caged in a 3-D ice lattice structure formed by the water molecules. The melting point of the structure is +18°C, somewhat higher than pure ice. **Clathrates form under conditions of high pressure and low temperature, such as are found in cold waters and under ocean sediments.** The methane was produced over thousands of years by bacteria that facilitated the anaerobic decomposition of organic matter in the sediments.

If seawater warmed by the enhanced greenhouse effect penetrates to the bottom of the oceans, the clathrate compounds could decompose and release their own methane, as well as reservoirs of pure methane currently trapped below them, to the air above. Methane trapped far below the permafrost in northern areas and in offshore areas in the Arctic also exists in the form of clathrates; it would be released eventually if the Arctic warmed sufficiently. Measurements made thus far do not indicate any significant emissions from these sources. It has been suggested by some scientists that CH_4 released from clathrates may be oxidized to CO_2 before it reaches the air, thereby greatly reducing the global warming potential.

Although the uncertainties concerning methane feedback are large, the stakes are higher than with any other gas. A few scientists believe that several positive climate feedback mechanisms, including those involving methane, could possibly combine to trigger an unstoppable warming of the globe. This worst-case scenario is called the *runaway greenhouse effect*. Such climate change would threaten all life on Earth, as the temperature would rise markedly, ocean currents would probably shift, and rainfall patterns would be very different from those we know. The possibility that the North Atlantic ocean current that brings warm water from the south and thereby warms Europe could cease to operate because of rapid global warming—induced by rapid increases in methane or carbon dioxide—is one of the most dramatic predictions about the possible consequences of the enhanced greenhouse effect.

PROBLEM 5-8

Calculate the mass of methane gas trapped within each kilogram of methane hydrate.

5.16 Nitrous Oxide

Another significant greenhouse trace gas is **nitrous oxide,** N_2O, "laughing gas," the molecular structure of which is linear NNO rather than the more symmetric NON. Its bending vibration absorbs IR light in a band at 8.6 μm, i.e., within the window region, and in addition, one of its bond-stretching vibrations is centered at 7.8 μm, on the shoulder of the window and at the same wavelength as one of the absorptions for methane.

Per molecule released, N_2O is 206 times as effective as CO_2 in causing an immediate increase in global warming, and about 114 times as effective when averaged over a century. Its atmospheric lifetime is about 120 years. Like that of methane, the atmospheric concentration of nitrous oxide was constant until about 300 years ago. It then began to increase, from 275 ppb (preindustrial) by just 17% to 323 ppb (2010). The yearly growth rate in the 1990s was erratic, falling at first and then rising, but since the turn of the century has resumed its annual linear growth rate of about 0.7 ppb. The increased amounts of nitrous oxide that have accumulated in the air since preindustrial times have produced about one-third of the amount of additional warming that methane has induced.

About 30–40% of nitrous oxide emissions, amounting to about 6.7 teragrams (Tg), currently arise from anthropogenic sources. In 1990 it was discovered that the traditional procedure, using *nitric acid,* HNO_3, to synthesize *adipic acid,* a raw material in the preparation of nylon, resulted in the formation and release of large amounts of nitrous oxide. Since that time, nylon producers have instituted a plan to phase-out N_2O emissions.

The greater part of the natural supply of nitrous oxide gas comes from release by the oceans and processes occurring in the soils, especially of tropical regions. The gas is a by-product of the biological denitrification process in aerobic (oxygen-rich) environments and in the biological nitrification process in anaerobic (oxygen-poor) environments; the chemistry of both

FIGURE 5-12 Nitrous
oxide production as a
by-product during the
biological cycling of
nitrogen.

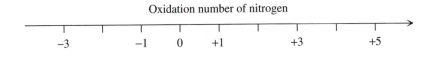

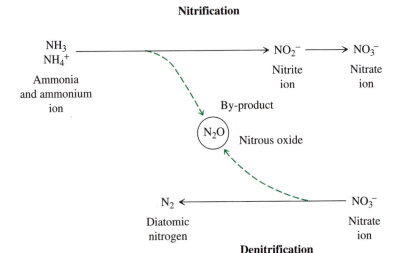

FIGURE 5-12 Nitrous oxide production as a by-product during the biological cycling of nitrogen.

processes is illustrated in Figure 5-12. **In denitrification, fully oxidized nitrogen in the form of the nitrate ion, NO_3^-, is reduced mostly to molecular nitrogen, N_2. In nitrification, reduced nitrogen in the form of ammonia or the ammonium ion is oxidized mostly to nitrite (NO_2^-) and nitrate ions.**

Chemically, the existence of the nitrous oxide by-product in both processes is simple to rationalize: nitrification (oxidation) under oxygen-limited conditions yields some N_2O, which has less oxygen than the "intended" nitrite ion, and denitrification (reduction) under oxygen-rich conditions yields some N_2O, which has more oxygen than the "intended" nitrogen molecule. Nitrification is more important than denitrification as a global source of N_2O. Normally, about 0.001 mol of nitrous oxide is emitted per mole of nitrogen oxidized, but this value increases substantially when the ammonia or ammonium concentration is high and relatively little oxygen is present.

Overall, the increased use of fertilizers—both synthetic and organic—for agricultural purposes probably accounts for the majority of anthropogenic emissions of nitrous oxide. The decomposition of livestock-produced manure under aerobic conditions, including its use as a fertilizer, contributes significantly to nitrous oxide emissions; manure produces very little N_2O if decomposed anaerobically.

Some portion of the nitrate and ammonium fertilizers used agriculturally, particularly in tropical areas, is similarly converted (an unintended effect, to be sure) to nitrous oxide and released into the air. Tropical forests

Soils emit about 6.6 Tg annually, oceans another 3.8 Tg.

in wet areas are probably a huge natural source of the gas. Apparently, nitrous oxide released from new grasslands is particularly significant in the years following the burning of a forest.

At one time it was believed that fossil-fuel combustion released nitrous oxide as a by-product of the chemical combination of the N_2 and O_2 in air, but this belief was based upon faulty experiments. Only when the fuel itself contains nitrogen, as does coal and biomass (but not gasoline or natural gas), does N_2O form; apparently N_2 from air does not enter into this process at all. However, some of the NO produced from atmospheric N_2 during fuel combustion in automobiles is unavoidably converted to N_2O rather than to N_2 in the three-way catalytic converters currently in use and is subsequently released into air. Some of the newer catalysts developed for use in automobiles do not suffer from this flaw of producing and releasing nitrous oxide during their operation.

As mentioned in Chapter 1, **there are no sinks for nitrous oxide in the troposphere.** Instead, all of it rises eventually to the stratosphere where each molecule absorbs UV light and decomposes, usually to N_2 and atomic oxygen, or reacts with atomic oxygen.

Fossil-fuel use and industrial processes collectively release about 0.7 Tg of nitrous oxide annually.

5.17 CFCs and Their Replacements

Gaseous compounds consisting of molecules with carbon atoms bonded exclusively to fluorine and/or chlorine atoms have perhaps the greatest potential among trace gases to induce global warming, since they are both very persistent and absorb strongly in the 8–13-μm atmospheric window region. Absorption due to the C—F bond stretch is centered at 9 μm. The C—Cl bond stretch, and various bond-angle-bending vibrations involving carbon atoms bonded to halogens, also occur at frequencies that lie within the window region.

As discussed in Chapter 2, the **chlorofluorocarbons** $CFCl_3$ and CF_2Cl_2 have already been released into the atmosphere in large quantities and have long **residence times, a term defined as the average amount of time one of its molecules exists in air before it is removed by one means or another.** Due to this persistence, and to their high efficiency in absorbing thermal IR in the window region, each CFC molecule has the potential to cause the same amount of global warming as do tens of thousands of CO_2 molecules. The *net* effect of CFCs on global temperature is small, however. The heating that the CFCs produce by the redirection of thermal infrared is partially cancelled by a separate effect, the cooling that they induce in the stratosphere due to their destruction of ozone. However, the decrease in stratospheric ozone allows more UV light to reach the lower atmosphere and the surface and to be absorbed there. The cooling and heating effects produced by CFCs occur at very different altitudes, so their net effect on the Earth's weather may be substantial.

Recall from Chapter 1 that the stratosphere is heated when oxygen atoms, recently detached photochemically from ozone molecules, collide with O_2 molecules to produce an exothermic reaction.

Ironically, the use of CFCs in insulating freezers, refrigerators, and air conditioners has reduced the energy requirements of this equipment and so has reduced CO_2 emissions resulting from electricity production.

The influence of CFCs on climate in the future will be reduced as a result of the requirements of the Montreal Protocols, which banned further production in developed countries after 1995, as discussed in Chapter 2. Most of the *HCFC* and *HFC* replacements for CFCs (with the notable exception of HFC-143a) have shorter atmospheric lifetimes and absorb less efficiently in the center of the atmospheric window region, and thus on a molecule-for-molecule basis they pose less of a greenhouse threat. However, if their levels of production and release become high in future decades because of expanding world population and increasing affluence, they will make significant contributions to global warming if they are released into the air.

For this reason, many people feel that these replacement substances must be used only in closed systems from which leakage to the atmosphere does not occur, and that they must be recovered from equipment before its eventual disposal. Indeed, prevention of the chronic release of long-lived gases of all types to the atmosphere is a principle now agreed to by many scientific, business, and governmental groups. The release of the refrigerant HFC-134a (see Chapter 2) from the air-conditioning units of modern cars has a global warming impact that is about 4–5% of that from the carbon dioxide emitted from them.

PROBLEM 5-9

Fully fluorinated compounds such as tetrafluoromethane and hexafluoroethane are released as by-product wastes into the air in the production of aluminum. They were also briefly considered as CFC replacements. Will such molecules have a sink in the troposphere? Will they act as greenhouse gases? Would your answers be the same for monofluoromethane and monofluoroethane? ●

5.18 Sulfur Hexafluoride

Sulfur hexafluoride, SF_6, is a little-known greenhouse gas. It has some importance, however, since it is such a good absorber of thermal IR—23,900 times greater than CO_2 in global warming potential—and because, like other fully fluorinated compounds, it is very long-lived in the atmosphere (3200 yr). It is used by electric utilities and in the semiconductor industry as an insulating gas. Formerly it was vented to the air during routine maintenance of equipment but now is mainly recycled instead. Currently its atmospheric concentration is 7.0 ppt (2010); its level has been increasing linearly since about 2006 at a rate significantly greater than the constant increases observed in the previous decade.

5.19 Tropospheric Ozone

Like methane and nitrous oxide, tropospheric ozone, O_3, is a "natural" greenhouse gas, but one which has a short tropospheric residence time.

Although the ozone molecule is homonuclear, in its bent structure the central oxygen is not equivalent to the terminal oxygen atoms, and consequently the O—O bonds are somewhat polar. For that reason, the dipole moment of O_3 molecules changes during the symmetric-stretching vibration, which occurs in the atmospheric window region between 9 and 10 μm, and the molecules can absorb this IR light. The dip near 9 μm in the window region of the outgoing thermal IR distribution (Figure 5-7) is due to absorption by this vibration in atmospheric ozone molecules. Ozone's bending vibration occurs at 14.2 μm, near that for CO_2, and thus it does not contribute much to the enhancement of the greenhouse effect since atmospheric carbon dioxide already removes much of the outgoing light at this frequency. The antisymmetric-stretching vibration of O_3 occurs at 5.7 μm, where there is very little outgoing IR.

As explained in Chapter 3, ozone is formed in the troposphere as a result of pollution from power plants and motor vehicles, and from forest fires and grass fires, as well as from natural processes. As a result of these anthropogenic activities, the levels of ozone in the troposphere probably have *increased* since preindustrial times. The best guess is that approximately 10% of the increased global warming potential of the atmosphere results from increases in tropospheric ozone, although this value is very uncertain. The amount of thermal IR absorbed by stratospheric ozone has probably dropped slightly due to its recent decline there.

Review Questions 11–18 are based upon the material in the preceding sections.

The Climate-Modifying Effects of Aerosols

In Chapter 2 we saw that the initial neglect by scientists of the effects of atmospheric aerosol particles, specifically ice crystals in the stratosphere, led to a large underestimation of the amount of ozone that would be destroyed by chlorine.

Similarly, neglect of aerosols led to misleading predictions about the extent of global warming to be expected. It is now realized that **aerosols offset and thereby mask a significant fraction of the atmospheric temperature increase that would have otherwise occurred due to the anthropogenic greenhouse gas emissions.**

The types of particulate matter of most importance in this context are those expelled by powerful volcanic eruptions into the upper atmosphere, *and* those produced by industrial processes and emitted into the lower troposphere. In order to understand how aerosols can affect global warming, it is necessary to understand how they interact with light.

5.20 The Interaction of Light with Particles

All solids and liquids—including atmospheric particles—have some ability to **reflect** light. **Atmospheric particles can reflect incoming sunlight, with**

FIGURE 5-13 Interaction of sunlight with suspended atmospheric particles. (a) Modes of interaction. (b) Illustration of the indirect effect of producing increased reflection by small water droplets as compared to larger ones having the same total volume.

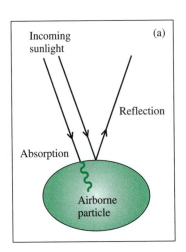

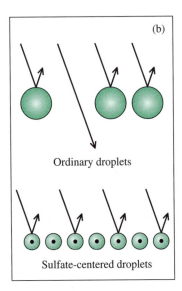

the consequence that some of it is directed back into space and so is unavailable later for absorption at the surface (see Figure 5-13a). The particles can also reflect outgoing infrared light (not shown), with the consequence that some of it is redirected back toward the surface rather than escaping from the atmosphere. **The redirection of light by a particle is sometimes called scattering;** reflection backward is "backscattering."

Recall that the fraction of sunlight reflected by a surface is called its **albedo**. For example, the albedos of fresh snow and of the top surfaces of many clouds are close to unity since they are highly reflective, whereas the albedo for bare ground is only about 0.1. **Certain types of suspended particulates in air reflect some of the sunlight which shines upon them back into space and so have a significant albedo value. This reflection of sunlight by the aerosol cools the air mass and the surface below it since none of the reflected light is subsequently absorbed and then converted to heat.**

Some types of aerosol particles can *absorb* **certain wavelengths of light (Figure 5-13a). Once absorbed, the energy that was associated with the light is rapidly converted into heat, which then becomes shared with the surrounding air molecules as a result of their collisions with the heated particle. Thus, absorption of light by a particle leads to warming of the air immediately surrounding it.** The absorption of sunlight, with consequent warming, is significant only for dark-colored particles such as those composed primarily of soot, often called **black carbon,** and of ash particles from volcanoes.

The contribution of black carbon to global warming has only recently been fully appreciated, and indeed was exceeded in magnitude in the past only by that of the greenhouse gases. Like other aerosols, the atmospheric lifetimes of black carbon particles are quite short—weeks, rather than years. However, black carbon still exerts its effects when it falls on snow, and consequently may be responsible for some of the unusually large amount of warming observed in Arctic regions.

The emission into the atmosphere of black carbon is largest in developing countries, where incomplete combustion of coal and biomass is widespread. Its effect globally is to increase air temperatures by its absorption of sunlight, with the subsequent export of this warmed tropospheric air to other areas. However, its *local* effect may be cooling, since it blocks sunlight from reaching the surface. Black carbon's effects on local climate may be substantial, increasing drought in some areas and flooding in others. However, recent research indicates that the relative importance of black carbon in adding to global warming has diminished in many areas in recent decades.

Recall from Chapter 3 that the *sulfur dioxide* gas predominantly released as a pollutant from the burning of fossil fuels—especially coal—and from the smelting of nonferrous metals creates a **sulfate aerosol. Pure sulfate aerosols do not absorb sunlight** since none of their constituents—water, sulfuric acid, and the ammonium salts thereof—absorb light in the visible or the UV-A regions. The sulfate aerosols are not particularly effective in trapping outgoing thermal IR emissions. Only if tropospheric sulfate aerosols incorporate some soot will absorption of sunlight by these particles be significant.

Overall, however, **anthropogenic sulfate-rich aerosols produced in abundance—especially in the Northern Hemisphere—reflect sunlight back into space much more effectively than they absorb it, so they significantly increase Earth's average albedo.** As a result, less sunlight is available to be absorbed and converted to heat in the lower troposphere and the surface. Thus the net **direct effect** of the sulfate aerosols—i.e., by reflection of sunlight by the particles themselves—is to cool the air near ground level and thereby to offset some of the global warming induced by greenhouse gases.

In addition to the direct effect of sulfate aerosols in reflecting sunlight, there are **indirect effects** that arise because the **sulfate particles act as nuclei for the formation of small water droplets.**

• Such small droplets are more effective in backscattering light than are an equal mass of larger ones (see Figure 5-13b).

• Also, small droplets are less likely to coalesce into raindrops, so their clouds are longer-lived than otherwise expected, and thus can reflect sunlight for longer periods.

Both these indirect effects result in more sunlight being reflected back into space, thereby cooling the Earth's surface. In addition, the "atmospheric brown cloud" (Chapter 4) formed by aerosol particles reduces the strength of the essential monsoon rains over India and Asia. These monsoons, and droughts elsewhere, are influenced by the nonuniform distribution of atmospheric aerosols.

A short-term, dramatic example of the effects of atmospheric aerosols upon climate occurred as a consequence of the massive eruption of substances into the troposphere and stratosphere by the Mount Pinatubo

A proposal to deliberately inject more sulfate particulates into the upper troposphere or lower stratosphere to counter global warming is discussed later in this chapter.

volcano in the Philippines in 1991. Initially, the lower stratosphere was warmed due to the dominant effect of the large volcanic ash particles, which absorbed some of the incoming sunlight and subsequently converted it to heat, and by the interception of outgoing infrared from the surface by them. Due to their relatively large size, the ash particles were not long-lived in the stratosphere. The longer-term effect of the Pinatubo eruption on air temperatures at ground level was a significant decrease. The *stratospheric* aerosol that remained suspended after a few months was formed by the oxidation of the 30 million tons of SO_2 that the volcano had blasted directly into the lower parts of this region.

The sulfate aerosol remained there for several years, during which time it efficiently reflected sunlight back into space. Inspection of Figure 5-2b indicates a drop in the globally averaged surface temperature of about 0.2°C in 1992 and 1993, due apparently to this "volcanic cooling." Many regions, including North America, experienced several cool summers in the early 1990s due to this effect. Because of the gradual sedimentation of the aerosol, the Earth returned to 1990–1991 temperatures by 1995.

5.21 Aerosols and Global Warming

The cooling effect of the sulfate aerosol is concentrated almost entirely in the Northern Hemisphere because most industrial activity takes place in that half of the globe, so it is there that most emissions occur. The relatively short lifetime of such sulfate aerosols precludes their spreading to the Southern Hemisphere; consequently the concentration of sulfate particulates is much higher over the Northern Hemisphere.

The short lifetime of the sulfate particles can be understood by considering the processes of their removal from air. The average diameter of the tropospheric sulfate aerosol particles is about 0.4 μm and their average altitude is about half a kilometer. For particles of this size at this altitude, the expected atmospheric lifetime before gravitational settling to the surface is several years. However, **since the sulfate aerosol droplets are also removed efficiently by rain, their actual lifetime in the lower troposphere is of the order of days rather than years.**

The increase of global SO_2 emissions from fossil-fuel combustion over the last century and a half was discussed in Section 3.20 and shown in Figure 3-17. Presumably the trend in anthropogenic sulfate aerosol production has approximately followed the pattern of SO_2 emissions. The amount of sunlight that was reflected into space by these aerosols in different parts of the world as of the late twentieth century is illustrated by the contours of the map of Figure 5-14. The bulk of the anthropogenically produced aerosol in North America is centered above the Ohio Valley and directly reflects sunlight mostly above that area. Equal or even larger effects are observed over southern Europe and portions of China. Indeed, according to some calculations, the cooling effect from aerosols outweighs the heating effect due to greenhouse gases for some regions in these areas (see Box 5-3).

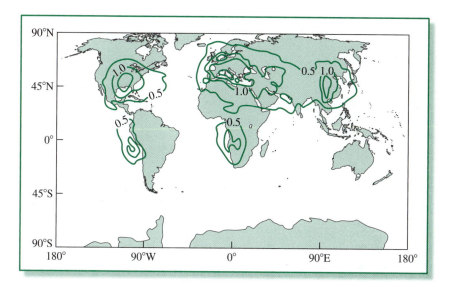

FIGURE 5-14 The amount of sunlight reflected into space by anthropogenic aerosols by the direct mechanism, in units of watts per square meter of the Earth's surface. [Source: J. T. Houghton et al., *Climate Change 1994— Radiative Forcing of Climate Change* (Intergovernmental Panel on Climate Change) (Cambridge: Cambridge University Press, 1995).]

It is not clear how the amount of tropospheric sulfate aerosol will change in the future. Emissions of sulfur dioxide from power production in North America and Western Europe are now more tightly controlled in order to combat acid rain. However, the anthropogenic sulfate aerosol concentrations over southern Europe and parts of Russia and China are considerably higher than the current maximum values in North America and will not be affected by legislative controls in some of these regions. The only substantial domestic energy source currently available to China for its rapid industrialization is coal, so the SO_2 emissions from this source and from India may well continue to rise, as discussed in Chapter 3.

Aerosols also result from the oxidation of the gas *dimethyl sulfide* (DMS), $(CH_3)_2S$, which is produced by marine phytoplankton and subsequently released into the air over oceans. Once in the troposphere, DMS undergoes oxidation, some of it to SO_2 which then can oxidize to sulfuric acid, and some to *methane-sulfonic acid*, CH_3SO_3H. Both of these acids form aerosol particles, which in turn lead to the formation of water droplets and hence of clouds over the oceans. The particles and droplets both deflect incoming light from the Sun. Some scientists believe that increased emissions of dimethyl sulfide by the oceans will occur when seawater warms as a result of the enhancement of the greenhouse effect and that this negative feedback will temper global warming.

Although the sulfate aerosol has a short lifetime, new supplies of it are constantly being formed from the sulfur dioxide pollution that pours into the atmosphere on a daily basis. Consequently there is a steady-state amount of the aerosol in the troposphere; **sulfur dioxide emissions keep postponing the full effects of global warming induced by the rise in greenhouse gas concentrations.**

BOX 5-3   Cooling over China from Haze

Measurements of the amount of sunshine reaching the surface in China indicate a significant decrease over the last half-century (Figure 1). The blockage of solar intensity results from the aerosols in the air above the region produced mainly from the sulfur dioxide emitted by burning coal. As a consequence of the increasing presence of the aerosols, maximum summer temperatures in heavily polluted eastern China have fallen by about 0.6°C per decade. Similar effects are observed in the Brazilian Amazon region, due to soot and ash emitted into the local air from forest and grass fires lit to clear land, and in the African country of Zambia, due to grass fires.

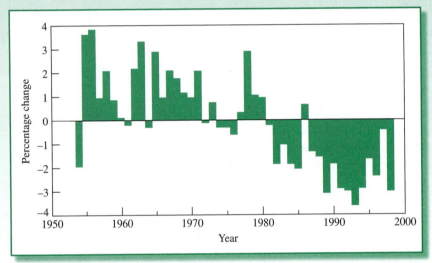

FIGURE 1 Change in the amount of sunlight reaching China relative to the average, over 50 years. [Source: F. Pearce, "Pollution Is Plunging Us into Darkness," *New Scientist* (14 December 2002):6.]

Global Warming to Date

5.22 Allocation of Warming to Natural and Anthropogenic Factors

The best estimates of global warming or cooling, as of 2005, arising from the various factors is summarized by the bar graphs in Figure 5-15; the effect of each factor is expressed as a percentage of the total anthropogenic effect. The order of the greenhouse gases in terms of the amount of extra warming they have produced is

$$CO_2 > CH_4 > O_3 > CFCs > N_2O$$

The value in Figure 5-15 for CFCs includes the cooling of the stratosphere induced by their destruction of ozone, and that for methane includes the

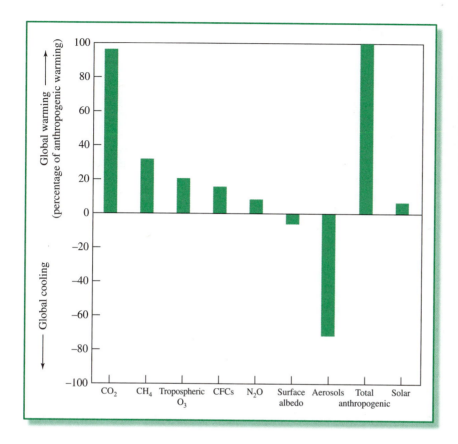

FIGURE 5-15 Contributions to global warming by 2005 and cooling produced by various factors, expressed as percentages of the total anthropogenic warming. [Data source: Intergovernmental Panel on Climate Change, *Climate Change 2007: The Physical Science Basis. Summary for Policymakers* (February, 2007).]

warming of the stratosphere produced from the additional water vapor formed there by its decomposition. The cooling labeled *Surface albedo* is the net result of cooling due to changes in land use minus warming arising from the deposition of sunlight-absorbing black soot on snow and ice.

Overall, the cooling from anthropogenic aerosols currently cancels about 40% of the net warming from all greenhouse gases. However, the aerosol effect—which is the sum of the direct effect and indirect effects affecting cloud albedo—has by far the largest uncertainty of any of the factors in Figure 5-15. There is some evidence that the sulfate aerosol level in the atmosphere gradually decreased in the 1990s and early 2000s, and was responsible for the larger-than-expected global warming observed in that period.

Greenhouse gas emissions from airplanes traveling long distances high in the troposphere are particularly effective in promoting global warming. They emit, into the low-density air in which they fly, the carbon dioxide and water vapor that result from the combustion of their hydrocarbon fuel. Because the air temperature there is so low, the IR absorbed by this CO_2 and H_2O is unlikely to be re-emitted, but instead warms the surrounding air and therefore enhances the greenhouse effect.

5.23 Global Warming: Geography

The year-to-year variations in the average worldwide surface temperature from 1880 to 2010 were shown in Figure 5-2b. The increases in temperature were not evenly spread around the globe, however, as indicated by Figure 5-16, which shows the changes in average global temperatures in the 2001–2005 period, relative to the 1951–1980 average. In Figure 5-16, the darker the shade of green, the greater the increase in temperature; the few areas shown in gray underwent decreases in temperature. **In general, air temperatures over land areas have experienced a greater increase than those over the seas.** Sulfate aerosols keep the eastern portions of the United States and Canada cooler than they would be otherwise.

The Arctic region has warmed most of all, with the consequence that its sea ice is disappearing, at a rate that is far greater than that expected from climate models. Indeed, the years 2007–2009 saw the smallest cover of sea ice in the Arctic ever recorded, with both the Northwest and Northeast passages ice-free in 2008 for the first time in living memory. The reduction in the minimum area covered by Arctic sea ice in 2007 is contrasted with its minimum extent in previous decades in the picture on the next page.

The melting of sea ice produces a positive feedback effect: since ice reflects sunlight more efficiently than does liquid water, the increasing amount of sunlight absorbed as the ice is replaced by open water raises surface water and surface air temperatures, thereby inducing even more melting. However, scientists have found that evaporation of the open water has produced more cloud cover in the region, which is a *negative* feedback effect that partially negates the positive one, since the clouds reflect sunlight and thereby

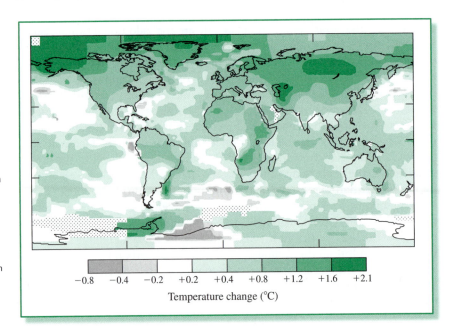

FIGURE 5-16 Changes, in degrees Celsius, in the mean surface temperature in 2001–2005 relative to the 1951–1980 mean. The dotted regions indicate areas for which data is insufficient. [Source: J. Hansen et al., "Global Temperature Change," *Proceedings of the National Academy of Science* 103 (2006): 14288.]

Temperature change (°C)

−0.8   −0.4   −0.2   +0.2   +0.4   +0.8   +1.2   +1.6   +2.1

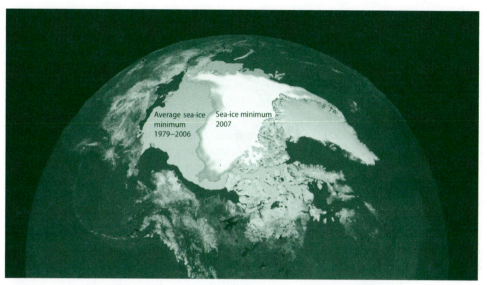

Minimum sea-ice area in 2007 contrasted with the average minimum in the 1979–2006 period.
[Source: *The Copenhagen Diagnosis Climate Science Report*, University of New South Wales Climate Change Research Centre]

increase the albedo. Up to 70% of the observed increase in Arctic temperatures may have arisen from the *decrease* in sulfate aerosol concentrations over northern areas due to lowered emissions of SO_2 from North America and Europe, and from *increases* in black carbon emissions in Asia which warm the air.

The increased cooling of air temperatures due to higher sulfate aerosol concentrations does not *permanently* cancel out all the warming owing to greenhouse gases because of the very different atmospheric lifetimes of the particles as compared to the gases. Since the tropospheric aerosols last only a few days, they do not accumulate with time, so their effect is short term. In contrast, today's emissions of carbon dioxide into the atmosphere will exert effects for decades or centuries to come: CO_2 emissions are cumulative over the medium term. This is also true for CFCs and nitrous oxide, which are also important greenhouse gases, but it is less so for methane, since its half-life is only about one decade. Thus, although sulfate-producing SO_2 emissions can temporarily cancel the effects of corresponding amounts of CO_2 emissions, eventually the cumulative effects of the carbon dioxide and the other greenhouse gases win out.

In summary, global warming has been experienced by most areas in the last half century. Most of the warming has arisen due to emissions of carbon dioxide into the atmosphere, with lesser amounts of warming arising from increased levels of methane, tropospheric ozone, and nitrous oxide, and the introduction of CFCs. The increased water vapor in the atmosphere that directly resulted from the warming due to these gases has itself produced at least as much additional warming. The increased emissions of sulfur dioxide that accompanied fossil-fuel combustion generated aerosols that countered some but not all of the warming produced by the greenhouse gases.

5.24 Global Circulation Models

In continuing research that began in the 1980s, scientists have attempted by computer modeling to predict the consequences of the increases in atmospheric gases and particles upon the past, present and future climate of the planet. There are some uncertainties in such an endeavor, including the fact that we don't yet fully understand all the sources and sinks of the greenhouse gases or the net effect of aerosols.

Probably the most important remaining problem with these **global circulation models** is their treatment of clouds. In particular, will the net feedback from the increased cloudiness expected for a warmed atmosphere be negative, positive, or zero? **Clouds operate both to cool and to heat the atmosphere.** They cool it by reflecting incoming light back into space; we experience this dramatically when the Sun goes behind the clouds on a warm, sunny day, and we immediately feel the air cooling off. We also know that the water droplets in clouds absorb infrared light emitted from below them.

• *Low-lying* clouds are warm, so they re-emit almost all this energy in random directions, rather than warm the air in their immediate surroundings. Since some of the IR emitted by these low clouds is directed downward, the surface is warmed—think of the general phenomenon that cloudy nights are generally warmer than cloudless ones. However, the fact that the IR is all re-emitted means that these clouds don't warm the atmosphere very much by this effect. Thus the *net* effect of low-lying clouds is to cool the Earth, because their reflection of incoming sunlight is dominant.

• In contrast, *high-lying* clouds absorb outgoing IR but don't emit much of it since they are cold. Thus all the absorbed IR is converted to heat, warming the nearby air. Therefore, the net effect of high-lying clouds is to warm the Earth, since their conversion of outgoing IR into heat is more important than their reflection of incoming sunlight.

Because we don't accurately know whether global warming will produce more additional low-lying or high-lying clouds, it is uncertain whether the feedback from this factor will be positive or negative.

Notwithstanding these problems, current global circulation models reproduce the climate of the last century and a half quite well. From the time variations of the input factors of the models, we can deduce the reasons why the trends in average air temperatures over the last century or so (Figure 5-2b) have arisen. In particular:

• The rise in temperature observed over the first half of the twentieth century was due to a combination of increased solar radiation intensity, a low level of volcanic activity, and increasing greenhouse gas concentration.

• The warming effects of increasing greenhouse gas concentrations were offset and masked by the sunlight-shielding effects of increased concentrations of sulfate aerosols during the 1940s–1970s, producing the observed temperature flattening.

- The temperature increases over the second half of the twentieth century were not due to natural effects, but to increases in greenhouse gas concentrations and to a lesser extent by black carbon, offset to some extent by cooling due to sulfate aerosols.

It is not *absolutely* certain that all or indeed that any part of the observed increases in the last 100 years are attributable to the anthropogenic-induced enhancement of the greenhouse effect. However, natural, century-long global warming or cooling trends as large as the 0.5°C we have recently experienced appear only once or twice a millennium. There is a 80–90% likelihood that the increase in the twentieth century was *not* a wholly natural climatic fluctuation. Furthermore, on the basis of computer simulations such as that discussed above, the UN-sponsored **Intergovernmental Panel on Climate Change** (IPCC) group concluded in its 2007 report that "most of the observed increase in globally averaged temperatures since the mid-twentieth century is *very likely* due to the observed increase in anthropogenic greenhouse gas concentrations."

http://www.ipcc.ch/ publications_and_data/ar4/ wg1/en/contents.html

5.25 Signs of Climate Change

In addition to a rise in average global surface air temperature, there have been a number of other shifts that indicate that our climate indeed is changing.

The likelihood that some of these changes have been caused by anthropogenic contributions is discussed in Chapter 6.

- *Precipitation has increased in most areas, but decreased in others.* An important aspect of climate is the amount of precipitation—rain and snow—that falls at various locations on Earth. In the twentieth century, the total annual precipitation in the Northern Hemisphere increased, with most mid- and high-latitude regions of eastern North and South America, of Europe, and of Asia becoming somewhat wetter. An overall increase in global precipitation is expected, since warming of the air warms the surface waters of lakes and oceans, and warmer water evaporates faster and thereby increases the water content of the atmosphere. Greenhouse gas increases have caused heavy precipitation events to have increased in frequency over land areas in the Northern Hemisphere over the last half-century. Flooding in many regions has increased as a consequence, including that in Western Australia and Queensland in 2010 and Sri Lanka in 2011. Indeed, 2010 globally was the wettest year on record. However, many areas just north and south of the Equator, especially in Africa and parts of southern Asia, became much drier, producing more intense and longer droughts, with disastrous consequences for food production there.

- *Extreme weather is becoming more common.* The phenomenon by which a relatively small increase in average air temperature can result in a

FIGURE 5-17 Schematic representation illustrating the effect on extreme hot and cold temperatures when the average temperature increases, based upon normal temperature distributions. [Source: Adapted from S. Solomon et al, editors, *Climate Change 2007: The Physical Science Basis*, Intergovernmental Panel on Climate Change, Cambridge University Press.]

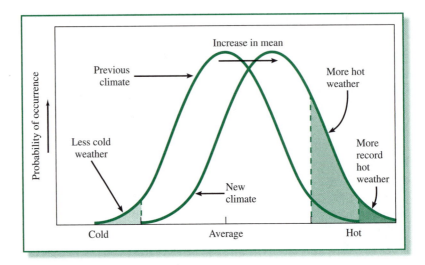

substantial increase in the number of hot-weather and record-hot days— and a decrease in very cold days, both especially at night—is illustrated in Figure 5-17. **The extreme events occur at the edges of the normal temperature distribution, which are greatly changed in size by a small shift of the curve.** Similar distributions presumably apply to precipitation as well.

Indeed, the frequency of extreme and violent weather events has recently increased in many areas of the world. Such events include stronger blizzards and storms with heavy snow and freezing rain in northern areas but record heat waves, hurricanes, and drought in others. For example, the number of heat waves lasting three days or longer almost doubled in the United States between 1949 and 1995. The heat wave in Europe in the summer of 2003 was a tragic example of the phenomenon, and was followed by another in 2010. Moreover, the frequency of storms with heavy or extreme precipitation increased in many nontropical regions in the last half-century. For example, over two-thirds of the extreme rainfall events over the last quarter of the twentieth century in the northeastern United States have been related to hurricane events, some of them 500 km away. The economic damage caused by storms, including hurricanes, has increased markedly over the last two decades.

• *Winters have become shorter, by about 11 days.* In the Northern Hemisphere, spring has been arriving sooner, and autumn has been starting later. Over the last three decades, the advent of spring—as observed by the appearance of buds, the unfolding of leaves, and the flowering of plants— has advanced by an average of six days in Europe, while the start of autumn—as defined by the date at which leaves turn color and begin to fall—has become delayed by about five days. In Alaska and northwest

Canada average temperatures have risen one degree per decade recently, resulting in a gradually increasing earlier date for the last frost and significant thawing of the permafrost. Consequently, in many regions there are fewer "frost days" now than there used to be—19 fewer in the western United States, and 3 less in the eastern part. The response of the plants has been driven by the increase in average daily air temperatures. Indeed, the range boundaries of some plants have increased toward the poles, and phenologies (season-dependent behaviors) have shown an earlier start for spring for the great majority of plants and animals. The earlier springtime disappearance of mountain snowpacks as well as higher temperatures in spring and summer have led to a substantial increase in the number, duration, and intensities of wildfires in the western United States since the mid-1980s.

• *The Earth's ice cover is shrinking fast.* Glaciers, polar icecaps, and polar sea ice are melting and disappearing at unprecedented rates as a consequence of global warming. For example, the remaining glaciers in Glacier National Park in the U.S. Rocky Mountains could disappear in 30 years if current melting rates continue. About 10% of the world's winter snow cover has disappeared since the late 1960s. Sea ice in the Arctic has not only decreased in area, by about 9% per decade recently, but it has also thinned dramatically. Warmer weather has also delayed the seasonal formation of sea ice. All these changes have caused a sharp decline in some populations of Antarctic penguins and Arctic caribou.

• *Warming water is killing much of the coral in ocean reefs and threatening sea life.* Coral reefs in tropical waters nurture and protect fish and attract scuba divers. As water warms, corals "bleach" themselves by expelling the algae that give them color and provide nutrition. Over 95% of the coral is already dead in some parts of the Seychelles Islands. Thus far, reefs in the central Pacific Ocean have escaped bleaching, and it is just beginning in the Caribbean. The death of the coral reefs not only affects the tourist trade but also threatens fishing for species that depend on the reefs for food. Beaches will erode if the reefs break up and no longer provide protection. The increased amount of CO_2 dissolved in ocean water has increased its acidity level, with potential harm to coral reefs, as discussed in more detail in Chapter 6.

• *Mosquito-borne diseases have reached higher altitudes.* Because of warmer temperatures, mosquitoes are now able to survive in regions where they formerly were not viable. As a result, mosquitoes have carried malaria to higher mountain regions in parts of Africa and *dengue fever* to new regions in Central America. Outbreaks of malaria have occurred in Texas, Florida, Michigan, New York, New Jersey, and even southern Ontario in the last decade. Warmer weather and changing precipitation patterns allowed the West Nile virus, another mosquito-borne disease, to become established in the New York City area in the late 1990s and southern Canada and virtually all the contiguous United States by the early 2000s.

FIGURE 5-18 Global sea-level rise in recent decades. The black curve represents a time-smoothed average of tide-gauge observations, whereas the green curve is obtained from satellite observations. [Source: Adapted from J. J. McCarthy, *Science* 326, 1646 (2009)]

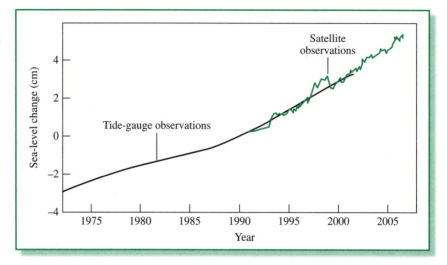

• *Rising seas are threatening to engulf Pacific islands.* Warming the air eventually leads to a rise in sea levels, for reasons that will be discussed in detail in Chapter 6. Acceleration in sea level rising began in the 1930s, and amounted to about 2 mm annually for some years. However, as illustrated in Figure 5-18, further increases in the rate have subsequently occurred. Indeed, the annual rate of increase in the 1993–2006 period amounted to 3.3 mm. Overall, the sea level rose about 60 mm (6 cm) in the last half-century. Some low-lying islands are threatened by extinction if sea-level rise continues unabated. The IPCC 2007 report concluded that it is "more likely than not" that there has been an anthropogenic contribution to the increased incidence of extreme high sea levels.

Review Questions 19–21 are based upon the material in the preceding section.

All of these signs of global warming from greenhouse gas emissions point toward the lack of sustainability of our current course of action. They reinforce the need for *systems thinking* to avoid additional unintended consequences and *external costs*, and are also examples of the *tragedy of the commons* (Table 0-1).

Geoengineering Earth's Climate to Combat Global Warming

Geoengineering refers to massive-scale schemes that intentionally attempt to alter the climate of the entire planet, usually by lowering its air temperature. Such schemes could be used in the future to counter the effects of global warming brought about by the increase in atmospheric greenhouse gas concentrations. To date, no geoengineering method has been attempted on a significant scale and relatively little research about them has been performed; as a consequence, the feasibility of none of them has been firmly established. However, some scientists and policymakers have called for their

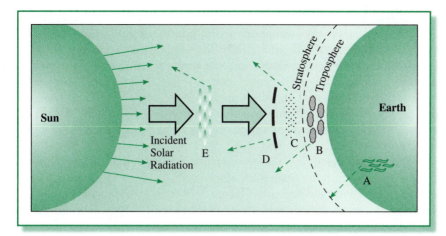

FIGURE 5-19 Schematic overview of solar radiation management schemes to deflect sunlight
Legend for areas: A, Increasing reflectivity of the ocean surface; B, Cloud whitening; C, Stratospheric aerosols; D, Orbital mirrors or reflectors; and E, Arrays of reflectors at Lagrange point.
[Source: Redrawn from J. J. Blackstock et al., *Climate Engineering Responses to Climate Emergencies* (Novim, 2009)]

further development in case they are required in the future, especially to counter a climate emergency.

Geoengineering schemes attempt to slow or halt global warming by one of two methods:

- By *reflecting* a small fraction of the sunlight headed toward Earth back into space, thereby reducing the amount of sunlight that reaches the atmosphere and the surface and is converted into heat. These schemes are discussed in detail below.

- By *removing* carbon dioxide from ambient air, and storing it. Such schemes will be discussed in Chapter 6.

Solar radiation management (SRM) methods all involve the placement of reflecting microscopic particles or macroscopic objects high in the sky. One advantage of such methods over those involving CO_2 removal is that **their effect on climate would occur relatively quickly,** requiring only a few years to take effect. The solar radiation management schemes proposed to date are summarized schematically in Figure 5-19. Their usual stated ambition is to reduce, by about 2%, the amount of sunlight reaching the Earth, an amount sufficient to counter the amount of greenhouse warming caused by doubling the CO_2 concentration in the air.

5.26 SRM: Using Metal Reflectors in Space

The SRM scheme that involves objects placed furthest out in space envisages tens of trillions of small silicon disks, each about 60 cm across, and collectively covering a total area of about 3 million km^2, in orbit around the Earth—about 1.85 million km away—where the gravitational effects of the Sun and Earth are equal, thus allowing the objects to remain in position indefinitely (Figure 5-19, area E).

This distance from Earth is called the Lagrangian or L_1 radius.

Each disk would be radio-controlled from the Earth, with mirrored fins and solar-cell power that would orient and position it, and would allow the units to be "turned off" if desired. Each disk would be studded with holes sized close to the wavelength of visible sunlight so that passing sunlight would be scattered. Collectively they would cast a weak shadow over the Earth, slightly dimming the sunlight reaching the atmosphere and surface. The cost of this *far-orbit reflector* scheme would be in the trillions of dollars, and would require decades to implement since so many units are required to be fabricated and launched.

Other space-based proposals would place sunlight reflectors much closer, in near-Earth orbit. One suggestion involves 55,000 mirrors, each 100 m² in area, positioned in random orbits (Figure 5-19, area D). The cost of launching reflectors of any type increases in proportion to their total mass; however, the lighter the individual reflector, the faster it will lose orbit due to pressure of the sunlight photons it reflects.

5.27 SRM by Increasing Sulfate Aerosols in the Stratosphere

The most discussed SRM method involves **the placement of an artificial layer of microscopic sulfate particles in the lower stratosphere.** There is naturally present in that region of the atmosphere a small concentration of particles consisting of liquid droplets of sulfuric acid (about 75%) and water. They originate from all the sulfur-containing gases (SO_2, H_2S, COS) that enter the stratosphere and are ultimately oxidized to sulfuric acid, H_2SO_4, which then condenses with water vapor molecules.

The stratospheric sulfate particles that exist naturally in time periods without significant volcanic activity have diameters in the submicron range ($\mu m = 10^{-6}$ m), with a size distribution that extends into the wavelength range of (ultraviolet and) visible light. As a consequence of the near match of wavelength with their diameter, these particles efficiently *scatter* incoming sunlight, as do sulfate particles in the troposphere, as we discussed previously. Thus some incoming sunlight is deflected back into space by stratospheric sulfate particles before it can reach the troposphere or the Earth's surface and be converted ultimately to heat if it is absorbed.

The geoengineering proposal is to artificially greatly increase the concentration of these sulfate aerosol particles (or indeed of any reflective particles of this size) **in the lower stratosphere and thereby deflect an additional significant fraction** (again about 2%) **of incoming sunlight from reaching the troposphere and the Earth's surface** (Figure 5-19, area C). Scientists have good experimental evidence that this scheme would work, based upon observations gathered after the eruption of Mount Pinatubo that we discussed earlier. Average global temperatures did show a significant decline for several years as a consequence of the decrease in sunlight reaching the surface caused by sunlight scattering by the Pinatubo sulfate aerosol.

The average diameter of the aerosol droplets produced by Pinatubo was somewhat larger than those normally present in the stratosphere. The larger particles in such a distribution not only affect incoming solar light, but also absorb outgoing infrared light from the Earth, thereby *heating* the air to some extent, although the overall effect of the particles was cooling rather than heating. It is clear from this evidence that the effectiveness in cooling the Earth of a sulfate aerosol depends to some extent on the sizes of the aerosol particles.

Based upon theoretical calculations and the experience with the Mt. Pinatubo aerosols, scientists have calculated that a few teragrams ($Tg = 10^{12}$, a million tonnes) of sulfur (in the form of one of its gases or an aerosol) per year would have to be injected into the stratosphere for the resultant sulfate aerosol to counter the global warming effect produced by a doubling of atmospheric carbon dioxide. The actual mass required would depend upon the average size of the particles. More would be required if their average size was similar to those of the Mt. Pinatubo aerosols, since, as discussed above, one of their effects would be to prevent the escape of some infrared light from the Earth. The injection would have to be repeated annually, since the aerosol is flushed from the stratosphere, in a matter of months over the polar regions and years over the tropics. Smaller particles also have the advantage of a longer lifetime.

A more-or-less uniform distribution of the aerosol in the lower stratosphere (15–25-km altitude) over the Earth could theoretically be achieved by emissions from special aircraft cruising at those altitudes. About a million flights per year (almost 3000 per day) would be required according to one estimate. Other methods that have been suggested for delivering the sulfate payload to the stratosphere include the use of artillery guns and balloons. Some such methods would initially produce localized pockets of air with high sulfate concentrations, rather than a smooth uniform distribution.

Recently it has been pointed out that continuously supplying new aerosols to the stratosphere would result in the growth of *existing* sulfate particles to sizes that are less effective in scattering incoming sunlight, and thus ever-larger loads of sulfate would be required to achieve the desired effect on climate. It has also been pointed out that local or small-scale global tests of sulfate injection would not produce a climatic result that would be distinguishable from normal climate variations. In other words, **if the SRM sulfate scheme is to be tried, a full-scale implementation would be required to see if it was going to be effective.** If it did accomplish its objective but was discontinued abruptly due to negative side effects, the climate would return relatively quickly to pre-injection temperatures. The economic and monetary costs of global warming depend significantly on its rate, so such a rapid change would have particularly negative consequences.

Although the details of the technology that would be used to deploy the sulfate shield have not yet been settled, its annual cost is estimated to be in the tens of billions of dollars, considerably less than other methods being considered. One advantage of sulfate over metal reflectors for solar radiation management is that it could be implemented much more rapidly and would take effect within a year or two.

5.28 Precipitation and Stratospheric Ozone in a Geoengineered World

All SRM geoengineering systems would not only reduce average global temperature, but also affect the climate in other, somewhat unpredictable and potentially disastrous ways. **A major concern is the reduction in regional rainfall that would accompany the artificial reduction in sunlight.** Precipitation would be affected because decreasing sunlight intensity reduces the heat available at the surface to evaporate the water that ultimately returns to Earth as rainfall (and snowfall).

The Asian monsoon provides most of the rainfall necessary to grow crops in the area.

For example, the eruption of Mt. Pinatubo led to a significant reduction in precipitation over land and consequently of river flow for 1 to 2 years. Computer projections predict that artificially enhancing the sulfate layer by geoengineering would seriously reduce the Asian monsoons and African rainfall patterns by its reduction of precipitation, with adverse effects on the food supply in those regions. Possible catastrophic drought in some regions is predicted by some researchers; there would be winner and loser countries from modifications of the hydrological system. However, it should be kept in mind that **unmitigated global warming will also affect rainfall patterns.**

Another negative impact of geoengineering by sulfate aerosols is likely to be some decrease in stratospheric ozone. The aerosol particles would provide surfaces on which ozone-destroying chemical reactions could occur, in a manner similar to that described in Chapter 2, Section 2.4 for the nitric acid/water particles found over the winter/spring Antarctic stratosphere. Indeed, there is evidence that the aerosols from Mount Pinatubo enhanced the destruction of ozone over the Antarctic for several years. One study predicted that geoengineered aerosols would strongly increase ozone depletion over the Arctic and would delay for many decades the recovery of the seasonal ozone hole over the Antarctic. However, the increase in UV-B at the surface due to this loss of ozone would be largely offset by the decrease brought about by the decrease in sunlight reaching the surface owing to scattering by the particles themselves.

5.29 Geoengineering by Ground-Level Systems

Several methods by which additional sunlight can be reflected back into space before it can be absorbed at ground level, and that operate from the Earth's surface, have been proposed. They include

- "whitening" clouds over the oceans (Figure 5-19, area B);

- painting roofs etc. a reflective white color;

- covering desert areas with metal reflectors; and

- spreading small reflecting spheres, such as white golf balls, across tropical seas (Figure 5-19, area A)

The basis of the cloud-whitening method is the significant increase in the albedo of low-level marine clouds when they consist of very tiny microdroplets compared to the same total volume of larger ones. The increase in cloud reflectivity occurs because the surface area—and hence the area presented to incoming sunlight—of water is greater the smaller are the individual particles (see Figure 5-13b and Additional Problem 9).

The sizes of the water droplets can be kept small by having them form on large numbers of suspended solid particles, the *cloud-condensation nuclei*. The cloud-whitening effect would be produced by artificially inserting such particles into the air above the oceans, which on average are 25% covered by clouds at any given time. Once the particles of suitable diameter and hydrophilicity have been manufactured, they could be spread by aircraft above the ocean area. One type of particle that would be effective and could be cheaply produced is sea salt, which could be distributed from the air, or alternatively from the sea's surface from a large number of ships. The ships would spray—perhaps using wind power—seawater droplets into the air, some of which would naturally travel upward due to turbulent mixing. Eventually, the water evaporates from the droplets, leaving sea salt particles which serve as condensation nuclei when they rise and reach supersaturated air masses located just below the bases of existing clouds. To be effective, several thousand such vessels would be needed.

The feasibility of producing sufficient nuclei and the overall effectiveness of cloud albedo enhancement methods remain uncertain. Because of the short lifetime (~ days) of aerosol particles at low altitudes, continual replenishment of the particles would be required. The restriction of the albedo change by these techniques to ocean areas could affect weather patterns in deleterious ways.

The proposal to paint roofs and paved roads a bright reflective shade of white would indeed increase the albedo of the built environment, and would be especially effective in the summertime in sunny regions. The reduction in the amount of sunlight absorbed by buildings would also reduce their need for air conditioning in the summer, thereby lowering the use of fossil-fuel-derived energy. However, the cost of materials and labor for painting (and repainting every decade or so) would be very high.

Deserts cover about 10% of the Earth's land surface. Many are hot and experience high levels of incident sunshine, of which only about one-third is naturally reflected. This fraction could be increased to at least three-quarters if desert surfaces were covered by an artificial highly reflective material such as aluminized polyethylene. Such a process would significantly increase the Earth's overall albedo and significantly combat global warming from CO_2 doubling if all major desert areas were covered. However, the highly localized nature of the change in sunlight absorption and conversion to heat would likely cause large-scale changes in precipitation patterns, including the

TABLE 5-1	Characteristics of SRM Schemes			
Method	Effectiveness	Affordability	Timeliness	Safety
Space disks	High	Low to very low	Very low	Medium
Stratospheric aerosols	High	High	High	Low (rainfall problems)
Cloud whitening	Low to medium	Medium	Medium	Low (weather patterns)
Roof & road whitening	Very low	Very low	Medium to high	Very high
Desert reflectors	Low to medium	Very low	High	Very low

Note: Adapted from "Geoengineering the Climate," *The Royal Society*, Sept. 2009.

yearly monsoon that brings rain to sub-Saharan Africa. The prospects of such deleterious climate change, and the very high cost to install and maintain the reflective materials—similar in magnitude to the "white roof method" discussed above—makes the desert reflector scheme unpalatable.

5.30 SRM Schemes Summary

A summary of the important characteristics of the geoengineering proposals discussed above is presented in Table 5-1. Note that monetary cost is the inverse of *Affordability*. The main safety issue causing several techniques to be classified as *Low* is that they would affect weather patterns, probably causing decreased rainfall in several inhabited areas. *Timeliness* refers to the speed at which the method could be implemented.

The sulfate aerosol scheme scores high on effectiveness, affordability, and speed of implementation, but rather low on safety owing to its likely effect on rainfall and monsoon patterns. The greater the area covered by sulfate or ground-based reflectors, the greater the (larger) area over which precipitation will be disturbed.

Given that some regions of the Earth would likely be affected negatively by any SRM geoengineering scheme, and that implementation would require not only the political agreement of most nations but a sharing of the gigantic cost of the scheme, it seems unlikely that any will be implemented short of a climate catastrophe.

Atmospheric Residence Time Analysis

The extent to which a substance, such as a greenhouse gas, accumulates in some compartment of the environment, such as the atmosphere, depends upon the rate, R, at which it is received from the source, and the mechanism by which it is eliminated, i.e., its sink. In this section, we explore the mathematics by which its concentration changes with time depending upon its rate of input and the efficiency of its sinks.

Commonly, the rate of elimination via the sink is directly proportional to the concentration, C, of the substance in the organism or environmental compartment. In chemistry, this is known as a *first-order* relationship, since the power to which the independent variable is raised is unity. If the rate constant for the elimination process is defined as k, the rate of elimination is kC.

$$\text{rate of intake} = R$$

$$\text{rate of elimination} = kC$$

In some cases, such as the atmospheric sink for methane, reaction involving a second substance is involved, and k incorporates the steady-state concentration of this other substance.

If none of the substance is initially present, i.e., if $C_0 = 0$, then initially the rate of elimination must be zero. The concentration then builds up solely due to its input or ingestion, as illustrated near the origin in Figure 5-20. However, as C rises, the rate of elimination rises and increases since it is proportional to C; eventually it matches the rate of intake if R is a constant. Once this equality is achieved, C does not vary thereafter. It is in a steady state, which, as we saw in Chapter 1, Section 1.10, is defined as one for which $dC/dt = 0$. Since under these steady-state conditions

$$\text{rate of elimination or loss} = \text{rate of intake}$$

$$kC = R$$

It follows that the steady-state value for the concentration—C_{SS}—is

$$C_{SS} = R/k$$

Often the speed of elimination or loss of a substance is discussed in terms of the **half-life period, $t_{0.5}$, the length of time required for half of it to decay under the circumstance that all new input has now ceased.** Under the latter condition, we know

$$dC/dt = -kC$$

Bringing to the left side of the equation all the terms involving C and putting to the right side the time dependence, we have

$$\int dC/C = -k \int dt$$

If we integrate both sides of the equation, we can find how C changes with time. Performing the integration, we obtain

$$\ln C = -kt + \text{constant}$$

Thus we see that the logarithm of the concentration will decline linearly with time for the substance. At time $t = 0$, we have $\ln C =$ a constant, so

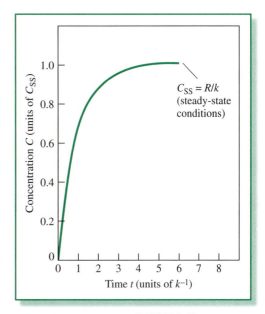

$C_{SS} = R/k$
(steady-state conditions)

FIGURE 5-20 Increase in concentration with time to eventually reach the steady-state value, C_{SS}.

we see that the (integration) constant equals the logarithm of the original concentration, C_0, the concentration at $t = 0$. Thus

$$\ln C - \ln C_0 = -kt$$

From the property of logarithms that $\log x - \log y = \log(x/y)$, we obtain the simpler equation

$$\ln(C/C_0) = -kt$$

or, in exponential form,

$$(C/C_0) = e^{-kt}$$

It is convenient to discuss the rate of decline of a substance in terms of its half-life. Substituting $C = 0.5\, C_0$ into the logarithmic equation above yields

$$\ln(0.5\, C_0/C_0) = -kt_{1/2}$$

But $\ln(0.5) = -0.693$, so we obtain $t_{1/2} = 0.693/k$.

We can use this final result to advantage to substitute for k in our equation for the steady-state concentration of a chemical achieves when it is being both created and destroyed:

$$C_{SS} = R/k$$

We obtain

$$C_{SS} = R\, t_{0.5}/0.693$$

or

$$C_{SS} = 1.44\, R\, t_{0.5}$$

Clearly, **the longer the half-life of a substance in its elimination process, the higher the steady-state accumulation level** it will achieve. The variations with time of concentrations and rates for systems of this type are illustrated in Figure 5-20 for the specific case where R, k, (and hence C) are expressed in units of C_{SS}, and time t is expressed in units of k.

Every atmospheric gas that is present at or near a steady state has its own characteristic **residence time** t_{avg}, which was defined earlier as the average amount of time one of its molecules exists in air before it is removed. **The average lifetime, or residence time, t_{avg}, of a substance is equal mathematically to the time required for its overall concentration to fall to $1/e$ times its initial value,** where e is the base of natural logarithms. Since at that time $C = C_0/e$, then

$$\ln[(C_0/e)/C_0] = -kt_{avg}$$

But $\ln(1/e) = -1$, so

$$t_{avg} = 1/k$$

Substituting this expression for k into C_{SS} gives us $C_{SS} = R/t_{avg}$. Sometimes this expression is more useful in the form

$$t_{avg} = C_{SS}/R$$

since it tells us the lifetime of a substance if we know its steady-state concentration and rate of input into the environment.

In order to assess the impact of any substance upon the enhanced greenhouse effect, it is necessary to know how long the substance is expected to remain in the atmosphere, since the longer its atmospheric lifetime, the greater will be its total effect. Thus, for example, if the steady-state atmospheric concentration of a gas is 6.0 ppm, and if its global rate of input, as determined by dividing the yearly amount of input by the volume of the atmosphere, is 2.0 ppm per year, then according to the last equation above, its average residence time is 6.0 ppm/2.0 ppm y^{-1} or 3.0 y.

The residence times of greenhouse gases such as nitrous oxide and the CFCs are many decades, so the influence of recent emissions of them will extend over long periods of time. In contrast, methane has a residence time of less than a decade.

The above analysis is applicable only to substances having a single first-order sink process. Thus it does not apply to carbon dioxide, for example, since this gas has many different sinks (dissolution in the oceans; absorption by plant matter, etc.) and sources.

PROBLEM 5-10

If the average, steady-state residence time of a trace atmospheric gas is 50 y and its input rate is 2.0×10^6 kg y^{-1}, what is the total amount of it in the atmosphere? ●

PROBLEM 5-11

The steady-state concentration of an atmospheric gas of molar mass 42 g mol^{-1} is 7.0 μg g of air, and its residence time is 14 y. What is the annual total release of the gas into the atmosphere as a whole? See Problem 5-6 for additional data. ●

Review Questions 22–26 are based upon the material in the preceding sections.

Review Questions

1. Sketch a plot showing the main trends in global air temperature over the last century and a half.

2. What is the wavelength range, in μm, for infrared light? In what portion of this range does the Earth receive IR from the Sun? What are the wavelength limits for the *thermal IR* range?

3. Explain in terms of the mechanism involved what is meant by the *greenhouse effect*. Explain what is meant by the *enhancement* of the greenhouse effect.

4. Explain what is meant by the terms *symmetric* and *antisymmetric bond-stretching* vibrations, and by *angle-bending* vibrations.

5. Explain the relationship between the frequency of vibrations in a molecule and the frequencies of light it will absorb.

6. Why don't N_2 and O_2 absorb thermal IR? Why don't we consider CO and NO to be trace gases which could contribute to enhancing the greenhouse effect?

7. What are the two main anthropogenic sources of carbon dioxide in the atmosphere? What is its main sink? What is *fixed carbon*?

8. Is water vapor a greenhouse gas? If so, why is it not usually present on lists of such substances?

9. Explain what is meant by *positive* and *negative feedback*. Give an example of each as it affects global warming.

10. What is meant by the term *atmospheric window* as applied to the emission of IR from the Earth's surface? What is the range of wavelengths of this window?

11. What reaction is the dominant tropospheric sink for methane?

12. What are four important trace gases that contribute to the greenhouse effect?

13. What are the six most important sources of methane?

14. What are the three most important sinks for methane in the atmosphere? Which one of them is dominant? What is meant by the term *clathrate compound*?

15. Is the enhancement of the greenhouse effect by release of methane from clathrates due to increased temperature an example of feedback? If so, is it positive or negative feedback? Would an increase in the rate and amount of photosynthesis with increasing temperatures and CO_2 levels be a case of positive or negative feedback?

16. Explain in chemical terms what is meant by *nitrification* and *denitrification*. What are the conditions under which nitrous oxide production

is enhanced as a by-product of these two processes?

17. What are the main sources and sinks for N_2O in the atmosphere?

18. Are the proposed CFC replacements themselves greenhouse gases? Why is their emission considered to be less of a problem in enhancing the greenhouse effect than was that of the CFCs themselves?

19. By which two mechanisms does light interact with atmospheric particles?

20. Explain how sulfate aerosols in the troposphere affect the air temperature at the Earth's surface, by both the direct and indirect mechanisms.

21. List four important signs, other than increases in average air temperature, that global warming is occurring.

22. Define the terms *geoengineering* and *solar radiation management*.

23. Describe how the use of sunlight reflectors in space could reduce global warming.

24. Describe how the scheme of increasing stratospheric sulfate concentration could reduce global warming. Why is particle size an important factor? What schemes have been suggested by which the sulfate could be delivered to the stratosphere?

25. What would the likely effects of solar radiation management be on (a) rainfall levels and (b) stratospheric ozone recovery?

26. How is the residence time of a substance related to its rate R of input/output and to its total concentration C?

Additional Problems

1. The tropospheric pollutant gases SO_2 and NO_2 have nonlinear molecular structures which, like that of CO_2, have the central atom connected to two oxygen atoms. The wavelengths for their

vibrations are given in the table shown on the next page. (a) Which of the vibrations are capable of absorbing infrared energy? (b) Based upon the wavelengths for the IR-absorbing vibrations and the spectrum in Figure 5-7, decide which if any vibrations could contribute much to global warming. (c) What lifetime characteristic of these gases would limit their role in global warming?

Gas	Symmetric stretch	Antisymmetric stretch	Bending
SO_2	8.7 μm	7.3 μm	19.3 μm
NO_2	7.6 μm	6.2 μm	13.3 μm

2. (a) How can the fact that nitrous oxide has three vibrations that absorb infrared light be used to prove that its structure is NNO rather than NON? (b) Would methane molecules absorb IR during the vibration in which all four C—H bonds stretched or contracted in phase?

3. Anthropogenic carbon dioxide emissions into the atmosphere amounted to 178 Gt from January 1990 to December 1997. Calculate the fraction of this emitted carbon dioxide that remained in the air, given that in that same eight- year period, the carbon dioxide concentration in air rose by 11.1 ppm. Note that the molar masses of C, O, and air, respectively, are 12.0, 16.0, and 29.0 g, that the mass of the atmosphere is 5.1×10^{21} g, and that 1 Gt is 10^{15} g.

4. The total amount of methane in the atmosphere in 1992 was about 5,000 Tg, and was increasing by about 0.6% annually due to the fact that the annual input rate exceeded the annual output rate of 530 Tg y^{-1}. Calculate the percentage by which anthropogenic releases of methane, which account for two-thirds of the total, had to be reduced if the atmospheric concentration of this gas was to be stabilized in 1992.

5. As mentioned in the text, the fraction F of light that is absorbed by any gas in air is logarithmically related to the concentration c of the gas and the distance d through which the light travels; this relationship is called the Beer–Lambert law:

$$\log_e (1 - F) = -Kcd$$

Here K is a proportionality constant. Show by simple trial calculations that for concentrations near zero (e.g., where $Kcd = 0.001$), that F is

related almost linearly to c, whereas for larger Kcd values (e.g., near 2), that doubling the concentration does not nearly double the light absorption.

6. The vapor pressure P of a liquid rises exponentially when it is heated according to the equation

$$\ln(P_2/P_1) = -\Delta H/R (1/T_2 - 1/T_1)$$

Here P_2 and P_1 are the vapor pressures of the liquid at the Kelvin temperatures T_2 and T_1 after and before the temperature increase, R is the gas constant 8.3 J K^{-1} mol^{-1}, and ΔH is the liquid's enthalpy of vaporization, which for water is 44 kJ mol^{-1}. Calculate the percentage increase in the vapor pressure of water that occurs if the temperature is raised from 15°C to 18°C. Give several reasons why the amount of outgoing thermal infrared in water's absorption bands may not be increased by exactly the percentage you calculate if the average air/surface temperature is increased to 18°C.

7. Suppose that some climatic crisis inspired the Earth's population to switch to energy systems that did not emit carbon dioxide, and that the transition occurred within a decade. What would be the predicted immediate effect to the Earth's average air temperature of this change, given that both carbon dioxide and sufur dioxide emissions from fossil fuels would have rapidly declined?

8. Given that the concentration of CH_4 in the atmosphere is 1.8 ppm, calculate the total mass of this gas that is present in the atmosphere. Note that the total mass of the atmosphere is 5.1×10^{18} kg and that its average molar mass is 29.0 g mol^{-1}.

9. Consider an area of space measuring $10D \times 10D$, in which there are ten aligned rows, each of ten spheres lined up one on top of the other. Calculate the total volume occupied by the spheres, and the total area they cover (of the $100D^2$ maximum if there were no "empty spaces"). [Hint: The volume V of a sphere equals $4\pi r^3/3$, where r is its radius.]

Energy Use, Fossil Fuels, CO₂ Emissions, and Global Climate Change

In this chapter, the following introductory chemistry
topics are used:

- ⮞ Percentage composition; stoichiometry
- ⮞ Combustion; heat of combustion
- ⮞ Structural chemistry of hydrocarbons (see online Appendix)
- ⮞ Acidity; weak acids; pH
- ⮞ Phase diagrams; condensation of liquids; sublimation of solids
- ⮞ Catalysis
- ⮞ Density
- ⮞ Polymerization

Background from previous chapters used in this chapter:

- ⮞ Greenhouse effect and greenhouse gases
- ⮞ Sinks
- ⮞ ppm concentration scale for gases
- ⮞ Clathrates
- ⮞ Albedo

Introduction

The usage of energy involves its transformation from one form to another, eventually resulting in its degradation to waste heat, and as such does not pose any global environmental problem *per se*. However, there are side effects associated with the production and/or consumption of energy that are serious environmental issues.

As we saw in Chapter 5, the Earth's climate has probably already been influenced by the enhancement of the greenhouse effect due to increasing atmospheric concentrations of carbon dioxide and other gases. A continuing buildup of CO_2 in the air leads to the conclusion that we are in store for further increases in global air temperatures and other changes to our climate.

In this chapter:

• We will investigate energy usage over the last several decades and inquire into predictions of the likely trends in energy usage over the next few decades.

• The nature of fossil fuels used in energy generation, and their role in producing CO_2, are then analyzed and the prospect of burying these emissions as they are generated is then explored.

• We finish by considering the predictions for future CO_2 emissions and the consequences for the climate, and of ramifications to civilization from these emissions if they continue unabated or with only weak controls.

Global Energy Usage

Ever since the Industrial Revolution, the worldwide use of *commercial energy*—that sold to users and usually derived on a large scale from fossil-fuel combustion, hydroelectricity, and nuclear power, as opposed to the biomass collected and used by individual families—has risen almost every year. In recent decades, the annual global growth rate had been about 1–2%. The period of the most rapid increase began after World War II, when global commercial energy consumption was only about one-tenth of the current level.

Because the amounts are so huge, it is useful to discuss global quantities of energy in terms of the large energy unit EJ, an *ekajoule*, which is 10^{18} joules. The total amount of commercial energy consumed currently amounts to almost 500 EJ annually, with the United States consuming about 20% of this total.

6.1 Global Energy Usage Trends and Relationships

The fantastic rise in global energy usage in the second half of the twentieth century was due mainly to industrial expansion and to increases in the standard of living in the *now-developed countries*, which collectively are known as the *Organization for Economic Cooperation and Development* (OECD). The energy consumption in these countries continued to expand, albeit slowly, until early into the current century, as the lower portion (light gray) of the graph in Figure 6-1 indicates. However, economic growth in the *developing countries*—which contain over three-quarters of the world's population—has risen quickly and with it their total energy consumption, as indicated by the top two shaded portions of Figure 6-1. The total global consumption of energy—dark green section at top of Figure 6-1—grew every year illustrated except 2008–2009, when the recession produced a small decrease in OECD countries at least. Energy usage increased by an annual average of 3% from 2000 to 2010, about double the rate of the preceding 15 years.

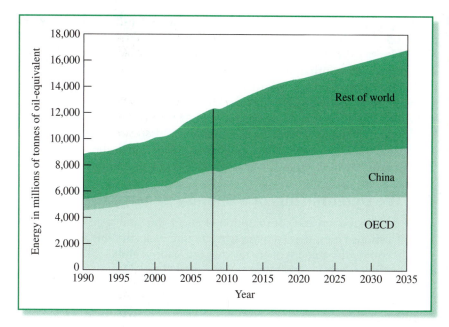

FIGURE 6-1 Global energy demand by region. Predictions after 2008 are from IEA's *New Policies Scenario*. Note that 1 million tonnes of oil are equivalent in energy to 42×10^{15} J, i.e., to 0.042 EJ. [Source: *World Energy Outlook 2010*, International Energy Agency]

Over the medium to long term, global energy usage is expected to increase exponentially with time, i.e., its yearly increases are compounded. Thus the amount of energy E used by the world after t years is related to the amount, E_0, used at the beginning of the time interval by the equation

$$E = E_0 \, e^{kt}$$

Here k is the fractional increase experienced in each time period. According to the *International Energy Agency* (IEA), between 2008 and 2035 energy use is likely to grow each year by an average of 1.4% of the previous year's value. The total increase over this 27-y period according to the above equation would be 46%, since if $t = 27$, $k = 0.014$, then $E/E_0 = e^{0.014 \times 27} = 1.46$.

Noncompounded growth of 1.4% over 27 y would produce an increase of only 38%.

PROBLEM 6-1

If the new energy policies discussed by countries at the Copenhagen 2009 meeting were to be implemented, the IEA estimates that annual global energy growth would be reduced to 1.2% over the 2008–2035 period. (a) What would then be the total percentage growth in energy by 2035? (b) In order to keep the total growth to 20% by 2035, to what percentage what would the average annual growth rate have to be reduced?

The recent prediction of energy demand growth over the next few decades is illustrated by the area to the right side of the vertical line in Figure 6-1. The scenario used by IEA in the projections is one that averages 1.2% annual growth (see Problem 6-1), notwithstanding the much larger rate observed over the first decade of this century. **The total energy used by developed countries is projected neither to rise nor to fall significantly. Almost all the growth in energy usage over the 2008–2035 time period is expected to occur in developing countries,** continuing the trend established in the last decade.

China's energy use was less than half that of the United States at the turn of the century, but overtook that of the United States by 2009. China's energy growth by 2035 over the 27-y period according to this IEA scenario amounts to 75%, and its share of global consumption could then reach 24% (compared to 16% in 2007). In contrast, the United States share will decline from 21% to 16% over that period, having been about 25% at the turn of the century.

6.2 The Determinants of a Country's Energy Use

Since the production and consumption of energy inevitably generates pollution of various types, it is of interest to inquire as to the important factors that determine the amount of energy that is produced and used. The usage of commercial energy by a country turns out to depend upon many factors, including its population, level of production of goods and services, geography, and climate, and also upon the cost of energy.

• *Population as a driving force*: The *total energy usage per capita* (in units of tonnes of oil-equivalent) is shown in descending order in Table 6-1, where

TABLE 6-1	Energy Supply per Capita for Various Countries
Energy Supply per Capita (tonnes of oil-equivalent per person, 2007)	**Countries (in descending order)**
6	Iceland, Luxemburg, Canada (8.17), USA (7.75), Finland
5–6	Australia (5.87), Norway, Sweden, Belgium
4–5	Netherlands, Russia, S. Korea, Czech Rep., Estonia, France, Germany (4.03), Japan (4.02), New Zealand
3–4	Austria, Slovenia, Denmark, United Kingdom (3.48), Ireland, Switzerland, Slovak Rep., Spain, Israel
2–3	Greece, S. Africa, Hungary, Poland, Portugal
1–2	Chile, Mexico (1.74), China (1.48), Turkey, Brazil (1.23)
0.5–1	Indonesia, India (0.53)

Source: OECD Factbook, 2010.

we have grouped countries within ranges. Per capita energy use in the United States currently amounts to about 10,000 J sec^{-1} (i.e., 10,000 W, the equivalent of one hundred 100-W light bulbs burning continuously), almost twice that in Germany, Great Britain, and Japan, and four times the world average of 1.82 tonnes.

Although the United States and China were essentially tied in *total* energy consumption, the population of China is more than four times as large, so their per capita energy use is less than a quarter as much. The energy use per capita of India is only about one-third that of China.

Since there are substantial differences—amounting to an order of magnitude at least—between countries in the energy per capita ratio, population cannot be the *only* or the most dominant factor in determining total energy use.

- *Economic activity as a driving force:* Rather than population, **the most important factor in total energy usage of a country turns out to be its gross domestic product (GDP),** i.e., the sum total value of all the goods and services its residents produce in a year. In industrialized societies, about 8 megajoules (8 million joules) of energy are needed on average to produce one (U.S.) dollar's worth of goods or services. **The ratio of energy consumption to GDP is called the energy intensity of a country.** Thus a higher than average energy intensity means that more energy than usual is being used to produce a given value of goods or services. The current energy intensity of various countries in order of their level of affluence, as measured by their GDP per capita, is shown in Table 6-2. The energy intensity values for almost all constituencies lie within 50% of the world average. From this relative constancy we conclude that GDP is indeed the dominant factor in determining the energy needs of countries.

The energy-to-GDP ratio usually rises when a country begins to industrialize, but then drops gradually as its infrastructure becomes more substantial and efficient. For example, the ratio for the United States has dropped, almost continuously, by a factor of about four over the last century. The variation with time of energy intensity for some large economies over the last two centuries is illustrated in Figure 6-2. In all cases illustrated, the energy intensity has peaked and now is declining; all seem to be veering toward the same value of about 0.1 tonnes of oil-equivalent per thousand dollars of GDP. China's energy intensity fell by 19% from 2005 to 2010, and according to their Green Plan of 2011, it is to be reduced by a further 16% over the subsequent five years. The majority of China's energy is used for manufacturing of iron, steel, and cement.

From the analyses above, we can conclude that **global energy usage will increase continuously in future decades as the world economy expands, driven especially by rapid growth in developing countries,** although the increase will be tempered to some extent by gradually increasing efficiency in energy use. It is on such bases that energy growth rates of 1.2–1.5% over the

1 tonne of oil-equivalent equals 41.9 GJ.

Per capita means per person.

TABLE 6-2	Energy Intensity and per Capita Gross Domestic Product for Various Countries and Regions	
Country or Region	Energy Intensity* Energy (GJ) ($ GDP)$^{-1}$	($ GDP) person^{-1}
USA	7.9	43.8
Canada	12.0	35.7
Australia	8.0	33.3
Japan	5.9	33.1
United Kingdom	4.5	31.8
Western Europe**	5.6	30.2
South Korea	10.3	24.5
Russia	15.4	12.2
Mexico	5.2	10.7
Brazil	6.3	8.8
China	11.7	7.8
India	6.6	3.8
World	7.9	

* 2008 energy and GDP data per U.S. 2005 $ at purchasing power parity values.
**Average for France, Germany, Italy, and Spain.

Data Source: U.S. Energy Information Administration.

next few decades have been projected by the IEA, as discussed above. In terms of an equation, for each country, region, or the world as a whole

$$\text{energy usage} = \text{population} \times (\text{GDP/population}) \times (\text{energy/GDP})$$

6.3 Energy Sources

The important energy sources that form the components of primary commercial energy will be discussed in this and the following three chapters. Some perspective regarding how the size of the five dominant sources—coal, oil, natural gas, renewables, and nuclear energy—have changed in the past few decades and are expected to change over the next few is shown in Figure 6-3. **Oil is the world's dominant energy source (35%), although coal does not lag far behind (27%)** and will likely catch up in the future. Natural gas (23%) had been growing in importance, and represents a larger source than either renewables (mainly hydroelectric) or nuclear power, even with the fast growth expected for the former.

"Liquids" in Figure 6-3 includes both crude oil and biofuels.

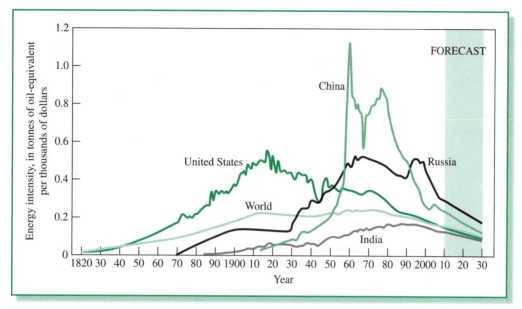

FIGURE 6-2 Historical and projected energy intensity of selected economies. GDP values are at 2009 purchasing-power-parity rates in U.S. dollars. Energy is in tonnes of oil-equivalents. Note that 1 million tonnes of oil are equivalent in energy to 42×10^{15} J, i.e., to 0.042 EJ. [Source: Adapted from *The Economist Online*, http://www.economist.com/blogs/dailychart/2011/01/energy_use]

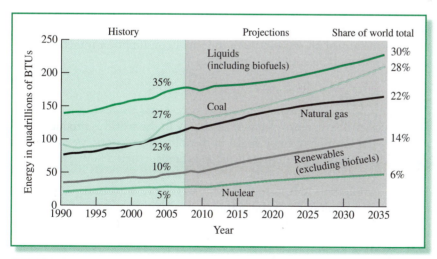

FIGURE 6-3 Global energy use by fuel type in the past and for the next few decades as projected by the U.S. Energy Information Administration. Liquids here include biofuels, whereas the Renewables category does not. Note that 1 quadrillion BTUs are equivalent to 1.06×10^{18} J, i.e., to 1.06 EJ. [Source: Adapted from *International Energy Outlook 2010*, U.S. Energy Information Administration.]

FIGURE 6-4 Consumption of energy types by geographic region, 2010. Coal is shown in light gray; renewables in dark gray; hydroelectricity in black; nuclear energy in light green; natural gas in medium green; and oil in dark green. [Source: Adapted from BP, *Statistical Review of World Energy 2011*.]

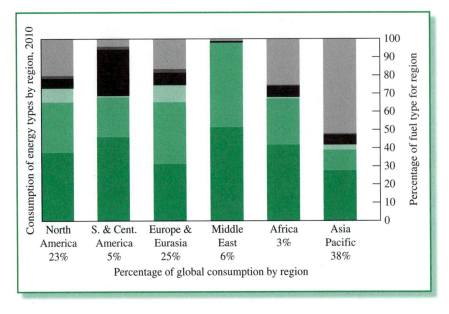

The magnitude of energy consumption, and the relative importance of the different energy sources consumed, varies widely with geographic region, as illustrated in Figure 6-4. Coal (shown in light gray) is the most prominent energy source in the Asia-Pacific region, which includes China and India, due to its ample supply and low cost. This region now leads the world in energy consumption. The North American and European/Euroasian regions are next highest in consumption, and their source distribution most closely parallels the global allocation. Note the small consumption totals in Africa and in South and Central America.

Review Questions 1–5 are based on the preceding sections.

Fossil Fuels

As mentioned above, **most commercial energy in the world is produced currently by the burning of fossil fuels.** In the following sections, we discuss the nature and future supplies of these fuels, and compare their differing abilities to produce the greenhouse gas carbon dioxide. Later in the chapter we shall see that the main problem of fossil-fuel usage in the present century is the CO_2 emissions that result from its combustion rather than a shortage in supply.

6.4 Coal

The main fossil-fuel reserve is coal, which is available in abundance in many regions of the world, including developing countries, and is cheap to mine and to transport. Five countries—the United States, Russia, China, India, and Australia—have 75% of the world's coal reserves. At today's rate of consumption, coal reserves are estimated to last another 200 years, much longer than oil or gas (see below). Coal use in absolute terms is projected to grow faster than any other energy source, according to the projections in Figure 6-3. Currently, coal produces about half the electric power in the United States and about 80% of that in Australia and China.

Although it is a mixture, **to a first approximation, coal is graphitic carbon, C.** It was formed from the tiny proportion of ancient plant matter that was covered over by water and could not be recycled back to CO_2 at that time. This also accounts for the buildup of O_2 in the atmosphere. Coal was formed from the highly aromatic, polymeric component of land-based woody plants called *lignin*.

An approximate empirical formula for lignin is C_3H_3O.

Over long periods of time during which the material was subjected to high pressures and temperatures, both water and carbon dioxide were lost. The material polymerized further in the process to yield the very carbon-rich, hard material known as coal.

Unfortunately, over the years the coal also incorporated during its formation measurable quantities of virtually every naturally occurring element, so that when it is burned, it emits not only CO_2 and H_2O but also substantial quantities of many air pollutants, notably sulfur dioxide, fluoride, uranium and other radioactive metals, and heavy metals including mercury. Thus coal has a reputation for being a "dirty fuel."

The burning of coal *domestically* in stoves and furnaces produces a great deal of soot, and it therefore has been largely discontinued in developed countries. However, coal is still used in most developed and developing countries for electric power production. When coal is burned in such plants, the soot problem is readily solved, although emissions of sulfur and nitrogen dioxides and of mercury require more sophisticated and expensive equipment to control, as discussed in Chapters 3 and 12.

The removal of some of these impurities, especially sulfur, by various modern technologies to produce "clean coal" was discussed in Chapter 3.

The heat that combustion of the fossil fuel produces is used to generate steam, which is then used to turn turbines and thereby produce electricity. As discussed below, however, the ratio of CO_2 to energy produced from coal is substantially greater than for the other fossil fuels. Coal can also be used to produce alternative fuels—as discussed in Chapter 7—but unfortunately the conversion processes are not energy efficient. Although the emission of carbon dioxide is not reduced by such conversions, they do allow the removal of sulfur dioxide and other pollutants and so are "clean" ways to use coal.

6.5 Natural Gas

Petroleum and natural gas are predominantly mixtures of hydrocarbons. They originated as the small fraction of microscopic marine organisms that became buried by sediment and therefore cut off from the oxygen that was required for complete oxidation. The high temperatures and pressures produced by further layers of sediment to which this buried material was later subjected decomposed it further, into liquid and gaseous hydrocarbons. Like

FIGURE 6-5 Schematic geology of natural gas resources. Associated and nonassociated refer to the presence or absence of oil. [Source: U.S. Energy Information Administration and U.S. Geological Survey, *Today in Energy*, February 14, 2011.]

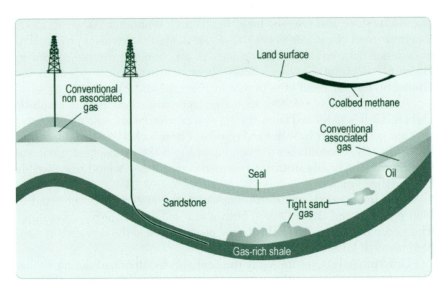

petroleum, natural gas deposits are found in geological formations in which the gas mixture has been trapped by a mass ("seal") of impermeable rock. Drilling a hole down through the rock seal releases the gas in a steady flow to the surface (Figure 6-5).

In terms of its hydrocarbon component, **natural gas as it exits from the ground consists predominantly (60–90%) of methane, CH$_4$.** The other component alkanes—*ethane, propane,* and the various *butane* isomers—are gases present to varying extents depending upon the geographic origin of the deposit.

Methane's boiling point is so low ($-164°C$) that it does not readily condense into a liquid, even at moderately high pressures. In contrast, the other gaseous alkanes possess substantially higher boiling points. This makes it possible to largely remove the other alkanes from natural gas by lowering the mixture's temperature and thereby condensing the other hydrocarbons to liquids.

Ethane, C$_2$H$_6$, which forms up to 10% of raw natural gas, is used primarily to make *ethene,* C$_2$H$_4$, which in turn is converted to polyethylene and other polymers, as will be discussed in more detail in Chapter 16.

$$C_2H_6 \longrightarrow H_2 + C_2H_4 \longrightarrow \text{polyethylene, } (C_2H_4)_n$$

Butane and the higher alkanes from the raw gas constitute most of the *natural gas liquids* used for components of gasoline and other petrochemical purposes.

Sulfur compounds are also important impurities in natural gas, as previously mentioned: some deposits contain more H$_2$S than CH$_4$! The **hydrogen sulfide** is removed from the gas by the Claus reaction, as discussed in Chapter 3. After processing to remove the other alkanes and the sulfur compounds, the natural gas—which now is mainly methane—is transported under pressure by pipeline to consumers.

See the online Appendix if you are unfamiliar with the terminology and numbering systems of organic molecules.

Raw natural gas also contains variable amounts of helium, nitrogen, and carbon dioxide.

Unfortunately, as we have discussed in Chapter 5, a small fraction of the methane being transported from its source to the consumer is lost to the atmosphere when natural gas pipelines leak; the greenhouse-enhancing effect of this methane could override some of the advantage methane has in producing less CO_2 per joule upon combustion compared to oil and especially compared to coal (see Additional Problem 4).

The enormous quantity of natural gas held in *methane hydrates* (clathrates) in ocean sediments and permafrost, as mentioned in Chapter 5, would double the fossil-fuel reserves if they could be tapped. The technology to extract the clathrates, most of which are in dilute form and mixed with sediments that lie far below the seabed, does not yet exist. However, recent experiments in offshore sand reservoirs have tapped natural gas from concentrated deposits of hydrates hundreds of meters below the sea floor. The methane is released from the icy clathrates by reducing the pressure and then collected. Coarse sands have been identified as the most likely initial sources of methane from clathrates.

Since the turn of the twenty-first century, natural gas held within deep shale deposits has begun to be exploited in the United States, significantly extending the domestic supply of the fuel. Such *shale gas* (Figure 6-5) may soon also be extracted in Canada, several European and Asian countries, and Australia. The water-intensive technique by which this gas can be obtained, and the environmental problems associated with it, are explored in Box 6-1.

BOX 6-1   Shale Gas

Like conventional natural gas, **shale gas** is a mixture of alkane hydrocarbons, with methane greatly preponderating. The shale rock naturally contained organic matter (up to 25%) that under the influence of high pressures and temperatures over geological time has been degraded anaerobically into methane. The rock must be fractured before the gas is released from tiny pores in the shale and can be brought to the surface. Only recently has the *fracking* technology developed sufficiently that the gas can be tapped economically.

The United States alone has at least a half dozen major shale deposits, which collectively could supply the nation with its natural gas requirements for many decades. The most important such deposit is the Marcellus Shale, centered in Pennsylvania, which covers a quarter of a million square kilometers (about 100,000 square miles), and lies about 1.6 km (1 mile) below the surface. The amount of gas that can be extracted from any given shale well is small compared to conventional ones, so many installations must be constructed—at a cost of several million dollars each—to obtain a large amount of gas. Generally, dozens of wells are drilled at a given site.

The most important environmental concern about shale gas is the large amount of water—amounting to tens of millions of liters (several million gallons) required to fracture the shale, and the disposal of this water after drilling has been completed and it has been returned to the surface. Water cools the drilling bit and carries the rock cuttings out of the hole and is also used to fracture the rock. Concerns have been raised about the withdrawal of large quantities of

(continued on p. 234)

water from local supplies for these purposes, especially in water-scarce areas.

Once vertical drilling has reached the shale deposit, horizontal drilling is used to maximize the area of fracture. The well is then cased and cemented, and a perforation gun is used to shoot holes at several spots along the horizontal component. Water-based fracturing fluids containing various additives are used to fill the horizontal well. It is then temporarily sealed at successive stages along its length and a sequence of very high pressure pulses are applied to it to fracture the rock at the various stage areas and thereby liberate the gas. About half the fluid is then recovered from the well, and gas flows through the perforations and up the well to the surface, where it is collected.

Sand or ceramic material is added to the water used during fracturing to keep the fractures in the cracked rock open, as are gels to increase the fluid's viscosity. Minor additives to the water include hydrochloric acid, biocides, oil to reduce friction, scale and corrosion inhibitors, and surfactants. One concern about shale-gas drilling is the possible contamination of drinking water supplies by these additives. Reinjecting the spent water at depths shallower than that of the horizontal well, a common procedure, also raises concerns about pollution of aquifers used for drinking water. Often the wastewater from the fracturing is simply allowed to evaporate and the residual solids are buried. Attempts to treat the wastewater by conventional methods or to reuse it at another site are usually precluded by the very high level (up to 20%) of dissolved solids it contains.

Among the innovations that are being tested to overcome these environmental problems is the use of nonaqueous fluids for the shale-fracturing stage. Both liquefied petroleum gas (LPG) and liquid carbon dioxide have been tested as possible replacements for water to transport the various additives into the rock.

A further environmental concern associated with gas-well drilling is the radioactivity of the rock cuttings and wastewater from some shales that contain uranium and its decay products, such as the radon gas discussed in Chapter 9. In addition, concerns have been raised that minor earthquakes have resulted from fracking in some areas.

Methane that is adsorbed onto coal is also an important source of natural gas. A hole is drilled from the surface into the underground coal seam and fitted with a steel-encased cylinder. The seam is then dewatered by pumping, after which the methane de-adsorbs from the coal, and rises through the well, driven by the pressure differential. The disposal of the water from the well can present an environmental problem if it has a high salt content. However, the collected gas itself is free of hydrogen sulfide and contains much less propane and higher alkanes than that from conventional sources. *Coalbed methane* is an important source of natural gas in the United States, Canada, Australia, England, and South Africa and is under development in Indonesia.

6.6 Natural Gas and Propane (LPG) as Fuels

In the developed world, **natural gas is used extensively as a fuel. It consists mainly of methane, but contains small amounts of ethane and propane.**

Normally the gas is transported by pipelines from its source to domestic consumers, who use it for cooking and heating, and to some utilities who burn it instead of coal or oil in power plants to produce electricity.

$$CH_4(g) + 2\,O_2(g) \longrightarrow CO_2(g) + 2\,H_2O(l) \quad \Delta H = -890 \text{ kJ mol}^{-1}$$

Unfortunately, where pipelines do not exist, the natural gas that is produced as a by-product of petroleum production at oil wells, etc., is often simply wasted by venting or flaring it off, thereby adding to the atmospheric burden of greenhouse gases.

Highly **compressed natural gas** (CNG) is used to power some vehicles, especially in Canada, Italy, Argentina, the United States, New Zealand, and Russia. Due to the cost of converting a gasoline or diesel engine to accept natural gas as the fuel, the current use of CNG in vehicles is mainly restricted to those such as taxis, buses, and commercial trucks that are in almost constant service. For such vehicles, the additional capital cost of converting the fuel system is much less in the long run than are the savings from the lower cost of the fuel. Because the compressed gas must be maintained at very high pressures in order to keep its storage volume reasonable, heavy fuel tanks with thick walls are required. In order to keep the weight and size of the tank to reasonable values, the driving range (before refilling) of CNG automobiles is usually considerably shorter than gasoline-powered vehicles. Given the size of fuel tanks, CNG is more suited to powering large vehicles such as trucks and buses, and an active program of switching from diesel to compressed natural gas is underway for both types of vehicles.

Compressed natural gas has both environmental advantages and disadvantages as a vehicular fuel when compared to gasoline.

• Since methane's molecules contain no carbon chains, neither organic particulates nor reactive hydrocarbons are formed and emitted into air as a result of its combustion; however, a small amount of each pollutant type is formed from the ethane and propane component of commercial natural gas. Overall, regional air quality is improved by the use of natural gas rather than gasoline or diesel oil.

• However, the release of methane gas from pipelines during its transmission or from tailpipes of vehicles due to its incomplete combustion could lead to increased global warming since methane is a potent greenhouse gas. A massive conversion in North America to CNG as a vehicular fuel would be limited by supply problems for the gas, which is now used extensively for domestic heating and cooking and increasingly as the fuel in new electric power plants.

Some interesting proposals have been made to improve the performance of natural gas as a vehicular fuel. More efficient burning of the methane results if a small amount—about 15% by volume—of hydrogen gas is added to it. Alternatively, a smaller volume for storage of methane results if it is liquefied rather than simply compressed; however, more energy is expended in the process.

Similar but somewhat less serious drawbacks apply to **propane,** C_3H_8, also a main component of **Liquified Petroleum Gas** (LPG), in its use as a gasoline replacement in vehicles. The heat energy produced per gram of propane combusted, 50.3 kJ, is not quite as high as that of 55.6 kJ for methane. The heat released per gram by burning gasoline depends upon the composition of the particular blend under consideration, but is generally slightly less than that for propane. Both LPG and propane are readily liquefied under pressure, so they can be stored much more efficiently than can natural gas.

6.7 Petroleum—Composition

Petroleum, or crude oil, is a complex mixture of thousands of compounds, most of which are hydrocarbons; the proportions of the compounds vary from one oil field to another. **The most abundant type of hydrocarbon usually is the alkane series,** which can be generically designated by the formula C_nH_{2n+2}. In petroleum, the alkane molecules vary greatly, from the simple methane, CH_4 (i.e., $n = 1$), to molecules having almost one hundred carbons. Most of the alkane molecules in crude oil are of two structural types: one type is simply a long, continuous chain of carbons; the other has one main chain and only short branches—e.g., 3-methylhexane.

Petroleum also contains substantial amounts of cycloalkanes, mainly those with five or six carbons per ring, such as the C_6H_{12} systems *methylcyclopentane* and *cyclohexane:*

methylcyclopentane   cyclohexane

Cycloalkanes in crude oil are (somewhat confusingly) called *naphthenes.*

Petroleum contains some aromatic hydrocarbons, principally benzene and its simple derivatives in which one or two hydrogen atoms have been replaced by methyl or ethyl groups. As discussed in Section 6.10, toluene is benzene with one hydrogen replaced by one methyl group, and the xylenes are the three isomers having two methyl groups.

PROBLEM 6-2

Deduce the structures of all the trimethylated benzenes. [*Hint:* For each of the three dimethylated benzenes, draw all the structures corresponding to placement of a third methyl group. Inspect each pair of structures you draw to eliminate duplicates.]

It is the component of petroleum that contains these aromatic hydrocarbons that is the most toxic to shellfish and other fish when an oil spill occurs in an ocean, whether from an oil tanker or from an off-shore oil well. Higher-molecular-weight hydrocarbons form sticky, tar-like blobs that adhere to birds, sea mammals, rocks, and other objects that the oil encounters.

Petroleum is found in certain rock formations in the ground and is pumped to the surface in oil wells. As it exits from the ground, crude oil is not a very useful substance because it is a mixture of so many compounds. To become useful, it must first be separated into components, each of which has several particular uses.

The liquid compounds that are present in crude oil consist of hydro-carbons containing from 5 to about 20 carbon atoms each. Although no attempt is usually made to isolate individual compounds from the mixture, crude oil is separated into a number of **fractions**—different liquid solutions whose components all boil within a relatively small temperature range. This separation of oil into fractions is accomplished by a process called **distillation,** which involves the vaporization by boiling of a liquid mixture, followed by the cooling of the vapor to cause its condensation back to the liquid state; it is described in Box 6-2. Because of the different boiling points of the compounds, it is possible to separate the mixture into components. Each day, a total of about fourteen billion liters of crude oil are distilled by this procedure in hundreds of *petroleum refineries* located around the world.

BOX 6-2   Petroleum Refining: Fractional Distillation

As we noted in the text, fractional distillation separates crude oil into a number of fractions having molecules of similar sizes. The crude oil mixture is first continuously fed through pipes that pass through a furnace that heats the oil to 360–400°C. Even higher temperatures are not employed because of the tendency of the oil to decompose under such conditions. At the temperatures that are used, most of the oil is converted to a gas.

The portion of the oil that is not vaporized is a hot liquid, called the *bottoms* or the *residuum,* which contains the heaviest molecules found in oil. It is drawn off and subsequently separated into components that are used as solid products such as waxes and asphalt, or it is used for making the form of carbon called coke that is employed in steel production. It is also possible, by reducing the pressure to almost a vacuum, to boil the bottoms fraction of crude oil and, by various techniques, to split the vaporized long molecules of this fraction into shorter ones, for use in gasoline and diesel fuel.

The vaporized oil is injected into a vertical distillation or fractionating tower, which is several meters in diameter and up to 30 m high. The temperatures in the tower decrease

(continued on p. 238)

as the hot gases move higher and higher; thus the vapor cools as it rises (see Figure 1). Since they correspond to compounds with high boiling points and hence high condensation temperatures, the first gaseous molecules to recondense to liquids as the vapor rises through the tower are those with 17 or 18 or more carbon atoms. By means of a series of collection trays situated in the tower at positions where the temperature falls to an average of about 350°C, this liquid fraction of the

oil can be collected and drained off, thereby separating it from the remainder, which continues to rise in the tower. This first fraction of the petroleum, called gas oil, is a rather viscous liquid when cooled to room temperatures and finds commercial use as lubricating oils.

Another series of collection trays and an output pipe are located somewhat higher in the tower, where the temperature has cooled sufficiently, to about 300°C, to allow hydrocarbons

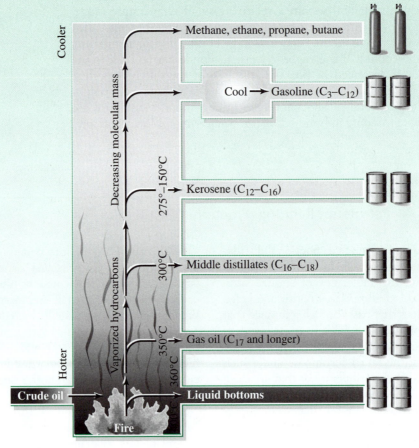

FIGURE 1 Petroleum fractions in a refining tower.

in the C_{16} to C_{18} range to condense and be collected. This second fraction, used as diesel fuel and industrial heating oil, is called middle distillates. The final fraction, called kerosene or heavy naphtha, which contains primarily hydrocarbons with 12 to 16 carbons, is collected by trays near the top of the tower, where temperatures have cooled to about 150–275°C. The kerosene fraction is used as diesel and jet fuels and as oil for home heating. There is no particular reason why the fractions described above, with these particular boiling point ranges, should be the only ones collected. In fact, different petroleum distillation towers collect different fractions by using different collection temperatures, not just those we have described. The decision as to exactly which fractions are to be collected is made by considering the end uses for the various products.

At the top of the tower, the remaining uncondensed vapor contains hydrocarbons consisting primarily of molecules having 1 to 12 carbon atoms each. This vapor is cooled almost to normal outdoor temperatures in a separate unit, a procedure which condenses the molecules with 5 to 12 carbons to the liquid known as straight-run gasoline or light naphtha. This fraction, which constitutes about one-fifth of the original oil, is the basis of the gasoline used to power motor vehicles.

Alkanes having more than about 12 carbons cannot be used in gasoline since they do not evaporate in the engine sufficiently to burn properly.

The C_1 to C_4 gases—namely methane, ethane, propane, and butane—that remain uncondensed at the top of the tower can be collected and used for the purposes previously described for components of natural gas. The C_4 alkanes (butanes) are used as components both of gasolines and of liquefied petroleum gas. Unfortunately, the C_1 to C_3 gases are sometimes ignited and simply "flared off" into the air if facilities for their condensation or transportation do not exist at the petroleum refining site.

In summary, the fractionating tower separates crude petroleum into a number of useful materials, each of which is a mixture of hydrocarbons in which the different constituents have approximately the same number of carbon atoms and which all boil within a small range relative to the larger range of the crude oil. For many applications, a further separation of a fraction into subfractions, each consisting of a smaller set of hydrocarbons, is accomplished subsequently.

In addition to hydrocarbons, crude oil also contains small quantities of compounds that contain other elements. The most predominant of these elements is sulfur, which occurs in oil to the extent of 0.5–4%, depending upon the origin of the material. Metals such as vanadium, nickel, and iron also are present, at a total concentration of more than 1000 ppm in some cases, but they are usually found mainly in the bottoms.

REVIEW QUESTION 1

Explain the process by which crude oil is converted to usable fractions.

REVIEW QUESTION 2

How are gasoline and kerosene different? How are they alike?

REVIEW QUESTION 3

Why can both gasoline and kerosene be used as fuels?

In addition to hydrocarbons, petroleum also contains some sulfur (up to 4%), in the form of compounds: hydrogen sulfide gas and organic sulfur compounds that are alcohol and ether analogs in which an S atom has replaced the oxygen. These substances are much more readily removed from oil than is the sulfur from coal, making petroleum products inherently cleaner. Petroleum that is low in sulfur is termed "sweet," and that higher in sulfur is called "sour." (The same adjectives are applied to natural gas.) Any sulfur that remains in fuels is converted during the fuel's combustion into *sulfur dioxide*, which is a serious pollutant if released into the air (Chapter 3). Some organic compounds of nitrogen also occur in petroleum, and are the source of the "fuel NO" (Chapter 3) formed when gasoline and diesel oil burn.

Petroleum and fuels made from it, such as gasoline and diesel oil, have the great advantage that they are energy-dense liquids that are convenient, relatively safe to use, and (currently) relatively cheap to produce. Virtually all nonelectrical transportation systems in both the developed and developing worlds are based upon cheap petroleum fuels. The possibility of switching to alternative fuels for transportation is discussed in detail in Chapter 7. It will be much more difficult to switch from oil to other chemical feedstocks for the production of pharmaceuticals and polymers once oil runs out.

> Diesel fuel made from petroleum is sometimes called *petrodiesel*.

6.8 Petroleum—Supply

Although it took nature about half a billion years to create the world's supply of petroleum, humans will probably will have used almost all of it during a 200-year period which started just before the end of the nineteenth century. Indeed, the production of petroleum in the lower 48 states of the United States has already peaked. In commerce, petroleum is measured in barrels, each of which is equivalent to 159 L, or 42 U.S. gal. Current annual world oil production amounts to about five *trillion* (5×10^{12}) L, or 32 billion barrels (2010). Much of the proven oil reserves occur in the Middle East.

> The world's top producers of crude oil are Russia, Saudi Arabia, and the United States.

Although it is commonly said that we are "running out" of oil and gas, this will probably not occur globally in the short-to-medium term. Improvements in the technology of petroleum extraction allow greater and greater proportions of the oil in a given deposit to be extracted. The initial extraction of oil from a reserve usually proceeds without difficulty. Later on, however, the remaining oil occurs mostly in pores as droplets that are larger than the connecting necks of the porous formation, and will not flow because of surface tension. In *secondary extraction*, surfactants, water, or pressurized carbon dioxide are used to lower the surface tension and drive additional oil from the reserve. In *tertiary extraction*, steam is injected to lower the viscosity of the remaining oil and allow it to be pumped. As the more accessible supplies dry up and the price of oil rises, such extraction processes are becoming more economically viable. Indeed, most major oilfields are now undergoing at least secondary extraction.

> Surfactants are *surface active* agents that contain both water-loving and water-repelling components.

The terms *light* and *heavy*, as used in the petroleum industry, refer to the density of the oil and consequently to its viscosity. Light oil is dominated by unbranched (straight-chain) alkanes (paraffins) and consequently requires little refining to produce gasoline and diesel. Heavy oil contains a much higher proportion of napthenes and aromatic molecules and so requires more refining. Generally speaking, heavy oils contain more sulfur than light ones.

Some petroleum geologists and analysts believe that global oil production will peak within the next few decades. Indeed, U.S. production already reached a maximum in about 1970. A recent estimation by the International Energy Agency of the manner in which oil production is expected to vary with time (under their *New Policies Scenario*) is shown in Figure 6-6. Notice that production of *conventional crude oil* (black line) peaked about 2006–2007, but it is predicted to stay constant at almost that level for the next quarter century at least. The future supply of conventional oil depends significantly on oil fields yet to be developed—and in some cases yet to be discovered—as indicated by the medium green region, since that from existing fields (darker green region) is in rapid decline.

The area at the top of the graph includes the *unconventional sources* of oil, namely, the *heavy oil* obtained from *oil/tar sands* and that from *oil shales* (light gray region at top of graph) and more heavily on *natural gas liquid* hydrocarbons (NGL). It is clear from the graph that unconventional sources are expected to play an increasingly important role in oil supplies in future decades.

Very-long-chain hydrocarbons obtained from *oil shales*, a type of carbon-rich sedimentary rock, and *oil sands*, which are sandstone or porous rock impregnated with very viscous crude oil ("extra heavy" or "ultraheavy oil"),

NGL: Natural gas under ground contains varying but small amounts of other alkanes—ethane through pentanes at least—that are liquid under pressure. They are almost entirely removed, then separated by distillation and used for purposes such as components of gasoline or petroleum feedstocks before the methane is pipelined to users.

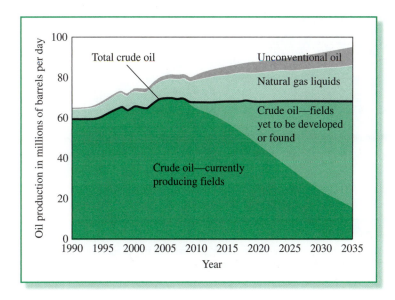

FIGURE 6-6 Global oil production by type according to IEA New Policies Scenario. Note that 1 million barrels of oil is equivalent in energy to 6.1×10^{15} J, i.e., to 0.0061 EJ. [Source: *World Energy Outlook 2010*, International Energy Agency.]

could extend the petroleum supply. The production of difficult-to-extract "tight oil" from shales, using techniques similar to that described in Box 6-1 for shale natural gas, is economically feasible in some locales such as Manitoba, in western Canada.

Chemical aspects of the 2010 *Deepwater Horizon* oil disaster in the Gulf of Mexico are discussed in Box 6-3.

BOX 6-3   The *Deepwater Horizon* Oil Spill Disaster

In the spring of 2010, the explosion and blowout of a deepwater drilling rig in the Gulf of Mexico, 66 km (41 miles) offshore and south of the coast of the state of Louisiana, produced the largest oil spill in the history of the United States and the second largest in world history. Before the exploratory well, located 1.5 km (5000 ft) below the sea's surface, was capped three months later, an estimated 780 million liters (200 million gallons) of crude oil had escaped into the sea. The oil distributed itself over all layers of the ocean, with some rising to the surface. Included in the release was about 500,000 tonnes of gas, mainly methane.

In the Gulf oil spill, additional environmental risk was evident in the two principal methods used to keep surface-level crude oil from reaching the beaches and wildlife habitats on the shores of the Gulf coast:

• setting fire to the oil in enclosed areas on the ocean surface (see Figure 1);

• spraying chemical "dispersants" onto the oil on the ocean surface and below it, near the leaking wellhead.

FIGURE 1 A fireboat response crew battling the blazing remnants of the offshore *Deepwater Horizon* oil rig on April 21, 2010. [AP Photo/U.S. Coast Guard]

The crude oil that contaminated the Gulf consisted of a complex mixture of thousands of organic compounds, most of which were hydrocarbons, although a minority of them also contained small amounts of oxygen, nitrogen, and sulfur. As in other conventional crude oils, the hydrocarbons were mainly *alkanes* (including cycloalkanes) and *aromatic* compounds containing one or more *benzene* rings.

One component of crude oil that was especially plentiful in that escaping from the sea bottom was *asphaltenes*. Compounds in this group consist of molecules that are very large, and generally contain one main region ("island") of several benzene rings in condensed combination, to which are bonded chains of saturated carbon atoms, as illustrated in Figure 2. Commonly the molecules also contain sulfur, nitrogen, and oxygen atoms, some of them within the aromatic region. Asphaltenes are much slower to biodegrade than are most other components of crude oil.

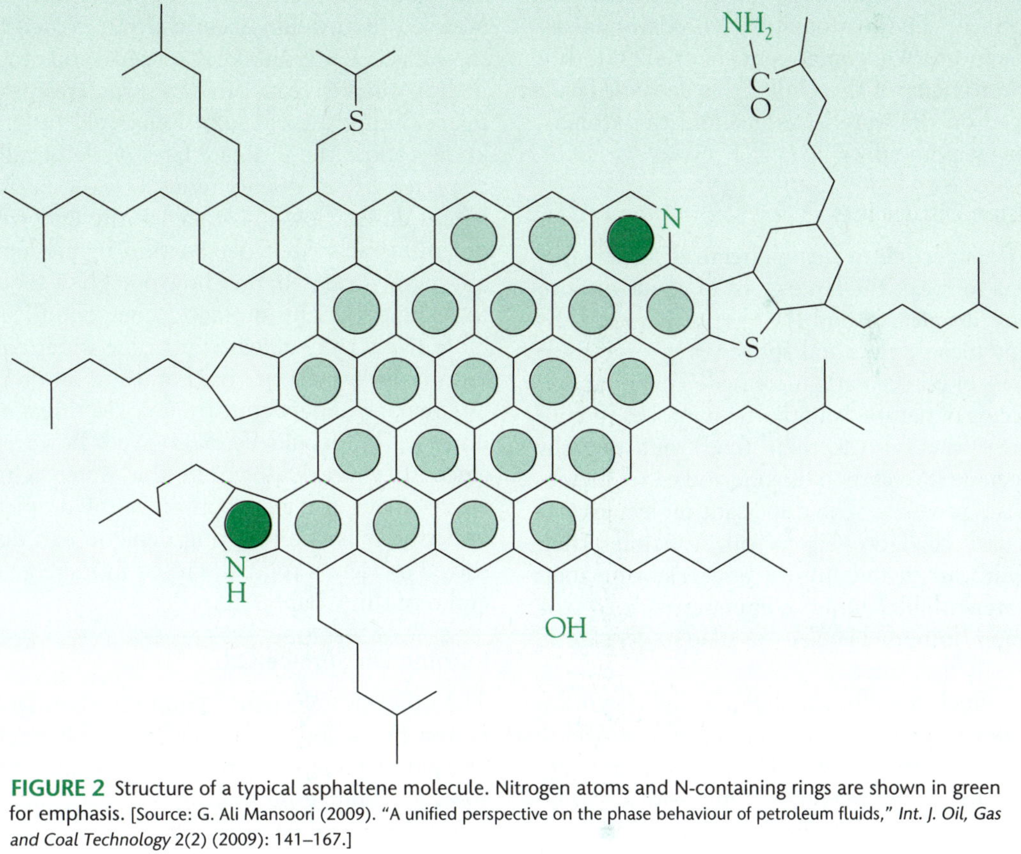

FIGURE 2 Structure of a typical asphaltene molecule. Nitrogen atoms and N-containing rings are shown in green for emphasis. [Source: G. Ali Mansoori (2009). "A unified perspective on the phase behaviour of petroleum fluids," *Int. J. Oil, Gas and Coal Technology* 2(2) (2009): 141–167.]

(continued on p. 244)

Hydrocarbons as a group are less dense than water, so the oil escaping from the sea bottom rose gradually to the surface. However, the composition of the discharged oil changed over a period of hours and days as it became exposed first to the seawater and then to the air. Although hydrocarbons are largely insoluble in water, benzene and related aromatic hydrocarbons such as toluene, xylenes, and ethylbenzene (i.e., BTEX) dissolve to some extent. In addition, the aromatics are much more volatile than are the majority of oil molecules, so they largely escape from the liquid and become a component of the air after the oil reaches the surface. These aromatic hydrocarbons are the most toxic components of crude oil; thus the toxicity of the Gulf crude *decreased* as it aged on the water's surface and the aromatics evaporated.

Using Dispersants

The objective of using chemical dispersants was to break the large globs of crude oil into tiny droplets about 10^{-5} m in diameter. By this means, the total surface area of a given mass of oil is greatly increased. This promotes access of natural waterborne microbes to consume the oil (as their food) and thereby degrade it to carbon dioxide and other metabolism products. The dispersant molecules surround small droplets of oil, attaching their hydrocarbon tails into it and orienting their water-soluble ionic components into the water, forming *micelles*—as soap molecules do in oily dishwater.

Such micelles are likely to fall from the water surface into the water column, where they are either degraded by microbes or swept out to sea by ocean currents—rather than invading coastal wetlands and beaches as oily blobs. The strategy of having microbes degrade wastes—sometimes accelerated by the addition of fertilizers—is called *bioremediation*, and is a standard technique in the treatment of toxic wastes, as we shall discuss in Chapter 16. The use of dispersants at the wellhead prevented large blobs of oil from rising directly upward, where many ships involved in the cleanup were present, and prevented large blobs from reaching the shoreline.

One environmental problem associated with dispersants is that **the compounds themselves are somewhat toxic,** though much less so than the oil. Ironically, by dispersing the oil into droplets that remain in the water column, they also expose waterborne organisms to more oil than would occur if it just floated to the surface. Large plumes of dispersed oil drops in deep water were reported by several research teams during the cleanup. Relatively little is known about the ultimate fates of chemically dispersed oil at sea in terms of how fast it breaks down, how it is taken in by undersea organisms and what degradation by-products are thereby created, and how quickly it binds to sediments. The aromatic compounds are more likely to be released into the water column if dispersants are used and the oil droplets remain underwater; their toxic threat to undersea life would be short-lived, however, since they break down in the water. One worry about the metabolism of oil droplets underwater is the resulting depletion of dissolved oxygen, a topic discussed in more general terms in Chapter 10.

Burning Off Surface Oil

The other strategy used to minimize contamination by the oil was to set fires deliberately on enclosed regions of the ocean to eliminate surface oil. Immense quantities of black smoke were thereby produced and became dispersed into the air. The blackness of the smoke indicated the presence of many particles of elemental graphitic carbon, onto whose surface toxic hydrocarbon molecules would be adsorbed. Also present in the smoke were large quantities of the poisonous gas carbon monoxide. Presumably the oil-burns at sea were restricted to days in which the smoke would have been swept out to sea and greatly diluted by air currents rather than migrating toward land and thereby threatening the health of Gulf Coast residents.

Because they are so large and nonvolatile, asphaltene molecules were less likely than aromatics and alkanes to burn in the fires set on the sea surface; thus they remained as residues after the fires burned out. The reddish-brown oil streaks on the sea surface seen from the air consisted largely of asphaltenes in the form of emulsions. **Emulsions consist of microscopic droplets of one liquid, in this case oil, dispersed in another, here water;** the combined particles are about 10^{-6} m in diameter. These combinations of oil and water are difficult to clean up, as the particles do not readily coagulate.

Overall, it is estimated that only about 8% of the oil released from the spill was chemically dispersed, about 5% of it was burned, about 3% of it was skimmed, and 17% was recovered directly from the wellhead. More than half the oil was naturally dispersed into the surrounding water by the current or is still present on the sediments near the wellhead and in either case is being naturally degraded, or it was evaporated.

The Gulf oil spill disaster provides an excellent example of the fact that the cleanup of any pollution site involves tradeoffs, for example, that between creating air pollution by burning surface oil and allowing it to reach the shore and thereby contaminate the wetlands and beaches. The avoidance of pollution in the first place is obviously the most desirable course of action in activities that may lead to environmental problems.

6.9 The Alberta Oil Sands

The potential for extracting oil from bitumen-covered sands in northeastern Alberta, Canada, exceeds the oil reserves of Saudi Arabia! Bitumen is a thick, sticky, tar-like viscous liquid that chemically consists mainly of highly condensed *polycyclic aromatic hydrocarbons*, PAHs (Chapter 15). The Alberta deposits are a 50-m-thick band that is a mixture of bitumen (about 10 %), sand, and water that lies up to 75 m below the surface. The individual tiny particles of sand are covered by a film of water, and a small collection of such particles are covered by a layer of bitumen. Large-scale extraction of the oil is capital-intensive, energy-intensive, and water-intensive; it results in the emission of more greenhouse gases than does the production of conventional crude and generates huge volumes of wastewater.

About 1.3 million barrels of oil per day were being produced from the resource in 2010, and this was projected to rise to 3 million barrels by 2018. The entire oil sands region in Alberta covers 140,000 km^2 (the size of New York State); recovery of oil from only a fraction of this area is economically feasible. The recoverable resource currently lies beneath 8000 km^2 of land, of which over 600 km^2 of forest and wetlands in northern Alberta have been stripped away to create the current mines.

About 2 tonnes of the asphalt-like mixture, and at least 2 barrels of water, are required to produce 1 barrel of oil. Energy in the form of natural gas must be expended to separate the oil from the sands and to split the long-chain hydrocarbon molecules of the tar into shorter ones for use in gasoline. Since condensed PAHs are hydrogen-poor, H$_2$ gas must be created from natural gas and used to hydrogenate them to create a hydrocarbon with an oil-like C:H ratio of approximately 2:1 (see the discussion of such processes in Chapter 7, Section 7.18). After the heavy oil is produced, it must be refined to remove sulfur (up to 4%) and convert the hydrocarbons into useful fractions, as discussed for petroleum in Section 6.7 and Box 6.2.

About half the current production uses large-scale open-pit mining of shallow reserves. The crushed bitumen is separated from the sand by heating it with several times as much hot water (50–80°C) that has been heated by burning natural gas, which results in high carbon dioxide emissions. The other half employs an *in situ* underground technique that mixes steam with the bitumen to extract the latter. The steam is injected into a long horizontal well. Eventually, the liquefied bitumen separates from the sand and flows down into a second well several meters below the first, where it is collected and pumped to the surface. This *steam injection* process uses much less water but more energy than near-surface mining, so it generates less wastewater but more CO$_2$ per barrel. Most of the water used in the *in situ* process is recycled. The use of hydrocarbon solvents can potentially reduce the *in situ* steam requirements and therefore of the amount of natural gas required and CO$_2$ generated, but this process is more expensive if the solvent is not recycled efficiently. Many more sites in Alberta could be exploited by the *in situ* process than by surface mining since most bitumen deposits lie many meters below the surface.

Recall that *in situ* means "in place."

The "tailings" from the open-pit mining of oil sands consist of a slurry of water, sand, silt, fine particles of clay, and some unrecovered bitumen, and currently cover an area of 130 km^2. The volume of tailings produced is about six times that of the bitumen. The tailings are sent into "ponds" where the components separate into layers. The bitumen fraction contains small amounts of many toxic organic and inorganic chemicals, and it is potentially a health problem to local birds as well as downstream human inhabitants if the ponds leak into groundwater. Unfortunately, the suspended clay particles take decades to settle out spontaneously. Some success has been achieved recently by using a polymer additive that greatly speeds up the settling process in the layer in which most of the suspended matter has settled after a few months, and can produce dry tailings. Although much of the water from the oil sands operations is reused, tailings ponds from oil sands operations covered over 100 km^2 of Alberta. One pond dam is second only to China's Three Gorges dam in terms of size. Were such ponds to be ruptured, the consequent water pollution would be widespread.

Recently, evidence has been uncovered that the extraction process pollutes the freshwater supply in the Athabasca River not only with hydrocarbons but also with heavy metals. Much of the releases originate with the process of upgrading the bitumen. However, the oil sands project is more controversial because of its high CO_2 emissions than because of the water, air, and land pollution that it causes. Because the original oil itself is of such poor quality and because it is attached to the sand, its mining and processing generates two-to-three times as much carbon dioxide per barrel as does conventional crude oil, although it is still less than that emitted into the air when the petroleum products are combusted.

6.10 Petroleum—Gasoline

Regular gasoline contains predominantly C_7 and C_8 hydrocarbons; diesel fuel contains mainly hydrocarbons with 9–11 carbon atoms. Generally speaking, the more carbon atoms in the alkane, the higher its boiling point and the lower its vapor pressure—and hence the lesser its tendency to vaporize—at a given temperature. For this reason, gasoline destined for warm summer conditions is formulated with smaller amounts of the smaller, more easily vaporized alkanes such as butanes and pentanes than is that prepared for winters in cold climates. The presence of volatile hydrocarbons in gasoline is vital in cold climates so that automobile engines can be started.

Gasoline that consists primarily of unbranched alkanes and cycloalkanes has poor combustion characteristics when burned in internal combustion engines. A mixture of air and vaporized gasoline of this type tends to ignite spontaneously in the engine's cylinder before it is completely compressed and sparked, so the engine "knocks," with a resulting loss of power. Consequently, all gasoline is formulated to contain substances that will prevent knocking.

In contrast to unbranched alkanes, highly branched ones such as the octane isomer *2,2,4-trimethylpentane*, "isooctane" (illustrated below) have excellent burning characteristics. Unfortunately, they do not occur naturally in significant amounts in crude oil. The ability of a gasoline to generate power without engine knocking is measured by its **octane number.** To define the scale, isooctane is given the octane number of 100, and *n*-heptane is arbitrarily assigned a value of zero.

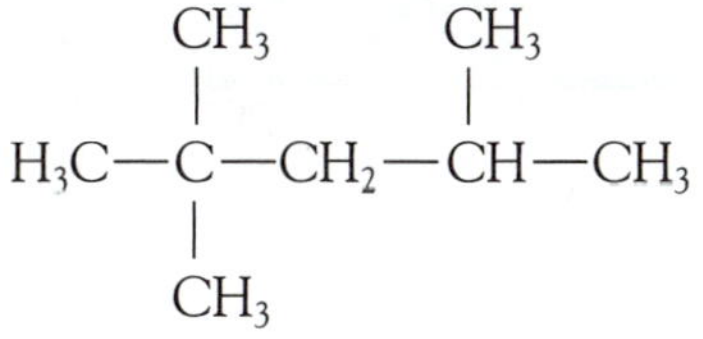

$$
\begin{array}{ccc}
CH_3 & & CH_3 \\
| & & | \\
H_3C-C-CH_2-CH-CH_3 \\
| \\
CH_3
\end{array}
$$

2,2,4-trimethylpentane ("isooctane")

Gasoline that has been produced simply by distillation of crude oil has an octane number of about 50, much too low for use in modern vehicles. However, when added to gasoline in small amounts, the compounds **tetramethyllead,** $Pb(CH_3)_4$, and its ethyl equivalent prevent engine knocking and hence greatly boost the octane number of gasoline. For decades, they were added worldwide to gasoline consisting predominantly of unbranched alkanes and cycloalkanes. However, these additives have largely been phased out now in most developed countries owing to environmental concerns about lead— especially the fact that it poisons catalytic converters, a topic discussed in Chapter 12.

In some European countries, Russia, Canada, and recently in Australia, lead compounds were replaced by small quantities of an organic compound of manganese called **MMT,** which stands for *Methylcyclopentadienyl Manganese Tricarbonyl.* The use of MMT has been controversial for health reasons since manganese concentrations rise in air and soil as a result, and for technological reasons since some car manufacturers claim it degrades emissions components on vehicles. MMT was banned in the United States until 1995 and is still used there only to a very small extent.

The alternative to using lead or manganese additives to boost octane ratings is to blend into gasoline significant quantities of highly branched alkanes or BTX or other organic substances such as MTBE (discussed later), which themselves have high octane numbers. A list of the common additives is shown in Table 6-3.

benzene   toluene   *p*-xylene

Collectively, the *benzene* + *toluene* + *xylene* component in gasoline is called **BTX.** Gasoline often also contains some trimethylated benzenes and *ethylbenzene*; the mixture is then called BTEX. Currently, most unleaded

TABLE 6-3	Octane Numbers of Common Gasoline Additives
Compound	Octane Number
Benzene	106
Toluene	118
p-Xylene (1,4-dimethylbenzene)	116
Methanol	116
Ethanol	112
MTBE	116

gasoline sold in the United States contains significant quantities of BTX (as high as 40% content in the past) or of ethanol (especially in the midwest) or of MTBE to boost the octane number. Unfortunately, the BTX hydrocarbons are more reactive than the alkanes that they replace in causing photochemical air pollution, so in a sense the lead pollution has been reduced at the price of producing more smog. In addition, the use of BTX in unleaded gasoline in countries such as Great Britain, in which few cars were equipped with catalytic converters, resulted in the past in an increase in the BTX concentrations in outdoor air. Benzene in particular is a worrisome air pollutant since at higher levels it has been linked to increases in the incidence of leukemia (see Chapter 4).

The **reformulated gasoline** introduced in the 1990s in North America limited the concentration of benzene and of aromatics in total, and set a minimum of 2% oxygen (by mass). The second phase of reformulated gasoline, which entered the U.S. market in 2000, reduced the benzene and BTX components even further, and lowered the sulfur content to 30 ppm. There are now many different sets of specifications for reformulated gasoline in the United States, depending upon region and season. The most restrictive are probably those in California, in which the maximum level of sulfur is 20 ppm, of benzene is 0.8 ppm, of aromatics is 25 ppm, and of olefins is 6 ppm. The maximum gasoline benzene level in Canada, Australia, India, and the European Union is 1 ppm.

The use of alcohols and compounds derived from them as additives or as "oxygenated" motor fuels in their own right is discussed in Chapter 7.

Olefins are nonaromatic hydrocarbons containing $C=C$ bonds.

Review Questions 6–14 are based on the preceding sections.

6.11 Green Chemistry: Polylactic Acid—The Production of Biodegradable Polymers from Renewable Resources; Reducing the Need for Petroleum and the Impact on the Environment

Our everyday lives are permeated by the chemicals in products such as pharmaceuticals, plastics, pesticides, personal hygiene products, cleaners, fibers, dyes, paints, clothes, building materials, computer chips, packaging, and food. The vast majority of these chemicals are ultimately made from oil, consuming approximately 2.7% of the production of this natural resource. The compounds that are isolated from oil and used to produce chemicals are known as *chemical feedstocks*. Approximately 60 billion kg of these feedstocks are employed to create 27 billion kg of polymers (many are loosely referred to as plastics) each year.

Some of the more familiar polymers (as will be discussed later, in Chapter 16) that are produced from crude oil include *polyethylene terephthalate* (PET), which is used to make plastic beverage bottles and fibers for clothes; polyethylene, which is used to produce plastic grocery and trash bags; and polystyrene, which we discussed in the green chemistry section in Chapter 2.

Trade names of polymers, such as *Dacron*, *Teflon*, *Styrofoam*, and *Kevlar*, represent polymers that are part of our everyday lexicon.

Approximately 2 billion kg of PET are produced each year. PET is one of the main targets for recycling of plastics, yet less than a quarter of this total is recycled in the United States, with the rest being landfilled or incinerated. Even when PET is recycled, it generally can't be reused as beverage bottles; it is downward recycled into polyester fiber products such as carpets, T-shirts, fleece jackets, sleeping bags, and car trunk linings or into thermoformed sheet products such as laundry scoops, nonfood containers, and containers for fruits.

When we use oil to produce items that we dispose of or incinerate (including the use of oil as a fuel), we are consuming a resource that has taken nature millions of years to produce. Petroleum is a finite, nonrenewable resource. Although there are still considerable oil reserves, at the rate of our current use we will deplete the supply of cheap, readily accessible oil within the next quarter century. We must learn to use renewable resources such as biomass rather than petroleum to produce chemical feedstocks.

Scientists at *NatureWorks LLC* (formerly Cargill Dow LLC) have developed a method for producing a polymer called **polylactic acid** (PLA) from renewable resources—such as corn (called maize in the United Kingdom and elsewhere) and sugar beets—for which they won a Presidential Green Chemistry Challenge Award in 2002. NatureWorks produces PLA at a plant in Blair, Nebraska. Ultimately, the goal is to utilize waste biomass as the source of this polymer. As in the steps shown in Figure 6-7, the corn is milled into starches, which are then reacted with water to yield glucose, which is then

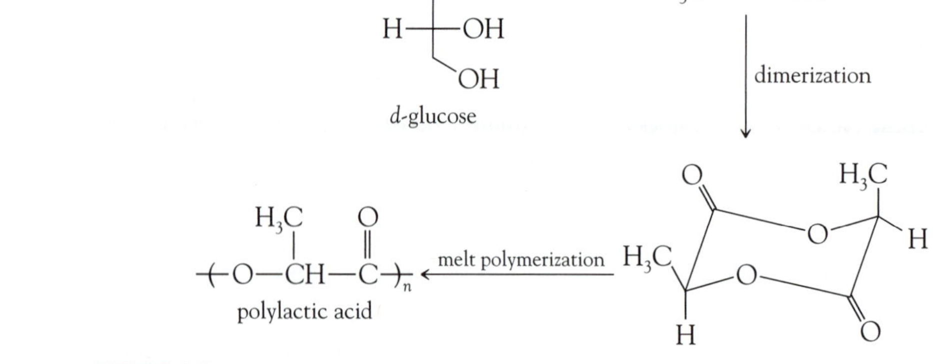

FIGURE 6-7 The synthesis of polylactic acid.

converted to lactic acid by natural fermentation. This naturally occurring compound is then converted to its dimer, followed by polymerization to PLA.

The environmental advantages of PLA over petroleum-based polymer include the following:

• It is made from annually renewable resources (corn, sugar beets, and eventually waste biomass).

• Production of PLA consumes 20–50% less fossil-fuel resources than petroleum-based polymers.

• It uses natural fermentation to produce lactic acid. No organic solvents or other hazardous substances are used.

• It uses catalysts, resulting in reduced energy and resource consumption.

• High yields of >95% are obtained.

• The use of recycle streams help to reduce waste.

• PLA can be recycled: converted back to its monomer via hydrolysis, then repolymerized to produce virgin polymer (i.e., closed-loop recycling).

• PLA can be composted (it is biodegradable); complete degradation occurs in a few weeks under normal composting conditions.

Another environmental consideration of PLA is that the plants, such as corn, used to produce this polymer consume atmospheric carbon dioxide, thus reducing the concentrations of this greenhouse gas. When PLA biodegrades it releases this carbon dioxide back into the atmosphere in amounts approximately equal to that absorbed by the plants used to produce it, thus making PLA, in theory, carbon neutral. However, fossil fuels are required during the production of PLA. Life cycle assessment studies indicate that PLA requires 25–55% less fossil-fuel-supplied energy than petrochemical polymers.

PLA can be used to produce products that are currently made from petroleum-based polymers, such as cups, rigid food containers, food wrappers/bags, bags for refuse, furnishings for homes and offices (carpet tile, upholstery, awnings, and industrial wall panels), as well as fibers for clothing, pillows, and diapers. Small purveyors of natural foods have been using PLA packaging for several years. In 2005 PLA received a significant boost when Walmart announced plans to use 114 million PLA containers a year. According to the company this will save 800,000 barrels of oil annually.

Biodegradable polymers produced from renewable resources help to reduce our consumption of oil, and have the potential to offer significant environmental and economic advantages over petroleum-based polymers. However, we must remember that even production of chemicals via annually renewable resources such as biomass does not offer a complete solution to energy and environmental problems. Growing crops, whether they are used to produce food or chemicals, requires fertilizers and pesticides. Energy is needed to plant, cultivate, and harvest; to produce, transport, and apply fertilizers and pesticides; to make and run tractors; to transport seeds, biomass, monomers, and polymers. Use of land to produce crops for chemicals and more significantly for bio-based fuels, also removes land that could be used to produce food and feed, and increases the price for food.

Sequestration of CO_2

Carbon dioxide can be extracted from the exhaust gases of major point sources that would otherwise release it into the atmosphere, especially power plants that burn fossil fuels and that collectively are responsible for one-third of total CO_2 emissions. The carbon dioxide gas so recovered would then be **sequestered—i.e., it would be deposited in an underground or ocean location that would prevent its release into the air.** For example, the CO_2 could be sequestered by burial in deep ocean waters, where it would dissolve, or in very deep aquifers under land or the seas, or in empty oil and natural gas wells or coal seams. The total amount of carbon dioxide that will be produced by fossil-fuel combustion over this century will amount to more than one trillion tonnes, so vast volumes for storage will be required to sequester even a fraction of it (Problem 6-3). The name given to the overall process is **carbon capture and storage,** CCS. Several small-scale operations of this process are underway, with more planned for the next few decades in order to determine whether CCS is a technically and economically feasible way to deal with CO_2 emissions.

PROBLEM 6-3

Calculate the length, in kilometers, of a cube of liquid or solid carbon dioxide whose density is the same as that of water, i.e., 1 g cm^{-3} and whose mass is a trillion (10^{12}) tonnes.   ●

Since it is not economically feasible to transport or store enormous volumes of emission gases from power plants, **the dilute carbon dioxide** in the gas mixture (about 4% from natural gas plants, up to 14% from coal plants) **must be captured and concentrated on site.** It can then be transported as a dense supercritical fluid (see Section 6.15) by pipeline to its storage destination. **The energy input required for the CO_2-capture-and-concentrating phase of carbon sequestration schemes is the major expense of CCS.** It is largest by far of all the steps involved; it will represent a substantial fraction, from one-third to one-half, of the output of the power plant to which it is connected.

As a consequence of the energy intensity of carbon capture, **substantially more fuel would be combusted, thus producing more CO_2 and other air pollutants.** The equipment required to capture and concentrate CO_2 is itself very large, since huge volumes of air—consisting mainly of nitrogen and some residual oxygen—are involved. The capture/concentration/compression of the gas accounts for about three-quarters of the cost of the entire sequestration process. Nevertheless, some observers feel it would still be cheaper to sequester carbon from fossil fuels than to convert to a renewable-fuel economy based on wind and solar energy.

6.12 Reversible Capture of CO$_2$

The capture and concentration of carbon dioxide from conventional power-plant emissions is accomplished by using an agent that will *selectively* remove the gas, collect it in ever-more-concentrated form over time, and later release pure CO$_2$ to be compressed, transported, and buried. The capture agent must be able to be recycled to make the operation economically viable. Although cooling the emission gases first to remove water and then carbon dioxide by condensation at a lower temperature ($-79°C$ at ambient pressure) would seem the simplest procedure, this process is too energy-intensive—and therefore too expensive—to be practical with dilute sources of CO$_2$.

Most commonly the removal agent is a chemical to which CO$_2$ weakly and selectively bonds, with later detachment of the gas achieved by heating or a drop in pressure. Since carbon dioxide is an *acid anhydride*—it reacts with water to form **carbonic acid,** H$_2$CO$_3$—the chemicals most convenient to capture it are bases since acid–base reactions are fast and largely proceed to completion.

The most commonly used technology for carbon dioxide capture from power plants uses amines. The cooled (to about 50°C) emission gases are passed upward through a packed tower, down which flows an aqueous solvent that selectively removes CO$_2$ (see Figure 6-8). The solvent (at about 38°C) contains 15–30% by mass of *amines*, R$_2$NH (or RNH$_2$), which are weak bases, B, that initially combine with the CO$_2$ to produce anions. In water the *overall reaction* is

$$H_2CO_3 + B \rightleftharpoons HCO_3^- + BH^+$$

The CO$_2$-laden liquid is then led to another tank, where heat is used to reverse this reaction (at about 120°C) and thereby produce a concentrated stream of carbon dioxide. The regenerated and cooled amine solution is then recycled for further use (Figure 6-8).

The amines commonly employed in this **chemical absorption** technology are *monoethanolamine* and *diethanolamine*, since they are highly water-soluble and thus can absorb high loads of the gas and require relatively little heating to later reverse the reaction to release the carbon dioxide. With these amines, over 95% recovery of CO$_2$ can be achieved.

Heating the solution to liberate the CO$_2$ and regenerate the amine solvent, and compressing the gas once it is produced, are the most energy-intensive steps of the process. In addition, enormous amounts of the solution

Monoethanolamine is
HO—CH$_2$—CH$_2$—NH$_2$.

FIGURE 6-8 Schematic
diagram of amine solvent
capture of CO_2 from fossil-
fuel power plant emissions.
See description in text for
details.

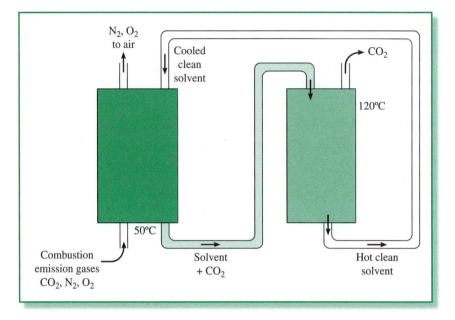

FIGURE 6-8 Schematic diagram of amine solvent capture of CO_2 from fossil-fuel power plant emissions. See description in text for details.

must be transported from place to place. Nitrogen and sulfur oxides must be removed from the emission gas *before* it enters the solution, since otherwise they would react with and degrade the amine. Nevertheless, some decomposition of the amine, producing ammonia and salts, inevitably occurs during the repurification cycles.

The amine-based technology described above has been used industrially in several contexts for many years. Newer technologies based on chemical techniques useful when the carbon dioxide is dilute in the emission gas and physical techniques for use with more concentrated sources are outlined in Table 6-4.

6.13 Oxycombustion

One way to circumvent the high expense and energy consumption required to isolate and concentrate the CO_2 from conventional power plants is by **oxycombustion.** In this technique, currently under development, **the fossil fuel is burned not in air but in oxygen gas, O_2.** If the correct stoichiometric ratio of oxygen to fuel is supplied, this process requires no expensive isolation step since the exhaust gas from oxycombustion would consist entirely of carbon dioxide and water vapor, which is readily separated from the carbon dioxide by condensation at lower temperatures. (By contrast, since air is only 19% oxygen by volume, the maximum level of CO_2 produced when it is used for combustion is also only 19%.) Incomplete combustion during oxycombustion results in some carbon monoxide and unburned fuel in the emission gases, but these can be catalytically oxidized to carbon dioxide prior to the water condensation step.

TABLE 6-4	Innovative Technologies for CO_2 Capture from Gas Streams
Technology	**Advantages/Disadvantages**
Carbonate Capture: Uses carbonate ion, CO_3^{2-}, as the weak base (formed in chilled ammonia solution) that captures the carbon dioxide after it forms carbonic acid in the water.	Less energy-intensive than capture by amines, and not susceptible to decomposition by the presence of SO_2 and other impurity gases.
Warming the solution (to 120°C) reverses the reaction and releases the carbon dioxide gas, and reforms the carbonate ion, which allows the solution to be reused.	
Carbonate Looping: Carbon dioxide can be chemically absorbed by certain basic metallic oxides, which will release the gas when heated. For example, calcium oxide, CaO, can quickly remove CO_2 from hot emission gases by formation of calcium carbonate, $CaCO_3$:	The $CaCO_3$ decomposition process is highly endothermic and requires high temperatures; however, schemes have been proposed to employ some of the corresponding exothermicity of the forward reaction to offset this energy requirement.
$$CaO(s) + CO_2(g) \rightleftharpoons CaCO_3(s)$$	
Subsequent heating of the solid to about 900°C, once it has been largely converted to the carbonate, reverses the reaction and produces concentrated CO_2 and regenerates CaO.	Unfortunately, calcium oxide deactivates relatively quickly over many absorption/desorption cycles, so fresh oxide must constantly be added to maintain the absorptive activity of the solid.
Lithium Compounds: Lithium silicate, Li_4SiO_4, reversibly reacts with CO_2, absorbing it at relatively low temperatures and releasing it at high temperatures. Carbon dioxide could be captured from gas streams and later regenerated in concentrated form by heating.	
Solvent Extraction: When it is present in relatively concentrated form (about 50% or more) in a gas stream, it is feasible to separate the CO_2 from the other components by bubbling it through an organic solvent such as methanol or a *glycol* in which it is selectively soluble until the solution is saturated, and then heating the solution to re-release the gas in pure form.	The heat required for desorption is significantly less than that required when CO_2 is chemically captured, since no chemical bond is formed in solution that must be broken before the gas is released. Some newly developed ionic liquids may also be suitable for their selective solubility of carbon dioxide.
Membrane Separation: Carbon dioxide at high concentration can be separated from other gases in some mixtures by use of polymeric membranes and molecular sieves that allow CO_2 to pass through them, while the other gases are excluded.	This technology has been used for many decades in the oil industry, and would be well suited to natural gas purification and to fermentation plants. Research and development is underway to devise membranes that would be efficient and economical for power-plant emissions.

(continued on p. 256)

TABLE 6-4	Innovative Technologies for CO$_2$ Capture from Gas Streams *(continued)*
Technology	Advantages/Disadvantages
Activated Charcoal Adsorption: Carbon dioxide will adsorb onto the large surface areas of activated charcoal. When present in relatively concentrated form, it can be partially separated from components that do not adsorb by passing the gas mixture under pressure over the surface of the solid. Later, the gas pressure is greatly reduced and the CO$_2$ desorbs and can be collected ("pressure-swing method"); alternatively, the solid can be heated to release the gas ("temperature-swing method").	The process is not completely selective for CO$_2$.

The pure oxygen needed for oxycombustion is produced by cryogenic (very low temperature) separation of the gas from nitrogen and argon in pure air.

However, **the original isolation of oxygen from air requires energy comparable in amount to that required to separate dilute CO$_2$ from emissions of conventional power plants.** In addition, the combustion facilities must be redesigned in order to use pure oxygen rather than air. In practice, since combustion in pure O$_2$ produces a flame too hot (3500°C) for power-plant materials, it is diluted with some of the CO$_2$ from the combustion to reduce the flame temperature.

6.14 Other Concentrated Sources of Carbon Dioxide

Another proposed scheme involves the conversion of a fossil fuel, either coal or natural gas, to **hydrogen gas,** H$_2$, which would be employed as the fuel, either in a power plant or in a vehicle, in a reaction that generates no additional carbon dioxide. In essence, the fuel value of the coal or natural gas is transferred to hydrogen by the gasification process. Such techniques for generating H$_2$ for use as a fuel are described in detail in Chapter 7; in general, the overall process corresponds to the reaction

$$\text{carbon-hydrogen fuel} + \text{water} \longrightarrow CO_2 + H_2$$

The high concentration of pressurized CO$_2$ in the mixture of gaseous products (in principle, 50% by volume) allows for a more economical separation of the carbon dioxide and hydrogen, in this case using a liquid glycol solvent, than the capture of carbon dioxide from emission gases in a conventional power plant. Alternatively, the separation may be accomplished via a membrane that allows only hydrogen to pass through, producing a gas stream largely composed of CO$_2$ (Table 6-4).

Other industrial processes in which carbon dioxide at relatively high concentrations can be separated by membrane techniques include natural gas purification and fermentation plants.

The cement manufacturing industry is responsible for 7% of global CO$_2$ emissions. The concentration of CO$_2$ in emission gases from cement plants, which produce the gas by heating calcium carbonate to release calcium oxide, reaches 15–30%. These emissions should thus be susceptible to more economical capture of carbon dioxide than the more diluted emissions from power plants. For example, carbonate looping (Table 6-4) is suited to the cement-manufacturing industry.

The Storage of Carbon Dioxide

A number of different methods and locations for storing carbon dioxide have been proposed and are currently under investigation, as discussed in the following sections.

6.15 The Physical States of CO_2

In order to understand the options for transporting and burying carbon dioxide, it is vital to know the physical phases in which it exists under various pressure and temperature conditions, and the complicating factor of the presence of water. Graphically, this information can be obtained from the *phase diagram* for pure CO_2 (Figure 6-9).

Under ambient conditions (about 1 atm or 1 bar, and 20°C), carbon dioxide exists as a gas. Unlike most substances, no amount of cooling produces a liquid state for it at atmospheric pressure—instead it turns directly into a solid ("dry ice"). This can be seen on the phase diagram in Figure 6-9, where an imaginary horizontal line through ambient conditions (intersection of 1 bar and mid-way between 0 and 50°C) does not intersect the liquid region that lies above it.

Proposals to store CO_2 underground or in deep seas would necessarily involve the very high pressures that are inherent in those environments. Thus the appropriate portion of the phase diagram lies well *above* that for

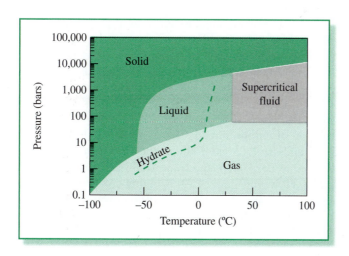

FIGURE 6-9 Phase diagram for carbon dioxide. Note that the pressure scale is logarithmic, and that 1 atm is approximately equal to 1 bar. [Source: Adapted from NOAA Pacific Marine Environmental Laboratory.]

ambient conditions. **At moderately high pressures,** we see the following from the phase diagram:

• **Carbon dioxide is a liquid, especially at cool temperatures a few dozen degrees on either side of 0°C.** This includes the temperature range for seawater.

• **Carbon dioxide becomes a supercritical fluid under very high pressures (>72 bar, 73 atm) and warm temperatures (>31°C),** as are encountered very deep underground.

Supercritical fluids have properties that lie between those of gases and those of liquids; the former are displayed at relatively low pressures and the latter at higher pressures. Thus the density of supercritical CO_2 varies over a considerable range. It is important to realize that **liquid carbon dioxide is much more compressible than is liquid water;** at less than about 270 bar, CO_2 is *less* dense than seawater, but at higher pressures it is *more* dense. In water bodies, pressure increases by 1 bar for every 10 m of depth, so **liquid carbon dioxide is denser than water below about 2700 m,** the exact depth being somewhat dependent on water temperature.

6.16 Direct Deep-Ocean Disposal of CO_2

Concentrated carbon dioxide destined for storage as such in the ocean could be transported by a pipeline, either originating from the shore or from a ship stationed above the disposal site, extending to the depth required (Figure 6-10). The average depth of the ocean is 3800 m and the oceans cover 70% of the Earth's surface, so there is a huge potential for CO_2 storage in the seas.

Shallow injection of the gas in the ocean, at 200–400-m depth, could produce a satisfactory result, provided that the seafloor at the site is slanted sufficiently to allow the dense, CO_2-rich water to be transported by gravity to greater depths. However, simulations show that most of the gas would return to the surface and enter the atmosphere within a few decades if the CO_2-rich water is simply diluted by mixing with surrounding water, rather than sinking. Over a period of centuries, excess carbon dioxide would eventually return to the atmosphere, but presumably by that time alternative energy sources would have replaced fossil fuels and the atmospheric CO_2 problem would then be less serious.

The fraction of carbon dioxide that is retained, at least for centuries, by the ocean tends to be greater the deeper it is placed. If it is deposited below about 500 m deep, water pressure would be sufficient to liquefy pure carbon dioxide. However, liquid CO_2 is highly compressible and much less dense than water, so any that did not dissolve would float upward to the surface. **Below about 2700 m deep, liquid carbon dioxide is compressed so much that it is denser than water and would sink.**

Between 2500- and 3000-m depth, liquid CO_2 has a density close to that of seawater and therefore is buoyant.

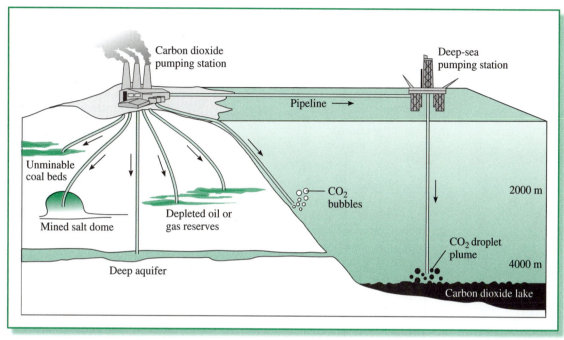

FIGURE 6-10 Potential sequestration sites for carbon dioxide. [Source: Redrawn from *Scientific American* (Feb. 2000): 72–79.]

However, since ocean temperatures are less than 9°C, the liquid or concentrated gas at depths below 500 m could combine with water to form the solid, ice-like *clathrate hydrate*, $CO_2 \cdot 6H_2O$ that, if fully formed, is denser than seawater and would sink to the deep ocean. The hydrate would subsequently slowly dissolve, and form CO_2-rich seawater.

Direct disposal to the deep ocean of CO_2 would require a pipeline to penetrate to a depth of 3000–5000 m, producing a pool of liquified carbon dioxide, which is denser than seawater at this depth. Some—perhaps just the surface—or all of the liquid carbon dioxide would convert to the solid clathrate. Figure 6-11 illustrates an experiment to test this hypothesis in which a beaker of liquid carbon dioxide was placed at almost 4000-m depth off Monterey Bay, California. Hydrate formed at the interface between water and liquid CO_2 and then sank to the bottom of the beaker since its density is greater than the liquid's.

Any pool of liquid carbon dioxide at the ocean bottom would, probably over centuries, dissolve into the surrounding water. Some of it is also expected to seep down into the sediments since it is denser than the seawater that it would displace. Unfortunately, sea life under this pool would be exterminated. There is also some fear that earthquakes or asteroid impact could destabilize

FIGURE 6-11 Liquid carbon dioxide overflowing from a beaker placed on the seafloor at 3650-m depth. A mass of transparent hydrate formed at the upper interface, sank to the bottom, and pushed out some of the liquid CO_2. [Source: P. G. Brewer et al., "Direct Experiments on the Ocean Disposal of Fossil Fuel CO_2," *Science* 284 (1999): 943.]

FIGURE 6-12 Capacities and storage times for various CO_2 sequestration technologies. [Source: Adapted from K. S. Lackner, "A Guide to CO_2 Sequestration," *Science* 300 (2003): 1677.]

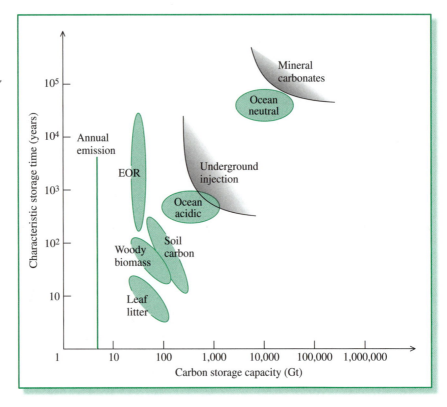

the pool, resulting in the release of massive amounts of carbon dioxide gas into the air above.

The various schemes for delivering carbon dioxide in massive amounts to the seas and depositing it there as such are collectively labelled "ocean acidic" in Figure 6-12, and have the potential to store many hundreds of gigatonnes of carbon dioxide for many hundreds of years. The schemes are referred to as acidic because dissolving carbon dioxide gas directly in seawater produces carbonic acid, H_2CO_3, a weak acid that would increase the acidity of the ocean in the near vicinity:

$$CO_2(g) + H_2O(aq) \rightleftharpoons H_2CO_3(aq)$$

The pH of ocean water would be lowered by tenths of a unit by the addition of large amounts of carbon dioxide, although much larger decreases of several pH units would occur near the points of injection.

Near the seafloor, dissolved carbon dioxide or carbonic acid could eventually react with the solid calcium carbonate, $CaCO_3$, in sediments, formed from seashells, etc. to produce soluble **calcium bicarbonate**, $Ca(HCO_3)_2$:

This reaction is discussed in detail in Chapter 10.

$$H_2CO(aq) + CaCO_3(s) \longrightarrow Ca(HCO_3)_2(aq)$$

For practical purposes, the CO_2, now chemically trapped in the bicarbonate form, would remain dissolved indefinitely.

An additional scheme proposed for carbon dioxide storage is by injection into sediments that lie under the seafloor. Because it is present under high pressure and low temperature, the CO_2 would exist in the sediment as a dense liquid or would combine with the water in the sediments and occur as the solid hydrate. Although expensive, this injection process could be useful for power plants in coastal locations.

6.17 Disposal of Neutralized CO_2

In the scheme labelled "ocean neutral" in Figure 6-12, calcium carbonate would first be reacted with the carbon dioxide to transform it to aqueous calcium bicarbonate by the same reaction shown above that occurs on the seafloor. The resulting solution would be then drained into ocean depths. **Acidity problems associated with direct carbon dioxide dissolution in seawater are avoided in this way.**

Huge amounts of calcium carbonate in the form of the mineral *limestone* would be required for this form of sequestration, but the potential for CO_2 storage by this method is very great and the storage time of the gas is many thousands of years (Figure 6-12). In addition, it may be possible to react power-plant emissions *directly* with the mineral, thereby avoiding the energy-expensive step of extracting and concentrating the carbon dioxide.

PROBLEM 6-4

Calculate the mass, in tonnes, of calcium carbonate that is required to react with each tonne of carbon dioxide.   ●

Alternatively, surface rocks containing alkaline silicates could be crushed and then reacted with carbon dioxide to produce insoluble solid carbonates that could simply be buried in the ground. These abundant minerals are substances whose empirical formulas can be written as combinations of SiO_2 and one of the alkaline oxides CaO, MgO, and FeO. Such minerals will react with CO_2 to displace *silicon dioxide* and form carbonates. For example, the silicate Mg_2SiO_4 would react as follows to produce *magnesium carbonate*, $MgCO_3$:

Silicate minerals do not actually contain the metal oxides as structural components, as shall be discussed Chapter 16.

$$(MgO)_2SiO_2 + 2CO_2 \longrightarrow 2\,MgCO_3 + SiO_2$$

Unfortunately, such direct *carbonation reactions* involving CO_2 are slow to occur unless the mineral is heated, a step that is costly in money and energy. In one indirect scheme, the magnesium silicate rock is first reacted with *hydrochloric acid*, HCl, to produce silicon dioxide and *magnesium chloride*, $MgCl_2$. Subsequent reaction of this salt with carbonic acid produces insoluble magnesium carbonate and reforms the hydrogen chloride that, in principle, can be recycled:

$$(MgO)_2SiO_2 + 4\ HCl \longrightarrow 2\ MgCl_2 + 2\ H_2O + SiO_2$$

$$MgCl_2 + H_2CO_3 \longrightarrow MgCO_3 + 2\ HCl$$

There are still energy costs and CO$_2$ production emissions associated with such procedures, however.

6.18 Deep Underground Storage of CO$_2$

A number of schemes by which massive quantities of carbon dioxide could be "buried" underground have been proposed and, in some cases, tested.

- Depleted oil and gas reservoirs could be used to store carbon dioxide (Figure 6-10). These underground caverns are known to be stable, since they have held their original materials for millions of years. A demonstration project using this technique is underway in Australia. Carbon dioxide storage in coal seams that lie too far underground to be mined may also be feasible. Pumping CO$_2$ into the coal helps it release adsorbed methane, which then can be pumped to the surface and used. Several hundred gigatonnes of carbon dioxide could be sequestered in coal.

- In the short run, the easiest route to begin sequestration of carbon dioxide is probably to inject it into reservoirs containing crude oil or natural gas. The total capacity for carbon dioxide storage by such **enhanced oil recovery,** labelled "EOR" in Figure 6-12, is less than 100 Gt, however. This technology is already used to enhance the recovery of oil in some fields, although currently most of the CO$_2$ is recovered and reused. In an interesting international project that began in 2000, 5000 tonnes a day of liquefied CO$_2$ is sent through a 325-km-long pipeline from a coal gasification plant in North Dakota to an oilfield in Weyburn, Saskatchewan. The gas is injected through pipes 1500 m underground, allowing more oil to be extracted from the source and sequestering most of the carbon dioxide in the brine of the oil reservoir. The injected CO$_2$ that accompanies the oil extracted from the field is captured and re-injected underground.

  Monitoring of the Weyburn site has found no significant amount of the gas to be escaping through the rocks and soil of the area. Any injected carbon dioxide escaping from the area could be identified by the characteristic low $^{13}C/^{12}C$ isotope ratio of the North Dakota fossil-fuel source; artificial tracer gases were also added to the injected gas to further monitor potential emissions from the site. Analysis of the fluid that accompanies oil produced in this way revealed that the carbon dioxide has been dissolving in the reservoir brine and been trapped by some of the minerals in the reservoir.

• The CO_2 output from power plants could be pumped deep underground, or deep under the seafloor, into regions where cracks and pores in common alkaline rocks such as *calcium aluminosilicates* could react with the gas, in microorganism-catalyzed processes, to produce calcium carbonate and thereby store the CO_2. Such carbonate minerals are known to be present in deep caves in Hawaii and elsewhere, so the process could well be feasible if the reactions occur quickly enough. In 1996, Norway began to pump concentrated CO_2 gas into sandstone rocks located a kilometer below the North Sea; the pores in the rock were left empty by extraction of natural gas from them in the past. The gas could potentially react with the rock and thus be immobilized.

The heading Underground injection in Figure 6-12 shows the large scope for storage of such types.

• Much larger in volume and capacity than oil and gas reservoirs are *saline aquifers*, large formations of porous rocks that are saturated with salty water and that lie far underground, well below fresh-water supplies. Carbon dioxide injected into such an aquifer initially remains as a compressed gas or a supercritical liquid, but it is somewhat soluble (about 5%), so much of it will slowly dissolve in the sometimes very alkaline brine (Figure 6-10). In aquifers that contain alkaline silicate minerals, the resultant carbonic acid can react with the minerals in the rocks to form stable, insoluble carbonate precipitates, as discussed above.

The brine is contained mainly in small pore spaces occupying about 10% of the volume in the soil. If carbon dioxide is to be stored underground, a depth of more than a kilometer is required so that its density is comparable to that of water. Even so, it is likely to rise to the top of the geological formation in which it is enclosed, and to spread laterally. Consequently, the natural *caprock* overlying the formation must be one that is secure if significant amounts of CO_2 are not to migrate upward through the soil and into the atmosphere over time. The stability of each aquifer to potential seismic instability and gas leakage would have to be individually assessed before it is used.

The Norwegian energy company *Statoil* has demonstrated the feasibility of this approach by storing annually, since 1996, about a million tonnes of carbon dioxide (a 9% impurity in the crude natural gas) in a saline aquifer that lies 1000 m under the floor of the North Sea. The North Sea aquifers are sufficiently large to absorb all emissions of carbon dioxide produced by Europe for many hundreds of years. A 2008 report by the company stated that no leakage of CO_2 from the site has occurred, and that, as expected, CO_2 is spreading through the structure.

Experiments to store carbon dioxide in saline aquifers are underway in several locations in the world, including Texas and Japan. In the United States, large saline aquifers are found in the states lying just below the lower Great Lakes, in southern Florida, northeastern Texas, and northern mid-west states.

Review Questions 15–22 are based on the material in the preceding sections.

ACTIVITY

Perform an internet search to discover which *carbon capture and sequestration* (CCS) projects are currently underway around the world. Classify each one in terms of the technique used to capture and to sequester the gas. Be sure to use only up-to-date information, as several of the CCS projects proposed and planned in the past have now been cancelled. As a starting guide, consider the summary given in Figure 4 of the review by R. S. Haszeldine, *Carbon Capture and Storage: How Green Can Black Be?*, Science 325, 1647–1652 (2009).

Other Schemes to Reduce Greenhouse Gases

Several other notable schemes to reduce the emissions of carbon dioxide and other greenhouse gases, or even to remove CO_2 directly from ambient air, have been proposed, and are outlined below.

6.19 Removing CO_2 from the Atmosphere

A potential technique for extracting some of the carbon dioxide that already is dispersed into the atmosphere and depositing it in ocean depths is by iron fertilization. Large portions of the seas, especially the tropical Pacific and the Southern oceans, lack plankton because they are very iron deficient. Artificially adding iron to these areas would result in massive blooms of plankton, some of which, in the Southern Ocean (which surrounds Antarctica) at least, would descend into the deep oceans, thereby locking away the carbon dioxide that it had used in photosynthetic activity.

Small-scale experiments have been undertaken to test the feasibility of oceanic iron fertilization. However, some scientists have warned that fertilization on a large scale would greatly disrupt the ocean ecosystem, perhaps in unpredictable ways. In addition, the decomposition of the phytoplankton consumes oxygen and encourages bacteria that produce methane and nitrous oxide, thereby potentially increasing the concentration of these greenhouse gases in air upon their release from the oceans. In any case, recent results indicate that iron fertilization does *not* have great potential for sequestering carbon.

Carbon dioxide can also be removed from the atmosphere by growing plants specifically for this purpose. Some utility companies and some countries have proposed a scheme by which they are given credit to offset some of their CO_2 emissions by planting forests that would absorb and temporarily sequester carbon dioxide as they grew. However, the crediting of carbon dioxide sequestering by growth of biomass is controversial. Of course the carbon stored in trees would be released back into the atmosphere if the wood burned or rotted.

A few scientists have proposed removing carbon dioxide chemically from ambient air—for example, from the high-speed wind already used to turn turbines—using chemicals such as the amines already discussed or some other chemical absorber. The practicality of chemically extracting carbon dioxide from air, in which its concentration is only 0.04%, compared to about 13% in power-plant emissions, is yet to be determined. Such *Direct Air Capture* schemes are discussed in Box 6.4.

BOX 6-4   Removing CO$_2$ from the Atmosphere: Direct Air Capture

Some scientists and policymakers have researched the feasibility of removing carbon dioxide from ambient air—rather than from a power-plant emission stack—and then sequestering it using one of the techniques described in Sections 6.16 and 6.17. Indeed, a number of capture schemes have been proposed and several are now being tested on a (very) small scale.

A principal difficulty in removing CO$_2$ from air is its very low concentration there—about 0.04% compared to the levels several hundred times greater in power-plant emissions. The thermodynamic limit of the minimum energy required to extract a dilute gas from a mixture is proportional to the *logarithm* of its partial pressure, so the difference in required energy amounts only to a factor of about 3, not several hundred. Nevertheless, if the energy required in the various stages of the extraction is generated mainly by the combustion of fossil fuels, there exists the possibility that the corresponding CO$_2$ emissions could exceed the amount of the gas extracted!

The proposed extraction processes operate by having ambient air blow over large-area vertical structures ("carbon trees") that are covered with substances that absorb CO$_2$ by reacting with it. Most schemes propose the use of electrically driven fans to blow the air over the absorber and hence use energy, but some have suggested placing the structures in windy areas where no fans would be needed. Once sufficient gas had been extracted, the absorber would be heated to release the CO$_2$ in concentrated form.

The prototype scheme suggested for Direct Air Capture of carbon dioxide employs an aqueous solution of sodium hydroxide, NaOH, as the absorber, as it is known that this base reacts efficiently with the gas:

$$2\,NaOH(aq) + CO_2(g) \longrightarrow Na_2CO_3(aq) + H_2O(aq)$$

It is not feasible to decompose even a concentrated sodium carbonate solution directly owing to the large amount of energy needed to evaporate the water and dehydrate the salt. Instead, the solution is reacted with calcium hydroxide, Ca(OH)$_2$, since its calcium ion combines with carbonate ion to produce insoluble calcium carbonate, CaCO$_3$, which then can be readily removed by filtration—rather than energy-intensive evaporation—from the aqueous solution. Note that NaOH is thereby also generated and the remaining solution can then be used again, to capture more carbon dioxide:

$$Na_2CO_3(aq) + Ca(OH)_2(aq) \longrightarrow CaCO_3(s) + 2\,NaOH(aq)$$

Unfortunately, the energy required to dehydrate and especially to decompose the calcium

(continued on p. 266)

carbonate is large, the latter as previously discussed in Table 6.4, though some is regained when the resulting calcium oxide, CaO, is rehydrated:

$$CaCO_3(s) \longrightarrow CaO(s) + CO_2(g)$$

The carbon dioxide produced from generating the 900°C heat required for this latter reaction by burning coal or natural gas would negate much of that captured in the process unless it was captured and sequestered by a CCS procedure.

Several scientists are currently researching alternative absorbers by which CO_2 can be captured and then released in concentrated form that do not have the high energy requirements of the NaOH scheme. Unless the energy requirements can be reduced substantially, however, it is unlikely that Direct Air Capture of carbon dioxide will be widely employed until more abundant emission sources of the gas have been eliminated by conversion to renewable energy and/or CCS is widely implemented.

6.20 Reducing CO_2 Emissions by Improving Energy Efficiency

Much of the current energy expended on both industrial and domestic scenes could be saved by adopting the most *efficient* technologies for each purpose. For example, the use of low-wattage compact fluorescent light bulbs instead of incandescent ones would significantly reduce the amount of electrical energy used for lighting in homes; the monetary *payback period*— the time required for the much higher capital cost of these bulbs to be matched by savings in electrical costs—is a few years. Similarly, automobiles could be made much more energy efficient and thus use less gasoline to travel a given distance.

However, improving energy efficiency would *not* necessarily lead to a reduction in the demand for energy and a reduction in carbon dioxide emissions. If energy-consuming equipment is made more efficient, the monetary cost for performing a given task drops, and there follows a natural tendency to use the equipment more, since it is so cheap to operate. For example, if you buy a very energy-efficient car, you will be able to afford to take more trips in it since each one would be cheaper than with a "gas guzzler." Thus some policymakers believe that because of this *rebound effect*, energy savings and CO_2 reductions from efficiency would not be achieved in the long run by making energy-consuming devices more efficient. Increased efficiency would have to be accompanied by price increases on the fuel, presumably in the form of taxes, for it to reduce overall consumption.

6.21 Reducing Methane Emissions

Although our focus so far has been on reducing emissions of carbon dioxide to the atmosphere, global warming can also be combatted by decreasing the amount of methane that is released. The major opportunities for methane reduction are

• better maintenance of natural gas pipelines so as to reduce their leakage, a practice already under way in Russia;

• capture and combustion of the methane released by landfills, underground coal mines, and oil production; and

• changes in the techniques of rice production, by draining the field a few days before the plants flower, the point at which maximum emissions begin.

Carbon Dioxide Emissions in the Future

The most serious long-range global environmental problem associated with energy use is the release into air of CO_2 when fossil fuels are combusted to produce heat. Carbon dioxide is produced when *any* carbon-containing substance undergoes combustion:

$$\text{C in substance} + O_2 \longrightarrow CO_2$$

As we saw in Chapter 5, CO_2 is the most important greenhouse gas, and the increase in its atmospheric concentration is responsible for the largest fraction of global warming of any anthropogenic factor.

6.22 Past Growth in CO_2 Emissions

Since the Industrial Revolution, the rate of emission of CO_2 to the atmosphere has climbed hand-in-hand with the expansion of commercial energy use, since so much of the latter (currently 88%) is obtained from fossil-fuel sources. Barring an unforeseen, massive development of nuclear energy or renewable fuels or carbon sequestration, CO_2 emission rates are expected to match commercial energy rate increases as the developing world undergoes rapid industrialization. Indeed, developing nations collectively now emit more carbon dioxide than do developed ones. However, if developing countries could implement the renewable energy technologies to be discussed in Chapters 7 and 8 while expanding their economies, they would avoid some of the heavy fossil-fuel dependence and intense carbon dioxide emissions characteristic of all currently developed countries.

By consulting government documents on the internet or in a library, determine how the use of oil, coal, and natural gas and other fossil fuels has varied over the last few decades in your country. If CO_2 emissions from these sources are also available, compare their trends to the global ones shown in Figure 6-3. Prepare a report summarizing your findings.

Developed countries accounted for about three-quarters of all carbon dioxide emissions from fossil-fuel combustion and cement manufacture from the beginning of the Industrial Revolution to the end of the millennium. The more recent emissions from the five top source countries, from 1990 to 2009, are shown in Figure 6-13. Notice that:

- The United States historically was the biggest emitter, but now has been surpassed by China, whose emissions took off sharply in the early 2000s.

- Emissions from two developed countries in the Top 5, the United States and Japan, rose slowly in the early years of the 2000s, but fell during the 2008–2009 recession.

- Emissions from Russia decreased significantly in the early 1990s, after the collapse of the economy of the Soviet Union.

- Emissions from India have risen significantly.

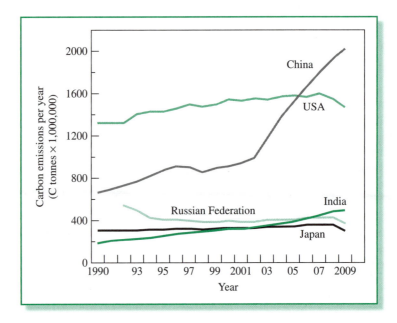

FIGURE 6-13 Carbon dioxide emissions for the Top 5 emitting countries. Note that emissions are given in terms of carbon content. [Source: Adapted from *Global Carbon Budget 2009*, Global Carbon Project.]

6.23 Carbon Intensity

Almost all the increase in CO_2 emissions has resulted from the expansion in the energy derived from fossil fuels, demand for which, as we have seen, is driven primarily by the production of goods and services. However, the ratio of carbon dioxide to energy varies among countries and over time because of differences in the fraction of energy produced by combustion of fossil fuels and because different fossil fuels produce quite different amounts of carbon dioxide per joule.

To a first approximation, the amount of heat released when a carbon-containing substance burns is directly proportional to the amount of oxygen consumed. From this principle, we can compare different fossil fuels by the amount of CO_2 released when they are burned to produce a given amount of energy.

Consider the reactions of coal (mainly carbon), oil (essentially chains of CH_2), and natural gas (essentially CH_4) with atmospheric oxygen, written so that the amount of carbon dioxide produced is identical in each case:

$$C + O_2 \longrightarrow CO_2$$

$$CH_2 + 1.5\ O_2 \longrightarrow CO_2 + H_2O$$

$$CH_4 + 2\ O_2 \longrightarrow CO_2 + 2\ H_2O$$

It follows from the stoichiometry of these reactions that, per mole of O_2 consumed and thus approximately **per joule of energy produced, natural gas generates less carbon dioxide than does oil, which in turn is superior to coal,** in a ratio of 1:1.33:2. (The actual ratio is computed in Problem 6-5.)

Overall then, the carbon dioxide produced per unit amount of energy from different sources varies with the source in the following order:

$$\text{coal} > \text{oil} > \text{natural gas} \gg \text{renewable and nuclear}$$

Some policymakers have promoted natural gas as a "bridge fuel" to replace coal and petroleum in the transition to a low-carbon economy—eventually using renewable energy sources—since it generates less CO_2 per unit energy. As we shall see in Chapters 7 to 9, there are minor emissions of CO_2 in the production of energy equipment (such as windmills and solar panels) that capture renewable energy and generate nuclear energy, although no emissions arise directly from their operation.

The replacement of diesel fuel by natural gas to fuel trucks is a current initiative.

As illustrated in Figure 6-3, the fraction of fossil-fuel energy obtained from coal is expected to *increase* in the future—due to higher and higher prices for oil and natural gas as they become more scarce—thereby increasing the carbon content of the fuel mix.

PROBLEM 6-5

Given the thermochemical data below, determine the actual ratio of CO_2 per joule of heat released upon the combustion of methane, CH_2, and elemental carbon (graphite). ΔH_f values in kJ mol^{-1} are CH_4: –74.9; $CO_2(g)$: –393.5; $H_2O(l)$: –285.8; C(graphite): 0.0; CH_2: –20.6.   ●

PROBLEM 6-6

The relative amounts of oxygen required to oxidize organic compounds to carbon dioxide and water can be deduced from calculating the *change* in the oxidation number (state) of the carbon atom in going from the fuel molecule to the product. Show that the ratio of oxygen required to combust C, CH_2, and CH_4 stands in the ratio of 2:3:4 according to such a calculation. What would the ratio be if the formula of coal was taken to be CH, rather than C?   ●

Because of the importance of CO_2 emissions to our climate, the term **carbon intensity** has become widely discussed by policymakers. Two versions of this parameter are in use:

1. The carbon intensity of the economy is defined as the ratio of CO_2 emitted per unit of gross domestic product, and thus is analogous to the concept of energy intensity discussed in Section 6.2. In shorthand, it equals CO_2/GDP.

2. The carbon intensity of the energy supply is defined as the ratio of CO_2 emitted per unit of energy consumed. In shorthand, it equals CO_2/energy.

The two definitions of carbon intensity are simply related. Since

carbon intensity (economy) = CO_2/GDP

= (CO_2/energy) × (energy/GDP)

therefore

carbon intensity (economy) = carbon intensity (energy) × energy intensity

The **Kaya identity** relates the carbon dioxide emissions of a country (or the world) to the intensity factors:

CO_2 emission = population × (GDP/population) × (energy use/GDP) × (CO_2 emitted/energy)

CO_2 emission = population × (GDP/population) × energy intensity × carbon intensity (of the energy)

The *carbon intensity of energy* varies significantly from one country to another, depending upon the proportion of the energy derived from renewable and nuclear versus fossil-fuel sources, and upon which fossil fuel is the dominant one. Globally, this carbon intensity of energy has not displayed much variation at all in the last few decades, but is predicted to decrease slightly in coming years before levelling off again as the use of all types of energy will increase.

Since energy intensity has declined, but the carbon intensity of energy has remained almost constant, the global *carbon intensity of the economy* declined over the last half of the twentieth century, especially over its last two decades when it fell by about one-quarter; however, it remained approximately constant from 2000 to at least 2005. Currently, the production of one dollar's worth of goods and services results on average in the emission of about 730 g of carbon dioxide. Carbon intensities are usually expressed as the carbon content alone of the emitted CO_2, or about 200 g of carbon per dollar in the present case.

PROBLEM 6-7

Show that the carbon content of 730 g of CO_2 is 199 g.   ●

The time evolution of carbon intensities for the world's largest two economies—the United States and China—are quite different. The American carbon intensity has fallen gradually and almost continuously, whereas that of China fell precipitously once major industrialization began, reached a minimum in about 2000, and rose significantly at least in the half-decade that followed. Presumably, much of the decrease in the U.S. intensity is due to the continuing conversion of its economy from the production of manufactured goods to a knowledge economy, whereas the increase in China arises from its development as an energy-intensive economy that produces large quantities of manufactured goods.

The driving forces behind the changes over the last few decades in global and regional CO_2 emissions can be understood by considering the various factors involved. As the world economy developed, the average GDP per person also rose, as did the global population (see the thin black line and dashed green lines in Figure 6-14).

The product of these two quantities is the global GDP, which rose in a faster-than-linear fashion since each of these factors rose more-or-less linearly with time. In the period until about 2000, the carbon intensity of the global GDP (green line in Figure 6-14) *declined* almost linearly with time, so the total carbon dioxide emission rate—which is a product of the three factors—rose only gradually and more-or-less linearly in that period (heavy black curve in Figure 6-14). However, since the carbon intensity did not fall from 2000 to 2005, the emission rate of carbon dioxide rose dramatically. The continuing strong rise in per capita GDP in China, combined with the increasing carbon intensity from 2000 to 2005 and again in 2010, resulted in sharply increasing emissions.

The term **carbon footprint** has become a popular way of assessing the total CO_2 emissions associated with a particular activity (such as an airplane flight) or manufactured product (such as an automobile). The term can also be applied to individual nations by considering, not the amount of gas emitted directly from a country, but rather the amount associated with goods and

Of course, the CO_2 emissions rose mainly because of the increase in energy use as economies expanded, as discussed in Section 6.2, without much change in the fuel mix.

Note that all values in Figure 6-14 are relative to those in 2000, each taken to be 1.0.

FIGURE 6-14 Global CO_2 emissions (heavy black curve; includes both fossil-fuel combustion and cement production) and important factors that influence it. Dashed green curve: population; thin black curve: per capita GDP; solid green curve: carbon intensity of GDP. [Source: M. Raupach (Global Carbon Project), *Carbon in the Earth System: Dynamics and Vulnerabilities*, Beijing, November, 2006.]

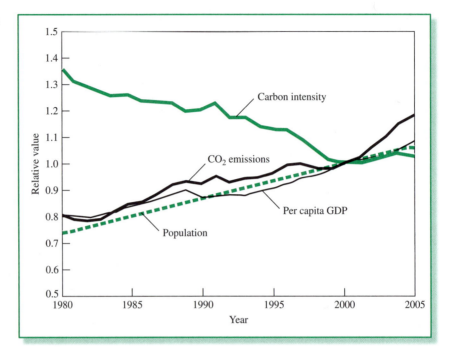

services that are consumed by people and organizations within them. Thus, for example, emissions resulting from the energy-intensive manufacture of a computer or TV in China for use in a developed country would be "charged to" the latter, not to China. Not unexpectedly, the resulting national carbon footprints are found to correlate more strongly with a country's *expenditure* rather than with its GDP. The footprint also shows some correlation with the need for heating in cold countries and for cooling in developed countries. Some economists believe that the cost of producing and upgrading energy ought to be charged to nations which ultimately use it; this would reduce the footprint value of resource-producing nations such as Australia, Russia, and Canada.

6.24 Per Capita Carbon Dioxide Emissions

A *tonne*, the metric ton, is 1000 kg, i.e., 2200 lb, whereas a conventional ton is 2000 lb.

On average, each person in the industrial countries is responsible for the release of about 11 tonnes of CO_2 from carbon-containing fuels each year!

Some of that carbon dioxide output is direct—for example, that released in the exhaust gases when vehicles are driven and homes are warmed by burning a fossil fuel. The remainder is indirect, arising when energy is used to produce and transport goods; to heat and cool factories, classrooms, and offices; to produce and refine oil; and in fact, to accomplish virtually any constructive economic purpose in an industrialized society.

Another way of considering the carbon dioxide emissions from different countries is to consider the per capita releases of this gas into the atmosphere. Currently, the emissions of carbon dioxide amount to about 4 tonnes per person per year when averaged over the global population; this factor is usually expressed as 1 tonne of *carbon*, and it is this reference to carbon that will be used henceforth. The global average carbon emission remained remarkably steady, at about 1.1 tonnes (C), from 1980 to 2000, but increased by about 10% shortly thereafter owing to the jump in global energy usage.

There is considerable variation in the per capita releases among different industrialized countries. People in developed counties have much higher annual average emissions than do those in developing countries: 3 versus 0.5 tonne of carbon per person. The bar graph in Figure 6-15 lists in order the Top 20 emitter countries of carbon dioxide as of 2009. The dark green bars indicate the country's percentage of total global emissions, and the light green bars show the per capita emissions. China now is the largest net emitter, exceeding the United States in part because a greater fraction of its energy is produced by burning coal.

Notice in Figure 6-15 that the United States, Canada, and Australia have the highest per capita CO_2 emission rates—aside from Saudi Arabia, where most emissions result from oil production—in part due to the high transportation requirements of these vast lands, compared to other

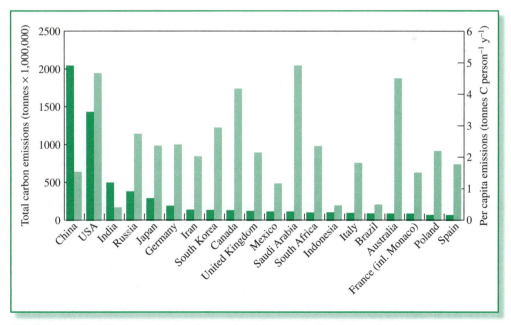

FIGURE 6-15 Total (dark green bars) and per capita (light green bars) emissions of carbon dioxide by Top 20 emitter countries in 2009. [Source: Adapted from *Global Carbon Budget 2009*, Global Carbon Project.]

developed countries that are compact and densely populated, e.g., European countries and Japan. In addition, in these three countries fossil-fuel energy is much cheaper than in European countries. Although Canada leads the United States and Australia in energy use per capita (Table 6-2), it follows them in CO_2 emissions because of its greater use of hydroelectric energy. The other developed countries have remarkably similar per capita annual carbon emissions—about 2 tonnes—which perhaps is the value currently developing countries will reach once they are fully developed. The per capita carbon emissions from the United States and the European Union, and from the world in total, remained remarkably constant over the last few decades of the twentieth century, their increases in GDP being matched by decreases in energy and carbon intensities.

Because populations of different countries vary so much, their greenhouse gas emissions per capita or per dollar of GDP are no guide to their *total* emissions. Thus in Figure 6-13 we see that China and India make substantial contributions to the total global emissions since their populations are so large, even though their per capita emission rates are still quite modest. Both countries generate most of their electricity by burning coal. China's CO_2 emissions growth since the turn of the millennium has exceeded even the rapid growth of the 1980s and early 1990s. Since 1980, India's annual emissions have been growing almost linearly, currently by about 3–4%, although it still has the lowest per capita emissions of any of the Top 20 emitter countries.

6.25 Patterns of Growth in CO₂ Concentrations

Because carbon dioxide has such a long lifetime in the atmosphere, a century or more on average, the gas *accumulates* in air. Thus almost all the CO_2 emissions from the 1990s, for example, that did not find a temporary sink will remain in the air for decades to come, adding to the bulk of the emissions from the 1980s, the 1970s, and previous decades. The actual carbon dioxide molecules that constitute this additional mass will change from year to year, as some CO_2 molecules leave temporary sinks and enter the atmosphere, while an equal number from the air enter one or another such sink.

The growth pattern of the CO_2 concentration in air is determined mainly by the pattern of CO_2 emissions. Suppose, for example, that the same amount of carbon dioxide emissions was added to the air each year and did not find a temporary sink (Figure 6-16a, black line). The total amount of CO_2 in air—and hence its concentration—would then annually *increase* by a constant amount, a linear increase (Figure 6-16a, green line). If the world could hold its carbon dioxide emissions constant at the year 2000 value, then the CO_2 concentration would increase linearly, by the current 2-ppm yearly increment, and would consequently become about 500 ppm in 2060. Although not constant, current global CO_2 emissions are only increasing by about 1% annually. Thus the atmospheric carbon dioxide concentration is currently increasing at nearly a linear rate (see insert in Figure 5-8).

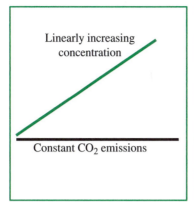

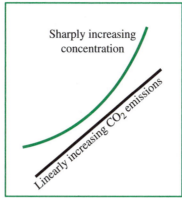

FIGURE 6-16 CO_2 concentration related to its emission level. (a) Constant emissions of CO_2 produce a linearly increasing concentration of the gas. (b) Linearly increasing emissions produce a quadratically increasing concentration of the gas.

Another scenario, which at some times in the past has been more realistic than the situation just described, is that the CO_2 *emissions* were not the same each year, but themselves *increased linearly*, i.e., by a constant amount k each year. Thus, if the emissions one year amounted to A, the next year they were $A + k$, and the following year $A + 2k$, etc. (Figure 6-16b, black line). In this case, if the fraction of the gas that enters the oceanic sink each year is constant, the growth in CO_2 *concentration* will be quadratic, much sharper than linear: the resulting plot of CO_2 concentration curves upward, as illustrated in Figure 6-16b (green curve).

Indeed, in the decades preceding the mid-1970s, the CO_2 concentration (see insert in Figure 5-8) did increase quadratically since carbon dioxide emissions were increasing rapidly. However, since then, the CO_2 concentration has increased in an approximately linear manner with time, reflecting the slower rate of increase in emissions in this period, among other factors—including the fact that the fraction of emissions that find a temporary sink varies with time (as was discussed in Chapter 5).

6.26 IPCC Scenarios for CO_2 Emissions and Concentrations

In its 2001 report, the IPCC described a number of very different scenarios for greenhouse gas emissions for the rest of the century. The magnitude of the emissions predicted at century's end vary dramatically: 5, 13.5, 20, and 29 Gt C annually, the range spanning from 0.6 to 3.5 times the current value of about 8 Gt Cy^{-1} . The carbon dioxide concentrations projected for 2100 for the IPCC scenarios range from 500 to more than 900 ppm, compared to today's 390 and the preindustrial 280-ppm levels.

Even with constant carbon dioxide emissions, at current levels or a few percent lower, the carbon dioxide concentration in the atmosphere will continue to grow. Some policymakers have promoted the idea that through international agreements or allocation schemes, the world should

Recall that Gt stands for gigatonnes, i.e., 10^9 tonnes.

FIGURE 6-17 Possible CO₂ emission scenarios to keep global warming from exceeding 2°C. The year that maximum emissions occur for each scenario are indicated by dots. [Source: *The Copenhagen Diagnosis Climate Science Report*, University of New South Wales Climate Change Research Centre.]

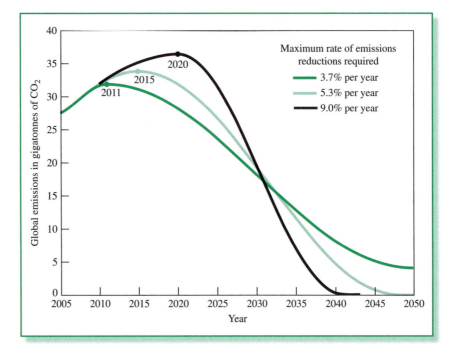

control future CO_2 emissions such that the atmospheric level of the gas never *exceeds* some specific *concentration*. Although there is no consensus as to what is the most appropriate target, for our discussion we shall use 550 ppm. This value is twice the preindustrial value—in other words, a situation in which human actions have doubled the natural atmospheric carbon dioxide concentration.

Several ways in which global CO_2 *emissions* could rise and fall with time and still probably limit global warming to 2 Celcius degrees are shown by the curves in Figure 6-17. The longer emissions are allowed to increase, the sharper the decrease that must follow their maximum levels since, as we have discussed, carbon dioxide accumulates in the air for many decades. It is *not* possible to defer emission reductions indefinitely if the concentration and/or temperature targets are to be achieved.

The Extent and Potential Consequences of Future Global Warming

As we saw in Chapter 5, the Earth's climate has already been affected by the enhancement of the greenhouse effect due to increasing atmospheric concentrations of carbon dioxide and other gases. The continuing buildup of CO_2 in the air leads to the conclusion that we are in store for further increases in global air temperatures and other changes to our climate.

In this section, we summarize what projections tell us on a qualitative basis about future climate changes and some of the consequences for human health.

Those of us who currently suffer through severe winters each year may look forward to the warmer climate associated with the enhanced greenhouse effect. After all, in the eleventh and twelfth centuries, an increase of a few tenths of a degree in the northern temperate zone was sufficient to allow farming on the coast of Greenland, for vineyards to flourish extensively in England, and for the Vikings to travel the North Atlantic and settle in Newfoundland.

However, the climate changes predicted for the twenty-first century and beyond do not present a uniformly pleasant prospect. The *rate* of change in our climate, which to date has been modest, will be dramatic by the middle of the century. Indeed, **the rapid rate of global change will probably be the greatest problem with which humanity will have to contend.** A more gradual transition, even to the same end result, would be much easier to handle, for it would allow humans and other living organisms more time to adapt to these changes.

It is very difficult for scientists to model the climate—even with the assistance of the fastest computers in the world—in order to make definitive statements about what changes will occur in particular regions in the future. We know that there will be substantial changes in the climate, but we are unable to specify exactly what they will be.

6.27 Predictions for Climate Change by 2100

By altering the concentrations of the gases and particles in the Earth's atmosphere that largely control the climate, we are conducting a giant experiment on our planet. As we discuss below, some of the predicted *external costs* of the experiment will be rather high, and invoking the *precautionary principle* to guide our actions might well be a wise course of action. Since the costs are oblivious to geopolitical boundaries, climate change is a global example of the *tragedy of the commons*.

The terms in italics here were defined in Table 0-1 on pages xxiii–xxiv.

The significant changes in the Earth's climate that have occurred in the past half-century and are predicted to continue over the twenty-first century, and that are likely to have been caused primarily by anthropogenic effects, as judged by the U.N.'s International Panel on Climate Change (IPCC) in its 2007 report, are summarized in Table 6-5.

Although the consequences of the global experiment involving greenhouse gases will only be revealed by the passage of time, scientists are employing ever increasingly sophisticated computer models to make predictions about how the climate will change. According to simulations of the future climate reported by the IPCC in 2007, the rise in the average global air temperature by 2100 (compared to 1990) could be as small as 1.4°C or as high as 4.0°C. As in the twentieth century, more of the warming will occur with night-time temperatures. The magnitude of the temperature increase will

TABLE 6-5	Recent Trends and Projections for Extreme Weather Events That Have Undergone Recent Change		
Phenomenon and Direction of Trend	Likelihood That Trend Occurred in Late 20th Century (Typically Post 1960)	Likelihood of a Human Contribution to Observed Trend	Likelihood of Future Trends Based on Projections for 21st Century
Warmer days and fewer cold days and nights over most land areas	Very likely	Likely	Virtually certain
Warmer and more frequent hot days and nights over most land areas	Very likely	Likely (nights)	Virtually certain
Warm spells/heat waves. Frequency increases over most land areas	Likely	More likely than not	Very likely
Heavy precipitation events. Frequency (or proportion of total rainfall from heavy falls) increases over most areas	Likely	More likely than not	Very likely
Area affected by droughts increases	Likely in many regions since 1970s	More likely than not	Likely
Intense tropical cyclone (hurricane) activity increases	Likely in some regions since 1970s	More likely than not	Likely
Increased incidence of extremely high sea level (excludes tsunamis)	Likely	More likely than not	Likely

Source: IPCC, *Climate Change 2007: The Physical Science Basis. Summary for Policymakers:* www.ipcc.ch

depend greatly on whether emissions (including those of sulfur dioxide) are controlled or not. At a minimum, however, **the world will warm more than twice as fast in this century as it did in the last.**

Part of the rather large range of the prediction is due to uncertainty about exactly how sensitive the climate is to carbon dioxide. Indeed, research reported in 2003 indicates that aerosols have countered more greenhouse warming in the past than previously thought; consequently, scientists may have significantly underestimated the sensitivity of temperature to CO_2, in which case the predicted increases will have to be revised upward.

An increase of a few degrees may seem small, but our current average air temperature is less than 6°C warmer than that in the coldest periods of the ice ages! Snow cover and sea-ice area will continue to decline in the future. There may well be enough melting of ice in the Arctic region for the Northwest Passage to be used for commercial transport, since the warming in winter of all Arctic regions is projected to be much greater than the global average.

Indeed, the Arctic regions of Alaska and western Canada warmed at the rate of 0.3–0.4°C per decade in the 1961–2004 period. The Arctic Ocean will probably become ice-free in the summer, a situation that has not occurred for at least one million years. A similar situation is occurring over the land, where global warming is now shortening the season when the land is snow-covered and has a much higher albedo than that of the soil and vegetation that becomes exposed in spring. In addition, most of the region of permafrost—land in northern Canada, Alaska, Siberia, and northern Scandinavia that stays frozen year-round—will likely melt to a depth of 3 m or more during this century.

The total amount of global rainfall is projected to increase, since more water will evaporate at the higher surface temperatures. The global average precipitation increases by about 2% for every centigrade-degree rise in temperature. Although the world overall will become more humid, some areas will become drier. To make matters worse, **most areas of the world that currently suffer from drought are predicted to become even drier.** Continental interiors at mid-latitudes will have continuing risk of drought in summer due to continued drying of the soil, the increased rate of evaporation from higher air temperatures being greater than the increase in the rate of precipitation. Subtropical areas will experience less precipitation, and equatorial regions and high-latitude regions will experience more, continuing the twentieth-century trends.

An increase in the average atmospheric temperature means that the air and water at the Earth's surface contain more energy and that more extreme weather disturbances could result, the effect of global warming that will affect many of us the most. **The number of days having intense rain showers or very high temperatures will both increase.** Storm wind intensities and heavy downpours will increase in some tropical areas.

6.28 Predictions Concerning Sea Levels

The sea-level rise that has been observed over the past decades (Figure 5-18), and the much more rapid increase predicted for the next century, result from a combination of several factors associated with global warming. The contributions from these factors over the last half-century are plotted in Figure 6-18 and are discussed individually below.

Although air and land surfaces are quick to warm with an increase in average global temperature, the same is *not* true of seawater. It takes many centuries for an increase in air temperature to gradually make its way down deeper and deeper into the ocean. For this reason, **the rise in sea levels resulting from any particular amount of global warming is largely delayed**

FIGURE 6-18
Contributions from various
factors to the sea-level rise
since 1961. [Source: Adapted
from Catia M. Domingues et al.,
"Improved estimates of upper-
ocean warming and multi-
decadal sea-level rise," *Nature*
453 (2008): 1090–1093.]

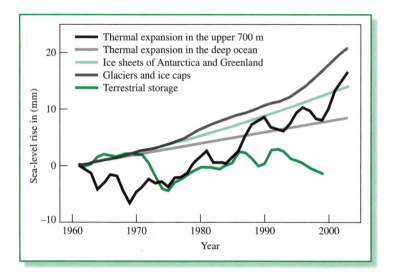

for many years. Consequently, even if atmospheric carbon dioxide levels did not increase at all beyond today's values and no further global warming occurred, sea levels would *continue* to rise for centuries yet to come, as deeper and deeper layers of the oceans became heated—and expanded—by the air that has already been warmed.

Sea levels are predicted to rise by another 80 cm (0.8 m) by 2100, although there is a large uncertainty in this value; it could be as high as 200 cm (2 m). Much of the predicted rise in sea level is due to the melting of glaciers and ice caps, with some contribution from the Antarctic and especially the Greenland ice sheets, and the rest arises mainly from the **thermal expansion** of seawater. The **thermal expansion occurs because the density of water decreases gradually as water warms beyond 4°C,** the temperature at which it reaches its maximum density, as illustrated in Figure 6-19. Since density is mass divided by volume, and since the mass of a given sample of water cannot change, the volume it occupies must increase if its density decreases. Thus, as seawater warms (beyond 4°C), the volume occupied by a gram or kilogram of it increases; the only way that this can occur is if the top of the water—the sea level—increases in height.

The thermal expansion that has occurred since 1960 in the top 700 m of the ocean is shown by the black curve in Figure 6-18, and that at deeper levels by the light gray line (which is based upon annual expansion that is assumed to be constant). Overall, thermal expansion accounts for about 40% of sea level rise in the last half century.

Global warming and other factors have led to the partial melting of many glaciers and ices sheets around the world. Once melted, the water flows from its location above land surfaces eventually into the oceans, raising their levels. Indeed, about one-third of the sea-level rise over the last

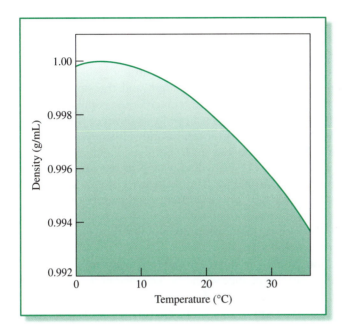

FIGURE 6-19 Density of liquid water versus temperature.

half-century has originated from this accelerating source, as documented by the dark gray curve in Figure 6-18. The shrinking of glaciers by rapid melting will eventually be problematic for the many people whose drinking and irrigation water supplies historically have been rivers supplied by the water from slowly melting glaciers, such as those in the Andes and Himalayan mountains.

Scientists predict that the Greenland Ice Sheet may eventually melt entirely due to global warming, raising ocean levels by about 7 m, but they are uncertain as to the fate of the Antarctic Ice Sheet. However, complete melting will require millennia to complete. Parts of both ice sheets sit above sea level on land. Consequently, the transfer by melting of their surface ice and the resulting draining of liquid water into the oceans causes an increase in sea levels, as does the transfer into the oceans of icebergs broken off from the ice sheets. The collapse of ice shelves, which are extensions into the sea of glaciers, such as the Larsen-B shelf in the Antarctic that collapsed in the early 2000s, allows glaciers that had been blocked to migrate more quickly to open water. The contributions of the ice sheets of Antarctica and Greenland to the sea-level rise over the last half-century is shown by the light green curve in Figure 6-18 and amounts to about 30% of the total increase. Recent research has found that the ice sheets in both Greenland and Antarctica are melting at a faster rate than previously realized.

The dark green curve in Figure 6-18 shows that contributions to sea-level rise from terrestrial water storage have been small over the past few

Figure 5-18 illustrated the total sea-level rise from all factors.

decades. The total sea-level rise (not shown in Figure 6-18) from all factors amounts to 60 mm (6 cm) over the half-century; the rate of the rise had increased to about 2.3 mm y^{-1} by the turn of the century.

Although an increase of half a meter in sea level does not seem very large, there are countries, like Bangladesh and the small island nation of Tuvalu in the South Pacific, in which much of the population currently lives on land that would be flooded by a rise in sea level of this amount. In addition, damage to low-lying coastal areas worldwide from tropical storms would increase because of these higher sea levels. Even if global temperatures eventually are stabilized, sea levels will continue to rise for centuries thereafter since it takes so much time for heat to melt ice sheets and to work its way down into ocean depths. Current estimates of the range of sea-level increases over the coming centuries are illustrated in Figure 6-20. The ranges are very wide due to both scientific uncertainty and whether and when effective climate control is achieved.

In the long term, the most dramatic—though unlikely—effect of substantial global climate change would be a change in the circulation patterns of water in the Atlantic Ocean. Currently, warm surface waters flow northward from the tropics into the North Atlantic, bringing heat to Europe and to a lesser extent to eastern North America. Some scientists have speculated that a rapid rise in temperature and rainfall levels could weaken or even eliminate this circulation pattern, as historical geological records indicate has happened in the past.

FIGURE 6-20 Global sea levels, relative to 1990: history and range of projections for future centuries. Notice that the uncertainty in the projection increases with time. [Source: Adapted from *The Copenhagen Diagnosis, 2009: Updating the world on the Latest Climate Science.* I. Allison et al., The University of New South Wales Climate Change Research Centre (CCRC), Sydney, Australia.]

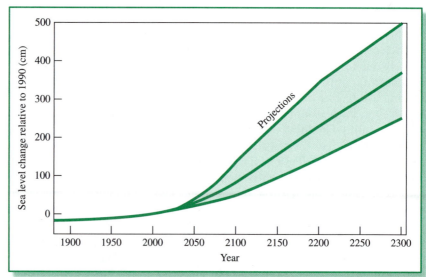

In addition to a rise in its levels, ocean water is also becoming more acidic—or more accurately, less alkaline— as a result of anthropogenic CO_2 input. Its pH at the upper levels has already fallen by 0.1 unit, to 8.07, and is predicted to decline by a further 0.15 units once the atmospheric CO_2 concentration reaches double its preindustrial-age value. Changes in ocean pH can lead to dramatic affects on marine life, particularly calcifying organisms such as mussels and coral. As acidity increases, the calcium carbonate in seashells begins to dissolve. Furthermore, the H^+ combines with dissolved CO_3^{2-}, decreasing the amount of carbonate available to sea creatures to build their shells and skeletons.

6.29 Climate Predictions for Specific Regions

It is much harder for scientists to make specific, reliable predictions for individual regions than for the globe as a whole. The climate changes that seem likely for the various continents, according to the 2007 IPCC report, are summarized in Table 6-6.

The higher latitudes of the Northern Hemisphere are expected to experience temperature increases substantially greater than the global average. Warming over some land areas, including the United States and Canada, should be noticeably faster than the average rate for the globe.

In the Midwest regions of the United States and the areas just north of them in Canada, as well as in southern Europe, the soil probably will become much less moist because of increased rates of evaporation due to the warmer air and ground. This could affect the continued suitability of these areas for the growing of grain. High latitude regions, however, could experience increased productivity, at least where the soil is suitable for agriculture. In areas that become drier, the positive CO_2 fertilization effect on plants will cancel some of the negative effects of decreased rainfall. There will be longer frost-free growing seasons at northern latitudes but increased chances that heat stress will affect crops grown there. Food production in temperate areas will probably also be affected by the attack of insects that in the past have been largely killed off during the winters but could survive and flourish under warmer conditions.

Temperature and moisture changes will occur quickly compared to those that have taken place in the past, and consequently some ecosystems will be destabilized. Coastal ecosystems such as coral reefs are particularly at risk. The species composition of forests is likely to change, especially in regions far removed from the Equator. For example, the hardwood forest in eastern North America may be at risk of extinction if climate zones shift faster than their migration can keep pace with. The boreal forest of central Canada could be eliminated by fire by 2050; indeed the frequency of fires in these woodlands is already climbing.

TABLE 6-6	Significant Impacts of Climate Change That Will Likely Occur in the Continents in the Twenty-First Century
Arctic	Significant retreat of ice; disrupted habitats of polar megafauna; accelerated loss of ice from Greenland Ice Sheet and mountain glaciers; shifting of fisheries; replacement of most tundra by boreal forest; greater exposure to UV radiation
North America	Reduced springtime snowpack; changing river flows; shifting ecosystems, with loss of niche environments; rising sea level and increased intensity and energy of Atlantic hurricanes increase coastal flooding and storm damage; more frequent and intense heat waves and wildfires; improved agriculture and forest productivity for a few decades
Europe	More intense winter precipitation, river flooding, and other hazards; increased summer heat waves and melting of mountain glaciers; greater water stress in southern regions; intensifying regional climatic differences; greater biotic stress, causing shifts in flora; tourism shift from Mediterranean region
Central and Northern Asia	Widespread melting of permafrost, disrupting transportation and buildings; greater swampiness and ecosystem stress from warming; increased release of methane; coastal erosion due to sea ice retreat
Central America and West Indies	Greater likelihood of intense rainfall and more powerful hurricanes; increased coral bleaching; some inundation from sea-level rise; biodiversity loss
Southern Asia	Sea-level rise and more intense cyclones increase flooding of deltas and coastal plains; major loss of mangroves and coral reefs; melting of mountain glaciers reduces vital river flows; increased pressure on water resources with rising population and need for irrigation; possible monsoon perturbation
Pacific and small islands	Inundation of low-lying coral islands as sea level rises; salinization of aquifers; widespread coral bleaching; more powerful typhoons and possible intensification of ENSC extremes
Global oceans	Made more acidic by increasing CO₂ concentration; deep overturning circulation possibly reduced by warming and freshening in North Atlantic
Africa	Declining agricultural yields and diminished food security; increased occurrence of drought and stresses on water supplies; disruption of ecosystems and loss of biodiversity, including some major species; some coastal inundation
South America	Disruption of tropical forests and significant loss of biodiversity; melting glaciers reduce water supplies; increased moisture stress in agricultural regions; more frequent occurrence of intense periods of rain, leading to more flash floods
Australia and New Zealand	Substantial loss of coral along Great Barrier Reef; significant diminishment of water resources; coastal inundation of some settled areas; increased fire risk; some early benefits to agriculture
Antarctica and Southern Ocean	Increasing risk of significant ice loss from West Antarctic Ice Sheet, risking much higher sea level in centuries ahead; accelerating loss of sea ice, disrupting marine life and penguins

Source: Scientific Expert Group Report on Climate Change and Sustainable Development, *Confronting Climate Change*, United Nations Foundation (2007): www.confrontingclimatechange.org.

6.30 Predicted Effects of Climate Change on Human Health

Some scientists have speculated that human health will be affected adversely by global warming. There probably will be more extreme heat waves in summers but fewer prolonged cold snaps in winters. The expected doubling in the annual number of very hot days in temperate zones will affect people who are especially vulnerable to extreme heat, such as the very young, the very old, and those with chronic respiratory diseases, heart disease, or high blood pressure. This stress will be particularly acute for poor people, who have less access to air conditioning. The heat wave in the summer of 2003 in Europe resulted in the death of at least ten thousand people in France alone.

Domestic violence and civil disturbances could also increase, as they tend to occur more frequently in hot weather. On the other hand, there would probably be a decrease in cold-related illnesses because of the milder winters. Air quality in summer will probably degrade further, as the background concentration of ground-level ozone will probably increase substantially. Higher CO_2 concentrations and warmer weather will increase the production and release by plants of pollen such as ragweed, thus exacerbating allergic responses.

Less directly, global warming may extend the range of insects carrying diseases such as malaria into regions where people have developed no immunity, and it may intensify transmission in regions where such diseases already are prevalent. It has been predicted that malaria could claim an additional million victims annually if the temperature rise is sufficient to allow parasite-bearing mosquitoes to spread into areas not now affected. A different type of mosquito carries the dengue and yellow fever viruses, and its range could increase with warming, thereby spreading these diseases. There also is evidence that cholera rates increase with warming of ocean surface waters, because coastal blooms of algae are breeding grounds for the disease, and they increase with water temperature. However, some experts in disease control discount these predictions, arguing that other effects such as increased rainfall could well negate or even reverse the effects on disease rates of increases in air temperature.

Animal health could also be affected by the spreading of disease by parasites. In addition, some species such as polar bears and caribou that live in very cold regions could be at risk of extinction from habitat changes that threaten their hunting practices.

6.31 International Agreements on Greenhouse Gas Emissions

Faced with the prospect that increased CO_2 emissions over the next century could result in a significant increase in global air temperature with its resultant modifications of climate, some national governments and organizations have been debating how future emissions can be minimized while still allowing economic growth to occur.

The first agreement on greenhouse gas emissions was reached at the Rio Environmental Summit meeting in 1992; each developed country was to voluntarily ensure that its CO$_2$ emission rate in 2000 would be no greater than that in 1990. This target was met, in fact, by very few countries; by 2000, most were emitting at levels well above their targets.

The second agreement was reached in negotiations held at Kyoto, Japan, in 1997. Thirty-nine industrialized nations agreed to decrease their collective CO$_2$-equivalent emissions by 5.2% by 2008–2012, compared to 1990 levels. Emissions by developing countries were not controlled by the Kyoto Accord, since they had not been significant players in emitting greenhouse gases in the past and therefore had not contributed much to climate change. By 2005, a global reduction greater than that required by the Kyoto agreement had been achieved, though rather inadvertently, because countries in the former Soviet bloc had reduced their emissions by a large amount as a result of the partial collapse of their economies. Japan and the European Union were expected to meet their reduction targets, but the United States, Canada, Australia, and several other countries have greatly exceeded their original targets.

The follow-up meeting to the Kyoto Accord was held in Copenhagen, Denmark, in 2009. There was general agreement that global temperature increases should be held to no more than 2°C (relative to the value in 1990 or later). Each country was to have submitted an emission reduction target for its own economy to be achieved by 2020. The *Copenhagen Accord* also called for countries to invest in clean energy technology and practices, and to assist in adaptations of climate change. The United States and Canada each have a 17% reduction target, relative to 2005 emissions, while Japan and the European Union aim for about 25% reductions relative to 1990 values.

The existing increase, by over one-third, of the atmospheric CO$_2$ level, and the temperature increase and climate modification that this probably caused resulted in large measure from the industrialization and the increase in standard of living that developed countries have achieved in the past two centuries. Without a significant change in the methods by which energy is produced and stored, and/or implementation of carbon sequestration on a massive scale, these same nations will continue to require about the same rate of CO$_2$ emissions in the future in order to maintain their economic growth, as discussed in Section 6.1. Interestingly, the infrastructure (power plants, vehicles, etc.) that would produce most of the 2-degree increase is yet to be built, so the world does have a good chance of meeting the new guidelines if there is a rapid switch to clean energy production.

Rather than utilizing a procedure in which countries have CO$_2$ emission targets that are negotiated at international meetings, schemes have also been discussed that are based upon allocations that could be traded between

The greenhouse gases affected by the Kyoto Accord agreement are carbon dioxide, methane, nitrous oxide, hydrofluorocarbons, perfluorocarbons, and sulfur hexafluoride.

(or within) countries on the open market. In a manner similar to the way in which SO_2 emission rights currently are traded in the United States, countries that need to emit more than their collective CO_2 allocations could purchase unused allocations from countries with an excess. Putting a price on carbon dioxide emissions internalizes the costs of this pollutant.

In the past, the cost of the emissions was borne by all that were negatively affected by the emissions—the tragedy of the commons. Putting a cost on the emissions (which becomes part of the cost of the energy that produced the emissions) forces those that benefit from the energy (that results in the emissions) to pay a cost for the emissions. A bonus of such a **cap-and-trade** scheme is that it provides an incentive to develop and invest in cleaner technologies, since avoiding CO_2 emissions could be cheaper than purchasing additional rights— especially in the future when few nations will have excess emission capacity and the price of emission rights will rise. The European Union began a cap-and-trade system for CO_2 emissions *within* its countries in 2005.

The question of how national CO_2 allocations can be made fairly in order to initiate the free-market CO_2 emission cap-and-trade scheme on a world-wide basis is a perplexing one. In the simplest scheme, each country would be assigned an allocation based strictly upon its (current) population. For example, if it was concluded that the current average annual emission of 1 tonne of carbon as CO_2 per capita could be sustained indefinitely, then this quantity would be allocated to a country for each of its residents. If it was decided to cut back current global emission levels, e.g., by one-quarter, then only 0.75 tonnes per capita per year would be allocated, etc.

An immediate consequence of a per capita allocation method would be the annual transfer of substantial funds from all developed countries to developing and undeveloped countries, since according to the data in Figure 6-15, the former nations all exceed the 1-tonne average, by factors ranging from two to five times. Although this method would provide external funding so that developing countries could establish efficient energy infrastructures, it would likely not prove popular in developed countries since their energy costs would rise.

One alternative allocation scheme is based upon how much energy is required for industrial production by a country and how efficiently it uses energy. Thus a country's carbon dioxide allocation would be directly proportional to its GDP. This allocation method rewards compact, energy-efficient developed countries at the expense of those—both developed and developing—which emit more CO_2 per unit of GDP. However, such a scheme would permit continued economic growth by developing countries, since their CO_2 allocations would track their economic growth. The global ratio of allowed carbon dioxide to dollar of GDP would have to decline with time if global emissions are to be controlled, since global GDP rises by several percent per year. As we have seen, the $CO_2/\$$ GDP ratio is more independent of the level of economic development than is the ratio based upon population.

Some policymakers believe that **carbon taxes,** i.e., taxes based on the amount of carbon contained in a fuel rather than upon its total mass, should be instituted as a disincentive to use fossil fuels, especially coal, since it generates more CO_2 per joule of energy produced than does natural gas. In fact, the hydrogen-to-carbon ratio of the average global fuel mix has been continuously increasing over the last century and a half, as we moved from economies whose energy source was dominated by wood (H/C ratio of about 0.1) to coal (~1.0 ratio) to oil (about 2.0) and now to natural gas (4.0); this is the same direction as moving to lower CO_2/energy ratios, as implied above.

We conclude by commenting upon the paradox that faces humanity today concerning the enhancement of the greenhouse effect. On the one hand, there exists the slight possibility that doubling or quadrupling the CO_2 concentration will have much less of an effect on climate than predicted, and that efforts taken to prevent such an increase not only would represent an economic burden for both the developed and the developing worlds, but would perhaps be wasted in the outcome.

On the other hand, if the predictions of scientists who model the Earth's climate turn out to be realistic, but we do nothing to prevent further buildup of the gases, both present and future generations will collectively suffer from rapid and perhaps cataclysmic changes to the Earth's climate. While the *precautionary principle* was invoked and action was taken to mitigate the destruction of the ozone layer (Chapter 2), the governments of the world have been unable to come to a consensus on action to mitigate climate change, even though the consequences or *external costs* of inaction will likely be extremely high.

Review Questions 23–29 are based on material in the preceding sections.

Review Questions

1. Define the term *commercial energy.*

2. What is the equation relating exponential growth to the annual increase in a quantity?

3. How does the rate of energy usage by a country depend on its population and its GDP?

4. Define the term *energy intensity.* Describe how the energy intensity has changed over the last few decades (a) globally, (b) for the United States, and (c) for China.

5. What are the five main global sources of primary commercial energy?

6. What are the ultimate origins of coal, oil, and natural gas? Which fuel is in greatest reserve abundance?

7. What is the main component of *natural gas?* Write out the balanced chemical equation illustrating its combustion.

8. Why is natural gas considered to be an environmentally superior fuel to oil or coal? What phenomenon involved in its transmission by pipeline might offset this advantage?

9. What is meant by the term *CNG?* What are the advantages and disadvantages to fueling vehicles with CNG? What is meant by *LPG?*

10. What are the three important classes of hydrocarbons present in crude oil?

11. What is meant by the term *heavy oil?* Why does it require a large amount of energy to extract petroleum from Alberta oil sands?

12. What is meant by *engine knocking*? How is the *octane number* rating scale for fuels defined?

13. What is meant by the *BTX fraction* of gasoline?

14. List several ways in which the octane number of fuels can be increased by the addition of other compounds to straight-chain alkane mixtures.

15. What is meant by the term *carbon sequestration*?

16. Name two techniques whereby carbon dioxide could be stripped from power-plant emission gases, and two techniques that are more appropriate to more concentrated sources of the gas. For each technique, specify whether a chemical or a physical process is involved.

17. Define *oxycombustion*, and state its advantages and limitations.

18. Sketch qualitatively the phase diagram for CO_2, showing the lack of a liquid state under ambient pressure.

19. What is mean by a *supercritical fluid*? What are some of its characteristics?

20. Explain the difference between *ocean acidic* and *ocean neutral* techniques of storing CO_2 in oceans. How does depth of disposal play an important role in sequestering carbon dioxide in seawater?

21. Define *enhanced oil recovery* and explain its relationship to the underground storage of carbon dioxide.

22. What are *saline aquifers*? Explain how they might be used to store carbon dioxide.

23. Define the two meanings of the term *carbon intensity*. What equation relates them to the Kaya identity?

24. Explain why the concentration of CO_2 in air is expected to rise linearly with time if its rate of emission remains a constant.

25. Explain the reasons why sea levels are expected to rise as global air temperatures increase.

26. List some of the consequences, including those affecting human health, that may occur as a result of climate change in the future. Why might soil in some areas be too dry for agriculture even though more rain falls on it?

27. What was the *Kyoto Protocol*? What gases are limited in their emissions under it?

28. Describe the *cap-and-trade* scheme whereby a nation's allocation of carbon dioxide emissions could be traded on a market. Describe two schemes by which initial allocations could be made.

29. What is a *carbon tax*, and what are the arguments in favor of it? Why do you think many people oppose it?

 # Green Chemistry Questions

1. The formation of polylactic acid (PLA) from biomass developed by Cargill Dow won a Presidential Green Chemistry Challenge Award.

(a) Which of the three focus areas (see page xxviii) for these awards does this award best fit into?

(b) List three of the twelve principles of green chemistry (see pages xxiii–xxiv) that are addressed by the green chemistry developed by Cargill Dow.

2. What are the environmental advantages of using PLA in place of petroleum-based polymers?

3. Why does the use of biodegradable polymers (to replace petroleum-based polymers) not offer complete solutions to energy and environmental problems?

Additional Problems

1. Over the time period 1980 to 2000, global energy use increased by 40%. What was the average annual increase, assuming growth was compounded? By what percentage would energy use have grown by 1990?

2. Assuming that energy usage grows exponentially with time, derive a general formula relating the number of years required for energy usage to double as a function of the annual fractional increase k. What is the doubling time when 4% annual growth (i.e., $k = 0.04$) is in operation? What about 3%, 1.5%, and 1.0% annual growth rates?

3. Why do you suppose that the concentration of olefins in gasoline is restricted to a small value, whereas that of aromatics is allowed to be much higher and alkanes are unrestricted? [*Hint:* Recall the relative reactivity of different classes of hydrocarbons to OH, as discussed in Chapter 3.]

4. The replacement by natural gas of oil or coal used in power plants has been proposed as a mechanism by which CO$_2$ emissions can be reduced. However, much of the advantage of switching to gas can be offset since methane escaping into the atmosphere from gas pipelines is 23 times as effective, on a molecule-per-molecule basis, in causing global warming as is carbon dioxide. Calculate the maximum percentage of CH$_4$ that can escape if replacement of oil by natural gas is to reduce the rate of global warming. [*Hint:* Recall that the heat energy output of the fuels is proportional to the amount of O$_2$ they consume.]

5. Given that the density of dry ice (solid CO$_2$) is 1.56 g cm^{-3}, calculate the diameter in meters of a dry-ice ball that would be produced from the 5 metric tonnes of CO$_2$ produced on average by each person in developed countries each year.

6. Write down a list of reasons that you think proponents of carbon dioxide allocations based strictly upon a country's population would advance in support of their position. What objections can be raised to their position? Repeat this exercise for an allocation scheme based upon current GDP.

Biofuels and Other Alternative Fuels

In this chapter, the following introductory chemistry topics are used:

- ⮕ Ideal gas law
- ⮕ Thermochemistry calculations
- ⮕ Oxidation numbers; redox half-reactions; batteries
- ⮕ Basic organic structures: Alcohols, ethers, carboxylic acids, esters, sugars, carbohydrates, aldehydes
- ⮕ Vapor pressure of liquids
- ⮕ Distillation
- ⮕ Viscosity

Background from previous chapters used in this chapter:

- ⮕ Greenhouse gases (carbon dioxide, nitrous oxide)
- ⮕ Anaerobic decomposition
- ⮕ Fossil fuels—gasoline and petrodiesel
- ⮕ Air pollution: Photochemical smog, particulates, gaseous pollutants
- ⮕ Catalytic converters
- ⮕ Light absorption as energy; photons

Introduction

In previous chapters, we have discussed the large-scale production and use of energy from fossil fuels, and the environmental costs—especially in terms of CO_2 emissions—that result. In this and the following chapter, we consider alternatives to these sources and fuels, especially ones that are **renewable in the sense that their production and use generate very low greenhouse gas emissions and whose production can be sustained indefinitely**.

In this chapter, we discuss in some detail the chemistry, the development, and the issues surrounding new vehicular fuels that are alternatives to gasoline and petrodiesel. Of special interest are *biofuels*, which use *contemporary* photosynthetic mass and thus are indirect forms of solar energy. The biofuels considered in this chapter have the inherent advantage over hydrogen, which we discuss in detail later in the chapter—and even over natural gas—that they are liquids under normal temperatures and pressures, and burn easily in air to produce considerable heat. Like the gasoline and diesel fuels with which they can be blended or can replace, they are *energy-dense* fuels. Since they all contain carbon, however, their combustion releases carbon dioxide.

In the sections that follow, we discuss the production of biofuels from biomass using mainly biological methods. Thermochemical methods of production of alternative fuels are discussed subsequently.

Biomass and Biofuels: Issues

The biomass produced by the worldwide operation of photosynthesis constitutes a form of solar energy. The annual amount of commercial energy currently generated from this source is about 55 EJ; a much larger amount is potentially available. Overall, **the power density of photosynthesis (about 0.6 W m^{-2}) is too low for it to be used to supply the majority of the world's energy needs.** The density is low because **the efficiency of conversion of sunlight to chemical energy by photosynthesis is very low,** no more than 1–2% even in the most productive areas. At today's consumption levels, the amount of land required to supply the world's energy needs *entirely* by biomass equals that of all agricultural land currently developed, which constitutes more than 10% of Earth's land surface.

7.1 Biofuels: The Major Issues

In recent decades, enormous attention, research, and development have been devoted to the conversion of living plant matter—**biomass**—into liquid fuels, especially those capable of powering vehicles. The incentives behind the production and use of such **biofuels** rather than fossil fuels are

• their *carbon-neutral* characteristic, i.e., the amount of CO_2 that they emit upon combustion is matched by that absorbed during photosynthesis by the next crop;

$$CO_2(g) + H_2O(g) \; \underset{\text{combustion}}{\overset{\text{photosynthesis}}{\rightleftharpoons}} \; \begin{array}{l}\text{plant matter}\\(\text{carbohydrates,}\\\text{oils, protein,}\\\text{lignin})\end{array} \; + \; O_2$$

• the fact they are *renewable*, in the sense that their production could be sustained indefinitely into the future even if fossil-fuel supplies run out;

- their oxygen content, which generally results in *smaller amounts of air pollutants* being produced and potentially released into the air; and

- their *replacement of imported fuel*, mainly petroleum, if they are grown in the country of their use.

Two important issues bedevil the production of biomass and its conversion to vehicular fuels as a means of counteracting global climate change from future greenhouse gas emissions:

- *The clearance of land* of its existing biomass in order to generate new areas to cultivate for biofuel production—or new areas to grow crops displaced by growing biofuel biomass on land currently used for agriculture—plus the fuel use consumed during these activities would produce a massive emission of carbon dioxide into the air. This *carbon debt* would take decades to be paid back by emission savings resulting from replacement of fossil fuels by the biofuels grown there.

- *The conversion of biomass into biofuels requires the expenditure of significant amounts of energy.* If this energy is provided by the combustion of fossil fuels, the savings in net CO_2 emissions by use of biofuels in vehicles are partially (or completely) cancelled.

The term *payback period* can also be used for the number of years that are required for the greenhouse gas emissions from biofuel use to repay the carbon debt.

The carbon debt issue was first raised quantitatively in 2008, and is a controversial one. **When forest is cleared to provide new cropland, the trees and brush are often burned on site, thereby generating much carbon dioxide without using the biomass energy of combustion in a productive manner.** In addition, cultivation by tilling of the soil provokes microbial action that oxidizes the organic carbon stored in the soil, releasing more CO_2 into the air. Moreover, if the forest was fairly young and still growing (rather than mature), it would have stored additional carbon over the next decades, and this "foregone sequestration" adds to the carbon debt. Conversion of grassland produces less carbon dioxide from above-ground sources. However, reforesting grassland would extract more carbon dioxide from the air than that saved by growing biofuels on it and displacing the equivalent amount of fossil fuels.

Initial estimates of the payback period required for greenhouse gas emission savings from the displacement of fossil fuel by biofuel to overcome the carbon debt ranged from several decades to centuries, depending upon the type and location of the new land. The use of nitrogen fertilizers to grow biofuel agricultural crops such as corn or soybeans also results in an increased emission of the powerful greenhouse gas *nitrous oxide*, as well as generating water pollution from unused fertilizer (about one-third of the applied nitrogen) in the drainage from the fields.

Research reported in 2011 indicates that some biofuel crops such as sugarcane reflect sunlight (have a higher albedo) better than do the food crops and pasture they displace, thereby generating a cooling effect on the Earth's climate and countering some of the fuel's carbon debt discussed above.

7.2 Burning Biomass Itself

The use of wood, crop residues, and dung (dried excrement from plant-consuming animals) was the world's first energy source and is still widely used in less-developed countries, but its domestic and small-scale usage is very polluting to the air and inefficient. Small-scale biomass burning is generally phased out in favor of commercial energy such as fossil fuels and electricity as the country's economy develops. Nevertheless, biomass was second only to hydroelectric power in the production of renewable energy even in the United States in the late twentieth century.

Recently, technology has been developed to burn biomass in large-scale installations that do not pollute the air. For example, wood-chip waste can be burned to produce steam. Alternatively, wood can be gasified, or digested by bacteria, and converted into alcohol fuels—see the methanol discussion in Section 7.19. However, simply burning the biomass—either on its own or *co-fired* with coal—to produce electricity ("bioelectricity") for battery-driven cars would yield more vehicle miles than converting it to any form of biofuel for gasoline- or diesel-powered vehicles. The wood must be dried of most or almost all its original water content—often exceeding 50%—and compressed into pellets, both processes requiring energy, before co-firing is practical. The greenhouse gas emissions from burning wood or straw pellets to produce electricity or just to heat buildings are quite low (see Figure 7-1).

Co-firing with wood has the advantage that some of the SO_2 from combustion of the coal reacts with alkali components in the wood to produce solid sulfates and consequently less of the gas is released into the air.

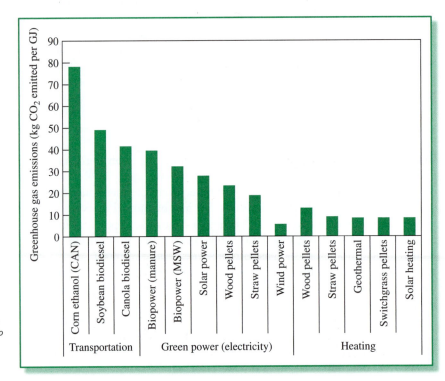

FIGURE 7-1 Life-cycle greenhouse gas emissions for various bioenergy technologies, by sector. (MSW = municipal solid waste) [Source: Adapted from REAP Canada, *Analysing Ontario Biofuel Options, Final Report* (2008), BIOCAP Foundation Canada.]

A significant problem for the development of large-scale biomass combustion is establishing a reliable, concentrated source of feedstock. Fast-growing trees in plantations could be used for this purpose, using land not suitable or needed for agriculture. In India, rice husk pellets—available in ample supply—are burned to produce steam for electricity generation for homes in some rural areas.

Ethanol

7.3 Ethanol as a Fuel

Ethanol, C_2H_5OH, also called *ethyl alcohol* or *grain alcohol*, is a colorless liquid that has been used as an automobile fuel as far back as the late 1800s; indeed, Henry Ford designed his original cars to run on ethanol.

As a fuel for vehicles, ethanol can be used "neat," i.e., in pure form, or as a component in a solution that includes gasoline. Often **these fuels are referred to by the letter E (for ethanol) combined with a subscript that indicates the percentage of alcohol in the gasoline–ethanol mix.** In North America, the "gasohol" currently sold is 10% ethanol and 90% gasoline, i.e., E_{10}. Ethanol and gasoline are freely soluble in each other, so all possible combinations can be produced. Currently in Brazil, E_{25} is used by all gasoline-powered vehicles. E_{100} is used mainly in Brazil, where some cars are designed with engines that can use the pure alcohol.

One of the difficulties in using pure ethanol (and also pure methanol) as vehicular fuel is its low vapor pressure: see Figure 7-2, in which the vapor pressure of gasoline–alcohol mixtures is plotted against their composition, with pure gasoline at the left side of the horizontal axis and

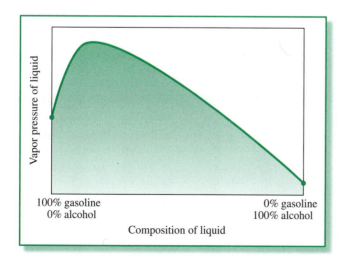

FIGURE 7-2 Variation in vapor pressure with composition for typical alcohol–gasoline mixtures.

pure alcohol (ethanol or methanol) at the right. Consequently, in cold climates, with pure ethanol there is very little vaporized fuel available with which to start a cold automobile engine. However, a blend of 85% ethanol and 15% gasoline has a high enough vapor pressure to overcome the *cold start* problem. In North America and Europe, E_{85}-fueled vehicles are now available, usually labeled *flex fuel*, because they can also burn gasoline. The E_{10} blend of ethanol in gasoline is now widely available in North America.

In order to reduce the evaporation of volatile organic compounds (VOCs) from gasoline—since they contribute to the ozone problem—the maximum vapor pressure of gasoline sold in the United States in the summer months is regulated. To achieve the lower overall volatility, the amount of (highly volatile) butane in gasoline is being reduced and replaced by substances that have low volatility. Unfortunately, as a minor additive, ethanol is quite volatile and actually *increases* the vapor pressure of gasoline. (The same phenomenon occurs for methanol–gasoline blends.) This behavior can be understood by reference to Figure 7-2. As an alcohol such as ethanol is added to gasoline, the vapor pressure of the mixture rises sharply since the C_2 compound behaves in this hydrocarbon environment much like a low-molecular-weight—and therefore volatile—hydrocarbon. In contrast, as a pure liquid, ethanol has a *lower* vapor pressure than gasoline since the extensive hydrogen bonding between ethanol molecules in this situation provides a "glue" that makes it difficult to break the molecules apart and vaporize them.

Another disadvantage of ethanol (which is even more true for methanol) as a fuel is that the energy it produces per liter combusted is somewhat less than is generated by an equal quantity of gasoline. To travel the same distance, fuel tanks for alcohols would have to be larger. In principle, about 1.25 gal of ethanol are needed to generate the same amount of energy as is obtained with one gallon of gasoline. In practice, however, the efficiency of the combustion is greater with the alcohols, so the volume penalty is not this large.

7.4 Air Pollution from Ethanol Combustion

One attractive feature of "oxygenated" transportation fuels such as ethanol is that they result in lower emissions of many pollutants—specifically carbon monoxide, alkenes, aromatics, and particulates—compared to pure gasoline or diesel fuel, particularly in older vehicles that do not have catalytic converters. In North America, however, the turnover of vehicle fleets means that very few cars still on the road emit much CO.

The reduction in urban ozone that would result from the lowered emissions of carbon monoxide and reactive hydrocarbons would be countered by increases due to the higher amounts of *acetaldehyde* and vaporized ethanol that would be emitted into the air.

The involvement of acetaldehyde in photochemical smog is discussed in Chapter 17.

This is particularly true for urban areas in which ozone formation is NO_X-limited rather than hydrocarbon-limited. However, NO_X emissions from engines burning ethanol are lower than from those burning gasoline since the combustion flame temperature is lower and therefore less thermal NO_X is generated.

Limitations on ozone were discussed in Section 3.9.

It is curious that some proponents of ethanol as a fuel point to its ability to reduce CO emissions, which in fact is important only for cars without catalytic converters, but downplay the effects of acetaldehyde emissions by stating that they can always be minimized by use of catalytic converters!

7.5 Bioethanol Production

Industrially, ethanol is made by catalytically adding water to petroleum-based *ethene*, $CH_2{=}CH_2$, to produce CH_3CH_2OH. In contrast, **ethanol for fuel is produced on a massive scale by the fermentation of carbohydrates in plants.** Such **bioethanol** is produced by the yeast-driven fermentation principally of **glucose**, $C_6H_{12}O_6$.

In North America, most carbohydrate used for bioethanol production is derived from the starch in kernels of corn, although some sugar beets, wheat, and other grains are also used. **In Brazil and some other semitropical countries, sucrose from sugarcane is widely employed.** A number of developing countries, including Thailand and China, produce ethanol from *cassava*, a woody shrub that produces a tuberous root high in starch content. If starch rather than sucrose is the starting material, it must first be converted to glucose. In the fermentation process, the sunlight-derived energy of the glucose becomes more concentrated in the ethanol product, since some of the carbon is released as carbon dioxide gas:

$$C_6H_{12}O_6 \xrightarrow{\text{yeast}} 2\,CO_2 + 2\,C_2H_5OH$$

PROBLEM 7-1

Show by comparison of oxidation numbers that the carbon atoms in ethanol are more reduced—and therefore better fuels upon their oxidation—than they were in the glucose molecules from which they originated before fermentation. Show also that there is no net change in oxidation number of carbon in going from reactants to products in the fermentation reaction.   ●

PROBLEM 7-2

Using the enthalpies of formation listed below, calculate the enthalpy change for the fermentation reaction of glucose into ethanol and carbon dioxide. Is the process exothermic or endothermic? From your answers, decide whether the fuel value of the ethanol product is slightly greater or slightly less than that of the glucose from which it is created.

$$\Delta H_f \text{ values in kJ mol}^{-1}: C_6H_{12}O_6(s) \quad -1273.2$$
$$C_2H_5OH(l) \quad -277.8$$
$$CO_2(g) \quad -393.5$$

Unfortunately, fermentation must be carried out in a large quantity of water in order to solubilize the starch from which the glucose is obtained, and because the yeast dies if it is present in concentrated alcohol. Indeed, the greater the percentage of alcohol in the mixture, the slower the rate of conversion. A total inhibition of fermentation occurs when the alcohol solution reaches about 8–11% ethanol by volume (i.e., about one-tenth of the aqueous solution). For this reason, **only dilute solutions of alcohol can be produced by fermentation.**

The dilute solution of ethanol in water (about equal in concentration to that of wine) produced by fermentation will not burn; **to be used as a vehicular fuel, almost all the water must be removed.** Consequently, the solution is distilled to separate the alcohol from the water. **Distillation is a very energy-intensive process,** since the watery mixture must be constantly kept at the boil. What ultimately results from repeated distillations is not pure ethanol, but a solution of 95.6% ethanol and 4.4% water (by volume). The last vestiges of water *cannot* be removed by more distillation, but this can be accomplished by a different process involving equipment (molecular sieves) that also uses heat energy when it is dried in order to allow its reuse. Many of Brazil's vehicles operate on *hydrous ethanol*, i.e., 95% C_2H_5OH, thereby avoiding the need for the final step of purification.

The rapid recent growth in world production of ethanol is illustrated by the top of the heavy green curve in Figure 7-3. By 2010 the production of bioethanol in the United States (light gray area) exceeded even the huge

When the *azeotropic mixture* of 95.6% ethanol boils, the vapor has the same composition as the liquid phase, and consequently separation by distillation is not effective.

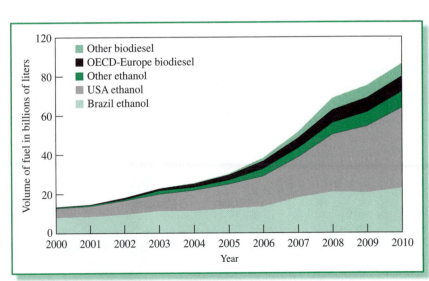

FIGURE 7-3 Global biofuel production in the first decade of the twenty-first century. [Source: Adapted from International Energy Agency, *Technology Roadmap: Biofuels for Transport 2011.*]

amounts are produced from sugarcane in Brazil (light green area). Smaller quantities of ethanol are obtained elsewhere in the world (dark green area)—from cane in Zimbabwe, and from corn and grain in Canada.

As of 2010, 45 billion liters (12 billion U.S. gallons) of ethanol for use as fuel were being produced and used annually from the starch of corn and grain in the midwestern United States; this was scheduled to increase to 57 billion liters (15 billion gallons) by 2015. One-third of the entire corn crop in the United States and almost half the sugarcane crop in Brazil were used to produce bioethanol in 2009–2010. Many farmers in the United States and Canada have provided major political support for the production and use of ethanol in gasoline, and in particular for the government subsidies required to make it economically competitive with petroleum. Unfortunately, the air and water pollution, and soil erosion, resulting from the Brazilian operation are massive.

7.6 Energy and CO_2 Balances in Bioethanol Production

The crux of the matter is that **heat generated by burning a large amount of fuel is needed to distil the ethanol from the water.** In the modern production of ethanol from corn in the United States and Canada, a nonrenewable fuel—either coal or natural gas—is burned to generate the heat required in the distillation process. As a result of this combustion, a large amount of carbon dioxide—a significant fraction of that produced when the alcohol is later burned as a fuel—is released into the atmosphere.

However, if, as is done in Brazil, biomass crop residues (called *bagasse* in the case of sugarcane) rather than a fossil fuel are used to power the distillation, the carbon dioxide that they release upon combustion is reabsorbed by the growth of such material the next season, so there is very little net addition of CO_2 into the air from this step. However, the particulate air pollution from the smoke that can accompany biomass combustion restricts its use in countries with strict regulations.

Many scientists and policymakers have attempted to add up the pluses and minuses of greenhouse gas release and CO_2 absorption, and of energy production and consumption, in the generation of ethanol for fuel from corn in North America, and have compared them to the comparable values for gasoline in order to determine whether or not corn bioethanol is truly a renewable fuel. Their conclusions about whether the overall balances for ethanol relative to gasoline are positive or negative depend largely on the assumptions they make, although all agree that the size of the difference is relatively small.

The analyses are complicated by the fact that commercial materials such as corn gluten feed, corn oil, and dried distiller grains, obtained

© FL Stock/Alamy

from the nonstarch component of the corn kernels, are co-produced with ethanol in the distillation step of the corn mash. Presumably, some of the fossil-energy usage and greenhouse gas emissions in the process ought to be associated with the *co-products*, rather than assigning it all to the alcohol since the co-products displace other substances on the market that would require energy to produce.

Most analyses conclude that **modern North American ethanol production from corn requires about two-thirds the amount of fossil fuel that would be required to generate the same amount of energy in the form of gasoline produced from conventional petroleum sources.** We can conclude then that the production and use of ethanol derived from carbohydrate biomass in North America reduces by about one-third the amount of petroleum per se that is required to produce energy for modern vehicles. **In essence, the energy of ethanol is derived from a combination of two parts fossil fuel and one part captured solar energy;** thus the production of ethanol from corn in North America is essentially the conversion of natural gas or coal into a convenient vehicular fuel, ethanol.

The preceding discussion of the merits of ethanol production from corn or other biomass gives us a glimpse of the types of material and energy flows that are part of what is known as a **life-cycle assessment** (LCA). Although at first glance ethanol from corn may seem like an environmentally sound idea, closer examination of the material and energy inputs and waste outputs shows that it is not a clear-cut proposition. An LCA allows one to assess the environmental impacts (in some cases, economic and social impacts are also considered) of competing products. LCA has been applied to problems such as paper versus plastic grocery bags, and cloth versus disposable diapers. An LCA embraces *systems thinking* and *cradle-to-grave* or *cradle-to-cradle* analysis.

LCAs, systems thinking, and cradle-to-grave analysis are terms defined in Table 0-1. LCAs are discussed in detail in Chapter 16.

Overall, bioethanol produced from corn causes about 86% of the greenhouse-effect enhancement of the gasoline that it displaces. The greenhouse gas percentage exceeds that for fossil-fuel use mainly because it includes the contribution from the nitrous oxide gas emitted as a by-product when fertilizers are used to grow the corn (see Chapter 5), as well as that from the carbon dioxide released when nitrogenous fertilizers are made synthetically. Long-distance transport of bioethanol from production to usage site also adds economic and environmental costs. Because ethanol readily absorbs water, gasohol cannot be shipped through pipelines; the alcohol is transported separately and then blended at gas stations. The increased consumption of E_{85} near production centers would alleviate some of this transport problem.

Although bioethanol produced from sugarcane has a more favorable energy balance than that produced from corn, it can require considerable water to grow and especially to process—the latter even exceeding that used to extract oil from the tar sand in Alberta. Although irrigation is not required in the parts of Brazil where it currently is produced, growing sugarcane in some other countries is placing a burden on their water resources.

Indeed, ethanol production from corn is also water-intensive in the United States for both irrigation and to a lesser extent for processing, and this can be a controversial issue if production plants are sited in water-poor areas. Only a small part of the corn grown in North America currently uses irrigated water, rainfall usually being sufficient, although this fraction could well grow if ethanol production is expanded or drought frequency increases.

7.7 Bioethanol from Cellulose

An emerging technology for bioethanol production uses the cellulose and hemicellulose components of plants, rather than starch, as the abundant feedstock from which sugars are produced and fermented. The hope is that such **cellulosic bioethanol** can be produced in the future in larger amounts and at cheaper prices in terms of fossil-energy consumption and dollars than that obtained from starch sources.

The main components in the woody plants used for cellulosic ethanol are *cellulose* (35–50%), *hemicellulose* (25–30%), and *lignin* (15–30%). Cellulose is a long polymer of the C_6 sugar glucose (see Figure 3-14), approximate formula $(CH_2O)_n$, whereas hemicellulose consists of shorter polymer chains consisting mainly of sugars such as xylose that contain five, rather than six, carbon atoms. Lignin is an unfermentable component of the biomass, and would be burned to provide process heat.

Cellulosic ethanol is currently the most important example of a *second-generation biofuel*. In contrast to *first-generation biofuels*—such as ethanol from corn and biodiesel from soybeans or rapeseed/canola (see Section 7.11)—the second-generation ones are *not* produced from edible food crops themselves but rather from waste materials from crops (stems, husks, leaves), from crops not used for food purposes, and from industrial waste biomass. **In general, the second-generation biofuels are more sustainable.** Some specific advantages of second- over first-generation biofuels include

> Second-generation biofuels are sometimes called *next-generation* or *advanced*.

- the much smaller need for fertilizers and irrigation water, and the much greater savings in greenhouse gas emissions required for their processing;

- the lesser competition with food crops, or land devoted to food crops, and hence the larger potential volume of production and the lesser tendency to drive up the costs of food; and

- the lower production cost.

In order to be converted into alcohol, the biomass must first be ground up and pretreated to break the seal of lignin, and to disrupt the crystalline cellulose structure to enable enzymes to reach and react with the cellulose and hemicellulose within. A number of different pretreatment techniques, some of them physical, some chemical (such as treatment with steam or dilute acid or ammonia) are available, but all those developed to date are relatively expensive to accomplish. Once the pretreatment has liberated the cellulose, enzymes are available to depolymerize it via hydrolysis to glucose, which then is fermented to (dilute) ethanol:

$$\text{biomass} \xrightarrow[\text{pre-treatment}]{} \underset{\text{cellulose}}{(CH_2O)_n} \xrightarrow[\substack{\text{enzyme} \\ \text{hydrolysis}}]{} \underset{\text{glucose}}{(CH_2O)_6} \xrightarrow[\text{fermentation}]{} \text{ethanol}$$

The use of ionic liquids to extract cellulose from biomass, as discussed in Section 3.17, may be another way to solve the pretreatment problem.

Until recently, the enzymes used to hydrolyze cellulose were expensive to produce, but this problem has now been overcome, leaving pretreatment as the most expensive step in the sequence. Unfortunately, the C_5 sugars that are the chief product of the depolymerization of hemicellulose are not fermented to ethanol by naturally occurring enzymes, although genetically engineered yeasts have now been produced that ferment both C_5 and C_6 sugars, albeit producing a very dilute (<5%) alcohol solution.

One widely discussed biomass source for cellulosic bioethanol production is *switchgrass*—a perennial wild grass that was once widespread in the Great Plains of the United States—grown for this purpose. Other sources could be agricultural residue, such as the stalks and other nonkernel parts of corn plants, or wood or waste paper or municipal solid waste. However, nitrogen fertilization—with its accompanying nitrous oxide release—would be required to maintain switchgrass production. Short-rotation hardwood crops—grown on land that is currently out of production or on marginal cropland—require substantially less fertilizers and pesticides than do corn and switchgrass, and together with waste from other forestry and agricultural operations can produce much more biomass per unit area than does corn.

A real advantage of the production of cellulose for bioethanol—and potentially for other second-generation biofuels—is the combustion of the mechanically dewatered lignin component of the biomass to fuel the distillation process, assuming it is burned cleanly. This greatly reduces the fossil fuel required to about 8%, and the net greenhouse gas emissions to 12%, of the amounts involved in the energy-equivalent gasoline cycle. Presumably substantial reductions would also be achieved in producing ethanol from corn itself if the stalks, cobs, etc. were used rather than coal and/or natural gas to fuel the distillation.

The United States plans to produce 16 billion gallons of biofuels from cellulosic sources by 2022, notwithstanding the fact that not a drop of commercial biofuel of this type was being produced in 2010.

7.8 Biobutanol and Other Liquid Fuels

The alcohol *n-butanol* (1-butanol), $CH_3-CH_2-CH_2-CH_2-OH$, has been proposed as a biofuel. Its feedstocks are the same as those for ethanol—corn, sugarcane, etc.—as a first-generation biofuel, and nonfood biomass for second-generation. Butanol has been produced commercially by fermentation of starch using a special bacterium for many decades. Its production for biofuel purposes is under active investigation by several large companies. One major problem is the small yield of butanol from fermentation, compared to ethanol.

The longer hydrocarbon chain in butanol compared to ethanol (and methanol) makes it more similar to the alkanes in gasoline: its energy density is higher, and it is less corrosive and less hygroscopic—making it able to be distributed via gasoline pipelines.

Some researchers have developed genetically altered microorganisms that can ferment sugars into straight-chain alkanes containing 12 to 16 carbon atoms, which, if production becomes economically feasible, could be used to blend with petrodiesel. No distillation step is required, since the hydrocarbons form a separate phase, insoluble in the aqueous sugar layer. One such process, *LS9*, is discussed in the Green Chemistry example of Section 7.15.

Biodiesel from Plant Oils and from Algae

7.9 Plant Oils as Vehicular Fuels

In addition to carbohydrates, plants also contain **oils,** which have in the past found some use as fuels. The oils that can be squeezed out of certain plants or their seeds consist of molecules that are mainly, though not completely, composed of carbon and hydrogen. As those who have experienced grease fires in the kitchen know only too well, plant-based cooking oils (whether corn oil, olive oil, or sunflower oil) can combust when spilled onto a hot stove or into a flame, and give off copious amounts of heat (and smoke) as they burn. Thus natural oils themselves are candidates for fuels in diesel engines. Historically, the first diesel engines could be powered using peanut oil, but it was more expensive than fuel derived from petroleum and this practice fell into disuse.

Modern diesel engines are designed to use fuel that is not as *viscous*—i.e., not as resistant to flow—as are natural oils, which owe their high viscosity to the fact that their component molecules are very large, each containing about 50 to 60 carbon atoms—about three times as many as those in petrodiesel. Unrefined vegetable oils also contain unwanted impurities—such as free fatty acids, water, and odorous substances.

The flow problem can be overcome by preheating the oil before it enters the engine, since viscosity decreases with increasing temperature. Thus some diesel vehicles have battery-driven heaters attached to the fuel tanks in order to be able to power them with natural oils. Other drivers employ two fuel tanks—one that contains petrodiesel fuel for use when the engine is cold (and also before it is switched off) and the second containing cooking oil, to which the driver switches after enough driving time has elapsed that the oil has been sufficiently warmed by heat from the engine.

However, the use of such so-called *straight vegetable oils* (SVOs) in unmodified modern diesel vehicles can lead to serious engine problems stemming from their incomplete combustion. The consequent polymerization of their unsaturated hydrocarbon components produces gums, which results in carbon deposits and thickening of lubricating oil in the engine. Indeed, gums form even during the storage of SVO, and they plug filters and injectors upon

So-called *grease cars* are fueled with grease from deep-frying.

use of the fuel in an engine. Used (waste) cooking oil has been substituted for diesel fuel by some owners of automobiles and trucks in the U.S. south. However, the oil becomes contaminated during its use for cooking and these constituents pose additional problems when it later is used to replace diesel fuel.

7.10 Biodiesel Fuel: Its Constituents

To overcome the practical difficulties in using SVOs themselves, the virgin vegetable oils commonly are transformed into a less viscous, less corrosive fuel called **biodiesel.** The molecules in biodiesel are simple derivatives of so-called *fatty acids*, which consist of carbon atoms (and their associated hydrogen atoms) joined together in chains averaging 16 to 18 units long, about the same size as those in petrodiesel fuel; in fatty acids, the chains terminate in a carboxylic acid group, as illustrated schematically below.

fatty acid

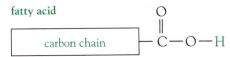

The carbon atoms are held together mainly by C—C bonds, but in contrast to the hydrocarbons in petrodiesel, some are of the C=C type.

Because there are technical problems in using the acids themselves as fuel, the fatty acids are reacted with **methanol,** CH_3OH, which converts them into *methyl esters*; thus the hydrogen atom (green) of the acid group is replaced by a *methyl group*, CH_3:

Recall that an ester corresponds to the replacement of the acidic H of the carboxylic acid, R—COO**H**, by an alkyl chain, R′, giving R—COO**R**′.

methyl ester

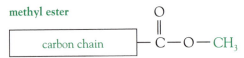

These *fatty acid methyl esters*, abbreviated *FAME*, are the constituents of biodiesel.

The vegetable oils themselves consist largely of triglycerides, molecules in which three separate fatty acid chains (not usually of the same length) are bonded by individual ester linkages to the trialcohol **glycerin** (sometimes called *glycerol*). The structure of the triglyceride is illustrated schematically below:

triglyceride

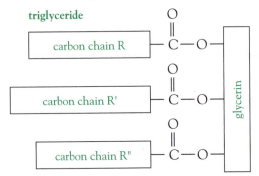

Glycerin has the formula CH_2OH—$CHOH$—CH_2OH. The total number of carbon atoms in a typical glyceride is about 60, and this large size accounts for its high viscosity. Note that the fatty acid and its ester, and the triglyceride itself, are dominated by their hydrocarbon chains, and thus are similar in many respects to hydrocarbons of the same size as far as their energy content upon combustion is concerned.

7.11 The Conversion of Plant Material to Biodiesel

The triglyceride oil is extracted from the original plant material by a mechanical pressing and/or by using a solvent. For example, in the United States, soybeans are de-hulled and crushed mechanically, and the oil extracted from the resulting flaked seeds using *hexane*, C_6H_{14}, a petroleum-derived solvent which is recovered (by evaporation) and reused.

The extracted oil then undergoes the *transesterification* chemical reaction to produce biodiesel. In particular, the triglyceride is reacted with methanol in the presence of the catalyst *sodium hydroxide*, NaOH. The reaction products are not mutually soluble and so form two separate phases that are easily separated; glycerin, being denser, forms the bottom layer. Some of the glycerin is sold separately for other uses, but its massive production in the biodiesel synthesis has produced a glut of it on the market. Most excess glycerin currently is incinerated and some useful heat recovered in this manner.

An excess of methanol is used to shift the position of equilibrium in the reaction toward the products, and thereby obtain a greater yield.

PROBLEM 7-3

Given the fatty acids RCOOH, R′COOH, and R″COOH, draw the molecular structure of a triglyceride that would be produced by them. The formula for glycerin is given above.   ●

The oil-bearing plants that grow best in any given country depend on the local climate and soil conditions. Thus, soybeans are the optimum oil-bearing crop in the United States, whereas palm and coconut oils are more suitable in semitropical areas. Almost all the biodiesel produced in the United States uses domestic soybeans (which are about 20% oil) as its raw material. The oil yield, about 40%, is even higher for rapeseed (canola), which is widely used in Europe.

In tropical areas, massive plantations of palm trees are being created to produce palm oil for biodiesel, since the yield there of oil per square kilometer greatly exceeds that from soybean or rapeseed crops. Unfortunately, in the rush to produce more palm oil destined to become biodiesel in Europe, huge areas of tropical rainforest in Malaysia, Brazil, and Borneo, and peat land in Indonesia, have been burned and cleared and thereby destroyed. This produces large amounts of greenhouse gas emissions, amounting to one-twelfth of all global CO_2 in the Indonesian case. The advantage of biofuels in producing lower greenhouse gases than conventional gasoline or diesel will be overcome by these emissions for many years, as discussed in Section 7.1.

7.12 Use of Biodiesel in Motor Vehicles

The fraction of biodiesel in diesel fuel is designated by a B labeling system analogous to that used for alcohols in gasoline. Thus B_5 symbolizes diesel fuel containing 5% biodiesel by volume, and B_{100} is pure biodiesel. In the past, the most common blend was B_{20}—indeed, the U.S. Navy, the largest single user of biodiesel in the world, uses that blend in all nontactical vehicles—but more dilute blends such as B_2, B_5, and B_7 are becoming popular.

The rapid rise in annual global biodiesel production began in the late 1990s. Its growth in volume over time is documented in Figure 7-3. Currently, the largest manufacturers and users of biodiesel in the European Union are Germany and France, who together account for about half the totals. The European Union collectively exceeds the United States in biodiesel production, and will have increased its biodiesel requirement in commercial diesel fuel to 10% by 2020.

7.13 Greenhouse Gas and Air Pollution Emissions from Biodiesel

The methanol currently used in transforming triglycerides from oil to biodiesel is obtained from natural gas, as described in detail in Section 7.19. **The use of methanol created from a fossil fuel makes biodiesel less than 100% renewable,** although the great majority of the carbon atoms in the fuel esters—and hence most of its fuel value—come from the original vegetable oil.

Overall, soybean-derived biodiesel generates over 90% more energy than is used to produce it, compared to about 25% for corn-based ethanol. Although there is both fossil energy and fossil-fuel-derived methanol used in its production, and nitrous oxide emissions are associated with fertilizers to grow the plants, biodiesel produced from soybeans on *existing* agricultural land overall reduces the CO_2 equivalent emissions by about 40%. The much greater decrease in CO_2 from biodiesel, compared to corn-based ethanol, is due primarily to the much smaller amount of process energy required: soybeans create oil that can be obtained readily from seed by *physical* methods that are not energy-intensive, whereas ethanol requires fuel-intensive distillation.

Soybean production also uses much less fertilizer, and releases much smaller amounts of nitrogen, phosphorus, and harmful pesticides into the watershed environment, than does corn production for ethanol. The greenhouse gas emissions emitted over the lifetime of biodiesel fuel are compared with that from corn-based ethanol and from other forms of renewable energy in Figure 7-1.

Biodiesel blends produce significantly fewer emissions of carbon monoxide, particulate matter (PM_{10}), and hydrocarbons—including polycyclic aromatic ones and those that promote photochemical smog—when it is combusted than the 100% petrodiesel fuel that it replaces; the reduction in soot and CO arises because biodiesel is an oxygen-containing fuel. The reductions increase with the percentage of biodiesel in the blend. There is

Research is under way to find an efficient method to convert glycerin into the methanol required in the reaction. If this process is feasible, biodiesel would be 100% renewable.

considerable controversy as to whether biodiesel blends produce more or less NO_X than does pure diesel. The presence of some $C=C$ bonds in biofuel molecules results in a slightly higher combustion temperature, which would produce more thermal NO_X (Chapter 3).

The advantages and disadvantages of producing and fueling vehicles with biodiesel and its blends as compared to petrodiesel are summarized in Table 7-1. The problems associated with biodiesel production's raising the price of food, and of its resulting in more greenhouse gas emissions from land cleared to grow it or to grow food on land its production displaces, are similar to those discussed for bioethanol. There are also concerns about the amount of water required to grow oil-producing plants in some areas.

TABLE 7-1	Comparing Biodiesel with Petrodiesel
Advantages of Biodiesel	**Disadvantages of Biodiesel**
Biodiesel produces significantly fewer air pollutants, other than NO_X, due to its lower aromatic content and to better combustion efficiency resulting from the presence of oxygen in its molecular structure.	Biodiesel has a slightly lower energy content (about 7–9%, depending on feedstock), though this is partially offset by its superior combustion efficiency.
Biodiesel has a significantly higher *flash point* so it is safer to handle.	Biodiesel is more viscous, though much less so than SVO, so its use in cold climate winters can be problematic.
Biodiesel is a better lubricant, and reduces long-term engine wear.	Biodiesel undergoes degradation by reaction with air during long-term storage due to the presence of $C=C$ bonds in the original oils, which make the carbon chain more susceptible to oxidation.
Biodiesel biodegrades faster (owing to its oxygen content, a point of enzyme attack) in freshwater and soil, and is much less toxic.	Biodiesel may attract water from atmospheric moisture owing to the presence of mono- and diglycerides not completely esterified by methanol. The water causes corrosion of the engine's fuel system, reduces the efficiency of combustion, can speed the gelling of the fluid at cool temperatures, and accelerates the growth of fuel-plugging microbe colonies.
Growth of plants for biofuel absorbs much of the CO_2 subsequently emitted during its production and combustion; thus it reduces by about half the greenhouse gas emissions from diesel fuel combustion.	The CO_2 reduction is more than offset for decades if new land must be cleared to grow the plants.

7.14 Algae as the Raw Material for Biodiesel

There has been much excitement in recent years over the prospect of using **microalgae** as a new renewable energy source. The basic idea is simple: to grow huge masses of algae in water contained in open ponds or in networks of clear-plastic closed *photo-bioreactors* in geographic regions having ample sunshine, and to convert this algae to a useful fuel. Carbon dioxide gas emitted from nearby fossil-fuel power plants or cement production facilities could be bubbled into the watery mix to accelerate algal growth and capture the CO_2 before it escapes into the air. However, the practical mass production of single-species algae has *not* yet been achieved, but is a very active field of research and development.

The open ponds would be about 0.3 m(1 ft)in depth; the plastic reactors would be about 0.1 m in diameter.

An appealing characteristic of algae is its oil content, with some strains consisting of over 50% triglycerides. This oil could presumably be converted into biodiesel as an alternative to using crops such as soybeans or palms grown for that purpose. The yield of biodiesel from algae per square kilometer could greatly exceed even that of tropical palm oil. Unfortunately, achieving a high yield of triglyerides requires the algae to be nutrient-deprived, which slows their production.

One of the great advantages of algae production over biodiesel crops is that no land need be cleared of its existing vegetation or no existing farmland converted from food production for the purpose. Indeed, the ponds or tubes in which the algae grow—the *bioreactors*—can in principle be located in deserts. As a consequence, algae production does not begin with the huge carbon deficit from clearing land for biodiesel or bioethanol production that is required directly or indirectly.

In addition, the algae grow quickly; about 1% of sunlight is absorbed and converted to biomass, which is small by comparison to ~10–15% capture by solar cells, but the initial capital costs are much less. In some instances, the yield of oil was reported to be about 0.6% of incident sunlight energy, which is much higher than that for oil-bearing plant crops used for biodiesel. The water used in the bioreactor or ponds need not be pure—wastewater can be used and may even be advantageous if it contains some of the nutrients required for algal growth (nitrogen and phosphorus). However, any commercial-scale production would be water-intensive as well as energy-intensive since large masses of water need to be pumped between ponds and the algae must be extracted. A commercial operation would have to be located close to a concentrated source of CO_2 since algal growth from atmospheric carbon dioxide would probably be too slow to be commercially viable.

One problem associated with the triglycerides derived from algae is that they are *polyunsaturated*: their fatty acid chains often contain four or more C═C bonds. This makes the resulting biodiesel fuel even more susceptible to oxidation during storage than biodiesel produced from crops such as soybeans. To prevent this deterioration, partial hydrogenation of the oil can be performed. In this catalytic process, **hydrogen gas, H_2,** is added to C=C bonds, thereby converting them to C—C linkages, which are much less reactive.

$$C=C + H_2 \longrightarrow H-C-C-H$$

Of course, the hydrogenation adds an element of nonrenewability to the biofuel since hydrogen usually is derived from natural gas (Section 7.22).

Other problems in algae systems are that in open reactors many strains of algae are present, and that the water tends to evaporate. The reactors must cover a wide area, since, if the algae layer is more than a few centimetres thick, insufficient light reaches underneath. Since the oil-rich types of algae grow more slowly than do the oil-poor ones, the open bioreactor is gradually taken over by the oil-poor strains, and consequently the yield of oil drops.

To overcome these problems, some researchers use enclosed tubular photo-bioreactors. However, they greatly increase the capital cost of setting up the system since expensive metal support structures are used to hold the tubes vertically. Troughs that lie on the ground would be a less expensive alternative, but a workable system has yet to be devised. In either case, the reactors need to be sprayed with water to keep them cool. A two-stage solution being tried is to grow the algae initially in the tubular reactors, where the strains can be limited to those with high oil content. Later, the mixture is transferred to an open system and allowed to continue growing. In any photo-bioreactor, significant amounts of energy must be used to continuously stir the watery mix and/or keep it flowing through the tubes, so that all the algae are continuously exposed to sunlight.

Once the algae have been separated from the water by centrifugation, their cell walls need to be broken and the oil can then be solvent-extracted using hexane, which as in conventional biofuel production would be recycled. Both the centrifugation and cell breakdown processes are energy-intensive.

After extraction, the oil would be partially hydrogenated (if necessary) and the triglycerides reacted with methanol and sodium hydroxide in the usual manner to produce biodiesel from plant oils, as previously described. The plant residue that remains after the oil is extracted consists of carbohydrates and protein. The former could potentially be converted to ethanol and the latter used for animal feed. In some proposals, the carbohydrates would undergo anaerobic decomposition to produce methane (biogas) that could be burned to generate the electricity required to power the mixing and flow of the growing algae solutions, the drying of the algae, and other aspects of the production process. Some authors maintain that without this replacement of the electrical power and heat that would be produced by burning fossil fuels, the net energy balance in generating biodiesel from algae would be negative.

Another biofuel option being explored is the production of ethanol from blue-green algae.

PROBLEM 7-4

When double bonds are hydrogenated, the process of H_2 addition is somewhat exothermic. Does hydrogenation therefore increase or decrease the heat energy obtained when the biofuel eventually is burned? Is this an advantage or a disadvantage?

ACTIVITY

By surveying the service stations in your area or researching on the web, determine which of the following alternative vehicular fuels are available and what is the price per unit volume of each: CNG, LPG (propane), M_{85}, M_{10}, E_{100} or E_{85}, B_5 or any other biodiesel blend, and hydrogen. For each available fuel, determine by web research or library resource its energy content per unit volume, and then calculate its price per megajoule. Repeat the research and calculations for conventional gasoline and diesel fuel in your area. Are any of the alternative fuels competitive in cost with gasoline and diesel?

7.15 Green Chemistry: Bio-based Liquid Fuels and Chemicals

The routes to liquid fuels, such as ethanol and biodiesel, from renewable resources have taken many paths with varying degrees of success. This will continue to be a vigorously pursued area of research and development for many years to come as these fuels will play a significant role in the transition to a sustainable world.

In 2010 a company called *LS9* was recognized with a Presidential Green Chemistry Challenge Award for such a development. The LS9 process has the capability of producing not only fuels such as bio-derived diesel, but also many other chemicals that can be used to produce such materials as surfactants, lubricants, polymers, and soaps (Figure 7-4). The transformations to

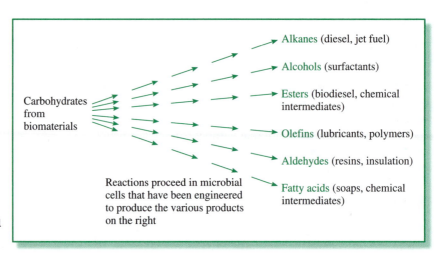

FIGURE 7-4 LS9 microbial production of bio-based liquid fuels and chemicals

useful products are accomplished via biochemical reactions in microbial cells that have been bioengineered to produce selected products from a wide array of biomaterial feedstocks, which include such carbohydrates and carbohydrate-rich materials as glucose, arabinose, xylose, mannose, sucrose, glycerin, and materials resulting from the hydrolysis of stover, wood chips, and bagasse. These substances are converted within the microbial cells by a series of complicated biochemical pathways to produce a multitude of products, as indicated in Figure 7-4.

What makes the LS9 method stand out from similar processes is that the feedstocks are converted to the desired compounds completely *within* the microbial cell in high yields, using non-food-based feedstocks. In addition to biochemical steps, other processes generally require additional chemical steps, which require significant quantities of chemicals (of particular note are heavy metals), energy, and isolation reactions, and which result in significant quantities of waste. In the LS9 process, not only are the compounds produced completely via the biochemical steps within the cell, but the products are easily isolated from the cells, which may then be reused.

While the LS9 process offers the potential to commercially produce fuels and other chemicals from bio-based feedstocks, the production of diesel fuel is the furthest along the road to development. Compared to petrodiesel, LS9 diesel is lower in nitrogen and sulfur, and contains no benzene, lead, manganese, mercury, or arsenic. In addition to being produced from non-food-based renewable resources, perhaps the most significant environmental advantage of LS9 diesel is the dramatic reduction in greenhouse gas emissions that results. Based on a lifecycle assessment, using Brazilian sugar cane as the feedstock to produce LS9 diesel, there is an 85% reduction in greenhouse gas emissions when compared to diesel produced from petroleum.

7.16 Green Chemistry: Recycling Carbon Dioxide—A Feedstock for the Production of Chemicals and Liquid Fuels

As has become apparent in the previous two chapters, the combustion of fossil fuels produces tremendous quantities of carbon dioxide that are altering the natural science of our planet. Thinking of this carbon dioxide as *waste* from the viewpoint of the cradle-to-grave paradigm, the fossil fuels are the "cradle" and carbon dioxide is the "grave." However, one popular saying reminds us that "one man's trash is another man's treasure." In the case of waste carbon dioxide, we saw (Chapter 2) that it can be used as a blowing agent and a solvent (Chapter 3), and in the example below, we see that it can be used as a feedstock for the production of liquid fuels (gasoline and diesel) and chemicals (polymers and pharmaceuticals). Thus this is a **cradle-to-cradle** approach (see Table 0-1 on pages xxiii–xxiv).

Carbon dioxide is not only readily available at no cost as a waste product, it is also nontoxic, and nonflammable. At first glance carbon dioxide might appear to be an ideal feedstock to produce liquid fuels and other organic compounds, especially from the viewpoint of carbon neutrality. The problem with this scenario is that carbon dioxide is very stable and fully oxidized, as evidenced by the energy given off by the burning of fossil fuels (see Chapter 6). However, we have seen that plants in fact do use it as a feedstock—through photosynthesis they use energy from the Sun to convert carbon dioxide to organic molecules. Some of these plant materials were buried, and, in the absence of oxygen, over the course of hundreds of millions of years, these compounds were converted into methane, crude oil, and coal.

The supplies of fossil fuels are finite and unfortunately we do not have hundreds of millions of years to wait for nature to produce additional supplies of them from carbon dioxide; hence we refer to fossil fuels as nonrenewable resources. So the question becomes, can we mimic nature but do so in much shorter periods of time? A team of investigators, led by James Liao at the University of California Los Angeles (UCLA), was awarded a Presidential Green Chemistry Challenge Award in 2010 for a process by which carbon dioxide can be converted into alcohols, which can then be used as liquid fuels and as feedstocks for chemicals.

Liao's group bioengineered a photosynthetic bacteria to convert carbon dioxide into **isobutanol,** $(CH_3)_2CHCH_2OH$, using the energy of the Sun. Like algae production, this process does not require agricultural land or compete with the production of food or result in high economic, energy, and environmental costs since it does not involve the deconstruction of biomass materials (cellulose and triglycerides) into simpler components and their subsequent refinement. Longer-chain alcohols such as *n*-butanol and isobutanol also offer higher energy density and greater compatibility with existing energy infrastructure (engines, pipeline, gas pumps, etc.) than does ethanol. Isobutanol also can be readily dehydrated to *isobutene*, which is a precursor to several different industrial chemicals such as polymethyl methacrylate (*Plexiglas*).

Although commercialization of this bioengineering process is 5–10 years down the road, Liao calculates that this process has the potential to produce about 55 billion gallons of isobutanol as a fuel to replace about 25% of the gasoline used in the United States, as well as 5 billion gallons of isobutanol to be used as a chemical feedstock. This would result in an 8.3% reduction of U.S. carbon dioxide emissions or 500 million tons. In 2010 Liao received a $4 million grant from the U.S. Department of Energy to further develop his work.

PROBLEM 7-5

Draw the molecular structure of the four alcohol isomers of butanol, $C_4H_{10}O$. Which ones have been proposed as a possible biofuel?

Thermochemical Production of Fuels, Including Methanol

Remarkably, the high-temperature decomposition of any carbon-containing substance—whether contemporary plant matter or fossil fuel—can ultimately lead to a variety of useful substances, especially when oxygen is excluded from the decomposing mixture. Indeed, such techniques were used in previous centuries to produce many organic compounds. The reactions that occur are strictly chemical in nature, in contrast with the ambient-temperature biological (fermentation) processes we discussed for bioethanol production. In this section, we discuss the production of various alternative fuels by such **thermochemical processes.** In each of them, heat is first used to decompose the original material into fragments, which then are catalytically reformed into the desired products.

7.17 Pyrolytic Production of Bio-oil

The least sophisticated thermochemical process for biomass utilization is **pyrolysis.** The dried cellulosic biomass is heated at ambient pressure *in the absence of air* to 300–600°C. The biomass substances decompose and rearrange their atoms to form other compounds. The final products of the process are

- a solid, which is mainly charcoal, and contains a significant fraction of the energy value of the original biomass;

- noncondensable gases, including hydrogen, methane, carbon monoxide, and carbon dioxide; and

- **bio-oil** (*pyrolysis oil*) formed from the aerosols and condensable gases, which contains a variety of oxygen-containing organic compounds.

This bio-oil is *not* the same as the product sold in beauty products.

The heat for the endothermic processes, including biomass drying, is often supplied by burning a portion of the original biomass.

The bio-oil liquid is quite acidic (pH about 3) owing especially to the presence of short-chain carboxylic acids. Its lignin content is also quite high. Although considered a biofuel, it has a fairly low energy content since the original potential of the biomass is split up between the solid charcoal phase, the gases, and the liquid. Bio-oil is not miscible with gasoline or diesel fuel. Although it will combust and can be used as heating oil, it is not suitable as a vehicular fuel due to its corrosiveness, as well as its poor energy content. It also slowly deteriorates over time with exposure to air.

The charcoal can be burned as fuel, or buried in soil to promote fertility and to store carbon. The gases could be used as synthesis gas, as described in the next section.

Research has centered upon upgrading bio-oil into a hydrocarbon-like liquid that would be compatible with gasoline or diesel fuel. One such method uses a catalyst that produces mainly the BTX aromatic compounds useful in gasoline. Alternatively, the oil can be catalytically hydrogenated and de-oxygenated using the same techniques used in refining crude petroleum (Box 6-2) to produce a liquid suitable for blending with gasoline or diesel fuel. Of course, the renewable aspect of bio-oil-derived fuels is dependent upon the source of any hydrogen gas used in its transformation: as we shall see later, there are renewable methods for H_2 production but currently most of it is derived from natural gas.

7.18 Synthesis Gas

The most widely used thermochemical method is **gasification.** Here, the carbon-containing material is subject to such high heat (700–1000°C) that it decomposes into the very stable gases **molecular hydrogen,** H_2, and **carbon monoxide,** CO, along with some residual tar which is cleaned out of the reactor. The decomposed material usually is a fossil fuel, as discussed in detail below, but it can also be biomass—e.g., low-grade wood or crop wastes. The high heat is supplied by burning some of the fuel, or potentially by concentrated *solar thermal energy,* as discussed in Section 8.14.

The gas mixture of H_2 and CO is called **synthesis gas,** since it can be used as the versatile reactant from which many different organic compounds can be created. Alternatively, it can be used as a gaseous fuel. Indeed, at the turn of the twentieth century and for several decades thereafter, the synthesis gas produced when white-hot coal reacts with steam was itself used as the fuel in many municipal street lighting systems around the world.

The approximate empirical formula for coal is simply C(s).

$$C(s) + H_2O(g) \longrightarrow CO(g) + H_2(g)$$

The production of synthesis gas from coal is the first step in the innovative **Integrated Gasification Combined Cycle** (IGCC) technology. The gas is then cleaned of impurities arising from the coal, such as SO_2, mercury, and particulates, and then burned to produce electricity via a steam turbine. The advantages of IGCC over a conventional coal-fired power plant are that CO_2 capture for subsequent sequestration is more straightforward since it and water are the only products of combustion, and that impurities in the coal are more readily and completely removed before combustion (Section 3.19). The very high cost of constructing such power plants has prevented their widespread adoption, but this situation could change if carbon capture (Chapter 6) is mandated in the future.

When used to create organic compounds, the synthesis gas is reacted under high pressure and temperature and in the presence of the specific catalyst that will hold the molecules in the correct positions such that the desired product is obtained. The hydrogen-to-carbon monoxide ratio in the synthesis gas must correspond to that in the product for efficient reaction to occur. One important such reaction was used to produce *synthetic gasoline,* the empirical formula for which we represent simply as CH_2. The synthesis reaction then is

$$H_2 + 2\,CO \longrightarrow \underset{\substack{\text{synthetic} \\ \text{gasoline}}}{CH_2} + CO_2$$

Hydrogen acts as a reducing agent in the reaction, since the *average* oxidation number of the carbon changes from $+2$ to $+1$. Note that carbon in the two carbon-containing products is both more reduced and more oxidized than it was in CO, a characteristic of all the synthetic fuels—including methanol, discussed below—produced from synthesis gas.

This process, generally called the *Fischer-Tropsch synthesis*, can be fine-tuned by variation in catalyst, temperature, and pressure, to yield alkane-like hydrocarbons of varying average lengths. It was used by Germany in World War II and by South Africa during the apartheid era to produce synthetic crude oil when their imports of petroleum were restricted.

Very different relative amounts of hydrogen and carbon monoxide result from different starting materials. In order to obtain the ratio appropriate to the desired end product, the **water-gas shift reaction** can be applied to the initial mixture. This equilibrium reaction requires the addition of carbon dioxide gas to the mixture if more CO than present initially is desired, or the addition of steam to increase the amount of hydrogen:

$$CO_2 + H_2 \xrightleftharpoons{\text{catalyst}} CO + H_2O$$

7.19 Methanol Production from Synthesis Gas

Methanol, CH_3OH, is a widely used chemical that can also be used as a fuel component in vehicles that are adapted for its use, as discussed in detail in Section 7.20. Most methanol is produced catalytically from synthesis gas that has a 2:1 molar ratio of H_2 to CO, reflecting the amounts of the three elements in the alcohol:

$$2\ H_2 + CO \xrightarrow{\text{catalyst (Cu/ZnO)}} CH_3OH$$

Unfortunately, existing catalysts allow only a partial conversion (about one-fifth) of the gases into methanol for each pass of the gas mixture over the catalyst, and the processes are energy-intensive and require relatively high temperatures. Research is under way to develop catalysts that will operate at lower temperatures and thereby allow higher product yields.

PROBLEM 7-6

The enthalpies of formation of CO(g) and CH_3OH(l), respectively, are -110.5 and -239.1 kJ mol^{-1}. Calculate the enthalpy of the reaction above that forms methanol from synthesis gas. From your answer, predict whether the equilibrium amount of methanol obtained will increase or decrease as the temperature is lowered. Given your result, comment on the interest in developing low-temperature catalysts. ●

The correct 2:1 molar ratio of H_2 to CO required for the methanol synthesis described above is rarely obtained initially from the raw materials. For example, a 3:1 ratio is produced by the reaction of methane with steam, as discussed below. **The water-gas shift reaction is used to reduce the ratio to obtain more CO at the sacrifice of some hydrogen.** To calculate the amount of adjustment required, let us denote the initial molar amount of CO initially available as the variable a; thus the initial amount of H_2 is $3a$. Since the hydrogen is initially in excess, the appropriate direction for the shift reaction is indeed the direction in which it is written above. When this reaction achieves equilibrium, a molar amount x of hydrogen will have been consumed and the same additional molar amount, x, of carbon monoxide will have been produced:

$$CO_2 + H_2 \longrightarrow CO + H_2O$$

from initial reaction $3a$ a

at new equilibrium $3a - x$ $a + x$

The value of x is obtained by requiring that the new, equilibrium ratio of H_2 to CO be 2:1:

$$\frac{3a - x}{a + x} = \frac{2}{1}$$

By algebraic manipulation of this equation, the ratio of x to a can be obtained:

$$x/a = 1/3$$

Thus the fraction $x/3a$ of the initial amount of H_2 from natural gas that must be sacrificed to produce sufficient CO is 1/9.

The two chemical reactions which, when combined, correspond to the conversion of methane into the correct 2:1 ratio are shown and added together below; it has been assumed for simplicity that $a = 1$ here

$$CH_4 + H_2O \longrightarrow CO + 3\,H_2$$

$$1/3\,H_2 + 1/3\,CO_2 \longrightarrow 1/3\,CO + 1/3\,H_2O$$

sum $CH_4 + 2/3\,H_2O + 1/3\,CO_2 \longrightarrow 4/3\,CO + 8/3\,H_2$

When combined by using a catalyst, the products from the sum of these reactions will yield 4/3 mol of CH_3OH according to the net overall reaction

$$CH_4 + 2/3\,H_2O + 1/3\,CO_2 \longrightarrow 4/3\,CH_3OH$$

PROBLEM 7-7

In order to synthesize a compound with the empirical formula CH_3O (and no other products), what ratio of H_2 to CO would be required? What fraction of the hydrogen gas originally present in a 3:1 ratio to CO would have to be sacrificed to produce carbon monoxide using the water-gas shift reaction in order to accomplish the transformation?   ●

Research is currently under way to find a method to directly convert the methane from natural gas into methanol in a much more efficient manner than described above. Most of the difficulty stems from the fact that CH_4 is a very unreactive molecule: its C—H bond-dissociation energy is the highest of all the alkanes. Once one C—H bond is broken, however, the molecule becomes highly reactive because the other C—H bonds are weakened, and in the presence of oxygen it tends to oxidize completely to carbon dioxide rather than partially to a useful intermediate stage such as methanol.

Although methanol can be used in some vehicles as a fuel, it can also be converted to ethanol or to synthetic gasoline, as indicated in Figure 7-5, where in schematic form the various manipulations for different starting materials and end-products, most involving synthesis gas, are summarized. The production of methanol, using biomass to generate synthesis gas, would produce a renewable biofuel, **biomethanol.** One difficulty in using wet biomass is the need to dry it before or during gasification, which consumes energy. Some progress has been made in developing catalysts that would allow gasification to occur using supercritical water, thereby avoiding the need for biomass drying.

As indicated in Figure 7-5, methanol can also be produced renewably if the hydrogen gas is produced renewably—e.g., from solar energy—rather than from a fossil fuel. Some of the CO_2 captured from fossil-fuel power plant emissions could be used as a reactant:

$$CO_2(g) + 3\ H_2(g) \xrightarrow{\text{catalyst}} CH_3OH(l) + H_2O(l)$$

Since this reaction is only slightly exothermic, most of the fuel energy of the hydrogen is present in the biomethanol product. Based upon equilibrium considerations, methanol formation would be favored by low temperatures and high pressures; research has centered upon finding a catalyst that will efficiently operate under such conditions without being deactivated.

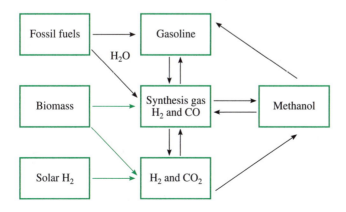

FIGURE 7-5 Scheme for the production of fuels in a hydrogen economy.

7.20 Methanol as an Alternative Fuel

Methanol, CH_3OH, is a colorless liquid that, like ethanol, is somewhat less dense than water. Although methanol was produced in the past from the destructive distillation of wood, giving rise to its historical name, *wood alcohol*, it is now produced mainly from fossil fuel via the synthesis gas process described in Section 7.19.

Methanol can be blended with gasoline to produce a fuel that burns more cleanly than gasoline. In a labeling scheme analogous to that used for ethanol–gasoline mixtures, blends of methanol are designated by an M rating; thus M_5 corresponds to 5% methanol and 95% gasoline. Although it is not widely available elsewhere, China has recently made the use of M_{15} methanol–gasoline blends a national priority, and has begun to produce methanol-fueled vehicles that will use M_{85}.

One disadvantage of methanol blends is that the pure alcohol is only soluble to the extent of about 15% in gasoline, corresponding to M_{15}; greater amounts of methanol form a second layer rather than dissolve. The inadvertent presence of water causes this unacceptable phase separation to occur at an even smaller percentage of methanol. Additives such as *tertiary-butyl alcohol* (i.e., 2-methyl-2-propanol) that are soluble in both methanol and gasoline prevent such separations from occurring. Looking at the solubility plot from the other direction, gasoline is moderately soluble in methanol, so fuel blends such as M_{85} have been tested and are now on sale in limited quantities. Another difficulty is that methanol cannot be used in conventional automobile engines because it reacts with and corrodes some engine and fuel-tank components.

Some concern has been expressed about the safety of methanol for use as a vehicular fuel, given the toxicity of the compound. Methanol–water solutions have been widely used as windshield-washer liquids in northern climates for many years without much environmental impact. However, the use of methanol as a fuel may be more dangerous as it would involve a very high concentration of the alcohol. Ethanol is much less toxic than is either methanol or gasoline.

However, alcohols also possess some advantages: they are inherently high-octane fuels and indeed methanol is used to power all the cars at the *Indy 500* races. Methanol has the added advantage that it does not produce a fireball when a tank-rupturing crash of racing cars occurs: it vaporizes less quickly than does gasoline, and once formed, the vapor disperses more quickly.

PROBLEM 7-8

Given that the enthalpies of combustion, per mole, of methanol and ethanol are -726 and -1367 kJ, and that the density of each is 0.79 g mL^{-1}, calculate the heat released by methanol and by ethanol (a) per gram and (b) per milliliter. From your results, comment on the superiority of one alcohol or the other with respect to energy intensity on a weight and on a volume basis. Are these alcohols superior or inferior to methane as fuels in terms of energy intensity per gram? How do they compare to gasoline, for which about 43 kJ are released per gram? ●

7.21 Ethers as Fuel Additives

Methanol can be used to produce **dimethyl ether,** CH_3OCH_3, which has been tested as a replacement for diesel fuel in trucks and buses:

$$2\ CH_3OH \longrightarrow CH_3OCH_3 + H_2O$$

This ether is nontoxic and degrades easily in the atmosphere, and, in fact, is used as a propellant in spray cans. Since its molecules contain no C–C bonds, soot-based particulate matter is produced in its combustion in only very small quantities (see Chapter 3) compared to diesel fuel. The NO_X emissions from dimethyl ether combustion are also lower than usually found for diesel engines.

Methanol is also used to produce the oxygenated gasoline additive MTBE (**methyl tertiary-butyl ether**), the structure of which is

$$\begin{array}{c} CH_3 \\ | \\ H_3C-O-C-CH_3 \\ | \\ CH_3 \end{array}$$

methyl *tert*-butyl ether (MTBE)

MTBE, octane number 116, is used in North American and European unleaded gasoline blends—up to 15%—to increase their overall octane number, and also to reduce carbon monoxide (and unburned hydrocarbon) air pollution since, like the alcohols, it is an "oxygenated" fuel and generates less CO during its combustion than would the hydrocarbons it replaces. The advantages of using MTBE rather than ethanol as an additive are that it has a higher octane number and that it does not evaporate as readily. However, as in the case with alcohols, its combustion can also produce more aldehydes and other oxygen-containing air pollutants than result from the hydrocarbons it replaces.

The usage of MTBE itself has become controversial. It has an objectionable odor resembling turpentine and ether. Another problem associated with MTBE is its contamination of well water, which has now occurred at many sites in the United States. The sources of MTBE in well water are leaking underground fuel tanks, leaking pipelines, and spillage of gasoline at gas stations and vehicle accidents and by homeowners. In contrast to the hydrocarbon components of gasoline, MTBE is rather soluble in water and therefore is quite mobile in soil and groundwater. It is also quite resistant to biological degradation because its carbon chains are very short; its half-life is of the order of years.

Various U.S. states and the EPA have set action levels for MTBE in drinking water at a few tens of parts per billion, values that are exceeded in some water supplies and at which the odor of the additive is apparent. Because of concerns about well-water contamination, California and many other states have now banned MTBE in gasoline; indeed, its use as a gasoline additive in the United States has dropped sharply, being replaced by ethanol, isooctane, and other high-octane substances.

Review Questions 1–19 are based on the material above.

Hydrogen—Fuel of the Future?

Hydrogen gas can be used as a fuel in the same way as carbon-containing compounds; some futurists believe that the world will eventually have a hydrogen-based economy. Hydrogen gas combines with oxygen gas to produce water, and in the process it releases a substantial quantity of energy:

$$H_2(g) + 1/2\ O_2(g) \xrightarrow{\text{spark}} H_2O(g) \quad \Delta H = -242 \text{ kJ mol}^{-1}$$

The idea that hydrogen would be the ultimate fuel of the future goes back at least as far as 1874, when it was mentioned by a character in the novel *Mysterious Island* by Jules Verne. Indeed, hydrogen has already found use as fuel in applications for which lightness is an important factor, namely, in powering the Saturn rockets to the moon and the U.S. Space Shuttles.

Hydrogen is superior even to electricity in some ways, since its transmission by pipelines over long distances consumes less energy than does the transmission through wires of the same amount of energy as electricity, and since batteries are not required for local storage of energy.

However, as we shall see in the material that follows, **the substantial technical problems in the production, storage, transportation, and usage of hydrogen—plus the need to create a new infrastructure for it—means that a hydrogen economy is probably still many decades away.** Indeed, almost all the hydrogen now produced is not used as a fuel, but rather to produce ammonia as fertilizer or to upgrade the quality of fossil fuels such as those from the Alberta oil sands, and in the gasification of coal and the production of synthetic liquid fuels.

The Haber process is used to produce ammonia:
$$N_2 + 3\ H_2 \longrightarrow 2\ NH_3$$

It is important to realize that hydrogen is *not* an energy *source*, since it does not occur as the free element on the Earth's crust. Hydrogen gas is an **energy vector** (*energy carrier*) only; it must be produced, usually from water and/or methane, by the *consumption* of large amounts of energy and/or of other fuels. The industrial infrastructure that would be required to produce enough hydrogen to fuel all the vehicles in the United States is enormous, since it would require about as much energy as the current electric power capacity.

7.22 Producing Hydrogen from Fossil Fuels

Virtually all (>95%) the hydrogen gas currently produced in the world is obtained by reactions of fossil fuels, i.e., nonsustainably. **Natural gas or**

petroleum or coal is reacted with steam to form hydrogen and carbon dioxide. The energy value of the fuel is transferred from carbon to water's hydrogen atoms; chemically speaking, the reduced status of the carbon is transferred to the hydrogen. The net reactions, assuming petroleum to have the empirical formula CH_2 and coal to be mainly graphite, are

$$CH_4 + 2\ H_2O \longrightarrow 4\ H_2 + CO_2$$

$$CH_2 + 2\ H_2O \longrightarrow 3\ H_2 + CO_2$$

$$C + 2\ H_2O \longrightarrow 2\ H_2 + CO_2$$

The relative importance of sources for hydrogen are natural gas > petroleum > coal.

Notice that as much carbon dioxide is produced in this way as would be obtained by combustion of the fossil fuels in oxygen.

The actual conversions occur in two steps:

• First the fossil fuel reacts with steam to yield carbon monoxide and some hydrogen, e.g.,

$$CH_4(g) + H_2O(g) \longrightarrow CO(g) + 3\ H_2(g)$$

In the case of natural gas, the steam is preheated to 1000°C.

Steam required for the process could be generated by concentrated solar process, or by burning some of the methane.

• Then the CO/H_2 synthesis gas mixture and additional steam are passed over a suitable catalyst to obtain additional hydrogen and complete the oxidation of the carbon by the water-gas shift reaction driven in the opposite sense to that shown previously, i.e.,

$$CO + H_2O \longrightarrow CO_2 + H_2$$

Hydrogen gas could be produced in a renewable way from biomass grown for this cycle. In addition, research indicates that aqueous solutions of both glucose and glycerin can be decomposed at moderate temperatures (225–265°C) and pressures (27–54 atm) with a platinum-based catalyst to produce hydrogen and carbon dioxide.

Associated with every conversion of one fuel to another are energy losses, mainly to waste heat. Some of these are dictated by the second law of thermodynamics (to be discussed in Section 8.16) and therefore cannot be avoided. For example, energy from natural gas can be transferred to hydrogen with an efficiency of about 72%, or from coal by 55–60%, by the processes described above. Thus, if the resulting fuel is used only to generate heat, significantly less CO_2 is emitted if the original fossil fuel is burned directly rather than being first converted to hydrogen.

7.23 Producing Hydrogen by Electrolysis

The most expensive commercial way to produce hydrogen is by **electrolysis of water,** using dc electricity generated by an energy source:

The water is made alkaline and conducting by the addition of potassium hydroxide, KOH.

$$2\ H_2O(l) \xrightarrow{\text{electricity}} 2\ H_2(g) + O_2(g)$$

Unfortunately, a quarter to one-half of the electrical energy is unavoidably converted to heat and therefore wasted in this process. A few percent of hydrogen currently is produced in this way, presumably using power that would otherwise be wasted.

There are prototype plants in Saudi Arabia and in Germany that use electricity from solar energy to produce hydrogen, a process now 13% efficient overall. The stored energy is later recovered by reacting the hydrogen with oxygen. Excess electricity from hydroelectric or nuclear or wind power installations—i.e., power generated but not required immediately for use—could likewise be used to produce hydrogen by electrolysis of water and later recovered. A hope for the future is that solar energy collected via photovoltaic collectors or wind power will be economically efficient in providing electricity to generate large quantities of hydrogen.

One catalyst that has been found to convert sunlight into hydrogen by electrolyzing water is **titanium dioxide,** TiO_2. A small potential is applied to the electrode in the cell's operation. Titanium dioxide is stable to sunlight (unlike many other potential light-absorbing materials) and cheap, but on its own absorbs only ultraviolet light. By blending carbon into it so that carbon atoms replace some of the oxide ions, the efficiency in producing hydrogen gas is increased eightfold, to more than 8% of the Sun's energy. The addition of carbon extends absorption into the visible region (to 535 nm), thereby greatly increasing the efficiency in producing hydrogen gas.

PROBLEM 7-9

Determine the longest wavelength of light that has photons that are capable of decomposing liquid water into H_2 and O_2 gases, given that for this process $\Delta H = +285.8$ kJ mol^{-1} of water. In which region of the spectrum does this wavelength lie? [*Hint:* Recall from Chapter 1 the relationship between reaction heat and light wavelength.]

Even better than the use of solar electricity to electrolyze water would be the **direct photochemical decomposition of water** into hydrogen and oxygen by absorbed sunlight, although no practical, efficient method has yet been devised to effect this transformation. One of the difficulties in using sunlight to decompose water is that H_2O does not absorb light in the visible or UV-A regions. Thus some substance must be found that can absorb sunlight, transfer the energy to the decomposition process, and finally be regenerated. The substances devised to date for this purpose are very inefficient in converting sunlight. In addition, since the light-absorbing substances and others required are not 100% recoverable at the end of the cycle, they must be continuously resupplied; thus the hydrogen that is produced is not really a renewable, sustainable fuel.

7.24 Hydrogen from Water: Thermochemical Cycles

In principle, the thermal conversion of sunlight into heat can produce temperatures hot enough to decompose water into hydrogen and oxygen. Research in Israel, using a *solar tower* of mirrors (Section 8.14) to concentrate sunlight by a factor of 10,000 and thereby produce temperatures of about 2200°C in a reactor, has succeeded in splitting about one-quarter of water vapor at low pressures into H_2 and O_2.

The process would produce a mixture of hydrogen and oxygen, which then would have to be separated.

Various **thermochemical cycles** by which water can indirectly be decomposed by heat into hydrogen and oxygen have been proposed. Ideally, such cycles should operate at moderate temperatures, be efficient in conversion of heat into hydrogen, and not degrade the reactants so they can be recycled.

Perhaps the most practical such scheme is the **sulfur–iodine cycle,** in which *elemental iodine* is first reduced by *sulfur dioxide*, SO_2, to *hydrogen iodide*, HI:

$$I_2 + SO_2 + 2\,H_2O \longrightarrow 2\,HI + H_2SO_4 \qquad \text{at } 120°C$$

The hydrogen iodide is then thermally decomposed to yield hydrogen gas, recovering the elemental iodine, and the *sulfuric acid*, H_2SO_4, is thermally decomposed to yield oxygen gas, recovering the sulfur dioxide:

$$2\,HI + \text{heat } (320°C) \longrightarrow H_2 + I_2$$

$$H_2SO_4 + \text{heat } (830°C) \longrightarrow SO_2 + H_2O + 1/2\,O_2$$

Since the reactants HI and SO_2 are recovered in high yield, the cycle can be repeated over and over. Heat produced by nuclear reactors could drive this cycle, which has a conversion efficiency of about 50%. However, designing materials to withstand high-temperature concentrated sulfuric acid remains a challenge.

Other thermochemical cycles requiring very high temperatures are discussed in Section 8.15 on solar thermal energy.

7.25 Storing Hydrogen: As a Liquid or Compressed Gas

In rocketry applications, hydrogen is stored as a liquid, as is the oxygen. Since hydrogen's boiling point of only 20 K (−253°C) at 1 atm is so low, **much energy must be expended in keeping liquid hydrogen very cold, after considerable energy has been used to liquefy it.** This drawback effectively limits the applications of liquid hydrogen to a few specialized situations in which its "lightness" (low density) is the most important factor. Evaporative losses during inactive periods and upon refueling probably prevent the use of liquid hydrogen in consumer vehicles; such losses are minimized when large volumes are involved, so the liquid remains the best option for storage at present.

Hydrogen can be stored as a highly **compressed gas,** in much the same way as is done for methane in the form of natural gas. However, compared to CH_4, it has a drawback: a much greater amount of H_2 gas needs to be stored in order to release the same amount of energy. Compared with methane, the combustion of one mole of hydrogen consumes only one-quarter the oxygen, and consequently generates about one-quarter the energy, even though both occupy equal volumes under the same pressure (ideal gas law). Thus the "bulky" nature of hydrogen gas limits its applications (see Problem 7-11).

Nevertheless, most automobile manufacturers plan to use compressed hydrogen to fuel their first set of hydrogen cars, since the other alternatives are even less viable for small vehicles. With current technology, the fuel tank can be refilled within three to five minutes. Most trial-scale hydrogen bus fleets use compressed hydrogen on the vehicle itself, though it may be stored in liquid form at the depots. A pressure of about 700 atm (70 MPa) of hydrogen will likely be used for vehicular tanks made from carbon fiber; even higher pressures do not store much more of the gas since it is much less compressible than an ideal gas under these extreme conditions (see Problem 7-12).

It is instructive to compare the volumes of hydrogen required to fuel a hydrogen-fuel-cell car (assuming 50% efficiency) to travel 400 km (240 miles), approximately the distance one can obtain in an efficient gasoline-powered car with a tank capacity of 40–50 L. The amount of hydrogen required is 4 kg, which occupies

- 45,000 L, or 45 m^3, i.e., a balloon having a 5-m diameter, or a cube 3.6 m on each side, if it existed as a gas at normal atmospheric pressure, or

- 100 L (about 25 gal, equal to several normal-sized gasoline tanks) as a gas compressed by a pressure of about 700 atm (now achievable), or

- 56 L as a liquid (or solid) maintained at −252°C (at 1 atm pressure), or

- 35–75 L if stored as a metal hydride, if effective systems can be developed, as discussed below.

Overall, the problem of devising a practical, economical, and safe way of storing hydrogen has not yet been achieved, and in the eyes of some analysts, "no breakthrough is yet in sight" despite much interest and research activity. The performances of different systems for storing hydrogen are compared with each other and to gasoline and diesel fuel in Figure 7-6. The significant mass of the storage containers is included in the values for the liquefied and compressed element. Liquid hydrogen comes closest to the fossil fuels in terms of the weight of hydrogen per total system mass.

Hydrides and other chemical storage methods for hydrogen are discussed in detail below. No practical system discovered to date has reached the target of the U.S. Department of Energy (Figure 7-6) in terms of combining high density with a high percentage of 6% of hydrogen in its total mass by 2010. The target for 2015 of 9% hydrogen by weight with a volumetric density of over 80 g L^{-1} will perhaps also prove overly optimistic.

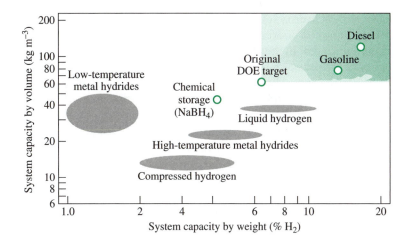

FIGURE 7-6 The performance of various hydrogen storage techniques. Note that both horizontal and vertical scales are logarithmic. [Source: R. F. Service, "The Hydrogen Backlash," *Science* 305 (2004): 958–961.]

PROBLEM 7-10

Assuming that the energy released by combustion of H_2 is proportional to the amount of oxygen it consumes, estimate the ratio of heat released by one mole of methane compared to one mole of hydrogen gas.

PROBLEM 7-11

Using the thermochemical information in the H_2 combustion equation above, calculate the enthalpy (heat) of combustion of hydrogen per gram, and by comparing it to that of methane (see page 235) decide which fuel is superior on a weight basis. By comparing the actual energy released by combustion per mole of gas—and hence per molar volume—decide which fuel is superior on a volume basis.

PROBLEM 7-12

For H_2 gas compressed to 700 atm at 27°C, calculate its density in kilograms per cubic meter, assuming (incorrectly) that it behaves as an ideal gas. Given that its actual density at that pressure is only 37 kg m^{-3}, calculate the percentage error in the calculated density due to the nonideal behavior of the gas under this extreme pressure. (Recall that for an ideal gas $PV = nRT$, and that $R = 0.082$ L atm mole^{-1} K^{-1}.)

7.26 Storing Hydrogen: Metal Alloys and Graphite

A practical and safe way to store hydrogen for use in small vehicles may be in the form of a **metal hydride.** Many metals, including alloys, chemically absorb large amounts of hydrogen gas reversibly—as a sponge absorbs water. The molecular form of hydrogen becomes dissociated into atoms at the surface of the metal as it is absorbed, and forms metal hydrides by incorporating the small atoms of hydrogen in "holes" in the crystalline structure of the metal. **Thus the hydrogen exists as atoms, not molecules, within the lattice,** which expands slightly to incorporate the atoms. For example, titanium metal absorbs hydrogen to form the hydride of formula TiH_2, a compound in which the density of hydrogen is twice that of liquid H_2! Heating the solid gradually releases the hydrogen as a molecular gas, which then can be burned in air or oxygen to power the vehicle.

Research continues in the effort to find a light metal alloy that can efficiently store hydrogen without making the vehicle excessively heavy. Even existing metal hydride systems are lighter than are the pressurized tanks needed to store liquid hydrogen. Most industrial research now centers on metal systems. Practical considerations require that an alloy to store hydrogen

- be capable of quickly and reversibly absorbing hydrogen,

- not become brittle after many repeated cycles of absorption and desorption,

- operate in the pressure and temperature ranges of 1–10 atm and 0–100°C,

- not be so dense as to weigh down the vehicle excessively, and

- not require a huge volume.

Lanthanum–nickel alloys derived from $LaNi_5$ have all of these characteristics except one: they are too heavy (hydrogen mass % < 2), a deficiency shared by all known metal hydrides that operate near ambient temperature. Many lighter hydrides and alloys, such as MgH_2 and Mg_2NiH_4, are known, but they do not operate reversibly under moderate conditions. Research on systems formed by the lighter metals continues, but as yet is unsuccessful in producing alloys that fulfill all five conditions listed above.

One of the practical difficulties in using hydrogen as a fuel is its tendency to react over time with the metal in pipelines or storage containers in which it is present. This reaction embrittles the metal and it deteriorates, eventually forming a powder. Recent progress has been made in overcoming this difficulty by using composite materials rather than simple metals as the structural materials for storage and transport facilities.

PROBLEM 7.13

Calculate the mass of titanium metal required to absorb each kilogram of hydrogen and form TiH_2 in a "tankful" of hydrogen. Repeat the calculation for magnesium if the hydride has the formula MgH_2. Which metal is superior for storage of hydrogen from a weight standpoint? ●

7.27 Storing Hydrogen as a Compound

For some applications, it may be feasible to transmit and store hydrogen in the form of an *energy-dense liquid* such as methanol and to generate it as required for power generation. *Toluene*, C_7H_8, has also been proposed as a hydrogen carrier over long distances; it could be dehydrogenated when the H_2 is required. The same is true for *formic acid*, HCOOH.

Boron compounds are attractive carriers of hydrogen since the element is so lightweight. One such system is an alkali salt (lithium or sodium) of the **borohydride ion,** BH_4^-. For their prototype minivan that ran on fuel cells, *Chrysler* used a 20% solution of sodium borohydride in water to "store" hydrogen. Hydrogen is released when the solution is pumped over a ruthenium catalyst, prompting the redox reaction of the H^- in borohydride with the H^+ in water to produce H_2:

$$BH_4^- + 3\ H_2O \longrightarrow 4\ H_2 + H_2BO_3^-$$

The density of hydrogen in the borohydride solution is comparable to that in liquid hydrogen. The analogous *alanate ion*, AlH_4^-, in the form of its sodium salt is also a candidate for hydrogen storage in vehicular fuel cells:

$$2\ NaAlH_4 \longrightarrow 2\ NaH + 2\ Al + 3\ H_2$$

Several molecular boron compounds, including **ammonia borane,** BH_3NH_3, and an organoboron–phosphorus system, have recently been proposed as hydrogen carriers. Ammonia borane releases hydrogen molecules sequentially in three stages, the first two of which occur at modest temperatures ($<100°C$), forming a polymer having the approximate empirical formula BNH:

$$\underset{\text{polymer}}{BH_3NH_3 + \text{heat} \longrightarrow [BNH]_x} + 2\ H_2$$

Extracting the last hydrogen from the BNH polymer would produce a BN polymer that cannot be reused.

None of the chemical systems described above could have the hydrogen source regenerated onboard a vehicle, however; the products would have to be returned to a central facility for recycling. For example, the $H_2BO_3^-$ product in the borohydride system would later be reacted with *magnesium hydride*, MgH_2, which itself is produced by reacting magnesium metal with hydrogen gas. Ammonia borane can be reformed from its BNH polymer by reaction with *hydrazine*, N_2H_4, obtained from ammonia, which in turn is formed by the Haber process from hydrogen and nitrogen gases.

PROBLEM 7-14

Calculate the percentage by mass of hydrogen in ammonia borane, and recalculate the mass percentage of hydrogen that is released when the BNH polymer is formed. Do both these values meet the DOE guideline for minimum percentage of releasable hydrogen? ●

7.28 Combusting Hydrogen

Hydrogen gas can be combined with oxygen to produce heat by conventional flame combustion, or via low-temperature combustion in catalytic heaters. The combustion efficiency, i.e., fraction of energy converted to useful energy rather than to waste heat, is approximately 25%, about the same as for gasoline. The main advantages to using hydrogen as a combustion fuel are its low mass per unit of energy produced, and the lesser (but not zero) quantity of polluting gases its combustion produces, when compared to other fuels.

Although it is sometimes stated that hydrogen combustion produces only water vapor and no pollutants, this in fact is not true. Of course, no carbon-containing pollutants, including carbon dioxide, are emitted with hydrogen. Since combustion involves a flame, however, some of the nitrogen from the air that is used as the source of oxygen reacts to form nitrogen oxides, NO_X. Some *hydrogen peroxide*, H_2O_2, is released as well. Thus, hydrogen-burning vehicles are not really zero-emission systems. It is true that the lower flame temperature for the $H_2 + O_2$ combustion, compared to that for fossil fuels with oxygen, inherently produces less NO_X, perhaps two-thirds less. The nitrogen oxide release can be eliminated by using pure oxygen rather than air to burn the hydrogen; alternatively, it can be reduced by passing the emission gases over a catalytic converter or by lowering the flame temperature as much as possible, which can be achieved by reducing the H_2/O_2 ratio to half the stoichiometric amount.

Hydrogen is sometimes considered to be a dangerous fuel due to its high flammability and explosiveness; it ignites more easily than do most conventional fuels. On the positive side, however, spills of liquid hydrogen rapidly evaporate and rise high into the air.

7.29 Generating Electricity by Powering Fuel Cells with Hydrogen

Hydrogen and oxygen can be combined in fuel cells in order to produce electricity (a hydrogen technology also used in space vehicles). Fuel cells are similar in operation to batteries except that the reactants are supplied *continuously*. In the hydrogen–oxygen fuel cell, the two gases are each passed over separate electrodes that are connected

• by an external electrical connection through which electrons travel

and also

• by an electrolyte through which ions travel.

The components of a fuel cell, then, are the same as an electrolysis operation in which water would be split into hydrogen and oxygen, but the chemical reaction that occurs is exactly the opposite. Instead of electricity being used to drive the electrolysis reaction, it is produced. Fuel cells have the advantage over combustion in that a more useful form of energy is produced (electricity rather than heat), and the process creates no polluting gases as by-products. In principle, the only product of the reaction is water. At the catalytic surface of the first electrode, the H_2 gas produces H^+ ions and electrons, which travel around the external circuit to the second

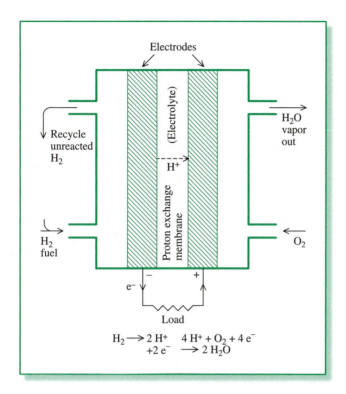

Electrodes

(Electrolyte)

H_2O
vapor
out

Recycle
unreacted
H_2

H^+

Proton exchange
membrane

H_2
fuel

O_2

e^- −   +

Load

$$H_2 \longrightarrow 2\,H^+ \qquad 4\,H^+ + O_2 + 4\,e^-$$
$$+2\,e^- \longrightarrow 2\,H_2O$$

FIGURE 7-7 Schematic diagram of a hydrogen–oxygen fuel cell (PEMFC version).

electrode, across which O_2 gas is bubbled (see Figure 7-7). Meanwhile the H^+ ions travel through the electrolyte, and recombine with the electrons and O_2 to produce water at the second electrode.

Although some of the reaction energy in a hydrogen fuel cell is necessarily released as heat—about 20%, due to requirements of the second law of thermodynamics—the majority of it is converted to electrical energy associated with the current that flows between the electrodes. Electric motors, whether in a fuel cell or vehicle battery, are 80–85% efficient in converting electrical to mechanical energy. Real fuel cells overall are now about 50–55% efficient; 70% efficiency may be attained eventually. By contrast, internal combustion engines using gasoline are 15–25% efficient, diesels 30–35%.

Although commercial electric cars are currently powered by batteries, future improvements in technology could lead to their replacement by fuel cells. *Mercedes-Benz* began to produce small numbers of fuel-cell cars in 2010. Several other car makers have prototype vehicles ready. Prototype buses running in Vancouver and Chicago use innovative fuel cells for their power source. The electrolyte used in the fuel cells of these vehicles is a hair-thin (about 100 mm) synthetic polymer that acts as a *proton-exchange membrane*. The membrane, when moist, conducts protons well by incorporating sulfonate groups. It also keeps the hydrogen and oxygen gases from mixing.

FIGURE 7-8 Operating characteristics of various fuel cells. [Source: B. C. H. Steele and A. Heinzel, "Materials for Fuel Cell Technologies," *Nature* 414 (2001): 345.]

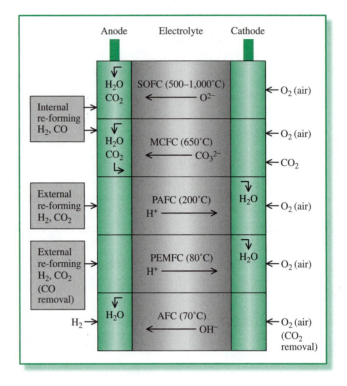

The electrodes of such **polymeric electrolyte membrane fuel cells** (PEMFC in Figure 7-8) are composed of graphite, with a small amount of platinum dispersed as nanoparticles in thin layers (about 50 μm thick) on its surface. Each cell, which operates at about 80°C, generates about 0.8 V of electricity, so many must be stacked together in order to provide sufficient power for the vehicle. In the current versions of the buses, compressed hydrogen is stored in tanks under the roof of the vehicle.

The incentives to develop vehicles that use fuel cells powered by hydrogen are

• to reduce urban smog, which is partially driven by emissions from gasoline and diesel engines,

• to reduce energy consumption, since fuel cells are much more efficient in producing motive power than are combustion engines, and

• to reduce carbon dioxide emissions, since fuel cells powered by hydrogen are carbon-free.

Some analysts point out that the cost of improving air quality and CO_2 emissions by switching the transportation system to such vehicles over the next few decades is much higher than alternative strategies, including scrapping old cars, improving vehicle fuel efficiency, reducing NO_X emissions from power plants, and capturing and sequestering CO_2 emissions from power plants. In addition, hydrogen leaked into the air acts indirectly as a greenhouse gas, since it reacts with, and decreases the concentration of, OH, a result that will slightly increase the atmospheric lifetime of methane and hence its concentration, since the main sink for CH_4 is its reaction with OH, as discussed in Chapter 3.

7.30 Obtaining Fuel-Cell Hydrogen from Liquid Fuels

Because of the limited practicality of transporting hydrogen around in individual cars and trucks, however, there is active research in designing systems that allow it to be extracted as needed from liquid fuels, which are much more convenient to transport. For example, in the near future, the hydrogen may instead be obtained as required from liquid methanol by on-board decomposition of the latter to hydrogen gas using the reverse of the methanol-formation reaction previously discussed:

$$CH_3OH \longrightarrow 2\,H_2 + CO$$

In the *General Motors* version of this process, the "reformer" unit operates at 275°C and uses a copper oxide/zinc oxide catalyst to promote the reaction. The water-gas shift reaction is subsequently used to react the CO in the synthesis gas with steam and provide additional H_2 gas, giving the overall reaction

$$CH_3OH + H_2O \longrightarrow 3\,H_2 + CO_2$$

Similar processes have been developed that convert gasoline, diesel fuel, octane, or aqueous ethanol into carbon dioxide and hydrogen. Hydrogen can be obtained from gasoline or methanol with about 75% efficiency, and this may be the source of hydrogen in fuel-cell vehicles. Unfortunately, the current PEMFC and alkaline fuel cells (see Problem 7-15), as well as the one based upon phosphoric acid, all require relatively pure hydrogen, free especially of carbon monoxide—a gas which is formed in the reformer process and difficult to eliminate completely. CO bonds to the sites of the catalyst (e.g., platinum) intended to promote the fuel-cell reaction and blocks the catalytic activity there.

Even concentrations of carbon monoxide greater than 20 ppm in the hydrogen gas slows down most fuel cells appreciably. Perhaps a CO-tolerant electrode catalyst, possibly one incorporating a second metal or a metal oxide as well as platinum, will be developed in the future to overcome this problem by oxidizing the adsorbed CO to carbon dioxide. Hydrogen that is virtually free of carbon monoxide could be produced from methanol by an oxidative steam reforming process at 230°C:

$$4\,CH_3OH + O_2 + 2\,H_2O \longrightarrow 10\,H_2 + 4\,CO_2$$

Since oxygen is involved as a reactant, however, not all the fuel value of methanol is captured in the hydrogen product.

An alkaline electrolyte can be used in the H_2—O_2 fuel cell to replace the acidic environment mentioned above (see Figure 7-8). Assuming that the reaction of O_2 with water and electrons produces hydroxide ions, OH^-, and that these hydroxides travel to the other electrode where they react with hydrogen to give up electrons and produce more water, deduce the two balanced half-reactions and the balanced overall reaction for such a fuel cell. (The *alkaline fuel cell*, labeled AFC in Figure 7-8, is used in space shuttles and Apollo spacecraft to provide electricity.)

7.31 Fuel Cells for Power Plants

Fuel cells may also be used in the near future in small electric power plants, partially because pollutant emissions from them are so small compared to those from fossil-fuel combustion (only about 1% of the NO_X, for example). Indeed, **phosphoric acid**-based fuel cells (PAFC in Figure 7-8) have been operating since the early 1990s in some hospitals and hotels to generate power. The most promising fuel cells for power plants involve a **molten carbonate** salt—potassium and lithium carbonates, plus additives, at 650°C for example—as the electrolyte. (The *molten-carbonate fuel cell* is labeled MCFC in Figure 7-8.) The hydrogen gas reacts with **carbonate ions,** CO_3^{2-}, to produce carbon dioxide, water, and electrons at the anode, whereas at the cathode, the carbon dioxide reacts with electrons and oxygen from air to reform the carbonate ions. The carbon dioxide must be recycled from the anode back to the cathode during the cell's operation. The hydrogen is produced on-site by reaction of methane with steam, so CH_4 is the actual source of the fuel energy here. By-product waste heat can be recovered and used in a cogeneration mode, and the dc produced by the fuel cell is converted to ac for distribution. MCFC efficiency approaches 50%.

Like the molten-carbonate fuel cell, that based upon **solid oxides** (see SOFC, Figure 7-8) is much more tolerant of carbon monoxide impurities in the fuel hydrogen gas than are the other fuel cells. The electrolyte in the solid-oxide fuel cell is a ceramic mixture of oxides of *zirconium* and *yttrium*. The **oxide anion,** O^{2-}, produced from oxygen gas carries the charge between the electrodes, travels through the solid from cathode to anode as the fuel is consumed, and forms water at the anode. The high operating temperature (up to 1000°C) of the solid-oxide cell allows fuel to be reformed internally and to form hydrogen ions without the use of expensive catalysts, so methane or other hydrocarbons can be used as the fuel instead of hydrogen. However, carbon deposits tend to form at the anode and stick to it at the high operating temperatures that are involved with these units.

The solid-oxide fuel cell, like the molten-carbonate one, is practical for central power plants, and its fuel efficiency is also high. Both types suffer from practical problems with electrodes, such as carbon deposition. This latter problem for solid-oxide fuel cells can be overcome by operating at a lower temperature.

PROBLEM 7-16

Obtain balanced half-reactions for the processes at the two electrodes, and add them together to determine the overall reaction in the molten-carbonate fuel cell.   ●

7.32 Other Uses for Fuel Cells

The first prototypes of consumer-electronics products powered by fuel cells began to enter the market in 2011. **Many observers believe that laptop computers will eventually be powered by fuel cells, rather than by rechargeable batteries.** These fuel cells will use methanol, rather than hydrogen, directly as fuel, and contain removable cartridges containing the alcohol. The advantage for the user of fuel cells over batteries is a much longer working time before the power runs out. Using methanol or natural gas, rather than H_2, directly as the fuel in fuel cells does avoid the problem of the generation and storage of hydrogen.

One problem with methanol is the generation of by-product carbon monoxide, which poisons the catalyst as discussed above. Diluting the methanol with water lessens this problem, but cuts the power output of the cell. In addition, neither methanol nor other liquid fuels that could, in principle, be used directly instead of H_2 as the fuel in a fuel cell, react fast enough to produce the electrical current required in a vehicle, although research on the use of dilute solutions of methanol indicates that these problems may be overcome eventually.

7.33 Electric Cars Powered by Batteries

An alternative to vehicles that use fuel cells are those powered, at least partially, by batteries. In previous decades, prototype "electric cars" used the lead-acid batteries that gasoline-powered vehicles have traditionally employed singly to operate the starter motor and were used to power golf carts, forklift trucks, etc. However, the energy density of lead-acid batteries is too low to make them practical in today's vehicles. The electrical power source used in most "hybrid" vehicles currently on the road is either the nickel–metal-hydride battery or the lithium ion battery.

In the **nickel–metal-hydride battery,** power is produced when **hydroxide ion,** OH⁻, produced by the reduction half-reaction of the *nickel oxyhydroxide*, NiO(OH), electrode with water, forming *nickel hydroxide*, $Ni(OH)_2$, migrates through the aqueous alkaline electrolyte to the electrode composed of a compound of an intermetallic alloy, M, in which *metal hydrides*, MH, are present. The hydrogen of the metal hydride reacts with the hydroxide, forming water and releasing an electron:

$$NiO(OH) + H_2O + e^- \longrightarrow Ni(OH)_2 + OH^-$$

$$MH + OH^- \longrightarrow M + H_2O + e^-$$

The **lithium ion battery** has one electrode composed of *lithium cobalt dioxide*, $LiCoO_2$, to which Li^+ ions migrate from a graphite electrode, through a nonaqueous electrolyte solvent containing complexed lithium ions, when power is drawn.

The turn of the century saw the introduction of hybrid combustion/electric-powered vehicles, such as the Toyota Prius, into the market. Nickel–metal-hydride batteries, which are lighter and more compact than lead ones of equivalent potential, are used in the Prius. Such vehicles overcome the requirement of long and frequent recharging of all-battery systems, since the battery is recharged continuously when the (small) gasoline engine is in operation and is developing excess power. During braking, recovered kinetic energy is channeled back to the battery. Motive power for the vehicle is supplied by both the gasoline and electric motors, the proportion depending upon the driving situation. Hybrid vehicles are highly fuel efficient so they emit much less carbon dioxide, and also emit much less nitrogen oxides, carbon monoxide, and VOCs than do conventional vehicles.

The Chevrolet Volt, a *plug-in hybrid* introduced in late 2010, uses power from a lithium ion battery for the first 80 km (50 miles) or less of driving, beyond which a gasoline motor takes over. The battery can be recharged overnight at the owner's residence, a process that requires about 12 h from regular 120-V outlet but only 4 h if a 240-V supply is available. Regenerative braking also helps recharge the battery during driving, as in the Prius.

The lithium ion battery in the Volt weighs 180 kg (400 lb).

The practical difficulties that discourage widespread adoption of electric vehicles are their high cost, the low driving range between battery recharges, the length of the battery recharge period, and the weight of the batteries. Like fuel-cell systems, they have the attraction of zero emissions and little operating noise during electrical operation. Of course, there is pollution emitted into the environment when the electricity required for these cars is generated in the first place.

Review Questions 20–28 are based upon the above material.

In the future, electric cars will probably use improved lithium-based batteries that have a much higher energy density, closer to that of gasoline, than the lithium-ion system. *Lithium–sulfur* and especially the *lithium–air* batteries currently under development promise energy densities many times as great.

Review Questions

1. Define the term *renewable energy*.

2. List three incentives supporting the development and use of biofuels.

3. Describe what is meant by the term *carbon debt* for biofuels.

4. Name some advantages and disadvantages of burning biomass to produce electricity.

5. What role does vapor pressure play in the composition of ethanol–gasoline mixtures

designed for automobiles? What is meant by E_{10} fuel?

6. What are the advantages and disadvantages of using ethanol blends in gasoline fuels with regard to air pollution?

7. Describe the method used in producing bioethanol in volume for use as a fuel. What are the potential feedstocks for this process?

8. What are the highly energy-intensive steps involved in the production of corn ethanol? What is mean by a *co-product*? Why isn't ethanol a fully "renewable" fuel?

9. What is meant by the term *cellulosic ethanol*? Describe one likely source of biomass for its production. How does a *second-generation biofuel* differ in nature from a *first-generation* one?

10. What is the nature of the viscosity problem in using straight vegetable oils as diesel fuels? Is the viscosity problem completely solved by biodiesel?

11. In terms of their chemical composition, describe what is meant by a *fatty acid* and how it differs from a *triglyceride* and the ester used in biofuel. What is *glycerin*?

12. What vegetable oil sources are commonly used as raw materials for biodiesel?

13. List some advantages and disadvantages of producing and using biodiesel rather than petrodiesel.

14. List some of the advantages and disadvantages of producing biofuel from algae compared with agricultural biomass. What is a *photo-bioreactor*?

15. What is meant by the *thermochemical production* of a fuel?

16. What are the three products of the *pyrolysis* of biomass?

17. What is meant by the term *synthesis gas*?

18. What is the *water-gas shift reaction*? Describe the methods by which methanol can be produced in volume for use as a fuel.

19. What does M_{85} mean? What are some advantages and disadvantages of using methanol as a vehicular fuel?

20. What is the difference between an *energy source* and an *energy carrier* (vector)? Into which category does H_2 fall?

21. Write the reaction scheme by which hydrogen is produced from natural gas.

22. Describe the production of hydrogen gas by electrolysis. Can solar energy be used as an energy source for hydrogen production? Why isn't water decomposed directly by absorption of sunlight?

23. Write the reactions by which hydrogen can be produced from water by a thermochemical method involving iodine.

24. Describe the three ways in which hydrogen can be stored in vehicles for use as a fuel, and discuss briefly the disadvantages of each method.

25. Describe two ways hydrogen can be stored in compounds of boron.

26. Does the burning of hydrogen really produce no pollutants? Under what conditions do no pollutants form?

27. Describe how a hydrogen fuel cell works. What other types of fuel cells exist?

28. Describe the chemical nature of the electrodes in (a) a nickel–metal-hydride battery and (b) a lithium ion battery.

🟢 Green Chemistry Questions

1. The development of bio-based fuels and chemicals by *LS9* won a Presidential Green Chemistry Challenge Award.

(a) Into which of the three focus areas (see page xxviii) for these awards does this award best fit?

(b) List four of the twelve principles of green chemistry (listed on pages xxiii–xxiv) that are addressed by this discovery. Explain each of your answers.

2. What are the environmental advantages of the bio-based fuels and chemicals developed by LS9?

3. The development of a biochemical process for the conversion of carbon dioxide into isobutanol won a Presidential Green Chemistry Challenge Award.

(a) Into which of the three focus areas (see page xxviii) for these awards does this award best fit?

(b) List three of the twelve principles of green chemistry (listed on pages xxiii–xxiv) that are

addressed by this discovery. Explain each of your answers.

4. What are the environmental advantages of the conversion of carbon dioxide to isobutanol developed by James Liao?

Additional Problems

1. The burning for energy production of scavenged wood and wood-chip waste has been proposed as a way of reducing CO_2 emissions, even though the combustion produces larger amounts of carbon dioxide per unit of heat produced than burning fossil fuels. Explain the rationale behind this proposal.

2. Consider the use of methanol, CH_3OH, as an oxygenated liquid fuel for suitably modified cars:

(a) By writing the balanced chemical equation for its combustion in air, determine whether it is more similar to coal, oil, or natural gas in terms of the joules of energy released per mole of CO_2 produced.

(b) Determine the balanced equation by which methanol can be produced by reacting elemental carbon (coal) with water vapor, given that CO_2 is the only other product in the reaction. Use the change in oxidation numbers to perform the balancing, rather than considering the intermediate synthesis gas reactions.

(c) Does the combined scheme of parts (a) and (b) represent a way of using coal but producing less carbon dioxide per joule than through its direct combustion? Explain your answer.

3. Deduce the fraction of the CO or H_2 produced by the reaction of coal with steam that must be converted to H_2 or CO, respectively, by the water-gas shift reaction in order to obtain a 2:1 ratio of hydrogen to carbon monoxide that is required to synthesize methanol. Deduce also the net reactions of conversion of coal to methanol, and

confirm that your answer is the same as that reached in the problem above.

4. Deduce the balanced reaction in which synthesis gas is formed by combining equal volumes of methane and carbon dioxide. From enthalpy of formation data given in this chapter, deduce the enthalpy change for this reaction. By applying LeChatelier's principle, deduce whether the conversion of the gases to carbon monoxide and hydrogen will be favored by low or by high pressures, and by low or by high temperatures. Combining these results with those obtained in Additional Problem 3, determine the fraction of the total carbon dioxide in the production and combustion of methanol synthesized from this synthesis gas that would be "renewable", i.e., recycled from the consumption of carbon dioxide in the process.

5. As discussed in Chapter 6, Canada has massive supplies of very heavy oil in oil sands, which are being used to make gasoline by combining them with natural gas. Assume that the empirical formulas of these three fuels are CH, CH_2, and CH_4, respectively, and that gasoline is made by hydrogenating the tar with hydrogen produced from natural gas in its reaction with water to produce CO_2 and H_2. Combine the hydrogenation and hydrogen production equations so as to use all the H_2, and thereby deduce the overall reaction of CH and CH_4 with steam to produce gasoline and carbon dioxide.

8

Renewable Energy Technologies
Hydroelectric, Wind, Solar, Geothermal, and Marine Energy and Their Storage

In this chapter, the following introductory chemistry topics are used:

- ➲ Electrochemistry: oxidation numbers; redox half-reactions; batteries; electrolysis; ac versus dc
- ➲ Crystalline versus amorphous solids
- ➲ Electronic structure of atoms; bonding versus antibonding electrons
- ➲ Ideal gas law; gas density
- ➲ Heat capacity

Background from previous chapters used in this chapter:

- ➲ Greenhouse gases (carbon dioxide, methane, nitrous oxide)
- ➲ Anaerobic decomposition; eutrophication
- ➲ Light absorption as energy; photons; solar spectrum; IR; albedo, excited states
- ➲ Hydrogen as a fuel

Introduction

In this chapter, the use of solar energy—whether by direct absorption or in indirect forms such as wind—to generate electricity and heat renewably and sustainably is discussed. We first consider *indirect* solar forms such as hydroelectric, wind, wave, and tidal energy. We also describe geothermal energy, which is largely renewable although not solar-based. *Direct* solar energy—whether used to generate electricity using solar cells or by first creating high-temperature heat sources—is then discussed, as are the thermodynamic limitations on the latter conversion processes.

A generic problem with solar, wind, wave, and tidal energy sources is their variability over short periods of time and total unavailability for hours or days; the consequent need to store electrical power using batteries and by other methods is also discussed and current advances in the area outlined.

Hydroelectric Power

8.1 Potential and Usage

Of all the forms of renewable energy, hydroelectric power is by far the most important, with a capacity of about 1000 GW. Worldwide, it constitutes over 80% of renewable energy (other than that based upon biomass), 16% of global electric capacity, and 3% of global commercial energy. Although its availability depends upon seasonal rainfall patterns in some areas, overall it is a more reliable resource than solar or wind energy.

Hydroelectric power is an indirect form of solar energy. In the hydrological cycle, the Sun's energy evaporates water from oceans, lakes, rivers, and the soil and transports the H_2O molecules upward in the atmosphere via winds. After the water molecules condense to raindrops, they still possess considerable potential energy owing to their elevation, only a portion of which is dissipated if they fall onto land or a water body that lies above sea level. We can harness some of its remaining potential energy by forcing the downward-flowing water to turn turbines and thereby generate electricity.

Although there are small hydroelectric installations that use the flow of a river, most large-scale facilities use dams and waterfalls where the water pressure—and hence the power yield—is much greater. In particular, the energy imparted to a turbine by falling water is directly proportional not only to the volume of the water but also to the height from which it falls. For this reason, new hydroelectric projects usually involve the construction of a high dam along the path of a flowing river. Water then collects behind the dam and its level rises to a considerable height. The water that is allowed to flow over the top of the dam falls a considerable distance before encountering the turbines positioned near the bottom. Unfortunately, the collection of water behind the dam floods considerable areas of land, creating a lake having environmental problems such as those discussed below.

Hydropower is a widespread resource. If all sites worldwide were exploited, the total amount of energy that could be obtained from hydroelectric sources is about 100 EJ per year; about 20% of this total is obtained at present. Most of the sites in developed countries that require little modification to use, and that are located within a reasonable distance of centers that use considerable electric power, have already been exploited. For example, about 75% of the feasible sites in Europe are already exploited.

To use a common expression, most of the "low-hanging fruit" has already been picked.

However, there are many river systems in developing countries, especially in Africa, South America, and Asia, where considerable new hydroelectric power is currently being developed by the construction of dams. An example is the Merowe Dam in northern Sudan, which dams the Nile River. Globally, more than 100 GW of hydropower capacity is under construction. The construction cost of hydroelectric power ranges from 1 to 5 million dollars per megawatt capacity, depending upon the size and site of the operation. Changes in rainfall patterns, as well as the melting of glaciers, resulting from climate change could reduce the future capacity of hydropower in some regions.

A multitude of new hydroelectric dams are planned for the Amazon River in South America and the Mekong River in Asia.

8.2 Environmental Problems

Although hydropower is often thought of as pollution-free, there are environmental and social costs associated with it, especially ones resulting from the creation of the reservoirs behind dams. The most important of these costs include:

- **the displacement of human populations** from lands flooded to create reservoirs;

- **the eutrophication of water in reservoirs;**

- **the release of greenhouse gases,** especially methane, from flooded areas;

- **the release of mercury** into reservoir water, and consequently into the fish that swim in the water and human populations that eat the fish; this topic is discussed in more detail in the Case Study at the website for this chapter entitled *Mercury Pollution and the James Bay Hydroelectric Project (Canada)* and in Chapter 12;

- **the devastation to fish populations,** such as salmon, from the blockage of their migratory routes by dams;

- **the buildup of silt behind dams,** with the result that less silt is carried to locations farther along the waterway.

Unfortunately, the construction of new hydroelectric projects involving the damming of river systems, especially in developing countries, often proceeds without adequate environmental assessment and planning ahead of time. The World Bank and several other large financiers of such hydroelectric facilities do insist on an independent assessment of the project's impacts before they provide financial assistance.

The largest hydroelectric project in the world is the 26-turbine *Three Gorges Dam* in China, completed in 2009, which provides 18 MW of power—equivalent to five large coal-fired power plants—and cost $25 billion to construct. Although more than a million people had to be relocated to avoid being flooded by the artificial lake, the dam also controls flooding on the Yangtze River, and thereby saves thousands of lives.

Side effects of the dam that have proven to be worse than anticipated include riverbank erosion and deteriorating water quality in the tributaries.

The expansion of wetlands that occurs by the deliberate flooding of land to produce a large, deep reservoir of water generally creates a long lake covering hundreds or thousands of square kilometers. The Three Gorges Dam resulted in a lake 660 km long! The deep water in such lakes is largely anaerobic, especially if the flooded land was not first cleared of vegetation. The anaerobic decomposition of the original trees, bushes, etc. on the land produces carbon dioxide and methane in almost equal volumes, both of which escape from the surface and enter the atmosphere. The emission especially of methane from such reservoirs is significant since it is such a powerful greenhouse gas.

Deep, small reservoirs produce and emit much less methane than do shallow ones that contain large areas of flooded biomass, such as those in the Brazilian Amazon. As pointed out in Section 5.14, the combined global warming effect of the methane and carbon dioxide produced by a large, shallow reservoir created to generate hydroelectric power can, for many years, exceed the carbon dioxide that would be emitted if a coal-fired power plant were used instead to generate the same amount of electrical power!

Even after the original vegetation has decayed, new plants that have grown on the shores of the lakes during the dry season when water levels recede are later engulfed by rising water when the levels rise in the wet season, and decompose, releasing more methane. In some new dams, water from different levels is mixed or released, minimizing the problems of anaerobic regions at lower depths.

Wind Energy

8.3 Introduction

Winds are air flows that result from the tendency of air masses that have undergone different amounts of heating, and that therefore have developed unequal pressures, to equalize those pressures. Air naturally flows from regions of high pressure to those of low pressure. The heating of air results directly or indirectly from the absorption of sunlight; indeed, about 1–2% of the Sun's energy received on the Earth is transformed into wind energy. A large quantity of such indirect solar energy is potentially available as **wind power,** although only 0.05% of it currently is being tapped.

Polar areas receive less sunlight—and therefore less heat—than do the tropics. To reduce the resulting temperature difference between tropical and polar regions, winds arise in the air as do currents in the oceans. Warm air and water are carried toward the poles, whereas cold air and water are transported in the opposite direction, toward the Equator. However, these flows do not follow simple trajectories, owing to factors such as the spinning motion of the Earth around its axis and the effects of local terrain.

The force of the wind can be exploited to do useful work or to generate electrical energy in the same way that the force of flowing water is used in hydroelectric power plants. Historically, the strong, sustained winds in central North America were exploited by windmills to pump water and later to generate small amounts of electricity on individual farms until the middle of the twentieth century. Of course, windmills have been in use in Europe—especially in the Netherlands—for centuries, following their invention in Persia more than a thousand years ago.

8.4 Large-Scale Wind Power

In recent decades, the large-scale generation of electricity by arrays of huge, high-tech windmills gathered in "wind farms" has become feasible. Wind power is currently the world's fastest-growing source of energy as a percentage of its capacity (though not in absolute terms), and has shown exceptional growth since the late 1990s. The cumulative global capacity of wind power had been growing almost exponentially for more than a decade until 2010 at least, as illustrated in Figure 8-1.

The global wind-power capacity in 2010 was almost 200 GW, about 2.5% of worldwide electricity capacity. As of 2010, China and the United States had the greatest amounts of installed capacity, with Germany, Spain, and India following them in that order. Although China had been responsible for about half of new installations in the late 2000s, Europe collectively had half the installed total. It is important to distinguish between installed *capacity* and actual wind energy *production*. For example, although only third in global capacity, Spain produced more wind power in 2010 than any other country, which accounted for about 16% of its total electrical needs. Denmark and Portugal exceeded those percentages.

Technically, several times the world electricity output could be produced from wind. A landmass the size of China would be needed to satisfy world

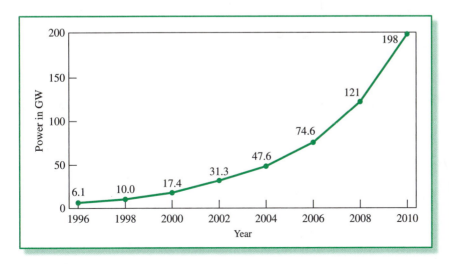

FIGURE 8-1 Cumulative global wind power capacity. [Source: Renewable Energy Policies Network for the 21st Century, *Renewables 2011 Global Market Report*.]

FIGURE 8-2 Percentage of land area estimated to have class 3 or higher wind power in the contiguous United States. [Source: "Wind Energy Resource Atlas of the United States," Chapter 2.]

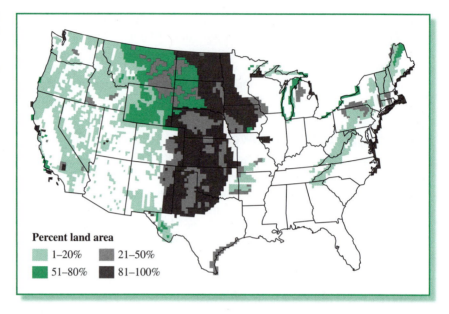

electricity demand from wind alone. More realistically, wind power could be expanded to provide up to perhaps one-fifth of the world's electricity.

If price is not taken into account, then the country with the highest potential for wind power is the United States. Indeed, the United States has enough potential wind power to generate all its electricity now and in the foreseeable future, although it only supplied 1% of it in 2008. About 90% of the U.S. potential for wind power lies in twelve states in the Midwest, ranging from North Dakota to northern Texas (see Figure 8-2), although the main demand for electricity is centered far from most of these areas.

8.5 Wind Speed and Windmill Size

As one would intuitively expect, the greater the velocity, v, of the wind, the greater the amount of energy a windmill or a wind turbine can produce, and this indeed proves to be the case. In fact, the energy increases very sharply with wind speed. **The energy yield from wind is proportional to v^3, i.e., to the third power of the wind speed.** Consequently, a small improvement in velocity produces a large increase in yield; e.g., an increase to 26 from 22 mph improves the energy yield by two-thirds!

The cubic dependence of energy on wind speed is the result of two factors.

• First, the *kinetic energy* of the motion of the air mass in the direction of the wind is proportional to the square of the air speed, since from physics we know that for any moving body, its kinetic energy is given by $mv^2/2$.

• Second, the amount of wind passing over the blades per unit time increases linearly in direct proportion to the wind speed.

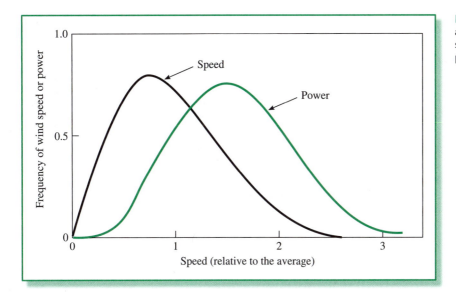

The energy available to the wind turbine is equal to the product of these two factors, so it is proportional to v^3.

Wind speeds vary over a wide range at most locations. A typical distribution is shown by the black curve in Figure 8-3, in which the frequency of occurrence of a given wind speed v is plotted against the speed, normalized to its average value v_{avg}. The peak in the distribution occurs at 0.8 v_{avg}. The resulting distribution of power from such a distribution is shown by the green curve in Figure 8-3, which is displaced strongly to the right of the wind speed distribution, since the power is proportional to the cube of the wind speed. The power distribution peaks at 1.6 v_{avg}. Clearly there is little power to be gained at low wind velocities compared to that at higher velocities.

PROBLEM 8-1

(a) Calculate the ratio of wind energy available for velocity of 8 meters per second compared to that of 5 mps.
(b) In general, by what factor does available wind energy increase for a doubling of wind speed?
(c) In one particular location, the wind blows at 12 mps about 10% of the time and at about 5 mps the rest of the time. What wind speed delivers the most energy overall?   ●

At very high wind speeds, windmills are shut down to prevent damage to their blades. Most windmills operate about three-quarters of the time, but the blades are often turning quite slowly, at considerably less than the value for optimum energy generation but above the "cut-in" speed, below which it is not worthwhile to operate the system. On average, a windmill's *capacity factor*—the ratio of what it actually produces divided by what would be obtained if it operated at its *rated value* 100% of the time—is about one-third according to modern experience. Thus, it would take about 300 MW of rated wind power capacity to replace 100 MW of fossil-fuel power capacity.

The energy the windmill can gather is proportional to the square of its blade length, since the area the blade sweeps out is proportional to the length squared. Since wind speeds increase with height above the ground, a tall turbine is also more efficient for this reason. However, the mass of the turbine rotor increases with the cube of the blade's length, adding disproportionately to the cost as size increases.

Each windmill in a wind farm extracts energy from the flow of air on it, so the individual windmills must be physically separated from each other to some extent. For technical reasons, no more than about one-third of the energy passing by a windmill can be extracted from the flow of air around it.

8.6 Potential Wind-Energy Sites

As a consequence of the local terrain, some geographical regions experience almost constant windy conditions. Geographical areas are commonly classified into seven classes of potential wind-power density, with *class 7* having the highest potential. Ideal locations for wind farms are those having almost constant flow of nonturbulent winds in all seasons. Although wind energy does rise steeply with wind velocity, locations with sudden gusts of high-speed wind are not considered favorably. Locations at less than 2 km altitude, with average wind speeds of at least 5 m s^{-1}, corresponding to 18 km h^{-1} (11 mph), are generally required for a location to be considered economically feasible. Other experts use the criterion of annual mean wind speeds $\geq$6.9 m s^{-1} (25 km h^{-1}, or 15 mph) measured at 80 m, the tip of the blade height of modern windmills, as those suitable for low-cost wind-power generation. Such sites are considered to be *class 3* or higher potential wind-energy sites (see Figure 8-2).

The regions of high wind-power potential at reasonable cost are the United States, Canada, South America, OECD Europe, and the former U.S.S.R. The areas with the lowest potential are in Africa, Eastern Europe, and Southeast Asia. In most areas, the potential exceeds the current electricity usage. Within a given country, the best locations are usually mountain passes, high-altitude plains, and coastal areas. In general, wind speed increases somewhat with altitude.

Many new wind farms are located on seacoasts. Coastal daytime summer breezes arise because of the difference in air density over the water and over the adjacent land. Since sunlight heats dry land more quickly than it does water, the air over the land also becomes warmer than that over the lake or sea. Since warm air rises—due to its lower density (according to the gas law, density is inversely proportional to Kelvin temperature)—and high above the surface moves out to sea, the remaining air over the land surface has a lower density and pressure than that over the sea. Consequently, to equalize pressures, surface air flows from over the sea to the land mass, creating a cooling sea breeze. At night, the situation is reversed, since the land cools more quickly than does the water, producing an outward breeze by the same mechanism.

Consistently breezy offshore shallow areas, such as the sandbanks off the coasts of Denmark and Ireland, are ideal sites and are now used extensively for wind farms. Indeed, offshore locations are popular in Europe, and most are anchored in water 8–10 m deep. Locations in New England, on Lake Erie, and off the coast of mid-Atlantic U.S. states could alone generate up to 20% of the U.S. electricity supply. However, the physical conditions at some potential offshore locations are rather harsh, and it is difficult to service broken turbines in open water. West-coast waters off North America are too deep for placing wind farms, and those in the southeastern United States are too prone to hurricanes.

8.7 Practical Considerations

In comparison to other forms of energy generation—fossil-fuel or nuclear or hydroelectric power plants—wind towers must be numerous and somewhat spread out since each captures a relatively small amount of energy. Indeed, a large "footprint" is characteristic of most forms of renewable energy, including solar cells, since solar energy is so diffuse, as opposed to its concentrated nature in fossil fuels or uranium. However, once they are constructed, the land between the wind towers can be used for other purposes.

In setting up a wind farm, the spacing between windmills must be considered. Air passing over a windmill will have its speed reduced since some of its energy is thereby extracted. In addition, the passage of air over a windmill imparts some turbulence to it. In general, a distance equal to at least five rotor diameters between adjacent windmills is necessary for the wind to come back up to its original speed after passing over one and before reaching the next windmill. An arrangement in which the rows are staggered with respect to each other (ABABAB . . .) gives the most efficient packing arrangement. As a rule of thumb, about one-tenth of a square kilometer of land is required per megawatt of power.

The most efficient and largest commercial wind turbines currently are the 5-MW units—four times the power of the models of the mid-1990s. They have huge, 120-m-long blades—more than ten times the length of those of the early 1990s. The rotor area of such windmills is that of a football field! By contrast, modern coal-fired power plants generate from 125 MW to 1000 MW, so hundreds of windmills would be required to replace the power generated by one coal-fired plant. About 1000 North American homes can be supplied with electricity from a 5-MW system on typical hot afternoons, when the power draw peaks due to air-conditioner usage.

Several companies are currently developing 10-MW turbines.

Of all the forms of renewable energy (other than hydroelectricity), wind power is the most economical. The cost per megawatt of large onshore windmills is about 2 million dollars, or almost 6 million per average delivered megawatt, since they operate on average at one-third their rated capacity. Offshore wind farms generally are somewhat more expensive to install. The cost of generating electricity using modern windmill technology—and feeding it into existing power grids—is now almost competitive with conventional energy sources. As mentioned above, however, there would also be significant transmission costs if wind energy were to be expanded in the U.S. midwestern states. If the world switches eventually to a hydrogen economy, hydrogen generation powered by wind in this area could generate much of the U.S. supply. Currently the electrolyzers required for such a process are expensive.

There exists considerable potential for wind energy to supply a significant fraction of the future electricity of many countries at lower cost to the environment than for any other feasible alternative. The price for new wind-power supplies is comparable to that of new conventional sources such coal- and nuclear-fired power plants, and will probably be lower if any realistic costs associated with the environmental impact of conventional sources is assessed in the future. However, there are unresolved problems of efficient energy storage for locations having intermittent winds that probably prevent the adoption of wind as the main source of electrical energy generation in many locations.

Indeed, many power grids are unwilling to rely on wind for more than a fraction of their energy supply due to its intermittent nature. The temporary storage of excess electrical power from wind and solar sources in order to provide a continuous supply is considered in Sections 8.22 and 8.23.

Some individual buildings, including households that are too remote to be connected to power lines, generate their own electricity using rooftop-mounted wind turbines. When all the power generated is not needed to run the 12-V direct current appliances, etc. within the building, the excess is stored in 12-V batteries, to be drawn upon in times of low or no wind.

8.8 Environmental Issues

In terms of **energy payback—the amount of time required to generate the energy used in constructing the unit**—that for wind is only 3–4 months and similar for the payback period for its CO_2 emissions, and it is about a year for air pollutants emitted. Carbon dioxide emissions from wind power are the least for any electric power source according to Figure 7-1. The CO_2 and air pollutant emissions arise mainly in the manufacture from ores and other raw materials of the steel and aluminum required for the blades, the copper needed for the electric generator and transformer, and the cement required for the windmill's foundations. If the electricity generated by large-scale wind turbines replaces that produced by the combustion of fossil fuels, there is a large net reduction in emissions of carbon dioxide and of air pollutants over the windmill's life span.

There is often public resistance to placing wind turbines in both urban and rural areas owing to their perceived visual unsightliness, although many people consider windmills to be attractive and a sign of the growing presence of renewable energy. To be efficient, wind turbine blades must be enormously long, and as such they constitute an artificial presence that is difficult to ignore. Some people are offended by the sight of their addition to an otherwise natural landscape, particularly when the location is one of natural beauty such as a mountaintop or the seashore.

Ironically, many such attractive locations are ideal from the viewpoints of ample wind and relative closeness to areas that consume substantial electricity. Thus wind turbines are a classic example of the *NIMBY* phenomenon ("not in my back yard"), in which a small set of individuals—in this case, those living near or frequenting the location of a proposed turbine—object to their construction since they are forced to absorb a "penalty" in order that society as a whole experience a small improvement in its overall environment, in this case a small reduction in air pollution and climate change.

Locations least likely to be objected to on aesthetic grounds are those remote from travel, such as large farms (where the only people to see the turbines are likely to be farmers who have a financial interest in their presence), or far offshore in large bodies of water such as the seas or the Great Lakes. For this reason, placing turbines on agricultural land or offshore has become popular. Indeed, many farmers supplement their income by leasing land sites for turbines. Once installed, individual turbines take up relatively little space, allowing almost all of the land around them to be used for farming activities such as cattle grazing.

However, using land that has not previously been developed requires new roads to be built, new power lines installed, and often trees to be cleared. A huge new 1000-MW wind farm, involving more than 300 turbines, has been proposed for central Labrador, in eastern Canada. One advantage of this project is its remote location, hundreds of kilometers from habitation, so any perceived unsightliness of the windmills is not an issue. Windmills in regions that experience very cold temperatures must be protected from ice formation on turbine blades by the use of internal heaters, which thereby consume a small fraction of the windmill's power output.

The pros and cons most often raised about wind power are summarized in Table 8-1. The most controversial issue in many areas is that of low-frequency noise resulting from the operation of the windmill's blades. Especially when the wind farm is located in a rural area—where noise levels at night are characteristically quite low—some people report sleep disturbance.

TABLE 8-1	Pros and Cons of Wind Power
Arguments Against Wind Power	**Arguments in Favor of Wind Power**
Many sites—including offshore ones—are far from centers of demand, requiring long transmission lines to be built.	This is also true of many potential new hydroelectric projects.
Wind power needs some tax incentives to compete with traditional forms of electricity production.	Conventional and nuclear power plants receive much larger, though indirect, subsidies.
The construction of windmills at some remote sites requires roads, forest clearing, and other destructive infrastructure.	
Windmills kill wildlife, especially bats and birds of prey.	Studies show that very few birds are killed by wind turbines, especially compared to the number killed by cars, cats, etc.
Huge areas of land, and therefore of habitat, are required to construct enough windmills to have a substantial effect on electricity supply.	
The continuous motion of the blades produces low-grade noise pollution nearby.	Noise level is comparable to traffic.
Onshore wind farms are a form of "visual pollution."	Sites remote from areas of dense population can be used.
Wind power is usually intermittent, with a lower annual load factor, and requires backup facilities using traditional resources to remain constantly on-call.	Excess wind energy can be stored mechanically by pumping water to elevated storage facilities or in batteries and then used when needed to produce electricity.
	Only small amounts of greenhouse gas emissions are associated with wind energy compared to fossil-fuel combustion. There is no nuclear waste to store or potential radiation problems as occur with nuclear power.

Marine Energy: Wave and Tidal Power

Wave power and **tidal power,** sometimes collectively known as **marine energy,** can be obtained in many coastal regions of the world and is competitive economically in niche markets. It is estimated that about 20 EJ of power is potentially recoverable annually from waves and tides.

The source of the energy of tides is the gravitational influence of the Sun and the Moon on the water mass. In some locations, coastal currents generated by tides can be exploited to turn submerged turbines mounted on pipes that fit into holes drilled in the seafloor. Because water is so much denser than air, slow currents—about 10 km h^{-1} are best—driving such "submerged windmills" efficiently generate electricity. Much slower speeds, as occur in the deep ocean where collectively most of the tidal energy is found, are too slow to be viable. Shallow seas surrounding the ocean perimeters are the best locations for tidal power.

Tides cause large masses of water to be lifted and then lowered twice a day. If the tides in a coastal basin are generally high, a gate that can be opened or closed can be built across the basin. When the tide is coming in, the gate is left open so the water behind it rises. At high tide, the gate is closed. The dammed water leaving the basin turns a turbine, generating electricity.

Tidal power plants are now in operation in France, Nova Scotia, China, Norway, Northern Ireland, and Russia. These installations had high capital costs and can operate only twice daily. Although the energy produced is renewable and pollution-free, sedimentation occurs behind the dam gates, and tidal mudflats are often destroyed as a result of the operation.

Wave power at the sea's surface can be exploited. The machines based upon an **oscillating water column** consist of a chamber located just above the water surface that contains trapped air. Wave power is generated by using the up-and-down motion of water that results from waves, which are caused by winds and thus are an indirect form of solar energy. The rising wave compresses air trapped in the chamber. The high-pressure air is then released through a valve, turning a turbine to produce electricity. As the wave recedes, air rushes back in through another valve, also spinning the turbine. Presently, there are thousands of oceanic navigation buoys whose 60-W light bulbs are powered by this mechanism.

Large-scale wave-power facilities are in the development phase. West-facing coasts at mid-latitudes—especially in Europe, North America, Australia, and South Africa—are considered the most promising locations. The world's first **wave farm** opened in 2008 off the coast of Portugal, but has closed down for technical and financial reasons. Several plants are now under development and construction off the coast of Scotland.

Geothermal Energy

8.9 Introduction

Geothermal energy, though not solar-based, is another form of renewable energy. It has proven particularly useful in countries that have no fossil-fuel resources, and currently accounts for slightly less than 0.1% of the world's energy supply.

Geothermal energy is heat that emanates from beneath the Earth's surface and results from the radioactive decay of elements and from conduction from the molten core ($>5000°C$) of the Earth. Because of the movement of crustal (tectonic) plates, there are volcanic zones in which the heat is brought closer than usual to the surface. An example of the heat gradient with

FIGURE 8-4 Underground temperature gradients in a typical area (dashed line) and in one having geothermal potential (solid green curve). [Source: Geothermal Education Office, at http://geothermal.marin.org.]

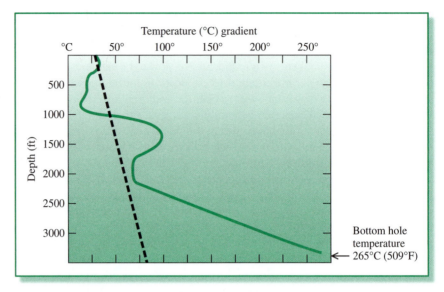

increasing depth for a geothermal zone is compared to that for a nongeothermal area in Figure 8-4. When deep groundwater circulates within a geothermal zone, the water is heated by contact with the hot rocks, and is sometimes vaporized. If the hot fluids are trapped in porous rocks under a layer of impermeable rock, a **geothermal reservoir** can form and potentially be tapped.

Geothermal energy is available as steam and/or hot water at temperatures ranging from 50 to 350°C from reservoirs. The fluid generally has to be piped 200–3000 m to the surface to be usable, although in a few locations it exits spontaneously from the ground as "hot springs." Production of hot fluid generally declines over time once a reservoir is tapped.

High-temperature (>180°C) geothermal energy in the form of steam or superheated water is usually found only in volcanic regions and island chains and is used to generate electrical energy. Geothermal energy in the form of moderately hot water (50–150°C) is most often used directly for space heating of buildings, including greenhouses, for spas, and for aquaculture. Intermediate-temperature hot water is used both for heating and to generate electricity.

The terms high, medium, and low *enthalpy* are sometimes used to describe these three types of geothermal energy.

8.10 Production of Electricity

Electricity is generated from high-temperature geothermal resources by turbines driven by the steam that exits at ground level. "Dry steam" exists as a gas underground in some locations, and exits the reservoir without accompanying liquid water. If the underground resource consists instead mainly of superheated water, then when it reaches the lower pressure conditions in a tank at the surface, most of it "flashes" into steam. In either case, the steam is used to turn a turbine and thereby produce electricity.

Geothermal electricity has the advantage over that generated by solar, wind, and marine sources that it is available 100% of the time at a uniform rate; it can be used as baseload power and need not be stored.

In accordance with the *second law of thermodynamics* discussed in Section 8.16, the efficiency of geothermal heat-to-electricity conversion is not great, however, since the temperature difference between the steam and the cooling water used to condense the steam is not as large as in fossil-fuel power plants. The warm water exiting from geothermal power generation is sometimes also used for space heating. The condensed steam is pumped down an injection well back into the source to sustain production.

If the geothermal energy is available only as an intermediate-temperature resource, the energy of the hot water can be first transferred to an organic fluid—such as *isobutane* or *isopentane*—whose boiling point is lower than 100°C, and the hot organic vapor is used to turn a turbine and generate electricity. The overall efficiency of such power conversion is less than 12%.

By 2010, 11,000 MW of geothermal electrical power had been installed globally, an increase of 20% since 2005. Geothermal energy provides some developing countries with a significant fraction of their electricity—more than 10% in Iceland, the Philippines, El Salvador, Costa Rica, and Kenya—in power plants that are as large as 100 MW. The United States (in California and Hawaii especially) is the largest net producer of geothermal electricity, followed by Mexico, the Philippines, Indonesia, Italy, and New Zealand. A map showing U.S. areas of potential geothermal energy sites for direct heating and for electrical power production is presented in Figure 8-5.

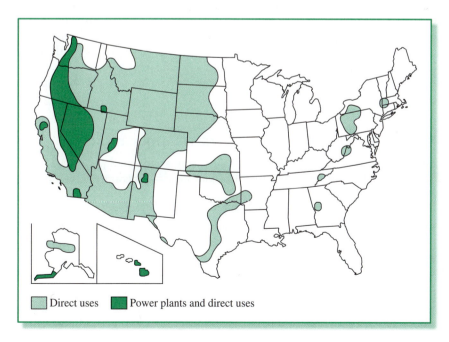

Direct uses ▪ Power plants and direct uses

FIGURE 8-5 Regions of potential sites for geothermal energy production in the United States. Direct uses correspond to heat. [Source: Geothermal Education Office, at http://geothermal.marin.org.]

Deep geothermal reservoirs, from which steam can be obtained, are tapped to generate some of Iceland's electricity. Eventually their plan is to tap more reservoirs and use the excess electricity to produce hydrogen gas by electrolysis of water. Iceland hopes to become the world's first economy based on hydrogen, and to have its economy totally free of fossil fuels by about 2030.

As expected, **the main costs associated with geothermal electrical power production are capital ones;** operating costs are low. High capital investment is needed for exploration, for drilling the wells, and for the instillation of electrical generating equipment. The direct capital cost per kilowatt of installed capacity is about $1500 for large power plants that have a high-quality steam resource available. Geothermal project costs vary considerably, however, depending especially on the depth and temperature of the resource.

In the past, geothermal energy has been exploited to produce electricity only where unusual geological features result in the existence of steam or hot water close to the Earth's surface. However, a number of *Enhanced Geothermal Systems* (EGS) projects are currently under development to overcome this limitation. In them, boreholes are sunk into the ground to levels where hot rocks are known to exist. In operation, cold water is sent down one pipe to contact the hot rocks, and the resulting hot water is collected from an adjacent well. The rocks may have to be fractured to facilitate the flow between the holes, analogous to the exploitation of shale gas (Box 6-1). About one-fifth of the electrical energy produced by the turbines is required to pump the water down one well and up the other.

A pilot power plant using EGS technology was constructed at Soultz, France in the early 2000s. A 25-MW demonstration plant is under construction in the Cooper Basin region of South Australia, where very hot rocks (270°C) are present relatively close to the surface. A recent report from MIT, funded by the U.S. Department of Energy, concluded that enough Enhanced Geothermal Energy can be obtained in the United States by EGS techniques to meet the nation's entire electricity needs!

8.11 Space Heating and Hot Water Applications

The use of geothermal energy for space heating and heated pools (spas) has a very long history, and currently is widespread—70 countries worldwide employ it to varying extents.

Per capita, Iceland leads the world in geothermal energy. As a result of the mid-Atlantic meeting of two tectonic plates deep under the island, geothermal energy is plentiful. For some years, over 80% of the hot water and heating in the major city, Reykjavik, has been supplied by hot water tapped from sources a few hundred meters below the ground and distributed by pipelines.

Another application of geothermal energy is the use of **heat pumps.** These devices draw heat energy from the soil under the ground, or from a shallow underground river, and pump it to the surface to supplement heating of buildings and water in winter. In summer, the direction of heat flow is reversed, with energy being transferred *from* buildings above the ground *to* the underground location. Thus, **the Earth is used as a heat source in winter and as a heat sink in summer.** Since the heat pumps extend only 100 m or so into the ground, the heat they extract is only in the 12–15°C range.

In a related project in Achen, Germany, a borehole was sunk 2.5 km (1.1 miles) into the Earth's crust, where rock temperatures are about 80°C. Cold water flows down the outer component of the pipe, being warmed as it descends, and then pumped back to the surface through an inner pipe, where it is used for heating and cooling.

8.12 Environmental Aspects

The greenhouse gas emissions from geothermal energy production stem from energy used in the construction phase and emissions of CO_2 when the underground fluid is pumped to the surface, but they are minimal when compared to the amount of electricity produced and fossil-fuel combustion thereby avoided (Figure 7-1). Exceptions to this generality can arise if electricity derived from fossil-fuel combustion is used to power extensive pumping of hot water from low-temperature resources destined for space heating.

One drawback to geothermal energy is the large quantity of *hydrogen sulfide* gas, H_2S, that is often present when the hot fluid is tapped from below the Earth's surface, especially from deep sites. Emissions of the gas into air when the fluid reaches the surface are minimized by scrubbing (Chapter 3). Some carbon dioxide that accompanies the hot fluid from the ground is emitted into the air, but is tiny compared to that which would be emitted if the same amount of energy were generated by burning fossil fuels.

At some sites, the hot groundwater contains significant amounts—often about 1%—of dissolved, sometimes acidic, minerals, which leads to problems of corrosion of equipment and of deposition of scale. In some cases, the minerals contain salable substances such as silica and zinc. Other minerals commonly present in the water contain carbonates, sulfates, and chlorides. The potential problem of the pollution of surface waters by brines is usually solved by re-injecting the geothermal liquids back into their source reservoirs where pressure permits. If, instead, the geothermal water is mixed with local fresh water, impact on aquatic life in the surface waters from both *thermal pollution* (Chapter 10) and from the dissolved minerals can occur.

Although most water mined from underground in geothermal systems is eventually returned to its source, it is impractical to return 100% of it. Consequently, there is danger of land subsidence eventually occurring in resources that are predominantly liquid when the aquifer's overburden of rocks is sufficiently destabilized. To overcome this problem, fresh water is often used to supplement the returning geothermal liquid. However, injecting cold water into the reservoir can present its own problems—namely the fracturing of the underground rocks by thermal contraction caused by contact with the cold water. Land subsidence occurred at the Wairakei geothermal power plant in New Zealand, which was the first wet-steam facility in the world.

Minor earthquakes, called *microseismic activity*, can occur as result of the tapping of geothermal resources. This is especially true if water is not returned to the source or to a different depth, and reservoir pressure is therefore not maintained. Injection of water much cooler than the reservoir can also contribute. Hydraulic fracturing of the rock can also cause minor earthquakes, as it does in shale gas projects (Box 6-1). A potential EGS project in Basel, Switzerland, was abandoned after water injection induced thousands of seismic events.

Review Questions 1–7 are based on the above material.

Direct Solar Energy

The direct absorption of energy from sunlight, and its subsequent conversion to useful forms of energy such as electricity, can occur by two mechanisms:

- **thermal conversion, in which sunlight (especially its infrared component, which accounts for half its energy content) is captured as heat energy by some absorbing material.** (An everyday example of such a material is a shiny metal surface, which we know from experience becomes very hot when left in sunlight.) Solar energy is an excellent source of heat at temperatures near or below the boiling point of water, a category that accounts for up to half of total energy usage and has low life-cycle greenhouse gas emissions (Figure 7-1).

- **photoconversion, in which the absorption of photons associated with the ultraviolet, visible, and near infrared components of sunlight by photovoltaic materials brings about the excitation of electrons in the absorbing material to higher energy levels.** The excitation subsequently causes a physical or chemical change (rather than a simple degradation to heat).

8.13 Low-Temperature Solar Energy

An example of *passive* solar thermal technology—systems that use no continuous active intervention or additional energy source to operate them—is the use of solar box cookers in developing countries. In temperate climates, the design of buildings to absorb and retain (by insulation) in winter a maximum fraction of the solar energy that falls on them is another example.

Solar water heaters are used extensively in Australia, Israel, the southern United States, and other hot areas that receive lots of sunshine, as well as in China, Germany, Turkey, and Japan. The water heaters represent the biggest use of *active* solar thermal technologies, which are defined as ones that employ an additional energy source to operate them. Solar collectors located on the rooftops of private homes and apartment buildings, as well as some commercial establishments such as car washes, contain water that is circulated around a closed system by an electrically driven pump. Sunlight is absorbed by a black flat-plate collector, which transfers the heat to the water that flows over it and which is bounded on the outside by glass or a plastic window. The hot water is pumped to an insulated storage tank until it is required for bathing or laundry purposes, or needed to supplement the heating of swimming pool water.

In more elaborate installations, the hot water is passed through a **heat exchanger,** which is a system of pipes over which air is passed and thereby warmed by transfer of heat. The hot air can be used immediately in winter to heat the rooms of the building. If not needed immediately, the heat can be stored in other media such as rocks. Usually a backup system, in which water can be heated electrically or by burning a fossil fuel, is incorporated in these systems in order to provide heat on cloudy days or in high-demand situations.

8.14 Concentrated Solar Thermal Power

By focusing the sunlight reflected by mirrors onto a receiver that contains a solid or a fluid, *very* high temperatures can be achieved. The hot fluid can be used to generate electricity by turning turbines, or for other uses as discussed later. The potential amount of electricity from such **Concentrated Solar Power** units is several times greater than that from solar photovoltaics, although the method relies on the absorption of direct sunlight whereas solar cells can operate to some extent on cloudy days.

As will be discussed in Section 8.16, **the fraction of thermal energy that can be extracted and converted to electricity from a mass of hot fluid at an absolute temperature T_h is limited by the second law of thermodynamics to be no greater than $(T_h - T_c)/T_h$,** where T_c is the final absolute temperature of the cooling water. Consequently, it is very advantageous to use a gas that has been heated to the highest possible temperature in order to maximize the amount of energy that is transformed to electricity rather than just degraded, wasting heat.

Indeed, temperatures of 1500°C have been achieved in steam heated by focusing sunlight. Generally, power plants need gas heated to 1200–1350°C at a pressure of 10–30 atm to operate. Simply focusing sunlight on tubes of air cannot achieve more than 700°C at 1 atm pressure. It should also be kept in mind that hot bodies, such as a solar receiver, continuously lose energy by radiating it outward. The rate of loss of radiation is proportional to T^4, where T is its Kelvin temperature (Section 5.3), and to the surface area of the collector. Consequently, the efficiency in harvesting energy does not increase linearly with temperature. The size of the receiver is kept as small as possible to minimize the radiative loss of energy.

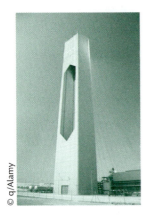

© q/Alamy

Solar towers are 80–200 m high, and the heliostat field covers a wide area around them. The first two commercial towers were erected in southern Spain.

Solar Towers, also known as *Power Towers*, have been constructed at several sunny locations around the world. They consist of a tall tower topped with a receiver onto which sunlight from hundreds of sun-tracking flat mirrors ("heliostats") on the ground are focused. The substance in the receiver is heated to a very high temperature, and its thermal energy is continuously removed for use in an electric power plant.

In one promising design, sunlight is focused by mirrors onto ceramic (silicon carbide) pins, which absorb the solar energy and heat up to 1800°C. Because they have a large surface area, the pins efficiently transfer the heat to air that flows around them. The hot, pressurized air (rather than steam) is then used to turn turbines and produce electricity.

Most solar towers use a *molten salt* to collect the energy and flow this liquid to a steam generator that then produces electricity. A liquid used for such purposes must possess the following characteristics:

- **high heat capacity,** so that much energy is stored in a relatively small volume and mass;

- **wide temperature range** over which it neither boils nor freezes, with a high upper limit so the initial hot liquid is as warm as possible;

- **stability** to decomposition; and

- **convenient and economical supply source.**

All these criteria are met by a mixture of the nitrate salts of sodium and potassium. Although **sodium nitrate,** $NaNO_3$, freezes at 307°C and **potassium nitrate,** KNO_3, at 337°C, as expected its mixtures have melting points depressed from these values. Indeed, the lowest-melting-point combination, called the *eutectic*, with a composition of 46 mol % $NaNO_3$ (60% by mass) and 54 mol % KNO_3, has the drastically lower melting point of 222°C. **The liquid range of this eutectic mixture extends over more than 300 degrees.** In addition, the liquid salt mixture is stable, has a high heat capacity, is available from natural supplies, and has a long history of being used in industry as a heat transport fluid. Heat loss from the hot tank amounts to only about 1% per day.

The storage tanks each hold tens of thousands of tonnes of salt.

In operation of a typical solar tower, the molten salt is heated to about 565°C by absorbed sunlight at the top of the power tower and continuously circulated to the steam generator, from which it exits at 288°C, ready to be heated again, as illustrated in Figure 8-6. Up to 18% of the sunlight energy is converted to electricity in tower systems. By using large storage tanks, excess hot liquid salt can be stored during the day and extracted at night when the Sun is not shining. In some installations, oil is used as the fluid heated in the receiver and to transmit energy to the turbines, with the nitrate salts used for storage of excess energy during the day to be used at night. Hot highly pressurized steam can also be used to store the energy. **Storing heat is more efficient and less expensive in general than storing the equivalent**

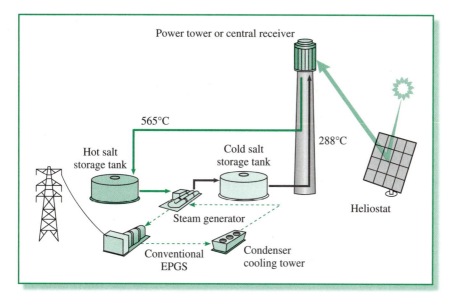

Power tower or central receiver

565°C

288°C

Hot salt
storage tank

Cold salt
storage tank

Heliostat

Steam generator

Conventional
EPGS

Condenser
cooling tower

FIGURE 8-6 Operation of a solar power tower using molten-salt heat storage. [Source: U.S. Department of Energy.]

amount of energy as electricity, an advantage solar thermal installations have over solar cells.

Several variants of solar electric systems are currently in development and at the demonstration level. In one, no tower is used. Instead, pipes containing a heat-absorbing fluid (usually oil) run the length along and above the center of a long system of curved mirrors that reflect sunlight onto it. The hot fluid in such *trough systems* is ultimately transported at about 400°C and processed through a heat exchanger to drive a steam turbine that produces electricity. During the day, the trough tilts so as to focus the maximum amount of sunlight on the collector tubes. The temperature of the collecting fluid in the trough systems is much lower than that in solar towers, so the efficiency in converting sunlight energy into electricity is lower (about 12%).

Likely locations for solar electric systems that meet the criteria of much sunny weather and open space for the heliostats and troughs include the southwestern United States, Australia, much of Africa, and southern European countries. In 2010, solar towers were operating in California, Spain, and Israel. Unfortunately many of the locations—such as deserts—with favorable sunlight and collection area conditions are deficient in cooling water required for the steam-driven turbines. The alternative of air cooling is feasible but makes the electricity production less efficient.

The first commercial trough system was constructed in the Mohave Desert of California in 1984, but is now closed.

Solar thermal electricity may become competitive in price with conventional sources, although this is not nearly the case at present. Solar thermal technology could become competitive, particularly if the waste heat from the process, e.g., steam near the boiling point of water, would also be used for some purpose. **The technique of using the waste heat from a heat-to-electricity conversion for a constructive purpose is called the cogeneration of**

energy. (It is a common feature of new power plants fuelled by natural gas.) Unfortunately, power plants based on steam require large amounts of cooling water in order to condense the steam back into the liquid state as part of the system's cycle; and as mentioned before, in many areas that have abundant land and sunlight there is little water available for this purpose. Scientists have also pointed out that if the absorption of sunlight by such systems occurred on a massive scale, the Earth's albedo would be altered, with consequent effects on the climate that are difficult to predict.

8.15 Thermochemical Applications of Concentrated Solar Energy

Another way to use the very-high-temperature concentrated solar heat is to drive an endothermic thermochemical process in order to produce a fuel such as hydrogen. An example is the thermal reduction of a metal oxide to oxygen gas and either the free metal or an oxide having less oxygen. The deoxygenated material is then either used as a fuel in its own right, or reacted with water or carbon dioxide to reduce them by oxygen atom removal and thereby produce a gaseous fuel.

An example of such a thermochemical cycle is that involving *zinc oxide*, ZnO, and metallic zinc. In the first step, the oxide is heated to 1600°C, which decomposes it into oxygen and the free metal:

$$ZnO(s) \longrightarrow Zn(s) + 1/2\ O_2(g)$$

The zinc metal could then be used to generate electricity in batteries or to react with water to form hydrogen fuel. In the latter case, steam is reacted with the oxide at about 700°C, releasing hydrogen gas and regenerating the oxide:

$$Zn(s) + H_2O(g) \longrightarrow ZnO(s) + H_2(g)$$

The overall reaction corresponds to the decomposition of water into hydrogen and oxygen gases that, because they are generated in different steps, are not mixed and do not have to be separated.

The *general reaction* by which water can be split into hydrogen and oxygen using the thermal decomposition of a metal oxide, MO, and water is:

Step 1: $MO + heat \longrightarrow M + 1/2\ O_2$

Step 2: $M + H_2O + heat \longrightarrow MO + H_2$

Fe$_3$O$_4$ is equivalent to a combination of iron oxides with Fe in the +2 and +3 ionic states, whereas FeO has only +2 iron, so this reaction reduces some of the iron at the expense of oxidizing some of the oxygen.

M need not be the pure metal but could be a lower oxide of the system. For example, a promising example of this technology is the very-high-temperature (~1500°C) partial de-oxygenation of *magnetite*, Fe_3O_4:

$$Fe_3O_4 \longrightarrow 3\ FeO + 1/2\ O_2$$

Ferrous oxide, FeO, is then used as a reducing agent, abstracting at high temperatures (~1100°C) an oxygen atom from water to produce hydrogen gas.

Alternatively, the oxygen source can be carbon dioxide, which would result in the production of carbon monoxide while regenerating the magnetite. Both CO and H_2 are required in synthesis gas to produce various fuels, as discussed in Section 7.18:

$$3 \ FeO(s) + H_2O(g) \longrightarrow Fe_3O_4(s) + H_2(g)$$

or

$$3 \ FeO(s) + CO_2(g) \longrightarrow Fe_3O_4(s) + CO(g)$$

All the solar thermochemical cycles suffer from practical difficulties resulting from the extreme temperatures required to carry out the reactions, such as the instability or melting of the oxides. The efficiency in converting sunlight into fuel is about 3% in the best of cases, so they are not competitive at this time with the conversion of sunlight into hydrogen by photovoltaic cells whose electrical output is used to electrolyze water with an overall efficiency of about 13%.

PROBLEM 8-2

A promising redox system for the thermochemical splitting of water involves the cerium oxides CeO_2 and Ce_2O_3. Write balanced equations concerning the thermal decomposition of CeO_2 to Ce_2O_3, and the reaction of the latter with steam to produce hydrogen. Combine the reactions to show that the overall process corresponds to the decomposition of water into hydrogen and oxygen. ●

Alternatively, high-temperature solar heat could be used to produce a combination of carbon monoxide and hydrogen (synthesis gas) from carbon dioxide and methane.

$$CO_2 + CH_4 \longrightarrow 2 \ H_2 + 2 \ CO \qquad \Delta H = +248 \ kJ \ mol^{-1}$$

Reversal of this highly endothermic reaction produces heat that can be used to generate electricity without the net emission of greenhouse gases, since the methane and carbon dioxide products are collected and reused. Alternatively, the mixture of H_2 and CO can be used as synthesis gas to make fuels such as methanol or synthetic gasoline (Sections 7.18, 7.30).

8.16 Limitations on the Conversion of Energy: The Second Law of Thermodynamics

In all processes that convert high-temperature heat into electricity, a portion of the original heat energy is inevitably lost as waste heat at a lower temperature. This loss is partially unavoidable as a consequence of the **second law of thermodynamics,** and applies to the production of solar thermal electricity as well as to other energy conversion processes.

According to the second law, **entropy** (or disorder) **must increase**—or **at the least be unchanged—when one type of energy is converted into another.** The law tells us that for any body at absolute temperature T that possesses an amount q of heat, its entropy S is a positive quantity given by the formula

$$S = q/T$$

Since high-quality (low-disorder) energy such as electricity has essentially zero entropy, clearly one cannot convert 100% of the heat into electricity, since the change ΔS in entropy associated with the conversion would be negative. However, if *some* of the initial heat energy q_h at the initial high temperature T_h is degraded to a smaller quantity q_c at a lower temperature T_c, then the entropy *change* ΔS for the process could be positive or zero:

$$\Delta S = \text{entropy of energy after conversion} - \text{entropy of energy}$$
$$\text{before conversion}$$

$$= q_c/T_c + 0 - q_h/T_h$$

The most complete conversion to electricity that is possible would correspond to the situation when $\Delta S = 0$; for this situation the above equation can be rearranged to give the new relationship

$$q_c/T_c = q_h/T_h \qquad \text{or} \qquad q_c = q_h \, T_c/T_h$$

The amount of heat converted to electricity is $q_h - q_c$, which upon substitution for q_c is equal to

$$\text{heat converted} = q_h - q_h \, T_c/T_h$$

$$= q_h \, (1 - T_c/T_h)$$

$$= q_h \, (T_h - T_c)/T_h$$

Thus the maximum fraction of the original heat that can be converted to electricity is

$$\frac{\text{heat converted}}{\text{initial heat}} = \frac{T_h - T_c}{T_h}$$

In other words, the *maximum* yield of electricity increases as the *difference* in temperatures between the original heat source and the temperature of the waste heat increases. Thus, if the sunlight can be converted to heat at 1500°C (1773 K), and if the temperature of the waste heat could be held to 27°C (300 K), the fraction of energy that could be converted to electricity would be

$$(1783 - 300)/1783 = 0.83$$

In fact, the efficiencies calculated by this formula are somewhat overestimated when more than one physical phase is involved. For example, associated with the conversion cycle in traditional power plants is a step that condenses steam back into liquid water, a process in which entropy is decreased. Consequently, additional energy beyond that calculated above must be degraded to waste heat to compensate.

PROBLEM 8-3

What is the maximum percentage of heat at 900°C that could be converted into electricity if the waste heat was produced as steam at 100°C?　　●

PROBLEM 8-4

Electricity could be obtained by exploiting the thermal gradient between the surface and the deep waters of the ocean. The maximum gradient, about 20 degrees, is achieved in tropical waters. What is the maximum percentage of the energy associated with this gradient that could be converted to electricity if the surface (cooling) water temperature is 25°C?　　●

PROBLEM 8-5

In order to reach a conversion efficiency of 50%, to what minimum Celsius temperature must a heat source be raised if the waste heat has a temperature of 57°C?　　●

8.17 Solar (PV) Cells

Electricity can be produced directly from solar energy by the photoconversion mechanism. This application exploits **the photovoltaic effect, which is the creation of separated positive and negative charges in a material as a result of the excitation by light of an electron within the solid from its normal energy level to a higher, excited state.** Both the excited electron and the *location* of the site of positive charge, the "hole," are mobile within the solid, so an electrical current could be made to flow in the material. The hole "moves" by means of the transfer of a (different) *bonding* electron from an atom adjacent to the initial hole to the atom on which the hole is located, thereby switching the position of the positive charge. Successive bonding electron transfers of this type allow the hole to move farther.

The two halves of a PV cell, usually designated **p** and **n,** are composed of different materials. This difference sets up an internal electrical potential that directs electrons in one direction and the holes in the opposite direction. The operation of a photovoltaic cell is illustrated in Figure 8-7. Once the electron has made its way through the **n** material to the first electrode, it travels through an external circuit to a second electrode, where it recombines with a hole that has migrated there through the **p** material. Placing electrical devices along the circuit allows them to be powered by extracting the electron's energy.

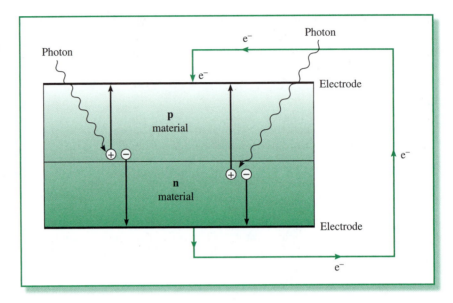

The material used for photovoltaic (PV) or solar cells is a **semiconductor, which is a solid that has a conducting behavior that is intermediate between that of a metal (freely conducting) and that of an insulator (nonconducting).** In semiconductors, the bonds linking the atoms are relatively weak, so the separation in energy between the bonding levels (called the valence band when applied to solids) and antibonding levels (called the conduction band) is relatively small (compared to that of an insulator). Consequently, the energy required to excite an electron from the least stable of the filled, bonding levels to the most stable of the empty, antibonding levels—called the **band gap**—is small but finite, and in many cases it lies in the energy range of photons of sunlight. The most common semiconductor used in solar cells is **elemental crystalline silicon,** for which the band gap separating the energy levels is 124 kJ mol^{-1}, which corresponds to infrared light.

> The two halves of a silicon solar cell are distinguished by having the element doped with small amounts of two different elements.

Silicon's light absorption ability extends from the band-gap energy of 124 kJ mol^{-1} up through to energies associated with the visible region, so it absorbs most of the photons of sunlight. However, **all the photon energy in excess of the 124 kJ mol^{-1} band gap is wasted by being converted into heat rather than promoting current flow.** When this loss of energy is combined with that wasted by immediate recombination of electrons and holes even in the purest single-crystal silicon—which prevents the pair of particles from contributing to charge flow *outside* the cell—a theoretical maximum of 29% of sunlight's energy can be converted into electricity. Commercial silicon cells now have efficiencies of 15–20%.

Semiconductor materials other than silicon can be used to construct PV cells. However, if one is chosen with a much smaller band gap, then, although more solar photons are absorbed, an even greater fraction of the energy of higher-energy photons is wasted as heat. Alternatively, if a much larger band gap is chosen, many fewer photons in sunlight are absorbed and produce electrons. Given these trade-offs and the actual solar spectrum received at the Earth's surface, the optimum band gap corresponds to a near-infrared wavelength of 920 nm, compared to that of 1100 nm for crystalline silicon. The theoretical maximum conversion for a single-junction solar cell of any type is 34%. A higher proportion of sunlight energy can be absorbed and utilized if several cell wafers having slightly different absorption characteristics are stacked on top of each other in a single cell, although such solar cells are expensive to manufacture.

Amorphous silicon has a maximum efficiency of only slightly more than half the value for pure crystalline silicon—because immediate electron–hole recombination occurs more readily, converting more of the captured energy into heat—but it is used extensively in solar cells because it is so much less expensive to manufacture, and is so much better at absorbing sunlight photons that it can be produced in **thin films** rather than the (relatively) thick wafers required for crystalline silicon.

Recall that an *amorphous solid* is one in which the atoms are packed in a much more random way than in a highly ordered crystal.

The film of amorphous silicon in a PV cell is deposited on a conducting surface by *chemical vapor deposition*. In this technique, the gas *silane*, SiH_4, at low pressure decomposes at 600–650°C into its elements, depositing a layer of amorphous silicon onto the surface:

$$SiH_4(g) \longrightarrow Si(s) + 2\,H_2(g)$$

The surface of the amorphous silicon used to form solar cells contains a small fraction of hydrogen atoms, which bond to any silicon atoms that do not form four bonds to their silicon neighbors and otherwise would promote spontaneous electron–hole recombination. The gradual loss over time of this hydrogen leads to degradation of the cells.

Amorphous silicon has a somewhat larger band gap than does the crystalline element—and indeed one that is larger than the optimum value—so it captures more of the visible energy of the solar spectrum but less of the infrared. Each crystalline silicon solar cell measures about 10 cm × 10 cm × 200 μm thick, and produces only about 1 W of electricity; so to generate electricity in useful quantities, many are joined together in a *solar array*. Amorphous silicon thin films are only about 2 μm thick, i.e., one-hundredth that of crystal silicon ones. Due to their thinness, they are transparent to visible light and can be used on windows, and indeed glass can be the surface on which amorphous silicon is deposited.

Another important thin-film PV cell uses *cadmium*: Cd–Te for one side of the cell and CdS for the other. The Cd–Te system starts absorbing light at 920 nm, which, as discussed above, is the optimum wavelength to collect the maximum number of photons and hence the number of electrons and maximum current available from the solar spectrum. Although they are rather inefficient collectors of light, the cadmium cells are cheap to produce, which accounts for their growing popularity.

Organic solar cells use two different polymeric organic semiconductors sandwiched together. Those devised to date have rather low efficiencies in converting sunlight into electric current, although they are inexpensive to manufacture and may achieve importance in the future.

8.18 Dye-Sensitized Solar Cells

In attempts to invent solar cells that are more efficient absorbers of sunlight across its spectrum, and to produce cells that are inexpensive to manufacture, chemists have investigated many dyes. In such cells, the various functions of the cell are performed by different substances, rather than essentially all by the same one, as occurs in silicon systems. **A thin film of a dye (the sensitizer) absorbs sunlight's photons. The resulting liberated electron quickly passes from the dye to a cheap, transparent semiconductor such as titanium dioxide, TiO_2, which is the n material of the cell, and is transported through it to an electrode.**

The most promising **dye-sensitized solar cell,** called the *Gratzel cell,* actually is a *photo-electrochemical system* since it incorporates an electrochemical reaction. A thin layer of the dye, an organometallic compound of *ruthenium,* is deposited on small particles of TiO_2, a layer of which covers one electrode. The cell's operation can be summarized as follows, and is illustrated in Figure 8-8:

- The electron liberated by the absorption of a photon by a dye molecule is quickly transferred to the **n** semiconductor, and travels through it to an electrode.

Michael Gratzel won the 2010 Millennium Technology Prize for invention of the cell that bears his name.

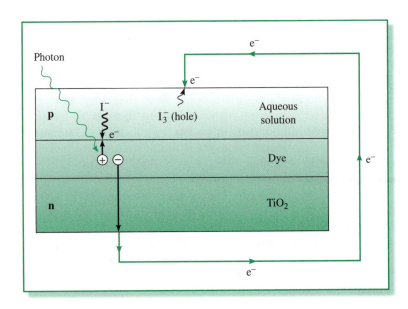

FIGURE 8-8 Schematic representation of the operation of a dye-sensitized solar cell. See text for details.

• The hole created by an absorbed photon travels through the dye and is transferred to a chemical in the electrolyte solution by the chemical's oxidation, which quickly liberates an electron to neutralize the dye's hole. This process occurs quite rapidly, and so avoids the possibility that the hole would instead recombine with a liberated electron.

• The oxidized chemical then travels to the second electrode, where it receives the electron from the external current and is thereby reduced, ready to react again with a hole.

• The reduced chemical in the electrolyte solution, which becomes oxidized in the process, is the **iodide ion,** I^-, present in a solution in contact with the dye. In this organic medium, oxidized iodide consists of the **triiodide ion,** I_3^-.

The expected oxidation product, *elemental iodine*, I_2, in the presence of excess iodide ion forms the triiodide ion: $I^- + I_2 \longrightarrow I_3^-$.

The chemistry of the cell's electrochemical reaction is shown schematically in the diagram below.

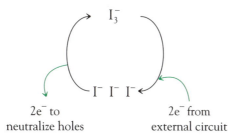

$$I_3^-$$

$I^- \quad I^- \quad I^-$

2e⁻ to
neutralize holes

2e⁻ from
external circuit

As shown in Figure 8-8 and by the chemistry in the above diagram, the liberated electron flows through the external circuit and subsequently re-enters the cell at an electrode, reducing the triiodide ion in the electrolyte solution back to iodide ion. In effect, the electrolyte—usually an organic solvent such as *acetonitrile*, CH_3CN—is the material that transports the hole from near its creation to the second electrode.

One of the advantages of dye-sensitized PV cells is that they operate in low-intensity light conditions, as occur on cloudy days, whereas silicon cells under these conditions are very inefficient due to electron–hole recombination. Sunlight-to-current conversions of greater than 11% have been achieved in dye cells. In addition, the purity of materials in dye-sensitized cells need not be nearly as high as in silicon systems because the various cell functions are separated, which lowers the cost.

8.19 Disadvantages of Solar Cells

One problem with the electricity that is generated using any PV cell is that it is *direct current* (dc) rather than the *alternating current* (ac) that is used in power grids and by most equipment and appliances. The dc electricity can be converted to ac, although with the loss of some power (as waste heat).

The cost of producing the solar cells and the problem of storing the electricity for use at night and on cloudy days are the greatest barriers to their increased use. As with other applications of solar energy, the capital cost in creating the infrastructure required to capture and use the "free" energy of the Sun by photocells is substantial. The costs of the encapsulation of cells, the wiring, and the construction of supporting structures is relatively high since so many cells are required, and collectively they add about as much to the total cost of a power system as do the cells themselves. Because the cells cannot operate at night or on very overcast days, the capacity factor overall for them is only about 14%, lower than that of any other form of renewable power.

The cost of manufacturing the solar cells has continued to fall with time, but electricity generated in this way is still not close to being competitive with conventional power-generation methods, nor is it likely to become so in the foreseeable future. As of 2010, the great majority of solar cells were still made from silicon rather than from any of the other materials.

Although the solar cells do not generate any carbon dioxide during their operation, their manufacture does consume significant amounts of energy and therefore causes substantial CO_2 emissions. Indeed, according to Figure 7-1, use of photovoltaic cells is considerably more CO_2-intensive than is wind power in generating electricity. The energy and carbon dioxide payback periods for solar cells and their infrastructure currently are about 3 years, but are expected to fall to 1–2 years when manufacturing techniques improve further. If manufacturing costs are not considered, however, use of solar cells to generate power in a home saves about as much carbon dioxide per year as is emitted from the family car.

8.20 Advantages of Solar Cells

Photovoltaic power may become attractive in hot, sunny locations such as the southwestern United States, where the peak power demand, driven by the need for air conditioning, coincides in time (summer afternoons) with the peak solar energy availability. Already solar-cell power (plus storage) is cheaper than extending power grid lines a kilometer or more away from an existing network into a remote region, and it is competitive in cost with the use of diesel generators for this purpose. Unlike other forms of renewable energy, it can be installed piecemeal rather than as one large project. The technology is also commonly used in water pumps, remote roadside telephones and signs, and satellites.

The use of solar cells in developing countries, most of which have sunshine in abundance, could obviate the need for the creation of power grids that carry electricity over long distances from source to user, and represents the greatest potential market for expansion of photovoltaic power. Indeed, the majority of solar cells made in the United States are exported abroad. Solar-cell electricity is already used to power water pumps, lights, refrigerators, and TVs in some developing countries. However, by 2000 more than 50% of new solar cells were connected to grid systems.

Like wind power, world production of solar cells for power production has soared since the mid-1990s. The global capacity of photovoltaic systems reached 40 GW in 2010, almost doubling that of the previous year and ten times that of 2004, but still considerably less than the total installed wind power. Europe leads the world in PV capacity, having 80% of the total, half of it in Germany alone. In 2010, for the first time, new installations of PV power in Europe exceeded those of new wind power.

Could solar cells be used to supply *all* of the electricity needs of developed countries? Using today's efficiencies and materials, the energy needs of the United States could be met by a continuous array of solar cells covering a square of land about 160 km × 160 km (100 miles by 100 miles). Although this area of 26,000 km² does not seem extraordinarily large, it should be realized that it is about a thousand times larger than the *total* area covered by all the solar cells manufactured through 1998, and would cost trillions of dollars to construct! An area about ten times larger—about 0.1% of the Earth's surface—could supply all the world's energy needs.

8.21 Conclusions about Renewable Energy

Although our survey of renewable energy sources reveals a huge potential and continuing rapid growth for power and fuel production, these systems (other than hydroelectric and biomass) have not as yet made a large dent in the global energy scene. As illustrated by data in Table 8-2 for electric power capacity, hydroelectric and biomass combustion continue to dominate renewable energy, followed now by a significant output from wind energy. Solar PV and geothermal power make some contributions, whereas that from concentrated solar and from marine sources are still negligible. Biomass combustion, solar collectors, and geothermal sources do make greater contributions to space and water heating than to electrical production (Table 8-2). However, renewable sources other than hydro collectively produced only 3.3% of global electrical energy in 2010.

An expert analysis in 2008 of the capital costs of *delivered* renewable energy (accounting for differences in capacity) found that wind energy is still much cheaper than either form of direct solar power:

Wind energy $4 per watt

Solar thermal energy $17 per watt

Photovoltaic energy $25 per watt

Although the greater use of these renewable energy technologies would indeed reduce greenhouse gas emissions that would arise from burning fossil fuels instead, we have seen that there are environmental implications associated with each replacement. There are no zero-impact solutions to any energy problem!

TABLE 8-2	Global Renewable Energy Capacity in 2008
Electric Power Generation	**Energy Capacity in GW**
Large-scale hydropower	860
Small-scale hydropower	85
Wind power	121 (198 by 2010)
Biomass combustion	52
Solar PV, grid-connected	13 (40 by 2010)
Geothermal	10
Concentrated solar thermal	0.5
Marine power (tidal plus wave)	0.3
Energy used for heating and hot water	
Biomass combustion	250
Solar collectors	145
Geothermal	50

Source: REN21-Renewable Energy Policy Network for the 21th Century, *Renewables Global Status Report: 2009 Update.*

Some general features concerning the use of direct solar energy, as opposed to fossil-fuel and nuclear energy, have emerged from the discussions above, and others have also been reached by energy analysts. Many of these conclusions apply to all forms of renewable energy.

The **advantages of direct solar energy** appear to be that it:

- is **free** and fantastically **abundant,**

- has **low environmental impact,**

- has **low operating costs,**

- does not necessarily require large, centralized suppliers and expensive distribution networks, and

- has **high public acceptance** as a "natural" form of energy.

The **disadvantages of direct solar energy** appear to be that it:

- is **intermittent** in its availability, and thus requires that efficient storage or backup systems be constructed so that power can be supplied continuously,

- is **diffuse**—it provides a low density of energy per unit of surface collection area, so large areas of solar collectors are required to harvest the energy, and

- requires high capital costs to construct the energy collection and storage systems; this offsets the "free" nature of the energy itself for many years until the investment is paid off.

The Storage of Renewable Energy—Electricity and Heat

One of the great advantages of fossil fuels is that they can be stored in relatively compact form until needed. Indeed, coal, oil, and natural gas are generally drawn from the ground as needed. The same is true of hydroelectric power—water can be kept behind dams until it is needed to produce electricity. Nuclear power and geothermal energy possess similar advantages.

However, direct solar energy in the form of photovoltaics or thermal solar, and wind energy and also of that derived from tides or waves, are dilute, dynamic forms of energy whose availability varies with time and which cannot be conveniently stored as such until they are required. To store wind or direct solar energy, they must be temporarily converted to other forms of energy, from which they are later reformed when needed.

Unfortunately, there is always some loss of energy, normally as waste heat, whenever a physical or chemical process is used to so alter a quantity of energy. Some of the waste heat is generated as a result of the second law of thermodynamics (Section 8.16) and some by practical inefficiencies. If the original conversion to temporary storage form retains say A% of the energy in useful form, and the subsequent return from temporary to useful form is B% efficient, then the **round-trip efficiency** is $A \times B$ percent, usually much less than 100%. In the next two sections we explore the technologies currently in use to temporarily store renewable energy from time-variable supplies.

The losses to waste heat in the conversions are $100 - A$ and $100 - B$ percents.

8.22 Storing Electricity in Batteries

The efficient large-scale storage of electricity in batteries is a technological problem which has not been solved and therefore is often avoided where possible. Photovoltaics produce electricity directly, so it must either be sent immediately to be used, or be stored, by forming a chemical fuel (Section 7.27), by converting it to mechanical energy, or by charging a battery or a capacitor. The same is true for wind energy.

A number of different battery types are used for the large-scale storage of dc electricity. Their important characteristics are:

- **Lead-acid storage battery:** Low energy density in terms of both volume and mass. Not reliable for long-term repeated charge–discharge cycling, although their round-trip efficiency is high: 80%.

- **Lithium ion battery:** High energy density and highly efficient in charging–discharging, but high manufacturing cost at present; potential fire hazard requires robust construction.

- **Sodium-sulfur battery:** Electricity stored by dissociation of *sodium polysulfide*, an ionic compound of Na^+ and S_n^{2-} ions into sulfur and molten sodium:

$$Na_2S_n + \text{electrical energy} \longrightarrow 2\ Na(l) + n\ S(l)$$

However, the molten electrode has to be kept in a hot chamber (at about 300°C) that is robust (and hence expensive) until it is used to reverse the reaction and thereby regenerate the electrical energy. The NaS batteries do have high energy density, can withstand thousands of charge–discharge cycles, operate efficiently in both directions, and are made from cheap, abundant materials.

- **Flow batteries:** In these systems, large quantities of aqueous solutions containing the substances to be oxidized and reduced are kept in separate tanks and poured into a reaction chamber when required. The most promising flow battery is based upon *vanadium* in its multiple ionic states.

The +4 and +5 states actually occur not as monatomic ions but as the oxyanions VO^{2+} and $VO_2{}^+$.

At one electrode, the redox half-reaction is $V^{3+} \longrightarrow V^{2+}$, whereas at the other it is $V^{4+} \longrightarrow V^{5+}$, when electrical energy is available. The solutions are separated by a thin membrane in the battery itself, and then pumped into tanks to be stored until electricity is required. As charge flows in the discharge mode, the potential between the two solutions *decreases*, so more electrolyte is added to each battery compartment in order to maintain the current. The batteries are inexpensive to build, readily scalable, and could therefore handle large amounts of stored electrical energy and have a round-trip efficiency of about 80%.

8.23 Other Methods of Storing Electrical Energy

There are several methods of storing the energy of generated electricity which involve its temporary conversion to a *nonelectrical* form:

- **Wind energy is sometimes stored as highly compressed air.** Electricity is used to compress air at higher and higher pressure in a sealed underground cavern, such as a disused salt mine. When energy is required, the compressed air is released gradually to turn a turbine and produce electricity, or more commonly, to supply precompressed air to a natural-gas-fired generator, thereby saving the energy that would have been used to pressurize ambient air before combustion. However, gas warms as it is compressed and this limits the amount of air that can be stored before it becomes too hot; the heat of compression is gradually lost over time during extended storage. Compressed-air energy storage has a round-trip efficiency of 60–70%.

- **Wind energy—and solar energy—can be stored by pumping water uphill from one mountain stage reservoir to a higher one.** The electricity is regenerated when needed by allowing the water to flow downhill and thereby spin turbines, as in hydroelectric power. This practice is used in Norway, China, Japan, and the United States, and has a round-trip efficiency of 65–85%.

Review Questions 8–17 are based on the material in the above sections.

The storage of energy as reduced chemicals was discussed in Section 8.15, and the storage of heat was outlined in Section 8.14.

ACTIVITY

Several physical techniques for the storage of large amounts of electrical energy are under development, using (a) flywheels, (b) supercapacitors, and (c) superconducting magnets. Using web resources, prepare a report on how each of the three would work and their current status as practical techniques for energy storage.

Review Questions

1. List four environmental/social problems associated with the expansion of hydroelectric power.

2. What is the relationship between the energy generated by a windmill and (a) the wind's speed and (b) the length of the windmill's blades? (c) Calculate the ratio of the energy collected by a windmill having a blade length of 100 m to one having a 6-m blade.

3. Explain the origin of coastal winds.

4. List four pros and cons of wind power.

5. Define *energy payback*, and state which form of renewable energy has the lowest payback period and the lowest cost at present.

6. Which two types of energy are collected under the title of *marine energy*?

7. What is meant by *geothermal energy*? List some examples of how and where it is tapped. Why are all geothermal resources not used to generate electricity? What are some environmental issues associated with geothermal energy?

8. Describe the difference between the two methods of absorbing energy from sunlight. What is the difference between *active* and *passive* systems?

9. What is meant by *solar thermal electricity*, and how is it generated? Describe the operation of a

solar power tower that uses molten salts. What is meant by the term *cogeneration*?

10. Write the general chemical reactions involving metal oxides by which a fuel can be produced by using concentrated solar energy.

11. State the *second law of thermodynamics*. According to this law, what formula gives the maximum fraction of heat that can be transformed into electricity?

12. Define the terms *photovoltaic effect* and *band gap*. Why is amorphous rather than crystalline silicon used in some solar cells? Describe the process of *chemical vapor deposition*.

13. Describe the workings of the dye-sensitized solar cell involving iodide ions.

14. What are some of the advantages and disadvantages of solar cells? What is the chief difficulty preventing their widespread use?

15. List four general advantages and four general disadvantages of direct solar energy.

16. List four types of batteries that can be used to temporarily store large amounts of electrical energy.

17. What are two physical methods by which surplus electrical energy from wind can be temporarily stored?

Additional Problems

1. Given that an average of 342 W of sunlight energy falls on each square meter of the Earth, and that the surface area of a sphere is $4\pi r^2$, and that the Earth's radius r is about 6400 km, calculate in joules the total amount of sunlight received annually by the Earth. What percentage of this quantity needs to be captured in order to provide our current commercial energy needs of about 500 EJ?

2. The distribution of energy with wind speed is proportional to the function $(v^4/c^2)\exp(-v^2/c^2)$, where the constant c is equal to $(2/\sqrt{\pi})$ times the time-averaged value of the wind speed v. Plot the magnitude of this function for v values from 0 to 25 for average wind speeds of (a) 5 m s^{-1} and (b) 10 m s^{-1} on the same graph. Is it clear from the difference between the plots why low values of average wind speed are not economical to harvest?

3. Using the actual annual capacity figures for global wind power listed in Figure 8-1, prepare a plot of the logarithm of capacity versus year to decide if the growth has indeed been exponential, as is claimed in the text. By extending your graph, or obtaining a simple equation for it from the slope and intercept, project global wind capacity in the years 2015 and 2020.

9

Radioactivity, Radon, and Nuclear Energy

In this chapter, the following introductory chemistry topics are used:

- Mass and atomic numbers; isotope symbolism; elementary particles
- Half-lives; exponential decay and first-order processes
- Oxidation numbers; redox reactions; solubility processes

Background from previous chapters used in this chapter:

- Concepts of free radicals and synergism
- Half-life equation
- Second law of thermodynamics

Introduction

In all other chapters of this book, we are concerned with chemicals and chemical processes. In this chapter, we consider *nuclear* processes and how they affect the environment, our health, and our energy supply. These concerns all center on the effects of radioactivity, and it is with this topic that we begin. This allows us to discuss radon, the most important radioactive indoor air pollutant, and depleted uranium. We then switch to nuclear energy and explore the ways in which electricity can be produced from it and the environmental consequences of the radioactive waste that these processes generate.

The San Onofre nuclear power generating station at San Diego, California. All current nuclear energy is generated by fission, though fusion plants may be viable in the future. [Corbis Images]

Radioactivity and Radon Gas

9.1 The Nature of Radioactivity

Although most atomic nuclei are stable indefinitely, some are not. The unstable, or **radioactive, nuclei spontaneously decompose by emitting a small particle that is very fast moving and therefore carries with it a great deal of energy.** In some types of nuclear decomposition processes, atoms are converted from those of one element to those of another as a consequence of this emission. Very heavy elements are particularly prone to this type of decomposition, which occurs by the emission of a small particle. The nuclei produced by emission of the particle may or may not themselves be radioactive; if they are, they will undergo another decomposition at a later time.

Recall from introductory chemistry that the **mass number** is the *number* of heavy particles—protons and neutrons—and not the actual mass of the nucleus. **An alpha (α) particle is a radioactively emitted particle that has a charge of +2 and a mass number of 4**—it has two neutrons and two protons—and it is identical to a common **helium** nucleus. Thus an α particle is written as ^4_2He, where 4 is its mass number and 2 refers to its nuclear charge (i.e., number of protons). The nucleus that remains behind after an atom has lost an α particle has a nuclear charge that is 2 units *less* than the original, and it is 4 units *lighter*. For example, when a $^{226}_{88}\text{Ra}$ (radium-226) nucleus emits an α particle, the resulting nucleus has a mass number of $226 - 4 = 222$ units and a nuclear charge of $88 - 2 = 86$; this is a wholly new element that is an isotope of the element radon. The process can be written as a *nuclear reaction:*

$$^{226}_{88}\text{Ra} \longrightarrow \, ^{222}_{86}\text{Rn} + \, ^4_2\text{He}$$

Notice that both the total mass number and the total nuclear charge individually balance in such equations.

A beta (β) particle is an electron. It is formed when a neutron splits into a proton and an electron in the nucleus. Since the proton remains behind in the nucleus when the electron leaves it, the nuclear charge (or atomic number) *increases* by 1 unit (you may imagine this effect as "subtracting a negative particle"). There is no change in mass number of the nucleus, since the total number of neutrons plus protons remains the same. For example, when an atom of the lead isotope $^{214}_{82}\text{Pb}$ (lead-214) decays radioactively by the emission of a β particle, the nuclear charge of the product is $82 + 1 = 83$, corresponding to the element bismuth; the mass number remains 214:

$$^{214}_{82}\text{Pb} \longrightarrow \, ^{214}_{83}\text{Bi} + \, ^{\,0}_{-1}\text{e}$$

Notice that the symbol $^{\,0}_{-1}\text{e}$ used here for the electron shows its mass number (zero) and its charge; in the equation the total mass numbers and nuclear charge numbers each balance.

TABLE 9-1	Summary of Small Particles Produced by Radioactivity		
Particle Symbol and Name	Chemical Symbol	Comment	Effect on Nucleus of Particle Emission
α (alpha)	$_2^4\text{He}$	Nucleus of a helium atom	Atomic number reduced by 2
β (beta)	$_{-1}^0\text{e}$	Fast-moving electron	Atomic number increased by 1
γ (gamma)	None	High-energy photon	None

One other important type of radioactivity is the emission of a **gamma (γ) particle** (also called a *ray*) by a nucleus. This is a huge amount of energy concentrated in one photon and possesses no particle mass. Neither the nuclear mass number nor the nuclear charge changes when a γ particle is emitted. The emission of a γ ray often accompanies the emission of an α or β particle from a radioactive nucleus. The properties of all three types of nuclear radiation are summarized in Table 9-1.

PROBLEM 9-1

Deduce the nature of the species that belongs in the blank for each of the following nuclear reactions:

(a) $_{86}^{222}\text{Rn} \longrightarrow {}_2^4\text{He} + \underline{\hspace{1.5cm}}$

(b) $_{83}^{214}\text{Bi} \longrightarrow \beta + \underline{\hspace{1.5cm}}$

(c) $_{-}^{214}\text{Po} \longrightarrow {}_{-}^{214}\text{Pb} + \underline{\hspace{1.5cm}}$

(d) $\underline{\hspace{1.5cm}} \longrightarrow {}_{90}^{234}\text{Th} + \alpha$

9.2 The Health Effects of Ionizing Radiation

The α and β particles that are produced in the radioactive decay of a nucleus are not in themselves harmful chemicals, since they are simply the nucleus of a helium atom and an electron. However, they are ejected from the nucleus with an incredible amount of energy of motion. **When this energy is absorbed by the matter encountered by the particle, it often ionizes atoms or molecules; for that reason, it is called ionizing radiation, or just** *radiation.* **This radiation is potentially dangerous if we absorb it, since the molecular components of our bodies can be ionized or otherwise damaged.**

Although α and β particles are energetic, they cannot travel far within the human body, since they lose more and more of their energy—and consequently slow down—as they collide with more and more atoms.

Alpha particles can travel only a few thousandths of a centimeter within the body, so they are not penetrating. This is true because they are relatively massive, and when they interact with matter they slow down, capture electrons from it, and are converted into harmless atoms of helium gas. If an α particle is emitted outside the body, it will usually be absorbed in the air or by the layer of dead skin, so it will do you no harm. However, **inhaled or ingested radioactive atoms can cause serious internal damage when they emit α particles.** The damage is particularly severe with α particles since their energy is concentrated in a small area of absorption located within about 0.05 mm of the point of emission. In their interaction with matter, α particles are highly damaging—the most highly damaging of all particles—since they can knock atoms out of molecules or ions out of crystal sites. If the molecules affected are DNA or its associated enzymes, cell death can result. A more serious consequence for the individual can be the creation of mutations that could lead to cancer.

Beta particles move much faster than α particles since they are much lighter and can travel about 1 m in air or about 3 cm in water or biological tissue before losing their excess energy. Like α particles, they can cause considerable damage to cells if they are emitted from particles that have been inhaled or ingested and the radioactive nucleus is consequently close to the cell when it decays.

Gamma rays easily pass through concrete walls—and our skin. A few centimeters of lead are required to shield us from γ rays. **Gamma particles are the most penetrating and therefore the most damaging of the three, traveling a few dozen centimeters into our bodies or even right through them.** They are generally the most dangerous type of radioactivity, since they can penetrate matter efficiently and do not have to be inhaled or ingested. Although they can pass through our bodies, γ rays lose some of their energy in the process, and cells can be damaged by this transferred energy, since it can ionize molecules. Ionized DNA and protein molecules cannot carry out their normal functions, potentially resulting in radiation sickness and cancer.

The ions produced by radiation when its energy is transferred to molecules are free radicals; hence they are highly reactive (Chapters 1, 3, and 17). For example, a water molecule can be ionized by an α, β, or γ ray or by an X-ray. The resulting H_2O^+ free-radical ion subsequently dissociates into a hydrogen ion and the *hydroxyl free radical*, OH:

$$H_2O + \text{radiation} \longrightarrow e^- + H_2O^+$$

$$H_2O^+ \longrightarrow H^+ + OH$$

If the affected water molecule is contained in a cell, the hydroxyl radical can engage in harmful reactions with biological molecules in the cell, such as

DNA and proteins. In some cases, radiation damage is sufficient to kill cells of living organisms. This is the basis of food irradiation, where the death of microorganisms helps prevent subsequent spoilage of the food.

If human beings are exposed to substantial, though sublethal, amounts of ionizing radiation, they can develop *radiation sickness*. The earliest effects of this malady to be observed occur in tissues containing cells that divide rapidly, because damage to the cell's DNA or protein can affect cell division. Such rapidly dividing cells are found in bone marrow, where white blood cells are produced, and in the lining of the stomach. It is not surprising, then, to find that early symptoms of radiation sickness include nausea and a drop in white blood cell count. Children are more susceptible to radiation than adults because their tissues undergo more cell division. On the other hand, radiation can be effectively used to kill cancer cells since they are dividing rapidly. Unfortunately, radiation therapy cannot be completely selective in terms of the cells it affects, so it has side effects such as nausea.

Long-term effects from radiation may show up in genetic damage, because chromosomes may have undergone damage or their DNA may have mutated. Such damage may lead to cancer in the person exposed or to effects in her or his offspring if the changes occurred in the ovaries or testes.

9.3 Quantifying the Amount of Radiation Energy Absorbed

The amount of radiation absorbed by the human body is measured in rad (radiation absorbed dose) units, where 1 rad is the quantity of radiation that deposits 0.01 joule of energy to 1 kilogram of body tissue. The rad is not a particularly useful quantity, however, since the damage inflicted by 1 rad of α particles is 10 to 20 times greater than that inflicted by 1 rad of β particles or γ rays. **The scale of absorbed radiation that incorporates this biological effectiveness factor is the rem** (roentgen equivalent man). A more modern unit than the rem is the *sievert*, Sv, which equals 100 rem.

On average, we each receive about 0.3 rem, i.e., 300 mrem or 3000 μSv, of radiation annually. Its origin on average is about

- 55% from radon in indoor and outdoor air;

- 8% from cosmic rays from outer space;

- 8% from rocks and soil;

- 11% from natural radioactive isotopes (e.g., ^{40}K, ^{14}C) of elements that are present in our own bodies; and

- 18% from anthropogenic sources, chiefly medical X-rays.

The average contribution from nuclear power production is negligible at present.

An *acute* exposure of more than 25 rem results in a measurable decrease in a person's white blood cell count; over 100 rem produces nausea and hair loss; and an exposure of over 500 rem results in a 50% chance of death within a few weeks.

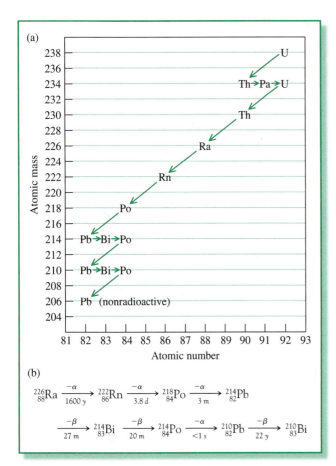

(a)

(b)

FIGURE 9-1 (a) The ^{238}U radioactive decay series. (b) The radium–radon portion of the ^{238}U radioactive decay series. The symbol above the arrow signifies the type of particle (α or β) emitted during the transition. The time period indicated below an arrow is the half-life of the unstable isotope.

9.4 Radioactive Nucleus Decay

The radioactive decay of the atoms in an isotope sample does not occur all at once. For example, in a sample of **uranium-238,** ^{238}U, just large enough to be visible, there are about 10^{20} atoms. Only about 10^7 of the ^{238}U nuclei in the sample decompose in a given second, so it requires billions of years for the decomposition process to be complete for the sample as a whole.

Since all radioactive nuclei disintegrate by processes that are kinetically first-order, it is convenient to express the decomposition rate as the time period required for half the nuclei in a sample to disintegrate—its half-life, $t_{1/2}$. (This is the same property that was used in earlier chapters to discuss the decomposition of substances by chemical reactions.) For example, the half-life of ^{238}U is about 4.5 billion years. Thus about half of this isotope of uranium existing when the Earth was formed (about 4.5 billion years ago) has now disintegrated; half the *remaining* ^{238}U, amounting to one-quarter of the original, will disintegrate over the next 4.5 billion years, leaving one-quarter of the original still intact. After three half-lives have passed, only one-eighth of the original will remain, and only one-sixteenth will be there after four half-lives.

PROBLEM 9-2

Recall from introductory chemistry that for first-order processes, the fraction F of reactant left after time t of reaction has elapsed is given by $F = e^{-kt}$, where k is the rate constant. Calculate the time required for 99% of a sample of ^{238}U to disintegrate. [*Hint:* The relationship between k and $t_{1/2}$ is $k = 0.693/t_{1/2}$.]

9.5 Radon from the Uranium-238 Decay Sequence

Many rocks and granite soils contain uranium, so the radioactive decay process takes place under our feet each day, with radon gas being one of its unwelcome products. Each $^{238}_{92}U$ nucleus eventually emits an α particle, and an atom of the *thorium* isotope $^{234}_{90}Th$ is formed:

$$^{238}_{92}U \longrightarrow {}^{234}_{90}Th + {}^{4}_{2}He$$

This is the first of 14 sequential radioactive decay processes that a ^{238}U nucleus undergoes, as illustrated in Figure 9-1a. The last of these reactions produces $^{206}_{82}Pb$, a nonradioactive (stable) isotope of *lead*; therefore, the sequence stops. The concentration of each member of the series can be determined using steady-state principles (see Box 9-1).

BOX 9-1   Steady-State Analysis of the Radioactive Decay Series

The various members, except the first and last, of the 14-step radioactive series of Figure 9-1a in a body of undisturbed uranium ore are each in a steady state (Chapter 1) with respect to their concentrations. Since all nuclear disintegration processes are kinetically first-order, the rate of production of each species, C, is proportional to the concentration of B, the species that precedes it in the chain A → B → C → D → · · · . The rate of disintegration of C is proportional to its own concentration. Thus

d [C]/dt = k_p [B] − k_d [C] = 0

at steady state

so

$$[C]_{SS}/[B]_{SS} = k_p/k_d$$

Since each first-order rate constant is inversely proportional to the half-life period for the process, it follows that

$$[C]_{SS}/[B]_{SS} = t_C/t_B$$

where t_C and t_B correspond to the half-life periods for decay by disintegration of C and B. Thus the steady-state concentration of any species along the series relative to the one that precedes it is directly proportional to its half-life decay period and inversely proportional to

the half-life decay period of the species that produces it. For example, the half-life of ^{222}Rn is 3.8 days and that of the ^{226}Ra that produces it is 1600 years, or 1.5×10^5 times as long, so the steady-state ratio of radon to radium is $1/1.5 \times 10^5$ or 6.5×10^{-6}. In qualitative terms, the radon concentration never comes close to reaching that of the radium since it decays so quickly after it is formed.

PROBLEM 1

Show that the ratio of the steady-state concentration of C relative to the first member, A, of the radioactive series A → B → C → D → · · · is equal to the ratio of their half-lives for disintegration, and thus that the steady-state concentration of each member of the series is directly proportional to its disintegration half-life.

PROBLEM 2

Using the equations developed in Box 11-4; deduce how long it will take for radon-222 to reach 80% of its steady-state concentration after a sample of radium-226 is purified of contamination by any other substance. [*Hint*: Recall that for first-order reactions, $t_{0.5} = 0.693/k$.]

Of particular interest is the portion of the 14-step sequence of ^{238}U radio-active decay that involves **radon,** since this element is the only one, other than the helium produced from the α particles, that is gaseous and therefore is mobile. Details concerning this portion of the radioactive decay series are shown in Figure 9-1b. The immediate precursor of the radon is radium-226, which has a half-life of 1600 years and decays by emission of an α particle:

$$^{226}_{88}Ra \longrightarrow {}^{222}_{86}Rn + {}^{4}_{2}He$$

The ^{222}Rn isotope has a half-life of 3.8 days, which can be long enough for it to diffuse through the solid rock or soil in which it is initially formed. Most radon escapes directly into outdoor air when the surface of the Earth where it appears is not covered, e.g., by a building. The very small back-ground concentration of radon in air that this produces nevertheless yields about half of our exposure to radioactivity, discussed in Section 9.3. Although the radon decays in a few days, it is constantly replaced by the decay of more radium.

In certain homes, radon becomes an important indoor air pollutant. Most radon that seeps into homes comes from the top meter of soil below and around the foundation; radon produced much deeper than this will probably decay to a nongaseous and therefore immobile element before it reaches the surface. Loose, sandy soil allows the maximum diffusion of radon gas, whereas frozen, compacted, or clay soil inhibits its flow. Radon enters the basements of homes through holes and cracks in their concrete foundations. The intake is increased significantly if the air pressure in the basement is low. The mate-rial used to construct the homes and water from artesian wells are other potential sources of radon in homes. Groundwater systems serving up to a few hundred people often have radon levels almost 10 times those of surface waters. When well water is heated and exposed to air, as occurs when it exits from a showerhead, radon is released to the air. However, radon from water usually represents only a small fraction of that arising from soil, although it represents a greater health hazard than that contributed from water disinfec-tion by-products and other dissolved chemicals.

Radon gas can accumulate to unhealthy levels in caves, including some that are used for recreational purposes.

9.6 Radiation Dose Scales

The rate of radioactive disintegrations in a sample of matter is usually mea-sured in bequerels, Bq, where 1 Bq corresponds to the disintegration of one atomic nucleus per second. The alternative unit used is the **curie,** Ci, which equals 3.7×10^{10} Bq and is the radioactivity produced by one gram of ^{226}Ra. Environmental regulations are usually expressed in the number of bequerels per unit volume or, in the United States, in terms of the number of picocuries, where 1 pCi = 10^{-12} Ci. For example, the U.S. EPA uses 4 pCi per liter of air as the upper limit for a safe radon level in houses.

We can calculate the amount of energy from radiation that is absorbed by a person's lungs in a year if he or she breathes air containing radon at the

4 pCi L^{-1} level, since the energy of each α particle emitted by a radon atom is known by measurement to be 9.0×10^{-13} J. Since 4 pCi L^{-1} is equivalent to $4 \times 10^{-12} \times 3.7 \times 10^{10} = 0.15$ disintegrations per liter per second, and since there are $60 \times 60 \times 24 \times 365$ sec in a year, the total annual number of disintegrations in 1 L of air is 4.7×10^6. Thus the total amount of energy liberated annually in the process is $4.7 \times 10^6 \times 9.0 \times 10^{-13}$ J $= 4.2 \times 10^{-6}$ J. If we assume that all this energy is absorbed by a person's lung tissues (rather than by the air in the lungs), that the exchangeable lung volume is about 1 L, and that the lung mass is about 3 kg, then since 1 rad $= 0.01$ J kg^{-1}, the energy absorbed is 1.4×10^{-4} rad or 0.14 mrad. Using a factor of 10 to convert rads to rems for α-particle radiation, we find that the annual radiation dose is about 1.4 mrem, i.e., about 0.5% of background exposure.

9.7 Radiation from the Daughters of Radon

Radon, the heaviest member of the noble gas group, is chemically inert under ambient conditions and remains a monatomic gas. As such, it becomes part of the air that we breathe when it enters our homes. Because of its inertness, physical state, and low solubility in body fluids, radon *itself* does not pose much of a danger; the chance that it will disintegrate during the short time it is present in our lungs is small and, as discussed above, the range of α particles in air before they lose most of their energy is less than 10 cm.

The danger arises instead from the radioactivity of the next three elements in the disintegration sequence of radon—namely, *polonium, lead,* and *bismuth* (see Figure 9-1b). These *descendants* are termed **daughters** of radon, which in turn is called the *parent* element. In macroscopic amounts, these particular daughter elements are solids, and when formed in the air from radon they all quickly adhere to dust particles. Some dust particles adhere to lung surfaces when inhaled, and it is under these conditions that the elements pose a health threat. In particular, both the ^{218}Po, which is formed directly from ^{222}Rn, and the ^{214}Po, which is formed later in the sequence (Figure 9-1b), emit energetic α particles that can cause radiation damage to the bronchial cells near which the dust particles reside. This damage can eventually lead to lung cancer. Indeed, as will be discussed, radon (or rather its daughters) is the second leading cause of such cancers, although it follows smoking by a wide margin.

Although some radon daughters in the sequence disintegrate by β-particle emission, the deleterious health effects of these particles are considered negligible because the α particles carry much more energy and, as discussed, it is the disruption of cell molecules by the burst of high energy that initiates cancer.

Notice that the sequence (see Figure 9-1b) of radon decay to ^{210}Pb formation takes less than a week on average. In contrast, disintegration of ^{210}Pb to ^{210}Bi has a half-life of 22 years, and, in fact, most of the lead will have been cleared from the body before this process occurs.

9.8 Evaluating the Health Danger from Indoor Radon

The greatest exposure to α particles from radon disintegration is experienced by miners who work in poorly ventilated underground uranium mines. Their rate of lung cancer is indeed higher than that of the general public, even after corrections to the data have been made for the effects of smoking. From statistical data relating their excess incidence of lung cancer to their cumulative level of exposure to radiation, a mathematical relationship between cancer incidence and radon exposure has been developed. Scientists have extrapolated this relationship to determine the risk to the general population from the generally lower levels of radon to which the public is exposed.

Based on linear extrapolation from the miners' data and other sources, the U.S. EPA estimates that radon currently causes about 21,000 excess lung cancer deaths annually; the estimate for the United Kingdom is 2000 cases per year. **Most of the excess deaths are associated with smokers, since radon and cigarette smoke are synergistic** (see Chapter 4) in causing lung cancer. In particular, the risk of lung cancer by age 75 is 4 in 1000 for a non-smoker living in a house with zero radon and is only increased to 7 in 1000 if he or she has constant exposure by inhalation to 400 Bq m^{-3} of radioactivity from the gas, for a net increase of 3 in 1000. However, a smoker's chances of lung cancer rise from 100 to 160 in 1000, an increase of 60, by exposure to the same level of radioactivity. **It is estimated that radon causes about 10% of all lung cancers,** which is about half the mortality rate from automobile accidents, for example.

The level of radioactivity in air is stated in units of either bequerels per cubic meter, Bq m^{-3}, or picocuries per liter, pCi L^{-1}. The radioactivity "action level" for indoor air, beyond which mitigation measures are recommended, is 150 Bq m^{-3}, i.e., 4 pCi L^{-1} in the United States, and lies in the 200–400 Bq m^{-3} range in most other developed countries.

Because radon dissolved in drinking water can escape into the air when it comes out of the tap, a maximum contaminant level of 150 Bq L^{-1} has been established by the U.S. EPA; the World Health Organization (WHO) guideline is 100 Bq L^{-1}. Radon concentrations at these levels are associated with groundwater that has passed through rock formations that contain natural uranium and radium.

Those skeptical of using uranium miners' data point out that the calculated estimates of radon-caused lung cancer may be too high, since miners work in much dustier conditions than are found in homes and their breathing during hard labor is much deeper than normal. Consequently, there is a much greater chance that radon daughters will find their way deep into the lungs of miners in comparison with the general population. The miners' exposure to arsenic and diesel exhaust may also contribute to an increase in the lung cancer rate that would have been counted as due to radon.

In order to establish whether or not radon gas buildup in homes causes lung cancer, several epidemiological studies were undertaken in the 1990s. These analyses, one from Sweden, one from Canada, and one from the United

States, reached contradictory conclusions about the risk of radon to householders. In the Swedish report the rate of lung cancer in nonsmokers and especially in smokers was found to increase with increasing levels of radon in their homes. The Canadian study focused on residents of Winnipeg, Manitoba, which has the highest average radon levels in Canada; no linkage between radon levels and lung cancer incidence was found. The U.S. study, conducted among nonsmoking women in Missouri, found little evidence for a trend of increasing lung cancer with increasing indoor radon concentration.

The environmental radon problem has received the greatest attention in the United States, where there are currently programs to test the air in the basements of a large number of homes for significantly elevated levels of the gas. Once radon is identified, the owners can then alter the air circulation patterns to reduce radon levels in living areas, thereby reducing the additional risk of contracting lung cancer. Gaps and cracks in basement walls are sealed, and venting pipes through basement floor slabs are installed.

9.9 The Health Effects of Very Low Levels of Radiation

A minority of scientists do *not* believe that radioactivity at very low levels causes any harm to humans. They are skeptical about the *linear-no-threshold* (LNT) assumption—that the observed effects of high doses of radioactivity can be extrapolated linearly to very low doses and that there is no *threshold* below which radioactivity causes no harmful effects such as cancer.

Indeed, some studies point to a threshold near 1000 Bq m^{-3} for lung cancer caused by residential radon. In addition, there is lack of evidence of increased cancer in many areas having high radon levels. The LNT theory is supported by evidence concerning Russian citizens who were exposed inadvertently to very low doses of radioactivity from nuclear weapons production. Recent reviews of epidemiological studies carried out in North America and Europe find no threshold for increase in lung cancer with exposure to residential radon.

A few scientists believe in a theory called *hormesis*, which states that exposure to radioactivity—and to some chemicals—at very low doses for short periods of time can actually be beneficial to human health. Although there are some animal and cell studies that support this idea, it is not widely accepted among scientists.

Review Questions 1–5 are based on the above material.

Nuclear Energy

Although most of the energy we use originates as heat generated by the combustion of carbon-containing fuels, **heat in commercial quantities can also be produced indirectly when certain processes involving atomic nuclei occur; this power source is called nuclear energy, used mainly to produce electricity.** Since nuclear forces are much stronger than chemical bond forces, the energy released per atom in nuclear reactions is immense compared to that obtained in combustion reactions. **One of the attractions of** nuclear power is that it does not generate *carbon dioxide* or other greenhouse gases during its operation. Some policymakers have promoted expansion of nuclear power as a way to combat global warming in the future.

There are two processes by which energy is obtained from atomic nuclei: fission and fusion.

• **In fission,** the collision of certain types of heavy nuclei with a neutron results in the splitting of the nucleus into two similarly sized fragments called **fission products.** Since the separated fragments are more stable energetically than was the original heavy nucleus, energy is released by the process.

• **The combination of two very light nuclei to form one combined nucleus is called fusion.** It also results in the release of huge amounts of energy, since the combined nucleus is more stable than the original, lighter ones.

9.10 Fission Power Reactors

Currently, there are about 440 fission-based nuclear power plants in operation in 30 countries around the world. Collectively, they generate 14% of global electricity demand. The fraction of electricity produced by nuclear energy in countries with the most reactors is listed in Table 9-2. Other

TABLE 9-2	Nuclear Power Around the World (2011)	
Country	Number of Operating Power Reactors	Proportion of Electricity Generated by Nuclear Power
United States	104	20%
France	58	74%
Japan	51	(29%)
Russia	32	17%
South Korea	21	32%
India	20	3%
Great Britain	19	16%
Canada	18	15%
Germany	17	28%
Ukraine	15	48%
China	13	2%
Sweden	10	38%
All others	63	

countries generating more than 25% of their electrical power from this source include Finland and Switzerland, plus many of the countries (Armenia, Bulgaria, Hungary, Slovakia, and Slovenia) formerly associated with the Eastern Communist Bloc, even though they individually have rather few reactors.

The most economically useful example of fission, and the one mainly used by power plants, is induced by the collision of a ^{235}U nucleus with a neutron. The combination of these two particles is unstable. When it decomposes, the products vary but are typically a nucleus of *barium*, ^{142}Ba, one of *krypton*, ^{91}Kr, and three neutrons:

$$\,^{1}_{0}n \;+\; \,^{235}_{92}U \;\longrightarrow\; \,^{142}_{56}Ba \;+\; \,^{91}_{36}Kr \;+\; 3\,^{1}_{0}n$$

$$\textcircled{n} + \left(\,^{235}_{92}U\right) \longrightarrow \left(\,^{142}_{56}Ba\right) + \left(\,^{91}_{36}Kr\right) + \begin{matrix}\textcircled{n}\\[2pt]\textcircled{n}\\[2pt]\textcircled{n}\end{matrix}$$

Not all the uranium nuclei that absorb a neutron form exactly the same fission products, but the process always produces two nuclei of about the sizes of Ba and Kr, together with several neutrons.

The two new nuclei produced in fission reactions are very fast-moving, as are the neutrons. It is heat energy from this excess kinetic energy that is used to produce electrical power. Indeed, the generation of electricity by nuclear energy and by the burning of fossil fuels both involve using the energy source to produce steam, which is then used to turn large turbines that produce the electricity.

An average of about three neutrons are produced per ^{235}U nucleus that reacts; one of these neutrons can produce the fission of another ^{235}U nucleus, and so on, yielding a **chain reaction.** In atomic bombs, the extra neutrons are used to induce a very rapid fission of all the uranium that is constrained to stay in a small volume, so energy is released explosively. In contrast, the **energy is released gradually in a nuclear power reactor by ensuring that, on average, only one neutron released from each ^{235}U fission event initiates the fission of another nucleus.** The extra neutrons produced in fission are absorbed by the **control rods** in a nuclear power reactor. These are bars made from neutron-absorbing elements such as *cadmium*. The position of the rods in the reactor can be varied to control the rate of fission.

The neutrons that leave ^{235}U nuclei upon fission are too fast-moving to be efficiently absorbed by other nuclei and cause further fission, so they must be slowed down if they are to be useful. This is accomplished by the **moderator,** which, depending upon the type of reactor, can be regular water, heavy water (i.e., water enriched in deuterium), or *graphite*. The **coolant** material that is used to carry off the heat energy produced by fission in most reactor types is water, but gaseous carbon dioxide is used in some. Ultimately, high-pressure, high-temperature steam is created—just as in a fossil-fuel

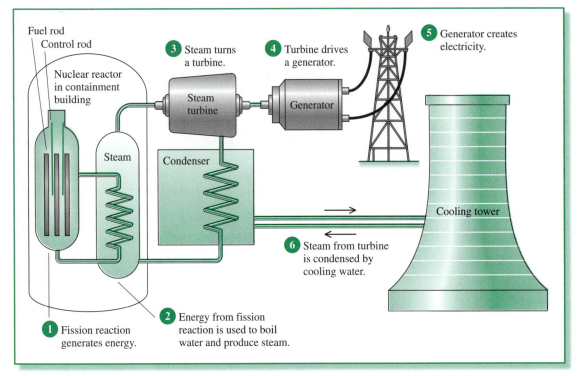

FIGURE 9-2 Step-by-step schematic diagram of the production of electrical power by a nuclear fission reactor.

power generating plant—and used to turn turbines and create electricity. The operating efficiency of nuclear power reactors is significantly lower than that of fossil-fuel ones since the core of the nuclear reactor, from which heat is drawn, is only a little above 300°C, whereas that of fossil-fuel plants may be as high as 550°C. The various steps in the production of electricity by a fission-based nuclear power plant are illustrated in Figure 9-2.

PROBLEM 9-3

Calculate the maximum percentage efficiency in converting heat at 300°C into electricity according to the second law of thermodynamics (see Section 8.16), assuming the cooling water is at 17°C. Repeat the calculation for 550°C heat. Do your calculated values lie above the average efficiencies of 30% and 40% observed for nuclear and coal-fired power plants, respectively?

Zirconium metal has only a small tendency to capture neutrons and is highly resistant to corrosion under normal conditions.

The uranium in reactors, usually in the form of the oxide UO_2, exists in the form of small pellets that are placed together in a series of *zirconium* alloy bars called **fuel rods.** When after several years of operation the uranium fuel in a rod is "spent," i.e., when its ^{235}U content is too low for it to be useful anymore as a fuel since the rate at which the uranium produces neutrons has become too low, it is removed from the reactor and replaced.

9.11 Actinide Products of Fission

The only naturally occurring uranium isotope that can undergo fission is ^{235}U, which constitutes only 0.7% of the native element. The remaining uranium is ^{238}U (99.3% natural abundance). However, a neutron produced by the fission of a ^{235}U nucleus can be absorbed on collision by a ^{238}U nucleus. The resulting ^{239}U nucleus is radioactive and emits a β particle, as does the heavy product (^{239}Np) of this process; consequently, a nucleus of **plutonium,** ^{239}Pu, is produced:

$$\ce{^{1}_{0}n} + \ce{^{238}_{92}U} \longrightarrow \ce{^{239}_{92}U} \xrightarrow{-\beta} \ce{^{239}_{93}Np} \xrightarrow{-\beta} \ce{^{239}_{94}Pu}$$

$$\text{(n)} + \left(\ce{^{238}_{92}U}\right) \longrightarrow \left(\ce{^{239}_{92}U}\right) \longrightarrow \beta + \left(\ce{^{239}_{93}Np}\right) \longrightarrow \beta + \left(\ce{^{239}_{94}Pu}\right)$$

Thus, ^{239}Pu **is produced as a by-product of the operation of nuclear power reactors.** Like ^{235}U, this isotope of plutonium can undergo nuclear fission—it is said to be "fissile." As its concentration in the fuel rods increases over time, it generates some of the neutrons and hence some of the power produced from the chain reaction. In addition, ^{239}Pu is a highly radioactive substance.

One or more successive neutron capture reactions, followed by one or more β-decay reactions, leads to the formation of other actinides in the fuel rods, as it does for plutonium. The important product neptunium-237 is formed from the small fraction of ^{235}U that does *not* fission after it absorbs a neutron. The resultant ^{236}U eventually absorbs another neutron, and then decays by β emission, producing *neptunium*:

$$\ce{^{235}_{92}U} \xrightarrow{+n} \ce{^{236}_{92}U} \xrightarrow{+n} \ce{^{237}_{92}U} \xrightarrow{-\beta} \ce{^{237}_{93}Np}$$

The neptunium-237 isotope has a half-life of 2 million years, so it is very slow to decay. Most of the α-particle decay from the spent fuel rods originates with it and with plutonium-239.

The concentration of $\ce{^{237}_{93}Np}$ in spent fuel rods actually *increases* over time since it is also formed *from* α-particle decay of the other important actinide formed in the fuel rod during its operation, $\ce{^{241}_{95}Am}$. This isotope of *americium* is created when some of the plutonium-239 successively absorbs two neutrons to produce $\ce{^{241}_{94}Pu}$, which decays by β emission with a half-life of 15 years to give $\ce{^{241}_{95}Am}$.

$$\ce{^{239}_{94}Pu} \xrightarrow{+n} \ce{^{240}_{94}Pu} \xrightarrow{+n} \ce{^{241}_{94}Pu} \xrightarrow{-\beta} \ce{^{241}_{95}Am} \xrightarrow{-\alpha} \ce{^{237}_{93}Np}$$

Americium-241 has a half-life of 432 years, so it will mainly have decayed to produce more neptunium within a few thousand years.

A very practical use has been found for americium-241. Because even a tiny sample of the isotope emits a constant stream of alpha particles over many years, it is used as the source of ionizing radiation in conventional smoke detectors. As long as the small electrical current comprised of charged particles passes between two electrodes in the ionization chamber of the detector, the alarm is not triggered. However, any smoke that is present will absorb the alpha particles, thereby reducing the current and setting off the alarm.

9.12 Radioactivity from Fission Products

At the time they are removed from the reactor, the spent fuel rods are almost a million times more radioactive than was the original uranium ore! The radioactivity arises both from the actinides produced from intact uranium nuclei, as discussed above, and from the fission products generated when the actinides undergo fission.

Each fission product has a nucleus about half that of the original uranium, with an atomic number in the 30 to 50 range rather than 92. Since the optimum neutron/proton ratio of nuclei increases with mass number, the fission products are *neutron-rich*. The excess neutrons split into a proton and an electron, with the latter expelled from the nucleus as an energetic β particle, thereby increasing the atomic number by 1 and lowering the neutron/proton ratio. Successive neutron decay continues until the product is no longer neutron-rich. For example, the krypton produced in the prototypical fission reaction previously discussed has a neutron/proton ratio of 1.53, typical of that of a nucleus the size of uranium. It has a half-life of only 10 seconds before it emits a β particle and becomes rubidium, which in turn (1-minute half-life) emits a β particle to become strontium, which in turn (10-hour half-life) emits a β particle to become yttrium, which in turn (59-day half-life) emits a β particle to become a stable, nonradioactive isotope of zirconium with a normal neutron/proton ratio (1.28) for elements of this size.

$$^{91}_{36}\text{Kr} \xrightarrow{-\beta} {}^{91}_{37}\text{Rb} \xrightarrow{-\beta} {}^{91}_{38}\text{Sr} \xrightarrow{-\beta} {}^{91}_{39}\text{Y} \xrightarrow{-\beta} {}^{91}_{40}\text{Zr}$$

9.13 Radioactivity in Spent Fuel Rods

Upon their removal from a reactor, the fuel rods are both radioactively "hot" and also thermally "hot" from heat produced by the β particles emitted from the fission products, as well as from α particles from actinides, even though fission has ceased. The spent fuel rod bundles initially generate about 6% of the heat that was produced in the reactor, though this decreases to only a few percent after a few hours and to about 0.1% after a year. Temperatures could reach 1500°C, destroying the tubes, if the rods were not immediately cooled after removal.

To cool the spent fuel rods, they are immersed in large pools of circulating water for several years. Water not only absorbs the heat but also captures most of the radiation that escapes as well. The large "swimming pools" that house the spent fuel rods are located at the power-plant site. Globally, there are more than 300,000 tonnes of spent fuel rods—about one-third of it in the United States—awaiting final disposition, much of it stored in pools.

The decline over time in the amount of radioactivity from a typical spent fuel bundle is illustrated by the black line in Figure 9-3; notice that both scales on the graph are logarithmic. For the first few hundred years, almost all the radioactivity originates from the medium-mass fission products (dark green curve, Figure 9-3). Their activity is essentially complete after a millennium. However, the radioactivity from actinides and their daughters is very much slower to decline, and becomes the dominant source after a few hundred years (light gray curve, Figure 9-3). Due to continuing emissions from the actinides, it is only after a few hundred thousand years have passed that the radioactivity level of the rods falls below that of uranium ore (dashed line in Figure 9-3).

After 10 years, most of the radioactivity from spent fuel rods is due to strontium-90, ^{90}Sr (half-life of 29 years), and to cesium-137, ^{137}Cs (half-life of 30 years), both of which are β-particle emitters. The dispersal of radioactive strontium and cesium into the environment would constitute a serious environmental problem. Ions of both metals can be readily incorporated into the body because strontium and cesium readily replace chemically similar elements that are integral parts of animal bodies, including those of humans. *Strontium*, a Group II metal, concentrates in the bones and teeth,

The dark green curve also includes emissions from *activation products,* which are nuclei in the zirconium casings that become radioactive from neutron bombardment.

The radioactivity level eventually falls below that of uranium ore, taken as 1.0 on the graph, as is that of yellowcake, because the ore contains other radioactive members of the decay chain as well as uranium itself.

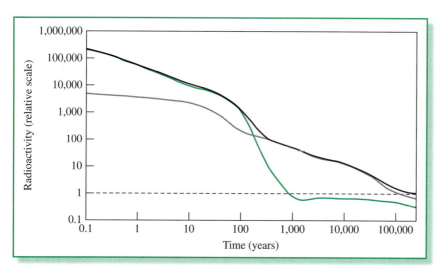

FIGURE 9-3 The radioactivity level—relative to that of uranium ore—emitted from spent fuel rods after their removal from the reactor. Note that both axes have logarithmic scales. The black curve corresponds to the activity from all sources; the dark green curve, from fission and activation products; and the light gray curve, from actinides and daughters they form after fuel rod removal. [Source: From J. Bruno and R. C. Ewing, "Spent Nuclear Fuel," Elements 2(6) (2006): 343–349.]

replacing ions of *calcium*, also a Group II metal that forms +2 ions. Ions of *cesium*, a Group I metal, can replace those of *potassium*, also from Group I and also +1 in charge, which is widespread in all body cells. Once these elements are in place in the body, their radioactive disintegration produces β particles that can damage the cells in which they are present or near. Radioactive waste from the spent fuel rods of nuclear power plants must therefore be carefully monitored and will eventually have to be deposited in a secure environment from which it cannot escape.

Environmental Problems of Uranium Fuel

Several of the steps in the *nuclear fuel cycle*, illustrated in Figure 9-4, generate environmental waste.

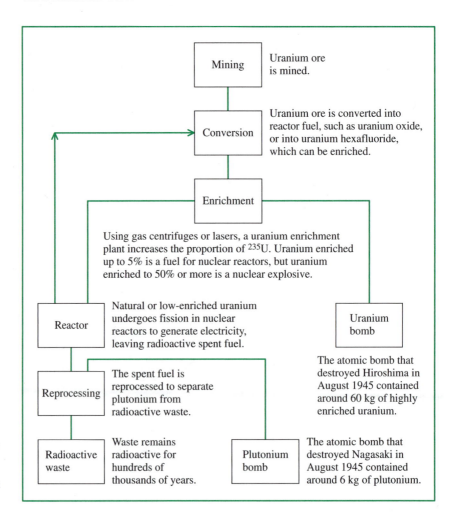

FIGURE 9-4 The nuclear fuel cycle. [Source: R. Edwards, "A Struggle for Nuclear Power," New Scientist (22 March 2003): 8.]

9.14 Uranium Ore

Major suppliers and processors of uranium ore are Canada, Australia, Russia, Niger, Ukraine, and Kazakstan. During the mining of uranium, contamination of the environment by radioactive substances commonly occurs. Since naturally occurring uranium slowly decays into other substances that are also radioactive, uranium ore contains a variety of radioactive elements. As a result, the very large volume of waste material that remains after the uranium is chemically extracted from the ore is itself radioactive.

The radioactive material that was immobilized in the original rock ore is in the form of liquid and powder *tailings* after mining. As in other mining operations, the liquid tailings are normally held in special ponds until the solids separate. Pollution of the local groundwater can occur if these ponds leak or overflow. In addition, when the solid tailings are exposed to the weather and are partially dissolved by rainfall, they can contaminate local water supplies. Gaseous emissions also escape from the wastes.

The use of the solid tailings as landfill on which buildings are constructed can also lead to problems, since the radon produced by the radioactive decay of the radium in the tailings is quite mobile. As previously noted, radon is a particular hazard for uranium ore miners, since the radioactive gas is always present in the ore and is released into the air of the mine. Indeed, the incidence of lung cancer among such miners was particularly high until mine ventilation was improved to permit more frequent changes of air and thus a more efficient clearing out of accumulated radon gas.

Uranium occurs as the ore **uranium dioxide,** UO_2, corresponding to the metal's +4 oxidation number. The ore contains other actinide metals and various impurities, so a multistage extraction process is required to isolate the uranium. In refining, it is first oxidized to the +6 state as UO_2^{2+} and $U_2O_7^{2-}$, then partially reduced by heating to *yellowcake*, U_3O_8, which can be thought of as a 1:2 combination of UO_2 and UO_3, the +4 and +6 oxides. To enrich the uranium (see below), the yellowcake is first oxidized completely to UO_3 and then fluorinated to UF_6. Finally, after reduction from +6 back to the +4 state, it is ready to be fabricated as UO_2 into fuel pellets for the reactor.

In most but not all nuclear power reactors (the Canadian *CANDU* system being the main exception), the uranium fuel must be enriched in the fissionable [235]U isotope; its abundance is increased to 3.0% from the 0.7% in the naturally occurring element. The extent of enrichment required for use in bombs is much greater. Uranium sufficiently enriched for this purpose, called *weapons-grade* material, is 90% or more [235]U. Enrichment is a very expensive, energy-intensive process since it requires physical rather than chemical means of separation, given the fact that all isotopes of a given element behave identically chemically. For separation, the uranium is temporarily converted to the gaseous compound *uranium hexafluoride*, UF_6.

9.15 Deuterium and Tritium and Heavy-Water Reactors

The CANDU reactor is able to use nonenriched uranium by employing **heavy water** rather than normal water as a moderator, since doing so decreases the probability that neutrons will be absorbed and therefore unavailable to continue the chain reaction. Heavy water contains a much-higher-than-normal fraction of **deuterium,** ^{2}H, the naturally occurring (0.02% natural abundance) nonradioactive isotope of hydrogen. Some of the neutrons produced by fission are absorbed by deuterium nuclei, producing **tritium,** ^{3}H.

$$^2H + {}^1n \longrightarrow {}^3H$$

Indeed, CANDU reactors produce about 30 times as much tritium as do light-water reactors. Tritium levels in moderator and coolant water increase with time, as the product of this reaction accumulates. Small losses of coolant from reactors result in the release of tritium into the aqueous environment.

Tritium is radioactive, a low-energy β-particle emitter with a half-life of 12.3 years. Less than 1% of the tritium in the present environment occurs naturally; the remainder is due to fallout from nuclear weapons testing in past decades and to releases from nuclear power reactors. The U.S. EPA classifies tritium as a human carcinogen; there is evidence that it is mutagenic and teratogenic (causes birth defects) as well. The EPA's maximum limit for tritium in drinking water is 740 Bq L^{-1}; Canada's limit is almost 10 times as high. The waters of Lake Huron and Lake Ontario, along whose shores several CANDU reactors are located, contain about 7 Bq L^{-1} of tritium, compared to 2 Bq L^{-1} for Lake Superior, which has no nuclear reactors nearby.

9.16 Depleted Uranium

The ^{235}U level in depleted uranium is 0.2–0.4%, about a third that of natural uranium.

Depleted uranium is what remains of natural uranium once most of the ^{235}U isotope, used in power reactors and for nuclear weapons, and the ^{234}U have been extracted from it. Indeed, about 200 kg of depleted uranium is produced for every kilogram of the highly enriched element, since uranium is only 0.7% ^{235}U. Because uranium is so dense (70% denser than lead), it is useful for making armor-piercing weapons, especially projectiles used against tanks. When the shells hit hard targets, the uranium ignites and the combustion creates clouds of dust containing uranium oxide. This dust settles and contaminates the soil in the area, but before that, it can be inhaled by people in the vicinity. Since most of the ^{235}U has been extracted from the uranium (for use in bombs and power production), and since ^{238}U has such a very long half-life, depleted uranium is less radioactive in terms of α-particle emissions (by almost half) than the naturally occurring element. However, the β emission from daughters of ^{238}U is still present. Concerns have been expressed about the effect of the residual radioactivity from depleted uranium on troops and civilians exposed to it during wartime.

9.17 Dirty Bombs

A dirty bomb is a conventional, chemical-based explosive mixed with radioactive material that would be distributed over a wide area as a result of the bomb's explosion. Dirty bombs could be made by terrorists, using radioactive material stolen from either hospitals or research institutes, or purchased on the black market from supplies originating in countries that have undergone disruption with a consequent loss of security at their nuclear energy facilities. Although the actual danger to human health from a dispersal of dirty bomb materials into the environment may well be quite small, the fear it would generate in the public and the expense of decontamination of wide areas would be substantial.

A similar terrorist threat would be the deliberate crashing of hijacked aircraft into nuclear power plants or into containers of nuclear fuel. However, most nuclear energy industry executives deny that the power plants or containers could be breached, and radioactivity released, by such an event.

9.18 Plutonium and the Reprocessing of Nuclear Fuel Rods

The plutonium-239 isotope that is produced during uranium fission is an α-particle emitter and has a long half-life of 24,000 years. After 1000 years, the main sources of radioactivity from spent fuel rods will be plutonium and other very heavy elements, since the medium-sized nuclei produced in fission, having much shorter half-lives than 1000 years, will have largely decayed by that time (see Figure 9-3). **Thus the long-term radioactivity of the spent fuel rods can be greatly reduced by chemically removing the very heavy elements from it.**

PROBLEM 9-4

Given that the half-life of ^{239}Pu is 24,000 years, how many years will it take for the level of radioactivity from plutonium in a sample to decrease to 1/128 (i.e., about 1%) of its original value? ●

The ^{239}Pu that forms in fuel rods is itself fissionable; once its concentration in the operating rods becomes high enough, some of it also undergoes fission and contributes to the power output of the reactor. However, many of the other fission by-products are efficient neutron absorbers, and their concentration eventually builds up to the point where the nuclear chain reaction cannot occur efficiently, and the fuel rod is removed. In an *open-cycle* system, such "spent" fuel rods are then stored as nuclear waste, as described in Sections 9.13 and 9.21.

However, in *closed-cycle* systems, the content of the fuel rods undergoes **reprocessing** to separate certain actinide elements and use some of them for nuclear fuel or for other purposes. The separation problem is not a trivial one, since the chemical properties of all the actinides are quite similar, so subtle differences between their characteristics must be exploited.

• To begin reprocessing, the fuel rods—having been cooled for 5–25 years in storage pools—are first chopped into pieces and their contents reacted with concentrated **nitric acid,** HNO_3, to dissolve them. This produces an aqueous solution containing oxidized cations of many metals, both actinides—with uranium as UO_2^{2+} and plutonium as Pu^{4+}—and medium-mass fission products.

• Uranium and plutonium can be separated from this aqueous solution by exploiting the fact that their oxidized cations—but not those of some other actinides and those of the lighter metals—will form complexes with an organic phosphate, namely *tri-n-butyl phosphate*, $O=P(-O-CH_2CH_2CH_2CH_3)_3$, that are much more soluble in hydrocarbon solvents such as kerosene than in the aqueous solution. Thus, mixing a neutralized aqueous solution with the phosphate and kerosene produces a two-phase system in which uranium and plutonium are selectively extracted into the organic layer:

$$Pu^{4+}(aq) + UO_2^{2+}(aq) + others(aq) \longrightarrow Pu^{4+}(org) + UO_2^{2+}(org)$$

• The aqueous layer, containing the lighter metals, must be stored as *high-level nuclear waste* since it contains many highly radioactive isotopes.

• The relatively small amount of plutonium is then isolated from the larger mass of uranium by selective reduction using a mild reducing agent:

$$Pu^{4+}(org) \xrightarrow{\text{reducing agent}} Pu^{3+}(org)$$

Since the +3 ion of plutonium is more soluble in aqueous solution than in the organic phase, it can be extracted by adding an acid solution, from which it can be precipitated, as *plutonium trifluoride*, PuF_3:

$$Pu^{3+}(org) \longrightarrow Pu^{3+}(aq) \longrightarrow PuF_3(s)$$

If, in order to guard against nuclear proliferation it is desired that plutonium *not* be isolated, it can be reduced *before* the organic phosphate layer is added; as the +3 ion, it will remain in the aqueous layer and become a component of the high-level waste.

• The uranium is recovered from the organic phase as the dioxide. If desired, it can be converted back to UF_6 and re-enriched in ^{235}U. It can also be combined as UO_2 with a small amount of plutonium that was re-oxidized to PuO_2 to produce *mixed oxide fuel* (MOX). This can be combined with a larger quantity of virgin UO_2 fuel and used in light-water reactors.

Plutonium obtained from the dismantling of nuclear weapons is also used to create MOX.

The highly radioactive liquid waste that remains after reprocessing, which contains both fusion products and some actinides, is solidified by heating to dehydrate it and to decompose its nitrate salts. The resulting powder, consisting mainly of the oxides of the metals, is then incorporated into borosilicate glass and stored in stainless-steel cylinders that are welded shut. Eventually these containers will be buried in long-term geological storage. Overall, reprocessing is costly, in part because of the need to dispose of the radioactive waste it liberates, and notwithstanding the fact that it is a chemical process and so is much less energy intensive than isotope separation to produce newly enriched uranium.

Spent fuel rods from civilian power reactors currently are reprocessed in France, India, Russia, China, and Japan but not in the United States or Canada. This reprocessing has resulted in a buildup of hundreds of tonnes of plutonium. Comparable quantities of the element are also available from the dismantling of nuclear weapons by the United States and Russia (see Box 9-2).

Aside from questions of health and safety, a major problem associated with the handling of plutonium from both civilian and military sources involves the security measures needed to prevent the material from slipping into the hands of terrorists and rogue governments who wish to fashion their own bombs. Only a few kilograms of weapons-grade plutonium, which consists of 93% or more of ^{239}Pu, is required to make an atomic bomb. A somewhat larger quantity of *reactor-grade* plutonium, which contains more of the other isotopes of plutonium and similar elements, is needed for a bomb. The world's current stockpile of plutonium exceeds 1000 metric tons and continues to grow.

BOX 9-2   Radioactive Contamination by Plutonium Production

Plutonium has been deliberately produced for more than 60 years to provide the readily fissionable material needed for nuclear weapons. In the United States, most plutonium production and processing were carried out at Hanford, Washington. Because huge quantities of radioactive waste were produced, stored, and disposed of at this facility, the surrounding environment is now so heavily polluted that it has been called the "dirtiest place on Earth." About 190,000 m³ of highly radioactive solid waste and 760 million liters of moderately radioactive liquid waste and toxic chemicals were deposited in the ground at this site. As much as a metric ton of plutonium may be contained within the masses of solid wastes buried there. The cleanup of the wastes will cost between $50 and 200 billion and will

not be completed earlier than 2020. One proposed—though expensive—technique for immobilizing the radioactive wastes involves passing a strong electrical current through the contaminated soil for a period of days; the electricity would fuse the soil and sand into a glassy rock from which the contaminants could not escape.

Plutonium was used as the actual explosive material in some atomic bombs and as a "trigger" for hydrogen (fusion) bombs, forcing the reactants together and thereby initiating the thermonuclear explosion. Consequently, about 100 metric tons of plutonium must be removed from nuclear weapons as tens of thousands of them are dismantled by the United States and Russia over the next few decades.

9.19 Breeder Reactors

Breeder reactors are nuclear power reactors that are designed specifically to maximize the production of by-product plutonium; such reactors actually produce more fissionable material than they consume. In particular, the special *fast breeder reactors* produce ^{239}Pu from ^{238}U by the reaction previously discussed (Section 9.11). The reactors are cooled by liquid sodium, since water is not adequate for the purpose and would absorb neutrons. They use **fast neutrons,** rather than those moderated to produce fission, since fast neutrons are more likely to be captured by ^{238}U and produce its daughters, which in turn quickly decay to plutonium. The plutonium must be separated from the residual uranium by reprocessing before it can be used as a fuel. Unfortunately, in an accident involving failure of the cooling system, the fast breeder reactor core could melt since plutonium's melting point is only 640°C. Another potential problem is the violent explosion that would occur if liquid sodium were to come into contact with water or air.

Alternatively, the **thermal breeder reactor** converts ^{232}Th by neutron absorption and β decay into ^{233}U, which can also be used as fuel in nuclear reactors since it will undergo fission; India has experimented with this design since it has substantial natural reserves of thorium.

Because of operational problems with the technologically sophisticated fast breeder reactors, the programs were abandoned in the United States, the United Kingdom, Germany, and France. Russia and China operate breeder reactors, and Japan is building a demonstration unit. Overall, the future of breeder reactors remains in doubt.

9.20 Disposing of Plutonium

The radiation from plutonium is weak and becomes a concern only if the material is ingested or inhaled, since its radiation (α particles) cannot pass through dead layers of skin or clothing or indeed through even a few centimeters of air. Elemental, metallic plutonium cannot be easily absorbed by the body. However, when exposed to air, plutonium forms the oxide PuO_2, a powdery dust that disperses readily and can be inhaled. Even microscopic amounts of plutonium oxide lodged in the lungs can induce lung cancer.

Two methods have been proposed to dispose of excess plutonium:

• Mix it with other highly radioactive liquid waste and then **vitrify** the mixture into durable glass logs that would be buried far underground in metal canisters, as will be discussed. **Vitrification** chemically bonds liquid waste into a stable and durable borosilicate glass in which fission product oxides make up about 20% of the mass.

• Convert it to plutonium oxide and mix it with uranium oxide to produce mixed oxide fuel, as previously described, that could be used in existing power plants. Indeed, some MOX is currently employed in reactors in

France, Germany, Switzerland, Belgium, and Japan. However, producing MOX fuel is more expensive at present than is generating low-enrichment uranium, so the incentive to use it is not economics.

9.21 Nuclear Reactor Waste Geological Storage

Although nuclear energy has been used to generate electricity now for many decades, there is still no consensus among scientists and policymakers concerning the best procedure for long-term underground storage of spent fuel rods or other high-level waste from these plants. Indeed, not a single gram of fuel rod waste has yet been deposited in a long-term storage location.

As discussed above, spent fuel rods are initially cooled in deep pools of water. After a few years, or perhaps a decade, the spent fuel rods have cooled enough thermally and radioactively to be placed in dry storage within cylinders in which air and water have been replaced by an inert gas. The sealed cylinders are then encased in steel-and-concrete canisters called *casks*. They are located outdoors, often on concrete pads. After a century has passed, their radioactivity level will have declined to only several hundred—rather than a million—times that of uranium ore (black curve, Figure 9-3).

Plans for the ultimate disposal of high-level nuclear waste, whether the fuel rods themselves or their residue after uranium and/or plutonium have been removed during reprocessing, all assume that their material would be encapsulated and then buried in a stable centralized location far underground, 300–1000 m below the surface. Ideally, the capsules would last for hundreds of thousands of years before any substantial leakage would occur, by which time the level of radioactivity of the waste would have declined almost to background values (Figure 9-3).

The most important optimum geological features of the burial sites would be:

• high stability from possible disruption by volcanic activity or earthquakes; and

• low permeability to water, to insure minimal interactions with groundwater and the biosphere.

Several countries that have many nuclear power reactors are pursuing individual burial plans for the rods. Sweden and Finland have the storage plans that currently are the most developed. They plan to bury the fuel rods several hundred meters underground in large iron canisters that are coated with 5 cm of corrosion-resistant copper and then covered in bentonite clay.

Bentonite clay swells when it absorbs water, which would prevent pooling of groundwater that could corrode the copper.

A 20-year test of this system is now underway in Sweden, with electrical heaters inside the canisters used to mimic the heat that would actually be produced from the fuel rods. Sensors monitor the temperature and the water movement, if any, within the chamber walls. Finland has already started excavations on their site.

By necessity or by design, most countries are planning on deep geological disposal sites that lie below the water saturation level, where conditions chemically are reducing rather than oxidizing. The Yucca site in the United States discussed below would have had disposal above the groundwater level. The advantage of storage *above* the water level is that any leakage of radioactive material out of the capsules would not then be mobilized in an aqueous environment.

The United States does have an underground storage facility, located in a kilometers-deep and thick underground salt deposit near Carlsbad, New Mexico, for cooled spent fuel from nuclear weapons.

The United States abandoned in 2010 its controversial plan—developed in the 1980s—to store spent fuel rods from power plants in a facility at Yucca Mountain in Nevada. The plan had been to transfer the rods, after they had cooled for a decade in water at local reactor facilities, into steel cylinders and place them in an underground tunnel to be cooled by fans for another 50 years before the facility was sealed.

Spent fuel rods are currently stored in aboveground casks at nine decommissioned nuclear plants in the United States.

A tentative new plan, proposed by a Blue Ribbon Commission, is to create a number of *interim storage sites* around the United States where spent fuel rods would be stored in outdoor steel-and-concrete casks for a number of decades—perhaps a century or more, but at least 50 years—after they had been cooled for a decade in ponds at the reactor sites. Eventually, a geologic repository would be created for the final disposition of the rods, one requiring no fans since they would have cooled sufficiently after that elapsed time.

Accidents and the Future of Nuclear Power

In developed countries, public opinion about fission-based nuclear power turned from positive to negative in the last two decades of the twentieth century, partly because of the accidents at the Three Mile Island power station in Harrisburg, Pennsylvania, in 1979, and at a power station at Chernobyl, Ukraine, in 1986, both of which are described below. As a consequence, no new reactors have been built in the United States or Canada or most other developed countries since that time. In contrast, many reactors have been constructed subsequently in developing countries where the need for electricity is great as their economies expand.

The public attitude in developed countries toward the construction of additional nuclear power plants —or at least to replacing reactors approaching the ends of their useful lives in coming decades—began to turn in a positive direction in the first decade of the twenty-first century. However, the major disaster in 2011 at the Fukushima Dai-ichi nuclear power plant in Japan—described in detail below—completely reversed this trend. Indeed, countries such as Germany decided to completely phase out nuclear power in the future. At the time of writing these words, it is not at all clear what the future of nuclear power will be.

9.22 The Catastrophe at Chernobyl

A much more realistic fear than that of a nuclear power plant running out of control and exploding like an atomic bomb is that highly radioactive fission products contained in operating fuel rods could be spread into the surrounding countryside if a nonnuclear explosion occurs in a power plant. This in fact did occur at one of the nuclear power plants in Chernobyl, Ukraine, in 1986. During a routine test, engineers at the plant lost control of the reactor by overriding the plant's safety mechanisms and withdrew most of the control rods from the reactor core. As a consequence, the reactor overheated and a fire broke out in the graphite moderator. This produced a huge explosion, which blew off the heavy plate covering the building, and several hundred million curies of radioactivity were released into the air and spread over a wide area. The dispersed radioactivity was mainly concentrated in the noble gas isotopes ^{131}Xe and ^{85}Kr, in ^{131}I, and in the cesium isotopes ^{134}Cs and ^{137}Cs, all of which are fission products.

About two dozen people, mainly plant operators and firefighters, died within a few months from high doses of radiation from the accident.

A concrete sarcophagus to entomb the reactor building was constructed within months of the accident, but it has begun to crumble. The current plan is to construct—by 2015—and slide into place an enormous steel arch to cover the building and allow robotic cranes to dismantle the sarcophagus and parts of the reactor. The cleanup work will take about 50 years to complete.

A major chronic health consequence to the public of the explosion at Chernobyl has been the great increase in thyroid cancer among children in the area. The thyroid cancer was initiated presumably by β radiation from radioactive iodine, ^{131}I, which has a half-life of only 8 days. The human body concentrates iodine in the thyroid gland. Thankfully, thyroid cancer has a high cure rate. In the most contaminated areas of Ukraine, about 4 times as many children as normal have been diagnosed with thyroid cancer, with very young children found to be the most susceptible to the radioactivity. Overall, more than 6000 children in the region have been diagnosed with thyroid cancer, although fewer than 20 have died from it thus far.

The probability of a Ukrainian child having thyroid cancer was found to increase in direct proportion to the dose of radioactive iodine ingested after the accident. Most of the iodine in the children's systems resulted from drinking milk from cows that had grazed on contaminated plants and from eating leafy vegetables, since the radioactive iodine was deposited from the air onto plant surfaces. The soil in the affected areas is chronically deficient in iodine, so the substance would readily have been taken up by the thyroid in children chronically deficient in the element. Indeed, tablets of *potassium iodide*, KI, designed to flood the body with iodine and thereby dilute the

134 emergency workers were diagnosed with acute radiation sickness.

radioactive form, were distributed to the inhabitants around Chernobyl—but not until a week after the explosion.

Radiation experts predict that about 4000–9000 premature deaths among people who were living in the heavily contaminated regions—Ukraine, Belarus, and Russia—have or will eventually be associated with radioactivity from the Chernobyl event, including not only ^{131}I but also the longer-lived cesium isotopes ^{134}Cs and ^{137}Cs (the latter having a half-life of 30 years). Cancers other than that of the thyroid have much longer development times. A much larger number of Europeans—perhaps as many as 60,000—living outside the most affected areas who ingested smaller doses may also be affected eventually if the dose–response relationship for radiation does not have a threshold. However, according to the World Health Organization, the largest public health problem for the people who were living in the highly contaminated regions at the time of the Chernobyl disaster is a long-term decrease in their mental health, stemming from the traumas of perceived health risks and from relocation out of areas declared uninhabitable, as was the case for a third of a million people.

9.23 The Disaster at the Fukushima Reactors in Japan

Japan's nuclear catastrophe of March 2011 began when a 9.0-magnitude earthquake occurred 25 km below the ocean surface about 100 km off the northeast coast of the country. The earthquake resulted in the automatic shutdown of all operating nuclear power stations along the coast. However, it generated a tsunami that was about 13 m (43 ft) in height when it reached the Fukushima Dai-ichi nuclear power plant site, which was protected by a seawall designed to cope with only 5.7-m (19-ft) tsunami waves.

Of the six boiling-water reactors at the site, three (numbers 1, 2, and 3) had been operating when the earthquake struck and shut down automatically, presumably by the insertion of the control rods, whereas the other three (4, 5, 6) had already been shut down for maintenance and refuelling. Electrical power connections from the plant to the outside world were lost as a result of the tsunami, which destroyed the power lines. It also flooded the buildings at the site, including emergency diesel-powered generators in the lower levels of the reactor buildings as well as pumps located outside them.

Less than an hour after the earthquake, the tsunami waves overwhelmed the defences at the power plants. Reactors 1 through 5 lost all electrical power, including that of their onsite diesel generators, whereas reactor 6's generator continued to function, and its power was subsequently connected to reactor 5 as well, ensuring that both could eventually be shut down completely.

However, notwithstanding the heroic attempts by plant operators working at times in near-darkness with only limited battery power, coolant water over the fuel rods was eventually lost at reactors 1, 2, and 3. In particular, **the lack of power and other problems prevented the continuing injection of** cooling water over the fuel rods. Due to leaks, the water levels began to fall, eventually exposing most or all of the fuel rods. As a result of the loss of coolant, the residual heat arising from radioactive decay of the fission products raised the temperature of the fuel rods in the core to thousands of degrees, melting them and allowing escape of the radioactive fission products from the rods. In addition, the fuel pellets would have fallen out of the zirconium tubes into the bottom of the vessel. The high temperatures also converted the remaining cooling water into high-pressure, very hot steam that raised the pressure inside the vessel.

The possibility of a core meltdown in a nuclear reactor was first brought to the public eye by the 1979 movie *The China Syndrome.*

As a consequence of the contact between the bare **zirconium metal,** Zr, of the fuel rods and the high-temperature (>1000°C) steam, a chemical reaction took place that produced **hydrogen gas,** H_2, and the dioxide of zirconium:

$$Zr(s) + 2\ H_2O(g) \longrightarrow ZrO_2(s) + 2\ H_2(g)$$

Later, after the increasing pressure in the inner primary containment vessel was relieved by allowing some hydrogen-laden air to enter the outer secondary containment area, apparently a spark randomly ignited the mixture, producing an explosion from the highly exothermic reaction of hydrogen and oxygen:

$$2\ H_2(g) + O_2(g) \longrightarrow 2\ H_2O(g) + heat$$

Explosions of this type eventually occurred at reactors 1, 2, and 3, resulting in the partial destruction of their buildings.

It is not clear whether or not the pooled uranium that had fallen out of the zirconium tubes ever reached critical mass and nuclear fission was restarted. The addition of boron salts to the seawater used later to cool the system was an attempt to reduce this possibility, since boron is a neutron absorber.

In addition to the fuel rods in the reactor itself, spent fuel rods of various ages were stored in pools of water in upper levels of the reactors at the Fukushima complex. The racks used to house the rods contain boron, to absorb α particles emitted from them and prevent any possibility of fission. Damage from the explosion at reactor 3 resulted in failure of the cooling-water system for the spent-fuel-rod assembly in shut-down reactor 4. An explosion occurred near the top level of reactor 4 from hydrogen either generated by its stored rods or migrated from reactor 3.

Spent fuel is not kept in water pools located above reactors in more modern designs.

As a result of the accident and cleanup operations, seven site personnel were exposed to more than the 250-milliSieverts (mSv) limit of radiation set retrospectively upward by the Japanese government. Radioactive material was released into the environment in the gas vented from the reactors to relieve pressure and in the coolant water discharged into the sea. The radioactivity in the gases would have consisted mainly of short-lived isotopes of xenon, Xe. As in the Chernobyl accident, the main fission products of concern would have been ^{131}I and the mass-134 and mass-137 isotopes of cesium, all three of which were detected in seawater near the reactor site at levels orders of magnitude greater than normal.

Milk and some vegetables in the Fukushima area were later found to be contaminated. However, the immediate large-scale and long-term evacuation of residents from a very large area surrounding the site presumably will prevent the development of childhood thyroid cancers that were characteristic of the Chernobyl event. At the time of writing this textbook, the cleanup of the disaster was incomplete. One wonders whether the most lasting public health problem for the evacuated residents will be similar to the type of trauma suffered by the Chernobyl evacuees discussed in the previous section.

9.24 The Accident at Three Mile Island

The other nuclear power plant accident was much less serious. It occurred in 1979 at the Three Mile Island plant in Pennsylvania. The problem originated with failures in the water-based cooling systems in the reactor. Although control rods were inserted into the core and the fission process stopped, the reactor kept heating. The source of the heat was the radioactive decay by γ-ray emission, not of uranium, but of fission products that had naturally built up over time in the fuel rods. Because of mechanical malfunction and operator error, the core became partially uncovered; it was therefore heated even further (>2200°C), and half of it melted. Hydrogen and oxygen gases were produced from the decomposition of the superheated water.

Fortunately, **there was no large explosion at Three Mile Island: The containment facility was not breached.** In contrast to the Chernobyl accident, the liquid and solid radioactive materials that escaped from the reactor did not escape from the containment building; the release of ^{131}I was about a million times less. Significant amounts of radioactive noble gas isotopes ^{131}Xe and ^{85}Kr did escape into the air but, as at Chernobyl, they were quickly diluted in the atmosphere.

Nuclear Fusion

9.25 Fusion Reactors

The optimum energetic stability per nuclear particle (protons and neutrons) occurs for nuclei of intermediate size, such as iron. That is why the fission of a heavy nucleus into two fragments of intermediate size releases energy. Similarly, **the fusion of two very light nuclei to produce a heavier one also releases substantial quantities of energy.** Indeed, fusion reactions are the source of the energy in stars, including our own Sun, and in hydrogen bombs.

Unfortunately, fusion reactions all have extremely large activation energies due to the huge electrostatic repulsion that exists between the positively charged nuclei when they are brought very close together, which must happen before fusion can occur. Consequently, it is difficult to initiate and sustain a controlled fusion reaction that provides more energy than it consumes.

The fusion reactions that have the greatest potential as producers of useful commercial energy involve the nuclei of the heavier isotopes of hydrogen, namely deuterium, ^{2}H, and tritium, ^{3}H. Note that two different sets of products can be produced from these sample reactions:

$$^2_1H \quad + \quad ^2_1H \quad \Bigg\langle \quad \begin{matrix} ^3_2He + {}^1_0n \\ \text{or} \\ ^3_1H + {}^1_1H \end{matrix}$$

$$\text{deuterium} \quad \text{deuterium}$$

The energy that is released when one of these reactions occurs is about 4×10^8 kJ mol^{-1}, which is about 1 million times the energy produced in a typical exothermic *chemical* reaction. An abundant supply of deuterium is available, since it is a nonradioactive, naturally occurring isotope (constituting 0.015% of hydrogen) and thus is a natural component of all water.

A somewhat lower activation energy is required for the reaction of deuterium with tritium:

$$^2_1H \quad + \quad ^3_1H \quad \rightarrow \quad ^4_2He \quad + \quad ^1_0n$$

$$\text{deuterium} \qquad \text{tritium}$$

However, because tritium is a radioactive element (a β emitter) with a short half-life (12 years), it is not a significant component of naturally occurring hydrogen and would have to be synthesized by the fission of the relatively scarce element *lithium*.

The environmental consequences of generating electrical power from fusion reactors should be less serious than those associated with fission reactors. The only radioactive waste produced *directly* in quantity would be tritium, although the neutrons emitted in the process could produce radioactive substances when they are absorbed by other atoms. Although the β particle that tritium emits is not sufficiently energetic to penetrate the outer layer of human skin, tritium is nevertheless dangerous since biological systems incorporate it as readily as they do normal hydrogen (^{1}H or ^{2}H)—by inhalation, absorption through the skin, or ingestion of water or food. Currently, tritium in drinking water (some of which results from artificial sources) constitutes the source of about 3% of our exposure to radioactivity.

About 80% of the energy emitted in the deuterium–tritium reaction is associated with the neutrons. The energy, captured by using neutron-heated coolants, would be used to create superheated steam to drive turbines. Unfortunately, the intense neutron bombardment will cause severe degradation of the structure used to confine the fusion reactants and will generate large amounts of radioactive nuclei. These problems would be largely overcome if so-called *advanced fuels* were to be used as reactants. Thus the fusion reaction of deuterium with helium-3 (^3_2He) releases only a few percent of its reaction energy as neutrons (the products of the dominant process being protons and ^{4}He nuclei), and that of two ^{3}He nuclei (to produce two protons and ^{4}He) is essentially neutron-free. These processes could operate with much higher energy conversion efficiencies but require even higher initiation temperatures and confinement conditions. The other serious problem is the lack of ^{3}He on Earth: The Moon is the best source for this material!

PROBLEM 9-5

Write and balance the two fusion reactions involving ^{3}He mentioned above.   ●

In 2006, a consortium consisting of the European Union, the United States, China, India, Russia, Japan, and South Korea agreed to finance the ITER (International Thermonuclear Experimental Reactor) project, the world's first nuclear fusion reactor. Construction of the facility in Provence, France, began in 2008, and it is expected to be fully operational by 2026. The reactor will not produce usable energy, but is being constructed to demonstrate that it is possible to build a fusion reactor that will generate more power than it consumes. The reactor design is based on the *tokamak*, a doughnut-shaped machine having a series of overlapping magnetic fields that can hold the plasma within reactor walls.

The ITER facility will use the deuterium-tritium reaction. The high-speed neutrons emitted by the reaction will be slowed down and absorbed by an enclosure "blanket" that contains lithium and that will breed more tritium required for the reaction. A major concern, however, is that the high-energy neutrons from fusion will damage the materials used to construct the reactor vessel.

A demonstration power plant, which could use the excess heat to boil water and turn turbines to generate electricity, is slated for about 2040. It will use the experience gained in the ITER project in dealing with the helium and tritium generated by the fusion reactions. Hopefully, a solution to the problem of radioactive reactor materials being generated by neutrons emitted by the fusion process will also have been overcome by that time.

PROBLEM 9-6

Write the nuclear reaction in which tritium and $^4_2$He are produced by bombarding $^6_3$Li with a neutron.   ●

9.26 The Energy Released in Fusion and Fission

The energy, E, released in fission and fusion processes comes from the conversion of a tiny fraction, m, of the masses of the atoms and other particles involved. According to Einstein's famous equation, the energy released is

$$E = mc^2$$

where c is the speed of light.

For example, in the conversion of one mole of deuterium atoms (mass 2.0140 g) and one mole of tritium atoms (mass 3.01605 g) into one mole of ^{4}He atoms (mass 4.00260 g) and one mole of neutrons (mass 1.008665 g), a total of 0.0188 g of matter is lost by conversion to energy. Since $m = 0.0188 \times 10^{-3}$ kg and $c = 2.99792 \times 10^8$ m s^{-1}, the energy released is

$$E = mc^2 = (0.0188 \times 10^{-3} \text{ kg}) \times (2.99792 \times 10^8 \text{ m s}^{-1})^2$$
$$= 1.69 \times 10^{12} \text{ kg m}^2 \text{ s}^{-2}$$
$$= 1.69 \times 10^{12} \text{ J}$$

This energy, 1690 million kilojoules per mole, is about 7 million times the energy released when the same quantity of hydrogen is burned in oxygen to produce water.

Review Questions 6–18 are based on the above material.

PROBLEM 9-7

Calculate the amount of energy released in the fission process described earlier in the chapter in which uranium-235 and a neutron are fissioned to barium-142, krypton-91, and 3 neutrons. Assume 1 mol of ^{235}U reacts and that the atomic masses of ^{235}U, ^{142}Ba, ^{91}Kr, and a neutron are, respectively, 235.044, 141.926, 91.923, and 1.008665. Is the energy released much greater than, much less than, or about the same per nucleus as that for the fusion reaction between deuterium and tritium?

Review Questions

1. What is the particulate nature of the radioactive emissions α and β particles? What is a γ ray?

2. Why are α particles dangerous to health only if ingested or inhaled?

3. What is meant by the terms *rad* and *rem*?

4. Explain the origin of radon gas in buildings.

5. Explain what is meant by the term *daughters of radon*. Why they are more dangerous to health than radon itself?

6. Define *fission* and write the reaction in which a ^{235}U nucleus is fissioned into typical products.

7. What are the functions in a fission power reactor of **(a)** the moderator and **(b)** the coolant?

8. Write the nuclear reaction that produces plutonium from ^{238}U in a fission reactor.

9. Explain why the spent fuel rods from fission reactors are more radioactive than the initial fuel.

10. Why would radioactive strontium and cesium be particularly harmful to human health?

11. Describe why the mining of uranium ore often pollutes the local environment.

12. What is a *breeder reactor*? Why is it useful to breed fissionable fuel? What is meant by *reprocessing* and why is it done?

13. Write the nuclear reaction that produces ^{233}U from ^{232}Th following the absorption of a neutron.

14. Describe the two main methods that have been proposed to dispose of excess plutonium.

15. What is *depleted uranium*? Is it radioactive at all?

16. Explain what is meant by a *dirty bomb*.

17. Define *fusion* and give two examples of fusion processes (i.e., reactions) that may be used in power reactors of the future.

18. Describe the nature of any radioactive by-products of the operation of fusion reactors. What damage could the neutrons do?

Additional Problems

1. Another possible radioactive decay mechanism involves the emission of a *positron*, a particle with the same mass as an electron but the opposite (i.e., positive) charge. The net nuclear effect of positron emission is the change of a proton into a neutron. Explain how the result of positron emission differs from β emission in terms of the product's position in the periodic table. Deduce the symbol for a positron to be used in balanced nuclear equations, by analogy with the symbol $_{-1}^{0}e$ used for a β particle (the symbol e for an electron is also used here to represent a positron). Predict the decay products and write the balanced equation for the positron decay of the radioactive isotopes $_{11}^{22}$Na and $_{7}^{13}$N.

2. Tablets of KI were distributed to people living around Chernobyl a week after the nuclear reactor explosion occurred, to help flush the radioactive ^{131}I isotope out of their bodies. What percentage of the ^{131}I released by the explosion would have remained by that time? How long would it have

taken for 99% of the released ^{131}I to undergo radioactive decay? Use the kinetics equation for the first-order decay of a species A as a function of time: $[A] = [A]_o\, e^{-kt}$, where k is the first-order rate constant for the decay. (The rate constant can be obtained from the value of the half-life given in the chapter: $t_{1/2} = 0.693/k$.)

3. One physical property that differs for ^{235}UF$_6$ and ^{238}UF$_6$ is the rate of effusion of these gases through porous membranes. The lighter ^{235}UF$_6$ will effuse faster, resulting in enrichment of this desired isotope after passing through the membrane. The effusion rate of gases can be described by Graham's law, which states that the effusion rate of a particle is inversely proportional to the square of its mass. Use this law to determine relative effusion rates of ^{235}UF$_6$ and ^{238}UF$_6$. Based on your result, explain why it is necessary in practice to perform this process through hundreds of membrane barriers to achieve useful enrichment.

PART III

WATER CHEMISTRY AND WATER POLLUTION

Contents of Part III

Introduction

Water is, of course, essential to life. Clean water is essential to healthy human life. One need not travel far outside developed countries to find that water that is safe to drink is in short supply, often unavailable except in bottles. Indeed, fresh water of any quality is becoming less and less available in the quantities needed for agriculture, industry, and humans in regions now increasingly affected by prolonged periods of drought.

In this part of the text, we concentrate our attention on natural waters, and specifically on the chemicals they contain—whether naturally occurring or a result of anthropogenic causes—and the chemical transformations that occur within them. These processes include redox and acid–base reactions, as well as some determined by solubility. The purification of drinking water, one of the great public health triumphs in the past few centuries, is discussed in detail. In addition, we examine five heavy metals that historically have been implicated in water pollution in many geographic regions. ●

10

The Chemistry of Natural Waters

In this chapter, the following introductory chemistry topics are used:

- ➲ Concepts of oxidation and reduction as electron loss/gain; half-reactions; redox reactions; oxidizing and reducing agents
- ➲ Oxidation numbers, and the balancing of redox reactions (reviewed in the Appendix)
- ➲ Acid–base and equilibrium concepts and calculations; pH
- ➲ Solubility and K_{sp}
- ➲ Titrations

Background from previous chapters used in this chapter:

- ➲ Equilibria involving gases dissolved in water: Henry's law

Introduction

10.1 The Global Supply and Use of Water

All life forms on Earth depend upon water. Each human being needs to consume several liters of fresh water daily to sustain life. Much more water is used for other domestic activities: typical daily usages in developed countries for showering or bathing, washing, and toilets each amount to about 50 L, in addition to about 20 L for dishwashing and 10 L for cooking.

The relationship between development and water consumption is illustrated by the fact that the rate of increase of water use over the last century was double the rate of population increase. The World Health Organization uses 20 L per capita per day available within a kilometer of residence as being reasonable access to water.

Vastly larger amounts of fresh water per capita are used by industry (about 20% globally) and especially for irrigation in agriculture (about 70%). The amount of water used to grow food and raise livestock is expected to almost double by 2050, compared to the year 2000 value of 7200 km³. This increase will occur partially due to population growth, but even more to the

A hose delivers 10 L or more of water per minute, so watering gardens and lawns can easily double average domestic consumption.

Domestic water consumption in the United States exceeds 400 L a day on average per person.

increase in consumption of animal protein that accompanies increasing affluence.

For example, the **water footprint—the volume of water required to produce one kilogram of product**—of beef is about 15,000 L kg^{-1}; in other words, mainly to grow the food consumed by cattle during their growth, about 15,000 kg (or 34,000 pounds) of water are needed for every kilogram of edible beef that results! The water footprints for lamb, pork, and chicken are only about one-third those of beef, but are several times the thousands of liters per kilogram to grow rice, wheat, and corn (maize). The term *virtual water* (or *embedded* water) has been coined to denote the amount of water used to produce a good or service, and is a useful concept in discussions of international trade.

However, fresh water is at a premium. Over 97% of the world's water is seawater, unsuitable for drinking and for most agricultural purposes. Three-quarters of the fresh water is trapped in glaciers and ice caps. Lakes and rivers are among the main sources of drinking water, even though taken together they constitute less than 0.01% of the total water on the planet. About half of drinking water is obtained from *groundwater*, the fresh water that lies underground and that is discussed in detail in Chapter 11. The annual fluxes of water between its global pools in the oceans, the air, and beneath the ground are shown in Figure 10-1, along with the sizes of each pool.

Humanity currently consumes, mostly for agriculture, about one-fifth of the accessible runoff water that travels through rivers to the seas; this fraction is predicted to rise to about three-quarters by 2025. Runoff water is highly variable in both location and time in terms of its availability unless

The growing use of biofuels is also adding to demand for water.

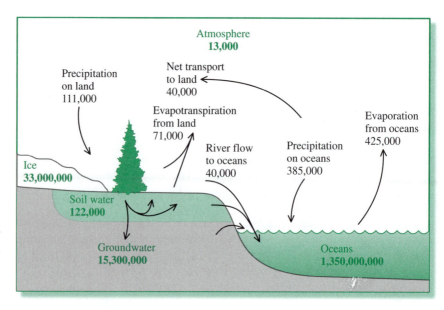

FIGURE 10-1 Global pools and fluxes of water on Earth, showing the size of groundwater storage relative to other major water sources and fluxes. All pool volumes (green) are in cubic kilometers, and all fluxes (black) are in cubic kilometers per year. [Source: W. H. Schlesinger, *Biogeochemistry—An Analysis of Global Change*, 2nd ed. (San Diego: Academic Press, 1997), Chapter 10.]

Atmosphere
13,000

Precipitation on land
111,000

Net transport to land
40,000

Evapotranspiration from land
71,000

River flow to oceans
40,000

Precipitation on oceans
385,000

Evaporation from oceans
425,000

Ice
33,000,000

Soil water
122,000

Groundwater
15,300,000

Oceans
1,350,000,000

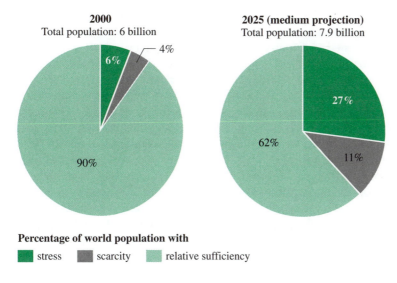

2000
Total population: 6 billion

4%
6%
90%

2025 (medium projection)
Total population: 7.9 billion

27%
62%
11%

Percentage of world population with

■ stress ■ scarcity ■ relative sufficiency

FIGURE 10-2 Global water supply in 2000 and projection for 2025. [Source: R. Engelman et al., *People in the Balance* (Washington, DC: Popular Action International, 2000) as reproduced in *Nature* 422 (2003): 252.]

storage and transport is available. In addition, the rapid melting of glaciers may produce intermittent flows to major Asian rivers in the future. Although only 10% of the world's population in 2000 lived under conditions of water stress or scarcity, this figure had been projected to rise to 38% by 2025 (Figure 10-2) but may be reached decades earlier, according to recent data. For example, some experts believe that water required for industry may be the ultimate limitation in China's growth rate in the future.

It is important to understand the types of chemical activity that prevail in natural waters, and how the science and application of chemistry can be employed to purify water intended for drinking purposes. Although some discussion of pollution problems is contained in this chapter, the remediation of contaminated water is considered in detail in Chapter 11.

10.2 Aspects and Concentration Units of Water Chemistry

It will be convenient to divide our considerations of water chemistry in this chapter into the two common reaction categories: acid–base and solubility reactions, and oxidation–reduction (redox) reactions. **Acid–base and solubility phenomena control the concentrations of dissolved inorganic ions such as carbonate in waters, whereas the organic content of water is dominated by redox reactions.** The pH and concentrations of the principal ions in most natural water systems are controlled by the dissolution of atmospheric carbon dioxide and soil-bound carbonate ions; such reactions are considered in detail later in the chapter. Before considering these acid–base processes, we consider some important redox processes, especially those involving dissolved oxygen. For clarity, the phase (aq) for ions and molecules dissolved in aqueous solution will not be shown in equations but simply assumed.

Before beginning these discussions, however, it is important to note that the solubilities of trace substances in water and in other liquids and solids are often expressed on a "parts per" scale, rather than on a mass or moles per unit volume basis.

In particular, **the "parts per" scales for condensed (nongaseous) media express the ratio of the mass of the solute to the mass of the solution or medium,** not the ratio of moles or molecules that is used for the gas phase. Since the mass of 1 L of a natural water sample is very close to 1 kg, the numerical solubility of a substance expressed in milligrams per liter of water *also* corresponds to the number of milligrams per 1,000 grams (kilogram) of solution. Multiplying both numerator and denominator of a solubility value by 1,000, we conclude that a solubility, e.g., of 0.0062 mg L^{-1} is equivalent to 0.0062 g of solute per 1 million grams of solution, i.e., to 0.0062 parts per million. **In general, the value for the ppm solubility of any trace substance in water is the same as its value expressed in units of milligrams per liter. Similarly, the ppb scale in aqueous solutions is equivalent to micrograms per liter.**

Another commonly used set of units for trace contaminants—particularly for substances dissolved in a medium such as soil or a biological specimen—is *micrograms* (of the contaminant) *per gram* (of the medium), $\mu g\ g^{-1}$. For example, the concentration of a pesticide in human fat could be listed as 2.0 $\mu g\ g^{-1}$. However, **the micrograms per gram scale is equivalent to that of parts per million,** which can be seen by multiplying the numerator and denominator of $\mu g\ g^{-1}$ by one million, giving grams per million grams. Similarly, **the nanograms per gram (ng g^{-1}) scale is equivalent to parts per billion.**

The concentration of 0.0062 ppm is one thousand times as large in parts per billion units, i.e., 6.2 ppb.

PROBLEM 10-1

For aqueous solutions, (a) convert 0.04 $\mu g\ L^{-1}$ to the ppm and ppb scales, (b) convert 3 ppb to the $\mu g\ L^{-1}$ scale, and (c) convert 0.30 $\mu g\ g^{-1}$ to the ppb scale.   ●

10.3 The Solubility of Gases and VOCs in Water

Volatile substances that are dissolved in natural waters can gradually evaporate from the liquid surface into the air, and similarly, the presence of such substances in air can lead to their partial dissolution in natural waters. If equilibrium is achieved for the substance between the two phases, Henry's law can be used to determine the concentration of the substance in water from its partial pressure in air, or vice-versa.

For example, consider *benzene*, C_6H_6, a volatile organic liquid which is present in many environments. Although hydrocarbons are usually considered to be insoluble in water, benzene dissolves to the extent of about one gram per liter, i.e., about 1000 ppm. The partial pressure $P_{benzene}$ of benzene vapor in air that would be present if equilibrium were to be established is related to its concentration in water by applying Henry's law:

$$[C_6H_6, aq] = K_H P_{benzene}$$

Upon rearrangement, we obtain

$$P_{benzene} = [C_6H_6, aq]/K_H$$

From tables of experimental data, for benzene $K_H = 0.18$ mol L^{-1} atm^{-1}. The molar concentration for benzene is readily obtained from the mass per volume value:

$$1.0 \text{ g } L^{-1}/78.12 \text{ g mol}^{-1} = 0.013 \text{ mol } L^{-1}$$

Thus the partial pressure of benzene above water is

$$P_{benzene} = [C_6H_6, aq]/K_H$$
$$= 0.013 \text{ mol } L^{-1}/0.18 \text{ mol } L^{-1} \text{ atm}^{-1}$$
$$= 0.07 \text{ atm}$$

Henry's law was discussed in Section 3.21.

This substantial vapor pressure for benzene would have occurred in air bubbles in water rising from the Gulf of Mexico oil disaster in 2010, since substantial amounts of benzene were present in the crude oil and partially migrated from it into the surrounding water.

Oxidation–Reduction Chemistry in Natural Waters

10.4 Dissolved Oxygen

By far the most important oxidizing agent in natural waters is dissolved **molecular oxygen**, O_2. Upon reaction, each of the oxygen atoms in O_2 is reduced from the zero oxidation number to -2, in H_2O or OH^-. The half-reaction that occurs in acidic solution is

$$O_2 + 4 \text{ H}^+ + 4e^- \longrightarrow 2 \text{ H}_2O$$

whereas that which occurs in basic aqueous solution is

$$O_2 + 2 \text{ H}_2O + 4e^- \longrightarrow 4 \text{ OH}^-$$

The concentration of dissolved oxygen in water is small and therefore precarious from the ecological point of view. For the dissolution of molecular oxygen gas in water

$$O_2(g) \rightleftharpoons O_2(aq)$$

the appropriate equilibrium constant is the Henry's law constant K_H, which for oxygen at 25°C has the value 1.3×10^{-3} mol L^{-1} atm^{-1}:

$$K_H = [O_2(aq)]/P_{O_2} = 1.3 \times 10^{-3} \text{ mol } L^{-1} \text{ atm}^{-1} \text{ at } 25°C$$

Since in dry air at sea level the partial pressure, P_{O_2}, of oxygen is 0.21 atmospheres (atm), it follows that the solubility of O_2 is 8.7 mg L^{-1} of water (see Problem 10-2). This value can also be stated as 8.7 ppm since, as discussed

Recall that oxidation is a loss of electrons, so oxidizing agents extract electrons from other species.

The concept and calculation of oxidation numbers is reviewed in the Appendix, as is the balancing of redox equations.

earlier in this chapter, ppm concentrations are based on mass rather than moles. (Note that for simplicity, molar concentrations, rather than activities, are used in all equilibrium calculations in this book since in general we are considering only very dilute solutions. In addition, we ignore the small correction for water vapor in determining atmospheric gas pressures.)

Because the solubilities of gases increase with decreasing temperature, the amount of O_2 that dissolves at 0°C (14.7 ppm) is greater than is the amount that dissolves at 35°C (7.0 ppm). The median concentration of oxygen found in natural, unpolluted surface waters in the United States is about 10 ppm.

> ### PROBLEM 10-2
>
> Confirm by calculation the value of 8.7 mg L^{-1} for the solubility of oxygen in water at 25°C. ●

> ### PROBLEM 10-3
>
> Given the solubility quoted above for O_2 at 0°C, calculate the value of K_H for it at this temperature. ●

River or lake water that has been artificially warmed can be considered to have undergone thermal pollution in the sense that, at equilibrium, it will contain less oxygen than colder water because of the decrease in gas solubility with increasing temperature. To sustain their lives, most fish species require water containing at least 5 ppm of dissolved oxygen; consequently, their survival in warmed water can be problematic. Thermal pollution often occurs in the region of electric power plants (whether fossil-fuelled, nuclear, or solar), since they draw cold water from a river or lake, use it for cooling purposes, and then return the warmed water to its source.

10.5 Oxygen Demand: Biological

The most common substance oxidized by dissolved oxygen in water is organic matter having a biological origin, such as dead plant matter and animal wastes. If, for the sake of simplicity, the organic matter is assumed to be entirely polymerized carbohydrate (e.g., plant fiber) having an approximate empirical formula of CH_2O, the oxidation reaction would be

C becomes oxidized from the 0 state in carbohydrate to +4 in CO_2, as oxygen is reduced from 0 in the molecular state to −2 in the products.

$$CH_2O(aq) + O_2(aq) \longrightarrow CO_2(g) + H_2O(aq)$$
$$\text{carbohydrate}$$

Dissolved oxygen in water is also consumed by the oxidation of dissolved **ammonia, NH_3,** and **ammonium ion, NH_4^+**—substances that like organic matter are present in water as a result of biological activity—eventually to **nitrate ion, NO_3^-** (see Problem 10-5).

PROBLEM 10-4

Show that 1 L of water saturated with oxygen at 25°C is capable of oxidizing 8.2 mg of polymeric CH_2O.

PROBLEM 10-5

Determine the balanced redox reaction for the oxidation of ammonia to nitrate ion by O_2 in alkaline (basic) solution. Does this reaction make the water more alkaline or less? [*Hint:* Recall the redox balancing procedure in the Appendix.]

Water that is aerated by flowing in shallow streams and rivers is constantly replenished with molecular oxygen. However, stagnant water or that near the bottom of a deep lake is usually almost completely depleted of oxygen because of its prior reaction with organic matter and the lack of any mechanism to replenish it quickly, diffusion being a slow process and turbulent mixing being absent.

The capacity of the organic and biological matter in a sample of natural water to consume oxygen, a process catalyzed by bacteria present, is called its biochemical oxygen demand, BOD. It is evaluated experimentally by determining the difference in concentration of dissolved O_2 at the beginning and at the end of a period in which a sealed water sample seeded with bacteria is maintained in the dark at a constant temperature, usually either 20°C or 25°C. A neutral pH is maintained by use of a buffer consisting of two ions of phosphoric acid, namely $H_2PO_4^-$ and HPO_4^{2-}:

$$H_2PO_4^- \rightleftharpoons HPO_4^{2-} + H^+$$

The BOD equals the amount of oxygen consumed as a result of the oxidation of dissolved organic matter in the sample. The oxidation reactions are catalyzed in the sample by the action of microorganisms present in the natural water. If it is suspected that the sample will have a high BOD, it first is diluted with pure, oxygen-saturated water so that sufficient O_2 will be available overall to oxidize all the organic matter; the results are corrected for this dilution. Usually the reaction is allowed to proceed for 5 days before the residual oxygen is determined. The oxygen demand determined from such a test, often designated BOD_5, corresponds to about 80% of that which would be determined if the experiment were allowed to proceed for a very long time—which, of course, is not very practical.

The median BOD for unpolluted surface water in the United States is about 0.7 milligrams of O_2 per liter, which is considerably less than the maximum solubility of O_2 in water (of 8.7 mg L^{-1} at 25°C). In contrast, the BOD values for sewage are typically several hundreds of milligrams of oxygen per liter. It is not uncommon for water polluted by organic substances associated with animal or food waste or sewage to have an oxygen demand that exceeds the maximum equilibrium solubility of dissolved oxygen. Under such circumstances, unless the water is continuously aerated, it will soon be depleted of its oxygen, and fish living in the water will die. The treatment of wastewater to reduce its BOD is discussed in Chapter 11.

10.6 Oxygen Demand: Chemical

A faster determination of oxygen demand of a water sample can be made by titrating it with a strong oxidizing agent and thereby determining its chemical oxygen demand, COD. Dichromate ion, $Cr_2O_7^{2-}$, can be dissolved as one of its salts, such as $K_2Cr_2O_7$, in sulfuric acid: the result is a powerful oxidizing agent, especially when hot. It is this solution, rather than O_2, that is used to ascertain COD values. The reduction half-reaction for dichromate when it oxidizes the organic matter is

$$Cr_2O_7^{2-} + 14\,H^+ + 6\,e^- \longrightarrow 2\,Cr^{3+} + 7\,H_2O$$

PROBLEM 10-6

Evaluate the oxidation number of the chromium in $Cr_2O_7^{2-}$ and show thereby that the reduction half-reaction above is indeed a six-electron process.   ●

(In practice, excess dichromate is added to the sample and the resulting solution is back-titrated with Fe^{2+} to the end point.) The number of moles of O_2 that the sample would have consumed in accomplishing the oxidation of the same material equals 6/4 (= 1.5) times the number of moles of dichromate, since the latter accepts 6 electrons per ion whereas O_2 accepts only 4:

$$O_2 + 4\,H^+ + 4e^- \longrightarrow 2\,H_2O$$

Thus the moles of O_2 required for the oxidation is 1.5 times the number of moles of dichromate actually used.

In working out the titration problems below, you could either use this factor of 1.5 or instead, work out the number of moles of electrons involved. For example, if 4.0 mL of a 0.0020 M solution of dichromate ion was required in a titration, then the moles of $Cr_2O_7^{2-}$ is equal to

$$0.0040\ L \times 0.0020\ mol\ L^{-1} = 8.0 \times 10^{-6}\ mol\ Cr_2O_7^{2-}$$

Since each mole of dichromate acquires 6 mol of electrons when it is reduced, then for this example

$$mol\ electrons = 6 \times (8.0 \times 10^{-6}\ mol) = 4.8 \times 10^{-5}$$

If the substance had instead been oxidized by O_2, then since each mole of O_2 requires 4 mol of electrons to be reduced, we divide the available moles of electrons by 4 to find moles of oxygen molecules that could react:

$$mol\ O_2 = mol\ electrons/4 = (4.8 \times 10^{-5})/4 = 1.2 \times 10^{-5}$$

We obtained the same result as if we had simply multiplied the moles of dichromate by 1.5, as mentioned above:

$$1.5 \times (8.0 \times 10^{-6} \text{ mol Cr}_2\text{O}_7^{2-}) = 1.2 \times 10^{-5} \text{ mol O}_2$$

PROBLEM 10-7

A 25-mL sample of river water was titrated with 0.0010 M $K_2Cr_2O_7$ and required 8.3 mL to reach the end point. What is the chemical oxygen demand, in milligrams of O_2 per liter, of the sample?

PROBLEM 10-8

The COD of a water sample is found to be 30 mg of O_2 per liter. What volume of 0.0020 M $K_2Cr_2O_7$ will be required to titrate a 50-mL sample of the water?

The difficulty with the COD index as a measure of oxygen demand is that **acidified dichromate is such a strong oxidant that it oxidizes substances that are very slow to consume oxygen in natural waters** and that therefore pose no real threat to their dissolved oxygen content. In other words, dichromate oxidizes substances that would not be oxidized by O_2 in the determination of the BOD. Because of this excess oxidation, namely of stable organic matter such as cellulose to CO_2, and of Cl^- to Cl_2, the COD value for a water sample as a rule is slightly greater than its BOD value. Neither method of analysis oxidizes aromatic hydrocarbons or many alkanes, which, in any event, resist degradation, and therefore oxygen consumption, in natural waters.

Finally, we note that there are two other measures used for the amount of organic substances present in natural waters. The **total organic carbon, TOC,** is used to characterize the dissolved *and* suspended organic matter in raw water. For example, the TOC usually has a value of approximately one milligram per liter, i.e., 1 ppm carbon, for groundwater. The parameter **dissolved organic carbon, DOC,** is used to characterize only organic material that is actually dissolved, not suspended. For surface waters, the DOC averages about 5 ppm, although bogs and swamps can have DOC values that are ten times this amount, and untreated sewage typically has a DOC value of hundreds of ppm.

The largest component of organic carbon in natural waters usually is carbohydrates, but many other types including proteins and low-molecular-weight aldehydes, ketones, and carboxylic acids are also present. Strong organic acids in waters are of interest because they can act in combination with strong inorganic ones to mobilize aluminum ion in lakes affected by acid rain (Section 4.7). The *humic* materials in water are discussed in Chapter 16.

10.7 Green Chemistry: Enzymatic Preparation of Cotton Textiles

Globally, more than 50 billion pounds (23 billion kilograms) of cotton are produced each year. Even with the invasion of synthetic fibers such as nylon, polyester, and acrylics, cotton still holds over 50% of the market share for apparel and home furnishings that are sold in the United States. In preparing raw cotton for use as a fiber, several steps, including desizing, scouring, and bleaching, are required, leaving a fiber which is 99% cellulose. These steps use copious amounts of chemicals, water, and energy, and produce millions of pounds of aqueous waste that is high in BOD and COD.

Raw cotton fibers are composed of several concentric layers. The outermost layer is composed of fats, waxes, and pectin, while the inner layers consist primarily of cellulose. The fats and waxes make the raw cotton fiber waterproof, and the pectin acts as a powerful glue to hold the layers together. In order to prepare cotton for use as a fiber for bleaching and dyeing, the outer layer must be removed. This process, which is known as scouring, has traditionally been done by immersing the cotton in 18–25% aqueous **sodium hydroxide** solution at elevated temperatures. This results in hydrolysis of the fats (saponification) and waxes, which solubilizes the components of the outer layer so that they can simply be rinsed away. Scouring results in fibers of even and high wettability. In addition to sodium hydroxide, chelating agents and emulsifiers are added during the scouring process. To end the process, the mixture is neutralized with acetic acid and rinsed several times.

The scouring process is estimated by the U.S. EPA to account for about half of the total BOD produced in the preparation of cotton fibers. The BOD in wastewater from cotton production generally exceeds 1100 mg L^{-1}, which is several times that of raw sewage. In addition to the large requirement of chemicals, energy, and water that are used, and the concomitant pollution that is produced, another disadvantage of this process is the unintended weakening and loss of some of the cellulose fibers.

An alternative to the traditional scouring process, known as BioPreparation, was developed by Novozymes-North America Inc. and won a Presidential Green Chemistry Challenge Award in 2001. BioPreparation employs an enzyme (a pectin *lyase*), which selectively degrades pectin at ambient temperatures. Since pectin acts as a glue to hold the outer layer of the cotton fiber together, destruction of the pectin results in disintegration of this layer. Because the *lyase* is selective for only pectin, its use is much less aggressive than the typical scouring process and removes much less organic material (including cellulose) from the cotton. Since the dissolved organic materials are what contribute to the high BOD and COD of the wastewater, this enzymatic treatment lowers the BOD by 20% and the COD by 50%.

In addition to these environmental advantages, BioPreparation eliminates the use of sodium hydroxide solutions at elevated temperatures, which in turn

- lowers the pH of the wastewater,

- eliminates the need for neutralization with acetic acid and the concomitant wastes,

- lowers energy requirements, and

- lessens rinsing requirements, reducing water consumption by 30–50%.

Elimination of the use of sodium hydroxide also provides for a reduction in risk to workers, and the reduced degradation of cellulose provides for more robust fibers and higher yield.

10.8 Decomposition of Organic Matter in Water

Dissolved organic matter will decompose in water under anaerobic (oxygen-free) conditions if appropriate bacteria are present. Anaerobic conditions occur naturally in stagnant water such as swamps and at the bottom of deep lakes. The bacteria operate on carbon to disproportionate it: i.e., some carbon is oxidized, to **carbon dioxide,** CO_2, and the rest is reduced, to **methane,** CH_4:

$$2\ CH_2O \xrightarrow{\text{bacteria}} CH_4 + CO_2$$
$$\text{organic matter}$$

C oxidation number   0   −4   +4

This is an example of a fermentation reaction, which in chemistry is defined as one in which both the oxidizing and the reducing agents are organic materials.
Since methane is almost insoluble in water, it forms bubbles that can be seen rising to the surface in swamps and it sometimes catches fire; indeed, methane was originally called *marsh* or *swamp gas*. The same chemical reaction shown above occurs in *digestor* units used by rural inhabitants in semi-tropical developing countries (India, for instance) to convert animal wastes into methane gas that can be used as a fuel. The reaction also occurs in landfills, as discussed in Chapter 5.
Since anaerobic conditions are reducing environments in the chemical sense, *insoluble* Fe^{3+} compounds that are present in sediments at the bottom of lakes are converted into *soluble* Fe^{2+} compounds, which then dissolve into the lake water:

$$Fe^{3+}\ +\ e^- \longrightarrow Fe^{2+}$$
$$\text{insoluble Fe(III)}\qquad\quad \text{soluble Fe(II)}$$

The general chemical name for reactions in which the same substance is oxidized and reduced is *disproportionation*.

Near neutral pH, most Fe(III) compounds are insoluble whereas most Fe(II) compounds are soluble.

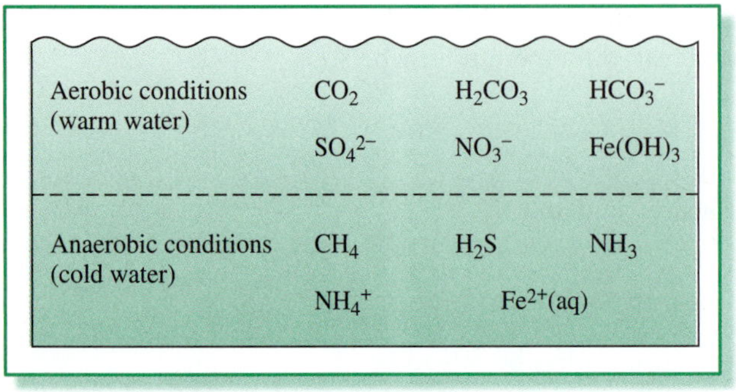

Aerobic conditions (warm water)	CO_2	H_2CO_3	HCO_3^-
	SO_4^{2-}	NO_3^-	$Fe(OH)_3$
Anaerobic conditions (cold water)	CH_4	H_2S	NH_3
	NH_4^+	$Fe^{2+}(aq)$	

FIGURE 10-3 The stratification of a lake in the summer, showing the typical forms of the major elements it contains at different levels.

Like most transition metals, iron loses its two valence-shell *s* electrons and forms a 2+ ion. It can also lose one 3*d* electron, giving a half-filled *d* shell in the 3+ ion.

It is not uncommon to find aerobic and anaerobic conditions in different parts of the same lake at the same time, particularly in the summertime when a stable stratification of distinct layers often occurs (see Figure 10-3). Water at the top of the lake is warmed by the absorption of sunshine by biological materials, while that below the level of penetration of sunlight remains cold. Since warm water is less dense than cold (at temperatures above 4°C), the warm upper layer "floats" on the cold layer below, and little transfer between them occurs. The top layer, called the *epilimnium*, usually contains near-saturation levels of dissolved oxygen, due both to its contact with air and to the O_2 produced in photosynthesis by algae.

Since conditions in the top layer of a lake are aerobic, elements exist there in their *most oxidized* forms:

- carbon, with an oxidation number (O.N.) of +4, as CO_2 or H_2CO_3 or HCO_3^-;

- sulfur, O.N. of +6, as SO_4^{2-};

- nitrogen, O.N. of +5, as NO_3^-; and

- iron, as Fe(III), in the form of insoluble $Fe(OH)_3$.

Near the lake bottom, in the *hypolimnium*, the water is oxygen-depleted since it has no contact with air and since O_2 is consumed when biological material, such as the dead algae that has sunk to these depths, decomposes. Under such anaerobic conditions, elements exist in their *most common reduced* forms:

- carbon, with an O.N. of −4, as CH_4;

- sulfur, O.N. of −2, as H_2S;

- nitrogen, O.N. of −3, as NH_3 and NH_4^+; and

- iron, as Fe(II), in the form of soluble Fe^{2+}.

Anaerobic conditions usually do not last indefinitely. In the fall and winter, the top layer of water is cooled by cold air passing over it, so that eventually the oxygen-rich water at the top becomes more dense than that below it, and gravity induces mixing between the layers. Thus in the winter and early spring the environment near the bottom of a lake usually is aerobic.

10.9 Sulfur Compounds in Natural Waters

The common inorganic oxidation numbers in which sulfur is encountered in the environment are illustrated in Table 10-1; they range from the highly reduced -2 form that is found in **hydrogen sulfide gas,** H_2S, and insoluble minerals containing the **sulfide ion,** S^{2-}, to the highly oxidized $+6$ form that is encountered in **sulfuric acid,** H_2SO_4, and in salts containing the **sulfate ion,** SO_4^{2-}.

In organic and bioorganic molecules such as amino acids, intermediate levels of sulfur oxidation are present. When such molecules decompose anaerobically, hydrogen sulfide and other gases such as CH_3SH and CH_3SSCH_3 containing sulfur in highly reduced forms are released, thereby giving swamps their unpleasant odor. The occurrence of such gases as air pollutants, collectives called *Total Reduced Sulfur,* was mentioned in Section 3.18.

As discussed in Chapter 3, hydrogen sulfide is oxidized in air first to **sulfur dioxide,** SO_2, and then fully to sulfuric acid or a salt containing the sulfate ion. Similarly, hydrogen sulfide dissolved in water can be oxidized by certain bacteria to **elemental sulfur** or more completely to sulfate. Overall the complete oxidation reactions correspond to

$$H_2S + 2\,O_2 \longrightarrow H_2SO_4$$

Some anaerobic bacteria are able to use sulfate ion as the oxidizing agent to convert organic matter, such as polymeric CH_2O, to carbon dioxide when the concentration of oxygen in the water is very low; the SO_4^{2-} ions are reduced in the process to elemental sulfur or even to hydrogen sulfide:

$$2\,SO_4^{2-} + 3\,CH_2O + 4\,H^+ \longrightarrow 2\,S + 3\,CO_2 + 5\,H_2O$$

Such reactions are especially important in seawater, for which the sulfate ion concentration is much higher than the average for fresh-water systems.

TABLE 10-1	Common Oxidation Numbers for Sulfur				
	Increasing Levels of Sulfur Oxidation $\longrightarrow$				
Oxidation Number of S	-2	-1	0	$+4$	$+6$
Aqueous solution and salts	H_2S			H_2SO_3	H_2SO_4
	HS^-			HSO_3^-	HSO_4^-
	S^{2-}	S_2^{2-}		SO_3^{2-}	SO_4^{2-}
Gas phase	H_2S			SO_2	SO_3
Molecular solids			S_8		

10.10 Acid Mine Drainage

One characteristic reaction of groundwater, which by definition is not well-aerated since it has spent much time not exposed to air, is that **when it reaches the surface and O_2 has an opportunity to dissolve in it, its rather high level of soluble Fe^{2+} is converted to insoluble Fe^{3+}, and an orange-brown deposit of $Fe(OH)_3$ is formed.**

$$4\ Fe^{2+} + O_2 + 2\ H_2O \longrightarrow 4\ Fe^{3+} + 4\ OH^-$$

$$4\ [Fe^{3+} + 3\ OH^- \longrightarrow Fe(OH)_3(s)]$$

The overall reaction is

$$4\ Fe^{2+} + O_2 + 2\ H_2O + 8\ OH^- \longrightarrow 4\ Fe(OH)_3(s)$$

An analogous reaction occurs in some underground coal and metal (especially copper) mines, especially abandoned ones, and in piles of mined coal left open to the environment. Normally FeS_2, called **iron pyrites,** or *fool's gold*, is a stable, insoluble component of underground rocks as long as it does not come into contact with air. However, as a result of the mining of coal and certain ores—and especially after underground mines have been abandoned and spontaneously fill with groundwater—some of it is exposed to water, oxygen, and certain bacteria, and becomes partially solubilized as a result of its oxidation. The **disulfide ion,** S_2^{2-}, in which sulfur has an average oxidation number of -1, is oxidized to sulfate ion, SO_4^{2-}, which contains sulfur in the $+6$ form:

$$S_2^{2-} + 8\ H_2O \longrightarrow 2\ SO_4^{2-} + 16\ H^+ + 14\ e^-$$

The main oxidizing agent acting on the sulfur is atmospheric O_2:

$$7\ [O_2 + 4\ H^+ + 4\ e^- \longrightarrow 2\ H_2O]$$

When this $(28e^-)$ reduction half-reaction is added to twice the $(14e^-)$ oxidation half-reaction, the *net redox reaction* is obtained:

$$2\ S_2^{2-} + 7\ O_2 + 2\ H_2O \longrightarrow 4\ SO_4^{2-} + 4\ H^+$$

Since the sulfate salt of the ferrous ion, Fe^{2+}, is soluble in water, the iron pyrites is effectively solubilized by the reaction. More importantly, the reaction produces a large amount of concentrated acid (note the H^+ product), only a portion of which is consumed by the air oxidation of Fe^{2+} to Fe^{3+} that accompanies the process:

$$4\ Fe^{2+} + O_2 + 4\ H^+ \longrightarrow 4\ Fe^{3+} + 2\ H_2O$$

In acidic environments, this reaction is catalyzed by bacteria (*Thiobacillus ferrooxidans*); the resulting Fe^{3+} can oxidize various metal sulfides, liberating the metal ions.

Combining the last two reactions in the correct ratio, i.e., 2:1, we obtain **the overall reaction for the oxidation of both the iron and the sulfur:**

$$4\, FeS_2 + 15\, O_2 + 2\, H_2O \longrightarrow 4\, Fe^{3+} + 8\, SO_4^{2-} + 4\, H^+$$
$$i.e.,\ 2\, Fe_2(SO_4)_3 + 2\, H_2SO_4$$

In summary, the oxidation of the fool's gold produces soluble iron(III) sulfate (also called ferric sulfate), $Fe_2(SO_4)_3$, and sulfuric acid. The Fe^{3+} ion is soluble in the highly acidic water that is first produced, the pH of which can be less than zero, with the usual range being 0–2. However, once the drainage from the highly acidic mine water becomes diluted and its pH consequently rises, a yellowish-brown precipitate of $Fe(OH)_3$ forms from Fe^{3+}, discoloring the water and waterway and smothering plant and animal life (including fish) in it (see Problem 10-9). **Thus the pollution associated with acid mine drainage is characterized in the first instance by the seeping from the mine of copious amounts of both acidified water and a rust-colored solid.** Unfortunately, the concentrated acid can liberate toxic heavy metals—especially zinc, copper, and nickel, but also lead, arsenic, manganese, and aluminum—from their ores in the rock of the mine, further adding to the pollution of the waterway.

The phenomenon of acid drainage is currently of particular importance in the many abandoned mines in the mountains of Colorado. However, the most acidic mine drainage water in the world comes from the Richmond Mine at Iron Mountain, California. There the pH reaches as low as -3.6 because the high temperatures (up to 47°C) of the mine water cause much of it to evaporate, thus concentrating the acid. By comparison, the most acidic natural waters occur near the Ebeko volcano in Russia, which have a pH as low as -1.7; the acidity is due to hydrochloric and sulfuric acids in the hot spring water.

Acid drainage is a serious problem in areas of the world where there are many mines that have been abandoned—examples arise in Ontario, in Cornwall and Wales in the United Kingdom, in Ireland, and in Tasmania, Australia. Thousands of shafts from abandoned gold mines near Johannesburg, South Africa, now generate so much acid that their rising water levels threaten to contaminate natural aquifers and even to reach the surface and pollute wetlands and rural areas near the city. Rising acidic water may even gradually corrode the foundations of buildings in Johannesburg itself.

The acid produced by acid mine drainage is spontaneously neutralized if the soil contains limestone, in the same way that we encountered for acid rain in Chapter 4. For example, some of the coal mines in Pennsylvania discharge water that is acidic (pH < 5), whereas others discharge alkaline water resulting from dissolution of limestone. Powdered or chipped limestone, or calcium oxide or hydroxide, can also be added to water exiting the mine to neutralize the acid, although the insoluble calcium sulfate that forms on the surface of limestone particles prevents their full reaction. Raising the pH precipitates most of the liberated heavy metals as their insoluble hydroxides in a sludge that can be removed from the water. An alternative method of remediation is

The oxidation of disulfide ion to sulfate ion in the above process is accomplished to some extent by the action of Fe^{3+} as the oxidizing agent, rather than by O_2: $S_2^{2-} + 14\, Fe^{3+} + 8\, H_2O \longrightarrow 2\, SO_4^{2-} + 14\, Fe^{2+} + 16\, H^+$

the introduction into the waters of anaerobic bacteria that reverse the oxidation of sulfate ion back to sulfide, thereby precipitating the heavy metals as insoluble sulfides as well as raising the pH. In some instances, the sulfides are rich enough in metals to be used as ores. Some abandoned mines have been sealed to prevent further intrusions of water and oxygen, but this means of preventing further production of acid is sometimes unsuccessful.

Review Questions 1–9 are based upon the material above.

PROBLEM 10-9

The K_{sp} values for $Fe(OH)_2$ and $Fe(OH)_3$ are 7.9×10^{-15} and 6.3×10^{-38}, respectively. Calculate the molar solubilities of Fe^{2+} and Fe^{3+} at a pH of 8, assuming they are controlled by their hydroxides. Also calculate the pH values at which the ion solubilities reach 100 ppm. ●

10.11 The pE Scale

Environmental scientists use the concept of pE to characterize the extent to which natural waters are chemically reducing or oxidizing in nature, by analogy to the way in which pH is used to characterize their acidity. In particular, **pE is defined as the negative base 10 logarithm of the effective concentration—i.e., of the activity—of electrons in water,** notwithstanding the fact that free electrons do not exist in solution (any more than do bare protons, H^+ ions). pE values are dimensionless numbers; like pH, they have no units.

• **Low pE values signify that electrons are readily available from substances dissolved in the water, so the medium is very reducing in nature.**

• **High pE values signify that the dominant dissolved substances are oxidizing agents, so that few electrons are available for reduction purposes.**

When several acids or bases are present in a water sample, it usually is the case that one of them makes a dominant contribution to the hydrogen or hydroxide ion concentration. In such situations, the position of equilibrium for the other, less dominant weak acids or bases is determined by the H^+ or OH^- level set by the dominant process. In a similar way, **in natural waters one or another redox equilibrium reaction is dominant and it determines the electron availability for the other redox reactions that occur simultaneously. If we know the position of equilibrium for the dominant process, we can calculate the pE of the water and from it the position of equilibrium—and hence the majority species—in the other reactions.**

When a significant amount of O_2 is dissolved in water, the reduction of the oxygen to water is the dominant reaction determining overall electron availability:

$$\tfrac{1}{4} O_2 + H^+ + e^- \rightleftharpoons \tfrac{1}{2} H_2O$$

In such circumstances, the pE of the water is related to the acidity and to the partial pressure of oxygen and its acidity by the following equation, the origin of which is discussed subsequently:

$$pE = 20.75 + \log([H^+] P_{O_2}{}^{1/4})$$
$$= 20.75 - pH + \tfrac{1}{4} \log(P_{O_2})$$

For a neutral sample of water that is saturated by oxygen from air, i.e., when $P_{O_2} = 0.21$ atm, and is free of carbon dioxide so that its pH = 7, the pE value is calculated from this equation to be 13.6. If the concentration of dissolved oxygen is *less* than the equilibrium amount, then the equivalent partial pressure of atmospheric oxygen is less than 0.21 atm, so the pE value is *smaller* than 13.6 and in some cases even negative.

The pE expression given above looks very much like the Nernst equations encountered in the study of electrochemistry. Indeed, the pE value for a water sample is simply the electrode potential, E, for whatever process determines electron availability, but divided by RT/F (where R is the gas constant, T the absolute temperature, and F the Faraday constant), which at 25°C has the value 0.0591 V:

$$pE = E/0.0591$$

Thus the pE expression for any half-reaction in water can be obtained from its standard electrode potential E^0, corrected by the usual concentration and/or pressure terms and evaluated for a *one-electron* reduction. For example, for the half-reaction linking the reduction of nitrate ion to ammonium ion, we first write the process as a (balanced) one-electron reduction:

$$\tfrac{1}{8} NO_3^- + \tfrac{5}{4} H^+ + e^- \rightleftharpoons \tfrac{1}{8} NH_4^+ + \tfrac{3}{8} H_2O$$

For this reaction, $E^0 = +0.836$ V (from standard tables), so $pE^0 = E^0/0.0591 = +14.15$.

The equation for pE involves the subtraction from the standard pE^0 of the logarithm of the ratio of concentrations of products to reactants, each raised to its coefficient in the one-electron half-reaction. Consequently, for the half-reaction above,

$$pE = pE_0 - \log([NH_4^+]^{1/8}/[NO_3^-]^{1/8} [H^+]^{5/4})$$

$$= 14.15 - \tfrac{5}{4} pH - \tfrac{1}{8} \log([NH_4^+]/[NO_3^-])$$

As usual, the concentration of water does not appear in the expression as its effect is already included in pE^0.

Here we have used the properties of logarithms that $\log a^x = x \log a$, and that $\log (1/b) = -\log b$.

PROBLEM 10-10

Deduce the equilibrium ratio of NH_4^+ to NO_3^- at a pH of 6.0 (a) for aerobic water having a pE value of +11, and (b) for anaerobic water for which the pE value is −3.   ●

Returning to the subject of the dominant reaction that determines pE in natural waters, we note that low values of dissolved oxygen usually are caused by the operation of microorganism-catalyzed organic decomposition reactions, and their dissolved products, rather than O_2, can determine electron availability. For example, in cases of low oxygen availability, the pE of water may be determined by ions such as nitrate or sulfate. In the extreme case of the anaerobic conditions found at the bottoms of lakes in the summer and in swamps and rice paddies, the electron availability is determined by the ratio of dissolved methane, a reducing agent, to dissolved carbon dioxide, an oxidizing agent, both of which are produced by the fermentation of organic matter discussed above. They are connected in the redox sense by the half-reaction

$$\tfrac{1}{8} CO_2 + H^+ + e^- \rightleftharpoons \tfrac{1}{8} CH_4 + \tfrac{1}{4} H_2O$$

The pE value for water controlled by this half-reaction is

$$pE = 2.87 - pH + \tfrac{1}{8} \log(P_{CO_2}/P_{CH_4})$$

For example, if the partial pressures of the two gases are equal and the water is neutral, the pE value is -4.1. **Thus the lower levels of a stratified lake are characterized by negative pE values, whereas the oxygenated top layer has a substantially positive pE.**

The pE concept is useful in predicting the ratio of oxidized to reduced forms of an element in a water body when we know how the electron availability is controlled by another species. Consider, for example, the equilibrium between the two common ions of iron:

$$Fe^{3+} + e^- \rightleftharpoons Fe^{2+}$$

For this reaction,

$$pE = 13.2 + \log([Fe^{3+}]/[Fe^{2+}])$$

If the pE is determined by another redox process and its value is known, the ratio of Fe^{3+} to Fe^{2+} can be deduced. Thus for the oxygen-depleted water discussed above that has a pE value of -4.1,

$$-4.1 = 13.2 + \log([Fe^{3+}]/[Fe^{2+}])$$

so

$$\log([Fe^{3+}]/[Fe^{2+}]) = -17.3$$

and hence

$$[Fe^{3+}]/[Fe^{2+}] = 5 \times 10^{-18}$$

In contrast, for a sample of aerobic water that has a pE of 13.6, the calculated ratio is 2.5:1 in favor of the Fe^{3+} ion. The transition between dominance of the two forms occurs when their concentrations are equal:

$$pE = 13.2 + \log(1) = 13.2 + 0 = 13.2$$

PROBLEM 10-11

Find the pE value for acidic water at which the ratio of concentrations of Fe^{3+} to Fe^{2+} is 100:1.   ●

PROBLEM 10-12

Assuming that dissolved oxygen determines the electron availability in an aqueous solution and that the partial pressure equivalent to the amount dissolved is 0.10 atm, deduce the ratio of dissolved CO_2 to dissolved CH_4 at a pH of 4.   ●

10.12 pE–pH Diagrams

A visual representation of the zones of dominance for the various oxidation states in water of an element can be displayed in a **pE–pH diagram,** as shown in Figure 10-4 for iron. It is clear from the diagram that the situation is more

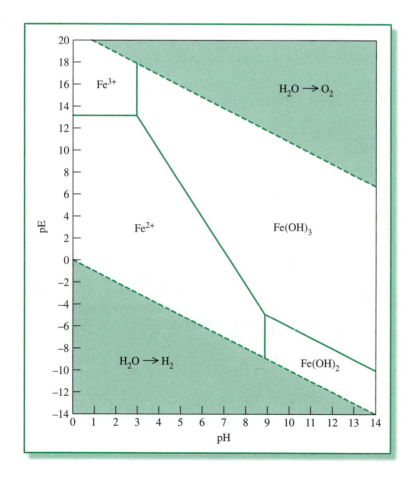

FIGURE 10-4 The pE–pH diagram for the iron system at 10^{-5} M concentration. [Source: Adapted from S. E. Manahan, *Environmental Chemistry,* 4th ed. (Boston, MA: Willard Grant Press/PWS Publishers, 1984).]

complicated than that described in our example above, since, in moderately acidic or alkaline environments, the solid hydroxides $Fe(OH)_2$ and $Fe(OH)_3$ also come into play in the equilibria. **The solid lines in the diagram indicate combinations of pE and pH values where the concentrations of the two species on either side of the line are equal.** Thus we see from the top-left side of Figure 10-4 that equilibrium between dissolved ions Fe^{2+} and Fe^{3+} is important only for pH < 3. Equality in the concentrations of these two dissolved forms corresponds to the short horizontal line in the figure. As expected from our calculations above, the transition occurs at pE $= 13.2$ regardless of pH; hence the line is horizontal.

If iron is in the 3+ state at higher pH, it exists predominantly as the solid $Fe(OH)_3$, whereas solutions containing iron in the 2+ state are generally not precipitated until the solution becomes basic (Figure 10-4).

The shaded regions at the top-right and bottom-left side of the pE–pH diagram represent extreme conditions, under which water itself is unstable to decomposition, being oxidized or reduced to yield O_2 or H_2, respectively:

$$2\,H_2O \longrightarrow O_2 + 4\,H^+ + 4\,e^-$$

and

$$2\,H_2O + 2\,e^- \longrightarrow H_2 + 2\,OH^-$$

10.13 Nitrogen Compounds in Natural Waters

In some natural waters, nitrogen occurs in inorganic and organic forms that are of concern with respect to human health. As discussed in Chapter 5, there are several environmentally important forms of nitrogen that differ in the extent of oxidation of the nitrogen atom.

The common oxidation numbers of nitrogen, along with the most important examples for each, are illustrated in Table 10-2. The most reduced forms have the -3 oxidation number, as occurs in ammonia, NH_3, and its conjugate acid, the ammonium ion, NH_4^+. The most oxidized form, having an oxidation number of $+5$, occurs as the nitrate ion, NO_3^-, which exists in salts, aqueous solutions, and in nitric acid, HNO_3. In solution, the most important intermediates between these extremes are the **nitrite ion, NO_2^-,** and **molecular nitrogen, N_2.**

TABLE 10-2	Common Oxidation Numbers for Nitrogen						
	Increasing Levels of Nitrogen Oxidation →						
Oxidation Number of N	-3	0	$+1$	$+2$	$+3$	$+4$	$+5$
Aqueous solution and salts	NH_4^+ NH_3				NO_2^-		NO_3^-
Gas phase	NH_3	N_2	N_2O	NO		NO_2	

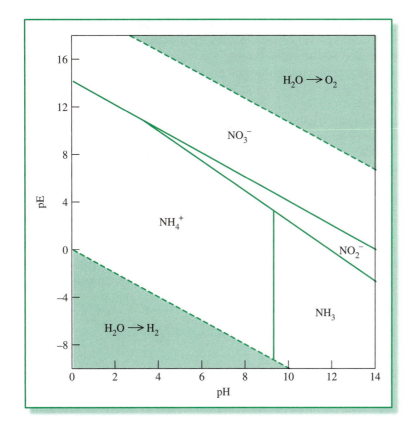

FIGURE 10-5 pE–pH diagram for inorganic nitrogen in an aqueous system. [Source: Adapted from C. N. Sawyer, P. L. McCarty, and G. F. Parkin, *Chemistry for Environmental Engineering,* 4th ed. (New York: McGraw-Hill, 1994).]

The pE–pH diagram for the existence of these forms in aqueous solution is shown in Figure 10-5. Notice the relatively small field of dominance (the small triangle on the right side of the diagram) for the nitrite ion, in which nitrogen's oxidation number has the intermediate value of +3. In particular, it is the predominant species only under alkaline conditions that are intermediate in oxygen content (small positive pE values).

The equilibrium between the most highly reduced and the most highly oxidized forms of nitrogen is given by the half-reaction

$$NH_4^+ + 3\,H_2O \rightleftharpoons NO_3^- + 10\,H^+ + 8\,e$$

Previously we derived the equation relating pE to pH for this reaction written as a reduction; this equation defines the diagonal line in Figure 10-4 that separates these two ions when their concentrations are equal, so that

$$pE = 14.15 - \tfrac{5}{4}\,pH - \tfrac{1}{8}\log(1) = 14.15 - \tfrac{5}{4}\,pH$$

We might conclude from this equation, and from the slope of the diagonal line in Figure 10-5 separating NH_4^+ and NO_3^-, that since the oxidation of

This equation applies only when the ion concentrations are equal, so their ratio is 1.0.

ammonium ion is highly pH dependent, it would not occur under highly acidic conditions. However, electron availability for the reduction of dissolved oxygen also decreases as the pH is lowered, so the pE value in such water is quite high, and nitrate still predominates. For example, the pE value calculated for oxygen in water at pH of 1 is about 20, so nitrate still predominates.

Recall from Chapter 5 that in the microorganism-catalyzed process of nitrification, ammonia and ammonium ion are oxidized to nitrate, whereas in the corresponding denitrification process, nitrate and nitrite are reduced to molecular nitrogen. Both processes are important in soils and in natural waters. In aerobic environments such as the surface of lakes, nitrogen exists as the fully oxidized nitrate, whereas in anaerobic environments such as the bottom of stratified lakes, nitrogen exists as the fully reduced forms ammonia and ammonium ion (Figure 10-3). Nitrite ion occurs in anaerobic environments such as waterlogged soils that are not sufficiently reducing to convert the nitrogen all the way to ammonia. Most plants can absorb nitrogen only in the form of nitrate ion, so any ammonia or ammonium ion used as fertilizer must first be oxidized by microorganisms before it is useful to such plant life.

Review Question 10 is based upon the material above.

PROBLEM 10-13

Consider the reduction of nitrate ion to nitrite ion in a natural water system.

(a) Write the balanced one-electron half-reaction for the process if it occurs in acidic media.
(b) Given that for this reaction, $E^0 = +0.881$ V, calculate pE^0.
(c) From your answer to (a), deduce the expression relating pE to pE^0 and ion concentrations.
(d) From your result in part (c), obtain an equation relating the pE and pH conditions under which the ratio of nitrate to nitrite is 100:1.
(e) From your result in part (c), deduce the ratio of nitrite to nitrate under conditions of pE = 12, pH = 5. ●

Acid–Base and Solubility Chemistry in Natural Waters: The Carbonate System

Natural waters, even when "pure," contain significant quantities of dissolved carbon dioxide and of the anions it produces, as well as cations of calcium and magnesium. In addition, the pH of such natural water is rarely equal to exactly 7.0, the value expected for pure water. In this section, the chemical processes that involve these substances in natural waters are analyzed.

10.14 Carbon Dioxide in Water

The acid–base chemistry of many natural water systems, including both rivers and lakes, is dominated by the interaction of the **carbonate ion, CO_3^{2-}**, a moderately strong base, with the weak acid H_2CO_3, **carbonic acid.** Loss of one hydrogen ion from the acid produces the **bicarbonate ion, HCO_3^-** (also called *hydrogen carbonate ion*):

$$H_2CO_3 \rightleftharpoons H^+ + HCO_3^- \tag{1}$$

The acid dissociation constant for this process, K_1, is numerically much greater than K_2, the constant for the second stage of ionization, which produces carbonate ion:

$$HCO_3^- \rightleftharpoons H^+ + CO_3^{2-} \tag{2}$$

In order to discover the dominant form at any given pH, it is instructive to consider a *species diagram* for the CO_2–bicarbonate–carbonate system in water, such as that shown in Figure 10-6. In it, **the fraction of the total inorganic carbon that is present in each of the three forms is shown as a function of the master variable, the pH of the solution.** Clearly, carbonic acid is the dominant species at low pH (< 5), carbonate is dominant at high pH (> 12), and bicarbonate is the predominant—but not the only—species present in the pH range of most natural waters, i.e., from 7 to 10. At the pH of natural rainwater, 5.6, most of the dissolved carbon dioxide exists as carbonic acid, but a measurable fraction is bicarbonate ion (Figure 10-6).

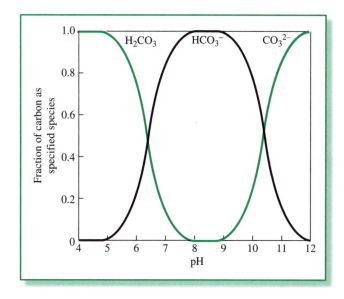

FIGURE 10-6 Species diagram for the aqueous carbon dioxide–bicarbonate ion–carbonate system. [Source: S. E. Manahan, *Environmental Chemistry*, 6th ed. (Boca Raton, FL: Lewis Publishers, 2000), Figure 3.3, p. 54.]

BOX 10-1   Derivation of the Equations for Species Diagram Curves

The final equations relating the concentrations of carbonic acid, bicarbonate ion, and carbonate ion to the pH, the equilibrium constants, and their total concentration C were obtained by algebraic manipulation of the three simultaneous equations.

Mass balance:

$$[H_2CO_3] + [HCO_3^-] + [CO_3^{2-}] = C$$

Acid dissociation constants:

$$K_1 = [HCO_3^-] [H^+]/[H_2CO_3]$$

and

$$K_2 = [CO_3^{2-}] [H^+]/[HCO_3^-]$$

From the second and third equations, we can express both $[H_2CO_3]$ and $[CO_3^{2-}]$ in terms of $[HCO_3^-]$ and $[H^+]$, and substitute these relationships solutions into the mass balance equation:

$$[H_2CO_3] = [HCO_3^-] [H^+]/K_1$$

$$[CO_3^{2-}] = K_2 [HCO_3^-]/[H^+]$$

Thus

$$([HCO_3^-] [H^+]/K_1) + ([HCO_3^-]) + (K_2 [HCO_3^-]/[H^+]) = C$$

Solving this equation for bicarbonate and substituting the solution into the preceding pair of equations yields the expressions given in the main text for the fraction of each species present at any pH.

The curves in Figure 10-6 were constructed by solving for the three unknowns $[H_2CO_3]$, $[HCO_3^-]$, and $[CO_3^{2-}]$, relative to their total concentration, C, from individual equilibrium constant expressions, as detailed in Box 10-1.

$$[H_2CO_3]/C = [H^+]^2/D$$

$$[HCO_3^-]/C = K_1[H^+]/D$$

and

$$[CO_3^{2-}]/C = K_1 K_2/D$$

where the common denominator $D = [H^+]^2 + K_1[H^+] + K_1 K_2$.

These expressions clearly show that **H_2CO_3 is the dominant species under conditions of high acidity, whereas carbonate ion is dominant at high pH, where the hydrogen ion concentration is negligible.**

The CO₂–Carbonate System

The carbonic acid in natural waters results from the dissolution of carbon dioxide gas in water, the gas originating either in air or from the decomposition of organic matter in the water. The gas in air and the acid in water in contact with the surface usually are at equilibrium:

$$CO_2(g) + H_2O(l) \rightleftharpoons H_2CO_3(aq) \qquad (3)$$

The relevant equilibrium constant for this reaction is the Henry's law constant, K_H, for CO$_2$. The pH of pure water in equilibrium with the current level of atmospheric CO$_2$ is 5.6, according to the methods discussed in Chapter 3 (see Problem 3-15).

The sequestration of anthropogenic carbon dioxide, as such, into oceans discussed in Chapter 6 would result in an increase in the acidity of the surrounding waters since the resulting increase in H$_2$CO$_3$ concentration (reaction 3) would give rise to further ionization (reaction 1); the process was classified as *ocean acidic* for that reason. The resulting decrease in pH could affect biological life in the vicinity. Indeed, the increase in the *atmospheric* concentration of CO$_2$ that has already occurred has resulted in a drop of about 0.1 pH units in surface ocean water worldwide.

The predominant source of the carbonate ion in natural waters is limestone rocks, which are largely made up of calcium carbonate, CaCO$_3$. Although this salt is almost insoluble, a small amount of it dissolves when water passes over it:

$$CaCO_3(s) \rightleftharpoons Ca^{2+} + CO_3^{2-} \tag{4}$$

Natural waters that are exposed to limestone are called calcareous waters. The dissolved carbonate ion acts as a base, producing its conjugate weak acid, the bicarbonate ion, as well as hydroxide ion in the water:

$$CO_3^{2-} + H_2O \rightleftharpoons HCO_3^- + OH^- \tag{5}$$

These reactions that occur in the natural three-phase (air, water, rock) system are summarized pictorially in Figure 10-7; the reactions of the carbon dioxide–carbonate system are summarized for convenience in Table 10-3.

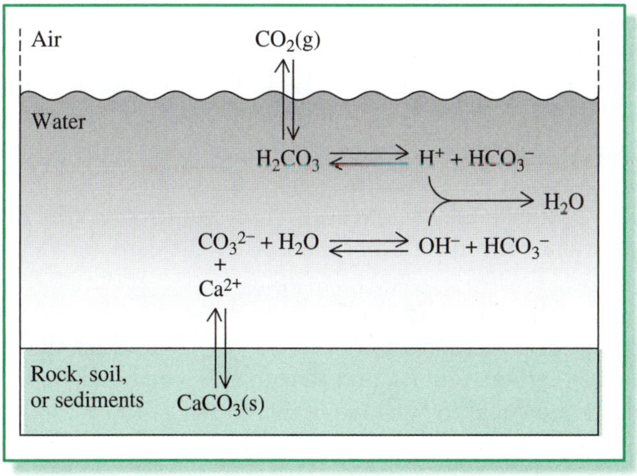

In fact, much of the dissolved carbon dioxide exists as CO$_2$(aq) rather than as H$_2$CO$_3$(aq), but following conventional practice we collect the two forms together and represent it all as the latter.

FIGURE 10-7 Reactions among the three phases (air, water, rocks) of the carbon dioxide–carbonate system.

TABLE 10-3	Reactions in the CO_2–Bicarbonate–Carbonate System		
Reaction Number	Reaction	Equilibrium Constant	K Value at 25°C
1	$H_2CO_3 \rightleftharpoons H^+ + HCO_3^-$	$K_1\ (H_2CO_3)$	4.5×10^{-7}
2	$HCO_3^- \rightleftharpoons H^+ + CO_3^{2-}$	$K_2\ (H_2CO_3)$	4.7×10^{-11}
3	$CO_2(g) + H_2O(l) \rightleftharpoons H_2CO_3(aq)$	K_H	3.4×10^{-2}
4	$CaCO_3(s) \rightleftharpoons Ca^{2+} + CO_3^{2-}$	K_{sp}	4.6×10^{-9}
5	$CO_3^{2-} + H_2O \rightleftharpoons HCO_3^- + OH^-$	$K_b\ (CO_3^{2-})$	2.1×10^{-4}
6	$CaCO_3(s) + H_2O(l) \rightleftharpoons Ca^{2+} + HCO_3^- + OH^-$		
7	$H^+ + OH^- \rightleftharpoons H_2O(l)$	$1/K_w$	1.0×10^{14}
8	$CaCO_3(s) + CO_2(g) + H_2O(l) \rightleftharpoons Ca^{2+} + 2\ HCO_3^-$		

In the discussions that follow, we analyze the effects on the composition of a body of water of the simultaneous presence of both carbonic acid and calcium carbonate. We shall see that **the presence of each of these substances increases the solubility of the other, and that the hydrogen ion and hydroxide ion produced indirectly from their dissolution largely neutralize each other, yielding water with almost neutral pH (reaction 7).**
In order to obtain a qualitative understanding of this rather complicated system, the effect of the carbonate ion alone is first considered.

10.15 Water in Equilibrium with Solid Calcium Carbonate: First Approximation

For simplicity, we first consider a (hypothetical) body of water that is in equilibrium with excess solid calcium carbonate and make the approximation that all other reactions are of negligible importance. The only process of interest in this case is reaction 4 (see Table 10-3). Recall from introductory chemistry that the appropriate equilibrium constant for processes that involve the dissolution of slightly soluble salts in water is the **solubility product**, K_{sp}, which equals the concentrations of the product ions, each raised to its coefficient of the balanced equation. Thus, for reaction (4), K_{sp} is related to the equilibrium concentration of the ions by the equation

$$K_{sp} = [Ca^{2+}]\ [CO_3^{2-}]$$

It follows from the stoichiometry of reaction (4) that as many calcium ions are produced as carbonate ions, and that in this simplified system both ion concentrations are equal to S, the solubility of the salt:

$$S = \text{solubility of } CaCO_3 = [Ca^{2+}] = [CO_3^{2-}]$$

After substitution of S for the ion concentrations in the K_{sp} equation and the K_{sp} value from Table 10-3, we obtain

$$S^2 = 4.6 \times 10^{-9}$$

Taking the square root of each side of this equation, a value for S can be extracted:

$$S = 6.8 \times 10^{-5} \, M$$

Thus the solubility in water of calcium carbonate is estimated to be 6.8×10^{-5} mol L^{-1}, assuming that all other reactions are negligible.

PROBLEM 10-14

Consider a body of water in equilibrium with solid calcium sulfate, CaSO$_4$, for which $K_{sp} = 3.0 \times 10^{-5}$ at 25°C. Calculate the solubility, in g L^{-1}, of calcium sulfate in water assuming that other reactions are negligible. ●

10.16 Water in Equilibrium with Solid Calcium Carbonate: Second Approximation

According to reaction (5) of Table 10-3, dissolved carbonate ion acts as a base in water. In this section, we incorporate the effects of this process on the solubility of calcium carbonate.

The relevant equilibrium constant for reaction (5) is the base ionization constant, K_b, where

$$K_b(CO_3^{2-}) = [HCO_3^-][OH^-]/[CO_3^{2-}]$$

Since the equilibrium in this reaction lies to the right in solutions that are not very alkaline, an approximation of the overall effect resulting from the simultaneous occurrence of reaction (4) and reaction (5) can be obtained by adding together the equations for the two individual reactions. The overall reaction is

$$CaCO_3(s) + H_2O(l) \rightleftharpoons Ca^{2+} + HCO_3^- + OH^- \tag{6}$$

We added reactions (4) and (5) together, and cancelled common terms to obtain this equation.

Thus the dissolution of calcium carbonate in neutral water results essentially in the production of calcium ion, bicarbonate ion, and hydroxide ion.

It is a principle of equilibrium that, **if several reactions are added together, the equilibrium constant K for the combined reaction is the product of the equilibrium constants for the individual processes.** Thus, since reaction (6) is the sum of reactions (4) and (5), its equilibrium constant K_6 must equal $K_{sp}K_b$, the product of the equilibrium constants for reactions (4) and (5). Since, from introductory chemistry we recall that the acid and base ionization constants for any acid–base conjugate pair such as HCO$_3^-$ and CO$_3^{2-}$ are simply related by the equation

$$K_a K_b = K_w = 1.0 \times 10^{-14} \text{ at } 25°C$$

it follows that for the conjugate base CO_3^{2-}

$$K_5 = K_b(CO_3^{2-}) = K_w/K_a \, (HCO_3^-)$$

Since K_a for HCO_3^- is the K_2 value for the carbonic acid system, then from Table 10-3,

$$K_b = 1.0 \times 10^{-14}/4.7 \times 10^{-11}$$

$$= 2.1 \times 10^{-4}$$

Thus, since K_6 for the overall reaction (6) is $K_{sp}K_b$, its value is $(4.6 \times 10^{-9}) \times (2.1 \times 10^{-4}) = 9.7 \times 10^{-13}$.

The equilibrium constant for reaction (6) is related to the ion concentrations by the equation

$$K_6 = [Ca^{2+}] \, [HCO_3^-] \, [OH^-]$$

If we make the approximation that reaction (6) is the *only* process of relevance in the system, then from its stoichiometry we have a new expression for the solubility, S, of $CaCO_3$, namely

$$S = [Ca^{2+}] = [HCO_3^-] = [OH^-]$$

Upon substitution of S for the concentrations, we obtain

$$S^3 = 9.7 \times 10^{-13}$$

Taking the cube root of both sides of this equation, we find

$$S = 9.9 \times 10^{-5} \, M$$

Thus the estimated solubility for $CaCO_3$ is 9.9×10^{-5} M, in contrast to the lesser value of 6.8×10^{-5} M that we obtained when the reaction of carbonate ion was ignored. **The CaCO₃ solubility here is greater than estimated from reaction (4) alone, since much of the carbonate ion it produces subsequently disappears by reacting with water molecules.** In other words, the equilibrium in reaction (4) is shifted to the right since a large fraction of its product reacts further (reaction 5).

From these results, it is clear that the saturated aqueous solution of calcium carbonate is moderately alkaline; its pH can be obtained from the hydroxide ion concentration of 9.9×10^{-5} M:

$$pH = 14 - pOH = 14 - log_{10}[OH^-] = 14 - log_{10}(9.9 \times 10^{-5}) = 10.0$$

Explicit inclusion of both equilibria in calculations predicts a slightly lower pH, 9.9, since only 70% of the carbonate is converted to bicarbonate.

That the solution is alkaline is not surprising, given that the carbonate ion, as weak bases go, is a moderately strong one.

The solubility of calcium carbonate in solutions whose pH is controlled by reactions that dominate over those considered in Table 10-3 is discussed in Box 10-2.

BOX 10-2 | Solubility of CaCO$_3$ in Buffered Solutions

In the analysis in Section 10.16, it was assumed that calcium carbonate dissolves in pure water, and that the reaction of carbonate with water determines the pH. In some real-world situations, the pH of the aqueous solution is predetermined by the presence of some other dominant source of H$^+$ or OH$^-$, and the contribution to the concentration of these ions from calcium carbonate is negligible. In the analysis which follows, we determine the solubility of CaCO$_3$ as a function of pH under such conditions.

The solubility, S, of calcium carbonate, and the resulting dissolved calcium ion concentration, are here equal to the sum of all the carbon-containing species, namely the carbonate and bicarbonate ion concentrations plus that of H$_2$CO$_3$, since we are neglecting any contribution from airborne carbon dioxide:

$$S = [Ca^{2+}] = [CO_3^{2-}] + [HCO_3^-] + [H_2CO_3]$$

In the analysis that follows, we use equilibrium constant expressions to relate [HCO$_3^-$] and [H$_2$CO$_3$] to the carbonate concentration and thereby express S in terms of it.

Since the pH of the solution is predetermined, the ratio of bicarbonate to carbonate ion can be determined from K_b (reaction (5)) and the known, fixed value of [OH$^-$]:

$$[HCO_3^-]/[CO_3^{2-}] = K_b/[OH^-]$$

and thus

$$[HCO_3^-] = K_b [CO_3^{2-}]/[OH^-] \quad (A)$$

Similarly, we can solve for [H$_2$CO$_3$] from the expression for K_1 (reaction (1)):

$$[H^+] [HCO_3^-] = K_1 [H_2CO_3]$$

$$[H_2CO_3] = [H^+] [HCO_3^-]/K_1$$

We can convert this equation into one involving the hydroxide ion concentration by using the water ionization constant K_w (reverse of equation (7) of Table 10-3)

$$[H^+] [OH^-] = K_w \quad \text{so} \quad [H^+] = K_w /[OH^-]$$

The preceding equation becomes

$$[H_2CO_3] = K_w [HCO_3^-]/K_1 [OH^-]$$

Finally, we can obtain an expression for [H$_2$CO$_3$] in terms of carbonate ion by substituting here for [HCO$_3^-$] by using equation (A). Thus

$$[H_2CO_3] = K_b K_w [CO_3^{2-}]/K_1 [OH^-]^2$$

Now we substitute these expressions for carbonic acid and bicarbonate ion into the above equation for the solubility S:

$$\begin{aligned} S = [Ca^{2+}] &= [CO_3^{2-}] + [HCO_3^-] + [H_2CO_3] \\ &= [CO_3^{2-}] + K_b [CO_3^{2-}]/[OH^-] \\ &\quad + K_b K_w [CO_3^{2-}]/K_1 [OH^-]^2 \\ &= [CO_3^{2-}] (1 + K_b/[OH^-] \\ &\quad + K_b K_w/K_1 [OH^-]^2) \end{aligned}$$

For convenience, we temporarily define $f = (1 + K_b/[OH^-] + K_b K_w/K_1 [OH^-]^2)$ and so $S = f [CO_3^{2-}]$.

Substituting this equation for [Ca^{2+}] into the K_{sp} expression gives

$$K_{sp} = f [CO_3^{2-}]^2$$

Solving for the carbonate concentration, we obtain

$$[CO_3^{2-}] = (K_{sp}/f)^{1/2}$$

and hence

$$S = f [CO_3^{2-}] = (f K_{sp})^{1/2} = \{K_{sp} (1 + K_b/[OH^-] + K_b K_w/K_1 [OH^-]^2)\}^{1/2}$$

(continued on p. 438)

Thus, as expected by application of Le Chatelier's principle to the reactions, **the solubility of calcium carbonate decreases as the (fixed) hydroxide concentration increases**, the limit at high levels of OH^- being $(K_{sp})^{1/2}$, the value we obtained assuming no reaction of carbonate ion with water. In contrast, for water that is neutral or acidic and therefore low in hydroxide ion, the $CaCO_3$ solubility is much larger than this value, as illustrated in Figure 10-8 where the logarithm of S is plotted against the pH of the water body.

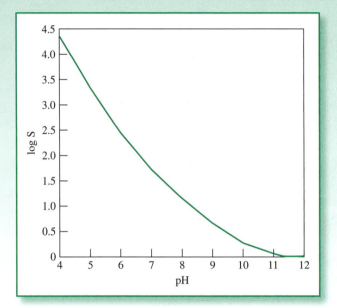

FIGURE 10-8 Molar solubility, S, in units of $K_{sp}^{0.5}$, of $CaCO_3$ in CO_2-free water versus pH.

PROBLEM 10-15

Repeat the calculation of the solubility of calcium carbonate by the approximate single equilibrium method using a realistic wintertime water temperature of 5°C; at that temperature, $K_{sp} = 8.1 \times 10^{-9}$ for $CaCO_3$, $K_a = 2.8 \times 10^{-11}$ for HCO_3^-, and $K_w = 0.2 \times 10^{-14}$. Does the solubility of calcium carbonate increase or decrease with increasing temperature?   ●

PROBLEM 10-16

What is the net reaction when reactions (1) and (3) are added together? How is the equilibrium constant for this combined process related to K_1 and K_3? Show that the pH of the aqueous solution resulting from a CO_2 partial pressure of 0.00039 atm in the combined process has the same value of 5.65 as is determined by considering the individual reactions consecutively, as in Problem 3-15.   ●

10.17 Water in Equilibrium with Both CaCO$_3$ and Atmospheric CO$_2$

The systems discussed above are somewhat unrealistic since they fail to consider the other important carbon species in water, namely carbon dioxide and carbonic acid, and the reactions that involve them. These reactions now will be considered in the context of a body of water that is also in equilibrium with solid calcium carbonate, i.e., the three-phase system illustrated in Figure 10-7.

At first sight, it might seem that since reaction (1) of Table 10-3 provides another source of bicarbonate ion, then by Le Chatelier's principle the production of bicarbonate from the reaction (5) of carbonate with water should be suppressed. However, a more important consideration is that **reaction (1) produces hydrogen ion, which combines with the hydroxide ion that is produced in reaction (4)** by the interaction of carbonate ion with water:

$$H^+ + OH^- \rightleftharpoons H_2O(l) \tag{7}$$

Consequently, the equilibrium positions of both reactions that produce bicarbonate ion are shifted to the right due to the disappearance of one of their products by the above reaction.

If reactions (1), (3), (4), (5), and (7) of Table 10-3 are all added together to deduce the net process, then after cancelling common terms the net result is

$$CaCO_3(s) + CO_2(g) + H_2O(l) \rightleftharpoons Ca^{2+} + 2\,HCO_3^- \tag{8}$$

In other words, combining equimolar amounts of solid calcium carbonate and atmospheric carbon dioxide yields aqueous calcium bicarbonate, Ca(HCO$_3$)$_2$, without any apparent production or consumption of acidity or alkalinity:

calcium carbonate (rock) + carbon dioxide (air) $\longrightarrow$ calcium bicarbonate
(in solution)

The salt calcium bicarbonate is stable as a dilute aqueous solution, but spontaneously decomposes by reversing this reaction if dried to a solid.

Natural waters in which this overall process occurs can be viewed as the site of a giant titration of an acid that originates with CO$_2$ from air with a base that originates with carbonate ion from rocks. (Note that we need not consider reaction 2 in this analysis, since reaction 5 of the conjugate base of bicarbonate with water was included.)

In the *ocean neutral* scheme of sequestration (Chapter 6), bulk carbon dioxide is reacted with solid calcium carbonate or with some other calcium-containing salt and the resulting slurry of calcium bicarbonate (or other salt) is then transported to and deposited in the ocean. This technique avoids the pH-lowering side effect of the *ocean acidic* scheme in which CO$_2$ is directly dissolved in the ocean water.

It should be noted that each of the individual reactions that were added together above is itself an equilibrium that does not lie *entirely* to the right. Since the reactions differ in their extent of completion, it is an approximation to state that the overall reaction shown above is the only resulting reaction. Nevertheless, it is the *dominant process*, and it is mathematically convenient to first consider this process alone in estimating the extent to which $CaCO_3$ and CO_2 dissolve in water when both are present.

Since reaction (8) equals the *sum* of reactions (1), (3), (4), (5), and (7) of Table 10-3, its equilibrium constant K_8 is the *product* of their equilibrium constants:

$$K_8 = K_{sp}K_bK_HK_1/K_w$$

- Here K_1 is the acid dissociation constant of 4.5×10^{-7} for carbonic acid.

- K_H is the Henry's law constant for reaction (3).

- Since K_w is the ion product for water, the equilibrium constant for reaction (7) is $1/K_w$.

- The other constants in the equation above for K_8 have been defined previously (see Table 10-3).

Thus at 25°C, for the overall reaction (8) it follows that

$$K_8 = 1.5 \times 10^{-6}$$

From the balanced equation for the reaction (8), it follows that the expression for K_8 in terms of chemical species is

$$K_8 = [Ca^{2+}] [HCO_3^-]^2/P_{CO_2}$$

If the calcium concentration again is called S, then from the stoichiometry of reaction (8) the bicarbonate concentration must be twice as large, or equal to 2S; after substitution for the concentrations in the equation for K_8 and rearrangement, we obtain

$$S (2S)^2 = K_8 P_{CO_2} \quad \text{or} \quad 4 S^3 = K_8 P_{CO_2}$$

Thus the solubility of calcium carbonate increases as the cube root of the partial pressure of carbon dioxide to which the water is exposed:

$$S = (K_8 P_{CO_2}/4)^{1/3}$$

Substituting the current partial pressure of CO_2 in the atmosphere, 0.00039 atm, corresponding to an atmospheric concentration of CO_2 of 390 ppm (Chapter 3), and the numerical value of K_8 into this equation yields

$$S = 5.3 \times 10^{-4} \text{ mol L}^{-1} = [Ca^{2+}]$$

and thus

$$[HCO_3^-] = 2S = 1.05 \times 10^{-3} \text{ M}$$

The amount of CO_2 dissolved is also equal to S, and is 34 times that which dissolves without the presence of calcium carbonate (see results of Problem 3-15). Furthermore, the calculated calcium concentration is five times that calculated without the involvement of carbon dioxide. **Thus the acid reaction of dissolved CO$_2$ and the base reaction of dissolved carbonate have a synergistic effect on each other that increases the solubilities of both the gas and the solid.** In other words, **water that contains carbon dioxide more readily dissolves calcium carbonate.** In fact, groundwater may become supersaturated with carbon dioxide as a result of biological decomposition processes; and in that case, the calcium carbonate solubility increases even more—at least until the water reaches the surface, when degassing of the excess CO$_2$ would occur.

PROBLEM 10-17

Repeat the above calculation for the solubility of CaCO$_3$ in water that is also in equilibrium with atmospheric CO$_2$ for a water temperature of 5°C. At this temperature, $K_H = 0.065$ for CO$_2$ and K_1 for H$_2$CO$_3$ is 3.0×10^{-7}; see Problem 10-15 for other necessary data.   ●

   Finally, the residual concentrations of CO$_3^{2-}$, of H$^+$, and of OH$^-$ in the system can be deduced from equilibrium constants for reactions (4), (5), and (7), since equilibria in these processes are in effect notwithstanding the overall reaction (8). Thus from reaction (4),

$$[CO_3^{2-}] = K_{sp}/[Ca^{2+}] = 4.6 \times 10^{-9}/5.3 \times 10^{-4} = 8.7 \times 10^{-6} \text{ M}$$

From reaction (5),

$$[OH^-] = K_b [CO_3^{2-}]/[HCO_3^-]$$
$$= (2.1 \times 10^{-4}) \times (8.7 \times 10^{-6})/1.05 \times 10^{-3} = 1.7 \times 10^{-6}$$

and finally from reaction (7)

$$[H^+] = K_w/[OH^-] = 1.0 \times 10^{-14}/1.7 \times 10^{-6} = 5.7 \times 10^{-9}$$

From this value for the hydrogen ion concentration, we conclude that, according to this calculation, **river and lake water at 25°C whose pH is determined by saturation with CO$_2$ and CaCO$_3$ should be slightly alkaline, with a pH of about 8.2.**

   Typically, the measured pH values of calcareous waters lie in the range from 7 to 9, in reasonable agreement with our calculations. Due to the smaller amount of bicarbonate in noncalcareous waters, their pH values are usually close to 7. Of course if such natural waters are subject to acid rain, the pHs can become substantially lower since there is little HCO$_3^-$ or CO$_3^{2-}$ readily available with which to neutralize the acid.

   About 80% of natural surface waters in the United States have pH values between 6.0 and 8.4. Lakes and rivers into which acid rain falls will have elevated levels of sulfate ion and perhaps of nitrate ion since the principal acids in the precipitation are H$_2$SO$_4$ and HNO$_3$ (see Section 4.3).

Review Questions 11–15 are based upon the material in the preceding section.

PROBLEM 10-18

In waters subject to acid rain, the pH is determined not by the CO_2–carbonate system but rather by the strong acid from the precipitation. Assuming that equilibrium with atmospheric carbon dioxide is in effect, calculate the concentration of HCO_3^- in natural waters with pH = 6, 5, and 4 at 25°C. (See Table 10-3 for data.) ●

PROBLEM 10-19

Using algebraic expressions and numerical values for K_1 and K_2 of H_2CO_3 (Table 10-3), calculate the pH values for which $[H_2CO_3] = [HCO_3^-]$ and for which $[HCO_3^-] = [CO_3^{2-}]$. ●

Ion Concentrations in Natural Waters and Drinking Water

10.18 The Abundant Ions in Fresh Water

The most abundant ions found in samples of unpolluted fresh calcareous water usually are calcium and bicarbonate, as expected from our previous analysis. The overall reaction (8) of carbon dioxide and calcium carbonate implies that the ratio of bicarbonate ion to calcium ion should be 2:1; the average ratio of concentrations is quite close to this predicted 2.0 value for river water in North America, Europe, and Africa, but significantly less than 2.0 in Asia, South America, and especially in Australia. The calculated calcium ion concentration, 5.1×10^{-4} M, agrees well with the North American river-water average value of 5.3×10^{-4} M, and similarly for the bicarbonate ion data. The close agreement between the calculated and the experimental results is somewhat fortuitous because river-water temperatures on average lie below 25°C—which results in a higher CO_2 solubility than has been assumed—and because several minor factors have been oversimplified in the calculation. In fact, even calcareous river water is usually unsaturated with respect to $CaCO_3$.

Water in rivers and lakes that is *not* in contact with carbonate salts generally contains substantially fewer dissolved ions than are present in calcareous waters. However, the concentrations of sodium and potassium ions may be as high as those of calcium, magnesium, and bicarbonate ions in these fresh waters. Even in areas with no limestone in the soil, the waters contain some bicarbonate ion due to the weathering of *aluminosilicates* in submerged soil and rock in the presence of atmospheric carbon dioxide. The weathering reaction can be written in general terms as

$$M^+(\text{Al-silicate}^-)(s) + CO_2(g) + H_2O \longrightarrow M^+ + HCO_3^- + H_4SiO_4$$

Here M is a metal such as potassium, and the anion is one of the many aluminosilicate ions found in rocks (see Chapter 16). The weathering of *potassium feldspar* is an example of one of the most important sources of potassium ion in natural waters:

$$3 \, KAlSi_3O_8(s) + 2 \, CO_2(g) + 14 \, H_2O \longrightarrow$$
$$2 \, K^+ + 2 \, HCO_3^- + 6 \, H_4SiO_4 + KAl_3Si_3O_{10}(OH)_2(s)$$

Thus bicarbonate normally is the predominant anion in both calcareous and noncalcareous waters since it is produced by the dissolution of limestone and aluminosilicates, respectively.

Commonly, river water also contains **magnesium ion, Mg^{2+}**, principally from the dissolution of $MgCO_3$; plus some sulfate ion, SO_4^{2-}; smaller amounts of **chloride ion, Cl^-**, and **sodium ion, Na^+**; and smaller levels of fluoride ion, F^-, and **potassium ion, K^+**.

Fresh water in which the concentration of ions is abnormally high is called **saline water,** and usually is unsuitable for drinking. Most surface saline water is the result of irrigation, in which water is transported onto land where little rainfall occurs. The water largely evaporates if the climate is hot and dry, leaving behind salts of the ions that were present in the irrigation water. The runoff water from rainfall and irrigation into water supplies consequently is saline. If the irrigation water is recycled, it becomes more and more saline as time goes on. The wintertime de-icing of roads in northern climates also contributes salinity to water bodies.

10.19 Fluoride Ion in Water

The level of **fluoride ion, F^-**, in natural waters displays substantial variations, from less than 0.01 ppm to as much as 50 ppm, in different regions of the world. The source of most F^- is weathering of the mineral **fluorapatite, $Ca_5(PO_4)_3F$**.

$$Ca_5(PO_4)_3F \longrightarrow 5 \, Ca^{2+}$$
$$+ \, 3 \, PO_4^{3-} + F^-$$

However, the maximum solubility of F^- in most natural waters is controlled by the solubility of **calcium difluoride, CaF_2**, which would precipitate excess fluoride ion if its level became too high—typically more than about 6 ppm in water having average calcium content (Problem 10-20). High levels of fluoride can exist only in water with an abnormally low calcium ion concentration, such as groundwater in some areas. Excessive levels of fluoride are typically removed from water destined for drinking by passing it over a solid source of calcium ions.

$$Ca^{2+} + 2 \, F^- \rightleftharpoons CaF_2$$

PROBLEM 10-20

Given that the average concentration of calcium ion in natural waters is 3.8×10^{-4} M, and that the solubility product K_{sp} for CaF_2 is 4×10^{-11}, calculate the maximum molar concentration of fluoride ion that can be dissolved before precipitation begins. Show that this molarity is equivalent to about 6 ppm. [*Hint*: Write the solubility product expression for CaF_2, and substitute the molar value for $[Ca^{2+}]$ into it.] ●

Tooth enamel consists largely of hydroxyapatite, $Ca_5(PO_4)_3OH$.

Increasing the public's daily exposure to fluoride ion in order to reduce the incidence of dental caries, especially among children, has been a widespread practice that began in the middle of the twentieth century. A veneer of fluorapatite forms over the enamel of the teeth, strengthening it to attack by acids.

In Mexico, Columbia, Jamaica, and some European countries including France and Germany, **sodium fluoride,** NaF, is added to table salt for this purpose. The addition of fluoride to milk or yogurt is practiced in some other countries. In many communities of English-speaking countries, including the United States, Canada, the United Kingdom, Ireland, Australia, and New Zealand, a soluble fluoride compound such as **fluorosilicic acid,** H_2SiF_6, or its sodium salt, both of which react with water to release fluoride ion, is added in order to increase the fluoride level to about 1 ppm in geographic regions in which its concentration in the drinking water source is low.

1 ppm = 1 mg L^{-1}, which for fluoride is 5×10^{-5} M.

$$H_2SiF_6 + 4 H_2O \longrightarrow H_4SiO_4 + 6 F^- + 6 H^+$$

The recommended range for F^- in the United States is 0.7–1.2 ppm.

The 1 ppm value was considered at least in the past to be optimum in strengthening children's teeth against decay while providing a margin of safety, although a slightly lower concentration is now recommended in many areas. If the fluoride level is much in excess of this value, as it is in some natural waters, deleterious effects on teeth, such as staining (*fluorosis*) and mottling, can occur.

Almost all brands of toothpaste sold in developed countries since the 1970s have contained an added fluoride ion source in the form of sodium fluoride, **stannous fluoride,** SnF_2, or **sodium monofluorophosphate,** Na_2FPO_3.

$$O$$
$$\parallel$$
$$F-P-O^-Na^+$$
$$|$$
$$O^-Na^+$$

Most children in developed countries receive topical fluoride not only from their toothpaste, but also in some cases by application from their dentists. Toothpaste containers warn children not to swallow the toothpaste, since this would deliver a high level of fluoride to their bodies. Some communities in Japan and several European countries have ceased fluoridation of water without their children experiencing a large increase in the rate of dental caries; the use of fluoride-containing toothpaste is apparently more effective than is delivery of the ion in drinking water in areas where regular brushing of teeth is practiced.

The addition of fluoride ion to public supplies of drinking water continues to be a controversial subject because at *high* concentrations fluoride is known to be poisonous and perhaps carcinogenic (bone cancer), and because some people feel that it is immoral to force everyone to drink water to which a substance has been added. In fact, for many people, the total amount of fluoride ion ingested from food and beverages (especially tea) exceeds that from water.

The U.S. Centers for Disease Control lists fluoridation of water as one of the ten top public health achievements of the twentieth century.

Suppliers of bottled water sometimes advertise their products as having "zero" concentrations of sodium and/or fluoride ions or as being *sodium-free* or *fluoride-free*. These are misleading statements since in reality the actual concentrations are not zero but below the level of detection in the analytical method used by the bottler, or below a threshold specified by government. "Zero" is not a meaningful chemical concept to answer the question of "how much" of a substance is present in a sample.

ACTIVITY

By visiting the websites of the brands, make a table of the ion concentrations for the types of bottled water that are commonly available where you live. Do any of the values exceed those recommended by your country for drinking water? In the case of brands of water that is purified, deduce from the names of the salts that are subsequently added which ions are being fortified. Does your research indicate why a *minimum* concentration of certain ions is desirable?

10.20 Seawater

The total concentration of ions in seawater is much higher than that in fresh water since it contains large quantities of dissolved salts. The predominant species in seawater are sodium and chloride ions, which occur at about one thousand times their average concentration in fresh water. Seawater also contains some Mg^{2+} and SO_4^{2-}, and lesser amounts of many other ions.

If seawater is gradually evaporated, the first salt to precipitate is $CaCO_3$ (present to the extent of 0.12 g L^{-1}), followed by $CaSO_4 \cdot H_2O$ (1.75 g L^{-1}), then NaCl (29.7 g L^{-1}), $MgSO_4$ (2.48 g L^{-1}), $MgCl_2$ (3.32 g L^{-1}), NaBr (0.55 g L^{-1}), and finally KCl (0.53 g L^{-1}). Thus "sea salt" is a mixture of all these salts, which together constitute about 3.5% of the mass of seawater.

Due primarily to the operation of the CO_2–bicarbonate–carbonate equilibrium system discussed previously for fresh water, the average pH of surface ocean water is about 8.1. Seawater has a low organic content, its DOC value being about 1 mg L^{-1}.

10.21 Alkalinity Indices for Natural Waters

The actual concentrations of the cations and anions in a real water sample cannot simply be assumed to be the theoretical values calculated above for calcium, carbonate, and bicarbonate for two reasons:

• the water may not be in equilibrium with either solid calcium carbonate or with atmospheric CO_2; and

• other acids or bases may also be present.

The index devised by analytical chemists to represent the *actual* concentration in water of the anions that are basic is provided by the **alkalinity** value for the sample. **Alkalinity is a measure of the ability of a water sample to**

act as a base by reacting with hydrogen ions. In practical use, the alkalinity of a body of water is a handy measure of the capacity of the water body to neutralize acids and hence to resist acidification when acid rain falls into it. (Alkalinity differs from [OH$^-$], or pOH, in that the latter equals only the hydroxide concentration a water sample has at a particular moment, and does not include its ability to generate additional OH$^-$ if acid is added to it.)

From an operational viewpoint, alkalinity, more properly termed **total alkalinity, is the number of moles of H$^+$ required to titrate one liter of a water sample to the (slightly acidic) end point.** For a solution containing carbonate and bicarbonate ions, as well as OH$^-$ and H$^+$, by definition

$$(\text{total) alkalinity} = 2\,[CO_3^{2-}] + [HCO_3^-] + [OH^-] - [H^+]$$

The factor of two appears in front of carbonate ion concentration since in the presence of H$^+$ it is first converted by the ion to bicarbonate ion, and then the latter is converted by a second hydrogen ion to carbonic acid:

$$CO_3^{2-} + H^+ \rightleftharpoons HCO_3^-$$

$$HCO_3^- + H^+ \rightleftharpoons H_2CO_3$$

Minor contributors to the alkalinity of fresh-water systems can include dissolved ammonia and the anions of *phosphoric, boric,* and *silicic* acids and of H$_2$S, as well as natural organic matter.

The alkalinities of natural waters range from less than 5×10^{-5} M (50 μM) to more than 2×10^{-3} M (2000 μM), compared to the value of about 3×10^{-4} M that corresponds to our estimate for water in contact with atmospheric CO$_2$ (see Problem 10-22). Lakes having alkalinities less than about 200 μM are considered to be "high" in sensitivity to acid rain; those from 200 to 400 μM are classified as "moderate" in sensitivity; and those with greater than 400 μM alkalinity are considered to have "low" sensitivity.

By convention in analytical chemistry, **methyl orange** is used as the indicator in titrations by which total alkalinity is determined. Methyl orange is chosen because it does not change color until the solution is slightly acidic (pH = 4); under such conditions, not only has all the carbonate ion in the sample been transformed to bicarbonate, but virtually all the bicarbonate ion has been transformed to carbonic acid (see Problem 10-21 and Figure 10-6).

Another index encountered in the analysis of natural waters is the **phenolphthalein alkalinity** (also called *carbonate alkalinity*), which **is a measure of the concentration of the carbonate ion and of other similarly basic anions.** In order to titrate only CO$_3^{2-}$ and not HCO$_3^-$ as well, the indicator **phenolphthalein** or one with similar characteristics is used. Phenolphthalein changes color in the pH range between 8 and 9, so it provides a fairly alkaline end point. At such pH values, only a negligible amount of the bicarbonate ion has been converted to carbonic acid, but the majority of CO$_3^{2-}$ has been converted to HCO$_3^-$ (Figure 10-6). Thus

$$\text{phenolphthalein alkalinity} = [CO_3^{2-}]$$

Alkalinity values are sometimes reported as milligrams of CaCO$_3$ equivalent, rather than moles of H$^+$, per liter in a manner similar to that explained below for hardness.

Significant H$^+$ to neutralize OH$^-$ would also be required at high pH values.

PROBLEM 10-21

Calculate the value of the ratios $[HCO_3^-]/[CO_3^{2-}]$ and $[H_2CO_3]/[HCO_3^-]$ at pH values of 4 and 8.5 to confirm the statements made above concerning the nature of the species present at the methyl orange and phenolphthalein end points of the titrations. [*Hint:* Use the equilibrium constant expressions and K values for reactions (1) and (2).] ●

PROBLEM 10-22

Calculate the values expected for the total alkalinity and for the phenolphthalein alkalinity of a 25°C saturated solution of calcium carbonate in water, and compare them to the values for a solution that also is in equilibrium with atmospheric carbon dioxide. Use the concentrations quoted in the last column of Table 10-4. ●

PROBLEM 10-23

Calculate the total alkalinity for a sample of river water whose phenolphthalein alkalinity is known to be 3.0×10^{-5} M, whose pH is 10.0, and whose bicarbonate ion concentration is 1.0×10^{-4} M. ●

The alkalinity value for a lake is sometimes used by biologists as a measure of its ability to support aquatic plant life, a high value indicating a high potential fertility. The reasons for such a situation are often the following ones. Algae extract the carbon dioxide they need for photosynthesis from bicarbonate ion, which is plentiful in calcareous waters, by a reversal of the $CO_2/CaCO_3$ reaction discussed previously:

$$Ca^{2+} + 2\,HCO_3^-\,(aq) \longrightarrow CO_2 + CaCO_3(s) + H_2O$$

TABLE 10-4	Calculated Ion Concentrations for Aqueous Equilibrium Systems	
Ion	CaCO₃ Only	CO₂ and CaCO₃
HCO_3^-	9.9×10^{-5} M	1.05×10^{-3} M
CO_3^{2-}	—	8.7×10^{-6} M
Ca^{2+}	9.9×10^{-5} M	5.3×10^{-4} M
OH^-	9.9×10^{-5} M	1.7×10^{-6} M
H^+	1.0×10^{-10} M	5.7×10^{-9} M
pH	10.0	8.2

Indeed, small crystals of calcium carbonate are sometimes observed in lakes undergoing active photosynthesis, i.e.,

$$CO_2 + H_2O + sunlight \longrightarrow CH_2O \ polymer + O_2$$
$$(as \ algae)$$

In noncalcareous waters, which have low alkalinity and low calcium content, dissociation of the bicarbonate ion in the water forms not only carbon dioxide but also hydroxide ion:

$$HCO_3^- \rightleftharpoons CO_2 + OH^-$$

The algae readily exploit this CO_2 for their photosynthetic needs, at the cost of allowing a buildup of hydroxide ion to such an extent that the lake water becomes quite basic—with a pH as high as 12.3 in some cases.

10.22 The Hardness Index for Natural Waters

As a measure of certain important cations present in samples of natural waters, analytical chemists often use the **hardness index, which measures the total concentration of the ions Ca^{2+} and Mg^{2+},** the two species that are principally responsible for hardness in water supplies. Chemically, the hardness index is defined in this way:

$$hardness = [Ca^{2+}] + [Mg^{2+}]$$

Experimentally, hardness can be determined by titrating a water sample with **ethylenediaminetetraacetic acid** (EDTA), a substance that forms very strong complexes with metal ions other than those of the alkali metals.

Traditionally, hardness is expressed not as a molar concentration of ions but **as the mass in milligrams (per liter) of calcium carbonate that contains the same total number of dipositive (2+) ions.** Thus, for example, a water sample that contains a total of 0.0010 mol of Ca^{2+} + Mg^{2+} per liter would possess a hardness value of 100 mg of $CaCO_3$, since the molar mass of $CaCO_3$ is 100 g and thus 0.0010 mol of it weighs 0.1 g, or 100 mg.

Most calcium enters water either from $CaCO_3$ in the form of limestone or from mineral deposits of $CaSO_4$. The source of much of the magnesium is *dolmitic limestone,* $CaMg(CO_3)_2$. Hardness is an important characteristic of natural waters, since calcium and magnesium ions form insoluble salts with the anions present in soaps, thereby forming a scum in wash water. Water is termed "hard" if it contains substantial concentrations of calcium and/or magnesium ions; thus calcareous water is "hard." Some scientists define water as being hard if its hardness index exceeds 150 mg L^{-1}.

$$CaMg(CO_3)_2 \longrightarrow Ca^{2+}$$
$$+ \ Mg^{2+} + 2 \ CO_3^{2-}$$

Many areas possess soils that contain little or no carbonate ion, so there is no reaction with CO_2 producing bicarbonate. Such "soft" water typically has a pH much closer to 7 than does hard water, since it contains few basic anions. However, there are lakes with little dissolved calcium or magnesium but relatively high concentrations of dissolved **sodium carbonate,**

Na_2CO_3; such lakes have a very low degree of hardness but are high in alkalinity.

Interestingly, people who live in hard-water areas may have a lower average death rate from ischemic heart disease than do people living in areas with very soft water. Research in rural Finland and in some other countries found the risk of heart attacks decreased as the magnesium concentration in the local water supply increased. However, even more recent research from the Netherlands failed to find any such relationship, so the effect of soft water upon health is still uncertain.

PROBLEM 10-24

What is the value of the hardness index for a 500-mL sample of water that contains 0.0040 g of calcium ion and 0.0012 g of magnesium ion? ●

PROBLEM 10-25

Calculate the hardness, in milligrams of $CaCO_3$ per liter, of water that in equilibrium at 25°C with carbon dioxide and calcium carbonate, using results in the last column of Table 10-4. Is the calculated value greater or less than the median hardness value found for surface waters in the United States of 37 mg L^{-1}? ●

10.23 Aluminum in Natural Waters

The concentration of **aluminum ion,** Al^{3+}, in natural waters normally is quite small, typically about 10^{-6} M. This low value is the consequence of the fact that in the typical pH range for natural waters (6 to 9), the solubility of the aluminum contained in rocks and soils to which the water is exposed is very small. The solubility of aluminum in water is controlled by the insolubility of **aluminum hydroxide,** $Al(OH)_3$. Given that the K_{sp} of the hydroxide is about 10^{-33} at usual water temperatures, then for the reaction

$$Al(OH)_3 \rightleftharpoons Al^{3+} + 3\ OH^-$$

it follows that

$$[Al^{3+}]\,[OH^-]^3 = 10^{-33}$$

Take, for instance, a sample of water whose pH is 6. Since the hydroxide concentration in such water is 10^{-8} M, it follows that

$$[Al^{3+}] = 10^{-33}/(10^{-8})^3 = 10^{-9}\,M$$

Although this value is very small, for every one-unit decrease of the pH, the concentration of aluminum ion increases by a factor of a thousand, so it reaches 10^{-6} M at pH = 5 and 10^{-3} M at a pH of 4. Thus **aluminum is much more soluble in highly acidified rivers and lakes than in those where pH**

values do not fall below 6 or 7. Indeed, Al^{3+} is usually the principal cation in waters whose pH is less than 4.5, exceeding even the concentrations of Ca^{2+} and Mg^{2+}, which are the dominant cations at pH values greater than 4.5.

In the past, fears arose that human ingestion of aluminum from drinking water and from the use of aluminum cooking pots was a major cause of Alzheimer's disease; however, the research upon which this conclusion was reached could not be reproduced. Today many neuroscientists do not believe that there is a strong connection between the disease and intake of the metal, since past epidemiological studies on this matter have not been definitive or consistent. However, Canadian and Australian research reported in the mid-1990s indicates that consumption of drinking water with greater than 100 ppb aluminum—not an uncommon level in drinking water purified by aluminum sulfate (see Chapter 11)—can lead to neurological damage such as memory loss and perhaps to a small increase in the incidence of Alzheimer's disease.

It is thought that the principal deleterious effect of acid waters upon fish arises from the solubilization of aluminum from soil and its subsequent existence as a free ion in the acidic water, as discussed in Chapter 4. Unfortunately, the $Al(OH)_3$ then precipitates as a gel on contact with the less acidic gills of the fish, and the gel prevents the normal intake of oxygen from water, thus suffocating the fish.

It is also believed that **aluminum mobilization in soils is one of the stresses that acid rain places on trees resulting in the dieback of forests.** Soils that contain limestone are usually considered to be buffered against much change in pH, due to the ability of carbonate and bicarbonate ion to neutralize H^+. But over a period of decades, surface soil may gradually lose its carbonate content due to a continual bombardment by acid rain. Thus soils receiving acid rain eventually become acidified. When the pH of the soil drops below about 4.2, aluminum leaching from soil and rocks becomes particularly appreciable. Such acidification has occurred in some regions in central Europe, including Poland, the former Czechoslovakia, and eastern Germany, and the resulting solubilization of aluminum may have contributed to the forest diebacks observed there in the 1980s (see Section 4.8).

PROBLEM 10-26

What is the concentration, in grams per liter, of dissolved aluminum in water having a pH of 5.5?

PROBLEM 10-27

Review Questions 16–20 are based on the material in the preceding section.

Calculate the pH value at which the aluminum ion concentration dissolved in water is 0.020 M, assuming that it is controlled by the equilibrium with solid aluminum hydroxide.

Review Questions

1. Write the balanced half-reaction involving O_2 that occurs in acidic waters when it oxidizes organic matter.

2. How does temperature affect the solubility of O_2 in water? Explain what is meant by *thermal pollution*.

3. Define *BOD* and *COD*, and explain why their values for the same water sample can differ slightly. Explain why natural waters can have a high BOD.

4. What do the acronyms *TOC* and *DOC* stand for, and how do they differ in terms of what they measure?

5. Write the half-reaction, used in the COD titration, which converts dichromate ion to Cr^{3+} ion, and balance it.

6. Write the balanced chemical reaction by which organic carbon, represented as CH_2O, is disproportionated by bacteria under anaerobic conditions.

7. Draw a labeled diagram classifying the top and bottom layers of a lake in summer as either oxidizing or reducing in character, and showing the stable forms of carbon, sulfur, nitrogen, and iron in the two layers.

8. What are some examples of highly reduced and of highly oxidized sulfur in environmentally important compounds? Write the balanced reaction by which sulfate can oxidize organic matter.

9. Explain the phenomenon of *acid mine drainage*, writing balanced chemical equations as appropriate.

10. What is meant by the *pE* of an aqueous solution? What does a low (negative) pE value imply about the solution? What species determines the pE value in aerated water?

11. What is the acid and what is the base that dominate the chemistry of most natural water systems, and whose interaction produces bicarbonate ion?

12. What is the source of most of the carbonate ion in natural waters? What name is given to waters that are exposed to this source?

13. Write the approximate net reaction between carbonate ion and water in a system that is *not* also exposed to atmospheric carbon dioxide. Is the resulting water acidic, alkaline, or neutral?

14. Write the approximate net reaction between carbonate ion and water in a system that *is* exposed to atmospheric carbon dioxide. Is the resulting water mildly acidic or mildly alkaline? Explain why the production of bicarbonate ion from carbonate ion does not inhibit its production from carbon dioxide, and vice-versa.

15. If two equilibrium reactions are added together, what is the relationship between the equilibrium constants for the individual reactions and that for the overall reaction?

16. Which are the most abundant ions in clean, fresh, calcareous water?

17. What is the natural source of fluoride ion in water? How and why is the fluoride level in drinking water artificially increased to about 1 ppm in many municipalities? How can fluoride be removed from natural water in which its concentration is too high?

18. Define the *total alkalinity index* and the *phenolphthalein alkalinity index* for water.

19. Define the *hardness index* for water.

20. Explain why aluminum ion concentrations in acidified waters are much greater than those in neutral water. How does the increased aluminum ion level affect fish and trees?

Green Chemistry Questions

1. What takes place during the scouring of cotton, and why is this process necessary for the production of finished cotton fibers?

2. Biopreparation (an enzymatic process) replaced the use of large amounts of sodium hydroxide in the scouring of cotton.

(a) Describe any environmental problems or worker hazards that would be associated with the use of sodium hydroxide solutions in the scouring of cotton.

(b) Would these same environmental problems or worker hazards be eliminated by the use of Biopreparation?

3. The development of Biopreparation by Novozymes won a Presidential Green Chemistry Challenge Award.

(a) Which of the three focus areas (see page xxviii) for these awards does this award best fit into?

(b) List at least three of the twelve principles of green chemistry (see pages xxiii–xxiv) that are addressed by the green chemistry developed by Novozymes.

Additional Problems

1. Over a period of several days, estimate your approximate daily water usages in the categories of showering/bathing, clothes washing, toilets, dishwashing, and cooking. (Many flush toilets display volume-per-flush data. Washing machine water volume capacity can be estimated from the dimensions of the washer cavity: 1 L = 10 cm × 10 cm × 10 cm. Using a measuring cup, discover how long it takes your shower to deliver one liter of water, and adjust the data accordingly for the length of your average shower.)

2. The TOC parameter for water samples is measured by oxidizing the organic material to carbon dioxide and then measuring the amount of this gas evolved from the solution. If a 5.0-L sample of wastewater produced 0.25 mL of carbon dioxide gas, measured at a pressure of 0.96 atm and a temperature of 22°C, calculate the TOC value for the sample. Assuming the average composition of the organic matter to be CH_2O, calculate what the Chemical Oxygen Demand value for the water sample would be due to its organic content. (The gas constant $R = 0.0821$ L atm mol^{-1} K^{-1}.)

3. (a) Balance the reduction half-reaction that converts SO_4^{2-} to H_2S under acidic conditions.

(b) Deduce the expressions relating pE to pH, the concentration of sulfate ion and the partial pressure of hydrogen sulfide gas, given that for the half-reaction, pE° = −3.50 V when the pH is 7.0.

(c) Deduce the partial pressure of hydrogen sulfide when the sulfate ion concentration is 10^{-5} M and the pH is 6.0 for water that is in equilibrium with atmospheric oxygen.

4. Calculate the solubility of lead(II) carbonate, $PbCO_3$ ($K_{sp} = 1.5 \times 10^{-13}$) in water, given that most of the carbonate ion it produces subsequently reacts with water to form bicarbonate ion. Recalculate the solubility assuming that none of the carbonate ion reacts to form bicarbonate; is your result significantly different from that calculated assuming complete reaction of carbonate with water?

5. The bicarbonate ion, HCO_3^-, can potentially act as an acid or as a base in water. Write the chemical equations for these two processes, and from the information given in this chapter, determine the corresponding acid and base dissociation constants. Given the relative magnitudes of the dissociation constants, decide whether the dominant reaction of bicarbonate in water will be as an acid or as a base. Calculate the pH of an aqueous 0.010 M solution of sodium bicarbonate in water using the dominant reaction alone, and assuming that the amounts of carbonate ion and carbonic acid from other sources are negligible in this case.

6. A sample of lake water at 25°C is analyzed and the following parameters are found for it:

total alkalinity = 6.2×10^{-4} M

phenolphthalein alkalinity = 1.0×10^{-5} M

pH = 7.6

hardness = 30.0 mg L^{-1}

$[Mg^{2+}] = 1.0 \times 10^{-4}$ M

Extract all possible single ion concentrations that you can by combining one or more of these data. Also determine whether or not the water is at equilibrium with respect to the carbonate–bicarbonate system and whether or not it is saturated with calcium carbonate.

7. The O_2 concentration of a water sample can be determined using the so-called Winkler titration method. In it, the oxygen in a small sample of the water is reacted with $MnSO_4$ in a basic solution. The reaction precipitates the manganese as MnO_2, which converts added I^- to I_2. Molecular iodine is then quantitatively determined by titrating it in acidic solution with a standardized solution of sodium thiosulfate, $Na_2S_2O_3$. The set of equations for the reactions is:

$$2\,Mn^{2+} + 4\,OH^- + O_2\,(aq) \longrightarrow 2\,MnO_2(s) + 2\,H_2O$$

$$MnO_2(s) + 4\,H^+ + 2\,I^- \longrightarrow Mn^{2+} + I_2 + 2\,H_2O$$

$$I_2(aq) + 2\,S_2O_3^{2-} \longrightarrow S_4O_6^{2-} + 2\,I^-$$

In determining the BOD of a sample of water, a chemist used two 10.00-mL samples of the water, one before and one after the five-day incubation period. They required 10.15 and 2.40 mL of a 0.00100 M standard solution of $K_2S_2O_3$. Calculate the BOD, in units of milligrams per liter, of this water sample. On the basis of these results, would you consider this water to be polluted?

The Pollution and Purification of Water

In this chapter, the following introductory chemistry topics are used:

- ⮕ Acid–base and equilibrium concepts and calculations; pH
- ⮕ Le Chatelier's principle
- ⮕ Basic structural organic chemistry (see online Appendix)
- ⮕ Oxidation numbers; redox half-reactions and their balancing
- ⮕ Catalysis
- ⮕ Distillation

Background from previous chapters used in this chapter:

- ⮕ BOD
- ⮕ VOCs
- ⮕ Photochemical reactions; UV light
- ⮕ Concentration scales in aqueous solutions
- ⮕ Free radicals
- ⮕ BTX hydrocarbons; MTBE

Introduction

The pollution of natural waters by both biological and chemical contaminants is a worldwide problem. There are few populated areas, whether in developed or developing countries, that do not suffer from one form of water pollution or another. In this chapter we shall survey the various methods—both traditional and innovative—by which water can be purified, including the desalination of seawater.

We begin by discussing techniques that are used to purify drinking water from relatively uncontaminated sources, and then consider the pollution and remediation of groundwater and of sewage and wastewater. Lastly, we investigate modern advanced techniques whereby polluted water—and air—can be cleansed.

Water Disinfection

The quality of "raw" (untreated) water, whether drawn from surface water or groundwater, that is intended eventually for drinking varies widely, from almost pristine to highly polluted. Because both the type and quantity of pollutants in raw water vary, the processes used in purification also vary from place to place. The most commonly used procedures are shown in schematic form in Figure 11-1. Before discussing the major topic of disinfection, we shall discuss the various non-disinfection steps that are often taken in the overall purification process.

11.1 Aeration of Water

Aeration is commonly used in the improvement of water quality. **Municipalities aerate drinking water that is drawn from underground aquifers in order to remove dissolved gases such as the foul-smelling H_2S and organosulfur compounds, as well as volatile organic compounds, some of which have a detectable odor.**

Aeration of drinking water also results in reactions that produce CO_2 from the most easily oxidized organic material. If necessary for reasons of odor, taste, or health, most of the remaining organics can be removed by subsequently passing the water over activated carbon, although this process is relatively expensive and so rather few communities use it (see Box 11-1). Another advantage to aeration is that the increased oxygen content of water oxidizes water-soluble Fe^{2+} to Fe^{3+}, which then forms insoluble hydroxides (and related species) that can be removed as solids.

$$Fe^{3+} + 3\,OH^- \longrightarrow Fe(OH)_3(s)$$

(Recall that ions listed in equations without a state specified are assumed to be in aqueous solution.)

After aeration, colloidal particles in the water are removed, as described below. If the water is *excessively* hard, calcium and magnesium are removed

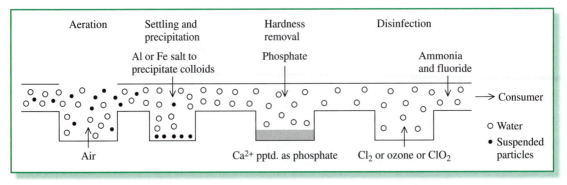

FIGURE 11-1 The common stages of purification of drinking water.

BOX 11-1	Activated Carbon

Activated carbon (activated charcoal) is a very useful solid for purifying water of small organic molecules present in low concentrations. The ability of this material to remove contaminants from water and to improve its taste, color, and odor has been known for a long time; indeed, the ancient Egyptians used charcoal-lined vessels to store water for drinking purposes.

Activated carbon is produced by anaerobically charring a high-carbon-content material such as peat, wood, or lignite (a soft brown coal) at temperatures below 600°C, followed by a partial oxidation process using carbon dioxide or steam at a slightly higher temperature.

The removal of contaminants by activated carbon is a physical adsorption process and therefore is reversible if sufficient energy is applied. The characteristic that makes activated carbon such an excellent adsorber is its huge surface area, about 1400 m^2 g^{-1}. This surface is internal to the individual carbon particles, so that crushing the material neither increases nor decreases the area. The internal structure of the solid involves series of channels (pores) of progressively decreasing size that are produced by the charring and partial oxidation processes. The internal sites where adsorption occurs are large enough only for small molecules, including chlorinated solvents.

At the typical ppm concentrations found for organic contaminants in water, each gram of activated carbon can adsorb a few percent of its mass in contaminants such as chloroform and the dichloroethenes, as well as much higher masses of TCE, PCE, and pesticides such as dieldrin, heptachlor, and DDT.

Once a sample of activated carbon has reached near-saturation in terms of adsorbed organics, three alternatives are available. It can be simply disposed of in a landfill, it can be incinerated to destroy it and the adsorbed contaminants, or it can be heated to rejuvenate the surface by driving off the organic pollutants, which can then be incinerated or catalytically oxidized.

from it, as described below, before the final stages of disinfection and the addition of fluoride, as discussed below (see Figure 11-1).

11.2 Removal of Calcium and Magnesium

If the water comes from wells in areas having limestone bedrock, it will contain significant levels of Ca^{2+} and Mg^{2+} ions, which are usually removed during the processing by precipitation reactions, since these ions can interfere with soaps and detergents used for washing.

- Calcium can be removed from water by addition of phosphate ion in a process analogous to that discussed later for phosphate removal; here, however, phosphate is *added* in order to precipitate the calcium ion.

- More commonly, calcium ion is removed by precipitation and filtering of the insoluble salt $CaCO_3$. The carbonate ion is either added as **sodium** carbonate, Na_2CO_3, or if sufficient HCO_3^- is naturally present in the water, **hydroxide ion, OH^-**, is added in order to convert dissolved bicarbonate ion to carbonate:

$$OH^- + HCO_3^- \longrightarrow CO_3^{2-} + H_2O$$

$$Ca^{2+} + CO_3^{2-} \longrightarrow CaCO_3(s)$$

• Magnesium ion precipitates as the insoluble **magnesium hydroxide, $Mg(OH)_2$**, when the water is made sufficiently alkaline, i.e., when the OH^- ion content is increased.

$$Mg^{2+} + 2\,OH^- \longrightarrow Mg(OH)_2$$

After removal by filtration of the solid $CaCO_3$ and $Mg(OH)_2$, the pH of the water is readjusted to near-neutrality by bubbling carbon dioxide into it.

PROBLEM 11-1

Ironically, calcium ion is often removed from water by adding hydroxide ion in the form of $Ca(OH)_2$. Deduce a balanced chemical equation for the reaction of calcium hydroxide with dissolved calcium bicarbonate, $Ca(HCO_3)_2$, to produce insoluble calcium carbonate. What molar ratio of $Ca(OH)_2$ to dissolved calcium should be added to ensure that almost all the calcium is precipitated?   ●

11.3 Disinfection to Prevent Illness

In terms of causing immediate sickness and even death, biological contaminants of water are almost always much more important than chemical ones. For that reason, we begin our discussion of the purification of water by extensively discussing its **disinfection, i.e., the elimination of microorganisms that can cause illness.**

Many of the microorganisms in raw water are present as a result of contamination by human and animal feces. The microorganisms are principally

• **bacteria,** including those of the *Salmonella* genus, one species of which causes typhoid. In this category is also *E. coli O157:H7*, whose transmission in water has caused a number of deaths in recent years, including those from an outbreak in Walkerton, Ontario, in 2000;

• **viruses,** including polio viruses, the hepatitis-A virus, and the Norwalk virus; and

• **protozoans** (single-celled animals), including *Cryptosporidium* and *Giardia lamblia*.

Because many microorganisms of these three types are pathogenic, causing mild to serious and sometimes fatal illnesses, they must be largely removed from water before it is suitable for drinking.

Notwithstanding well-known techniques for water disinfection, many of which have been used extensively for more than a century in developed countries, there are still more than one billion people in the world who do not yet have access to safe drinking water. According to the World Health Organization, about 4500 children die *daily* from the consequences of polluted water and inadequate sanitation.

11.4 Filtering of Water

In addition to dissolved chemicals, **the raw water that is obtained from rivers, lakes, or streams contains a multitude of tiny particles,** some of which consist of or contain microorganisms. Many of the small suspended particles are composed of clay, and resulted from the erosion of soil and rock, whether by natural forces or due to plowing of land for agriculture, mining, or commercial or housing development. The suspended particles increase water's turbidity, and thereby reduce the ability of light to penetrate deeply enough to support photosynthesis.

The larger of the particles suspended in water are often removed from the water by simply filtering it. Indeed, the filtration of water by passing it through a bed of sand is the oldest form of water purification known, dating back to ancient times. The sand retains suspended solids of all types, including microorganisms, down to about 10 μm in size.

11.5 Removal of Colloidal Particles by Precipitation

Most municipalities allow raw water to settle, since this permits large particles to settle out or to be readily separated. However, much of the insoluble matter—which originates from rocks and soil, and from the disintegration and decomposition of water-based plants and animals—will not precipitate spontaneously since it is suspended in water in the form of **colloidal particles.** These are **particles that have diameters ranging from 0.001 to 1 μm, and consist of groups of molecules or ions that are weakly bound together.** These groups dissolve as a unit, rather than breaking up and dissolving as individual ions or molecules. In many cases the individual units within a colloidal particle are spatially organized such that the surface of the particles contains ionic groups. The ionic charges on the surface of one particle repel those of like charge on neighboring particles, preventing their aggregation and subsequent precipitation.

Colloidal particles must be removed from drinking water for both aesthetic and health-safety reasons. To capture the colloidal particles, a small amount of either **iron(III) sulfate,** $Fe_2(SO_4)_3$, or **aluminum sulfate,** $Al_2(SO_4)_3$ ("alum"), is deliberately dissolved in the water. By subsequently making the water neutral or alkaline in pH (7 and up), both the Fe^{3+} and Al^{3+} ions produced from the salts form gelatinous hydroxides that physically incorporate the colloidal particles and form a removable precipitate. The

$M_2(SO_4)_3 \rightarrow 2\ M^{3+} + 3\ SO_4^{2-}$
for both M = Al and Fe

water is greatly clarified once this precipitate has been removed. Commonly, after the removal of the colloidal particles, the water is filtered through sand and/or some other granular material (see Figure 11-1).

Although the idealized formulas of the precipitates are $Fe(OH)_3$ and $Al(OH)_3$, the actual situation is much more complex. For example, aluminum actually forms a polymeric cation $Al_{13}O_4(OH)_{24}^{7+}$, which forms a loose network structure that is held together by hydrogen bonds. This network entraps the colloidal particles and forms the precipitate. Only if the pH rises to quite a high value does the aluminum in solution form the expected hydroxide $Al(OH)_3$. Since the concentration of aluminum sulfate added to the water is only about 10 μmol L^{-1}, very little residual aluminum ion is left in the treated water.

PROBLEM 11-2

Calculate the approximate number of atoms contained in colloidal particles of (a) 1-μm and (b) 0.01-μm diameters, assuming that their densities are similar to that of water and that the atomic mass of the atoms averages 10 g mol^{-1}. [*Hint:* Find the mass of the particles and the mass of each atom.] ●

11.6 Disinfection of Water by Membrane Technology

Water can be purified of most contaminant ions, molecules, and small particles including viruses and bacteria by passing or forcing it through a membrane in which the individual holes, called pores, are of uniform and microscopic size. The range of sizes for the various contaminants in raw water are summarized in Figure 11-2. Clearly, **for a technique to be effective in providing a barrier, the pore size of the membrane must be smaller than the contaminant size.** Generally speaking, before water is treated using membranes with even small pores (see below), it must be *pretreated* to remove the larger particles—especially colloids—which would otherwise clog and foul the finer membrane by leaving deposits, and since it is much less expensive to remove large particles by simple techniques than energy-intensive ones.

In the processes of **microfiltration** and **ultrafiltration,** a membrane or some other analogous barrier containing pores of 0.002- to 10-μm diameter (2–10,000 nm) is employed to remove constituents larger than these values from water. The water can be forced through the barrier by pressure or can be drawn through it by suction, leaving behind the larger impurities. In one modern version of this technology, the barrier is composed of thousands of strands of plastic tubing having walls that are pierced with thousands of tiny pores of similar size.

Some bacteria and colloid particles are as small as 0.1 μm and so can pass through conventional filters and even some microfilters (Figure 11-2). Viruses can be as small as 0.01 μm, and therefore require at least the ultrafiltration level to eliminate them all. However, filtration using membranes

Some authors use 0.05 and 0.5 μm as the upper limits to pore sizes for ultrafiltration and microfiltration, respectively.

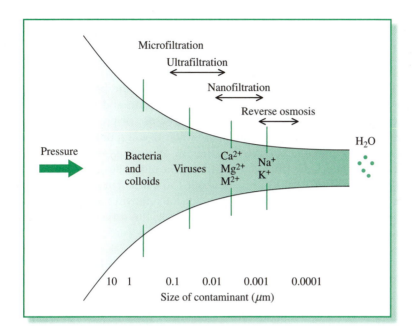

FIGURE 11-2 Filtration of contaminants by various methods.

can be used to disinfect water if a sufficiently small pore size is used and if the water is later irradiated with ultraviolet light to eliminate any microbes that have passed through the filtration stage.

Neither microfiltering nor ultrafiltering removes dissolved ions or small organic molecules. Membrane systems have been developed recently that purify water of virtually *all* contaminants by **nanofiltration.** Water is pumped under pressure through fine membranes that have pores only about 1 nm wide, which therefore remove not only bacteria and viruses, but also any larger organic molecules that would nourish the regrowth of bacteria. These **nanofilters** still allow water molecules to pass through the filter, since the molecules are only a few tenths of a nanometer in size. Unlike ultrafiltration, nanofiltration can be used to soften water, since hydrated divalent ions such as Ca^{2+} and Mg^{2+} are larger than the pores and so do not pass through. Hydrated monovalent ions such as sodium and chloride pass through some nanofilters, but not through ones with subnanometer pore sizes. As a consequence, some nanofilter membrane systems can be used to desalinate seawater and to help purify wastewater, as discussed later in this chapter.

11.7 Reverse Osmosis

The ultimate in membrane filtration occurs in the widely used technique called **reverse osmosis,** sometimes called *hyperfiltration.* Here, **water is forced under high pressure to pass through a semipermeable membrane,** composed of an organic polymeric material such as cellulose acetate or triacetate or a polyamide. **Since only water (and other molecules of its small size) can pass efficiently through the pores, the liquid on the other side of the membrane is purified water (Figure 11-2).**

The solution on the impact side of the membrane becomes more and more concentrated in contaminants as time goes on and is discarded. The procedure is called *reverse* osmosis because, by use of pressure, the natural phenomenon of osmosis—by which pure water would spontaneously migrate through the membrane *into* solution, thereby diluting it—is reversed.

Particles, molecules (including small organic molecules), and ions down to less than 1 nm (0.001 μm) in size, or about 150 g mol^{-1} in mass, are removed by reverse osmosis. It is particularly useful for removing alkali and alkaline earth metal ions, as well as salts of heavy metals. Thus it is employed in hospitals and renal dialysis units to produce water that is particularly free of ions. Reverse osmosis is used on a large scale for the *desalination*, i.e., the removal of salts, of seawater and brackish water, a topic considered in Box 11-2.

Water destined for drinking purposes is commonly pretreated, e.g., by filtering it through sand and gravel, and passing it over activated carbon, to remove the larger particles such as bacteria, and chlorine before subjecting it to reverse osmosis, in order to minimize fouling and degradation of the membrane.

Because of the high pressures needed to force water through the small pores in the membrane, reverse osmosis is an energy-intensive process. A pressure of about 2 atm is sufficient for portable and domestic units, but greater pressures must be used for brackish or salty water.

Some domestic consumers of drinking water have installed small under-the-sink reverse osmosis units to further purify their water by removing unwanted contaminants such as heavy metal cations (such as lead), hard water cations (calcium and magnesium), anions (such as nitrate and fluoride), and organic molecules from water obtained from domestic supplies. Small reverse osmosis units are also used in medical facilities for producing water that is particularly ion-free. However, **reverse osmosis tends to be wasteful in water, since so much of it—a third to a half—is discarded, even in large commercial installations.**

Some bottled water is purified and deionized by reverse osmosis, but small amounts of salts are reintroduced into it before it is sold to consumers. Drinking large amounts of deionized water is not healthy, since the ion balance in the body can be upset as a consequence.

11.8 Disinfection by Ultraviolet Irradiation

Ultraviolet light can also be used to disinfect and purify water. Powerful lamps containing mercury vapor whose excited atoms emit UV-C light (see Chapter 1) centered at a wavelength of 254 nm are immersed in the

BOX 11-2   The Desalination of Salty Water

Desalination is the production of fresh water from salty water, often seawater, by the removal of its dissolved ions. There are more than 15,000 large-scale desalination plants in operation, in more than 125 countries, providing about 1% of the world's drinking water. Desalinated water is also used in some locales to irrigate high-value crops, such as greenhouse vegetables and flowers. However, all large-scale facilities of desalination are both cost-intensive to construct and energy-intensive to operate.

Reverse osmosis (RO) is the most widely used method of desalination. As of 2010, the largest RO desalination plants in the world were located in Israel and produced drinking water from seawater for about $ 0.70 per cubic meter. State-of-the-art reverse osmosis systems produce a liter of fresh water from salt water using about 0.002 kWh of electricity, less than half the amount required in the 1980s. The reduction in cost has been due largely to efficient energy recovery. Large, new RO desalination plants have been built in several coastal cities in Australia, where water demand is rising but rainfall has declined. The desalination unit in Perth is powered in part by wind energy; those in Sydney and Queensland will be powered entirely from renewable energy sources.

Membrane fouling is a chronic problem with RO desalination systems and prevents its cost from falling even more. Producing membranes with more uniform pore-hole sizes will help increase the water flux and further decrease the cost. Microfiltering the seawater to remove particles that escape conventional pretreatment adds to the expense but allows the RO membrane to operate more efficiently. Low-grade waste heat, where available, can also help increase the flux of water through the membrane. Water produced by reverse osmosis and intended for irrigation suffers from lack of nutrients such as calcium, magnesium, and sulfate, and still contains boron at levels toxic to most crops (though not a hazard to human health).

The other main commercial desalination process is the thermal distillation—evaporation—of seawater or brackish water. Desalination of seawater by evaporation is a technique that goes back to ancient times, and is especially suited even today for seawater sources that contain particularly high levels of dissolved salts and suspended solids, such as those in the Persian Gulf. Indeed, the world's largest desalination plant is located in the United Arab Emirates and uses the thermal technique.

The evaporation method is even more energy-intensive than is reverse osmosis. Modern, large-scale thermal distillation plants use energy to raise salty water to the boiling point, then reduce the air pressure above the liquid to create a partial vacuum into which the liquid readily "flash" evaporates, leaving the salt behind in the remaining liquid. The vapor is removed and condensed as desalted water. Thermal distillation plants are often incorporated within electrical generating plants to use the low-grade waste steam from the latter as their energy source. Corrosion from the hot seawater is a chronic problem in thermal distillation plants.

A related but novel technology developed in India that can be employed in hot climates is **low-temperature thermal desalination.** Warm water (at about 26–30°C) from the

(continued on p. 464)

ocean surface is subjected to low pressure in a vacuum chamber, vaporizing much of it. The vapor is then channeled to a second chamber, where it is cooled and condensed to a salt-free liquid by using cold water (about 13°C), drawn from hundreds of meters deep in the sea, in condensers. Creation of the vacuum and pumping up the deep water are energy-intensive processes, so to date the installations using this technology have consumed more energy than does reverse osmosis per liter of drinking water.

A desalination technology still at the prototype stage is **forward osmosis.** As in reverse osmosis, pure water flows across a membrane. However, the receiving pool here initially is a relatively concentrated aqueous solution of **ammonium bicarbonate, NH_4HCO_3.** Because this solution has a greater concentration of ions (here NH_4^+ and HCO_3^-) than seawater, the direction of spontaneous water flow is *from* seawater, not into it. Once the ammonium bicarbonate solution has been sufficiently diluted, it can be cycled away and moderately heated (to about 40–60° C), which forces the salt to decompose to ammonia and carbon dioxide, which then bubble out of solution, leaving pure water:

$$NH_4HCO_3(aq) \longrightarrow NH_3(g) + CO_2(g) + H_2O(aq)$$

The captured gases can be bubbled into water again to recreate the ions and provide a new solution for the process.

To be feasible, this technology requires the development of suitable thin, porous membranes that can withstand the alkaline environment produced by the bicarbonate. Ideally, the moderate heat required would be supplied by waste heat from a power plant. In addition to requiring less energy than reverse

osmosis, this process would not produce concentrated brines.

Desalination of water is also sometimes accomplished using the technique of electrodialysis, which is described later in this chapter. Techniques based on filtering through carbon nanotubes or proteins are in the development phase.

A low-energy technique researched recently involves mixing seawater with a fatty acid solvent such as n-decanoic acid, $CH_3(CH_2)_8COOH$, at 40°C or higher. The salt does not dissolve in the solvent but remains in the residual brine; the solvent incorporates a few percent of water in a homogenous solution. The decanted solvent–water solution is then cooled below 34°C, at which point much of the pure water separates, and the much larger mass of solvent is reused. The residual solubility of the decanoic acid in the separated water is very small, amounting to a few dozen parts per million.

One major problem with desalination, especially of brackish inland waters, lies is the disposal of the accumulated discharges of brine—sometimes called *concentrate*—that is produced. In some locations, the brine is injected into an underground saltwater aquifer. Combining the brine output with spent cooling water from a power plant or with treated wastewater can serve to lower the salt concentration before its disposal in the ocean. Alternatively, the output can be dispersed over a large ocean area by having many small openings in the disposal pipes. The brine could cause cumulative environmental problems, such as harming fish populations, in the immediate area of the seacoast into which it is deposited if it is not first treated. In some installations, brine is left to evaporate in large outdoor pools, and the resulting salts later disposed of.

water flow. About ten seconds of irradiation are usually sufficient to eliminate the toxic microorganisms, including *Cryptosporidium*. **The germicidal action of the light disrupts the DNA in microorganisms, preventing their subsequent replication and thereby inactivating the cells.** At the molecular level, absorption of UV-C light results in the formation of new covalent bonds between nearby thymine units on the same strand of DNA. If sufficient thymine dimers are formed, the DNA molecule becomes so distorted that subsequent replication of the organism is prevented.

The use of ultraviolet light to purify water is complicated by the presence of dissolved iron and humic substances, both of which absorb the UV light and thus reduce the amount available for disinfection. Small solid particles suspended in the water also inhibit the action of the UV light since they can shade or absorb bacteria and also scatter or absorb the light. An advantage of UV disinfection technology is that small units can be employed to serve small population bases, whether in the developed or developing world, so the continuous monitoring activity of chemical systems is avoided. As discussed later, UV light can also be used to purify water of dissolved organic compounds, but via a different mechanism.

11.9 Disinfection by Chemical Methods: Ozone and Chlorine Dioxide

To rid drinking water of harmful bacteria and viruses, especially those arising from fecal matter from both humans and animals, by use of a chemical agent **requires an oxidizing agent more powerful than O_2.** In some localities, particularly in France and other parts of western Europe but also in some North American cities—Montreal and Los Angeles are examples—**ozone, O_3,** is used for this purpose.

Since O_3 cannot be stored or shipped because of its very short lifetime, it must be generated on-site by a relatively expensive process involving electrical discharge (20,000 V) in dry air. The resulting ozone-laden air is bubbled through the raw water; about 10 minutes of contact is usually sufficient for disinfection. Since the lifetime of ozone molecules is short, there is no residual protection in the purified water to ensure it will not be subject to future contamination. Some pollutants in water react with the ozone itself and others with free radicals such as hydroxyl and hydroperoxy (Chapter 3) that are produced when ozone reacts with water.

Unfortunately, **the reaction of ozone with bromine in water leads to the formation of oxygen-containing organic compounds,** particularly those containing the *carbonyl group*, C=O, such as *formaldehyde* and other low-molecular-weight aldehydes and various other compounds, some of which are toxic. In addition, **ozone reacts with bromide ion, Br⁻, present in the water to produce the bromate ion,** BrO_3^-, a carcinogen in test animals, which is also probably carcinogenic in humans. The reaction of ozone with bromide, a natural constituent of water that often is present at ppm concentrations, occurs in several steps; the overall reaction is

$$Br^- + 3\,O_3 \longrightarrow BrO_3^- + 3\,O_2$$

The bromate ion produced by ozonation may subsequently react with organic matter in the water to generate toxic organobromine compounds, although experiments have shown that the only brominated product under water treatment conditions is *dibromoacetonitrile*, $CHBr_2CN$, produced by the reaction of bromate ion with acetonitrile. Substances such as bromate ion that are generated during water purification are called **disinfection by-products, or DBPs. All chemical methods of disinfecting water produce DBPs;** more than 600 of them are now known.

The concentrations of toxic substances in drinking-water supplies are monitored and controlled, at least in developed countries. **In the United States, the limits are called maximum contaminant levels (MCL) and are the maximum permissible concentrations of substances dissolved in the water of any public system, based on an annual average.** Thus, an MCL of bromate ion in drinking water has been set at 10 ppb (0.010 ppm) by the U.S. EPA.

Similarly, **chlorine dioxide** gas, ClO_2, is used in more than 300 North American and in several thousand European communities to disinfect water. The ClO_2 molecules, themselves free radicals, operate to oxidize organic molecules by extracting electrons from them:

$$ClO_2 + 4\,H^+ + 5\,e^- \longrightarrow Cl^- + 2\,H_2O$$

The organic cations created in the accompanying oxidation half-reaction subsequently react further and eventually become more fully oxidized:

$$\text{organic cation} + O_2 \longrightarrow \longrightarrow CO_2 + H_2O$$

Since chlorine dioxide (unlike molecular chlorine) is *not* a chlorinating agent—it does not generally introduce chlorine atoms into the substances with which it reacts—and since it oxidizes the dissolved organic matter, much smaller amounts of toxic organic chemical by-products are formed than if molecular chlorine is used (see below).

As is the case with ozone, ClO_2 cannot be stored since it is explosive in the high concentrations that its practical use calls for, and so it must be generated on-site. This is accomplished by oxidizing its reduced form, the **chlorite ion,** ClO_2^-, from the salt **sodium chlorite,** $NaClO_2$:

$$ClO_2^- \longrightarrow ClO_2 + e^-$$

Some of the chlorine dioxide in these processes is converted to **chlorate ions,** ClO_3^-. The presence of chlorite and chlorate ions as residuals in the final water has raised health concerns due to their potential toxicity. The U.S. EPA has set an MCL of 1.0 ppm for chlorite ion, and a MRDL (*maximum residual disinfectant level*) of 0.8 ppm for chlorine dioxide, in drinking water.

11.10 Disinfection by Chlorination: Background

The most common water purification agent used in North America is hypochlorous acid, HOCl. This neutral, covalent compound kills microorganisms, as it readily passes through their cell membranes. Its action at the molecular level is to deactivate essential enzymes by oxidizing some of their side chains, particularly those involving sulfur:

$$R\text{—}S\text{—}H \xrightarrow{\text{HOCl}} \longrightarrow R\text{—}SO_3H$$

In addition to being effective, disinfection by **chlorination** is relatively inexpensive. **Incorporating a small excess of the chemical in the treated water provides it with residual disinfection power during subsequent storage and transmission to the consumer.** About half the U.S. population uses surface water, and one-quarter of the population uses groundwater that is disinfected by HOCl. Chlorination is more common than ozonation in North America because generally the raw water is less polluted. Chlorination of public water supplies in the United States, Canada, and Great Britain began in the early years of the twentieth century. For the previous 50 years, chlorination had been practiced on an emergency basis during epidemics.

11.11 Disinfection by Chlorination: Production of Hypochlorous Acid

Like ozone, HOCl is not stable in concentrated form and so cannot be stored. For large-scale installations, e.g., municipal water treatment plants, it is generated by dissolving **molecular chlorine** gas, Cl_2, in water. At moderate pH values and low concentrations, the equilibrium in the reaction of chlorine with water lies far to the right and is achieved in a few seconds:

$$Cl_2(g) + H_2O(aq) \rightleftharpoons HOCl(aq) + H^+ + Cl^-$$

Thus **a very dilute aqueous solution of chlorine in water contains very little aqueous Cl_2 itself.** If the pH of the reaction water were allowed to become too high, the result would be the ionization of the weak acid HOCl to the **hypochlorite ion,** OCl^-, which is less able to penetrate bacteria on account of its electrical charge. Once chlorination is complete, the pH of the water is adjusted upward, if necessary, by the addition of lime.

In small-scale applications of chlorination, as in swimming pools, the handling of cylinders of Cl_2 is inconvenient and dangerous. The chlorine can be produced as needed on the spot by the electrolysis of chloride ion in salt water swimming pools. More commonly, hypochlorous acid instead is generated from the salt **calcium hypochlorite,** $Ca(OCl)_2$, or is supplied as an aqueous solution of **sodium hypochlorite,** $NaOCl$. In water, an acid–base reaction occurs to convert most of the OCl^- in these substances to $HOCl$:

$NaOCl \longrightarrow Na^+ + OCl^-$

$$OCl^- + H_2O \rightleftharpoons HOCl + OH^-$$

Close control of the pH in an environment like a swimming pool is necessary to avoid the shift to the left side of the equilibrium for this reaction that occurs if a very alkaline condition is permitted to prevail. On the other hand, corrosion of pool construction materials—such as calcium carbonate in the cement—can occur in acidic water, so the pH usually is maintained above 7 to prevent such deterioration from occurring.

Sources of the ammonia include swimmers' sweat and urine. Urea can react directly with Cl_2 to produce NCl_3.

Maintenance of an alkaline pH also prevents the conversion of dissolved **ammonia,** NH_3, to the **mono-** and **dichloramines,** NH_2Cl, $NHCl_2$, and especially to **nitrogen trichloride,** NCl_3, which is a powerful eye irritant:

$$NH_3 + 3\,HOCl \longrightarrow NCl_3 + 3\,H_2O$$

Significant respiratory and eye irritation problems from exposure to chloramines in the air around indoor swimming pools have been reported when appropriate ventilation is not available.

The characteristic smell of swimming pools that chlorinate the water is due mainly to NH_2Cl.

If the level of chloramines becomes too high in pools, the operators "shock" the water by dramatically increasing the level of chlorine; this oxidizes the mono- and dichloramines to nitrogen trichloride, which in turn is oxidized by chlorine to gaseous nitrogen. The overall reaction from ammonia through to nitrogen is given by the equation

The -3 nitrogen is oxidized to the zero state in the reaction.

$$2\,NH_3 + 3\,Cl_2 \longrightarrow N_2 + 6\,HCl$$

It is desirable to adjust the equilibrium point in the $OCl^- \longrightarrow HOCl$ reaction so as to ensure a good supply of the superior disinfectant molecular species, $HOCl$. Since the equilibrium between $HOCl$ and OCl^- shifts rapidly in favor of the ion between pH values of 7 and 9, the acidity level must be meticulously controlled. A compromise pH of about 7.5 is usually chosen. At this level, half the $HOCl$ is ionized, but is available to be protonated later as molecular $HOCl$ gradually degasses from the water surface. Swimming pool acidity can be adjusted by the addition of acid (in the form of solid *sodium bisulfate,* $NaHSO_4$, which contains the acid HSO_4^-) or a base (sodium carbonate, Na_2CO_3). **Sodium bicarbonate,** $NaHCO_3$, is the buffer that maintains the pH in the appropriate range. Recall from Figure 10-6 that bicarbonate ion is the predominant carbonate-system ion at pH values just above 7.

$HSO_4^- \longrightarrow H^+ + SO_4^{2-}$

Chlorine must be constantly replenished in outdoor pools since UV-B and the short-wavelength components of UV-A in sunshine are absorbed by and decompose both hypochlorous acid and especially the hypochlorite ion, thereby affecting the equilibrium in the $OCl^- \longrightarrow HOCl$ process toward forming the chloride ion:

$$2\,ClO^- \xrightarrow{\text{UV}} 2\,Cl^- + O_2$$

Hypochlorous acid can also be generated in swimming pools by the reaction with water of the chlorine-containing compound **isocyanuric acid, $C_3N_3O_3H_3$**:

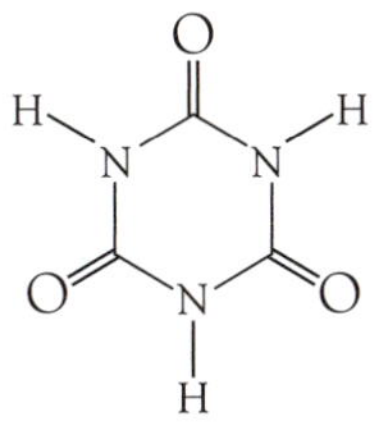

Either the trichloro derivative, in which each hydrogen is replaced by Cl to give $C_3N_3O_3Cl_3$, or the sodium dichloro derivative, $C_3N_3O_3Cl_2Na$, are used. In either case, the OH group from water combines with the chlorine to produce HOCl and the hydrogen of H_2O becomes bonded to the nitrogen, giving isocyanuric acid, $C_3N_3O_3H_3$:

$$C_3N_3O_3Cl_3 + 3\ H_2O \rightleftharpoons C_3N_3O_3H_3 + 3\ \textbf{HOCl}$$

Since this process is an equilibrium, not all the compound is immediately converted to hypochlorous acid. As HOCl is used up, both by its use as a disinfectant and through dissociation in sunlight of its ionic form, the equilibrium shifts to the right and more HOCl is produced (Le Chatelier's principle). None of the various forms of isocyanuric acid absorb UV light, so its chlorine is "protected" against decomposition by sunlight. Since the bulk chlorinated forms of isocyanuric acid are expensive, it is common to supply hypochlorite from a cheaper source and to add isocyanuric acid as a stabilizer, temporarily reversing the above reaction to "store" the chlorine until it is needed.

11.12 Disinfection by Chlorine: By-Products and Their Health Effects

An important drawback to the use of chlorination in disinfecting water is the concomitant production of chlorinated organic substances, some of which are toxic, since HOCl is not only an oxidizing agent but also a chlorinating agent. Examples of these important by-products are the group of *halogenated acetic acids* (haloacetic acids), such as $CH_2Cl—COOH$. The U.S. EPA restricts their collective annual average concentration to 60 ppb maximum; *haloacetonitriles*, such as $CH_2Cl—CN$; and *haloacetaldehydes* such as dichloroacetaldehyde, $CHCl_2—CHO$. **Dichloroacetic acid,** $CHCl_2—COOH$, is a more potent carcinogen than is chloroform. As would be expected, bromine-containing DBPs are more common when the raw water contains relatively high levels of bromide ion.

If the water to be disinfected contains **phenol,** C_6H_5OH, or a derivative thereof, chlorine readily substitutes for the hydrogen atoms on the ring to

Some chlorination of organic matter occurs by its reaction with chlorine monoxide, Cl_2O, a reactive species present at very low concentration in chlorinated water. See Additional Problem 5.

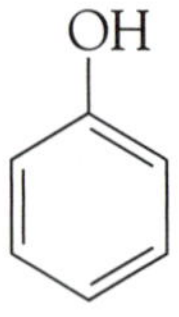

phenol

give rise to chlorinated phenols: these compounds have an offensive odor and taste and are toxic. Some communities switch from chlorine to chlorine dioxide for disinfection when their supply of raw water is temporarily contaminated with phenols to avoid the formation of chlorinated phenols.

A more general problem of the **unintended consequences** of the chlorination of water lies in the production of **trihalomethanes,** THMs. Their general formula is CHX_3, where the three X atoms can be chlorine or bromine or a combination of the two. **The THM of principal concern is chloroform, $CHCl_3$, which is produced when hypochlorous acid reacts with organic matter dissolved in the water** (see Box 11-3).

$$\text{organic matter} + HOCl \longrightarrow \longrightarrow CHCl_3 \text{ (overall reaction)}$$

Chloroform is a suspected liver carcinogen in humans, and it may also give rise to negative reproductive and developmental effects. Its presence, even at very low levels of approximately 30 ppb, raises the specter that chlorinated drinking water may pose a health hazard, though one that pales by comparison with the benefits that it confers in the elimination of fatal

---

BOX 11-3   The Mechanism of Chloroform Production in Drinking Water

Humic acids, with which HOCl reacts to form chloroform, are water-soluble, non-biodegradable components of decayed plant matter. Of particular importance are humic acids that contain 1,3-dihydroxybenzene rings. The carbon atom (#2) located between those carrying the —OH groups is readily chlorinated by HOCl, as in this elementary case:

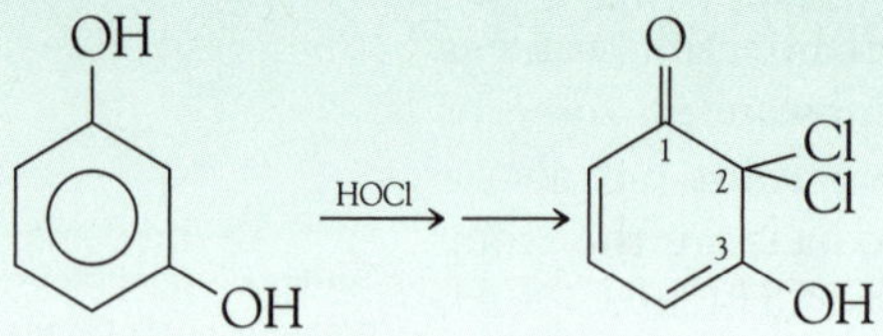

Subsequently the ring cleaves between C-2 and C-3 to yield a chain:

In the presence of the HOCl, the terminal carbon becomes trichlorinated, and the —CCl_3 group is readily displaced by the OH⁻ in water to yield chloroform:

Analogous sequences of reactions produce bromoform, $CHBr_3$, and mixed chlorine–bromine trihalomethanes from the action on humic materials of hypobromous acid, HOBr, which is formed when bromide ion in water displaces chlorine from HOCl:

$$HOCl + Br^- \rightleftharpoons HOBr + Cl^-$$

waterborne diseases. The annual average limit of total THMs in drinking water in the United States has been reduced to 80 ppb. The previous limit of 100 ppb is still used in Canada and the European Union. In fact, these 80–100-ppb limits are set not only to regulate the THM chemicals themselves but also as an indicator that the production of *other* chlorinated organic DBPs (see below) is not excessive.

The U.S. EPA has set a **maximum contaminant level *goal*, MCLG,** of 70 ppb for THMs in drinking water. An MCLG is the **maximum level at which the contaminant is believed to be safe,** allowing for adequate margins of safety, but unlike the MCL, it is not an enforceable standard. A nonzero goal is considered by some scientists and policymakers to be appropriate to substances, like chloroform, which are believed to operate indirectly as carcinogens. They do not damage DNA directly, but cause tissue damage that leads to rapid cell proliferation, which in turn increases the likelihood that cancer will form in the damaged tissue. **A threshold below which no effects are likely to be observed is expected for carcinogens that operate in this indirect manner.**

The level of trihalomethanes formed in water depends sharply on the organic content of the raw water, since they are formed from the reaction of the organics with HOCl (see Box 11-3). THM levels can reach 250 ppb in areas of Scotland and Northern Ireland that have peat moorlands. Water exposed to bogs in Newfoundland, Canada, has generated THM levels in excess of 400 ppb. As of the early 1990s, about 1% of the larger U.S. drinking-water utilities that used surface waters, but none that used groundwater, had average THM levels exceeding 100 ppb. The THM content of chlorinated water could be decreased by using activated carbon either to remove dissolved organic compounds before the water is chlorinated or to remove THMs and other chlorinated organics after the process, although THMs are not very efficiently adsorbed by the carbon and it is an expensive process.

An analysis of epidemiological studies relating the chlorination of water to cancer rates in various communities in the United States led to the conclusion that the risk of bladder cancer in humans increased by 21%, and that of rectal cancer by 38%, for Americans who drank chlorinated surface water in the past. A similar study in Ontario found even higher bladder cancer risk factors for people who drank water for 35 years or more that had THM levels greater than 50 ppb, and for colon cancer when the concentration exceeded 75 ppb, but found no correlations of THMs with rectal cancer rates. A recent study found no increased risk for pancreatic cancer from lifetime exposure to the by-products of chlorinated water.

Given that slightly more than half the population of the United States drinks surface water, one effect of chlorination would be to increase bladder cancer incidence by about 4,200 cases and rectal cancer incidence by about 6,500 cases annually. Because of these risks, some communities are

Canada has set a separate limit of 16 ppb for the THM $CHBrCl_2$.

considering a switch, or have already switched, to water disinfection by ozone or chlorine dioxide, since these agents produce little or no chloroform. The extent of chlorination has already been reduced in most American communities relative to the levels that led to these statistics. An interesting recent study from Spain found that the risk of bladder cancer from THMs is reduced for persons with high fluid intake, perhaps because the carcinogens are flushed more quickly through the body.

Recently public health officials have expressed concern about the possible link between THMs and adverse human reproductive outcomes, including first-term miscarriages, stillbirths, impaired fetal growth, and certain birth defects. Even though the existing research in this area is not yet definitive, some officials suggest that women drink bottled water rather than chlorinated tap water during their first three months of pregnancy.

Several other mutagenic chlorinated organic DBPs formed during chlorination have been detected in water, in addition to chloroform. It is not clear whether the main carcinogens in the chlorinated drinking water are the THMs themselves or some nonvolatile, higher-molecular-weight mutagenic by-product present at still lower concentrations but whose concentration would presumably be proportional to THM. The same risks do not usually apply to chlorinated well water, since its organochlorine content is much less (only 0.8 ppb on average, versus 51 ppb for surface water) because it contains much smaller amounts of organic matter that could become chlorinated. Brands of carbonated water that use municipal drinking water as their source also contain dissolved chloroform.

Exposure to chloroform by dermal contact, and inhalation of the gases desorbed from the hot water during showers, baths, and hand-washing of dishes contribute more to one's intake of THMs, as measured by the blood levels of these compounds, than does drinking the water itself. Swimming in pools in which the water is chlorinated for disinfection also contributes significantly to THM exposure. Various organic materials such as lotion, mucus, etc. can raise the THM levels in swimming pool water beyond that formed in the water chlorination process itself by their reaction with $HOCl$ and Cl_2.

A recent study in Spain found that chloroform levels in swimmers' urine tripled over the course of a 1-hour swim, and took several hours to recover. Chloroacetic acids were also found in their urine for a few hours after swimming, the major route of exposure being the inadvertent ingestion of the pool water. Chlorine-based chemicals in pool water or in the air above the pool surface may increase the prevalence of asthma among frequent swimmers.

The use of the disinfectant *triclosan*, a phenol-based compound present in some hand soaps, in chlorinated water can produce additional chloroform and chlorinated phenols, to which the user is then exposed. Indeed, the reactions of sodium hypochlorite with organic cleaners or air fresheners that some bleach products contain generates not only chloroform, but also carbon tetrachloride.

11.13 Disinfection by Chlorine: Advantages over Other Methods

Notwithstanding the preceding discussion of chlorination by-products, it is important to point out that the disinfection of water is extremely important in protecting public health, and saves many more lives—by a very wide margin—than are affected negatively. For example, both typhoid and cholera were widespread in both Europe and North America a century ago but have been almost completely eradicated in the developed world, thanks to chlorination and the other disinfection methods for drinking water and to improved sanitation in general. The same is not true in many developing countries; e.g., there were more than half a million cases of cholera in Peru in the early 1990s. A cholera outbreak a year after the 2009 earthquake in Haiti killed thousands. Overall, about 20 million people, most of them infants, die from waterborne diseases annually worldwide in underdeveloped countries, where water purification is often erratic or even nonexistent. Under no circumstances should effective disinfection of water be abandoned because of concern for the by-products of chlorination!

An advantage chlorination has over disinfection by chlorine dioxide or ozone or UV is that some chlorine remains dissolved in water once it has left the purification plant, so that the water is protected from subsequent bacterial contamination before it is consumed. Indeed, some chlorine is usually added to water purified by the other methods to provide this protection. There is very little danger of significant chloroform production in the purified water since its organic content has been virtually eliminated before the chlorine is introduced. If the chlorine level in water purified by chlorination is too high, it can be lowered by the addition of sulfur dioxide, which reduces it to chloride ion.

The residual chlorine in water often exists in the form of the chloramines NH_2Cl, $NHCl_2$, and NCl_3, which are produced from reaction with dissolved ammonia gas. Although not as fast as HOCl in disinfecting water, the mono- and dichloroamines especially are good disinfectants. The mixture of chloramines, called **combined chlorine,** is longer-lived in water than is hypochlorous acid and thus provides longer residual protection.

Indeed, ammonia is often added to purified drinking water in order to convert the residual chlorine to the combined form (Figure 11-1). Chloramines are increasingly used, rather than chlorine or ozone or chlorine dioxide, as the main disinfectant in the purification of drinking water. They have the advantage over chlorine of producing little (though not zero) amounts of THMs and haloacetic acids, although they react with some nitrogen-containing organic compounds to produce toxic nitrosamines (compounds discussed later in the chapter). Overall, however, chloramines are less reactive than the mixture of compounds in chlorination. The EPA has set MRDLs of 4.0 ppm for both chlorine and chloramine in drinking water.

Bromine rather than chlorine is sometimes used as the disinfectant in swimming pools. The main disinfecting agent in bromination is **hypobromous acid,** HOBr, in analogy with the role of hypochlorous acid in chlorination. HOBr reacts more rapidly with dissolved ammonia than does HOCl, producing mainly NH_2Br, which is also a good disinfectant. Ozone is used for disinfection, replacing most or all of the chlorination required, in some swimming pools. Silver is used to kill bacteria, and copper to control algae, in some ozone-based pool disinfection systems.

PROBLEM 11-3

Assuming that the nitrogen atom in monochloramine, NH_2Cl, has an oxidation number of -3, calculate that of the chlorine. Using the principle that unlike charges attract, predict whether it will be the hydrogen ion or the hydroxide ion from dissociated water molecules that will extract the Cl from NH_2Cl; from your result, predict the products of the decomposition reaction of chloramine in water. ●

A drinking-water quality issue of current concern involves the pathogenic protozoa called *Cryptosporidium*, which was responsible for the death of 100 people and for about 400,000 cases of watery diarrhea in Milwaukee in 1993. Less serious *Cryptosporidium* outbreaks occurred in Oxford, England, in 1989, and in Saskatchewan, Canada, in 2001. This deadly parasite is resistant to standard methods of disinfection such as chlorination at normal levels, and is so small (about 5-μm diameter) that it easily passes through the standard filters used to separate sediments. Several possible solutions have been advanced, including ozonation or the use of ultrafiltration or UV irradiation or the application of monochloramine following chlorination. A longer-than-usual exposure of water containing *Cryptosporidium* is necessary with ozonation, since the activation energy for inactivation of protozoa by ozone is about twice as large as that for bacteria (80 vs. about 40 kJ mol^{-1}).

Another protozoa, *Giardia lamblia*, also causes many instances of waterborne disease. Like *Cryptosporidium*, it is somewhat resistant to chlorination, but since it is larger (about 10-μm diameter), it is more easily removed by filtration through sand.

11.14 Point-of-Use Water Disinfection

In some instances, it is more practical or safer to disinfect water meant for drinking and cooking on-site domestically rather than at a central facility.

For example, to disinfect water for drinking purposes, hikers either boil raw water or treat it chemically with either chlorine, in the form of bleach solution or *calcium hypochlorite* crystals, which provide HOCl, or with iodine, as **elemental I_2** (supplied as a crystal or dilute solution) or **hypoiodous acid,** HIO, itself. Concerns have been expressed about chronic health problems such as thyroid dysfunction associated with long-term usage of iodine, however. Treating the water with elemental iodine tends to make it unpalatable as well. In emergency situations such as natural disasters, when traditional infrastructure has failed, pouring water through porous silver-impregnated paper is effective in killing much of its bacteria content. The silver is present in nanoparticle form, deposited on the paper by reducing silver nitrate.

The antibacterial activity of silver has been known since ancient times.

In the rural areas especially of many developing countries, the infrastructure for centrally supplied drinking water does not exist or does not reliably produce disinfected water, and many people instead use local surface water and groundwater. Because such sources often are contaminated by fecal matter, diarrheal disease especially among young children is widespread, resulting in the death of 1.6 million people yearly from the disease. To overcome such problems, several point-of-use water disinfection technologies have been developed and tested in an attempt to develop sustainable methods appropriate to these locations.

The characteristic requirements for any such technology to be successful are:

• the ability to consistently provide adequate quantities of disinfected water for daily household needs;

• effectiveness in treating different qualities of raw water, including those that are turbid and have high organic content;

• requirement for only a short treatment time by the user to produce a sufficient quantity of water;

• low cost;

• existence of a reliable, low-cost supply chain for materials or replacement parts;

• continued implementation following the introduction and education phase; and

• ease of use.

Chlorination of water is achieved by using concentrated liquid hypochorite bleach or sodium or calcium hypochlorite tablets. It is easy to use and its cost is low, but unfortunately the treatment does not clear the water of suspended particles, which may also react with the chlorine to produce objectionable odor and taste and reduce the concentration of hypochlorite available to inactivate viruses. The esthetic problems may help explain why many people do not continue to use the technology.

Sodium isocyanurate tablets may be a better alternative to HOCl solutions since they have a longer shelf life.

Chlorination plus coagulation adds a dry coagulant to a solid chlorine source, resulting in aesthetically improving the water and improving disinfection effectiveness. Issues with product availability and cost are found to limit its sustained usage.

SODIS (*solar disinfection*) works in two ways: by exposing transparent plastic bottles of raw water to sunlight, it allows destruction of pathogens by their absorption of UV-A and by absorption of the heat reflected back into the water from aluminum plates (see picture below). Some UV-A may also be absorbed by chemical contaminants, leading to the production of reactive free-radical species that can oxidize the pathogens. Although very low cost, many bottles are required per household and many hours of exposure are required. Very turbid or colored water may not become fully disinfected, since UV may not be able to penetrate sufficiently. The use of PET plastic bottles unfortunately prevents transmission of UV-B—which would be more effective in killing pathogens than UV-A alone—into the water. Although now widely used, questions concerning the effectiveness of SODIS alone in combating contracting diarrhea have been raised.

SODIS Eawag

Ceramic filters are made from clay and are effective if the pore size is controlled to filter out microbes by size exclusion. Water is also clarified as a result of the filtration. Slow flow-through rates limit its popularity, as does need of replacement of parts if a local supply chain is not available. The modest start-up cost is an advantage. Locally produced ceramic filters made by firing a combination of clay and waste agricultural products such as rice husks can allow a suitable water flow while trapping large particles containing pathogens. Such filters are often coated with a solution of silver to improve disinfection of the water. The filters are found to be more effective in decreasing bacteria and protozoa concentrations than virus concentrations.

Biosand filters flow raw water vertically through a bed of fine sand (see Figure 11-3). Provided that some water is continuously maintained above the sand, a biological zone (of a few centimeters depth) consisting of a dense population of microorganisms develops in the top of the sand layer. These species consume (as food) the water-borne pathogens that become adsorbed onto and trapped by the sand particles. The sand and gravel remove larger particles, clarifying the water and improving its taste. Although the unit has a significant initial cost (unless supplied by an aid agency), continuing costs are negligible.

Boiling the water is the most widely used and traditional method of disinfection in developing countries, and is still used in parts of developed countries, especially when temporary contamination of a water supply has occurred. It is highly effective in killing pathogens, even if temperatures of only 66°C or higher are employed. Unfortunately, large amounts of wood or other fuel must be gathered or purchased to produce the fire required to raise the required amounts of water to such temperatures. In addition, recontamination of the water can occur as it cools.

The relative scores on the important sustainability criteria for the five modern point-of-use technologies are listed in Table 11-1. Biosand filtration is the superior technology, and once installed, it tends to be used more successfully over a longer period of time than are the alternatives. Other experts believe that ceramic filters are the best choice.

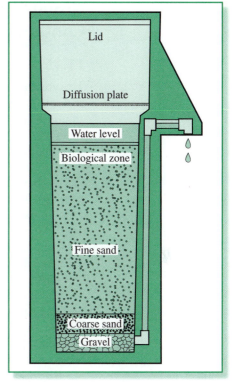

FIGURE 11-3 Biosand filter system. [Source: Redrawn from Wikimedia Commons, Free Software Foundation]

Review Questions 1–10 are based upon the material above.

TABLE 11-1	**Ratings of POU Technologies in Developing Countries**					
Technology	Water Quantity	Water Quality	Ease of Use	Cost	Supply Chain	Overall Score
Biosand filters	3	3	2	2	3	13
Ceramic filter	2	3	2	3	2	12
Chlorination	3	1	3	3 (liquid) 2 (tablet)	1	11 (liquid) 10 (tablet)
Solar + heat (SODIS)	1	1	1	3	3	9
Coagulation + chlorination	2	3	1	1	1	8

Note: Score range is 3 (highest) to 1 (lowest) for individual categories.

Source: Adapted from Table 3 of M. D. Sobsey et al., "Point of Use Household Drinking Water Filtration," *Environmental Science and Technology* 42 (2008):4261–4267.

Groundwater: Its Supply, Chemical Contamination, and Remediation

11.15 The Nature and Supply of Groundwater

The great majority (99%) of the available fresh water on Earth lies underground, half of it at depths exceeding a kilometer. As one digs into the ground below the initial belt of soil moisture, the **aeration** or **unsaturated zone,** where the particles of soil are covered with a film of water but in which air is present between the particles, is next encountered. At lower depths is the **saturated zone,** in which water has displaced all the air from these **pore spaces. Groundwater is the name given to the fresh water in the saturated zone** (see Figure 11-4); it makes up 0.6% of the world's total water supply. The ultimate source of groundwater is precipitation that falls onto the surface; a small fraction of it eventually filters down to the saturated zone. Underground water ranges in "age" from a few years to millions of years. For example, in zones that now are arid, much of the groundwater currently being accessed has been present since the wetter conditions of the last ice age and will not be quickly replaced.

The top of the groundwater (saturated) region is called the **water table.** In some places it occurs right at the surface of the soil, a phenomenon that gives rise to swamps. Where the water table lies above the soil, we encounter lakes and streams.

If groundwater is contained in soil that is composed of porous rocks such as sandstone, or in highly fractured rock such as gravel or sand, and if the water is bounded at its lower depths by a layer of clay or impervious rocks, then it constitutes a permanent reservoir—a sort of underground lake— called an **aquifer.** This groundwater can be extracted by wells, and is the main supply of drinking water for almost half the population of North America and over 1.5 billion people worldwide.

In the United States, groundwater constitutes about 33% of the water used for public supplies and 98% of that withdrawn for individual domestic systems, the latter being very common in rural homes. In Europe the proportion of public drinking water extracted from aquifers ranges from nearly 100% for Denmark, Austria, and Italy, to about two-thirds in Germany, Switzerland, and the Netherlands, to less than one-third in Great Britain and Spain. Some aquifers lie below several layers of impermeable rock or soil; these are called *confined* or *artesian* aquifers.

About two-thirds of groundwater usage in the United States is for irrigation purposes, almost all of it in the western states. The massive extraction of

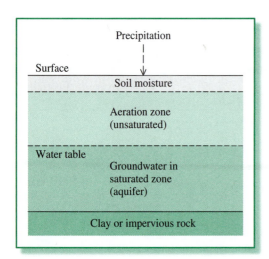

FIGURE 11-4 Groundwater location in relation to regions in the soil.

water from American aquifers—amounting to over 300 billion liters a day— has given rise to fears about future supplies of fresh water (and about the sinking of land above the aquifers), since such aquifers are replenished only very slowly. In the High Plains of the central United States, more than half the groundwater in storage has been depleted in some areas.

Globally, about 2 trillion liters of groundwater are extracted daily! In northern China the depletion of shallow aquifers is forcing the sinking of wells more than one kilometer deep in order to reach a new supply of groundwater. Indeed, groundwater depletion—along with the buildup of salts in the soil—is now the dominant threat to irrigated agriculture. In addition, the contamination of groundwater by chemicals is becoming a serious concern in many areas. A side effect of the rise in sea levels that will accompany global warming is the intrusion of salt water into aquifers near coasts.

11.16 The Contamination of Groundwater

Groundwater has been traditionally considered to be a pure form of water. Because of the filtration through soil and its long residence time underground, it contains much less natural organic matter and many fewer disease-causing microorganisms than water from lakes or rivers, although the latter point may be a misconception according to recent evidence. Some groundwater is naturally too salty or too acidic for either drinking or irrigation purposes, and may contain too much sodium, sulfide, or iron for many uses.

Humans have been concerned about the pollution of surface water in rivers and lakes for a long time. Indeed, a recent survey indicated that stream water in both agricultural and urban areas of the United States contains pesticide concentrations that exceed human health benchmark standards. In contrast, the contamination by chemicals of groundwater was not recognized as a serious environmental problem until the 1980s, notwithstanding the fact that it had been occurring for half a century. To a large extent, groundwater contamination was neglected because it was not immediately visible—it was "out of sight, out of mind"—even though groundwater is a major source of drinking water. We were ignorant of the long-range consequences of our waste disposal practices. Ironically, surface water can be cleaned up relatively easily and quickly, whereas groundwater pollution is a much harder, much more expensive long-range problem to solve.

Because we are now aware of the consequences—including high remediation costs—of the uncontrolled disposal of organic chemical wastes, most large corporations in developed countries have become much more responsible in their disposal of chemicals. Unfortunately, the collective discharges from smaller sources, including many municipalities, small industries, and farms, have not yet been controlled in like manner. Similarly, the huge number of septic tanks that exist are collectively a major contributor of nitrate, bacteria, viruses, detergents, and household cleaners to groundwater.

11.17 Nitrate Contamination of Groundwater

The inorganic contaminant of greatest concern in groundwater is the nitrate ion, NO_3^-, which commonly occurs in both rural and suburban aquifers. In the United States, Canada, and some other countries, **the MCL (or equivalent) of nitrate ion is cited not of the ion itself, but of the amount of nitrogen in it.** Thus the U.S. MCL of nitrate is stated as 10 ppm of "nitrate nitrogen." Since NO_3^- contains one nitrogen, molar mass 14.01, in a total molar mass of $14.01 + (3 \times 16.00) = 62.01$, the maximum concentration permitted expressed as nitrate ion itself is $(62.01/14.01) \times 10 = 44$ ppm.

Although uncontaminated groundwater generally has nitrate nitrogen levels of 4–9 ppm, about 9% of shallow aquifers—from which water often is extracted via privately owned wells—in the United States now have nitrate levels that exceed the 10-ppm nitrogen MCL value. Indeed, elevated levels of about 100 ppm can result from agricultural activity, with values as high as 400-ppm N found at some sites in India. The location of areas in the United States that have a high risk of nitrate contamination of groundwater is shown in Figure 11-5. Exceeding the 10-ppm MCL limit is much rarer (1%)

O⁻
\
N⁺=O
/
O⁻

nitrate ion
NO_3^-

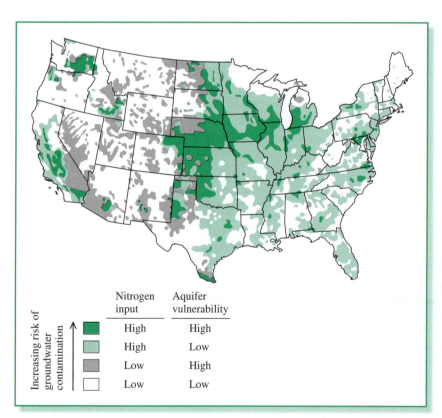

FIGURE 11-5 The risk of nitrate contamination of the groundwaters in the United States. [Source: B. T. Nolan et al., "Risk of Nitrate in Groundwaters of the United States—A National Perspective," *Environmental Science and Technology* 31 (1997): 2229.]

Increasing risk of groundwater contamination

	Nitrogen input	Aquifer vulnerability
■ (dark green)	High	High
■ (light green)	High	Low
■ (gray)	Low	High
□ (white)	Low	Low

for public U.S. groundwater supplies, partially because they are drawn from deeper aquifers; these are generally less contaminated because of their depth, because their locations are remote from large sources of contamination, and because natural remediation by denitrification of nitrate in the low-oxygen conditions can occur. Surveys in Nova Scotia and many sites in the European Union (including parts of the Netherlands, Belgium, England, France, Italy, and Spain) found that about one-fifth of groundwater sources in agricultural areas had nitrate levels exceeding the MCL.

The expenditure of public money on nitrate level reductions in drinking water has become a controversial subject. In Great Britain, in particular, hundreds of millions of dollars have been spent on achieving the 50-ppm maximum level of nitrate ion set by the European Union. Because nitrate removal from well water is very expensive, water contaminated with high levels of the ion is not normally used for human consumption, at least in public supplies.

PROBLEM 11-4

Convert the EU nitrate standard of 50 ppm to its nitrogen content alone. Is the EU standard more or less stringent than the United States regulatory limit of 10 ppm nitrogen as nitrate?

Nitrate in groundwater originates mainly from four sources:

- application of nitrogen fertilizers, both inorganic and animal manure, to cropland;

- cultivation of the soil;

- human sewage deposited in septic systems; and

- atmospheric deposition.

Concern has been expressed about the increasing levels of nitrate ion in drinking water, particularly in well water in rural locations; the main source of this NO_3^- is runoff from agricultural lands into rivers and streams. By the early 2000s, 12 million tonnes (12 Tg) of synthetic nitrogen fertilizer was being applied annually for agriculture in the United States; globally about 80 million tonnes was used. Manure production contributed almost 7 million tonnes more nitrogen in the United States at the time. Initially, oxidized animal wastes (manure), unabsorbed **ammonium nitrate,** NH_4NO_3, and other nitrogen fertilizers were thought to be the culprits in nitrate contamination of groundwater, since even reduced nitrogen unused by plants is converted naturally to nitrate, which is highly soluble in water and can easily leach down into groundwater.

It now *also* appears that intensive cultivation of land, even without the application of fertilizer or manure, facilitates the oxidation of reduced nitrogen to nitrate in decomposed organic matter in the soil by providing aeration and moisture. The original, reduced forms of nitrogen become oxidized in the soil to nitrate, which, being mobile, then migrates down to the groundwater where it dissolves in water and is diluted.

Denitrification of nitrate to nitrogen gas (see Chapter 5) and uptake of nitrate by plants can occur in forested areas that separate agricultural farms from streams, thereby lowering the risk of contamination in areas with significant woodland. However, rural areas with high nitrogen input, well-drained soil, and little woodland are at particular risk for nitrate contamination of groundwater.

A survey of nitrate levels in wells across the United States found that **redox conditions in the water had the greatest influence on their nitrate levels.** High levels were found where there not only was high nitrogen input, but also high levels of dissolved oxygen and low levels of iron and manganese. Low levels of nitrate occurred where for geological reasons, the water was reducing in character and contained high levels of iron. The Fe^{2+} present would act as a reducing agent, converting the $+5$ nitrogen in NO_3^- by electron transfer to less oxidized forms of nitrogen in nitrite ion ($+3$ oxidation number) and ammonium ion (-3) and molecular N_2 (0). Another recent survey found that the concentration of nitrate in U.S. aquifers that had been recently recharged averaged 15 mg L^{-1}, compared to 2 mg L^{-1} in the 1940s. The greatest amount of leaching occurred where there was high agricultural use of nitrogen and sandy well-drained irrigated soils, and the least in poorly drained soils with shallow groundwater shunted quickly by tile drains and ditches to surface waters.

The atmospheric deposition of nitrate results from its production in the atmosphere when NO_X emissions from vehicles and power plants, and its natural sources in thunderstorms, are oxidized in air to nitric acid and then neutralized to ammonium nitrate (see Section 3.25). In urban areas, the use of nitrogen fertilizers on domestic lawns and on golf courses, parks, etc. contribute nitrate to groundwater. Septic tanks and cesspools also are significant contributors where they exist.

Excess nitrate ion in wastewater flowing into seawater, e.g., the Baltic Sea, has resulted in algal blooms that pollute the water after they die. The algae consume the dissolved oxygen as they decompose, producing "dead zones" where fish and other organisms that require oxygen cannot survive. Nitrate ion normally does not cause this effect in bodies of fresh water, where phosphorus rather than nitrogen is usually the *limiting nutrient* in plant growth. Increasing the nitrate concentration there without an increase in phosphate levels does not lead to an increase in growth. There are, however, instances where nitrogen rather than phosphorus temporarily becomes the limiting nutrient even in fresh waters.

Algal blooms caused by excess nitrogen are, however, a problem in many coastal areas of the world, including the eastern, Gulf of Mexico, and Pacific northwest coasts of the United States, the coasts of Japan, and other areas in East Asia. The Black Sea, the Mediterranean Sea, and the Baltic Sea in

The production of algal blooms is a manifestation of *eutrophication*, a topic discussed in more detail in Section 11.28.

Europe are also affected. The main sources of the nitrogen are runoff from agricultural fields and other sources such as sewage plants, all of which eventually flows into rivers that empty into the seas. Estuaries are a particular problem area, since river flows there enter confined areas.

PROBLEM 11-5

The nitrate concentration in an aquifer is 20 ppm, and its volume is ten million liters. What mass of ammonia upon oxidation would have produced this mass of nitrate? [*Hint:* The moles of nitrogen in reactant and product are identical.] ●

11.18 Health Hazards of Nitrates in Drinking Water

Excess nitrate ion in drinking water is a potential health hazard since it can result in **methemoglobinemia** in newborn infants as well as in adults with a specific enzyme deficiency. The pathological process, in brief, runs as follows.

Bacteria, e.g., in unsterilized milk-feeding bottles or in the baby's stomach, reduce some of the nitrate to **nitrite ion,** NO_2^-:

$$NO_3^- + 2\,H^+ + 2\,e^- \longrightarrow NO_2^- + H_2O$$

The nitrite combines with and oxidizes the iron ions of the hemoglobin in blood from Fe^{2+} to Fe^{3+} and thereby prevents the proper absorption and transfer of oxygen to cells. The baby turns blue and suffers respiratory failure. (In almost all adults, the oxidized hemoglobin is readily reduced back to its oxygen-carrying form, and the nitrite is readily oxidized back to nitrate; also, nitrate is mainly absorbed in the digestive tract of adults before reduction to nitrite can occur.)

This occurrence of methemoglobinemia, or *blue-baby syndrome,* is now relatively rare in industrialized countries. It was a serious problem in Hungary until the late 1980s, and in Romania. The U.S. EPA MCL of 10 ppm of nitrate nitrogen was set in order to avoid blue-baby syndrome. The MCL for *nitrite* nitrogen of 1 ppm has also been adopted.

Recently, an increase in the risk of acquiring non-Hodgkin's lymphoma has been found for persons consuming drinking water having the highest levels (long-term average of 4 ppm or more of nitrogen as nitrate) of nitrate in drinking water for some communities in Nebraska. As discussed in the next section, excess nitrate ion in drinking water is also of concern because of its potential link with stomach cancer. Recent epidemiological investigations have, however, failed to establish any positive, statistically significant relationship between nitrate levels in drinking water and the incidence of stomach cancer.

A 2001 study found that older women in Iowa who drank water from municipal supplies having elevated nitrate levels (> 2.46 ppm) were almost three times as likely to be diagnosed with bladder cancer as those least

Nitrogen is $+5$ in nitrate ion and $+3$ in nitrite.

exposed (< 0.36 ppm in their drinking water). However, a recent large-scale study from the Netherlands failed to find an association between nitrate exposure and the risk of bladder cancer. A new study found a link between high nitrate in drinking water and the incidence of thyroid cancer. A review of the current literature concluded that there is also *no association* between nitrate exposure from drinking water and adverse reproductive effects.

Write balanced redox half-reactions (assuming acidic conditions) for the conversion of NH_4^+ to NO_3^-, and of NO_2^- to N_2.   ●

11.19 Nitrosamines in Food and Water

Some scientists believe that excess nitrate ion in drinking water and foods could lead to an increase in the incidence of stomach cancer in humans, since some of it is converted in the stomach to nitrite ion. **The nitrites could subsequently react with amines to produce *N*-nitrosamines, compounds that are known to be carcinogenic in animals.** N-nitrosamines are amines in which two organic groups and an —N=O unit are bonded to the central nitrogen:

N-nitrosamines   NDMA

Of concern not only with respect to its production in the stomach and its occurrence in foods and beverages (e.g., cheeses, fried bacon, smoked and/or cured meat and fish, and beer), but also as an environmental pollutant in drinking water, is the compound in which R is the methyl group, CH_3; it is called **N-nitrosodimethylamine,** or NDMA for short. This organic liquid is somewhat soluble in water (about 4 g L^{-1}). It is a *probable human carcinogen,* and a potent one if extrapolation from animal studies is a reliable guide. It can transfer a methyl group to a nitrogen or oxygen of a DNA base, and thereby alters the instructional code for protein synthesis in the cell.

Even though the commercial production of NDMA has been phased out, it can be formed as a by-product due to the use of amines in industrial processes such as rubber tire manufacturing, leather tanning, and pesticide production.

The levels of NDMA in drinking water drawn from groundwater are of concern in some localities that have industrial sources of the compound. Following the discovery that the water supply of one town had been contaminated by up to 100 ppt NDMA from a tire factory, the province of Ontario adopted a guideline maximum of 9 ppt of NDMA in drinking water, which corresponds to a lifetime cancer risk from it of 1 in 100,000. California established a notification level of 10 ppt in drinking water for NDMA and its ethyl and n-propyl analogs after nitrosamines were detected in some wells in that state.

Research in Canada has established that N-nitrosamines can be *formed* during the disinfection treatment of surface water by an unknown mechanism probably involving organic carbon sources in the raw water. The concentrations actually increased during transport of the water, presumably because of continuing reaction of the residual chlorine disinfectant with organic matter. N-nitrosamines now are considered to also be disinfection by-products of the water-purification process, especially methods using chloramines as the disinfectant.

Large quantities of nitrate are used to "cure" pork products such as bacon and hot dogs. In these foods, some of the nitrate ion is biochemically reduced to nitrite ion, which prevents the growth of the organism responsible for botulism. Nitrite ion also gives these meats their characteristic taste and color by combining with hemoproteins in blood. **Nitrosamines are produced from excess nitrite** during frying (e.g., of bacon) and in the stomach, as discussed. Government agencies have instituted programs to decrease the residual nitrite levels in cured meats. Some manufacturers of these foods now add vitamins C or E to the meat in order to block the formation of nitrosamines. Based upon average levels of NDMA in various foods and the average daily intake for each of them, most of us now ingest more NDMA from consumption of cheese (which is often treated with nitrates) than from any other source.

> Levels of NDMA in beer now are kept to about 70 ppt, compared to more than 3000 ppt in the 1980s.

11.20 Perchlorates

Perchlorate ion, ClO_4^-, is analogous to nitrate ion in that both involve nonmetals in their highest common oxidation states bonded to oxygen. For that reason, both ions are oxidizing agents, and both have been used in explosives and propellants. Both have both natural and anthropogenic sources. Because it too is highly water-soluble, perchloate does not accumulate in soil unless the area has little or no rainfall and is arid, conditions that occur in deserts.

perchlorate ion
ClO_4^-

Large quantities of **ammonium perchlorate,** NH_4ClO_4, are manufactured for use as oxidizing agents in solid rocket propellants, fireworks, batteries, and automobile air bags. Because rocket fuel has a limited shelf life, it must be replaced regularly. Large amounts of perchlorates were washed out of missiles and rocket boosters onto the ground or into holding lagoons in the second half of the twentieth century.

The map (Figure 11-6) of perchlorate releases in the United States indicates that most is found in the south-central and western states, especially in California. Perchlorate has also been detected in garden fertilizers, at concentrations approaching 1%. It occurs naturally in some Chilean deposits of nitrate, which in the past were exported to the United States and elsewhere

FIGURE 11-6 Regions of perchlorate use and contamination in the United States. [Source: B. E. Logan, "Assessing the Outlook for Perchlorate Remediation," *Environmental Science and Technology* (1 December 2001): 484A.]

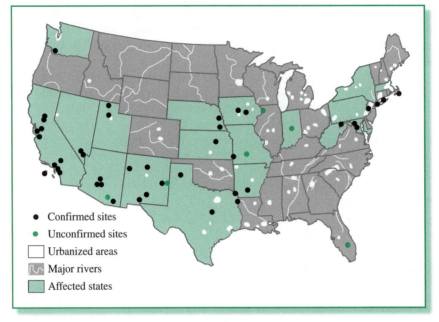

- Confirmed sites
- Unconfirmed sites
- ☐ Urbanized areas
- 〰 Major rivers
- ▮ Affected states

as fertilizers. Perchlorate also exists naturally in some minerals found in the U.S. Southwest. Its use as an oxidizer, in fireworks, rockets, etc., as well as its prior presence in fertilizer, make important contributions to its contamination of foodstuffs in the United States.

Perchlorate is a newly discovered (late 1990s) pollutant in the drinking-water supply of about 15 million Americans, though it has yet to become a problem in other countries. Perchlorate contamination has been established in 23 U.S. states, including much of the Colorado River and aquifers in the deserts of the southwest (see Figure 11-6). Concentrations of perchlorate in drinking water in the U.S. Southwest range from 5 to 20 ppb.

At high doses, perchlorate affects human health by reducing hormone production in the thyroid, where it competes with iodide ion. The hazard at low concentrations, if any, is not known, and makes the development of a drinking-water standard for the ion a difficult problem. No MCL for the ion has yet been set in the United States, but in 2009 the EPA issued an Interim Drinking Water Health Advisory Level of 15 ppb. Previously, several states had set their own limits, ranging from 1 to 18 ppb. For example, California's limit is 6 ppb, which also is the guidance level recommended by Health Canada.

Like nitrate, perchlorate is a difficult ion to remove from water supplies since it is a highly water-soluble anion that is very inert and does not adsorb readily either to mineral surfaces or to activated carbon. Barrier methods, using elemental iron, etc. are not successful because the anion is so unreactive. **The primary technologies currently in use to remediate**

perchlorate in water are ion exchange and biological treatment. Some ion exchange resins successfully remove perchlorate, though it tends to remain in solution until all other anions have been adsorbed. Ion exchange is used especially when perchlorate concentrations are low to begin with.

Certain bacteria found naturally in many soils, sediments, and natural waters biodegrade perchlorate by reduction to chloride ion:

$$ClO_4^- \longrightarrow \longrightarrow Cl^- + 2\,O_2$$

Perchlorate can be biodegraded in bioreactors, large vats that are engineered to maintain a high concentration of the appropriate bacteria in contact with the water.

11.21 Groundwater Contamination by Organic Chemicals

The contamination of groundwater by organic chemicals is a major concern. However, many organic substances decay rapidly or are immobilized in the soil, and so the number of compounds that are sufficiently persistent and mobile to travel to the water table and to contaminate groundwater there is relatively small.

The compounds that are most often detected in groundwater-based U.S. community public water supplies, including those near hazardous waste sites, are summarized in Table 11-2. Municipal landfills as well as industrial waste disposal sites are often the source of the contaminants. Liquid that contains dissolved matter that drains from a terrestrial source, such as a landfill, is called a **leachate.** In rural areas, the contamination of shallow aquifers by organic pesticides, such as *atrazine* (Chapter 13), leached from the surface, has become a concern. The insecticide *dieldrin* (Chapter 13), which has been banned since 1992, is the pesticide found most often to exceed human-health guideline levels in U.S. groundwater. Ironically, shallow groundwater aquifers used to supply drinking water are often more polluted by pesticides at greater than acceptable levels than are those in agricultural areas in the United States.

The typical organic contaminants in most major groundwater supplies are:

• Chlorinated solvents, especially **trichloroethene** (TCE, also called *trichloroethylene*), C_2HCl_3, and **perchloroethene** (PCE, also called *perchloroethylene* or *tetrachloroethene*), C_2Cl_4. These molecules contain a $C{=}C$ bond, with three or four of the four hydrogen atoms of ethene (ethylene) replaced by chlorine:

TCE   PCE

By a large margin, chlorinated solvents are the most prevalent organic pollutants in groundwater.

The technique of ion exchange in described in Section 11.30.

TABLE 11-2	Organic Compounds Commonly Found in U.S. Groundwater-Based Community Water Supplies and Their Properties		
Chemical		Density (g mL^{-1})	Water Solubility (g L^{-1})
Present at 25–50% of sites:			
Chloroform (trichloromethane)		1.48	8.2
Bromodichloromethane		1.98	4.4
Dibromochloromethane		2.45	2.7
Bromoform (tribromomethane)		2.89	3.0
Present at a smaller fraction of sites:			
Trichloroethene		1.46	1.1
Tetrachloroethene (perchloroethene)		1.62	0.15
1,1,1-Trichloroethane		1.34	1.5
1,2-Dichloroethenes		1.26, 1.28	3.5, 6.3
1,1-Dichloroethane		1.18	5.5
Carbon tetrachloride		1.46	0.76
Dichloroiodomethane		1.58	
Xylenes		0.86–0.88	0.18 (o)
1,2-Dichloropropane		1.16	2.8
Benzene		0.88	1.8
Toluene		0.87	0.54
Also commonly present at wells close to hazardous waste sites:			
Methylene chloride		1.33	20
Ethylbenzene		0.87	0.15
Acetone		0.79	sol
1,1-Dichloroethene		1.22	2.3
1,2-Dichloroethane		1.24	8.5
Vinyl chloride (chloroethene)		gas	8.8
Methyl ethyl ketone		0.80	268
Chlorobenzene		1.11	0.47
1,1,2-Trichloroethane		1.44	4.5
Chloroethane		0.90	5.7
Fluorotrichloromethane		gas	1.1
1,1,2,2-Tetrachloroethane		1.60	2.7
Methyl isobutyl ketone		0.80	19

Source: Based on U.S. EPA surveys of about 2% of U.S. water supplies.

- Hydrocarbons from the BTX component of gasoline and other petroleum products: **benzene,** C_6H_6, and its methylated derivatives **toluene,** $C_6H_5(CH_3)$, and the three isomers of **xylene,** $C_6H_4(CH_3)_2$. (See Chapter 6 for structures.)

- MTBE from gasoline (see Chapter 6).

The chemicals in the chlorinated solvent and hydrocarbon groups mentioned above occur commonly in groundwater at sites where manufacturing and/or waste disposal occurred, especially from 1940 to 1980. In that period, little attention was paid to the ultimate fate and residence in the ground of injected chemicals. The sources of these organic substances also include leaking chemical waste dumps, leaking underground gasoline storage tanks, leaking municipal landfills, and accidental spills of chemicals on land.

An extensive study published in 2007 of groundwater resources—both aquifers and shallow groundwater—across the entire United States found that TCE and PCE, along with the other chlorinated solvents **methylene chloride,** CH_2Cl_2, and **1,1,1,-trichloroethane** (TCA; methyl chloroform), $CCl_3\text{-}CH_3$, were among the most frequently detected. The concentrations of TCE, PCE, and methylene chloride were frequently found to be close to or exceed their MCLs, all of which are 5 ppb. The solvents were most frequently found at sites in the Northeast, in Iowa, and in California, and, with the exception of methylene chloride, were especially prevalent in shallow groundwater beneath urban areas rather than under agricultural regions.

Trichloroethene is an industrial solvent, used to dissolve grease on metal, as is perchloroethene. A 2006 report by the U.S. National Academy of Sciences concluded that TCE is a possible cause of kidney cancer, can impair neurological function, and causes reproductive and developmental damage. A link between TCE exposure and an abnormally low sperm count in males has been established. The International Agency for Research on Cancer has classified TCE as "probably carcinogenic to humans."

PCE is not only used in metal degreasing but also finds wide application as the solvent in dry-cleaning operations in developed countries, so it is released from a large number of small sources. A group of women in Cape Cod, Massachusetts, who were inadvertently exposed over several decades to high levels of PCE in their drinking water, were found to have small to moderate increases in their risk of contracting breast cancer.

Gasoline enters the soil via surface spills, leakage from underground storage tanks, and pipeline ruptures. Before 1980, underground gasoline storage tanks were made from steel; almost half of them were sufficiently corroded to leak by the time they were 15 years old. Once they descend to groundwater, the water-soluble components of the gasoline are preferentially leached into the water and can migrate rapidly in the dissolved state. The BTX component is the most soluble of the hydrocarbons, and often occurs at concentrations of 1–50 ppb in groundwater. However, the alkylated benzenes are rapidly degraded by aerobic bacteria and consequently are not long-lasting.

The MTBE component of gasoline is more water-soluble than the hydrocarbons, but unlike them, it is not readily biodegraded. It is not highly toxic. The main problem is the odor and taste that it gives to water: as little as 15 ppb of it in water can be tasted or smelled. MTBE contamination, albeit at low levels, of well water is becoming of concern in the United States since it has occurred at about a quarter of a million sites.

An analysis by the U.S. Geological Survey of a number of raw drinking-water sources—both surface and groundwater—that were suspected of having some human or animal contamination, mainly through wastewater, produced the results depicted by bar graphs in Figure 11-7. Clearly more surface water sites were contaminated for most categories of compounds, mainly because of the more direct pathway to them from wastewater discharges. The steroids detected, including cholesterol, all had natural sources. The nonprescription drugs included caffeine (and its metabolite and analog) and the common pain-killers acetaminophen and ibuprofen. TCE was the most frequently detected contaminant in groundwater. The insect repellant commonly detected was DEET. None of the drug compounds had concentrations exceeding 0.5 ppb in any of the samples.

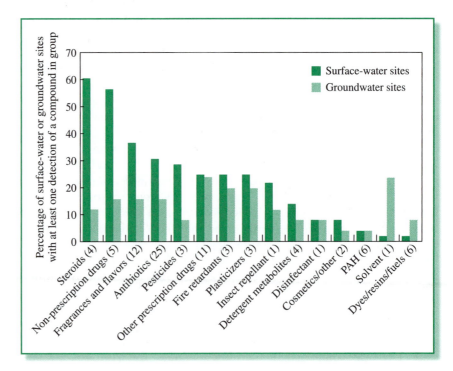

FIGURE 11-7 Detection of organic wastewater compounds, by use category, at selected U.S. surface and groundwater sites. [Source: Adapted from M. J. Focazio et al., *Science of the Total Environment* 402 (2008): 201–216.]

11.22 Drugs in Water

In recent years, trace concentrations of various drugs—prescription, over-the-counter, illegal, and veterinary—have been detected in the waters leading *from* sewage treatment plants, and in rivers and streams into which this water then flows, at concentrations up to the ppb level. About 100 substances have been detected in various rivers, lakes, and coastal waters. The substances—commonly including *estradiol*, *ibuprofen*, the antidepressant drug Prozac (*fluoxetine*), the anti-epileptic drug *carbamazepine*, and degradation products associated with cholesterol-reducing pharmaceuticals—are present in raw sewage after their excretion in urine or feces from humans and animals, since most drugs are poorly absorbed and metabolized by the body. They also result from the disposal of unused or expired medication down the toilet.

Chlorination is more effective in degrading certain pharmaceuticals, including some antibiotics and beta-blockers, than is UV treatment.

Most commonly, concentrations of drugs detected in *drinking water* are in the parts-per-trillion level, so their risk to human health is probably small. Research is underway to determine whether there could be effects on human health from sustained exposure to a combination of these substances. The synthetic hormones are thought to be the greatest risk to aquatic species. Certain fish have been found to undergo some skewed sexual development due to exposure to sewage effluent containing the synthetic estrogen in birth control pills.

The most frequently detected (legal) drugs in a mid-2000s survey of drinking water produced from surface raw waters in Ontario were

- *carbamazepine*, a widely prescribed anticonvulsant and mood stabilizer,

- *gemfibrozil*, one of the fibrate drugs used to control cholesterol levels, and

- *ibuprofen*, the nonprescription painkiller, which was also detected in a groundwater sample.

The survey did not analyze for metabolites of drugs. Antimicrobial drugs were commonly detected in untreated water. The survey also commonly detected bisphenol-A (see Chapter 15) in both treated and untreated water.

ACTIVITY

Make a list of the names of any prescription drugs taken in the recent past by yourself, members of your family, and close friends. Search the internet to discover which of the drugs have been detected in raw surface water and drinking water. What property explains why some common drugs are never detected in water?

11.23 The Ultimate Sink for Organic Contaminants in Groundwater

The subsequent behavior of the organic compounds that do migrate to the water table depends significantly upon their density relative to that of water, 1.0 g mL^{-1}. Liquids that are *less* dense (lighter) than water and have low solubility in it form a mass that floats on the top of the water table. All hydrocarbons having a small or medium molecular mass belong to this group, including the BTX fraction of gasolines and other petroleum products (see Table 11-2). In contrast, polychlorinated solvents are *more* dense (heavier) than water and insoluble in it, so they tend to sink deeply into aquifers; important examples are methylene chloride, *chloroform*, *carbon tetrachloride*, 1,1,1-trichloroethane, TCE, and PCE. Nonchlorinated but insoluble high-molecular-weight organic materials, such as *creosote* and *coal tar*, also belong to the heavier-than-water group. These substances are sometimes referred to as **dense nonaqueous-phase liquids**, DNAPL.

Although the oily liquid blobs that these organic compounds form generally are found in an aquifer at a position either directly below their original point of entry into the soil or close to it, the conclusion that they are horizontally immobile is misleading. Very slowly—in a process that often takes decades or centuries to complete—these low-solubility compounds gradually dissolve in the water that passes over the blob, and so provide a continuous supply of contaminants to the groundwater. The complete removal of such deposits usually is not feasible, since they may exist as several blobs whose exact locations are difficult to pinpoint. In addition, disturbance of the deposit during removal or treatment may increase its net exposure to the water phase. Even removing 90% of the substance does not necessarily result in reduction of its groundwater concentration. Thus, **plumes** of polluted water grow in the direction of the water's flow and thereby contaminate the bulk of the aquifer (see Figure 11-8). Because of such contamination, many drinking-water wells have had to be closed.

11.24 Decontamination of Groundwater: Physical and Chemical Processes

In the last few decades, considerable energy and money have been spent in the United States on attempts to control aquifer pollution by the oily liquids discussed above. Dense organic leachates, especially PCE and TCE, have contaminated the groundwater that lies below the waste sites associated with the U.S. Superfund remediation initiative (see Chapter 16). Unfortunately, no easy cure for the problem of contamination has been found. Control usually consists of **pump-and-treat** systems that pump contaminated water from the aquifer, treat it to remove its organic contaminants (using methods of the type described later in this chapter), and return the cleaned water to the aquifer or to some other water body. Alternatively, a fine mist of the contaminated groundwater is sprayed into the air above agricultural land using a

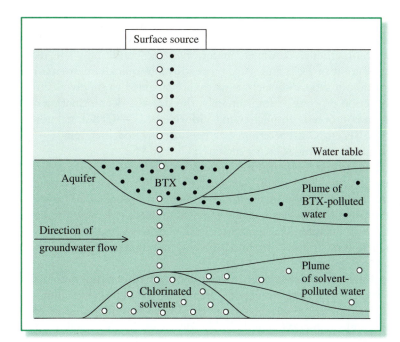

FIGURE 11-8 The contamination of groundwater by organic chemicals.

long, movable sprinkler system; the contaminant volatile organic compounds evaporate into the air and the cleansed water is used for irrigation.

The volume of water that must be pumped and treated in a given aquifer is huge. For organic contaminants with low water solubility, recontamination of water returned to the aquifer by additional dissolution from the blob will occur. Consequently, the treatment systems must operate in perpetuity, and there are already thousands of them spread across the United States.

Both *in situ* heating, to vaporize the organic liquids so their vapors rise to the soil surface, and the addition of oxidants to convert the substances to products such as carbon dioxide, have been tried in some locations. Typically, temperatures close to 100°C are used in heating, though it is not known if this is optimal in most cases. Heating and/or the production of gases or precipitates by oxidation may inadvertently change the geologic and biological conditions in the immediate vicinity of the treatment, with unforeseen effects on the distribution and mobility of the remaining pollutant.

11.25 Decontamination of Groundwater: Bioremediation and Natural Attenuation

Bioremediation is the term applied to the decontamination of water or soil using biochemical rather than chemical or physical processes. Recently, there has been interesting progress reported on using bioremediation to cleanse water of chlorinated ethene solvent contamination.

The biodegradation of chloroethenes by *aerobic* bacteria becomes less and less efficient as the extent of chlorination increases, so it is ineffective for perchloroethene. However, under *anaerobic* conditions, the reductive biodegradation of PCE and TCE proceeds more quickly, particularly if an easily oxidized substance such as methanol is added to supply electrons for the reduction processes. Unfortunately, the stepwise dechlorination of these compounds proceeds through **vinyl chloride,** $CH_2{=}CHCl$, a known carcinogen. Recently, a bacterium has been discovered that removes all the chlorine from organic solvents such as TCE and PCE.

Owing to the high cost and limited effectiveness of many groundwater cleanup technologies, the inexpensive process of *natural attenuation*—allowing natural biological, chemical, and physical processes to treat groundwater contaminants—has become popular. Indeed, it is now used at more than 25% of Superfund program sites in the United States and is the leading method to remedy the contamination of groundwater from leaking underground storage sites.

However, there is great controversy about whether or not natural attenuation is an appropriate strategy for managing groundwater contamination: many environmentalists feel that it is a cheap way for industry to avoid expensive cleanup costs. The U.S. National Research Council appointed a committee to determine which pollutants could be treated successfully by this technique. Table 11-3 summarizes their results. Only three pollutants are highly likely to be successfully treated by natural attenuation:

- BTEX hydrocarbons (i.e., BTX hydrocarbons plus ethylbenzene),

- low-molecular-weight oxygen-containing organics, and

- methylene chloride, CH_2Cl_2.

In all three cases, biotransformation is the dominant process by which attenuation occurs. Notice that neither highly chlorinated organics, including TCE and PCE, nor MTBE are usually successfully treated in this way, nor is mercury or perchlorate ion.

11.26 Decontamination of Groundwater: *In Situ* Remediation

Scientists have developed a promising *in situ* technique for treating groundwater contaminated by volatile (mainly C_1 and C_2) chlorinated organics. They construct an underground permeable "wall" of material (mostly coarse sand) along the path of the water. **The water becomes cleansed as a result of its passage through the wall** and never has to be pumped out of the ground (see Figure 11-9).

The ingredient that is placed within the sand bed and which chemically cleans the water is metallic iron, Fe⁰, in the form of small filings,

TABLE 11-3	Likelihood of Success of Groundwater Remediation by Natural Attenuation for Various Substances		
Chemical Class	Dominant Attenuation Processes	Likelihood of Success Given Current Level of Understanding	

	Organic Compounds		
Hydrocarbons			
BTEX	Biotransformation	High	
Gasoline, fuel oil	Biotransformation	Moderate	
Nonvolatile aliphatic compounds	Biotransformation, immobilization	Low	
PAHs	Biotransformation, immobilization	Low	
Creosote	Biotransformation, immobilization	Low	
Oxygenated hydrocarbons			
Low-molecular-weight alcohols, ketones, esters	Biotransformation	High	
MTBE	Biotransformation	Low	
Halogenated aliphatics			
PCE, TCE, carbon tetrachloride	Biotransformation	Low	
1,1,1,-trichloroethane (TCA)	Biotransformation, abiotic transformation	Low	
Methylene chloride	Biotransformation	High	
Vinyl chloride	Biotransformation	Low	
Dichloroethylene	Biotransformation	Low	
Halogenated aromatics			
Highly chlorinated PCBs, tetrachlorodibenzofuran, pentachlorophenol, multichlorinated benzenes	Biotransformation immobilization	Low	
Less chlorinated PCBs, dioxins	Biotransformation	Low	
Monochlorobenzene	Biotransformation	Moderate	
	Inorganic Substances		
Metals			
Ni	Immobilization	Moderate	
Cu, Zn	Immobilization	Moderate	
Cd	Immobilization	Low	
Pb	Immobilization	Moderate	

(continued on p. 496)

TABLE 11-3	Likelihood of Success of Groundwater Remediation by Natural Attenuation for Various Substances *(continued)*	
Chemical Class	**Dominant Attenuation Processes**	**Likelihood of Success Given Current Level of Understanding**
Cr	Biotransformation, immobilization	Low to moderate
Hg	Biotransformation, immobilization	Low
Nonmetals		
As	Biotransformation, immobilization	Low
Se	Biotransformation, immobilization	Low
Oxyanions		
Nitrate	Biotransformation	Moderate
Perchlorate	Biotransformation	Low

Source: Adapted from J. A. Macdonald, "Evaluating Natural Attenuation for Groundwater Cleanup," *Environmental Science and Technology* (1 August 2000): 346A.

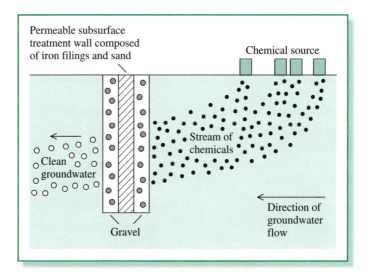

FIGURE 11-9 *In situ* purification of groundwater using an "iron wall."

a common waste product of manufacturing processes. When placed in contact with certain chlorinated organics dissolved in water, the iron acts as an reducing agent, giving up electrons to form the **ferrous,** or Fe(II), ion, Fe^{2+}, which becomes dissolved in the water:

$$Fe(s) \longrightarrow Fe^{2+}(aq) + 2\ e^-$$

Usually, these electrons are donated to chloroorganic molecules that are temporarily adsorbed onto the metal's surface; the chlorine atoms contained in these molecules are consequently reduced to **chloride ions,** Cl^-, which are released into aqueous solution. This technique is an example of **reductive degradation.** For example, the reduction of trichloroethene to its completely dechlorinated form, ethene, can be written in unbalanced form as

$$C_2HCl_3 \longrightarrow C_2H_4 + 3\ Cl^-$$

Upon application of a standard redox balancing technique for alkaline solution, we obtain the balanced half-reaction

$$C_2HCl_3 + 3\ H_2O + 6\ e^- \longrightarrow C_2H_4 + 3\ Cl^- + 3\ OH^-$$

Combination of the half-reactions (after tripling the oxidation step to ensure the number of electrons lost and gained is the same) yields the overall reaction

$$3\ Fe(s) + C_2HCl_3 + 3\ H_2O \longrightarrow 3\ Fe^{2+}(aq) + C_2H_4 + 3\ Cl^- + 3\ OH^-$$

One of the by-products of the reaction is hydroxide ion, OH^-. Recall from Chapter 10 that in limestone-rich areas, the groundwater contains significant concentrations of dissolved **calcium bicarbonate,** $Ca(HCO_3)_2$. The hydroxide ions produced in the groundwater remediation reaction react with bicarbonate to produce **carbonate ion,** CO_3^{2-}, which combines with dissolved calcium ions to produce insoluble **calcium carbonate,** $CaCO_3$, which then precipitates in the sand–iron mixture.

Field trials indicate that this new iron wall technology can work successfully for several years at least, and may replace the pump-and-treat methods for many applications involving chlorinated methanes and ethanes dissolved in underground water.

Coating the iron filings with nickel speeds up the rate of degradation of the organic compounds by a factor of ten; with this modification, the technique may be even more useful than first imagined. In addition, it has been discovered that the elemental iron filings in the barriers will reduce soluble Cr^{6+} ions to insoluble Cr^{3+} oxides, and can therefore remediate groundwater contaminated by Cr^{6+}. A technique for the *in situ* creation of elemental iron from its ions (Fe^{2+} and Fe^{3+}) by the injection of aqueous reducing agents has also been tested.

An *in situ* technique of treating TCE and PCE by hydrogenation has been developed recently. The process uses dissolved H_2 gas to rapidly dechlorinate these two organics, eventually forming ethane and HCl. The reaction uses a palladium catalyst, and can be done within a well bore so that the water need not be brought up to the surface.

PROBLEM 11-7

The dissolution of iron in the process described above produces some molecular hydrogen gas, H_2. Show by balanced equation(s) how the hydrogen could arise from the reduction of water rather than of TCE.   ●

PROBLEM 11-8

Deduce the overall reaction of excess hydroxide ion with dissolved calcium bicarbonate to produce insoluble calcium carbonate. Are there any reaction products in addition to $CaCO_3$?   ●

PROBLEM 11-9

Suppose that this "iron wall" technology reduced an appreciable fraction of TCE to vinyl chloride rather than completely to ethene. Why would this be an unacceptable result environmentally? (Note that in practice a sufficiently thick wall of iron is used to convert any vinyl chloride by-product to ethene.)   ●

PROBLEM 11-10

Deduce the overall reaction by which perchloroethene is converted to ethene by metallic iron.   ●

PROBLEM 11-11

Review Questions 11–17 are based upon the material in the preceding section.

At one test site for this remediation process, the water contained 270 ppm TCE and 53 ppm perchloroethene. Calculate the mass of iron required to remediate one liter of this groundwater.   ●

The Chemical Contamination and Treatment of Wastewater and Sewage

Most municipalities treat the raw sewage collected from homes, buildings, and industries (including food processing plants) through a *sanitary sewer* system before the liquid residue is deposited into a nearby source of natural waters, whether a river, lake, or ocean. Since the rainwater and melted snow that drains from streets and other paved surfaces is usually not highly contaminated, it is often collected separately by *storm sewers* and deposited directly into a body of natural water. Unfortunately, in some municipalities, storm-driven overflow occurs in the sanitary sewer system and this excess is combined with storm water and deposited, untreated, into waterways.

The main component of sewage—other than water—is organic matter of biological origin. It occurs mainly as particles—ranging from those of macroscopic size large enough to be trapped (together with such *objets d'art* as facial tissues, stones, socks, tree branches, condoms, and tampon applicators) by mesh screens to those which are microscopic in size and are suspended in the water as large colloids.

11.27 Sewage Treatment

In the primary (or mechanical) treatment stage of wastewater (see the schematic diagram in Figure 11-10), **the larger particles—including sand and silt—are removed by allowing the water to flow across screens and slowly along a lagoon or settling basin.** A **sludge** of insoluble particles forms at the bottom of the lagoon, while "liquid grease" (a term which here includes not only fat, oils, and waxes but also the products formed by the reaction of soap with calcium and magnesium ions) forms a lighter-than-water layer at the top and is skimmed off. About 30% of the Biochemical Oxygen Demand (BOD, Chapter 10) of the wastewater is removed by the primary treatment process, even though this stage of the process is entirely mechanical in nature. The treatment and disposal of the sludge is discussed later.

After passing through conventional primary treatment, the sewage water has been much clarified but still has a very high BOD—typically several hundred milligrams per liter (ppm)—and is detrimental to fish life if released at this stage (as occurs in some jurisdictions that discharge into the ocean). The high BOD is due mainly to organic colloidal particles. **In the secondary (biological) treatment stage, most of this suspended organic matter as well as that actually dissolved in the water is biologically oxidized by microorganisms to carbon dioxide and water or converted to additional sludge, which can readily be removed from the water.** Either the water is sprinkled

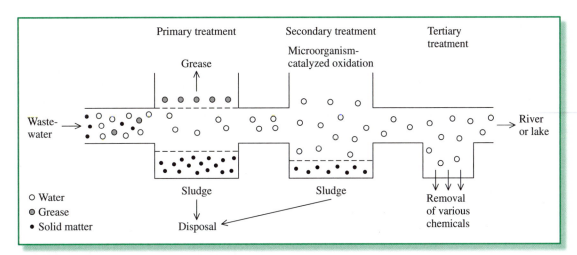

FIGURE 11-10 The common stages in the treatment of wastewater.

onto a bed of sand and gravel or plastic covered with the aerobic bacteria (*trickling filters*) or it is agitated in an aeration reactor (*activated sludge process*) in order to effect the microorganism-driven reaction. The system is kept well-aerated to speed the oxidation. In essence, by deliberately maintaining in the system a high concentration of aerobic microorganisms, especially bacteria, the same biological degradation processes that would require weeks to occur in open waters are accomplished in several hours.

In the new technology of *membrane bioreactors*, the slow sedimentation of sludge by gravity is replaced by particle removal using microfiltration or ultrafiltration membranes. The development of suitable membranes resistant to fouling is needed before this method can become widespread in use.

The biological oxidation processes of **secondary treatment** reduce the BOD of the polluted water to less than 100 mg L^{-1}, which is about 10% of the original concentration in the untreated sewage. For example, Canadian waste-water quality standards require that the BOD in treated water be 20 mg L^{-1} or less. The process of nitrification also occurs to some extent, converting organic nitrogen compounds to nitrate ion and carbon dioxide (see below). **In summary, the secondary treatment of wastewater involves biochemical reactions that oxidize much of the oxidizable organic material that was not removed in the first stage.** Treated water diluted with a greater amount of natural water can support aquatic life. Normally, municipalities take the water produced by secondary treatment and disinfect it by chlorination or irradiation with UV light before pumping it into a local waterway. Recent research in Japan has shown that chlorination of the effluent before its release produces mutagenic compounds, presumably by interaction of chlorine-containing substances with the organic matter that remains in the water.

A few municipalities employ **tertiary** (or **advanced** or chemical) waste-water treatment as well as primary and secondary. **In the tertiary phase, specific substances are removed from the partially purified water before its final disinfection.** In some cases, the water produced by tertiary treatment is of sufficiently high quality to use as drinking water. Alternatively, river water into which the effluent from sewage treatment plants has been deposited is used by municipalities downstream as drinking water. The re-use of water after it has been cleansed is particularly prevalent in Europe, where the supplies of fresh water are less available than in North America and the consuming population density is high.

Depending upon locale, tertiary treatment can include some or all the following chemical processes:

- **further reduction of BOD** by removal of most remaining colloidal material using an aluminum salt in a process that operates in the same manner as described previously for the purification of drinking water;

- **removal of dissolved organic compounds** (including chloroform) and some heavy metals by their adsorption onto activated carbon, over which the water is allowed to flow (see Box 11-1);

- **phosphate removal** (as discussed in the next section); some phosphorus is removed in the secondary treatment stage since microbes incorporate it as a nutrient for their growth;

- **heavy-metal removal** by the addition of hydroxide or **sulfide ions,** S^{2-}, to form insoluble metal hydroxides or sulfides, respectively;

- **iron removal** by aeration at a high pH to oxidize it to its insoluble Fe^{3+} state, possibly in combination with use of a strong oxidizing agent to destroy organic ligands bound strongly to the Fe^{2+} ion, which would otherwise prevent its oxidation; and

- **removal of inorganic ions present in excess,** as discussed below.

In some wastewaters, the further removal of **nitrogen compounds**—usually either ammonia or organic nitrogen compounds—is deemed necessary. Ammonia removal can be achieved by raising the pH to about 11 (with lime) so as to convert most ammonium ion to its molecular form, ammonia, followed by bubbling air through the water to air-strip it of its dissolved ammonia gas. This process is relatively expensive, however, because it is energy-intensive. Ammonium ion can also be removed by ion exchange using certain resins that have their exchange sites initially populated by sodium or calcium ions.

Alternatively, both organic nitrogen and ammonia can be removed by first using nitrifying bacteria to oxidize all the nitrogen to nitrate ion. Then the nitrate is subjected to denitrification by bacteria to produce molecular nitrogen, which bubbles out of the water. Since this reduction step requires a substance to be oxidized, **methanol,** CH_3OH, is added if necessary to the water and is converted to carbon dioxide in the process:

$$5\ CH_3OH + 6\ NO_3^- + 6\ H^+ \xrightarrow{\text{bacteria}} 5\ CO_2 + 3\ N_2 + 13\ H_2O$$

Of course, water contaminated by nitrate ion can also be treated by this latter step. A mathematical analysis of the kinetics of the transformations is given in Box 11-4.

PROBLEM 11-12

Given that the K_b value for ammonia is 1.8×10^{-5}, deduce a formula giving the ratio of ammonia to ammonium ion as a function of the pH of water. What is the value of this ratio at pH values of 5, 7, 9, and 11?

11.28 The Origin and Removal of Excess Phosphate

One of the world's most famous cases of water pollution involves Lake Erie, which in the 1960s was said to be dying. Indeed, one of the authors of this book can recall visiting a once-popular beach on its north shore in the early

BOX 11-4   Time Dependence of Concentrations in the Two-Step Oxidation of Ammonia

The bacteria-catalyzed oxidation of ammonia (or of other reduced organic nitrogen compounds) to nitrate is a reaction with two main steps, with nitrite ion, NO_2^-, an intermediate:

Step 1 $NH_3 + \frac{3}{2}O_2 \longrightarrow NO_2^- + H^+ + H_2O$

Step 2 $NO_2^- + \frac{1}{2}O_2 \longrightarrow NO_3^-$

If sufficient oxygen is available, the rate of each reaction is first-order only in the concentration of the nitrogen reactant, so the sequence can be represented as

$$A \xrightarrow{k_2} B \xrightarrow{k_2} C$$

where A stands for ammonia, B for nitrite ion, and C for nitrate ion, and k_1 and k_2 are the pseudo-first-order rate constants. Since the rate of step 1 depends on the first power of the ammonia concentration, then the rate of disappearance of this species is

$$\frac{d[A]}{dt} = -k_1[A]$$

Since B (nitrite) is produced at this rate by step 1, but is consumed in step 2 by a process whose rate is proportional to the first power of its concentration, we can write

$$\frac{d[B]}{dt} = +k_1[A] - k_2[B]$$

These differential equations can be coupled and integrated to yield the following expressions for the evolution of [A] and [B] with time, relative to $[A]_0$, the original concentration of A:

$$[A]/[A]_0 = e^{-k_1 t}$$

$$[B]/[A]_0 = k_1(e^{-k_1 t} - e^{-k_2 t})/(k_2 - k_1)$$

As can be seen from the solution to Problem 1, the concentration of B (nitrite) rises exponentially at first, reaches a peak value, then declines slowly. Thus significant concentrations of nitrite ion occur in water undergoing the two-step conversion of ammonia to nitrate.

PROBLEM 1

(a) Derive a general expression relating the time at which [B] reaches a peak to k_1 and k_2.
(b) Draw a graph showing the evolution of [A] and of [B] with time for the values $k_1 = 1$ and $k_2 = 2$.

1970s and being repulsed by the sight and smell of dead, rotting fish on the shoreline. Lake Erie's problems stemmed primarily from an excess input of **phosphate ion**, PO_4^{3-}, in the waters of its tributaries. The phosphate sources were the polyphosphates in detergents (as explained in detail later), raw sewage, and the runoff from farms that used phosphate fertilizers.

Since there is commonly an excess of other dissolved nutrients in lakes, phosphate ion usually functions as the limiting (or controlling) nutrient for algal growth: the larger the supply of the ion, the more abundant the growth of algae, and its growth can be quite abundant indeed. When the vast mass of excess algae eventually dies and starts to decompose by oxidation, the water becomes depleted of dissolved oxygen, with the result that fish life is adversely affected. The lake water also becomes foul-tasting, green, and slimy, and masses of dead fish and aquatic weeds rot on the beaches. The series of changes, including the rapid degradation and ageing, that occur when lakes receive excess plant nutrients from their surroundings is called **eutrophication.** When the enrichment arises from human activities, it is called **cultural eutrophication.**

To correct the eutrophication problem in the region, the United States and Canada in 1972 signed the *Great Lakes Water Quality Agreement.* Since that time, over eight billion dollars have been spent in building sewage treatment plants to remove phosphates from wastewater before it reaches the tributaries and the lakes themselves. In addition, the levels of phosphates in laundry detergents were restricted in Ontario and in many of the states that border the Great Lakes. The total amount of phosphorus entering Lake Erie has now decreased by more than two-thirds. As a result, Lake Erie has sprung back to life: its once-fouled beaches are regaining popularity with tourists and its commercial fisheries have been revived. However, phosphate previously stored on the lake bottom in sediments may be liberated in the low-oxygen "dead zone" still present when the ferric hydroxides on which it had been adsorbed are reduced and solubilized. Indeed, Lake Erie is now over-enriched in both phosphorus and nitrogen.

As we have pointed out, the presence of excess phosphate ion in natural waters can have a devastating effect on an aquatic ecology because it over-fertilizes plant life. Formerly, one of the largest sources of phosphate as a pollutant was detergents, and in the material that follows, the role of such phosphates is discussed.

The reaction of synthetic detergents with calcium and magnesium ions, to form complex ions, diminishes the cleansing potential of the detergent. **Polyphosphate ions,** which are anions containing several phosphate units linked by shared oxygens, are added to detergents as *builders* that preferentially form soluble complexes with these metal ions and thereby allow the molecules of the detergent to operate as cleansing agents rather than being complexed with Ca^{2+} and Mg^{2+} present in the water. Another role of the builder is to make the wash water somewhat alkaline, which helps remove the dirt from certain fabrics.

In particular, great quantities of **sodium tripolyphosphate** (STP), $Na_5P_3O_{10}$, were formerly added as the builder in most synthetic detergent formulations. As shown in Figure 11-11a, STP contains a chain of alternating phosphorus and oxygen atoms, with one or two additional oxygens attached to each phosphorus. In solution, one tripolyphosphate ion can form a complex with one calcium ion by forming interactions between three of its oxygen atoms and the metal ion (Figure 11-11b).

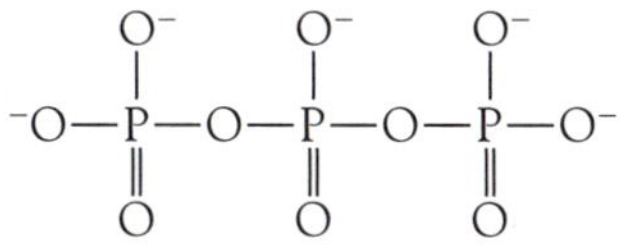

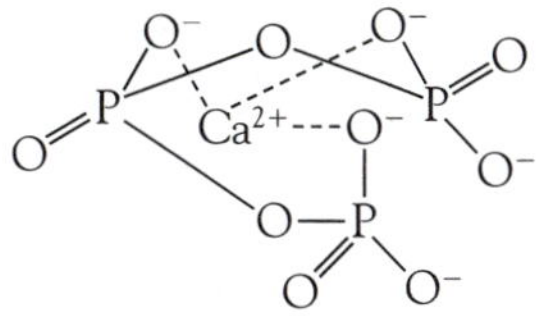

FIGURE 11-11 Structure of the polyphosphate ion: (a) uncomplexed and (b) complexed with calcium ion.

Substances like STP that have more than one site of attachment to the metal ion, and thereby produce ring structures that each incorporate the metal, are called **chelating agents** (from the Greek word for claw). Because several bonds are formed, the resulting chelates are very stable and do not normally release their metal ions back into a free form.

Tripolyphosphate ion, $P_3O_{10}^{5-}$, like phosphate ion itself, is a weak base in aqueous solution and thus provides the alkaline environment that is required for effective cleaning:

$$P_3O_{10}^{5-} + H_2O \longrightarrow P_3O_{10}H^{4-} + OH^-$$

Phosphate is sometimes called orthophosphate.

Unfortunately, when wash water containing STP is discarded, the excess tripolyphosphate enters waterways, where it slowly reacts with water and is transformed into phosphate ion:

$$P_3O_{10}^{5-} + 2\, H_2O \longrightarrow 3\, PO_4^{3-} + 4\, H^+$$

Note that when tripolyphosphate decomposes, STP behaves as an acid rather than a base (since H^+ is formed in the reaction).

Because of environmental concerns, phosphates are now used only sparingly as builders in detergents in many areas of the world. In Canada and parts of Europe, STP was replaced largely by **sodium nitrilotriacetate** (NTA) (see Figure 11-12a). The anion of NTA acts in a similar fashion to that of STP, chelating calcium and magnesium ions using three of its oxygen atoms and the nitrogen atom (Figure 11-12b). NTA is not used as a builder in the United States because of concerns that its slow rate of degradation might lead to health hazards in drinking water. However, the early experiments with test animals that led to this concern are open to question, as are fears about its persistence and its tendency to solubilize heavy metals into water supplies.

Other builders now used include sodium citrate, sodium carbonate (*washing soda*), and sodium silicate. Currently, substances called **zeolites** are also employed as detergent builders. Zeolites are abundant aluminosilicate minerals (see Chapter 16) consisting of sodium, aluminum, silicon, and oxygen. The latter three elements are bonded together to form cages, which the sodium ions can enter. In the presence of calcium ion, zeolites exchange

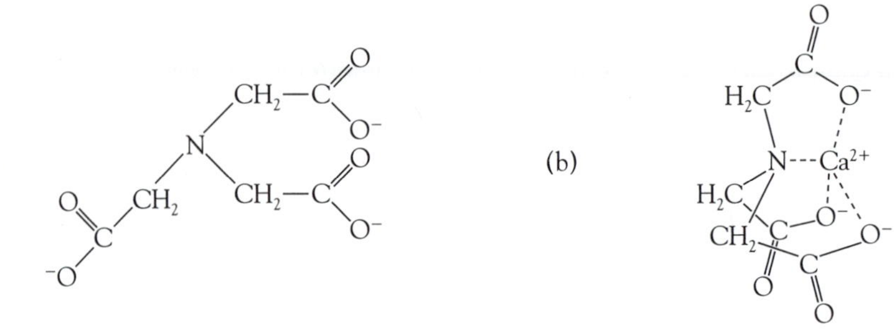

FIGURE 11-12 Structure of the nitrilotriacetate ion: (a) uncomplexed and (b) complexed with calcium ion.

their sodium ions for Ca^{2+} (though not for Mg^{2+}), thereby sequestering it in a manner similar to polyphosphates. Like polyphosphates, they also control pH. One disadvantage to the use of zeolites is that they are insoluble, so their use increases the amount of sludge that must be removed at wastewater treatment plants.

Phosphate ion can be removed from municipal and industrial wastewater by the addition of sufficient calcium as the hydroxide, $Ca(OH)_2$, so that insoluble calcium phosphates such as $Ca_3(PO_4)_2$ and $Ca_5(PO_4)_3OH$ are formed as precipitates that can then be readily removed. Phosphate removal could be a standard practice in the treatment of wastewater, but it is not yet practiced in all cities. Some policymakers believe that the optimum environmental solution is to use phosphates, rather than some other builder, in detergents and then to efficiently remove them at wastewater treatment plants.

Geographically, phosphate ion enters waterways from both point and nonpoint sources. **Point sources** are specific locations such as factories, landfills, and sewage treatment plants that discharge pollutants. **Nonpoint sources** are numerous large land areas such as farms, logged forests, and golf courses; individual domestic lawns and septic tanks; storm water runoff; and atmospheric deposition. Although each nonpoint source may provide a small amount of pollution, on account of the large number of them that are involved, they can generate larger *total* quantities than do point sources. For example, now that sewage treatment plants and detergent controls have been instituted, much of the remaining phosphate arises from nonpoint agricultural sources in many areas.

11.29 Green Chemistry: Sodium Iminodisuccinate, a Biodegradable Chelating Agent

Because most chelating agents are not biodegradable or are only slowly biodegradable, they not only place a load on the environment (e.g., phosphates acts as nutrients), but it may be necessary to remove them during treatment of wastewater in a wastewater treatment plant.

Unlike many chelating agents, **sodium iminodisuccinate** (IDS, see Figure 11-13) [also known as *D,L-aspartic-N-(1,2-dicarboxylethyl) tetrasodium salt*] readily degrades in the environment. Not only is IDS biodegradable, it is also nontoxic.

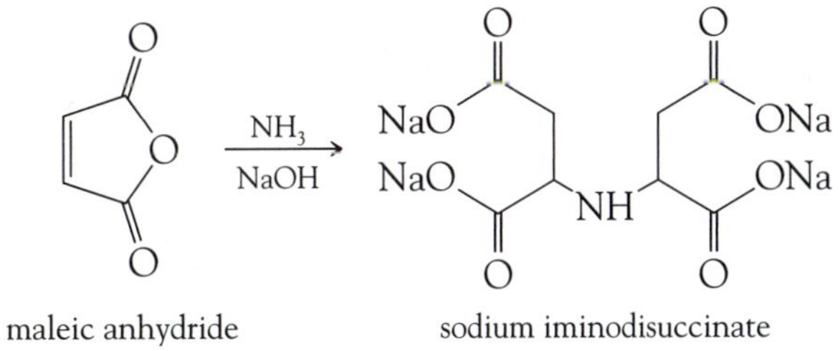

IDS can be used as an effective chelating agent for absorption of agricultural nutrients, metal ion scavenging in photographic processing, groundwater remediation, as a builder in detergents and household and industrial cleaners. The *Bayer Corporation* won a *Presidential Green Chemistry Challenge Award* in 2001 for the development of IDS as a chelating agent and for its synthesis from maleic

maleic anhydride   sodium iminodisuccinate

FIGURE 11-13 Synthesis and structure of sodium iminodisuccinate, a biodegradable cleaning agent.

anhydride (Figure 11-13). This synthesis is accomplished under mild conditions, in water as the only solvent. The excess ammonia that is used is recycled back into the production of more IDS. This synthesis stands in stark contrast to typical syntheses of aminocarboxylate chelating agents, which employ hydrogen cyanide as a reagent. Bayer markets IDS as a chelating agent under the name Baypure.

11.30 Reducing the Salt Concentration in Water

The decomposition of organic and biological substances during the secondary phase of wastewater treatment usually results in the production of inorganic salts, many of which remain in the water even after the techniques listed above have been applied. Water can also become salty due to its use in irrigation, or because water softener units have been recharged and their discharge disposed of as sewage. Inorganic ions can be removed from water by desalination by using one of the techniques listed below, or by using the precipitation methods mentioned above.

• **Reverse Osmosis:** As previously mentioned, this technique is also used to produce drinking water from salty water, such as seawater.

• **Electrodialysis:** Here a series of membranes permeable either only to small inorganic cations or only to small inorganic anions are set up vertically in an alternating fashion (see Figure 11-14) within an electrical cell. Direct current is applied across the water, so cations migrate toward the cathode and anions toward the anode. The liquid in alternating zones becomes more concentrated (enriched) or less concentrated (purified) in ions; eventually the ion-concentrated water can be disposed of as brine and the purified water released into the environment. This technology is also used to desalinate seawater for drinking purposes.

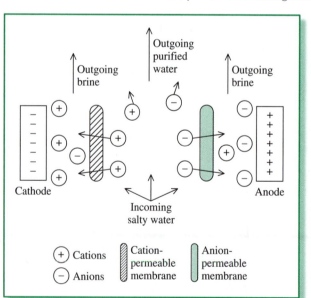

FIGURE 11-14
Electrodialysis unit
(schematic) for the
desalination of water.

• **Ion Exchange:** Some polymeric solids contain sites that hold ions relatively weakly, and so one type of ion can be exchanged for another of the same charge that happens to pass by it. Ion exchange resins can be formulated to possess either cationic or anionic sites that function in this manner. The exchange sites of a cationic resin of this type are initially occupied by H^+ ions, and the exchange sites of an anionic exchange resin are occupied by OH^- ions. When water polluted by M^+ and X^- ions is passed sequentially through the two resins, the H^+ ions on the first are replaced by M^+, and then the OH^- on the second resin are replaced by X^-. Thus the water that has passed through contains H^+ and OH^- ions, rather than those of the salt, which remains behind in the resins; of course, these two ions immediately combine to form more water molecules. Thus ion exchange can be used to remove salts, including those of heavy metals, from wastewater.

PROBLEM 11-13

Water polluted by inorganic ions could be purified by distilling it or by freezing it. Why do you think that such techniques are not generally used on a mass scale to purify water? ●

Transition metal cations can be removed from water using either precipitation or reduction techniques, in either case to form insoluble solids. Precipitation of sulfides or hydroxides has already been mentioned; a disadvantage of the latter is the production of a voluminous sludge that must be disposed of in an acceptable manner. Electrolytic reduction of metals leads to their deposition on the cathode. If, instead of the elemental metal, a concentrated aqueous solution of it is desired, the deposited metal can be reoxidized chemically by addition of hydrogen peroxide or electrolytically by reversing the polarity of the cell.

11.31 The Biological Treatment of Wastewater and Sewage

An alternative to the processing of wastewater through a conventional treatment plant in small communities is biological treatment in an **artificial marsh** (also called a *constructed wetland*) that contains plants such as bulrushes and reeds. **The decontamination of the water is accomplished by the bacteria and other microbes that live among the plants' roots and rhizomes.** The plants themselves take up metals through their root systems and concentrate contaminants within their cells.

In facilities that have been constructed to deal with sewage, primary treatment to filter out solids, etc. in a lagoon is usually implemented before the wastewater is pumped to the marsh, where the equivalent of secondary and tertiary treatment occurs. The plant growth uses up the pollutants, and increases the pH—which serves to destroy some harmful microorganisms. Artificial marshes also provide some water purification by its exposure to sunlight, by the ability of volatile contaminants to evaporate, and by the precipitation of its insoluble contaminants. Ammonia is removed from water by nitrification in the marsh's aerobic zones, followed by denitrification of the resulting nitrate to nitrogen gas in anaerobic zones near the water's bottom.

Artificial marshes are sometimes also used to process storm water runoff.

One advantage to biological treatment is that great amounts of sludge are not generated, in contrast to conventional treatments. In addition, it requires neither the addition of synthetic chemicals nor the input of commercial energy. Among the problems in such facilities is decaying vegetation, which must be limited so that the BOD of the processed water does not rise too much, and the fact that the marshes usually require a great deal of land unless they are constructed so that part of the routing is vertical.

Underground constructed wetlands have been found useful in secondary treatment of municipal wastewater in some areas, especially for storm water. The wastewater is fed through an inlet, and passes slowly through the porous subsurface under a bed of planted vegetation before being collected and released. By necessity, most of the flow is under anaerobic conditions, and therefore nitrification is limited.

In many rural and small communities, **septic tanks** are used to decontaminate sewage since central sewage facilities are not available. These underground concrete tanks receive the wastewater, often from only one home. Although solids settle in the tank, grease and oil rise to the top, from which they are periodically removed. The bacteria in the wastewater feed on the bottom sludge, thereby liquefying the waste. Partially purified water flows out of the tank into an underground drain, where further decontamination takes place. The system is relatively passive, compared to central facilities, and, as in the case of artificial marshes, time is required for the processes to occur. In addition, nitrogen compounds are converted to nitrate but the latter is *not* then reduced to molecular nitrogen, so groundwater under the septic system can become contaminated by nitrate.

11.32 The Treatment of Cyanides in Wastewater

The **cyanide ion,** CN^-, binds strongly to many metals, especially those of the transition series, and is often used to extract them from mixtures. Consequently, cyanide is widely used in mining, refining, and electroplating metals such as gold, cadmium, and nickel. Unfortunately, cyanide ion is very poisonous to animal life since it binds strongly to metal ions in living matter, e.g., to the iron in proteins that are necessary for molecular oxygen to be utilized by cells.

Cyanide is a very stable species and does not quickly decompose on its own or in the environment. Thus it is an important water pollutant, and should be destroyed chemically rather than simply disposed of in a waterway.

We can deduce the type of treatment that will be effective for cyanide by considering its acid–base and redox characteristics. Cyanide ion is the conjugate base of the weak acid HCN, **hydrocyanic acid,** which has limited solubility in water. Thus acidification of a cyanide solution will result in the release of poisonous HCN gas from it and therefore is not a good solution to the problem of cyanide contamination.

The redox chemistry of cyanide ion can be predicted by considering the oxidation numbers of the two atoms involved. If nitrogen, the more electronegative atom, is considered to be in its fully reduced –3 oxidation form, then the carbon must be +2. Thus one way to destroy cyanide ion is to oxidize the carbon more fully, to +4 as it is in CO_2 or HCO_3^-. This oxidation can be accomplished by dissolved molecular oxygen if high temperatures and elevated air pressures are used:

$$2\ \overset{(+2)}{C}\overset{(-3)}{N^-} + \overset{(0)}{O_2} + 4\ H_2O \longrightarrow 2\ \overset{(+4)(-2)}{HCO_3^-} + 2\ NH_3$$

The use of stronger oxidizing agents, such as Cl_2 or ClO^-, not only oxidizes the carbon from +2 to +4, but can also oxidize the nitrogen from the –3 state to the zero oxidation number of molecular nitrogen:

$$2\ \overset{(+2)(-3)}{CN^-} + 5\ \overset{(0)}{Cl_2} + 8\ OH^- \longrightarrow 2\ \overset{(+4)}{CO_2} + \overset{(0)}{N_2} + 10\ \overset{(-1)}{Cl^-} + 4\ H_2O$$

Other oxidizing agents that are used in cyanide treatment include **hydrogen peroxide,** H_2O_2, and/or oxygen, both with a copper salt added as a catalyst. The process can also be carried out electrochemically for high cyanide concentrations; the resulting low concentration can be subsequently oxidized by ClO^-.

Sodium cyanide is now used in some shallow tropical waters such as those in Indonesia to stun reef fish so that they can be captured and sold live as seafood or pets. Unfortunately, the cyanide kills smaller fish and destroys the coral.

Four of the ten electrons gained collectively by the chlorines were located on the carbons and six were on the nitrogens in this overall process.

PROBLEM 11-14

Deduce the balanced half-reaction when cyanide ion is oxidized in acid media by hypochlorite ion to bicarbonate ion and molecular nitrogen. ●

PROBLEM 11-15

If an oxidizing agent even more powerful than chlorine or hypochlorite were to be used in the treatment of cyanide, what other possibilities for the nitrogen-containing product would there be? ●

PROBLEM 11-16

For HCN in water, $K_a = 6.0 \times 10^{-10}$. Calculate the fraction of cyanide that exists as the anion rather than in the molecular form at pH values of 4, 7, and 10. ●

11.33 The Disposal of Sewage Sludge

The sludge from both the primary and secondary treatment stages of sewage is principally water and organic matter. It can be digested anaerobically, in a process that takes several weeks to complete. Bacteria levels in the sludge are not thereby completely eliminated, but the levels are reduced about a thousand-fold. The sludge that remains after this further organic decomposition has occurred and after the supernatant water is removed sometimes is then incinerated or used as landfill or simply dumped into a landfill or into a water body such as the ocean. However, **sludge is high in plant nutrients,** so about half the sewage sludge in North America and Europe is spread on farm fields, golf courses, and even residential lawns as low-grade fertilizer sometimes called *biosolid.*

Unfortunately, **sewage sludge may contain toxic substances,** which potentially could be incorporated into food grown on the land or could contaminate groundwater under the fields. In particular, **heavy-metal concentrations often are higher in sewage sludge than in soil,** principally because industrial wastes are sometimes released directly into sewage lines shared by households. For example, the lead level in municipal sludge can range from several hundred to several thousand parts per million, compared to an average of about 10 ppm in the Earth's crust. In a few communities, an attempt is made to eliminate these toxic materials before final disposal occurs.

Some scientists have worried that food crops grown in soil fertilized by sewage sludge may incorporate some of the increased amounts of heavy metals. Control experiments indicate that vegetables vary greatly in the extent to which they will absorb increased amounts of the metals; e.g., the uptake of lead by lettuce is particularly large, but that by cucumbers is negligible. The concentration of arsenic in such agricultural soils is also greatly increased if arsenic pesticides are applied to them; crops planted on these soils subsequently absorb some of the adsorbed arsenic. Other substances of concern in using sewage sludge as fertilizer for food are alkylphenols from detergents, brominated fire retardants, and pharmaceuticals—especially antibiotics given to farm animals.

Review Questions 25–30 are based upon the preceding section.

Modern Wastewater and Air Purification Techniques

The most important chemical (as opposed to biological) pollutants dissolved in wastewater usually are chloroorganics, phenols, cyanides, and heavy metals. Below we describe some of the high-tech methods that have recently been developed and put into practice to purify wastewater, particularly for removal of chloroorganics. Some of these same techniques are also used to cleanse the compounds from contaminated air. It should be realized that treating large volumes of water to remove trace contaminants, especially without producing toxic by-products, can be an expensive undertaking.

Chloroorganics are organic compounds containing chlorine, and are discussed in detail in Chapters 13 and 14.

11.34 The Destruction of Volatile Organic Compounds

The major stationary sources in North America of VOCs are the evaporation of organic solvents, the manufacture of chemicals, and the petroleum industry and its storage activities. Wastewater effluent that is contaminated with

VOCs, e.g., the water emanating from chemical or petrochemical plants, is commonly treated by a two-step process:

1. The VOCs are removed from the wastewater by air stripping. In this process, air is passed upward into a downward stream of the water, and the volatile materials are transferred from the liquid to the gas phase. This technique does not work well for compounds that are highly water-soluble.

2. The resulting VOCs, now present in low concentration in a contained mass of humid air, **are destroyed by a process of catalytic oxidation.** For example, the air, heated to 300–500°C, is passed for a short time over platinum, or, depending upon the VOC, some other precious metal, that is supported on alumina. The energy costs of this step are very high since it involves heating a large volume of humid air. Note that the outlet air from such processes contains HCl if the VOCs originally contained chlorine; this compound must be removed by scrubbing with a basic substance before the air is released into the atmosphere.

The removal of VOCs from *gaseous* emissions from industries usually operates by this same catalytic oxidation process; typically the concentration of VOCs in the air stream is thereby reduced by 95%. A primary heat exchanger recovers and reuses the VOCs' heat of combustion to warm incoming gases to the operating temperature.

The **adsorption** of compounds onto activated carbon (see Box 11-1) or onto synthetic carbonaceous adsorbents is a cost-effective technology used for the removal of low-level VOC concentrations from both liquid and vapor streams, and is also useful for nonvolatile organic compounds. These adsorbents can be easily regenerated by treatment with steam or by other thermal techniques as well as by solvents; the concentrated pollutants can be subsequently destroyed by catalytic oxidation.

11.35 Advanced Oxidation Methods for Water Purification

Conventional water purification methods often do not successfully deal with synthetic organic compounds such as chloroorganics that are dissolved at low concentrations; examples include the common groundwater pollutants trichloroethene and perchloroethene. The conventional method for the treatment of water containing such pollutants is adsorption of the chloroorganics onto activated carbon; this removes the compounds but doesn't destroy them. The wastewater from pulp and paper mills also contains organochlorines that are resistant to conventional treatments.

In order to cleanse water of these extra-stable organics, so-called **advanced oxidation methods** (AOMs) have been developed and deployed. The aim of these methods is to **mineralize** the pollutants, i.e., to convert them entirely to CO_2, H_2O, and mineral acids such as HCl. Most AOMs are ambient-temperature processes that use energy to produce highly reactive intermediates of high oxidizing or reducing potential, which then attack and destroy the target compounds. The majority of the AOMs involve the generation of significant amounts of the **hydroxyl free radical,** OH, which in aqueous solution is a very effective oxidizing agent, as it is in air (see Chapter 3). The hydroxyl radical can initiate the oxidation of a molecule by extraction of a hydrogen atom or addition to one atom of a multiple bond as it does in air (Chapter 3); in water, as an additional alternative, it can also extract an electron from an anion.

Since the generation of OH in solution is a relatively expensive process, it is economical to use AOMs to treat only the wastes that are resistant to the cheaper, conventional treatment processes. Thus, integrating an AOM with pretreatment of the wastewater by biological or other processes to first dispose of the easily oxidized materials is often appropriate.

Ultraviolet (UV) light is often used to initiate the production of hydroxyl radicals and thus to begin the oxidations. Commonly, **hydrogen peroxide,** H_2O_2, is added to the polluted water and UV light from a strong source in the 200–300-nm range is shone on the solution. The hydrogen peroxide absorbs the ultraviolet light (especially that closer to 200 nm than to 300 nm), and uses the energy so obtained to split the O—O bond, resulting in the formation of two OH radicals:

$$H_2O_2 \xrightarrow{\text{UV}} 2 \text{ OH}$$

Alternatively and less commonly, ozone is produced and then photochemically decomposed by UV. The resulting oxygen atom reacts with water to efficiently produce OH via the intermediate production of hydrogen peroxide, which is photolyzed:

$$O_3 \xrightarrow{\text{UV}} O_2 + O$$

$$O + H_2O \longrightarrow H_2O_2 \xrightarrow{\text{UV}} 2 \text{ OH}$$

A fraction of the oxygen atoms produced by ozone photolysis are electronically excited, and these react with water to directly produce hydroxyl radicals, as discussed in Chapters 1 and 3.

PROBLEM 11-17

Given that the enthalpies of formation for H_2O_2 and OH are, respectively, -136.3 and $+39.0$ kJ mol^{-1}, calculate the heat energy required to dissociate one mole of hydrogen peroxide into hydroxyl free radicals. What is the maximum wavelength of light that could bring about this transformation? [*Hint:* see Chapter 1]. Given that light of 254-nm wavelength is usually used, and that all the energy of each photon that is in excess of that required to dissociate one molecule is lost as waste heat, calculate the maximum percentage of the input light energy that can be used for dissociation itself.

Hydroxyl radicals for wastewater treatment can also be efficiently produced without the use of UV light by combining hydrogen peroxide with ozone. The chemistry of the intermediate processes is complex, but the overall reaction between these two species is

$$H_2O_2 + 2\ O_3 \longrightarrow 2\ OH + 3\ O_2$$

This *ozone/H$_2$O$_2$ method* is more cost-effective and is easier to adapt to existing water treatment systems than is any other AOM system.

It is also possible to generate the hydroxyl radical electrolytically. In most such applications, a metal ion (such as Ag^+ or Ce^{3+}) is first oxidized to a more positively charged ion (Ag^{2+} or Ce^{4+} in our examples) that will subsequently oxidize water to H^+ and OH.

The biggest liability associated with advanced oxidation processes is that their action produces toxic chemical by-products. For example, in the ozone/peroxide and peroxide/UV treatments of groundwater contaminated with trichloroethene and perchloroethene, the toxic intermediates trichloroacetic acid and dichloroacetic acid are formed in about 1% yield.

$$CCl_3 - C \overset{\displaystyle O}{\underset{\displaystyle OH}{\big<}}$$

trichloroacetic acid

11.36 Photocatalytic Processes

Another innovative technology for wastewater treatment involves the irradiation by UV light of solid semiconductor photocatalysts such as **titanium dioxide,** TiO$_2$, small particles of which are suspended in solution. Titanium dioxide is chosen as the semiconductor for such applications since it is nontoxic, is very resistant to photocorrosion, is cheap and plentiful, absorbs light efficiently in the UV-A region, and can be used at room temperature. Irradiation at wavelengths less than 385 nm produces electrons, e^-, in the conduction band and *holes,* h^+, in the valence band of the metal oxide. The holes in the valence band of the semiconductor can react with surface-bound hydroxide ions or with water molecules, thereby producing hydroxyl radicals in both cases:

The concepts of holes and electron bands were previously discussed in Section 8.17.

$$h^+ + OH^- \longrightarrow OH$$

$$h^+ + H_2O \longrightarrow OH + H^+$$

The holes can also react directly with adsorbed pollutants, producing radical cations that readily engage in subsequent degradation reactions.

Normally, O_2 molecules dissolved in the water react with the electron produced at the semiconductor surface, a process which eventually produces more reactive free radicals, but which is relatively slow. If hydrogen peroxide is added to the water instead, it will react with the electron to form the anion radical and generate reactive radicals more quickly.

The cost of the electrical energy required to generate the needed UV light is usually the major expense in operation of AOM systems. On this basis, the titanium dioxide methods are even less cost-effective than those described previously, since considerably more electricity is required per pollutant molecule destroyed. Sunlight could be used to supply the UV light, but only about 3% of its light lies in the appropriate UV-A range and can be absorbed by the solid. Another problem with the TiO_2 processes is the difficulty in separating the various reactants and products from the TiO_2 particles if the metal oxide has been used in the form of a fine powder. However, there now are closed systems in which the titanium dioxide slurry is efficiently separated from the purified water and recycled back to the inlet stream.

Some scientists have experimented with immobilizing TiO_2 as a thin film (1 μm thick) on a solid surface such as glass, tile, or alumina. Indeed, TiO_2-coated tiles are now used on walls and floors in some buildings. The low-level UV light from fluorescent lighting in such rooms is sufficient to allow the destruction of gaseous and liquid-phase pollutants that touch the oxide on the tiles! For example, odors that upset people are usually present in air at concentrations of only about 10 ppm; at such levels, the UV from normal fluorescent lighting should be sufficient to remove them with TiO_2 photocatalysts. Bacterial infections such as those that cause many secondary infections in hospitals can also be eliminated by spraying walls and floors (in rooms lit by fluorescent bulbs) to give them a titanium dioxide film.

Photocatalysts are quiet, unobtrusive cleansing materials. Since irradiated TiO_2 surfaces can inactivate viruses not fully affected by UV light, a combination of the two may be used in the future to disinfect water.

11.37 Other Advanced Oxidation Methods

A process called **direct chemical oxidation** has been proposed for the destruction of solid and liquid organic wastes in the aqueous phase, particularly in environments such as those under buildings, where the light required for UV processes cannot conveniently be supplied. It uses one or another of the strongest known chemical oxidants—e.g., acidified **peroxydisulfate anion,** $S_2O_8^{2-}$, under ambient pressure and moderate temperatures to oxidize the wastes. Such a process needs no catalysts, and produces no secondary wastes of concern. The sulfate ion that results from the peroxydisulfate can be recycled back to the oxidant. Other very strong oxidizing agents that have been tested are the **peroxymonosulfate anion,** HSO_5^- and the **ferrate ion,** FeO_4^{2-}; in the latter, iron has an oxidation number of +6, so it is not surprising that it is a strong oxidizing agent. Unfortunately, ferrate ion suffers from the problem of instability.

PROBLEM 11-18

Deduce the balanced half-reaction (acidic media) in which the peroxydisulfate ion is converted into sulfate ion. Repeat the exercise for the conversion of oxalic acid, $C_2H_2O_4$, into carbon dioxide. Combine these half-reactions into a balanced equation, and calculate the volume of 0.010 M peroxydisulfate that is required to oxidize 1 kg of oxalic acid.   ●

Review Questions 31–34 are based upon the material in the preceding section.

Review Questions

1. Describe the function of *aeration* in the purification of drinking water.

2. Describe the chemistry underlying the removal of excess calcium and magnesium ions from drinking water.

3. What are *colloids*? How are they removed from drinking water?

4. Describe how water can be disinfected by (a) membrane filtration, (b) reverse osmosis, and (c) ultraviolet irradiation.

5. Describe the two main methods used to desalinate seawater. Why is the drinking water produced by desalination relatively expensive?

6. What two other chemical methods, other than chlorination, are used to disinfect water? What are some advantages and disadvantages to these alternatives? What does DBP stand for?

7. Explain the chemistry underlying the disinfection of water by chlorination. What is the active agent in the destruction of the pathogens? What are the practical sources of the active ingredient?

8. Explain why pH control of water in swimming pools is important. What compounds are formed when the chlorinated water reacts with ammonia?

9. Discuss the advantages and disadvantages of using chlorination to disinfect water, including the nature of the THM compounds. What is "combined chlorine"?

10. Describe three important *point-of-use* techniques by which water can be disinfected.

11. What is meant by the terms *groundwater* and *aquifer*? How does the *saturated zone* of soil differ from the *unsaturated zone*?

12. Why did concern about groundwater pollution lag far behind that about surface water?

13. Name three important sources of nitrate ion to groundwater.

14. Construct a table that shows the common oxidation numbers for nitrogen. Deduce in which column the following environmentally important compounds belong: HNO_2, NO, NH_3, N_2O, N_2, HNO_3, NO_3^-. Which of the species become prevalent in aerobic conditions in a lake? Under anaerobic conditions? What is the oxidation number of nitrogen in NH_2OH?

15. Explain why excess nitrate in drinking water or food products can be a health hazard; include the relevant balanced chemical reaction showing how nitrate becomes reduced.

16. What is an *N-nitrosamine*? Write the structure and the full name for NDMA.

17. What is the formula for the perchlorate ion? What is the origin of perchlorate ion in U.S. drinking water?

18. Define *leachate*.

19. Name two types of organic contaminants found in groundwater, and give two examples of each type.

20. Explain the difference in vertical location in an aquifer between compounds such as chloroform and those such as toluene.

21. Define the term *plume* and describe how it forms in an aquifer.

22. Why are the BTX and MTBE components of gasoline the ones that are most often found in groundwater? Are both components easily biodegraded?

23. Define what is meant by the terms (a) DNAPL and (b) bioremediation.

24. What is meant by *reductive degradation*? Describe the *in situ* technique by which chloroorganics in aquifers can be destroyed by reductive dechlorination.

25. What procedures are involved in primary wastewater treatment? In secondary treatment?

26. List five possible water purification processes that are associated with the tertiary treatment of wastewater, including one that removes phosphate ion.

27. What polyphosphate was commonly used in detergents, and why did its use lead to environmental problems? What are the other main sources of phosphate to natural waters? What other builders are used in detergents?

28. Describe two important methods that are used to desalinate wastewater.

29. Describe the chemical processes by which cyanide ion can be removed from wastewater.

30. Describe how *artificial marshes* work.

31. Describe how VOCs dissolved in wastewater are usually removed and destroyed.

32. Describe what AOM stands for, and state the most common reactive agent in such processes. Describe three methods by which this reactive species can be generated.

33. Describe two photocatalytic methods that can destroy organic wastes.

34. What is *direct chemical oxidation*? What are two of the strong oxidizing agents that can be used for such procedures?

 ## Green Chemistry Questions

1. What are the environmental advantages of using iminodisuccinate compared to most chelating agents?

2. The development of iminodisuccinate by Bayer won a Presidential Green Chemistry Challenge Award.

(a) Which of the three focus areas (see page xxviii) for these awards does this award best fit into?

(b) List at least three of the twelve principles of green chemistry (see the list on pages xxiii–xxiv) that are addressed by the green chemistry developed by Bayer.

Additional Problems

1. Given that for HOCl, $K_a = 2.7 \times 10^{-8}$, deduce the fraction of a sample of the acid in water that exists in the molecular form at pH values (predetermined by the presence of other species) of 7.0, 7.5, 8.0, and 8.5. [*Hint:* Derive an expression that relates the fraction of HOCl that is ionized to the concentration of hydrogen ions.] Would it be a good idea to allow the pH of swimming pool water to rise to 8.5?

2. For efficient disinfection to occur, treated drinking water should contain about 0.5 mg L^{-1} of Cl_2 that remains after most of the chlorine has

been converted to HOCl. What pressure of $Cl_2(g)$ is required to maintain this concentration? The value of Henry's law constant for Cl_2 in water is 8.0×10^{-3} M atm^{-1}.

3. At a particular temperature, K_a for HOCl is 3.5×10^{-8}. What would be the pH values for 1.00 M and 0.100 M concentrations of HOCl at this temperature? What percentages of the HOCl is unionized at these two concentrations?

4. The equilibrium constant for the reaction of dissolved molecular chlorine with water to give hydrogen ions, chloride ions, and HOCl is 4.5×10^{-4}, where as usual the concentration of water is included in the K value. If the pH of the solution is determined by other processes so that the amount of hydrogen ion contributed by the chlorine reaction is negligible, calculate the fraction of the original 50 ppm chlorine that remains as Cl_2 at pH values of 0, 1, and 2. [Notes: (1) The dissociation of HOCl into ions is negligible at these low pH values. (2) Approximate solutions to the quadratic equation involved in these calculations will not be accurate due to the high percentage of reaction.]

5. As mentioned in the margin in the text, chlorinated water contains a small concentration of chlorine monoxide, Cl_2O. This chlorinating agent is produced when two molecules of HOCl react together:

$$2\ HOCl \rightleftharpoons Cl_2O + H_2O$$

(a) By calculating oxidation numbers, decide whether or not this is a redox reaction.

(b) Given that for this reaction, $K = 9 \times 10^{-3}$ and remembering that water does not appear in the equilibrium constant expression, calculate the equilibrium concentration of Cl_2O that exists in a typical water chlorination treatment plant, where $[HOCl] = 3 \times 10^{-5}$ M.

6. Calculate the oxidation number of the chlorine in molecular chlorine, HOCl, chlorine dioxide, monochloramine (see Problem 11-3), and sodium chloride. Given that the last item is the most stable form of chlorine, predict whether the other substances mentioned are likely to be oxidizing agents or reducing agents, and rank them in likely order of this redox behavior. Using this analysis and the section on chlorine compounds in your introductory chemistry textbook, suggest other compounds that might be useful to disinfect water.

7. Calculate the volume of $Ca_5(PO_4)_3OH$, the density of which is 3.1 g mL^{-1}, which is produced

for each gram of sodium tripolyphosphate present in a detergent, when it is removed in tertiary wastewater treatment. Estimate the annual mass of detergent used for laundry purposes for a typical household of four persons, and assuming that the phosphate levels in laundry detergents used to be about 50%, calculate the volume that was required annually to dispose of the family's waste laundry phosphate.

8. To desalinate seawater by reverse osmosis, pressure in excess of a solution's osmotic pressure, π, must be applied across the membrane. The total osmotic pressure exerted in a solution is determined by the total molar concentration, M, of its solutes, and is given by the equation $\pi = MRT$. Using the composition of seawater listed on page 445 of Chapter 10, determine the minimum pressure that must be exerted on seawater to desalinate it using reverse osmosis at 20°C.

9. What could be done to dispose of the solvents that are used to extract VOCs from adsorbents? Which of the methods you suggest would least contaminate the environment?

10. Given their names, can you deduce the nature of the similarity in molecular structure between hydrogen peroxide and the sulfur compounds that are used in direct chemical oxidation methods? By calculating the oxidation state of the atoms in hydrogen peroxide, deduce why it can act as an oxidizing agent.

11. Water samples from three wastewater streams were analyzed, and the important pollutants were determined to be those listed below. In each case, devise economical, practical processes (other than activated carbon treatment) for purifying the water of the three pollutants:

(a) phosphate ion, ammonium ion, and salt (in water containing bicarbonate ion)

(b) nitrite ion, PCE, and Fe^{2+}

(c) cadmium ion, carbon tetrachloride, and glucose

12. Chlorine-containing substances covering a wide variety of oxidation numbers have been encountered, some within this chapter and others previously in the book. For each substance in the list below, write out its formula, deduce the oxidation number of its chlorine, and fill in the appropriate row of the table below, as in the example shown:

chlorite ion   perchlorate ion

molecular chlorine   hypochlorous acid

chlorine dioxide   chlorine monoxide

chloride ion   chlorate ion

Oxidation Number	Formula of Example	Name of Example
−1		
0		
+1		
+2		
+3		
+4		
+5		
+6	ClO_3	Chlorine trioxide
+7		

12

Toxic Heavy Metals

In this chapter, the following introductory chemistry topics are used:

- ⟳ Redox half-reactions
- ⟳ Electrolysis
- ⟳ Basic structural organic chemistry (see online Appendix)
- ⟳ Half-life calculations
- ⟳ Solubility product and acid–base equilibrium constant calculations
- ⟳ Electron configurations of atoms and atomic ions

Background concepts from previous chapters used in this chapter:

- ⟳ Steady state; UV and visible light wavelengths
- ⟳ Synergism
- ⟳ Concentration scales in aqueous solutions
- ⟳ LD_{50}; maximum contaminant level
- ⟳ Aerobic, anaerobic, and calcareous waters
- ⟳ Radioactive decay series

Introduction

In chemistry, **heavy metal** refers not to a type of rock music but rather to a type of chemical element, many examples of which are poisonous to humans. The five main ones discussed here—**mercury** (Hg), **lead** (Pb), **cadmium** (Cd), **chromium** (Cr), and **arsenic** (As)—present the greatest environmental hazards due to their extensive use, their toxicity, and their widespread distribution. None has yet pervaded the environment to such an extent as to constitute a widespread danger. However, each one has been found at toxic levels in certain locales in recent times. Metals differ from the toxic organic compounds to be discussed in Chapters 13, 14, and 15 in that they are totally nondegradable to nontoxic forms; they ultimately may be transformed to insoluble forms, which therefore are biologically unavailable unless they are again converted into more soluble substances. The ultimate sinks for heavy metals are soils and sediments.

The heavy metals occur near the middle and bottom of the periodic table. Their densities are high compared to those of other common materials. For example, the densities of mercury and lead are more than ten times that of water.

Although we commonly think of heavy metals as water pollutants, they are for the most part transported from place to place via the air, either as gases or as species adsorbed on, or absorbed in, suspended particulate matter.

12.1 Speciation and the Toxicity of Heavy Metals

Although mercury *vapor* is highly toxic, the heavy metals Hg, Pb, Cd, Cr, and As are not particularly toxic as the *condensed* free elements. However, **all five are dangerous in the form of their cations and most are also highly toxic when bonded to short chains of carbon atoms.** Biochemically, the mechanism of the toxic action usually arises from the strong affinity of the cations for sulfur. Thus, *sulfhydryl groups*, —SH, which occur commonly in the enzymes that control the speed of critical metabolic reactions in the human body, readily attach themselves to ingested heavy metal cations or to molecules that contain the metals.

Because the resultant metal–sulfur bonding affects the entire enzyme, it cannot act normally; as a result, human health is adversely affected, sometimes fatally. The reaction of heavy metal cations M^{2+} (where M is Hg, Pb, or Cd) with the sulfhydryl units of enzymes R—S—H to produce stable systems such as R—S—M—S—R is analogous to their reaction with the simple inorganic chemical **hydrogen sulfide,** H_2S, with which they yield the insoluble solid sulfide MS.

PROBLEM 12-1

Write the balanced chemical reactions that correspond to the reaction of an M^{2+} ion (a) with H_2S, and (b) with R—S—H, to produce hydrogen ions and the products mentioned above.   ●

The toxicity for all four heavy metals depends very much on the chemical form of the element, i.e., upon its speciation. Substances that are almost totally insoluble pass through the human body without doing much harm. The most devastating forms of the metals are

● those that cause immediate sickness or death (e.g., a sufficiently large dose of arsenic oxide) so that therapy cannot exert its effects in time, and

● those that can pass through the membrane protecting the brain—the blood–brain barrier—or the placental barrier that protects the developing fetus.

For mercury and lead, the forms that have alkyl groups attached to the metal are highly toxic. Because such compounds are covalent molecules, they are soluble in animal tissue, and they can pass through biological membranes, whereas charged ions are less able to do so; e.g., the toxicities of lead as the ion Pb^{2+} and in covalent molecules differ substantially.

The toxicity of a given concentration of a heavy metal present in a natural waterway depends not only on its speciation, but also on the water's pH and on the amounts of dissolved and suspended carbon in it, since interactions such as complexation and adsorption may well remove some of the metal ions from potential biological activity.

Mercury

12.2 Elemental Mercury

Elemental mercury was employed in hundreds of applications, many of which took advantage of its unusual property of being a liquid that conducts electricity well. In automobiles built before 2000, electrical switches that operate convenience and trunk lighting contained mercury, as did instrument panels and antilock brakes; all of this mercury is lost to the environment when the cars are recycled for their steel unless the element is specifically collected.

Elemental mercury is still used in fluorescent light bulbs, including the small ones now used domestically to replace incandescent bulbs, and in the mercury lamps employed for street lighting. Although energized mercury atoms emit light in the ultraviolet rather than the visible wavelength region of the spectrum (Chapter 1), the bulbs are coated with a material that absorbs the UV and re-emits it from the bulb as visible light. The metal is released into the environment if such light bulbs are broken; however, the mercury content of fluorescent lamps has been reduced by about 80% since the mid-1980s, down to 5–10 mg each today. This amount of mercury is less than the additional quantity of the metal that would have been emitted into the air by a coal-fired power plant if an incandescent light bulb, with its much lower efficiency in converting electricity to light, had been used instead. Fluorescent bulbs are virtually the only use of mercury for which a suitable alternative has not been found. For street lighting, there has been a shift toward the use of sodium vapor lamps, since such bulbs present a lower toxicity hazard and are even more efficient light sources than mercury bulbs.

Mercury is the most volatile of metals, and its vapor is highly toxic. Adequate ventilation is required whenever mercury is used in closed quarters, since the equilibrium vapor pressure of mercury is hundreds of times the maximum recommended exposure. **Mercury vapor consists of free, neutral atoms, $Hg^0(g)$. If inhaled, the atoms diffuse from the lungs into the bloodstream, and then, because they are electrically neutral they readily cross the blood–brain barrier to enter the brain.** The result is serious damage to

Thermometers containing liquid mercury used to be quite common, but have largely been phased out.

the central nervous system, which is manifested by difficulties with coordination, eyesight, and tactile senses. *Liquid* mercury itself is not highly toxic, and most of that ingested is excreted. Nevertheless, children should not be allowed to play with droplets of the metal because of the danger from breathing the vapor.

The latter part of the twentieth century saw a substantial decline in anthropogenic emissions of mercury into the water and land environments from many sources in *developed* countries, resulting from governmental attempts to reduce its uses and emissions. Emissions of mercury from large industrial operations in developed countries have been successfully curtailed. The overall use of mercury has decreased in the United States by more than 95%. For example, mercury has been eliminated from almost all batteries. In the United States, the reduction of mercury emissions arising from disposal of mercury-containing products has resulted mainly

- from emission controls on municipal and medical-waste incinerators, and

- by removing batteries and paint from the waste stream.

In Canada, emission reductions have come from controls on metal smelting, the near-complete closure of the chlor-alkali industry (which had used mercury electrodes), and controls on waste incineration.

12.3 Mercury Amalgams

Mercury readily forms amalgams, which are solutions or alloys with almost any other metal or combination of metals. The dental amalgam that has been used to fill cavities in teeth for more than 150 years initially has a putty-like consistency and is prepared by combining approximately equal proportions of liquid mercury and a solid mixture that is mainly silver with variable amounts of copper, tin, and zinc. The slight expansion of volume that accompanies its solidification ensures that the final amalgam fills the cavity.

When a filling is first placed in a tooth, and whenever the filling is involved in the chewing of food, a tiny amount of the mercury is vaporized. Some scientists believe that mercury exposure from this source causes long-term health problems in some individuals, but an expert panel of the U.S. National Institutes of Health concluded that dental amalgams do not pose a health risk. A recent study of adults found no measure of exposure to dental mercury—neither the concentration of the element in the urine nor the number of dental fillings—correlated with any measure of mental functioning or fine-motor control; another study found no IQ or other neurological problems correlated with the use of amalgams for filling teeth in children.

Some countries in Europe, such as Germany, have banned the use of mercury in fillings, at least for pregnant women and small children. Mercury-free "amalgams" for use in dentistry are under development; porcelain fillings already are common, though expensive. Some fears have been expressed about the release of elemental mercury vapor into the atmosphere when dead human bodies that have amalgam-filled teeth are cremated, since the amalgam decomposes at high temperatures. In countries such as Sweden, crematoria are fitted with selenium filters that remove most mercury from emissions by forming mercury selenide crystals.

In some areas, dentists are now required to install a separator to capture mercury from their wastewater rather than have it flow down drains and become part of municipal sewage. On average, each dentist produces about one kilogram of mercury waste per year. Dentists collectively release about the same amount of the metal as emitted by coal-fired power plants. By means of the control of emissions by dentists, the sewage sludge used by farmers as fertilizer (Chapter 16) will have a much lower concentration of mercury in the future.

12.4 Mercury Emissions from Gold and Silver Production

In working some ore deposits, tiny amounts of elemental gold or silver are extracted from much larger amounts of the denser particles of soil or sediment by adding elemental liquid mercury to the mixture. The mercury extracts the gold or silver by forming an amalgam, which is then roasted to distill off the mercury. From 1570 until about 1900, this process was used to extract silver from ores in Central and South America. About one gram of mercury was lost to the environment for every gram of silver so produced, resulting in the release of almost 200,000 tonnes of mercury. The mercury was shipped to these regions from Almaden, Spain, and from Peru. Until recently, the corresponding process of extracting gold by amalgamation with mercury was carried out in China in both large-scale and small-scale operations. The ratio of mercury-to-gold in the workshops, some of which continue illegally today in remote regions, averaged 15:1.

Photo by H. John Maier Jr./Image Works/Time Life Pictures/Getty Images

Today, the gold extraction procedure using mercury is still carried out on a large scale in Brazil to obtain gold from muddy sediments, and results in substantial mercury pollution both in the air and, because of careless handling practices, in the Amazon River itself. The health hazards to workers using processes that involve the vaporization of mercury are significant, since the element is so toxic in its gaseous form. Indeed, mercury vaporized from such operations currently makes up more than 10% of the anthropogenic emissions of mercury in air. People who live in the mining regions often inhale air in which the concentration of elemental mercury exceeds 50 μg m^{-3}, which is 50 times the WHO public exposure guideline.

As a consequence, many "amalgam burner" workers exhibit tremors and other signs of mercury poisoning. In addition, mercury in surface sediments disturbed by slash-and-burn deforestation and agriculture in the region enters the aquatic environment, where some of it enters the food chain. Initiatives have been undertaken by the European Union to incorporate inexpensive technology into the process to prevent the massive release of mercury into the air and to the Amazon River during the extraction of gold.

12.5 Mercury and the Industrial Chlor-Alkali Process

An amalgam of sodium and mercury is used in some industrial **chlor-alkali** plants in the process that converts aqueous sodium chloride into the commercial products **chlorine, Cl_2, and sodium hydroxide, NaOH,** (and gaseous hydrogen) by electrolysis. In order to form a concentrated, pure solution of NaOH, flowing mercury is used as the negative electrode (cathode) of the electrochemical cell. The metallic sodium that is produced by reduction in the electrolysis immediately combines with the mercury and is removed from the NaCl solution without having reacted with the aqueous medium:

$$Na^+(aq) + e^- \xrightarrow{\text{Hg}} Na \text{ (in an Na–Hg amalgam)}$$

When metals such as sodium are dissolved in amalgams, their reactivity is greatly lessened compared to that for the free state, so that the otherwise highly reactive elemental sodium in the Na–Hg amalgam does not react with the water in the original solution. Instead, the amalgam is removed, and later induced by the application of a small electrical current to react with water in a separate chamber, thereby producing sodium hydroxide that is free of salt.

The mercury is recovered after NaOH production, and is recycled back to the original cell. The recycling of mercury is not complete, however, and some finds its way into the air and into the water body from which the plant's cooling water is obtained and to which it is returned. Although liquid mercury is neither soluble in water nor in dilute acid, it can be oxidized to soluble form by the intervention of bacteria that are present in natural waters. By this means the mercury becomes accessible to fish.

The mass of mercury lost to the environment from the average chlor-alkali plant has decreased enormously since the problem was identified in the 1960s. Nevertheless, installations in North America that use mercury electrodes have largely been phased out. They were replaced by those that use a fluorocarbon membrane that separates the NaCl solution from the chloride-free solution at the negative electrode. The membrane is designed such that Na$^+$, but not anions, can pass through it. In both types of cells, the overall reaction is

$$2\,NaCl(aq) + 2\,H_2O(l) \longrightarrow 2\,NaOH(aq) + Cl_2(g) + H_2(g)$$

12.6 The 2+ Ion of Mercury

Like its partners zinc and cadmium in the same subgroup of the periodic table, the element has an s^2p^0 valence-shell electron configuration in the free atom, and consequently **the common ion of mercury is the 2+ species, Hg^{2+}, the mercuric or mercury(II) ion.** An example of a compound containing the mercuric ion is the red ore *cinnabar*, HgS, i.e., $Hg^{2+}S^{2-}$. Like most sulfides, this salt is very insoluble in water; indeed, the wastewater at chloralkali plants is sometimes treated by adding a soluble salt such as Na_2S that contains the sulfide ion, since this action precipitates ionic mercury as HgS.

$$Hg^{2+} + S^{2-} \longrightarrow HgS(s)$$

The other inorganic ion of mercury, Hg_2^{2+}, is not very toxic since it combines in the stomach with chloride ion to produce insoluble Hg_2Cl_2. In terms of electron configurations, the ion Hg_2^{2+} corresponds to a single covalent bond formed by two s^1p^0 ions of Hg^+.

PROBLEM 12-2

The solubility product, K_{sp}, for HgS is 3.0×10^{-53}. Calculate the solubility of HgS in water in moles per liter, and transform your answer into the number of mercuric ions per liter. According to this calculation, what volume of water in equilibrium with solid HgS contains a single Hg^{2+} ion?

PROBLEM 12-3

A quantity of a mercury–chlorine compound is included in a shipment of waste to a toxic-waste disposal dump. Before it can be disposed of properly, the owners of the dump need to know whether the compound is $HgCl_2$, or Hg_2Cl_2, or some other compound. They send a sample of it for analysis, and find that it contains 26.1% chlorine by mass. What is the empirical formula of the compound?

Most of the mercury in the environment is inorganic, in the form of the Hg^{2+} ion, or complex inorganic ions formed by it. The levels of ionic mercury even in remote areas are two to five times as great as preindustrial values, with local polluted sites having levels ten times greater or more. In natural waters, much of the Hg^{2+} is attached to suspended particulates, and so is eventually deposited in sediments. This topic is considered in further detail when soil and sediment chemistry is discussed (Chapter 16).

Mercuric oxide, HgO, was present in a paste in *mercury cell* batteries such as those that were used in hearing aids. If the discarded spent batteries are subsequently incinerated as garbage, the volatile mercury can be released into the air. The amount of mercury used in ordinary flashlight batteries, added as a minor constituent in the zinc electrode to prevent its corrosion and to thereby extend the shelf life of the product, was first drastically

HgO consists of Hg^{2+} and O^{2-} ions.

curtailed—typically from about 10,000 ppm to about 300 ppm in alkaline batteries—and in many cases is now completely eliminated, thereby reducing by half the mercury in domestic garbage.

Very small amounts of mercury are still permitted in the "button batteries" used in hearing aids, watches, etc. The mercury prevents the release and buildup of hydrogen gas, which would otherwise occur when the zinc electrode gradually became corroded over time by the alkaline electrolyte:

$$Zn + 2\,H_2O \longrightarrow Zn^{2+} + H_2 + 2\,OH^-$$

By modifying the silver oxide electrode in such batteries to absorb any H_2 produced and by modifying the zinc electrode to reduce corrosion, Sony in 2009 introduced completely mercury-free button batteries.

Large amounts of mercury vapor currently are released into the air as a result of the unregulated burning of coal and fuel oil, both of which always contain trace amounts of the element, and of incinerating municipal waste that contains mercury in products such as batteries. Coal-fired power plants and municipal and medical waste incinerators are the biggest sources of mercury emissions to the atmosphere in North America. In Canada, primary base metal and iron and steel industries also contribute significantly to the load. Globally, about one-fifth of emissions arise from the production of gold and other nonferrous metals. The vaporized mercury is eventually oxidized and returns in rain, often falling far from the site of the original emissions—another example of the **tragedy of the commons** phenomenon discussed in the introductory chapter.

In air, the great majority of elemental mercury is in the vapor (gaseous) state, with only a fraction of it bound to airborne particles. Airborne gaseous mercury can travel long distances before being oxidized and then dissolving in rain and subsequently being deposited on land or in water bodies. This global cycling of mercury results in its being distributed even to remote parts of the planet. The concentration of Hg^0 in air is so small—about 2 ng m^{-3} on average, though wintertime levels in some Chinese cities are more than ten times that high—that it usually will not directly affect human health, even though mercury vapor is a very toxic form of the element. However, once mercury is deposited and enters the aquatic environment, it can be transformed by microorganisms to the very toxic and bioavailable *methylmercury*, as will be discussed in Section 12.10.

12.7 Mercury Emissions from Power Plants

Overall, about one-third of airborne mercury results from natural sources on land and water, especially wildfires. Of the other two-thirds, half are new emissions from point sources such as power plants, and the other half are recycled, legacy anthropogenic emissions that were temporarily deposited on land and water. Because the direct emission of mercury-containing wastes into natural waters and onto land has largely ceased in developed countries, environmental concentrations of mercury have fallen substantially there from their levels in the 1960s and 1970s. However, the levels are apparently not *continuing* to fall rapidly toward natural concentrations because one important source of anthropogenic mercury has not yet been tackled globally: the deposition onto water and soil from air into which the element has been released, largely from the burning of coal to produce electricity and the incineration of wastes.

Mercury occurs in coal as the Hg^{2+} ion in the inorganic, *pyritic* component of the fuel at a level of about 0.2 parts per million on average, the content varying from 0.01 to 1.5 ppm. During combustion of the coal, much of this mercury is liberated into the air rather than becoming part of the ash. About 4 billion tonnes of coal are currently combusted globally per year, corresponding to a potential mercury release of 800 tonnes, about 40% of global air emissions of the element.

In the mid-1990s, atmospheric mercury emissions in North America from incineration of waste were also large, due to the presence of mercury in products such as batteries, and amounted to more than 50% of emissions from fossil-fuel combustion. Since then, however, the use of mercury in batteries has been almost completely eliminated in developed countries, so incineration should now be a smaller source. Ironically, because there are restrictions on the placement of mercury-containing hazardous waste in landfills in the United States, much mercury waste now is burned, releasing most of the element into the air!

The global anthropogenic emissions into the air of mercury from all other sources combined are of about the same magnitude as that from coal combustion, yielding a total output of about 2000 tonnes per year. Eastern North America, western and central Europe, central and South Africa, China, Japan, India, and parts of Russia all have many high-emission mercury sites.

Asia is now the largest source of mercury to the environment, contributing more than half the total. Emissions from developing countries have risen as those from developed ones have fallen. Indeed, China is now the world's largest consumer of coal, accounting for 47% of the global total in 2009. Europe and North America have decreased their mercury emissions over the past few decades, and now each accounts for about 10% of the total. Overall, however, the reductions in mercury emissions achieved by the United States, Canada, and other developed nations have been offset by increased airborne emissions from developing countries, especially in Asia, so there has been little change in the total emission rate of the element over the last three decades. However, the global atmospheric concentration of mercury continuously *declined* by about 1.5% annually, amounting to about a 30% reduction in total, in the period from 1995 to 2010. This decline probably arose primarily from decreasing re-emissions from past deposits of mercury in soil and water.

Projections of the release of mercury into the air over the next few decades foresee almost the same annual amounts arising from the Americas,

Europe, the Middle East, and Africa, but significant increases from Asia. However, the totals depend heavily on whether economic expansion in Asia continues to be fueled mainly by coal combustion and on whether mercury capture from power-plant emissions is instituted.

12.8 The Nature of Airborne Mercury

The question of how far airborne mercury from power plants and incinerators travels before it is deposited on land or in water—and in particular, whether it is deposited locally or regionally or is globally dispersed—is an important and controversial issue in North America. To understand this issue, we must first consider the different forms of mercury that exit from these sources.

One form of emitted mercury is the gaseous, free, neutral atoms, Hg^0. This elemental mercury is formed in the combustion process by the chemical reduction of the Hg^{2+} that was present in the fuel. **The element is emitted mainly as a monatomic gas—rather than condensing to a liquid—since the temperatures in the combustion process and in the exhaust gases greatly exceed 357°C, the boiling point of the element.** Some of the Hg^0 is adsorbed onto particulates in the stack emissions.

In contrast with the behavior of organic pollutants that undergo long-range atmospheric transport, mercury vapor does not condense out in cool climates. In addition, it is not water-soluble. Rather, it continues to circulate in air as Hg^0 until it is eventually re-oxidized by OH or O_3 to Hg^{2+}, after which it is rained out relatively soon since the ion is water soluble. The average lifetime of **Gaseous Elemental Mercury**, GEM, $Hg^0(g)$, in air is about a year. Consequently, **no significant fraction of the GEM emitted by a power plant or incinerator is deposited in the region of its emission.**

A fraction (roughly 50%) of the mercury emitted from power plants and incinerators occurs in an oxidized form, since some of the atomic mercury vapor released during coal combustion is oxidized in the exhaust gas. Some of the oxidized mercury consists of Hg^{2+} ions that are bound to particles in the emissions, and is called **Particulate Mercury,** or PM (or TPM, for Total Particulate Mercury). The actual compound may be mercuric oxide, HgO. Since airborne particles tend to deposit within a few weeks, most particulate mercury is deposited regionally, downwind of the power plant.

Do not confuse this use of the term PM with the air pollution index discussed in Chapter 4.

The other emitted form of oxidized mercury is called Reactive Gaseous Mercury, or RGM. It is probably gaseous **mercuric chloride,** $HgCl_2$, the molecular compound of oxidized mercury that is formed when Hg^0 reacts with the various forms of chlorine-containing gases (Cl, Cl_2, HCl) released from the coal upon its combustion:

$$Hg^{2+}(coal) \xrightarrow{\text{flame}} Hg^0(g) \xrightarrow{Cl_x} HgCl_2(g)$$

However, much of the reaction to form oxidized mercury may be heterogeneous, and occur on surfaces, again to produce $HgCl_2$ ultimately. **The fraction** of mercury oxidized to Hg^{2+} depends significantly upon the chlorine content of the coal being combusted.

The lifetime of RGM in air is relatively short (about 1–10 days) since mercuric chloride is highly water-soluble and so is readily deposited by wet precipitation. For this reason, RGM is mainly deposited locally and regionally downwind of its source.

The various forms of airborne mercury, and their lifetimes, are summarized in the accompanying table:

Form of Mercury	Chemical Formula	Estimated Atmospheric Lifetime
Gaseous elemental mercury (GEM)	$Hg^0(g)$	Months–year
Particulate mercury (TPM)	Hg^{2+} (adsorbed, possibly as HgO, on solid); some adsorbed Hg^0	Few weeks
Reactive gaseous mercury (RGM)	$HgCl_2(g)$	Days–week

12.9 Mercury Emission Control by Power Plants

Particulate mercury can readily be captured by power plants—such as many of those in the United States and Europe but few of those in developing countries—that have electrostatic precipitators or fabric filters installed to efficiently remove particles. Since the $HgCl_2$ of RGM is water-soluble, it is efficiently removed from exhaust gases if they are scrubbed to remove other pollutants such as sulfur dioxide. Most of the mercury captured by these technologies would deposit locally and regionally if not removed. The various forms of mercury emitted from power plants and other sources and their destinies are illustrated schematically in Figure 12-1.

To date it has proven to be virtually impossible to keep all or most elemental mercury vapor from being emitted from coal-fired power plants, since the gas Hg^0 is inert. It has resisted chemical attempts—such as adding catalysts to exhaust gases to make it react—to trap it completely. Powdered activated carbon injection into the exhaust gas stream can be used to adsorb and thereby trap mercury vapor; the powder subsequently is captured by a fabric filter or an electrostatic precipitator. Partially brominating the carbon leads to higher capture efficiency. Indeed, adding elemental *bromine gas*, Br_2, or a chlorine-containing gas such as SCl_2 or S_2Cl_2 to the flue gas from the power plant oxidizes much of the elemental mercury to the covalent, water-soluble compounds $HgBr_2$, *mercuric bromide*, or $HgCl_2$, mercuric chloride, both of which can be removed efficiently during water-based scrubbing for other pollutants, thereby converting them to aerosols that can be collected

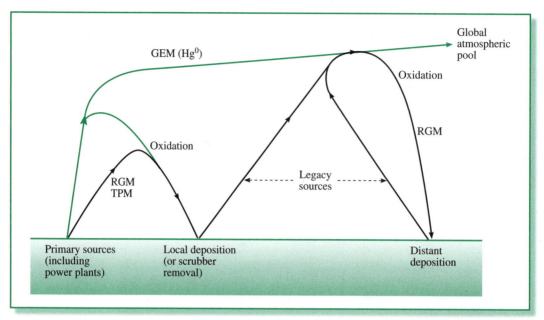

FIGURE 12-1 Schematic representation of the emissions of various forms of mercury into the atmosphere and their eventual destiny. Notice that the processes depicted at the left of the diagram could occur in the open air, resulting in deposition within a few kilometers, or within the power-plant scrubber unit, resulting in immediate collection. [Source: Adapted in part from M. Gustin and D. Jaffe, *Environmental Science and Technology*, 44 (2010): 2222–2227.]

by electrostatic precipitators. The initial oxidation of the mercury may occur on surfaces and be mediated by SO_2 in the exhaust gases, producing $HgSO_4$ which then reacts with HCl to produce the dichloride.

Dozens of new technologies are under development for mercury emission control. In one, called *Chem-Mod*, Hg^0 is captured by a liquid adsorbent material that either oxidizes it to Hg^{2+} or traps it on its surface. Another new technology irradiates the exhaust gases with UV light (ironically, using a mercury lamp), thereby ionizing the mercury atoms into the oxidized Hg^{2+} state, which then reacts with HCl already present in the exhaust gases to produce $HgCl_2$.

The 1200 coal-burning power plants in the United States are its last unregulated major emitters of mercury into the environment. As a result of legal pressure from environmental groups, by court order the U.S. EPA was forced in 2005 to issue regulations that reduce these emissions. The reductions are scheduled to occur in two phases:

• By 2010, total mercury emissions were to have been reduced from the current 48 tons a year to 38 tons. This was accomplished as a "co-benefit" of the new reductions in sulfur dioxide and nitrogen oxide emissions from the plants that would have been implemented by that date, as discussed in Chapter 3, since it mainly involved capture of oxidized mercury.

- By 2018, a further reduction of annual emissions to 15 tons total of mercury was to have occurred. For this to be accomplished, some capture of GEM will be required.

Each state has been assigned an emissions budget. States and power plants can trade reductions among themselves, as long as overall goals are achieved. Critics of the new program point out that averaging of reductions between years is allowed, so that higher emissions could occur in later years if greater-than-regulated reductions occur early. They also point out that "hot spots" of mercury deposition may be perpetuated because of the trading provisions, and that the overall reduction in emissions will be quite small for many years to come.

The EPA rules were further modified in 2011.

ACTIVITY

Using a search engine (e.g., Google Scholar) designed to search the technical scientific literature, explore current research ideas concerning the removal of Hg^0 from power-plant emissions. Prepare a one-page report or a few PowerPoint slides summarizing the technique you find which seems most promising.

12.10 Methylmercury Toxicity

When in combination with anions that are capable of forming covalent bonds, the mercuric ion, Hg^{2+}, forms covalent molecules rather than an ionic solid. For example, $HgCl_2$ is a molecular compound, *not* a salt of Hg^{2+} and Cl^-. In terms of electron configurations, in such covalent molecules, one of the s electrons in the s^2p^0 free mercury atom form is promoted to the empty p orbital, giving the s^1p^1 configuration and a divalent atom.

Just as chloride ion forms a covalent compound with Hg^{2+}, so does the *methyl anion*, CH_3^-, yielding the volatile molecular liquid **dimethylmercury,** $Hg(CH_3)_2$. **The process of dimethylmercury formation occurs in the muddy sediments of rivers and lakes, especially under anaerobic conditions, when anaerobic bacteria and microorganisms convert Hg^{2+} into** $Hg(CH_3)_2$. The active agent in the biomethylation process is a common constituent of microorganisms; it is a derivative of Vitamin B_{12} with a CH_3^- anion bound to cobalt and is called *methylcobalamin*.

The less volatile mixed compounds CH_3HgCl and CH_3HgOH, collectively called **methylmercury** (or *monomethylmercury*), are often written as CH_3HgX, or somewhat misleadingly as $CH_3Hg^+X^-$, since these substances, like most of those written as Hg^{2+}, consist of covalent molecules, not ionic lattices. In fact, the methylmercury ion CH_3Hg^+ exists as such only in compounds with anions such as nitrate or sulfate.

Methylmercury compounds are even more readily formed in the same way as dimethylmercury at the surface of sediments in anaerobic water.

FIGURE 12-2 The cycling of mercury in fresh-water lakes. Note that most CH_3Hg^+ species are in fact CH_3HgX molecules, not actual ions. [Source: Adapted from M. R. Winfrey and J. W. M. Rudd, "Environmental Factors Affecting the Formation of Methylmercury in Low pH Lakes," *Environmental Toxicology and Chemistry* 9 (1990): 853–869.]

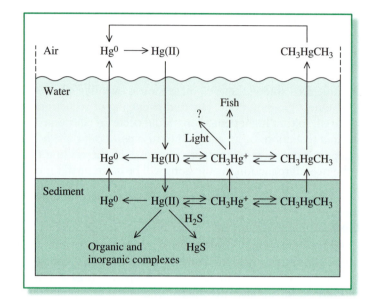

Methylmercury production predominates over dimethylmercury formation in acidic or neutral aqueous systems. Sulfate ion stimulates the sulfate-reducing bacteria that methylate the mercury; in contrast, the presence of sulfide ion results in formation of mercury sulfide complexes that do not undergo methylation.

Due to its volatility, dimethylmercury evaporates from water relatively quickly unless it is transformed by acidic conditions into the monomethyl form. The pathways for the production and fate of dimethylmercury and of other mercury species in a body of water are illustrated in Figure 12-2. Methylation of inorganic mercury does occur in anaerobic regions of lakes, especially near the interface of the epilimnion and the hypolimnion, and at the interface of the latter with sediments, but not in aerobic water. **Organic-rich sediments at the bottom of warm, shallow lakes are important sites of methylmercury production.** Wetlands are also active areas of methylmercury production. Methylmercury in surface water is photodegraded (to as yet unknown products) and is the most important sink for this substance in some lakes.

Mercuric ion itself is not readily directly transported across biological membranes. Methylmercury is a more potent toxin than are salts of Hg^{2+} because it is soluble in fatty tissue in animals and accumulates there, and is more mobile. Once ingested, the original covalent CH_3HgX compound is converted to substances in which X is a sulfur-containing amino acid; in some of these forms, it is soluble in biological tissue, and can cross both the blood–brain barrier and the human placental barrier, presenting a twofold hazard. **Methylmercury is, in fact, the most hazardous form of mercury, more dangerous even than the vapor form of the element.** The main toxicity of methylmercury occurs in the central nervous system. In the brain, methylmercury is converted to Hg^{2+}, which probably is responsible for the brain damage. Mercury vapor is also oxidized to this ion once it has entered the cell. Thus the usual barriers in the cell to Hg^{2+} are circumvented by Hg^0 and CH_3HgX, which, by their electric neutrality can penetrate the cell's defenses, and then be converted to the highly toxic $+2$ ionic form.

12.11 Mercury in the Human Diet

Most of the mercury present in humans is in the form of methylmercury. Almost all methylmercury originates from the fish in our food supply: mercury in fish is usually at least 80% methylmercury. Mercury contamination is the reason behind about 97% of the advisories against eating fish caught in various regions of North America. Indeed, the number of water bodies—both rivers and lakes—in North America having fish advisories due to mercury contamination is actually increasing, not falling.

In contrast to organochlorines, which predominate in the fatty portions of fish, **methylmercury can bind to the sulfhydryl group in proteins and so is distributed throughout the fish.** Consequently, the mercury-containing part cannot be cut away before the fish is eaten. At its current levels in the environment, mercury poses its greatest health risk to people whose diet is largely local fish and game, to others whose fish consumption exceeds the average, and to young children and pregnant women. It is estimated that hundreds of thousands of babies born annually in the United States could have some degree of neurological deficit owing to their exposure to mercury in the womb.

Organochlorines are organic compounds containing chlorine, and are discussed in detail in Chapters 13 and 14.

Fish absorb methylmercury that is dissolved in water as it passes across their gills, and they also absorb it from their food supply. The ratio between methylmercury in fish muscle and that dissolved in the water in which the fish swims is often about one million to one, and can exceed ten million to one. The highest methylmercury concentrations (over 1 ppm) usually are found in large, long-lived predatory marine species such as shark, king mackerel, tilefish, swordfish, and large tuna (sold as steaks and sushi), and in freshwater species such as bass, trout, and pike. Indeed, the U.S. Food and Drug Administration warns women of childbearing age not to eat the first four types of marine fish in the list above.

On average, the older the fish, the more methylmercury it will have accumulated. On average, most Americans consume almost half their methylmercury from tuna (mostly of the canned variety), followed by swordfish, pollock, shrimp, and cod. Canada has limited the maximum concentration of mercury in six species of ocean fish to 1.0 ppm; this concentration is not uncommon in swordfish, though most fish have levels of 0.10–0.15 ppm. The U.S. EPA has set a criterion of 0.3 ppm maximum for methylmercury in fish tissue.

The phenomenon of
bioaccumulation is
discussed in detail in
Chapter 13.

Noncarnivorous species such as whitefish do not accumulate very much mercury, since magnification along their food chain operates to a much lesser extent than in carnivorous fish. Most fish in Chinese reservoirs are noncarnivorous, with low mercury concentrations. Indeed, rice grown in areas of China contaminated by mercury from metal smelting is the dominant source of methylmercury to the local residents.

In lakes, the mercury content in fish is generally greater in acidic water, probably because both the solubility of mercury is greater and the methylation of mercury is faster at lower pH. In this way, the acidification of natural waters indirectly increases the exposure of fish-eaters to methylmercury. The relationship between mercury levels in small fish and the pH of the water is illustrated in Figure 12-3. The data in this figure is from a collection of lakes in Wisconsin, eastern Ontario, and Nova Scotia; most of the acidic lakes are in Nova Scotia.

The half-life of methylmercury compounds in humans, about 70 days, is much longer than that of Hg^{2+} salts, due in part to its greater solubility in a lipid environment. Consequently, methylmercury can accumulate in the body to a much higher steady-state concentration, even if on a daily basis a person consumes doses that individually would not be harmful.

PROBLEM 12-4

If the half-life in the human body of methylmercury is 70 days, what is its steady-state accumulation in a person who consumes daily 1 kg of fish containing 0.5 ppm methylmercury? [*Hint:* Recall the discussion of steady-state concentrations in Chapter 5.]   ●

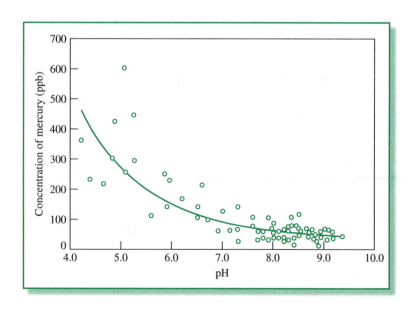

FIGURE 12-3 The relationship between pH and mercury concentration in a fish for a standard fish length for 48 lakes from Ontario, Nova Scotia, and Wisconsin.
[Source: D. Lean, "Mercury Pollution," *Canadian Chemical News* (January 2003): 23.]

12.12 Safe Level of Mercury in the Body

It is somewhat reassuring that both the direct effects of methylmercury on humans and the prenatal effects probably have thresholds below which no effects are observed. Currently the daily methylmercury intake of 99.9% of Americans lies below the World Health Organization's "safe limit." Nonetheless, some effects of methylmercury consumption on human vision are observed even when the concentration of (total) mercury in hair lies below the generally recognized threshold of 50 ppm.

However, if prenatal health is the main consideration and if a safety factor of ten is applied, a substantial fraction of the population of the United States would exceed the safe limit. The World Health Organization has concluded that levels of 10–20 ppm of methylmercury in hair indicate that a pregnant woman has sufficient methylmercury in her blood to represent a threat to a developing fetus. This places at risk the developing fetus of more than 30% of the women in some native communities in northern Canada, for example, in which fish play a large part in the diet. Although it is clear that high levels of methylmercury can result in developmental disabilities, there is continuing controversy whether methylmercury acquired through a diet high in fish and marine mammals can cause significant neurological damage to an adult or a developing fetus. The epidemiological studies of this question have, to this point, produced inconsistent results.

PROBLEM 12-5

What is the mass, in milligrams, of mercury in a 1.00-kg lake trout which just meets the North American standard of 0.50 ppm Hg? What mass of fish, each at the 0.50-ppm Hg level, would you have to eat in order to ingest a total of 100 mg of mercury?   ●

PROBLEM 12-6

The U.S. EPA recommends a maximum dietary intake for methylmercury of 0.1 μg per day per kilogram of the human's body weight. What mass of fish can a 60-kg woman safely eat each week if the average methylmercury level in the fish is 0.30 ppm? Approximately how many average servings of fish does this correspond to?   ●

12.13 The Minamata Disaster

At the fishing village of Minamata, Japan, a chemical plant employing Hg^{2+} as a catalyst in a process that produced polyvinyl chloride discharged mercury-containing residues into Minamata Bay. The methylmercury compounds, mainly CH_3Hg—SCH_3, that subsequently formed from the inorganic mercury through biomethylation by microorganisms in the bay's sediments then bioaccumulated to concentrations as high as 100 ppm in the fish, which were the main component of the diet for many local residents.

By way of contrast, the current North American recommended limit for mercury in fish to be consumed by humans is 0.5 ppm.

Thousands of people in Minamata were affected in the 1950s by mercury poisoning from this source, and hundreds died from it. Because the onset of symptoms in humans is delayed, the first signs of *Minamata disease* were observed in cats which ate discarded fish: they began jumping around and twitching, ran in circles, and finally threw themselves into the water and drowned. Symptoms in humans arise from dysfunctions of the central nervous system, since the target organ for methylmercury is the brain. They include numbness in arms and legs, blurring and even loss of vision, loss of hearing and muscle coordination, and lethargy and irritability.

Since methylmercury can be passed to the fetus, children born to Minamata mothers poisoned even slightly by mercury showed severe brain damage, some to a fatal extent. The infants showed symptoms similar to those of cerebral palsy: mental retardation, seizures, and motor disturbance, even paralysis. Just as in the case of high PCB levels, to be discussed in Chapter 14, the developing fetuses were much more affected by methylmercury than were the mothers themselves. The poisonings at Minamata must surely rank as one of the major environmental disasters of modern times.

12.14 Other Sources of Methylmercury

Organic compounds of mercury have been used as fungicides in agriculture and in industry and enter the environment as a side effect of these applications. However, as a result of contact with soil, the compounds are eventually broken down and the mercury becomes trapped as insoluble compounds by attachment to sulfur present in clays and organic matter.

Mercury is leached from rocks and soil into water systems by natural processes, some of which are accelerated by human activities. Flooding of vegetated areas can release mercury into water. For example, after the flooding of huge areas of northern Quebec and Manitoba in Canada to produce hydroelectric power dams, the newly submerged surface soils (and to a lesser extent the vegetation) released a considerable quantity of soluble methylmercury, formed from the "natural" mercury content of these media. The additional methylmercury resulted from contact of soil-bound Hg^{2+} with anaerobic bacteria produced by the decomposition of the immersed organic matter. In this way, previously insoluble inorganic mercury was converted to methylmercury, which readily dissolved in the water. The methylmercury subsequently entered the food chain through its absorption by fish, and native persons who ate fish from these flooded areas now have substantially elevated levels of mercury in their bodies. Indeed, the methylmercury concentration in fish from these areas, 5 ppm or more, approaches that previously associated only with regions of industrial mercury pollution.

Recent studies of new reservoirs in China created for hydroelectric power show little initial production of methylmercury because there is little organic matter in the freshly flooded soil. However, methylmercury levels increase there over time—rather than decrease as in North America—because

This issue was explored in detail in the online Chapter 8 Case Study *Mercury Pollution and the James Bay Hydroelectric Project (Canada)*.

the reservoirs gradually accumulate sediment, which facilitates conversion of the Hg^{2+} in the water to methylmercury.

Review Questions 1–13 are based on the above sections.

Lead

12.15 Lead Through History

Although the environmental concentration of lead, Pb, is still increasing in some parts of the world, the uses that result in its uncontrolled dispersion have been greatly reduced in the last few decades in many developed countries. Consequently, **lead's concentration in the soil, water, and air there has decreased substantially.**

Lead's relatively low melting point of 327°C allows it to be readily worked—it was the first metal to be extracted from its ores—and shaped. Lead was used as a structural metal in ancient times, and for weatherproofing buildings, in water pipes and ducts, and for cooking vessels. Lead is still used for roofing and flashing and for soundproofing in buildings. When combined with tin, it forms *solder*, the low-melting alloy used in electronics and in other applications (e.g., tin cans) to connect solid metals.

Analysis of ice-core samples from Greenland indicate that atmospheric lead concentration reached a peak in Roman times that was not equaled again until the Renaissance. The history of lead's presence in the environment can be seen in Figure 12-4, in which the ratio of two stable lead isotopes in samples taken from a peat bog in Switzerland is plotted against the depth in the bog at which the sample was taken. The layers of the bog were laid down gradually over millennia, and each layer incorporated lead-containing dust particles deposited from the air at the time.

Lead consists of a mixture of several stable isotopes, each produced originally from different radioactive decay series (Chapter 9) with different decay times. Consequently, the ratio of lead isotopes in a given sample is dependent upon the rock type and the lead ore or soil from which the element was derived, as well as upon the geographical location of the source. Thus **we can deduce the origin of atmospheric lead before it became airborne, knowing the isotope ratios of different sources.** The $^{206}Pb/^{207}Pb$ ratio in the peat samples in Figure 12-4 is close to 1.20 for dust laid down more than 3000 years ago; the sediment ages are determined by ^{14}C dating of their peat content. Beginning at about that time, however, the ratio in the dust fell to 1.18 due to the dominance of large emissions—first from mining lead-contaminated silver and later from European lead ores mined during the Roman era and thereafter, both of which had this lower $^{206}Pb/^{207}Pb$ ratio. Previous to that, in Greek times, silver was first mass-produced for use in coins; apparently the substantial amount of lead contaminant in the crude silver escaped into the air during the refining of the metal.

In about 1860, the isotope ratio of the lead deposited in Europe began a continuing decrease, with the rate of change increasing with the

FIGURE 12-4 Isotopic composition of lead in a Swiss peat bog and the chronology of atmospheric lead deposition. Notice the change in depth scale at 100 cm. [Source: W. Shotyk et al., "History of Atmospheric Lead Deposition Since 12,370 BP from a Peat Bog, Jura Mountain, Switzerland," *Science* 281 (1998): 1635, Figure 3B.]

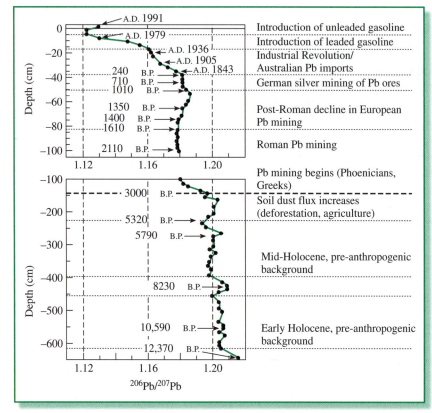

introduction of leaded gasoline in about 1940, probably as a consequence of the extensive use of lead derived first from Australia (ratio of 1.04) and then also from Canada in the gasoline. Recently, the ratio has begun to increase due to the decreased use of lead in European gasoline.

12.16 Elemental Lead in Ammunition

Elemental lead is found in ammunition (lead shot) used in huge amounts by hunters, especially of waterfowl. Many ducks and geese are injured or die from chronic lead poisoning after ingestion of lead shot, which dissolves in the acidic environment inside them. In addition, ducks consume the pellets left lying on the ground or at the bottom of ponds, since they look like food or grit. When birds that prey on some ducks and other waterfowl that have been shot by hunters but not harvested by them, or that have eaten lead shot to help grind food in their gizzard, these predators (such as bald eagles) became victims of lead poisoning. For these reasons, lead shot has been partially banned for shooting waterfowl in the United States, Canada, and some European countries. However, in North America, many loons die because they swallow and are subsequently poisoned by lead sinkers and jigs still used in sport fishing.

Lead ammunition in the form of bullets and shotgun shells that are used for shooting wild game also poses an environmental problem. Condors in California suffer from lead poisoning, sometimes fatally, when they eat deer that have been shot and then abandoned by hunters; the lead bullets explode into many fragments on impact and contaminate the meat. Humans too may be at risk when they eat game meat that is shot with lead ammunition.

Lead contamination of meat from livestock is regulated in the European Union, but game meat is exempt.

12.17 Ionic 2+ Lead and the Element

Although the elemental form is not an environmental problem to most life forms, lead does become a real concern when it dissolves to yield an ionic species. As expected from its s^2p^2 valence-shell electron configuration, **the common ions of lead are the +2 (loss of both p electrons) and +4 (loss of all s and p electrons).** The most common ion of lead is Pb^{2+}, i.e., Pb(II). Under highly oxidizing conditions, and in some covalent organometallic compounds, Pb(IV) as Pb^{4+} and PbR_4, respectively, are encountered.

Lead does not react on its own with dilute acids. Indeed, elemental lead is stable as an electrode in the *lead storage battery*, even though it is in contact with fairly concentrated **sulfuric acid,** H_2SO_4. However, some lead in the solder that was used commonly in the past to seal tin cans *will dissolve* in the dilute acid of fruit juices and other acidic foods *if air is present*—i.e., once the can has been opened—since **lead is oxidized by oxygen in acidic environments:**

$$2\ Pb(s) + O_2 + 4\ H^+ \longrightarrow 2\ Pb^{2+}(aq) + 2\ H_2O$$

The Pb^{2+} produced by this reaction contaminates the contents of the can; for this reason lead solder is not often used anymore for food containers in North America. Partially as a result of this change, the average daily intake of lead for two-year-old children dropped from about 30 μg in 1982 to about 2 μg in 1991.

The 1845 Franklin Expedition to find a Northwest Passage across the Arctic is thought to have failed because the members all died from lead poisoning from the solder in the tin cans that held their food. Canadian writer Margaret Atwood has written eloquently about the incident in her short story "The Age of Lead":

> . . . it was the tin cans that did it, a new invention back then, a new technology, the ultimate defence against starvation and scurvy. The Franklin Expedition was excellently provisioned with tin cans, stuffed full of meat and soup and soldered together with lead. The whole expedition got lead poisoning. Nobody knew it. Nobody could taste it. It invaded their bones, their lungs, their brains, weakening them and confusing their thinking, so that at the end those that had not yet died in the ships set out in an

Margaret Atwood, "The Age of Lead," in *Wilderness Tips,* copyright 1991 by O. W. Toad Limited.

idiotic trek across the stony, icy ground, pulling a lifeboat laden down with toothbrushes, soap, handkerchiefs and slippers, useless pieces of junk. When they were found (ten years later, skeletons in tattered coats, lying where they'd collapsed) they were headed back toward the ships. It was what they'd been eating that had killed them.

The effects of lead poisoning were known to the ancient Greeks, who realized that drinking acidic beverages from containers coated with lead-containing substances could result in illness. This information was not available to the Romans. Indeed, they sometimes deliberately adulterated overly acidic wine with sweet lead salts to improve the flavor. They also made use of lead sweeteners in cooking. The concentration of lead in the bones of Romans is almost 100 times that found for modern North Americans. Some historians have hypothesized that chronic lead poisoning, from wine and other sources, of upper-class Romans contributed to the eventual downfall of the Roman Empire because of the metal's detrimental effects on the neurological and reproductive systems. The latter effects include dysfunctional sperm in males and an inability to bring the fetus to term in females. Owing primarily to the contamination of beverages by lead from the distillation of alcohol in lead vessels, episodes of colic and gout due to lead poisoning were recorded through the Middle Ages and even until recent times.

Lower-class Romans could not afford lead-sweetened wine, and so did not suffer the same problems as did the upper classes.

12.18 Lead in Drinking-Water Systems

The recommended maximum levels for important heavy metal ions in drinking water are summarized in Table 12-1. The limit for lead, 10–15 ppb, is sometimes exceeded in water piped to the consumer even though it was sufficiently pure when it left the water treatment plant. Lead used in the solder in the joints of domestic copper water pipes, and lead used in previous decades and centuries to construct the pipes themselves, can dissolve in

TABLE 12-1	Drinking-Water Guidelines for Heavy Metals		
Metal	U.S. EPA Maximum Contaminant Level (ppb)	Canadian Maximum Acceptable Concentration (ppb)	World Health Organization Guideline (ppb)
As	10	10	10
Cd	5	5	3
Cr	100	50	50
Hg (inorganic)	2	1	6
Pb	15	10	10

drinking water during its transport to the point of consumption, particularly if the water is quite acidic or if it is particularly soft.

The contamination of water by lead is less of a problem in areas of calcareous water, since an insoluble layer containing compounds such as $PbCO_3$ forms on the surface of the lead by reaction of the metal with dissolved oxygen and the **carbonate ion,** CO_3^{2-}, in the water (Chapter 10). This layer prevents the metal underneath from dissolving in the water that passes over it. The Pb^{2+} in the insoluble carbonates deposited as scale can be oxidized further to the Pb^{4+} ion of the insoluble oxide PbO_2 by the action of residual hypochlorous acid from the disinfection process (Chapter 11):

$$PbCO_3(s) + HOCl + H_2O \rightleftharpoons PbO_2(s) + HCl + H_2CO_3$$

The scale then contains both $PbCO_3$ and PbO_2.

Release of lead from scale does not normally occur provided that no change occurs to the chemical properties of the water being transported through the pipes. However, if subsequently a change is made to a less powerful oxidizing agent—such as chloramines—to purify the water, the $+4$ lead in some of the PbO_2 present in the scale of pipes can be reduced to the soluble $+2$ form and contaminate the purified drinking water, particularly if the water is not alkaline. The half-reaction that releases the lead is

$$PbO_2(s) + 4 H^+ + 2 e^- \rightleftharpoons Pb^{2+} + 2 H_2O$$

Lead ion can also be released at the junction of lead and copper piping by the oxidation of Pb metal by the copper ion, Cu^{2+}, that itself was formed from its metal by the combination of chlorine and ammonia present in the water:

$$Pb(s) + Cu^{2+} \longrightarrow Pb^{2+} + Cu(s)$$

Thus, water purified and lead-free when it leaves the purification facility can become contaminated by dissolved lead before it reaches the consumer. Examples of lead release upon switching to chloramines for disinfection occurred in Washington, D.C., in 2003, and Greensville, North Carolina, in 2005. A similar result happened in London, Ontario, in 2007 when the acidity of the water was increased slightly.

PROBLEM 12-7

The scale on lead piping consists mainly of a mixture of Pb^{2+} compounds containing both carbonate, CO_3^{2-}, and OH^- ions. What are possible formulas, i.e., possible combinations of x and y, for compounds of the general formula $Pb_3(CO_3)_x(OH)_y$?

In some regions of England and in some cities in the northeast United States that have soft water and networks of old lead pipes, phosphates are added to drinking water in order to form an insoluble protective coating of *lead phosphate* on the inside of lead pipes and so reduce the concentration of dissolved lead.

In general, it is a good idea not to drink water that has been standing overnight in older drinking fountains or in the pipes of older dwellings; water in such plumbing systems should be allowed to run for a minute or so. Lead in water is more fully absorbed by the body than is lead in food. Now that many other sources of lead have been phased out, drinking water accounts for about one-fifth of the collective lead intake of Americans, whose major lead source is food. Many domestic water treatment systems successfully remove the great majority of lead from drinking water. Bottled water sold in plastic containers usually has very low levels of lead, averaging 16 ppt in one recent survey, which is not much higher than groundwater taken from pristine deep aquifers. Bottled water in glass containers has more lead, up to about 1 ppb, since tiny amounts of the metal are leached from the glass. However, some PET plastic water bottles leach small amounts of antimony, used as a catalyst in the manufacture of the plastic, into the liquid.

PET plastics are discussed in Chapter 16.

PROBLEM 12-8

According to an informal 1992 survey, the drinking water in about one-third of the homes in Chicago had lead levels of about 10 ppb. Assuming that an adult drinks about 2 L of water a day, calculate the total lead that residents of these Chicago homes obtain daily from their drinking water.   ●

12.19 Lead Salts as Glazes and Pigments

One form of the oxide PbO is a yellow solid that has been used at least as far back in history as ancient Egypt to glaze pottery. In glazing, the material is fused as a thin film to the surface of the pottery in order to make it waterproof and to give it a brilliant high gloss. The oxide becomes a hazard if applied incorrectly; some of it will dissolve over a period of hours and days if acidic foods and acidic liquids, such as cider, are stored in pottery containers, giving dissolved Pb^{2+} up to hundreds or even thousands of parts per million in the food:

$$PbO(s) + 2\ H^+(aq) \longrightarrow Pb^{2+}(aq) + H_2O$$

Indeed, lead-glazed dishware is still a major source of dietary lead, especially, but not exclusively, in developing countries. The leaching of lead from glazed ceramics used to prepare food is one of the major sources of the element for children in Mexico, where lead contamination is a leading public health problem. Nowadays, *lead silicate* rather than the oxide or sulfate is used for glazing in most countries since it is almost insoluble and thus much safer.

Various salts of lead have been used as pigments for millennia, since they give stable, brilliant colors. **Lead chromate,** $PbCrO_4$, is the yellow pigment used in paints applied to school buses and for the yellow stripes on roads. **Red lead,** Pb_3O_4, is a mixture of Pb(II) and Pb(IV) used in corrosion-resistant paints and has a bright red color. It was used in great quantities in the past to produce a rust-resistant surface coating for iron and steel.

Lead pigments have been used to produce the colors used in glossy magazines and food wrappers. In past centuries, lead salts were used as coloring agents in various foods. **White lead,** $Pb_3(CO_3)_2(OH)_2$, was extensively used until the middle of the twentieth century as a major component of white indoor paint. Since it was more durable than unleaded paint, it was often used on surfaces subject to punishment such as kitchen cabinets and window trim. However, when the paint peels off, small children might eat the paint flecks because Pb^{2+} has a sweet taste. Persons who renovate old homes are urged to ensure that dust from layers of old paint is properly contained. Children in inner-city slums, in which old coats of paint continue to peel, are often found to have elevated blood levels of lead.

In indoor paint, white lead has now been replaced by the pigment titanium dioxide, TiO_2.

Although now banned from use in indoor paints, lead pigments continue to be used in exterior paints, with the result that soil around houses may eventually become contaminated. Some of this lead-contaminated soil may be ingested by small children because of its sweet taste. Lead is still widely used in indoor paint sold in China, India, and some other Asian countries, sometimes at levels exceeding 180,000 ppm (as compared to the United States standard of 600 ppm maximum for new paints). Indeed, some children's toys manufactured in China were found to be covered by lead-containing paint and have been taken off store shelves and returned. A limit of 90 ppm in paint and surface coatings on any consumer product is in effect in the United States, and in Canada for any item used by children. A limit of 100 ppm for lead in all children's products was scheduled for 2011 in the United States.

An additional source of sweet lead-containing dust was the surface of some types of PVC miniblinds that had lead incorporated as a stabilizer in the plastic and that underwent partial decomposition from exposure to UV in sunlight. Lead is used as a stabilizer in other PVC products as well, including children's toys.

Lead dust, which originates as soil containing tiny particles of lead compounds, is now the biggest source of the element for children in inner cities. The lead collectively originates from individually small but numerous contributions from many of the sources mentioned above—paint flakes, ceramics, plastics, gasoline, recycling plants, and even lead salts used in hair coloring preparations for people with graying hair. The use of **lead arsenate,** $Pb_3(AsO_4)_2$, as a pesticide was another former source of Pb^{2+} to soil.

White lead acetate was probably used by Roman women as face paint.

12.20 Green Chemistry: Replacement of Lead in Electrodeposition Coatings

Sheet metal surfaces made of steel undergo corrosion very rapidly unless they are covered with a protective coating. Since the 1960s a technique called *electrodeposition* has competed with spray painting for coating steel.

In 1976, the first automobiles were treated by electrodeposition. In this technique, the surface to be treated is dipped in a bath, with the surface acting as a cathode or anode, and the coating is deposited electrophoretically. Electrodeposition has many advantages over spray painting, including

• lower air pollution, due to decreased solvent emissions,

• better corrosion protection, due to better coverage of poorly accessible areas,

• reduced waste, due to high transfer efficiency, and

• more uniform coating thickness.

Virtually all primer coats for automobiles are done by this method. Red lead, mentioned previously, offers significant corrosion resistance and primer coats use large amounts of this material. Although lead has been banned from house paints in the United States since 1972, the demand for corrosion resistance for motor vehicles has resulted in exemptions from environmental regulations regarding lead in automobile and truck paints.

PPG Industries has discovered that *yttrium oxide* serves as an excellent replacement for lead as a corrosion inhibitor and won a Presidential Green Chemistry Challenge Award in 2001. On a weight basis, yttrium is twice as effective in inhibiting corrosion as red lead but is only 1/120th as toxic. An additional consideration is the pretreatment process that is used to assist in adhesion and corrosion resistance prior to the application of the electrocoat. The use of yttrium eliminates chromium from metal pretreatments and reduces the amount of nickel compared to the lead process. It is estimated that employing yttrium in automobile electrodeposition will eliminate not only the use of 1 million pounds of lead but also 25,000 lb of chromium and 50,000 lb of nickel on an annual basis. As of 2006, more than 38 million motor vehicles had been coated with the yittrium-containing product since its introduction in 2001. According to PPG Industries, no customers in either the United States or Europe purchase any lead-containing coating product for any application, including automotive.

12.21 Dissolution of Otherwise-Insoluble Lead Salts

The presence of significant concentrations of lead in natural waters is seemingly paradoxical, given that both its sulfide, PbS, and its carbonate, $PbCO_3$, are highly insoluble in water:

$$PbS(s) \rightleftharpoons Pb^{2+} + S^{2-} \qquad K_{sp} = 8.4 \times 10^{-28}$$

$$PbCO_3(s) \rightleftharpoons Pb^{2+} + CO_3^{2-} \qquad K_{sp} = 1.5 \times 10^{-13}$$

However, the anions in both salts are fairly strong bases. Thus both of the above dissolution reactions are followed by the reaction of the anions with water:

$$S^{2-} + H_2O \rightleftharpoons HS^- + OH^-$$

$$CO_3^{2-} + H_2O \rightleftharpoons HCO_3^- + OH^-$$

Because these reactions reduce the concentrations of the original anions produced by dissolution of the salt PbS or PbCO$_3$, the position of equilibrium in the original reactions shifts to the right side, thereby dissolving more of the salt, in analogy with the process involving CaCO$_3$ that we analyzed in Chapter 10. **Thus the solubilities of PbS and PbCO$_3$ in water are substantially increased by the reaction of the anion with water** (see Additional Problem 2).

If highly acidic water comes into contact with minerals such as PbS, the "insoluble" solid dissolves to a much greater extent than in neutral waters. This occurs because the sulfide ion initially produced is subsequently converted almost entirely to **bisulfide ion, HS$^-$**, which in turn is converted by the acid to dissolved hydrogen sulfide gas, H$_2$S, since both S^{2-} and HS$^-$ act as bases in the presence of acid:

$$S^{2-} + H^+ \rightleftharpoons HS^- \qquad K = 1/K_a\,(HS^-) = 7.7 \times 10^{12}$$

$$HS^- + H^+ \rightleftharpoons H_2S \qquad K' = 1/K_a\,(H_2S) = 1.0 \times 10^7$$

When these two reactions are added to that for the dissolution of PbS into Pb^{2+} and S^{2-}, the overall reaction is seen to be

$$PbS(s) + 2\,H^+ \rightleftharpoons Pb^{2+} + H_2S(aq)$$

Since the equilibrium constant $K_{overall}$ for an overall process which is the sum of several others is the product of their equilibrium constants, in this case $K_{overall} = K_{sp}KK' = 6.5 \times 10^{-8}$. The expression for the equilibrium constant in terms of concentrations for this reaction is

$$K_{overall} = [Pb^{2+}]\,[H_2S]/[H^+]^2$$

Under conditions in which no significant amount of hydrogen sulfide gas is vaporized, but which are sufficiently acidic that almost all the sulfur exists as H$_2$S rather than as S^{2-} or HS$^-$, the stoichiometry of the reaction allows us to write that [Pb^{2+}] = [H$_2$S]. By substitution of this relationship into the above equation, we obtain

$$[Pb^{2+}]^2 = 6.5 \times 10^{-8}\,[H^+]^2 \quad \text{or}$$

$$[Pb^{2+}] = 2.5 \times 10^{-4}\,[H^+]$$

Thus the solubility of PbS increases linearly with the H$^+$ concentration in acidic water. At pH = 4, the solubility of PbS and the concentration of Pb^{2+} ion in water is calculated to be 2.5×10^{-8} M, whereas at pH = 2, the solubility is 2.5×10^{-6} M. We conclude that **dangerous concentrations of lead ion can occur in highly acidic bodies of water that are in contact with "insoluble" lead minerals.**

PROBLEM 12-9

By calculations similar to those for PbS above, deduce the relationship between the solubility of mercuric sulfide, HgS ($K_{sp} = 3.0 \times 10^{-53}$), and the hydrogen ion concentration in acidic water. Is the solubility of HgS increased substantially by exposure to acid? ●

12.22 Ionic 4+ Lead in Automobile Batteries

The elemental lead and the lead(IV) oxide, PbO_2, employed as the two electrodes in storage batteries in vehicles together now constitute the major use of the element. Storage batteries that are not recycled provide the main source of lead in municipal waste; some states and countries have banned the discarding of such batteries. However, the majority of used lead storage batteries are recycled for their lead content. During the recycling operation, lead can be expelled into the environment if careful controls are not maintained. Indeed, such recycling operations often constitute urban "hot spots" of lead emission into the surrounding communities.

Although lead-recycling operations in developed countries are carried out under strict control, this is not necessarily the case in developing countries, where batteries are often shipped for recycling. For example, recycling lead from car batteries had to be discontinued in Dakar, Senegal, after 18 children were found to have died from hand-to-mouth contact with lead-contaminated soil and sand.

12.23 Tetravalent Organic Lead Compounds as Gasoline Additives

Whereas the compounds of Pb(II) are ionic, **most Pb(IV) compounds are covalent molecules rather than ionic compounds of Pb^{4+}.** In this respect, tetravalent lead is similar to the corresponding behavior of the other elements (C, Si, Ge, Sn) in its group of the periodic table.

Commercially and environmentally, the most important covalent compounds of lead(IV) are *tetraalkyl* compounds, PbR_4, especially those formed with the methyl group, CH_3, and the ethyl group, CH_2CH_3, namely **tetramethyllead,** $Pb(CH_3)_4$, and **tetraethyllead,** $Pb(C_2H_5)_4$. In the past, both compounds found widespread use as additives to gasoline—about a gram per liter—to produce leaded gasoline. As discussed in Chapter 6, this practice now has been phased out in North America and in many other developed countries, except in aviation fuel, for which no acceptable substitute for lead has yet been found.

Since tetraalkyl lead compounds are volatile, they evaporate to some extent from gasoline and enter the environment in gaseous form. They are not water-soluble, but they are readily absorbed through the skin. In the human liver, PbR_4 molecules are converted into the more toxic compounds of PbR_3^+, which are neurotoxins because they can cross the blood–brain barrier. In substantial doses, these organic compounds of lead cause symptoms which mimic psychosis. It is not clear what the effects may be, if any, of chronic low-level exposure to them. At very high exposures, tetraalkyl lead compounds are fatal, as was discovered many years ago when several employees of the companies that originally produced these compounds died. In contrast to mercury, little or no methylation of inorganic lead occurs in nature. Thus almost all the tetraalkylated lead in the environment probably originated from leaded gasoline.

12.24 Environmental Lead from Leaded Gasoline

When these additives are used in gasoline, the atoms of lead that are liberated by the combustion of the tetraalkyl compounds must be removed before they form metallic deposits and damage the vehicle's engine. In order to convert the combustion products into volatile forms that can leave the engine in the exhaust gases, small quantities of *ethylene dibromide* and *ethylene dichloride* are also added to the leaded gasoline.

CH_2Br—CH_2Br and
CH_2Cl—CH_2Cl

As a result, the lead is removed from the engine and enters the atmosphere from the tailpipe in the form of a mixture of the mixed dihalide $PbBrCl$ and the dihalides $PbBr_2$ and $PbCl_2$. Subsequently, under the influence of sunlight, these compounds form PbO, which then exists in particulate form as an aerosol in the atmosphere for hours or days. Consequently, not all of it is deposited in the immediate surroundings of the roadway. The **unintended consequence** is that PbO can enter the food chain at more distant sites if it is deposited upon vegetables or on fields used by grazing animals.

A high proportion of environmental lead in many parts of the world is that emitted from vehicles, and occurs in the environment mainly in inorganic form. The conversion to nonleaded gasoline in North America and Europe, the initial impetus for which was the interference of lead in exhaust gases with the proper functioning of catalytic converters, has had the welcome side effect of greatly decreasing the average amount of lead ingested by urban inhabitants. Indeed, the noted environmentalist Barry Commoner has called the elimination of lead from gasoline "one of the (few) environmental success stories."

European scientists have traced the rise and fall of atmospheric alkylated lead by analyzing different vintages of a French red wine (*Chateauneuf-du-Pape*) that used grapes grown near two busy autoroutes. They found that the concentration of trimethyllead, PbR_3^+—the degradation product of the tetramethyl compound—rose steadily to a maximum in the mid-1970s,

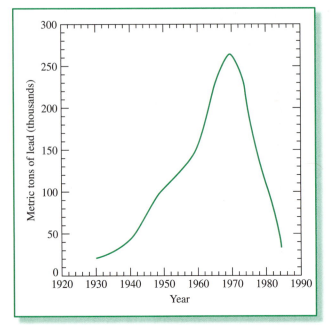

FIGURE 12-5 The
historical consumption of
lead in gasoline in the
United States. [Source: C. E.
Dunlop et al., "Past Leaded
Gasoline Emissions as a
Nonpoint Source Tracer in
Riparian Systems,"
*Environmental Science and
Technology* 34 (2000): 1211.]

which was followed by a steady decline to about one-tenth of the peak concentration by the early 1990s as the compound was phased out of gasoline. This pattern of usage is consistent with the variation in the U.S. consumption of lead for use in gasoline that is plotted in Figure 12-5, which shows a sharp rise from 1930 to 1970, followed by an even sharper decline thereafter.

In many countries of the world, the use of leaded gasoline still continues. In these areas, the air is the major source of lead ingested by humans, as it was in the past in North America and Europe. For example, in Mexico airborne lead from vehicular emissions was a major source of the element in the blood of children. Some of the gasoline-based lead enters the body directly from inhaled air, and some enters indirectly from food into which lead has been incorporated.

12.25 Lead's Effects on Human Reproduction and Intelligence

Most ingested lead in humans is initially present in the blood, but that amount eventually reaches a plateau. Any excess enters the soft tissues, including the organs, particularly the brain. Eventually, lead becomes deposited in bone, where it replaces calcium, because Pb^{2+} and Ca^{2+} ions are similar in size. Indeed, lead absorption by the body increases in persons having a calcium (or iron) deficiency and is much higher in children than in adults. A study in Mexico indicated that pregnant women can decrease the lead levels in their blood—and presumably in the blood of their developing fetus—by taking calcium supplements.

At high levels, inorganic lead (Pb^{2+}) is a general metabolic poison. **The toxicity of lead is proportional to the amount present in the soft tissues,** not to that in blood or bone. Lead remains in human bones for decades; thus it can accumulate in the body. The dissolving of bone, as occurs with old age or illness such as osteoarthritis and advanced periodontal disease, or in times of stress such as pregnancy and menopause, results in the remobilization of bone-stored lead back into the bloodstream where it can produce toxic effects. Excess lead may lead to the deterioration of bones in adults. A correlation has been found between periodontal bone loss and blood lead levels in U.S. adults, particularly in those who smoke. Children exposed to environmental lead also have more dental cavities.

Although there is some evidence that too much lead can slightly increase the blood pressure of adults, **the humans most at risk from Pb²⁺ even at relatively low levels are fetuses and children under the age of about seven years.** Both these groups are more sensitive to lead than are adults, partly because they absorb a greater percentage of dietary lead and partly because their brains are growing rapidly. The metal readily crosses the placenta and thus is passed from mother to unborn child. Because of the immaturity of the fetus's blood–brain barrier, there is little to prevent the entry of lead into its brain. Indeed, in the past women who worked in the lead industry suffered higher-than-average rates of miscarriages and still-births. In addition, lead is transferred postnatally from the mother in her breast milk and/or from the tap water used to prepare formula for bottle-fed babies.

The principal risk to children from lead is interference with the normal development of their brains. A number of studies have found small but consistent and significant neuropsychological impairment in young children due to environmental lead absorbed either before or after birth. Lead appears to have deleterious effects on children's behavior and attentiveness, and possibly also on their IQs. Children in a lead-smelting community (Port Pirie) in Australia that had a blood lead level of 300 ppb had an average IQ 4–5 points lower than those whose level was 100 ppb. This result is consistent with studies that indicate that there is an IQ deficit of about 2–3 points for each increase by 100 ppb of blood lead. Some studies indicate that prenatal exposure to lead—especially during the first trimester of pregnancy—has the greatest detrimental effect on the IQs measured in children in primary grades; some indicate it is the lead level at the age of two (when blood concentrations usually peak) that is predominant; and others that it is the concurrent lead level—even if lower than that in early childhood—that is the dominant factor. No threshold for the effects of lead upon IQ is apparent in the studies.

Surveys of American children aged 1–5 years, the population at highest risk from lead poisoning, indicate a factor-of-ten decrease in the average lead level in their blood from 1976–1980 to 2007–2008. The percentage of these young children with high levels, greater than the intervention level of 10 μg dL^{-1}, dropped by an amazing two orders of magnitude:

Survey Period	Mean Blood Lead	Fraction with > 10 μg dL^{-1}
1976–1980	14.9 μg dL^{-1}	88%
1991–1994	2.7 μg dL^{-1}	4.4%
2007–2008	1.5 μg dL^{-1}	0.9%

The average blood level for Americans one year and older in the 2007–2008 period was 1.3 μg dL^{-1}, also about one-tenth that in the 1970s.

FIGURE 12-6 The distribution of blood lead levels in U.S. children aged one to five years in 1988–1991. [Source: R. A. Goyer, "Results of Lead Research: Prenatal Exposure and Neurological Consequences," *Environmental Health Perspectives* 104 (1996): 1050–1054.]

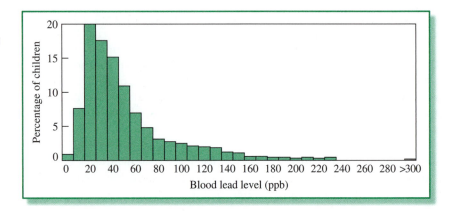

A graphical representation of the distribution of blood lead levels in American children in the 1988–1991 period is shown in Figure 12-6. Here the height of each block indicates the percentage of young children having levels within successive 10-ppb blocks; note the scale here is parts-per-billion, where 10 ppb approximately equals 1 μg dL^{-1}. The distribution peaks in the 20–30 ppb segment, and falls away exponentially thereafter.

A survey reported in 2010 found that the average level in Canadians aged 6 to 79 years was 1.3 μg dL^{-1}, about one-third the concentration found in the late 1970s, with less than 1% having more than 10 μg dL^{-1}. The graphs in Figure 12-7 are another way of showing the distribution of blood

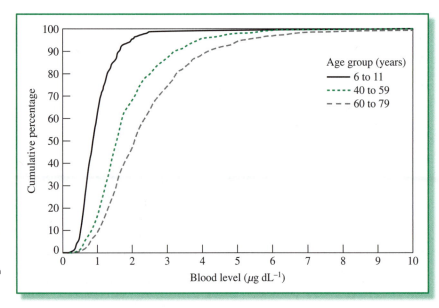

FIGURE 12-7 Cumulative distribution of blood lead concentrations in Canadians by age group, 2007–2009. [Source: "Lead and Bisphenol A Concentrations in the Canadian Population," Statistics Canada Document 82-003-X, 2010.]

levels among a cohort of people in a given age range. A point on any of the curves indicates the percentage (y axis) of Canadians in a particular age group with blood lead levels *less than* any particular value. For example, about 50% of those represented by the long-dashed curve had concentrations of $2~\mu g~dL^{-1}$ or less, as you can tell by seeing where the vertical axis for this value intersects the horizontal line for 50%. **In essence, the curves in such figures represent the sum (integration) of the subjects in graphs of the type in Figure 12-6 having levels from zero up to the chosen value.**

The curve in Figure 12-7 for the oldest (60–79 years, long dashes) sample of Canadians lies farthest to the right side in the graph, indicating that they had on average the highest levels. Although not shown, there is progressively less lead in the 40–59 age group, then the 20–39 group, then the 12–19 cohort, although the values for the latter were slightly greater than those for the 6–11 age group.

This trend of increasing lead levels with age (at least after age 5) is not unexpected, since the older the person, the greater his or her exposure to lead before it was phased out of gasoline. Lead becomes deposited for long periods in bones, which continually supply it in small amounts to the bloodstream.

$$Pb(blood) \rightleftharpoons Pb(bones)$$

PROBLEM 12-10

The concentrations of lead in blood samples are often reported in units of micrograms of Pb per deciliter of blood, or in micromoles of lead per liter of blood. Calculate the value of the concentration in these units of a blood sample containing 60 ppb lead, assuming that the density of blood is one gram per milliliter. ●

ACTIVITY

(a) According to Figure 12-7, do any significant number of Canadians have zero lead in their blood? If not, what appears to be the minimum value?
(b) Using Figure 12-7, for points corresponding to gradations of lead concentrations from 0.5 to 6.0 $\mu g~dL^{-1}$, read off as well as you can the percentage of Canadians in the 60–79 age bracket who have blood levels equal to or less than each of these values. Calculate the number of people with concentrations between each of these successive values by taking the difference between the two percentages. Using your results, make a bar graph of percentage of population versus blood lead of the type shown in Figure 12-6.
(c) Using Figure 12-6, for each of the bars from zero to 140 ppb, estimate the percentage of children in each 10-ppb range. By summing the values in turn, calculate the number of children with lead levels of less than 10 ppb, less than 20 ppb, . . . , less than 140 ppb from your graphical results, and use them to make an integration plot of the type in Figure 12-7.

(d) What do you consider to be the advantages and disadvantages in displaying results by each of the two types of graphical representation? Which do you prefer, and why?

In summary, on an atom-for-atom basis, lead is not as dangerous as mercury. However, the general population is exposed to lead from a greater variety of sources and generally at higher levels than those associated with mercury. Overall, more people are adversely affected by lead, though on average to a lesser extent, than those fewer individuals exposed to mercury. Both metals are more toxic in the form of their organic compounds than as the simple inorganic cations. In terms of its environmental concentration, lead is much closer–within a factor of ten–to the level at which overt signs of poisoning become manifest than is any other substance, including mercury. Thus it is appropriate that society continue to take steps to further reduce human exposure to lead.

Review Questions 14–22 are based on material in the above section.

Cadmium

Cadmium, Cd, lies in the same subgroup of the periodic table as zinc and mercury, but it is more similar to the former. Like zinc, the only common ion of cadmium is the 2+ species. In contrast to mercury, cadmium's compounds with simple anions such as chloride are ionic salts rather than covalent molecules.

12.26 Environmental Sources of Cadmium

Most cadmium is produced as a by-product of zinc smelting, since the two metals usually occur together. Some environmental contamination by cadmium often occurs in the areas surrounding zinc, lead, and copper smelters. As is the case for the other heavy metals, burning coal introduces cadmium into the environment. The disposal by incineration of waste materials that contain cadmium is also an important source of the metal to the environment.

A major use of cadmium is as an electrode in rechargeable *nicad* (nickel–cadmium) batteries used in calculators and similar devices. When current is drawn from the battery, the solid elemental metal cadmium electrode partially disintegrates to form insoluble **cadmium hydroxide,** $Cd(OH)_2$, by incorporating hydroxide ions from the medium into which it dips. When the battery is being recharged, the solid hydroxide, which was deposited on the metal electrode, is converted back to cadmium metal:

$$Cd(s) + 2\,OH^- \rightleftharpoons Cd(OH)_2(s) + 2\,e^-$$

Each nicad battery contains about 5 g of cadmium, much of which is volatilized and released into the environment if the spent batteries are incinerated in garbage. The metallic cadmium preferentially condenses on the smallest particles in the incinerator smoke stream, which are precisely the ones that are difficult to capture by pollution-control devices inserted in the gas stack. In order to avoid releasing airborne cadmium into the environment upon combustion, some municipalities require nicad batteries to be separated from other garbage. The recycling of metals from such batteries has also begun in some areas. However, the European Union has banned the use of nicad batteries, except in cordless power tools and systems used for safety and medical purposes. Some U.S. states have banned the disposal of nicad batteries. Battery manufacturers hope to replace nicad batteries soon with those that do not contain cadmium.

In ionic form, the main use of cadmium is as a pigment. Because the color of **cadmium sulfide,** CdS, depends on the size of the particles, cadmium pigments of many hues can be prepared. Both CdS and CdSe have been used extensively to color plastics. For several centuries, painters have used cadmium sulfide pigments in paints to produce brilliant yellow colors and they oppose any ban on them, since at present there are no suitable replacements. Van Gogh could not have painted his famous *Sunflowers* canvas without cadmium yellows, although it is speculated that cadmium poisoning may have contributed to the painter's anguished mental state. Some countries, including Sweden, have set limits to the amount of cadmium allowed in consumer products.

Cadmium is also released into the environment during the incineration of plastics and other materials that contain it as a pigment or as a stabilizer. Its release to the atmosphere occurs as well when cadmium-plated steel is recycled, since the element is fairly volatile when heated (its boiling point is 765°C).

12.27 Human Intake of Cadmium

Although Cd^{2+} is rather soluble in water, unless sulfide ions also are present to precipitate the metal as CdS, humans usually receive only a small proportion of their cadmium directly from drinking water or from air, except for individuals who live near mines and smelters, particularly those that process zinc. The MCL for cadmium in drinking water is 5 ppb in the United States and Canada (Table 12-1).

Smokers are also exposed to cadmium that is absorbed from soil and irrigation water by tobacco leaves and then released into the smoke-stream when a cigarette is burned. Heavy smokers have approximately double the net cadmium intake of nonsmokers.

Owing to its chemical similarity to zinc, plants absorb cadmium from irrigation water. The use on agricultural fields of phosphate fertilizers, which contain ionic cadmium as a natural contaminant, and of sewage sludge contaminated with cadmium from industrial releases, increases the cadmium level in soil and subsequently in plants grown in it. In the future, cadmium may be removed from phosphate fertilizer before it is sold to the consumer (see also Chapter 16). Soil also receives cadmium from atmospheric deposition. Since cadmium uptake in plants increases with decreasing soil pH, one effect of acid rain is to increase cadmium levels in food.

For most of us, **the greatest proportion of exposure to cadmium comes from our food supply.** Seafood and organ meats, particularly kidneys, have higher cadmium levels than do most other foods. However, the majority of cadmium in the diet usually comes from potatoes, wheat, rice, and other grains, since most people consume so much more of them than seafood and kidneys. An exception is the Inuit people of Canada's Northwest Territories; a prized component of their diet is caribou kidneys, organs which are highly contaminated by cadmium that has reached the Arctic regions on the wind from industrial regions in Europe and North America.

Historically, all episodes of serious cadmium contamination resulted from pollution from nonferrous mining and smelting. The most acute environmental problem involving cadmium occurred in the Jintsu River Valley region of Japan, where rice for local consumption was grown with the aid of irrigation water drawn from a river that was chronically contaminated with dissolved cadmium from a zinc mining and smelting operation upstream. Hundreds of people in this area, particularly older women who had borne many children and who had poor diets, contracted a degenerative bone disease called *itai-itai* or "ouch-ouch," so named because it causes severe pain in the joints. In this disease, some of the Ca^{2+} ions in the bones are replaced by Cd^{2+} ions since they have the same charge and are virtually the same size. The bones slowly become porous, and can subsequently fracture and collapse. The intake of cadmium by itai-itai sufferers was estimated at about 600 μg per day, about ten times the average ingestion of North Americans.

12.28 Human Protection Against Low Levels of Cadmium

Cadmium ion is acutely toxic: the lethal dose is about 1 g. Humans are protected against chronic exposure to low levels of cadmium by the presence in our bodies of the sulfur-rich protein **metallothionein,** the usual function of which is the regulation of zinc metabolism. Because it has many sulfhydryl groups, metallothionein can complex almost all ingested Cd^{2+}; the complex is subsequently eliminated in the urine. If the amount of cadmium absorbed by the body exceeds the capacity of metallothionein to complex it, the metal is stored mainly in the liver and kidneys. Indeed, there is evidence that chronic exposure to cadmium eventually leads to an increased chance of acquiring kidney diseases.

The average cadmium burden in humans is increasing. Cadmium is a cumulative poison; if it is not eliminated quickly (by metallothionein, as discussed above), its lifetime in the body is several decades. The geographic areas at greatest risk from cadmium are Japan and central Europe; in both regions the pollution of the soil by cadmium is particularly high due to contamination from industrial operations. As discussed above, rice grown in many areas of Japan is often contaminated with rather high cadmium levels. As a consequence, the dietary intake of cadmium by residents of Japan is substantially greater than for peoples of other developed countries. Indeed, in Japan the average daily amount of ingested cadmium is beginning to approach the maximum level recommended by health authorities, although this limit has a large built-in safety factor relative to levels at which health effects would occur.

Review Questions 23–26 are based on the material in the above section.

Arsenic

Arsenic is not actually a metal; it is a metalloid—its properties are intermediate between those of metals and nonmetals. However, for convenience we discuss it in this chapter.

Arsenic compounds such as the oxide As_2O_3, **white arsenic,** were common poisons used for murder and suicide from Roman times through the Middle Ages. In the seventeenth century, arsenic was believed in some European societies to be not only a poison but also a magical substance that was a cure for certain ailments, including impotence, and to be a prophylactic against the plague. Indeed, arsenic compounds have been used therapeutically for 2000 years, and even today about 50 Chinese drugs contain the element. There are small background levels of arsenic in many foods, and indeed a trace amount of this element apparently is essential to good human health.

12.29 Arsenic(III) Versus Arsenic(V) Toxicity

Arsenic occurs in the same group of the periodic table as phosphorus, and so also has an s^2p^3 electron configuration in its valence shell. Loss of all three p electrons gives the 3+ ion, whereas sharing of the three electrons gives trivalent arsenic; collectively these two forms are designated As(III). Arsenic (III) commonly exists in aqueous solution and in solids as the **arsenite ion,** AsO_3^{3-} (which can be considered to be As^{3+} bonded to three surrounding O^{2-} ions), or one of its successively protonated forms $HAsO_3^{2-}$, $H_2AsO_3^{-}$, or H_3AsO_3.

Alternatively, loss of all five valence-shell electrons gives the 5+ ion, and sharing them all gives pentavalent arsenic; collectively these two forms are designated As(V). Arsenic(V) also commonly exists as an oxyanion, the **arsenate ion** AsO_4^{3-} (equivalent to As^{5+} bound to four O^{2-} ions), or one of its successively protonated forms $HAsO_4^{2-}$, $H_2AsO_4^{-}$, or H_3AsO_4.

arsenite
AsO_3^{3-}

arsenate
AsO_4^{3-}

These phosphorus ions are
called *phosphite* and
phosphate, respectively.

Overall, arsenic acts much as does phosphorus, which commonly exists in the analogous oxyanion forms PO_3^{3-} and PO_4^{3-}.

However, arsenic has more of a tendency than phosphorus to form ionic rather than covalent bonds, since it is more metal-like. Owing to the similarity in properties, arsenic compounds coexist with those of phosphorus in nature. Consequently, arsenic often contaminates phosphate deposits and commercial phosphates.

Arsenic's lethal effect when consumed in an acute dose is due to gastro-intestinal damage, resulting in severe vomiting and diarrhea. Inorganic As(III) is more toxic than As(V), although some of the latter is converted by reduction to the former in the human body. It is thought that the greater toxicity of As(III) is due to its ability to be retained in the body longer since it becomes bound to sulfhydryl groups in one of several enzymes. Due to the subsequent inactivity of the enzymes, energy production in the cell declines and the cell is damaged. Once the arsenic becomes methylated in the liver, it does not bind tightly to enzymes and hence is largely detoxified.

12.30 Anthropogenic Sources of Arsenic to the Environment

Anthropogenic environmental sources of arsenic stem

- from the continuing use of its compounds as pesticides;

- from its unintended release during the mining and smelting of gold, lead, copper, and nickel, in whose ores it commonly occurs (the leachate from abandoned gold mines of previous decades and centuries can still be a significant source of arsenic pollution in water systems);

- from the production of iron and steel;

- from the combustion of coal, of which it is a contaminant; and

- from arsenic-contaminated underground water brought to ground level by wells.

The arsenic present in raw coal can become a serious pollutant, especially around areas where the fossil fuel is burned. The total pollution from arsenic can be substantial where the coal is burned in small, unventilated stoves rather than in large power plants. In the former cases, which occur in some developing countries, the arsenic becomes not only an indoor air pollutant, but also contaminates the food and water stored indoors. A particularly acute example occurs in the Chinese province of Guizhou, where arsenic levels in the coal are extraordinarily high, exceeding 1% (i.e., 10,000 ppm) in some cases. Many of the residents of Guizhou suffer arsenic-related health problems, since they use this coal for domestic cooking and heating. By contrast, the level of arsenic in U.S. coal averages about 22 ppm, and most coal worldwide has arsenic levels of less than 5 ppm.

Arsenic compounds found widespread use as pesticides before the modern era of organic chemicals. Although its use in these applications has decreased, arsenic contamination from pesticides remains an environmental problem in some areas of the world. The common arsenic-based pesticides include

- the herbicides sodium arsenite, Na_3AsO_3, and Paris Green, $Cu_3(AsO_3)_2$, both of which contain As(III) as AsO_3^{3-}; some methylated derivatives of arsenic acids are still used as herbicides, even in developed countries;

- the insecticide lead arsenate, $Pb_3(AsO_4)_2$, and the herbicide calcium arsenate, $Ca_3(AsO_4)_2$, both of which contain As(V) as AsO_4^{3-};

- the sodium salt of the $O{=}As(OH)_2O^-$ ion in which one —OH has been replaced by a methyl group, producing the methanearsonate ion, $O{=}As(OH)(CH_3)O^-$, which was an herbicide widely used on golf courses and cotton fields in the United States; such As(V) compounds act as weed killers because they enter into plant metabolism in place of phosphate ion.

An organic compound containing arsenic(V) is routinely used in chicken feed to stimulate growth and prevent disease; some scientists have worried about the contamination of soil and water by arsenic leached from chicken litter.

The environmental consequences of using another heavy metal, tin, in a pesticide are discussed in Box 12-1.

Since the 1970s, arsenic has been used in the form of the compound chromated copper arsenate, CCA, to pressure-treat lumber in order to prevent rot and termite damage. Unfortunately, some of the arsenic leaches out of the wood over time. U.S. and Canadian producers of CCA-treated wood voluntarily phased out use of the arsenic compound at the end of 2003 for wood destined for residential structures such as decks, picnic tables, fences, and playground equipment. CCA is discussed in further detail in Section 12.38 on chromium. The U.S. EPA has banned arsenic in all other pesticides.

12.31 Arsenic's Effect on Human Health

Arsenic—much of it from natural sources—is one of the most serious environmental health hazards. The presence of significant levels of arsenic in drinking-water supplies is a significant and controversial environmental issue. Although arsenic has been used for millennia as a poison, the major health problem stemming from its presence at low levels in drinking water is cancer. Drinking arsenic-contaminated water has also been linked to diabetes and cardiovascular disease, perhaps by disrupting a hormonal process associated with both conditions. Natural levels of arsenic in water can be quite high, and it is more common for health problems to arise from this source than from anthropogenic arsenic.

BOX 12-1   Organotin Compounds

Although inorganic compounds of tin (Sn) are relatively nontoxic, the bonding of one or more carbon chains to the metal results in substances that are toxic. Such organotin compounds have some common uses, such as additives to stabilize PVC plastics and fungicides to preserve wood, and therefore are of environmental concern.

Tin forms a series of compounds of general formula R_3SnX, which are molecular substances, though often shown in formulas as if they were ionic, e.g., $(R_3Sn^+)(X^-)$, where R is a hydrocarbon group and X is a monatomic anion; corresponding compounds such as $(R_3Sn)_2O$ also occur. All these compounds are toxic to mammals when R is a very short alkyl chain; maximum toxicity occurs when R is the ethyl group, C_2H_5, and decreases progressively with increasing chain length.

For fungi the greatest toxic activity is attained when each hydrocarbon chain has four carbons in an unbranched chain, i.e., if R is the n-butyl group, $—CH_2CH_2CH_2CH_3$ (or simply $n\text{-}C_4H_9$). *Tributyltin oxide*, $(R_3Sn)_2O$, where $R = n\text{-}C_4H_9$, and the corresponding fluoride have both been used as fungicides; commonly they are incorporated as antifouling agents in the paint applied to docks, to the hulls of boats, to lobster pots, to fishing nets, etc., to prevent the accumulation of slimy marine organisms such as the larvae of barnacles. In recent years, tributyltin has been incorporated into polymeric coatings for boat hulls; a thin layer of the compound subsequently forms around the hull. The tin compounds replaced copper(I) oxide, Cu_2O, in such applications since their effectiveness lasts longer than a single season.

Unfortunately, some of the tributyltin compound leaches into the surface waters in contact with the coatings or paint, particularly in harbors where the boats are moored, and subsequently enters the food chain via the microorganisms that live near the surface. This can lead to sterility or death for fish and some types of oysters and clams that feed on these microorganisms. Some countries have restricted the use of tributyltin compounds to large ships. Thus, although the concentration of tributyltin has decreased in the waters of small harbors and marinas, the pollutant still tends to concentrate in marine coastal regions due to its use on large vessels. Scientists are worried that the presence of tributyltin compounds in these waters could affect fish reproduction.

For this reason, the International Maritime Organization banned new applications of tributyltin to ships of any size, effective 2003, and required that this material be removed from all old applications by 2008. Ironically, the triazine herbicide (Chapter 13) added to copper-based antifoulant paints that were introduced to replace those based on tributyltin degrades only slowly in water and has now begun to accumulate there.

Higher organisms have enzymes that break down tributyltin fairly rapidly, so it is not very toxic to humans. However, most humans now have detectable levels of tributyltin in their blood.

Arsenic is carcinogenic in humans. Lung cancer results from the inhalation of arsenic and probably also from its ingestion. Cancers of the lung, bladder, and skin, and perhaps also of the kidney and prostate, arise from ingested arsenic, including that in water. The mechanism by which arsenic causes cancer is not clear; indeed, no animal model exists for it. Evidence suggests that it acts as a *cocarcinogen*, inhibiting the DNA repair mechanism and thereby enhancing the cancer-causing abilities of other carcinogens.

The term *cocarcinogen* is explained further in Section 12.33.

There is evidence from Chile that smoking and simultaneous exposure to high levels of arsenic in drinking water act synergistically in causing lung cancer; i.e., their effect when taken together is greater than the sum of their individual effects if each acted independently, as discussed in Chapter 4. Other data from Chile show that exposure to arsenic during early childhood or even *in utero* increases subsequent mortality in young adulthood from both malignant and nonmalignant lung diseases. Arsenic seems to act synergistically with several *cofactors*—i.e., factors the presence of which negatively affect the health of an individual to an extent greater than that of it or the arsenic operating independently.

Exposure to excessive levels of UV from sunlight and a lack of selenium in the diet (stemming from malnutrition and/or low selenium levels in local foods) are other cofactors with arsenic. The protective effect of selenium in reducing the amount of active arsenic in the body may arise from the formation of a biomolecule containing an $As\!=\!Se$ bond. Research is underway to determine if selenium supplementation of the diet would be effective in countering the negative health effects of excess arsenic in the drinking water of Bangladesh and Bengal India.

12.32 Arsenic in Groundwater

Drinking water, especially that derived from groundwater, is a major source of arsenic for many people. Although anthropogenic uses of arsenic can result in its presence in water, by far the greatest problems occur with the contamination produced by natural processes. **Groundwater in several parts of the world is highly contaminated with inorganic arsenic.** Unfortunately, the arsenic is tasteless, odorless, and invisible, so its presence is not easily detected.

Major problems from high arsenic levels in groundwater occur in the Bengal Delta, with the result that tens of millions of people in Bangladesh and in the West Bengal region of India drink arsenic-laced water. The World Health Organization has called this the "largest mass poisoning of a population in history." The problem arose from the creation of tens of millions of tube wells, which mine groundwater that was previously inaccessible. The concrete tube wells extend 20 m (60 ft) or more into the ground. Ironically, the wells were constructed by UNICEF in the 1970s and early 1980s in an otherwise highly successful project to eliminate epidemics of diarrhea, cholera, and other water-borne diseases, and to reduce the high child-mortality rate caused by use of microbially unsafe water drawn from streams, ponds, and shallow-dug wells used in the past. About half the tube wells—affecting about 50 million people in Bangladesh—produce water with arsenic levels as high as 500–1000 ppb, greatly exceeding the 10-ppb WHO guideline for drinking water (Table 12-1). Generally, the deeper the well beyond about 20 m, the lower the concentration of arsenic. The sediments through which the groundwater travels contain the arsenic which, as discussed below, can be solubilized into the water under certain conditions.

Several million people living in the Bengal Delta region will probably contract skin disorders from drinking arsenic-laced groundwater if remedial action is not taken; a fraction of them will also suffer from the more serious ailment of *arsenicosis*, which can cause cancer of the skin, bladder, kidneys, and lungs. Skin lesions appear after 5–15 years of exposure to high levels of arsenic in drinking water. A large number of residents of West Bengal, India, have already developed such lesions—the usual outward sign of chronic exposure to arsenic—that may develop into skin cancer because they consumed arsenic-laced groundwater from underground wells. The main cause of arsenic-related deaths among these people is lung cancer. However, it has also been established that rice and vegetables grown in Bangladesh using irrigation water from tube wells are also contaminated by arsenic, and this may be the dominant source of the element for some people. Grains and beans absorb additional arsenic from the water they absorb when they are cooked.

Research from Bangladesh indicates that increasing levels of arsenic and/or of manganese in drinking water confers progressively more and more negative effects on the intellectual levels of six- and ten-year-old children. The Mn levels in one such study averaged 1.4 ppb, compared to the WHO standard of 0.5 ppb. Elevated manganese levels are also present even in the United States: approximately 6% of domestic wells exceed the U.S. EPA lifetime health advisory concentration of 0.3 ppb Mn in drinking water. High manganese levels are found in almost half the groundwater wells in Vietnam. A recent study of children in rural Quebec who drink water from wells with high manganese levels found significantly lower IQ scores—averaging a deficit of 6 points—between highest and lowest exposed groups.

The origin of the dissolved arsenic in the water in Bangladesh and India is somewhat controversial. Normally the element, as arsenate ion, is coprecipitated with and strongly adsorbed onto the surface of iron(III) oxides in the soil, as would have occurred in ancient times when sediments were being laid down. However, the iron dissolves when insoluble Fe(III) is reduced to the more soluble Fe(II) if organic carbon consumes the oxygen and produces anaerobic, reducing conditions. **The arsenic previously bound to the solid iron oxide dissolves in the water along with the Fe(II).** Indeed, usually the higher the concentration of dissolved iron, the higher the arsenic concentration found in the water. Reduction of arsenic from As(V), as it exists when adsorbed onto the iron mineral, to the more soluble As(III) form is also believed to be a factor in solubilizing the element.

Manganese is also a heavy metal.

Oxidizing Conditions		Reducing Conditions
Fe as insoluble Fe(III) oxides and hydroxides to which is bound As(V) as AsO_4^{3-}	$\longrightarrow$	Fe as soluble Fe(II) dissolved in water and releasing As(V), and As(III) as AsO_3^{3-}

The controversy centers around whether the dominant reduction process

- is the natural one, by which buried peat acts as the reducing agent and has been doing so for millennia, or

- whether the release has been greatly accelerated in recent years as an indirect effect of annually lowering the water table by extracting massive amounts of water for crop irrigation.

In the latter mechanism, the subsequent recharge of the depleted aquifer below excavated irrigation ponds transports carbon in the water drawn down from the surface, resulting in further reduction of iron oxides and solubilization of the arsenic.

The water obtained from adjacent wells separated even by only tens of meters from each other can differ enormously in arsenic content, apparently as a result of being drawn from sediments initially laid down in ancient times by different streams that had different sources of organic carbon being deposited simultaneously. Widespread testing in 1999 of tube wells in Bangladesh identified those delivering high arsenic, and the handles on such wells were painted red to warn people of the danger. Thousands of larger, deeper wells, that draw water from less-contaminated aquifers, have subsequently been installed as centralized facilities in many villages. However, although the use of deep aquifers generates water that is very low in arsenic, experience covering more than a century of such exploitation in Vietnam shows that arsenic-containing water from shallower aquifers can be drawn down to the deep ones as a result.

Arsenic-contaminated drinking water is also a major problem in Chile, Argentina, Mexico, Nepal, Taiwan, Cambodia, in both the Red River and Mekong delta regions of Vietnam, and in large areas of China. Indeed, 8% of the deaths of Chilean adults over the age of 30 are attributable to arsenic poisoning. In a study of residents of Taiwan who were exposed to high levels of the element in their well water, a relationship between arsenic exposure and skin cancer incidence has been established. As in Bangladesh, arsenic only became a problem when people began to drink groundwater, which was touted as being purer than surface water, since the latter is often contaminated by sewage.

12.33 Drinking-Water Standards for Arsenic

Drinking water, especially groundwater, is a major source of inorganic arsenic for most people. The global average inorganic arsenic content of drinking water is about 2.5 ppb. The World Health Organization has set

10 ppb as the acceptable limit for arsenic in drinking water, and the United States, Canada, and the European Union have all adopted this standard (Table 12-1), although the EU value refers to total arsenic whereas the others refer to inorganic arsenic. The standard in many developing countries is still 50 ppb, which is no longer considered to be protective of human health. One argument against lowering the WHO standard to less than 10 ppb is the lack of inexpensive analytical technology for determining lower concentrations.

The shape of the dose–response curve for cancer at such low concentrations of arsenic is unknown. Assuming that no threshold exists, linear extrapolations of human cancer incidence from populations that were exposed to high levels of arsenic leads to the conclusion that there is a 1-in-1,000 lifetime risk of dying from cancer induced by normal background levels of arsenic. This estimate makes arsenic almost equivalent to environmental tobacco smoke and radon exposure as an environmental carcinogen. Drinking water over a lifetime at the 50-ppb level, the old U.S. standard, would have caused bladder or lung cancer in about 1% of the population, a much greater risk than continuously consuming any other water-based contaminant at its MCL.

Some environmentalists argue that the arsenic standard should be lowered still further, since at 10 ppm the extrapolated lifetime lung or bladder cancer risk is 12–23 per 10,000 people, compared to the usual target risk maximum of 1 in 10,000. About 57 million Americans currently drink water containing more than 1 ppb of arsenic; areas of the contiguous United States whose groundwater sometimes contains more than 10 ppb As are shown in dark green in Figure 12-8. Most affected systems lie in the West, Midwest, Southwest, and New England, and use groundwater having naturally occurring arsenic.

One of the difficulties in setting a standard for arsenic levels in drinking water is deciding the manner in which the element operates as a carcinogen. For carcinogens that induce cancer *directly*—by damaging DNA—the assumption is made that no amount of exposure to the substance is safe, since the risk from it rises from zero in proportion to exposure. However, as mentioned previously, there is evidence that arsenic does not act directly but indirectly, as a co-carcinogen, by inducing cell damage and regrowth or by inhibiting repair of DNA damage caused by *other* carcinogens such as UV light or tobacco smoke. For carcinogens that act *indirectly*, there *can* be a *threshold*, i.e., a level below which the substance can be considered to be safe and not cause damage. If such a threshold exists for arsenic, the cancer risk estimates discussed above are probably much too high.

Some scientists are not convinced that these estimates of cancer risk are at all realistic, since the extrapolation of the cancer incidence from high arsenic levels to the low environmental concentrations may not be valid if arsenic acts indirectly as a carcinogen. It will be difficult to resolve this issue by analyzing cancer trends in different parts of the United States, however,

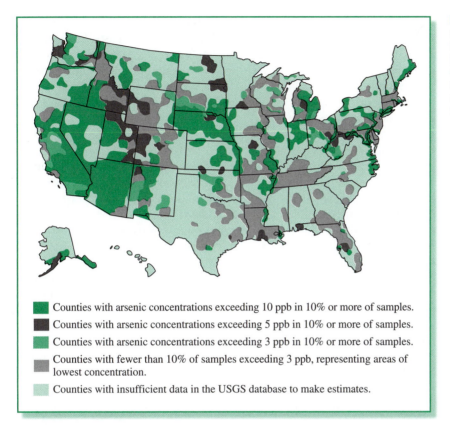

FIGURE 12-8 Average arsenic concentrations in U.S. drinking water. [Source: "Pressure to Set Controversial Arsenic Standard Increases," *Environmental Science and Technology* 34 (2000): 208A.]

■ Counties with arsenic concentrations exceeding 10 ppb in 10% or more of samples.

■ Counties with arsenic concentrations exceeding 5 ppb in 10% or more of samples.

■ Counties with arsenic concentrations exceeding 3 ppb in 10% or more of samples.

■ Counties with fewer than 10% of samples exceeding 3 ppb, representing areas of lowest concentration.

■ Counties with insufficient data in the USGS database to make estimates.

since the predicted fraction of bladder and lung cancers caused by arsenic is still a small percentage of the total for these diseases.

One argument that was advanced against making the arsenic standard even as low as 10 ppb in the United States is that it forces some small-scale suppliers of drinking water to shut down since they cannot afford the cleanup costs associated with introducing equipment to remove the element. Such shutdowns may lead consumers to turn to water supplies that are even more unsafe in other respects. Indeed, lowering the standard to 10 ppb is estimated to cost users of small water utilities, many in rural areas, several hundred dollars per year, whereas it costs users of large facilities only a few dollars annually.

12.34 Removal of Arsenic from Water

In Asia, the most widely used process for removing arsenic is to flow the drinking water from local wells over a surface such as a column of activated *alumina* (aluminum oxide). *Ferrous iron*, Fe^{2+}, in the water is readily oxidized by air to the insoluble ferric form, which gathers over a period of minutes onto the alumina surface as particles of hydrated Fe(III) oxide. **Once in place, the iron particles adsorb arsenic from the water subsequently flowing over them,** thereby efficiently filtering the element in a reversal of the reaction that liberated the arsenic into water from soil-bound iron in the first place!

In some areas of the world, such as the Red River delta in Vietnam, there is insufficient Fe^{2+} in the raw water for this purpose, so it must be added artificially. Abundant iron or chlorine can also oxidize As(III) to the more readily adsorbed As(V); alternatively, sunlight could be used to photochemically oxidize As(III).

An alternative system for arsenic removal used in parts of the United States and Bangladesh uses nanoparticles of *magnetite*, Fe_3O_4, which contains both Fe(II) and Fe(III). The high surface-area-to-volume ratio of nanoparticles allows a much smaller amount of oxide to be used to trap the arsenic in both its forms from the water. Phosphate ion adsorbs even more strongly than arsenates to the iron particles, so its presence in high concentrations can interfere with the process.

Some villages in India, Bangladesh, and Vietnam use the alumina method described above, or filter the water through sand. For any of these filters, the surface requires periodic cleaning of adsorbed species to remain effective. The arsenic can be flushed from the iron-based system using an alkaline solution, since the ferric oxides do not adsorb arsenic at high pH. The primary motivation in rural areas for the installation of sand filters is usually to improve the taste of the water by removing its high visible iron content, with simultaneous arsenic removal an unplanned bonus.

All removal techniques require regular maintenance of the equipment and the proper periodic disposal of the arsenic-laden wastes. In some cases, however, people weakened by arsenic poisoning found the cleaning process of individual filters to be onerous. In addition, eventual replacement of filters is too expensive for many individuals; community-based filters overcome some of these difficulties. In contrast, sand filters are regularly and easily cleaned to maintain the improved taste of the water. However, some analysts believe that none of the arsenic-removal techniques work reliably in many areas, partly due to poor maintenance, and that instead, people should be directed to deeper wells with low arsenic contamination rather than trying to clean arsenic out of water from shallower wells that contains high levels of the element. Centralized water-treatment plants using surface water are being constructed in some areas to overcome dependence on groundwater.

Like calcium and magnesium, **arsenic can be removed from drinking water at large treatment facilities by precipitating it in the form of one of its insoluble salts.** The arsenic in surface water normally exists as As(V). Since the salt formed between the *ferric ion*, Fe^{3+}, and arsenate ion is insoluble, the soluble salt *ferric chloride*, $FeCl_3$, is dissolved in the water, and the precipitated *ferric arsenate*, $FeAsO_4$, is filtered from the resulting mixture:

$$Fe^{3+} + AsO_4^{3-} \longrightarrow FeAsO_4(s)$$

> Recall that phosphate ion is PO_4^{3-}.

Arsenic in groundwater often exists as As(III) since reducing conditions occur underground. Such arsenic must be oxidized to As(V) before this removal process can operate.

Arsenic *cannot* be removed from water by cation exchange, since it occurs as an anion, not as a cation. However, anion exchange can be used to remove arsenic from drinking water. Anion exchange also works better for As(V) than for As(III), since the latter exists partially as the neutral H_3AsO_3 rather than as an anion, at normal water pH values (6.5–8.5), whereas As(V) is completely ionic in that range (see Additional Problem 5). Anion exchange is problematic if appreciable amounts of *sulfate ion*, SO_4^{2-}, are also present in the water, as these are exchanged preferentially to arsenate, thereby tying up many sites and leaving fewer at which arsenic can exchange.

Reverse osmosis can also be used to remove arsenic, although, as previously discussed (Chapter 11), the process is expensive.

12.35 Arsenic in Organic and Other Molecular Forms

The common environmental organic forms of arsenic are not simple methyl derivatives, as occur with mercury and lead. Rather, they are water-soluble oxyacid derivatives that can be excreted by the body and thus are less toxic than some inorganic forms. As previously mentioned, in water, arsenic occurs most commonly as the As(V) acid H_3AsO_4, i.e., $O=As(OH)_3$, or one of its deprotonated anions. Biological methylation in the environment by methylcobalamin initially involves the replacement of one or more —OH groups of the acid by —CH_3 groups. Monomethylation by the human liver and kidneys converts most but not all ingested inorganic arsenic to $O=As(CH_3)(OH)_2$ and then to the corresponding dimethyl acids, which are then readily excreted.

Although most daily exposure to *total* arsenic by North American adults is from food intake, especially meat and seafood, much of the arsenic present in food sources occurs in the organic form and is readily excreted. In seafood, the common forms of arsenic are either the $(CH_3)_4As^+$ ion, a type of As(III), or the compound with one methyl group replaced by —CH_2CH_2OH or —CH_2COOH. The organic forms of arsenic found in seafood are probably noncarcinogenic and are much less toxic than inorganic ones, as illustrated in dramatic fashion by their high LD_{50} values, which lie in the thousands of milligrams per kilogram, compared with those for inorganic arsenic, whose LD_{50}'s are about 1% of these values (see Table 12-2). However, the main exposure to inorganic arsenic by Europeans and North Americans, as well as many Asians, is through the rice and other cereals they eat, rather than from drinking water or from seafood. The arsenic in rice arises from both the water in which it was grown and that in which it is cooked.

In contrast to the compounds discussed above, neutral As(III) compounds such as **arsine, AsH_3,** and **trimethylarsine, $As(CH_3)_3$,** are the most acutely toxic forms of arsenic. Curiously, the trimethyl compound is

LD_{50} is the lethal dose of a toxic substance, measured in milligrams per kilogram of animal, and is discussed in detail in Section 13.13.

TABLE 12-2	LD$_{50}$ Values for Some Common Forms of Arsenic	
Name	Formula	LD$_{50}$ (mg kg^{-1})
Arsenous acid	H_3AsO_3	14
Arsenic acid	H_3AsO_4	20
Methylarsonic acid	$CH_3AsO(OH)_2$	700–1800
Dimethylarsonic acid	$(CH_3)_2AsO(OH)$	700–1800
Arsenocholine	$(CH_3)_3As^+CH_2CH_2OH$	6500
Arsenobetaine	$(CH_3)_3As^+CH_2COO^-$	>10,000

Source: X. C. Le, "Arsenic Speciation in the Environment," Canadian Chemical News (September 1999): 18.

produced by the action under humid conditions of molds in wallpaper paste with the arsenic-containing green pigment $CuHAsO_3$ in wallpaper. Instances of mysterious illnesses and even of human "death by wallpaper" have been reported due to chronic exposure to the $As(CH_3)_3$ gas released into rooms by this mechanism. There have been episodes of human poisoning from gaseous arsine that was accidentally generated and released when aqueous solutions of As(III) in the form of $HAsO_2$ came into contact with an easily oxidized metal such as aluminum or zinc and the arsenic was further reduced to As($-$III):

Review Questions 27–31 are based on the material in the above section.

$$2\ Al(s) + HAsO_2 + 6\ H^+ \longrightarrow 2\ Al^{3+} + AsH_3 + 2\ H_2O$$

Chromium

12.36 The Oxidation States of Chromium

Chromium normally occurs in the form of inorganic ions. Its common oxidation states are +3 and +6, i.e., Cr(III) and Cr(VI), known as *trivalent* and *hexavalent* chromium, respectively.

Under oxidizing, i.e., aerobic, conditions, chromium exists in the VI state, usually as the **chromate ion,** CrO_4^{2-}, though under even slightly acidic conditions this oxyanion protonates to $HCrO_4^-$.

At high concentrations not encountered in the environment, chromate dimerizes to give the orange *dichromate ion*, $Cr_2O_7^{2-}$, familiar as a strong oxidizing agent in the laboratory and used in the determination of the COD of water samples, as discussed in Chapter 11.

$$H^+ + CrO_4^{2-} \rightleftharpoons HCrO_4^-$$

The oxyanions of chromium(VI) are highly soluble in water. Both the Cr(VI) ions mentioned above are yellow, and impart a yellowish tinge to water even at chromium levels as low as 1 ppm.

Under reducing, i.e., anaerobic, conditions chromium exists in the III state. In aqueous solution, this state occurs as the +3 ion, i.e., Cr^{3+}. However, the aqueous solubility of this ion is not high, and Cr(III) is often precipitated as its hydroxide, $Cr(OH)_3$ under alkaline, neutral, or even slightly acidic conditions:

$$Cr^{3+} + 3\ OH^- \rightleftharpoons Cr(OH)_3(s)$$

Thus, **whether chromium occurs as an ion dissolved in water or as a precipitate depends upon whether the aqueous environment is oxidizing or reducing.** The difference is important, since hexavalent Cr(VI) is toxic and a suspected carcinogen, whereas trivalent Cr(III) is much less toxic and even acts as a trace nutrient. Chromate ion readily enters biological cells, apparently because of its structural similarity to sulfate ion, SO_4^{2-}. Inside the cell, chromate can oxidize DNA and RNA bases. Because hexavalent chromium is more toxic, more soluble, and more mobile than trivalent chromium, it is considered to pose a greater health risk.

The term *hexavalent chromium* was made famous in the movie *Erin Brockovich*, the story of how a legal assistant battled successfully against pollution of local groundwater by this substance.

12.37 Chromium Contamination of Water

Chromium is widely used for electroplating, corrosion protection, and leather tanning. In tanning, Cr(III) binds to protein in animal skin to form leather that is resistant to water, heat, and bacteria. As a consequence of industrial emissions, **chromium is a common water pollutant, especially of groundwater beneath areas with metal-plating industries.** It is also the second most abundant inorganic contaminant of groundwater under hazardous waste sites. The MCL for total chromium in U.S. drinking water is 100 ppb (Table 12-1).

Most dissolved heavy metals can be removed from wastewater by simply increasing the pH, since their hydroxides are insoluble. However, **Cr(VI) does not precipitate out at any pH since it does not exist as a cation but rather as an oxyanion in water.** Owing to the low solubility and hence the low mobility of Cr(III), however, **the usual way to extract chromium(VI) from water is to first use a reducing agent to convert Cr(VI) to Cr(III):**

$$\underset{\text{(soluble)}}{CrO_4^{2-}} + 3\ e^- + 8\ H^+ \rightleftharpoons \underset{\text{(insoluble)}}{Cr^{3+}} + 4\ H_2O$$

Reducing agents commonly employed for this conversion are gaseous SO_2 or a solution of *sodium sulfite*, Na_2SO_3.

The SO_2 or sulfite ions are oxidized to sulfate ion in the process.

In addition, reducing the Cr(VI) to Cr(III) by adding iron in the form of Fe(II), and then adding base to precipitate Cr(III), is a common practice in purifying Cr-contaminated wastewater. Fine-grained elemental iron placed in permeable underground walls positioned in the path of flowing polluted groundwater is another application of this technique. The iron reduces the chromium, and then as Fe^{3+} it forms an insoluble Fe(III)-Cr(III) compound. This reduction process can occur spontaneously in soils with, for example, Fe^{2+} or organic carbon as the reducing agent.

The use of iron walls to decontaminate groundwater was discussed in Section 11.26.

Hexavalent chromium is quite mobile in soils, since it is not strongly absorbed by many types of soil. However, it can be reduced to the less-mobile trivalent form by the humic substances in soils that are rich in organic matter.

12.38 The Wood Preservative CCA

Another potentially significant source of chromium to the environment stems from its presence in *chromated copper arsenate* (CCA), the widely used wood preservative previously mentioned. CCA is a waterborne mixture of metal oxides with which wood is treated using a vacuum-pressure impregnation process. The amount of CCA forced into the wood is almost 10% of the mass of the lumber. The chromium used here originally is hexavalent. However, during a period of *fixation*, which lasts for several weeks after treatment, almost all the Cr(VI) is reduced to Cr(III) by reaction with carbon in the wood. This process produces insoluble complexes that are slow to leach from the treated wood over its lifetime, since the copper and chromium at least are bound to the wood. Leaching of heavy metals from the wood becomes very slow a few months after treatment, with more copper and arsenic than chromium being lost.

One use of CCA is to protect wooden structures, such as residential docks, that are destined to be used in aquatic environments. For environmental and human health reasons, CCA had largely replaced organic preservatives such as creosote and pentachlorophenol in such applications. However, not only chromium but also arsenic and copper leach from the structures into the water over time.

Review Questions 32–35 are based on material in the above sections.

12.39 Green Chemistry: Removing the Arsenic and Chromium from Pressure-Treated Wood

Wood that is used for exterior construction decays in about three to twelve years unless it is treated with pesticides that prevent destruction from termites, fungi, and other wood-destroying agents. Most of the preserved exterior wood that is presently used is commonly called *pressure-treated wood*. Pressure-treated wood is found in over 50% of homes in the United States. It is also used in decks, fences, retaining walls, piers, docks, wooden bridges, picnic tables, and playground equipment, and lasts 20 to 50 years. Treatment of wood thus results in the conservation of millions of trees each year and limits the use of scarce woods that contain natural preservatives, such as redwoods.

Pressure-treated wood is produced by placing the wood in a horizontal cylinder and evacuating the cylinder, drawing out much of the moisture from the wood cells. An aqueous preservative solution is then pumped into the cylinder and the pressure is raised, forcing the preservative solution into the wood cells. Since the 1930s in the United States, the preservative solution used in 95% of pressure-treated wood was chromated copper arsenate (CCA) discussed in the previous section.

Although the exact composition varies, the most common formulation of relative percentages for the preservative solution is 35.3% CrO_3, 19.6% CuO, and 45.1% As_2O_3. Treatment with CCA results in wood with concentrations of 0.1–2.0% copper, 0.25–4.0% chromium, and 0.15–4.0% arsenic. For example, in 2001, 7 billion board feet of pressure-treated wood (enough to build 450,000 homes) was produced in the United States, utilizing 150 million pounds of CCA. The CCA contained 64 million pounds of hexavalent chromium and 40 million pounds of arsenic. A 12-ft-long 2 × 6 board of CCA-treated wood contains from 16 to 300 g of arsenic. If all this arsenic were ingested, it would be enough to kill many people. Although the preservatives are "locked" into the wood, health officials and environmentalists have long been concerned with the potential for leaching of arsenic and chromium from pressure-treated wood and the ingestion of these elements by infants and children by direct contact with the wood. Studies of the soils beneath decks made of pressure-treated wood gave copper, chromium, and arsenic concentrations averaging 75, 43, and 76 mg kg^{-1}, while control soils averaged 17, 20, and 4 mg kg^{-1}. Studies also indicate that measurable amounts of arsenic can be dislodged from the surfaces of pressure-treated wood by direct contact.

Because of the environmental and human health concerns associated with CCA, the U.S. EPA announced that wood producers voluntarily ceased production of CCA-treated wood on December 31, 2003, for products intended for residential use. *Chemical Specialties, Inc* (CSI) in 1996 introduced a new wood preservative called *Preserve* to replace CCA, for which they earned a Presidential Green Chemistry Challenge Award in 2002. Preserve is formulated with an alkaline quaternary (ACQ) wood preservative. The active ingredients in the preparation are copper and a *quaternary ammonium salt*, $R_4N^+Cl^-$ (either didecyldimethylammonium chloride, or alkyldimethylbenzylammonium chloride). According to the World Health Organization, none of these ingredients are mammalian or human carcinogens.

Because there are no environmental and health concerns for ACQ, under the EPA system ACQ is registered as a *nonrestricted pesticide* for treatment of wood products. Analogous formulations of copper and ACQ are used as algaecides and fungicides in lakes, rivers, and streams, as well as fish hatcheries and potable water supplies. Quaternary ammonium salts are also used as surfactants in typical household and industrial detergents and disinfectants, and unlike arsenic, they have low toxicity to mammals. It is also noteworthy that the copper that is used in the ACQ formulations is obtained from scrap copper. ACQ-treated wood not only eliminates the cancer and toxicity concerns associated with CCA, but it offers the advantages of simplified disposal of treated wood and elimination of hazardous waste generation at the approximately 450 treatment sites across the United States.

Review Questions

1. What is a *sulfhydryl group*, and how does it interact biochemically with heavy metals? How does the interaction affect processes in the body?

2. What is meant by *speciation*? What is its significance in environmental problems?

3. What accounts for the recent decline in emissions of mercury onto water and land?

4. Is the liquid or the vapor of mercury more toxic? Describe the mechanism by which mercury vapor affects the human body.

5. What is an *amalgam*? Give two examples, and explain how they are used.

6. Explain how the *chlor-alkali process* led to the release of mercury to the environment.

7. Name two uses for mercury in batteries.

8. What are some important sources of airborne mercury?

9. What three chemical forms of mercury are emitted into the air from coal-fired power plants? What is the chemical formula and approximate atmospheric lifetime of each form? How can emissions of each form be controlled?

10. Write the formulas for the methylmercury ion, for two of its common molecular forms, and for dimethylmercury. What is the principal source of exposure of humans to methylmercury?

11. Explain why mercury vapor and methylmercury compounds are much more toxic than other forms of the element.

12. Discuss the sources and relative importance of different dietary sources of mercury.

13. What is meant by *Minamata disease*? Explain its symptoms and how it first arose.

14. Explain how lead in ammunition can be a danger to wildlife.

15. What are the two common ionic forms of lead?

16. Explain how lead can dissolve—e.g., in canned fruit juice—even though it is insoluble in mineral acids.

17. Explain why lead contamination of drinking water by lead pipes is less common in hard-water areas than in soft-water areas. What can result in the release of lead from old water pipes into the drinking water they carry?

18. Why were lead compounds used in paints? Why were mercury compounds used in paints?

19. Explain why heavy metal compounds such as PbS and $PbCO_3$ are much more soluble in acidic water.

20. In what forms does lead exist in the lead storage battery?

21. What are the formulas and names of the two organic compounds of lead that were used as gasoline additives? What was their function? How did their presence in gasoline lead to environmental contamination?

22. Discuss the toxicity of lead, especially with respect to its effects on reproduction and intelligence. Which subgroups of the population are at particular risk from lead?

23. What are the main sources of cadmium in the environment?

24. What is the main source of cadmium to humans?

25. Describe what is meant by *itai-itai* disease and discuss where it arose and why.

26. What is *metallothionein*? What is its significance with respect to cadmium in the body?

27. What are some uses of arsenic that result in contamination of the environment? What lead compounds were used as pesticides?

28. What are the main health concerns about arsenic in drinking water? Why is the drinking water in many regions of Bangladesh heavily polluted with arsenic?

29. Describe the difficulties in setting health-based drinking-water standards for arsenic.

30. What organic compounds of arsenic are of environmental significance? Why is arsenic in organic acid forms not very toxic to humans?

31. Describe methods by which arsenic can be removed from water.

32. What are the two important oxidation states of chromium? Which one is the more toxic?

33. Explain how Cr(VI) can be removed from wastewater.

34. What is CCA? Which two toxic heavy metals does it contain?

35. Complete the chart shown in outline below:

Element	Common ionic forms	Common metal-organic forms	Most toxic forms
Mercury			
Lead			
Cadmium			
Arsenic			
Chromium			

 ## Green Chemistry Questions

1. The replacement of lead with yttrium in electrodeposition coatings won PPG a Presidential Green Chemistry Challenge Award.

(a) Which of the three focus areas (see page xxviii) for these awards does this award best fit into?

(b) List one of the twelve principles of green chemistry (see pages xxiii–xxiv) that are addressed by the green chemistry developed by PPG.

2. What environmental advantages does the use of yttrium oxide have over the use of lead oxide in electrodeposition coatings?

3. What environmental advantages does electrodeposition offer over spray painting?

4. The removal of arsenic and chromium from pressure-treated wood won Chemical Specialties, Inc. a Presidential Green Chemistry Challenge Award.

(a) Which of the three focus areas (see page xxviii) for these awards does this award best fit into?

(b) List one of the twelve principles of green chemistry (see pages xxiii–xxiv) that are addressed by the green chemistry developed by Chemical Specialties, Inc.

Additional Problems

1. The equilibrium vapor pressure of mercury at room temperature is about 1.6×10^{-6} atm. Imagine an old chemistry lab that over the years has experienced enough liquid mercury spills and thereby accumulated enough mercury in cracks in the floor, etc. that mercury liquid–vapor equilibrium has been established. What is the concentration, in units of milligrams per cubic meter, of Hg^0 in the air of the room? Does this value exceed the limit of 0.05 mg m^{-3} established by the American Conference of Governmental Industrial Hygienists for safe exposure based upon a 40-hour work week?

2. By adding the dissolving reaction for PbS(s) to that for reaction of S^{2-} with water, determine the overall reaction when PbS dissolves and most of the resulting sulfide ion reacts with water. Calculate the solubility of PbS with and without the subsequent reaction of sulfide, given that for HS^-, $K_a = 1.3 \times 10^{-13}$.

3. (a) Fit an approximately exponential decay curve to the blood-lead distribution among children based on the portion of the curve in Figure 12-6 from 20 ppb (the zero point of the function) to higher levels. By integration, determine the total percentage of children having levels in excess of 100 ppb that your function predicts.

(b) What does the fact that the curve in Figure 12-6 does not continue to rise as the blood level comes close to zero tell us about the background level of lead in the environment?

4. The object of this problem is to estimate the mass of lead that would have been deposited annually on each square meter of land near a typical, busy, six-lane freeway from the lead compounds emitted by cars using the roadway. Use reasonable estimates for the number of cars passing a point each day and their average mileage per liter or gallon of gasoline. Assume that the gasoline contained about one gram of lead per gallon, or 0.2 g L^{-1}, and make the approximation that half the lead was evenly deposited within 1000 m on each side of the freeway.

5. (a) Since the ion AsO_4^{3-} is basic, the forms $HAsO_4^{2-}$, $H_2AsO_4^-$, and H_3AsO_4 will all be present in aqueous solutions of its salts. Given that for H_3AsO_4, the successive acid dissociation constants are 6.3×10^{-3}, 1.3×10^{-7}, and 3.2×10^{-12}, deduce the predominant form of arsenic in waters of pH = 4, 6, 8, and 10.

(b) Arsenic in its As(III) form exists in solution as arsenious acid, H_3AsO_3, or one of its ionized forms. Given that the acid dissociation constant for H_3AsO_3 is 6×10^{-10}, calculate the ratio of its unionized molecular form to the ionized form $H_2AsO_3^-$ at pH values of 8 and 10.

6. The 2+ ions of mercury, lead, and cadmium each form a series of complexes by attaching to themselves in successive equilibrium reactions, up to four chloride ions. Deduce the formulas for the species for one of these metals, including the net charges on the complexes. Would the complexes having three or four chlorines more likely be found in fresh water or in seawater?

7. How does the phenomenon of acid rain indirectly affect the risk to human health from mercury, lead, and cadmium?

8. By reference to reliable websites or textbooks in your library on heavy metals and/or water pollution, determine why copper is considered to be toxic, and find out what types of organisms are at risk from elevated levels of copper in the environment. Does speciation affect the toxicity of copper?

9. Based upon the material in this chapter, write a paragraph supporting your choice of which of the five metals you consider still requires the most regulatory control for environmental reasons.

PART IV

TOXIC ORGANIC COMPOUNDS

Contents of Part IV

Introduction

When people in the general public discuss environmental problems, the topics most often raised these days are global warming and "chemicals." By the latter they are referring to synthetic organic compounds that are produced and used in great volume and that may consequently contaminate—albeit usually in small concentrations—the air, water, soil, food, and other materials to which we all are exposed.

In this portion of the textbook, we survey the most important classes of these contaminants with respect to their origin and uses, as well as to the potential health hazards that arise from their presence in the environment. First we consider *pesticides*, the substances used to control troublesome plants and insects. In the second chapter of this part, the infamous compounds called *dioxins* and *PCBs*—and the furan contaminants of the latter—are studied in detail. In the third chapter, a variety of other organic substances, ranging from toxic by-products of combustion to flame retardants and environmental contaminants that mimic human hormones, are documented, and the mechanism by which they can be transported over long distances from their original site of use is explained. ●

Pesticides

In this chapter, the following introductory chemistry topics are used:

- Elementary organic chemistry (see online Appendix)
- Concepts of vapor pressure; solubility; half-life; chemical versus physical change; enzymes; acids and bases
- Molarity

Background from previous chapters used in this chapter:

- Photochemical reactions
- Concentration units in water
- Maximum Contaminant Level (MCL)

Introduction

The term synthetic chemical is used to describe substances that generally do not occur in nature but that have been synthesized by chemists from simpler substances. The great majority of commercial synthetic chemicals are organic compounds, and most use petroleum or natural gas as the original source of their carbon.

In this chapter, we discuss common pesticides—most of which are synthetic organic chemicals. In addition to surveying their structures and uses, we also consider the environmental problems associated with their use and some general principles of toxicology that apply to their effects on human health.

13.1 Types of Pesticides

Pesticides are substances that kill or otherwise control an unwanted organism. The most common categories of pesticides are listed in Table 13-1. All chemical pesticides share the common property of blocking a vital metabolic process of the organisms to which they are toxic. In this chapter, we discuss first **insecticides**—substances that kill insects—and then **herbicides**—compounds that kill plants. Collectively, these two categories, together with *fungicides*, represent the great bulk of the 1 *billion* kilograms of pesticides that are used annually in North America.

Fungicides are substances that are used to control the growth of various types of fungus, especially to protect stored seeds before planting.

TABLE 13-1	Pesticides and Their Targets	
Pesticide Type	**Target Organism**
Acaricide	Mites
Algicide	Algae
Avicide	Birds
Bactericide	Bacteria
Disinfectant	Microorganisms
Fungicide	Fungi
Herbicide	Plants
Insecticide	Insects
Larvicide	Insect larvae
Molluscicide	Snails, slugs
Nematicide	Nematodes
Piscicide	Fish
Rodenticide	Rodents
Termiticides	Termites

About half of the usage of pesticides in North America involves agriculture; worldwide the figure rises to 85%. Almost all commercial food crops are now produced with the use of synthetic insecticides, herbicides, and fungicides, except, of course, for organic farming where some natural pesticides are used. Indeed, the current ability to produce and harvest large amounts of food on relatively small amounts of land with a relatively small input of human labor has been made possible by the use of pesticides. Depending upon the main crops they produce, countries vary in what types of pesticides they consume in large amounts. For example, herbicides account for three-quarters of pesticide use in Malaysia, whereas insecticides are the most widely used pesticides in India and the Philippines, as are fungicides in Columbia.

Currently, the greatest U.S. use of insecticides occurs in the growing of cotton, whereas the majority of herbicides are used in the growing of corn and soybeans. The application of insecticides to cotton has been significantly reduced by the introduction of cotton that has been genetically modified to incorporate resistance to insects, in particular the bollworm.

Some 80–90% of American domestic households contain at least one synthetic pesticide; typical examples are weed killers for the lawn and garden, algae controls for the swimming pool, flea powders for use on pets, and sprays to kill insects such as cockroaches.

13.2 Concerns about Pesticides

Almost since their introduction, synthetic pesticides have been a concern because of the potential impact on human health of eating food contaminated with these chemicals. Half the foods eaten in the United States contain measurable levels of at least one pesticide. For that reason, many have been banned or restricted in their use. Nevertheless, the U.S. National Academy of Science has pointed out that pesticide regulation to date has not paid enough attention to the protection of human health, especially that of infants and children, whose growth and development are at stake. Children, kilogram for kilogram, eat more food than adults and tend to eat more food (such as apples, grapes, and carrots) with higher pesticide levels than do adults. The internal organs—including the brain—of children are still developing and maturing, which makes them more vulnerable to any negative effects these chemicals may have. In addition, small children play on floors and lawns and put many objects in their mouths, increasing their exposure to pesticides used in homes and yards.

Some scientists dismiss these concerns by emphasizing that living plants themselves manufacture insecticides in order to discourage insects and fungi from consuming them, and consequently we are exposed in our food supply to much higher concentrations of these "natural" pesticides than to synthetic ones. Natural pesticides are not necessarily less toxic than are synthetic ones, as we shall see.

13.3 Traditional Insecticides

Insecticides of one type or another have been used by society for thousands of years. One principal motivation for using insecticides is to control disease: human deaths due to insect-borne diseases through the ages have greatly exceeded those attributable to warfare. The use of insecticides has greatly reduced the incidence of diseases transmitted by insects and the rodents which bear them: malaria, yellow fever, bubonic plague, sleeping sickness, and recently, illness from the *West Nile virus* scarcely exhaust the list of these scourges. The other principal motivation for insecticide usage is to prevent insects from attacking food crops: nevertheless, even with extensive use of pesticides, about one-third of the world's total crop yield is destroyed by pests or weeds during growth, harvesting, and storage. People also try to control insects such as the mosquito and the common fly simply because their presence is annoying.

The earliest recorded usage of pesticides was the burning of sulfur to fumigate Greek homes around 1000 B.C. Fumigants are pesticides that enter the insect as an inhaled gas. The use of sulfur dioxide, SO_2, from the burning of solid *sulfur*, sometimes by incorporating the element in candles, continued at least into the nineteenth century. Sulfur itself, in the form of dusts and sprays, was also used as an insecticide and as a fungicide; it is still employed as a fumigant against *powdery mildew* on plants.

Inorganic fluorides, such as **sodium fluoride,** NaF, were used domestically to control ant populations. Both sodium fluoride and *boric acid* were used to kill cockroaches in infested buildings. Various oils, whether derived from petroleum or from living sources such as fish and whales, have been utilized for hundreds of years as insecticides and as *dormant sprays* to kill insect eggs.

Pesticides containing
mercury and lead were
discussed in Chapter 12.

13.4 Organochlorine Insecticides

Many organic insecticides developed during and after World War II have largely displaced inorganic and metal-organic substances. Usually only small amounts of the organic compounds are required to be effective against the target pests, so smaller amounts of chemicals enter the environment. Given the dosage of each large enough to act as a pesticide, **the organic substances are generally much less toxic to humans than are the inorganic and metal-organic pesticides.** Pesticide formulations containing only organic compounds were initially thought to be readily biodegradable, though as we shall see, this has certainly not been found to be true in many cases.

In the 1940s and the 1950s, the chemical industries in North America and western Europe produced large quantities of many new pesticides, especially insecticides. The active ingredients in most of these pesticides were **organochlorines, which are organic compounds that contain chlorine.** Many organochlorines share several notable properties:

• stability against decomposition or degradation in the environment;

• very low solubility in water, unless oxygen or nitrogen is also present in the molecules;

• high solubility in hydrocarbon-like environments, such as the fatty material in living matter; and

• relatively high toxicity to insects, but low toxicity to humans.

The nonpesticidal POPs,
including three new ones,
are discussed in Chapters 14
and 15.

Twelve organochlorine substances constituted the original "dirty dozen," listed by the *United Nations Environmental Program* (UNEP) as **Persistent Organic Pollutants,** or POPs, which are now banned or are being phased-out by international agreement (see Table 13-2 for the insecticides on the list). Each country signing the *Stockholm Treaty* that bans these POPs develops its own implementation plan.

Many of the POP insecticides were synthesized from *1,3-cyclopentadiene.* Six other organochlorine insecticides were added to the list in 2009 or later (Table 13-2, second column). As of the writing of this text, *chlordane* and *mirex* were still being used for termite control in a few countries. *Endosulfan* is banned as of mid-2012, with specific uses exempted until 2017.

1,3-cyclopentadiene

TABLE 13-2	Organochlorine Pesticides on the U.N.'s Persistent Organic Pollutants (POP) List	
Pesticides on Original POP List	**Recently Added Pesticides**	
DDT	Three isomers of hexachlorocyclohexane, one of which is called lindane	
Aldrin		
Dieldrin	Chlordecone (Kepone)	
Endrin	Pentachlorobenzene.	
Chlordane	Endosulfan*	
Heptachlor		
Hexachlorobenzene (HCB)		
Mirex		
Toxaphene		

*Although formally banned as of mid-2012, some uses are exempted for a further five years.

The reasons for the persistence and accumulation of such substances in the environment, and the small-scale continuing use of DDT, are topics discussed below.

ACTIVITY

Using the internet or printed resources in the library, determine the structural formula for the following POPs, all of which were used as organochlorine insecticides at one time: *aldrin, dieldrin, endrin, chlordane, heptachlor, mirex, chlordecone (Kepone)*, and *endosulfan*. Circle the common structural cyclopentadiene component incorporated into each of them. What is the current legal status of each of these compounds in your country?

Although many organochlorine compounds have been banned, a few remain in widespread use. For example, the 1,4 isomer of **dichlorobenzene** is still employed as an insecticidal fumigant.

1,4-Dichlorobenzene is sometimes called *para*-dichlorobenzene.

It is one type of domestic *moth repellant* and is also used as a deodorizer in restroom urinals, garbage pails, etc. Although it is a crystalline solid, it has an appreciable vapor pressure. Consequently, enough of it will vaporize to act as an effective insecticide in the immediate area around the solid. The same compound has also been used as a soil fumigant. However, it is an animal carcinogen and accumulates to some extent in the environment. It may well be the chemical responsible for the greatest carcinogenic risk of all indoor VOCs (volatile organic compounds). Recent research indicates it is present in the blood of most U.S. residents; its presence at high levels associated with reduced pulmonary function.

13.5 Pesticides in Water

The pollution of aquatic environments is not merely a question of the concentration of pollutant actually dissolved, i.e., in *solution*, and the small values for the water solubilities of organochlorines can be deceptive on this score. **Most organochlorine compounds are much more soluble in organic media than in water.** In bodies of water such as rivers and lakes, **organochlorines are much more likely to be adsorbed on the surfaces of organic particulate matter suspended in the water and on the muddy sediments at the bottom than to be dissolved in the water itself.** From these surfaces, they enter living organisms such as fish.

For reasons discussed in detail later, the concentration of pesticides in fish is often thousands or millions of times greater than that dissolved in polluted drinking water. It is due to this phenomenon that concentrations of organochlorines have often reached dangerous levels in many species. Many organochlorine insecticides have been removed from use as a consequence. For humans, the amount of organochlorines ingested by eating a single lake fish is generally greater than the total organochlorine content acquired in a lifetime of drinking water from the same lake!

Review Questions 1–5 are based on the material in the preceding sections.

DDT

DDT, or *para-dichlorodiphenyltrichloroethane,* has had a tumultuous history as an insecticide, as discussed in detail in the online Case Study: *To Ban or Not to Ban DDT? Its History and Future.* As discussed there, DDT went from the status of "savior," since it saved so many lives during the World War II and in years following, to "pariah" when its ecological effects were later uncovered. This turnaround is an example of how *systems thinking* is necessary if we are to evade negative *unintended consequences.* Although our current knowledge of toxicity and persistence of chemicals was lacking in the middle of the twentieth century, today our knowledge of these subjects has greatly increased and needs to be applied in a systematic manner before bringing new chemicals to market. One such program that does this in the U.S. EPA's *Premanufacture Notice.*

The two terms in italics were defined and discussed in Table 0-1.

Unfortunately, DDT was widely overused in the 1950s and 1960s, particularly in agriculture, which consumed 70–80% of its production in the United States, and in forestry. As a result, its environmental concentration rose rapidly and it began to affect the reproductive abilities of birds that indirectly incorporated it into their bodies. By 1962, DDT was being called an "elixir of death" by the writer Rachel Carson in her influential book *Silent Spring* because of its role in decreasing the populations of birds such as the bald eagle, whose intake of the chemical in their diet was very high.

13.6 DDT's Structure and Characteristics

Structurally, a DDT molecule is a substituted ethane. At one carbon, all three hydrogens are replaced by chlorine atoms, while at the other, two of the three hydrogens are replaced by a benzene ring. Each ring contains a chlorine atom at the *para* position, i.e., directly opposite the ring carbon that is joined to the (shaded) ethane unit:

(DDT): *para*-dichlorodiphenyltrichloroethane

Many animal species metabolize (convert by biochemical reactions into other substances) DDT by the elimination of HCl; a hydrogen atom is removed from one ethane carbon and a chlorine atom from the other, thereby creating a derivative of ethene called **DDE, dichlorodiphenyldichloroethene:**

DDE ·

Substances that are produced by the metabolism of a chemical are called **metabolites;** thus DDE is a metabolite of DDT. The chemical DDE is also produced slowly in the environment by the degradation of DDT under alkaline conditions, and by DDT-resistant insects that detoxify DDT by this transformation. Unfortunately, in some birds DDE interferes with the enzyme that regulates the distribution of calcium, so contaminated birds produce eggs that have shells (calcium carbonate) too thin to withstand the weight of the parents who sit on them to make them hatch.

PROBLEM 13-1

The structures shown above for DDT and DDE have both the ring chlorines in the *para* position, and are sometimes labeled *p,p'*-DDT and *p,p'*-DDE, where the *p* and *p'* prefixes refer to the para chlorine positions in the first and second rings. Deduce the structures and the appropriate labels for all the other unique isomers of both DDT and DDE in which there is one chlorine atom on each ring. Note that the two rings are equivalent, so that, e.g., *o,m'*-DDT is the same compound as is *m,o'*-DDT, and that there is free rotation about the C—C bonds.   ●

DDT's persistence made it an ideal insecticide: one spraying gave protection from insects for weeks to years, depending upon the method of application. Its persistence is due to

• its low vapor pressure and its consequent slow rate of evaporation,

• its low reactivity with respect to light and to chemicals and microorganisms in the environment, and

• its very low solubility in water.

Recall that a half-life is the time required for a substance to decline in concentration by a factor of 2, assuming its rate of decay is directly proportional to its concentration.

Its rate of evaporation from the upper layer of soil in southern Canada, for example, is so slow that its volatilization half-life is about 200 years. Like other organochlorine insecticides, DDT is soluble in organic solvents and therefore in the fat of animal tissue. DDT and/or its degradation products have been found in all birds and fish that have been analyzed, even those living in deserts or the ocean depths.

We all have some "DDT" (to the extent of about 3 ppm for North American adults) stored in our body fat. In humans, most ingested DDT is slowly but eventually eliminated. Most of the "DDT" stored in human fat is actually the DDE that was present in the food we have eaten, having previously been converted from DDT that was originally in the environment. Unfortunately, DDE is almost nondegradable biologically and is very fat-soluble, so it remains in our bodies for a long time.

13.7 DDT Levels in Modern Times

For environmental reasons, DDT is now banned from use in most industrialized countries. However, its use had been declining as resistant insect populations evolved that metabolized DDT to the non-insecticidal DDE and thus rendered it inactive. The United Nations includes DDT on its list of Persistent Organic Pollutants (Table 13-2). Of the pesticides, only DDT will not be totally banned.

A controversial exception to the ban on DDT in the control of disease is for its use in controlling mosquitoes involved in the transmission of malaria. The issues in this debate are explored in the online Case Study *To Ban or Not to Ban DDT? Its History and Future.*

DDT and DDE still enter the environment everywhere as a result of long-range air transport—a topic discussed at length in Chapter 15—from developing countries where DDT is still in use to control malaria and typhus and for some agricultural purposes. DDT also continues to be degassed from soils in developed countries where it was used for agricultural purposes decades ago. Such *secondary ("legacy") sources* now represent a significant fraction of POPs present in the air, water, and biosphere and prevent their concentrations there from decreasing to zero, even long after their use has been discontinued.

Thanks to restrictions and bans, the environmental concentrations of DDT and DDE in developed countries dropped substantially in the early and middle years of the 1970s and have now become stabilized at low levels. As an example, the decline in DDT levels in lake trout from Lake Michigan over a third of a century is illustrated in Figure 13-1a. From this graph it appears that the concentration has leveled out to a nonzero amount. The functional dependence of the overall decline is clarified by the logarithmic concentration plot in Figure 13-1b.

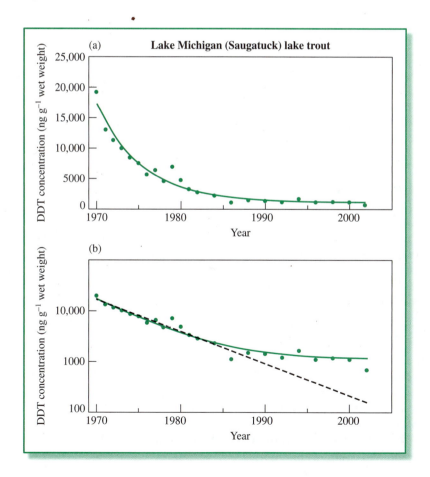

FIGURE 13-1 DDT concentrations in Lake Michigan lake trout using (a) a linear concentration scale and (b) a logarithmic scale. The dashed line in (b) represents a simple exponential fit to the early data, whereas the solid curve is an exponential that converges to a nonzero value. [Source: Modified from D. L. Carlson et al., *Environmental Science and Technology* 44 (2010): 2004–2010.]

Clearly, the simple exponential function representing the rapid decline in DDT levels seen from 1970 to the mid-1980s (dashed line), with a half-life of 5 years, has not been valid since that period. It is difficult to tell from the scatter in the experimental data whether the DDT level in the fish has been in very slow decline since then (with a half-life of about 17 years) or has been at a constant level. The latter explanation is the more likely, and probably arises because there remain small input sources of DDT—perhaps mainly from secondary (legacy) sources—to the lake water whose input rate now approximately equals the rate at which it continues to be removed. Overall, the data seem to fit well the mathematical function $y_0 + ae^{-bx}$, where the constant level y_0 is about 1000 ng g^{-1} (solid curves in Figure 13-1).

The same functional behavior is seen for the other Great Lakes and for other POPs there.

The concentration of DDT itself declined more rapidly than that of its metabolite DDE. Nevertheless, as a result of the substantial decline in DDE levels that has occurred, bald eagles have made a comeback around Lake Erie and elsewhere. Similarly, the population of Arctic peregrine falcons, a bird that was driven to near-extinction due to the effects of DDE, has now recovered to such an extent that it has been removed from the list of endangered species in the United States.

The concentration of DDT in humans has also declined drastically; for example, its 1997 level in human breast milk of Swedish women is only 1% of what it was in 1967. Data for women in North America show similar declines over this period. The concentration of DDE in breast milk did not drop as quickly as did that of DDT, its half-life being somewhat longer, about 6 years compared to 4 years for DDT. However, DDE levels remain high in regions such as Central and South America, Mexico, and Africa, where DDT has been used more recently.

Review Questions 6–9 and Additional Problem 1 are based upon the material in the preceding sections.

The Accumulation of Organochlorines in Biological Systems

Many organochlorine compounds are found in the tissues of fish in concentrations that are orders of magnitude higher than are those in the waters in which they swim. *Hydrophobic* (water-hating) substances like DDT are particularly liable to exhibit this phenomenon. There are several reasons for this **bioaccumulation** of chemicals in biological systems.

13.8 Bioconcentration

In the first place, many organochlorines are inherently much more soluble in hydrocarbon-like media, such as the fatty tissue in fish, than they are in water. Thus, when water passes through a fish's gills, the compounds selectively diffuse from the water into the fish's fatty flesh and become more concentrated there: this process (which also affects other organisms besides fish) is called **bioconcentration. The bioconcentration factor, BCF, represents the equilibrium ratio of the concentration of a specific chemical in** a fish relative to its concentration in the surrounding water, provided that the diffusion mechanism represents the only source of the substance to the fish. BCF values occur over a very wide range, and vary not only from chemical to chemical but also, to a certain extent, from one type of fish to another, particularly because of variations in the abilities of different fish to metabolize a given substance.

The BCF of a chemical can be predicted, to within about a factor of ten, for a typical fish from a simple laboratory experiment: the chemical is allowed to equilibrate between the liquid layers in a two-phase system made up of water and **1-octanol,** $CH_3(CH_2)_6CH_2OH$, an alcohol that has been found experimentally to be a suitable surrogate for the fatty portions of fish.

1-Octanol is also a good surrogate for humic materials and colloids found in the environment.

The **partition coefficient**, K_{ow}, for a substance S is defined as

$$K_{ow} = [S]_{octanol}/[S]_{water}$$

where the square brackets denote concentrations in molarity units. (Since water and fat have approximately the same densities, the ratio of molarities in the two phases is identical to the ratio of their masses; consequently, K_{ow} can also be taken as the ratio of ppm or ppb concentrations.) Largely as a matter of convenience, the value of K_{ow} is often reported as its base 10 logarithm, since its magnitude sometimes is quite large. For example, for DDT (see Table 13-3), K_{ow} is about 1,000,000, i.e., 10^6, and so $\log K_{ow} = 6$. Experimentally, the bioconcentration factor for DDT lies in the range of about 20,000 to 400,000, depending upon the type of fish. The K_{ow} value of a compound is a fairly reliable approximation to the BCF values found for fish. The approximation that BCF $= K_{ow}$ typically breaks down when the molecules are large to diffuse into the fish.

In general, the higher its octanol–water partition coefficient K_{ow}, the more likely a chemical is to be adsorbed on organic matter in soils and sediment and ultimately to migrate to fat tissues of living organisms.

TABLE 13-3	Selected Data for Some Pesticides	
Pesticide	Solubility in H_2O (ppm)	$\log K_{ow}$
Hexachlorobenzene (HCB)	0.0062	5.5–6.2
DDT	0.0034	6.2
Toxaphene	3	5.3
Dieldrin	0.1	6.2
Mirex	0.20	6.9–7.5
Malathion	145	2.9
Parathion	24	3.8
Atrazine	35–70	2.2–2.7

Data from K. Verschueren, *Handbook of Environmental Data on Organic Chemicals* (New York: Van Nostrand Reinhold, 1996).

However, log K_{ow} values of 7 or 8 or higher are indicative of chemicals with such strong adsorption to sediments that they are actually unlikely to be mobile enough to enter living tissue. Thus, **it is chemicals with log K_{ow} values in the 4–7 range that bioconcentrate to the greatest degree.**

In analogy to the K_{ow} parameter, scientists also use the *octanol–air partition coefficient, K_{oa}.* Organic chemicals with K_{ow} values as low as 100 but K_{oa} values greater than 10^6 (which includes about two-thirds of all commercial organic compounds) have a potential for bioaccumulation in food webs that contain air-breathing animals.

PROBLEM 13-2

For HCB, log K_{ow} = 5.3. What would be the predicted concentration of HCB due to bioconcentration in the fat of fish that swim in waters containing 0.000010 ppm of the chemical?   ●

13.9 Biomagnification

Fish also accumulate organic chemicals from the food they eat and from their intake of particulates in water and sediments onto which the chemicals have

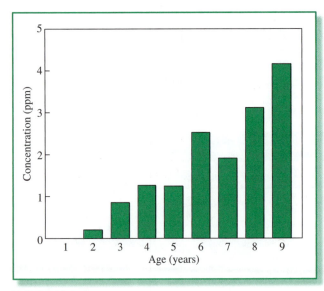

been adsorbed. In many such cases, the chemicals are not metabolized by the fish: the substance simply accumulates in the fatty tissue of the fish, where its concentration there increases with time. For example, the concentration of DDT in trout from Lake Ontario increases almost linearly as the fish ages, as illustrated in Figure 13-2. The average concentration of many chemicals also increases dramatically as one proceeds up a **food chain,** which is a sequence of species, each one of which feeds mainly upon the one preceding it in the chain. Over a lifetime, a fish eats many times its weight in foods from the lower levels of the food chain, but retains rather than eliminates or metabolizes most organochlorine chemicals from these meals.

FIGURE 13-2 Variation with age of DDT concentration in Lake Ontario trout caught in the same year. [Source: "Toxic Chemicals in the Great Lakes and Associated Effects" (Ottawa: Government of Canada, 1991).]

A chemical whose concentration **increases along a food chain is said to be biomagnified.** In essence, the biomagnification results from a sequence of bioaccumulation steps that occur along the chain. The difference between bioconcentration from water and biomagnification along a food chain is illustrated symbolically in Figure 13-3.

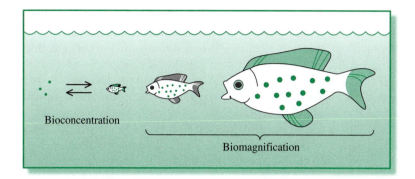

FIGURE 13-3 Schematic representation of the two modes of bioaccumulation that operate in biological matter present in a body of water.

Fish at the top of the aquatic part of the food chain bioaccumulate organochlorines such as DDT rather effectively, so that even higher concentrations are found in the birds of prey that feed on them. As an example of such biomagnification, consider that the DDT + DDE concentration in seawater in Long Island Sound and the protected waters of its southern shore at one time was as high as 0.000003 ppm, but it reached 0.04 ppm in the plankton, 0.5 ppm in the fat of minnows, 2 ppm in the needlefish that swim in these waters, and 25 ppm in the fat of the cormorants and osprey that feed on the fish, for a total biomagnification factor of about 10 million. It is by such mechanisms that DDE levels in some birds of prey became so great that their ability to reproduce successfully was impaired. **The bioaccumulation of organochlorines in fish and other animals is the reason why most of the human daily intake of such chemicals enters via our food supply rather than from the water we drink.**

The biomagnification up a Great Lakes food chain of a nonpesticidal family of organochlorine compounds is illustrated in the next chapter, in Figure 14-2.

13.10 Less Persistent Analogs of DDT

A number of compounds having the same general molecular structure as DDT are found to display similar insecticidal properties. This similarity arises from **the mechanism of DDT action, which is due more to its molecular *shape* than to its chemical interactions.** The shape of a DDT molecule is determined by the two tetrahedral carbons in the ethane unit and by the two flat benzene rings. In insects, DDT and other molecules with the same general size and 3-D shape become wedged in the nerve channel that leads out from the nerve cell. Normally, this channel transmits impulses only as needed via sodium ions. But a continuous series of Na^+-initiated nerve impulses is produced when the DDT molecule holds open the channel. As a consequence, the muscles of the insect twitch constantly, eventually exhausting it with convulsions that lead to death. The same process does not occur in humans and other warm-blooded animals since DDT molecules do not exhibit such binding action in nerve channels.

Examples of other molecules with DDT-like action include **DDD** (sometimes called TDE), **para-dichlorodiphenyldichloroethane,** which is an environmental degradation product of DDT: it differs only in that one chlorine from the —CCl$_3$ group is replaced by a hydrogen. Since the overall shapes and sizes of DDT and DDD are similar, their toxicity to insects is as well. In the past, DDD was itself sold as an insecticide, but it has also been discontinued because it bioaccumulates.

Notice that DDE, unlike DDT and DDD, is based upon a *planar* C=C unit rather than a C—C linkage which has tetrahedral groups at each end. Thus, whereas DDD is a DDT-like insecticide, DDE is not, since its three-dimensional shape is very different: DDE is flat rather than propeller-shaped, and so it does not become wedged in the insect's nerve channels.

Scientists have devised analogs to DDT that have its same general size and shape, and consequently possess the same insecticidal properties, but are more biodegradeable and thus do not present the bioaccumulation problem associated with DDT. The most important of these analogs is **methoxychlor:**

methoxychlor

The *para*-chlorines of DDT are replaced in methoxychlor by *methoxy* groups, —OCH$_3$, which are approximately the same size as chlorines but which react much more readily. In particular, under reducing conditions, the O—CH$_3$ bonds in methoxychlor are converted to O—H bonds. The hydroxylated products are water-soluble compounds, and not only do they degrade in the environment but they are excreted rather than accumulated by organisms. Methoxychlor is still used both domestically and agriculturally to control flies and mosquitoes.

Review Questions 10–13 and Additional Problem 2 are based upon the material in the preceding sections.

PROBLEM 13-3

Draw the molecular structure of DDD.

PROBLEM 13-4

Methyl groups are approximately the same size as chlorine atoms, but hydrogen atoms are significantly smaller. Would you expect insecticidal properties for DDT analogs in which (a) the —CCl$_3$ group is replaced by —C(CH$_3$)$_3$, and (b) the *para*-chlorines are replaced by hydrogens?

Principles of Toxicology

Toxicology is the study of the harmful effects to living organisms of substances that are foreign to them. The substances of interest include both synthetic chemicals and those that exist naturally in the environment. In toxicology, the effects are normally determined by injecting or feeding animals with the substance of interest, and observing how the health of the animal is affected. By contrast, in **epidemiology**, scientists do not run experiments in a lab but instead determine the health history of a selected group of human beings and attempt to relate differences in disease rates, etc. to differences in the substances to which they have been accidently exposed.

13.11 Types of Toxicity

Toxicological data concerning the harmfulness of a substance, such as an organochlorine pesticide or a heavy metal, to an organism are gathered most easily by determining its **acute toxicity, which is the rapid onset of symptoms—including death at the extreme limit—immediately following the intake of a dose of the substance.** For example, experiments show that it takes only a few tenths of 1 μg of the most acutely toxic synthetic compound—the "dioxin" to be discussed in Chapter 14—to kill most rodents within a few hours after it is administered orally to them.

Although the acute toxicity of a substance is of interest when we are exposed accidently to pure chemicals, in **environmental toxicology** we are usually more concerned about **chronic** (continuous, long-term) **exposures** at relatively low individual doses of a toxic chemical that is present in the air that we breathe, the water we drink, or the food we eat. Generally speaking, any effects of such continuing exposures are also long-lasting and therefore also classified as chronic.

The same chemical may give rise to both acute and chronic effects in the same organism, although usually by different physiological mechanisms. For example, a symptom of acute toxicity in humans of exposure to many organochlorines is a skin irritation that leads to **chloracne,** a persistent, disfiguring, and painful analog to common acne. There is the fear that persistent exposure to much lower individual doses than those which produce the skin disease could eventually lead to cancer.

The three types of chemicals that produce the detrimental effects on long-term human health of most concern are

- **mutagens,** substances that cause mutations in DNA, most of which are harmful and that can produce inheritable traits;

- **carcinogens,** substances that cause cancer; and

- **teratogens,** substances in the mother that cause birth defects in the fetus.

Some carcinogens operate in the *initiation step*, in which the substance—sometimes after itself having been transformed in the body—reacts directly with a strand of DNA. This alteration in DNA can lead to the growth of cancer cells. Others, called *promoters*, act only after cancer has been initiated, but speed up and enlarge the process of tumor formation.

13.12 Dose–Response Relationships

Most of the *quantitative* information concerning the toxicity of substances is obtained from experiments performed by administering doses of the substances to animals. Owing to practical considerations including cost and time, most experiments involve acute rather than chronic toxicity, even though it is the latter that usually is of primary interest in environmental science. To determine directly the effects of continuous, low-level exposures over long periods would require a very large number of test animals and a long project time. The practical alternative is to evaluate the effects using high doses—at which point the effects are substantial and clear-cut—and then extrapolate the results down to environmental exposures. Unfortunately, there is no assurance that such extrapolations are always reliable, since the cellular mechanisms that produce the effects at high and low doses could differ.

The dose of the substance administered in toxicity tests is usually expressed as the mass of the chemical, usually in milligrams, per unit of the test animal's body weight, usually expressed in kilograms, thus giving units of milligrams per kilogram, mg kg^{-1}. The division by body weight is necessary because the toxicity of a given amount of a substance usually decreases as the size of the individual increases. It is also assumed that toxicity values obtained from experiments on small test animals are approximately transferable to humans, provided the differences in body weight are taken into account. Normally the toxicity of a substance increases with increasing dose, although exceptions are known.

> The maximum recommended doses for medicines such as headache remedies are smaller for children than for their adults, primarily because of the difference in body masses.

PROBLEM 13-5

If a dose of a few tenths of a microgram of a certain substance is sufficient to kill a mouse, approximately what mass of the substance would be fatal to you? What average ppb concentration of the substance would have to be present in the water you drink if you were to receive a fatal dose from this source in a week? Note that your weight in kilograms is that in pounds divided by 2.2, and the weight of a mouse is about 30 g.   ●

Individuals differ significantly in their susceptibility to a given chemical: some respond to it even at very low doses whereas others require a much higher dose before they respond. It is for this reason that scientists created **dose–response relationships** for toxic substances, including environmental agents. A typical dose–response curve for acute toxicity is illustrated in Figure 13-4a. The dose is plotted on the (horizontal) x-axis, and the cumulative percentage of test animals that display the measured effect (e.g., death) when administered a particular dose is shown on the (vertical) y-axis. For example, in Figure 13-4a, about 60% of the test animals were affected by a dose of about 4 mg kg^{-1}.

Because the range of doses on such graphs often exceeds an order of magnitude, and because the effects at the low end of the concentration scale are often important in environmental decision making and cannot be seen clearly using linear scales, the dose–response plot is usually recast by using a logarithmic scale for doses. Usually an S-shaped or sigmoidal-shaped curve results from this transformation; see Figure 13-4b.

This type of plot was explained in Chapter 12, Figure 12-7.

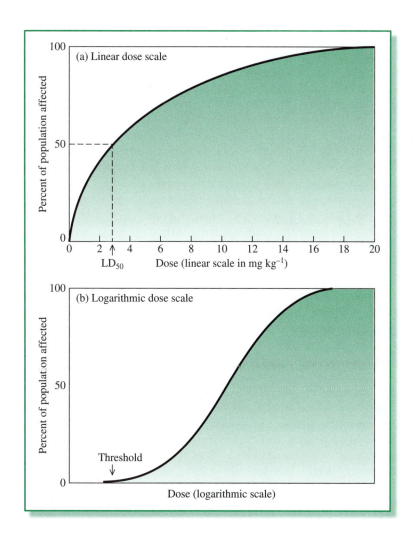

FIGURE 13-4 Dose–response curves for (a) linear dose scale; (b) logarithmic dose scale.

13.13 Lethal Doses and Concentrations

Most often, the response effect on test animals that is used to construct dose–response curves is death. **The dose that proves to be lethal to 50% of the population of test animals is called the LD_{50} value of the substance;** its determination from a dose–response curve is illustrated by the dashed lines in Figure 13-4a. **The *smaller* the value of LD_{50}, the more potent (i.e., more toxic) is the chemical, since less of it is required to affect the animal.** A chemical much less toxic than that illustrated in Figure 13-4b would have a sigmoidal curve shifted to the right of the one shown.

Many sources discuss values for the **LOD_{50},** the *lethal oral dose*, when the chemical has been administered orally to the test animals, as opposed dermally or by some other means. For example, the LOD_{50} value for DDT for rats is about 110 mg kg^{-1}. As mentioned previously, the presumption is usually made that LD_{50} and LOD_{50} values are approximately transferable between species. In the case of DDT, for example, humans are known to have survived doses of about 10 mg kg^{-1}, so presumably the LOD_{50} value for humans is greater than 10 mg kg^{-1}. However, we have no *direct* evidence that the 110-mg kg^{-1} value for rats is also valid for humans.

Of much more concern than the acute toxicity of DDT is its ability to cause chronic effects in humans, such as cancer. Although DDT is not traditionally considered to be a human carcinogen, some small-scale epidemiological studies found that the higher the concentration of DDE in a woman's blood, the more likely she was to have contracted breast cancer. However, later full-scale studies in the United States have failed to confirm this association between breast cancer and DDE. Thus it does not seem that lifetime exposure to DDT is an important cause of breast cancer. However, these studies do not address the issue raised recently of whether exposure during the teenage years, when breasts are developing rapidly, might not be a factor in developing breast cancer decades later.

The range of LD_{50} and LOD_{50} values for acute toxicity of various chemical and biological substances is enormous, and spans about ten powers of 10. Indeed, all substances are toxic at sufficiently high doses; as the Renaissance-era German philosopher Paracelsus observed, all things are poison, and it is the dose that differentiates a poison from a remedy.

The World Health Organization (WHO) has devised descriptors for four broad levels of toxicity for substances, especially pesticides; the properties and examples of pesticides and common substances for each category are shown in Table 13-4. All the pesticides in the *extremely toxic* category, Ia, are synthetic, but nicotine—which as a solution of its sulfate salt has been used as an organic insecticide in gardens for many decades—falls in category Ib, *highly hazardous*. The U.S. EPA classifies pesticides in a similar fashion to the WHO, but does not distinguish between the two sublevels of the WHO's category I. Category II substances, *moderately hazardous*, include many pesticides—both synthetic and natural—still on the market.

Most nicotine is destroyed by incineration in cigarettes before the smoke is inhaled.

TABLE 13-4		WHO and U.S. EPA Pesticide Hazard Categorization			
WHO Category Number	U.S. EPA Category*	WHO Description	LOD_{50}† $(mg\ kg^{-1})$	Examples	
				Synthetic Pesticide	"Natural" Pesticide
Ia	I	Extremely hazardous	<5	aldicarb; parathion; methyl parathion; turbufos	
Ib	I	Highly hazardous	5–50	azinphos-methyl; carbofuran; dichlorvos	nicotine
II	II	Moderately hazardous	50–500	carbaryl; chlorpyrifos; diazinon; dimethoate; endosulfan; fenitrothion; lindane; paraquat; propoxur	permethrin; pyrethrins; rotenone
III	III	Slightly hazardous	500–5000	alachlor; malathion; metolachlor; 2,4-D family; glyphosate	allethrin
III	IV		>5000		

*The United States EPA does not distinguish between WHO classes Ia and Ib but uses a single category I. The EPA also defines a fourth category, IV, for substances with LOD_{50} values greater than 5000 mg kg^{-1}.

†The LOD_{50} values quoted are for the solid form and are based upon experiments with rats; LD_{50} ranges are a factor of 2 higher. Lethal dose ranges for liquids are a factor of 4 larger than their respective LD_{50} and LOD_{50} ranges.

The WHO's category III, *slightly hazardous*, is subdivided into III and IV by the EPA.

In the dose–response curves for *some* substances, there exists a dose below which none of the animals are affected; this is called the threshold, illustrated in Figure 13-4b. The highest dose at which no effects are seen lies slightly below it and is called the *no observable effects level* (NOEL), although sometimes the two terms are used interchangeably. It is difficult to determine the threshold or NOEL level: it may be that if more animals were involved in a particular study, effects at low doses might be uncovered that are not apparent with only a small number of test animals. Most toxicologists believe that for toxic effects *other* than carcinogenesis there is probably a nonzero threshold for each chemical. A few scientists hold the controversial view that for some substances the curve in Figure 13-4b actually falls *below* the zero or NOEL value for very low concentrations before returning to zero at zero dose, indicating that tiny amounts of these substances could have a positive rather than a negative effect on health.

Experiments involving test animals are also used to determine how carcinogenic a compound is. However, the simple **Ames test** that uses bacteria can be used fairly reliably to distinguish compounds likely to be human carcinogens from those that are not.

A parameter that is useful in judging whether a specific chemical is present in an environmental sample in dangerous amounts or not is the **lethal concentration, LC,** of the substance. Usually this is listed as its **LC_{50}, the concentration of the substance that is lethal to 50% of a specified organism within a fixed exposure period.** LC_{50} values may refer to the concentration of a substance in air or in aqueous solution, and usually has units of milligrams per liter. For example, the LC_{50} for rainbow trout for a 4-day exposure to the insecticide endosulfan in water is only 0.001 mg L^{-1}; indeed, this insecticide is "supertoxic" to many fish species. The LC_{50} for a shorter exposure would be a value greater than 0.001.

LC ratings for many pesticides are available for exposure in air, or more precisely, as aerosols in air. The EPA categories of high through very low toxicity are then applied to those ratings, as they are to the **Dermal LD_{50}** ratings, which measure the sensitivity of skin to the pesticide. Categories of all three types of toxicity are listed for some of the common organophosphate insecticides in Table 13-5.

13.14 Risk Assessment

Once toxicological and/or epidemiological information concerning a chemical is available, a **risk assessment** analysis can be performed. This analysis tries to answer quantitatively the questions "What are the likely types of toxicity expected for the human population exposed to a chemical?" and "What is the probability of each effect occurring to the population?" Where necessary, risk assessment also tries to determine permissible exposures to the substance in question.

In order to perform a risk assessment on a chemical, it is necessary to know

• **hazard evaluation** information; i.e., the types of toxicity (acute? cancer? birth defects?) that are expected from it;

• quantitative **dose–response** information concerning the various possible modes of exposure (oral, dermal, inhalation) for it; and

• an estimate of the potential **human exposure** to the chemical.

For chronic exposures, threshold or NOEL dose information is normally expressed as milligrams of the chemical per kilogram of body weight per day. In determining the threshold level for the most sensitive members of the human population, it is common to divide the NOEL from animal studies by

TABLE 13-5	Structural Type, Toxicity, Regulatory Status, and Uses of Common Organophosphate Insecticides			
Organophosphate Name and Type[*]	Acute Oral Hazard Category (Other Routes)[†]	U.S. EPA Agricultural Status	Uses	Comments
Azinphosmethyl C	High	2012 phase-out	Fruits, vegetables	Usage is risk to farm workers
Chlorpyrifos B	Moderate; (moderate inhalation; moderate to very low dermal)	Restricted	Indoor insects[††], crop insects	
Diazinon B	Low-moderate; (low dermal; very low inhalation)	Restricted	Domestic insects[††], soil & foliage crop insects	Was widely used in pet collars
Dichlorvos (ddvp) A	High	Restricted	Fly strips, animal worms	
Dimethoate C	Moderate	Restricted	Crops insects, including domestic[††]	
Malathion C	Very low; (very low inhalation; low dermal)	Restricted	Mosquito and fly control; crop insects	Head lice shampoo ingredient
Parathion (ethyl-) B	Extremely high	2006 cancellation	Crop insect spray	Widely banned; fatal to many farm workers
Parathion (methyl-) B	Extremely high	2013 cancellation	Cotton insects	

[*]See Figure 13-5 for type classification meanings.
[†]Acute toxicity levels according to WHO Descriptions in Table 13-4. Inhalation and dermal toxicity are shown in parentheses, where available.
[††]Now withdrawn in United States for residential applications, but still used in developing countries for such purposes.

a *safety factor,* typically 100. The resulting value is called the maximum **acceptable daily intake,** ADI, or maximum daily dose; the U.S. EPA instead refers to it as the **toxicity reference dose,** RfD. Note that the ADI or RfD value does *not* represent a sharp dividing line separating absolutely safe from absolutely unsafe exposures, since the transferability of toxicity information from animals to humans is not exact, and in most cases the safety factor presumably is quite generous. Some scientists have suggested dividing the

NOEL by a further factor of ten in order to protect very susceptible groups such as children. Indeed, the Food Quality Protection Act in the United States requires that the EPA set limits for residues of pesticides on foods ten times lower than what is considered to be safe for an adult.

PROBLEM 13-6

The NOEL for a chemical is found to be 0.010 mg kg^{-1} day^{-1}. What would its ADI or RfD value for adults be set at? What mass of the compound is the maximum that a 55-kg woman should ingest daily?   ●

As mentioned, in a risk assessment, an attempt is made to estimate the exposure of the affected population. For example, for chemicals whose mode of exposure is primarily through drinking water, regulatory agencies such as the U.S. EPA consider a hypothetical average person who drinks about 2 L of water daily and whose body weight averages 70 kg (154 lb) through life. If the ADI (or RfD) of a substance is 0.0020 mg kg^{-1} day^{-1}, then for the 70-kg person, the mass of it that can be consumed per day is $0.0020 \times 70 = 0.14$ mg day^{-1}. Thus the maximum allowable concentration of the chemical in water would be 0.14 mg/2 L = 0.07 mg L^{-1} = 0.07 ppm. Of course, if there are other significant sources for the substance, they must be taken into account in determining the drinking-water standard. Also, exposure to several chemicals of the same type (e.g., several organochlorines) could produce additive effects, so the standard for each one should presumably be lowered to take this into consideration.

The U.S. EPA regulates exposure to most carcinogens by assuming that the dose–response relationship has no threshold and can be linearly extrapolated from zero dose to the area for which experimental results are available. The maximum daily doses are then determined by assuming that each person receives the dose every day over his or her lifetime and that this exposure should not increase the likelihood of cancer to more than one person in every million.

In determining regulations to control risk, usually no consideration is given to the economic costs involved. Many economists believe that, because regulations cost money to implement and because such resources are limited, a cost-benefit analysis should be used to help decide which substances to regulate. Associated with this line of thinking is the idea that regulations should show a positive payback: the benefits from regulation should exceed the costs. Of course, it is difficult to place a specific monetary value on environmental benefits in many cases. For example, is the $200,000 average cost for saving a life by regulating the chloroform content in water (see Chapter 11) a reasonable amount to pay? If so, would it still be reasonable if the cost instead were 10 or 100 or 1000 times this amount?

Review Questions 14–16 and Additional Problems 3 and 4 are based upon the material in the preceding sections.

Organophosphate and Carbamate Insecticides

13.15 Organophosphate Insecticides

Structurally, all **organophosphate** pesticide molecules can be considered as derivatives of **phosphoric acid,** $O=P(OH)_3$, and consist of a central, pentavalent phosphorus atom to which are connected

- an oxygen or sulfur atom doubly bonded to the P atom;

- two methoxy ($-OCH_3$) or ethoxy ($-OCH_2CH_3$) groups singly bonded to the P atom; and

- a longer, more complicated, characteristic R group singly bonded to phosphorus, usually through an oxygen or sulfur atom, that differentiates one organophosphate from another.

phosphoric acid

For example, R is $-CH=CCl_2$ for dichlorvos.

The organophosphate insecticides are toxic because they inhibit enzymes in the nervous system; thus they function as nerve poisons. In particular, organophosphates disrupt the communication that is carried between cells by the **acetylcholine** molecule. This cell-to-cell transmission cannot operate properly unless the bound acetylcholine molecule is destroyed after it has executed its function. Organophosphates block the action of the enzymes whose job it is to destroy the acetylcholine, by selectively bonding to them. (At the atomic level, it is the phosphorus atom of the organophosphate molecule that attaches to the enzyme and stays bound to it for many hours.) The presence of the insecticide molecule has the effect of suppressing the dissociation of the bound acetylcholine. Consequently, continuous stimulation of the receptor cell and its target muscle occurs.

Normally the enzyme occurs in large excess and therefore some exposure to OPs occurs without immediate effects, but symptoms begin to appear if a majority of them are inactivated. Since the affected nerves include those controlling gastrointestinal activities and bronchial secretions, massive gastric and respiratory secretions occur, along with involuntary motions. Death ensues if 80–90% or more of the enzyme sites are inactivated.

The largest use of organophosphate (OP) insecticides is in agriculture, but they also find many uses domestically. Organophosphates generally are nonpersistent; in this respect, they represent an environmental advance over organochlorines since they do not bioaccumulate in the food chain, presenting a chronic exposure and health problem to us later. However, **organophosphate insecticides are generally much more acutely toxic to humans and other mammals than are organochlorines.** Many organophosphates represent an acute danger to the health of those who apply them and to others who may come into contact with them. Exposure to these chemicals by inhalation, swallowing, or absorption through the skin can lead to immediate health problems. Organophosphates however, metabolize relatively quickly and are excreted in the urine.

General structure

Type A:

$$R-O-\overset{\overset{\displaystyle O}{\|}}{\underset{\underset{\displaystyle O-CH_3}{|}}{P}}-O-CH_3$$

Type B:

$$R-O-\overset{\overset{\displaystyle S}{\|}}{\underset{\underset{\displaystyle O-CH_3}{|}}{P}}-O-CH_3$$

Type C:

$$R-S-\overset{\overset{\displaystyle S}{\|}}{\underset{\underset{\displaystyle O-CH_3}{|}}{P}}-O-CH_3$$

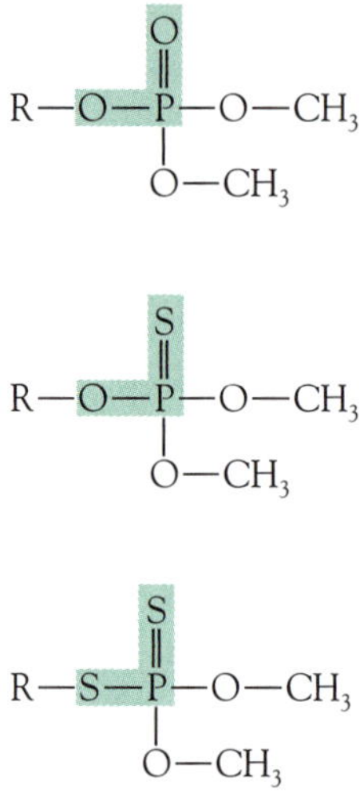

FIGURE 13-5 General structures of organophosphate insecticides. In some examples, the methoxy groups are replaced by ethoxy groups.

The three main subclasses of the organophosphates (for convenience called Types A, B, and C), each falling under one of these headings, are illustrated in Figure 13-5. Those containing a P=S unit are less acutely toxic, and therefore less dangerous to humans, but are converted within insects—but much less so within animals—to the corresponding molecules with a P=O unit, producing more toxic substances. The P=S form is also used initially because it penetrates the insect more readily. Such organophosphates decompose fairly rapidly in the environment because oxygen in air alters P=S bonds to P=O.

13.16 Organophosphate Insecticides in the Environment

After application, most organophosphates decompose within days or weeks, and thus are seldom found to bioconcentrate appreciably in food chains. However, because they have a wide range of uses in homes, on lawns, in commercial buildings, and in agriculture, almost everyone is regularly exposed to them. There is some evidence that OPs cause chronic as well as acute health problems. Children in the United States generally have their greatest exposure to organophosphates through the food they eat, with exposure from drinking water only 1–10% of that from food.

A 2003 report found that preschool children (in Seattle) who consumed mainly organic fruits, vegetables, and juices had much lower exposure to organophosphates, as measured by metabolites in their urine, than children with conventional diets. A number of studies, none of them large enough to be statistically definitive, have found links between indoor use of insecticides—organophosphates especially—and childhood leukemia and brain cancer. A recent study of seven-year-old children found that high exposure to organophosphates in the womb—though not after birth—can lower the IQ by amounts comparable to those from prebirth exposure to PCBs (as will be discussed in Section 14.15).

Most organophosphate insecticides decompose in the environment by *hydrolysis* reactions, i.e., reaction with H_2O. Water molecules split P—S and P—O bonds in organophosphates by adding H to the sulfur or oxygen atom and OH to the phosphorus and also add to the C—O bonds in the phosphoric acid esters formed as intermediates, ultimately yielding nontoxic substances such as phosphoric acid and alcohols and thiols. For example:

$$R'-S-\overset{\overset{\displaystyle O}{\|}}{P}(OR)_2 + H-OH \longrightarrow R'-S-H + HO-\overset{\overset{\displaystyle O}{\|}}{P}(OR)_2$$

$$\longrightarrow \longrightarrow R'SH + 2\,ROH + O{=}P(OH)_3$$

The best-known organophosphate insecticides are listed in Table 13-5, along with their structural category (as in Figure 13-5), toxicity level, legal status for agricultural applications in the United States, and common uses. The U.S. EPA recently re-evaluated individual organophosphates and took action to restrict or ban some of them (Table 13-5), and have eliminated domestic use of most of them. They have now undertaken a cumulative risk assessment encompassing all organophsphates insecticides, since they all act by a similar mechanism.

ACTIVITY

Using internet or library resources related to pesticides, add a column to a photocopy of Table 13-5 showing the structural formulas for the R groups of the organophosphate insecticides listed.

13.17 Malathion

The most important remaining organophosphate insecticide is **malathion.** Upon exposure to oxygen, it is converted to *malaoxon*, which has its doubly bonded sulfur atom replaced by oxygen and is more toxic than malathion. Introduced in 1950, malathion is not particularly toxic to mammals since their enzymes convert it to a harmless form faster than their livers convert the P$=$S form to the activated P$=$O form, but insects do not have the requisite enzyme to deactivate it and hence it is fatal to them. However, if improperly stored, malathion can be converted to an isomer that is almost 100 times as toxic; it was responsible for the death of five spray workers and the sickness of thousands more during a malaria eradication program in Pakistan in 1976.

Malathion is still used in domestic fly sprays and to protect agricultural crops. In combination with a protein bait, low concentrations of malathion have been sprayed from helicopters over several areas in the United States (California, Florida, Texas) to combat infestations of the Mediterranean fruit fly, a dangerously destructive pest. As in California, aerial spraying of malathion in Chile to combat the fruit fly has also proven to be controversial. It was also sprayed on parts of New York City (1999) and Florida (1990) to protect against *St. Louis encephalitis*, carried by mosquitoes. For decades, the entire city of Winnipeg, Manitoba, has been sprayed with malathion several times each summer to keep down the mosquito population and now also to protect against the West Nile virus. A recent analysis concluded that the health risks from this virus exceed those of the malathion used to control mosquitoes that carry it.

Malathion is a Type C insecticide in the terminology used in Figure 13-5.

13.18 Carbamate Insecticides

The mode of action of **carbamate** insecticides is similar to that of the organophosphates; they differ in that it is a carbon atom rather than a

phosphorus atom that attacks the acetylcholine-destroying enzyme. The carbamates, introduced as insecticides in 1951, are derivatives of **carbamic acid,** H_2NCOOH. One of the hydrogens attached to the nitrogen is replaced by an alkyl group, usually methyl, and the hydrogen attached to the oxygen is replaced by a longer, more complicated organic group symbolized below simply as R:

$$
\underset{\text{carbamic acid}}{H_2N-\overset{\overset{\displaystyle O}{\|}}{C}-OH}
\qquad\qquad
\underset{\text{the general formula of a carbamate}}{CH_3-\overset{\overset{\displaystyle H}{|}}{N}-\overset{\overset{\displaystyle O}{\|}}{C}-O-R}
$$

Like the organophosphates, the carbamates are short-lived in the environment since they undergo hydrolysis reactions and decompose to simple, nontoxic products. The reaction with water involves the addition of H—OH to the N—C bond; the species HO—C(=O)—OR decomposes to release CO_2 and the alcohol R—OH.

$$
CH_3-NH-\overset{\overset{\displaystyle O}{\|}}{C}-O-R + H-OH \longrightarrow CH_3-NH_2 + [HO-\overset{\overset{\displaystyle O}{\|}}{C}-O-R]
$$
$$
\longrightarrow HO-R + CO_2
$$

1-naphthyl

Carbaryl, the carbamate in which R is the naphthalene ring system, is the most widely used insecticide for domestic (lawn and garden pests) and agricultural purposes in the United States. Its oral toxicity is moderate, dermal toxicity low, and inhalation toxicity very low. However, it is particularly toxic to honeybees.

The other two well-known carbamate insecticides, **carbofuran** and **aldicarb,** are both highly toxic. Aldicarb is licensed only to be applied by professionals in the United States, and is banned in some European countries. Carbofuran likewise was used mainly by professional applicators, but has recently been effectively banned by the U.S. EPA. It is so toxic that it has been used to kill wildlife in some parts of the world, and was responsible for the inadvertent killing of tens of millions of birds in North America annually.

13.19 Health Problems of Organophosphates and Carbamates

Organophosphates and carbamates solved the problem of environmental persistence and accumulation associated with organochlorine insecticides, but sometimes at the expense of dramatically increased acute toxicity to the humans and animals who encounter them while the chemicals are still in the active form. These less persistent insecticides—together with the pyrethroids mentioned below—largely replaced organochlorines in residential uses.

Organophosphates are a particular problem in developing countries, where widespread ignorance about their hazards and failure to use protective clothing—due to lack of information or to the heat—has led to many deaths among agricultural workers. The types of pesticides used in developing countries are also more likely to be highly toxic, even banned elsewhere for that reason.

Estimates by the United Nations and the World Health Organization put the number of persons who suffer acute illnesses from short-term exposure to pesticides in the millions annually; 10,000–40,000 die each year from poisoning, about three-quarters of these in developing countries. Although most of the deaths from pesticide poisonings occur in developing countries, about 20,000 people receive emergency medical care in the United States annually for actual or suspected poisoning from pesticides, and about 30 Americans annually die from it.

Review Problems 17–19 and Additional Problem 5 are based upon the material in the preceding sections.

ACTIVITY

Fipronil is a relatively new residential insecticide, residues of which have been detected in many homes. Using web resources and perhaps a visit to a local pet store, write a report discussing its chemical nature, its uses as an insecticide, toxicity, and other advantages and disadvantages of it.

Natural and Green Insecticides, and Integrated Pest Management

13.20 Pesticides from Natural Sources

As pointed out earlier, many plants can themselves manufacture certain molecules for their own self-protection that either kill or disable insects. Chemists have isolated some of these compounds so that they can be used to control insects in other contexts. Examples are nicotine, rotenone, the pheromones, and juvenile hormones.

One group of natural pesticides that has been used by humans for centuries is the **pyrethrins.** The original compounds, the general structure for which is illustrated below, were obtained from the flowers of a species of chrysanthemum.

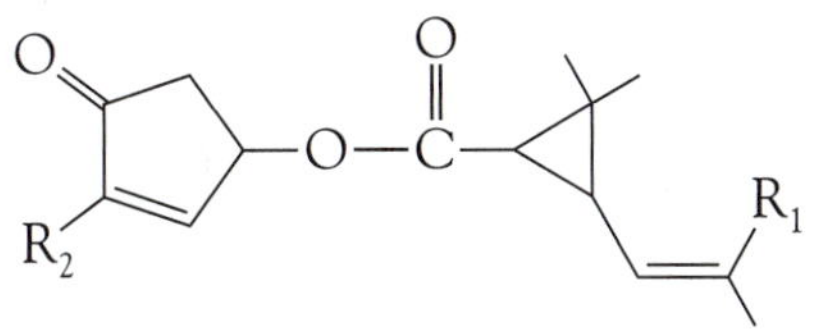

general pyrethrin structure

In the form of dried, ground-up flower heads, pyrethrins were used in Napoleonic times to control body lice; they are still used in flea sprays for animals.

Pyrethrins are generally considered to be safe to use. Like organophosphates, they paralyze insects, though they usually do not kill them. Unfortunately, these compounds are unstable in sunlight. For that reason, several synthetic pyrethrin-like insecticides that are stable outdoors—and therefore can be used in agricultural and garden applications—have been developed by chemists. Semisynthetic pyrethrin derivatives are called **pyrethroids** and are usually given names ending in *-thrin* to denote their nature (e.g., *permethrin*).

Pyrethroids are now also a common ingredient in domestic insecticides, as any visit to a garden center will testify. They have also been used in Mexico to spray houses in which malaria control is still necessary, and to spray neighborhoods in New York City to reduce mosquito populations carrying the West Nile virus. In order to make them more effective as insecticides, pyrethroid formulations are usually mixed with **piperonyl butoxide,** a semisynthetic derivative of the natural product *safrole,* which can be extracted from sassafras plants. The piperonyl butoxide inhibits the enzyme that insects employ to detoxify the pyrethroids, and thus makes them more potent in destroying the insects.

Pyrethroids are now so widely used that their metabolites were detected in the urine of most elementary schoolchildren from Seattle, Washington, who participated in a study in 2003–2004. Interestingly, the pyrethroid levels did not decrease when the children adopted an organic diet. However, the most important contribution to high pyrethroid levels was not dietary but instead correlated with the use of pyrethrin insecticides by their parents.

Rotenone, a complex natural product derived from the roots of certain tropical bean plants, has been used as a crop insecticide for over 150 years and for centuries to paralyze and/or kill fish. The compound enters fish via their gills and disrupts their respiratory system. It is also highly efficient against insects, but is decomposed by sunlight. Rotenone is widely use in hundreds of commercial products, including flea-and-tick powders and sprays for tomato plants. Notice in Table 13-4 that the "natural" insecticides pyrethrins and rotenone are classified as *moderately hazardous,* since they have about the same acute toxicity as some synthetic ones such as malathion, even though they are often marketed as safer, natural pesticides. There is epidemiological and toxicological evidence that chronic exposure to rotenone can contribute to the onset of Parkinson's disease. Indeed, there is some evidence that exposure to pesticides in general contributes to the incidence.

13.21 Integrated Pest Management

In recent years, **integrated pest management** (IPM) strategies have been developed. They combine the best features of various feasible methods of pest control—not just the use of chemicals—into a long-range, ecologically sound plan to control pests so that they do not cause economic damage.

> Pyrethroids commonly detected in U.S. households include cypermethrin and tetrametrin as well as permethrin.

Generally speaking, a unique plan is developed for each area and crop, with chemicals used only as a last resort and when the monetary cost of their use will be more than recovered by increased crop yield. The six pest control methods that can be combined are

• chemical control—the use of chemical pesticides, both synthetic and natural;

• biological control—reducing pest populations by the introduction of predators, parasites, or pathogens;

• cultural control—introducing farm practices that prevent pests from flourishing;

• host-plant resistance—using plants that are resistant to attack, including plants adapted by genetic engineering to have greater resistance;

• physical control—using nonchemical methods to reduce pest population; and

• regulatory control—preventing the invasion of an area by new species.

Review Questions 20–21 are based upon the material in the preceding sections.

13.22 Green Chemistry: Insecticides That Target Only Certain Insects

Insecticides such as the organophosphates and carbamates interrupt the function of specific enzymes that are common to most insects (and to humans). They are thus toxic to a wide array of insect species and are known as **broad-spectrum insecticides.** Although it may be an advantage to kill more than one species with a single pesticide, this often leads to the demise of beneficial insects such as pollinators (bees) and natural enemies (lady bugs and praying mantises) of insects that are pests.

An approach to limiting the environmental effect of an insecticide is to develop insecticides that are toxic to only certain species, i.e., the target organism. One way to accomplish this is to find a biological function that can be interrupted in the target organism but not in other organisms. The Rohm and Haas Company of Philadelphia, Pennsylvania, (whose agricultural business was purchased in 2001 by Dow AgroSciences) won a Presidential Green Chemistry Challenge Award in 1998 for the development of *Confirm, Mach 2,* and *Intrepid.* The active ingredients in these insecticides are members of the *diacylhydrazine* family of compounds (Figure 13-6a) and are effective in controlling caterpillars.

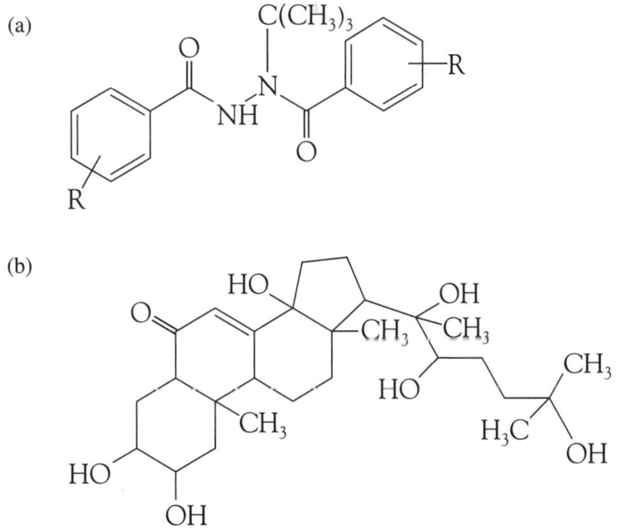

(a)

(b)

FIGURE 13-6
(a) Diacylhydrazine pesticides; (b) 20-hydroxyecdysone.

Caterpillars are the larval stage of insects such as moths and butterflies and during the larval stage they must shed their cuticle to grow. The concentration of *20-hydroxyecdysone* (Figure 13-6b), which is produced by the caterpillar and is a member of the steroid family of compounds, increases during the molting process. As a result of its presence, the caterpillar ceases to feed and sheds its cuticle. The concentration of this natural compound then decreases and the caterpillar resumes feeding. The diacylhydrazines present in the commercial products Confirm, Mach 2, and Intrepid mimic 20-hydroxyecdysone; however, their concentrations do not decline, and consequently the insect never resumes feeding. The insect thus dies of starvation or dehydration. These insecticides are primarily effective only on caterpillars, thus leaving most insects unaffected.

Confirm and Intrepid are classified as **reduced-risk pesticides** by the U.S. EPA. This classification program was started in 1993. To be placed in this category, a pesticide must meet one or more of the following requirements:

- it reduces pesticide risks to human health;

- it reduces pesticide risks to nontarget organisms;

- it reduces the potential for contamination of valued environmental resources; or

- it broadens the adoption of IPM (integrated pest management, discussed in the preceding section) or makes it more effective.

The diacylhydrazene insecticides certainly meet the first two of these requirements. In order to encourage the development of lower-risk pesticides, the EPA rewards the developers of pesticides that contain active ingredients that meet the EPA reduced-risk criteria with expedited review.

13.23 Green Chemistry: A New Method for Controlling Termites

Termites invade over 1.5 million homes in the United States annually and cause about $1.5 billion in damage. Traditional treatments for termites involve treating the soil around the affected structure with 100–200 gal of pesticide solution to create an impenetrable barrier. This process may result in groundwater contamination, accidental worker exposure, and detrimental effects to beneficial insects.

Dow AgroSciences in Indianapolis won a Presidential Green Chemistry Challenge Award in 2000 for their development of Sentricon. In contrast to the traditional control of termites, Sentricon employs monitoring stations to first detect the presence of termites prior to the use of any insecticide. The monitoring stations (Figure 13-7a) consist of pieces of wood (1) contained in perforated plastic tubes (2), which are placed in the ground around the

(a)

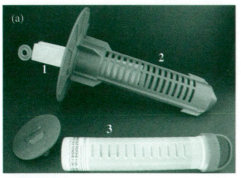

(b)

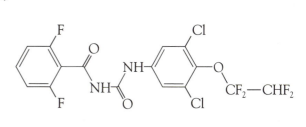

FIGURE 13-7 (a) Sentricon monitoring/baiting station; (b) hexaflumuron structure.
[Source: Photo by Michael Cann.]

structure. If termites are detected in any of the monitoring stations, the wood pieces are replaced by a perforated plastic tube (3) containing the bait. Bait stations may also be placed in the structure. The bait consists of a mixture of cellulose and the pesticide **hexaflumuron** (Figure 13-7b). Hexaflumuron interrupts the molting process of termites and thus is not harmful to most beneficial insects. Termites that have ingested the bait return to their nests and share the bait by trophallaxis, thus spreading the insecticide throughout the colony. Once the colony has been decimated, the bait is replaced with wood and monitoring resumes.

Hexaflumuron was the first pesticide to be classified as a reduced-risk pesticide by the U.S. EPA. It is significantly less toxic than traditional pesticides that are used for termite control and is used in quantities 100 to 1000 times less.

13.24 Green Chemistry: Spinetoram, an Improvement on a Green Pesticide

Throughout this book you have seen several examples of green chemistry. Perhaps a better description would be "greener chemistry," as there is always the potential for improving on the green characteristics of a product or process. One particular example of this is the development of pesticides by Dow AgroSciences. In 1999 this company won a Presidential Green Chemistry Challenge Award for the development of Spinosad and then another award in 2008 for Spinetoram. Dow AgroSciences has distinguished itself by becoming the only company to win three Presidential Green Chemistry Challenge awards (see the previous example of termite control).

Spinosad is a mixture of two naturally occurring macrocyclic lactones (Figure 13-8a) isolated from a fermentation broth of the soil bacteria *Saccaropolyspora Spinosa*. Spinosad is effective against many chewing insects

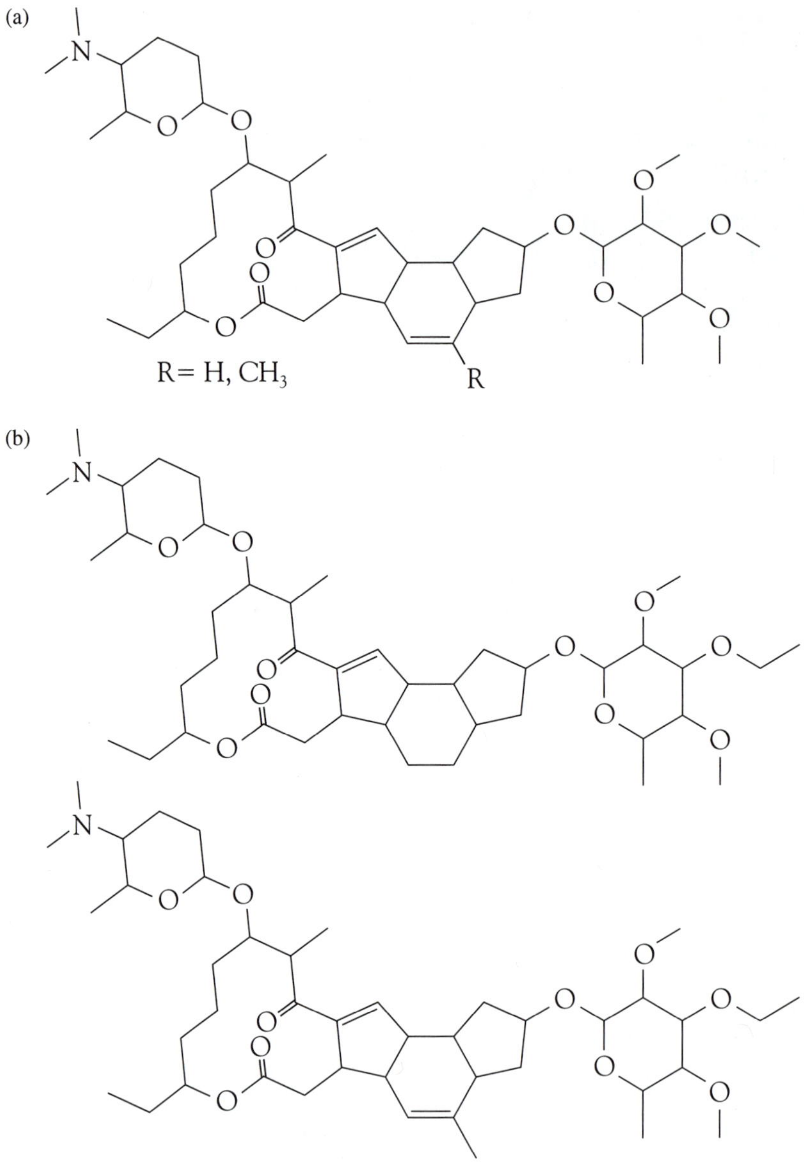

FIGURE 13-8
(a) Structures of the
spinosyns that are the
major components of the
pesticide spinosad;
(b) Structures of the
spinosyns that are the
major components of the
pesticide spinetoram.

(a)

$R = H, CH_3$

(b)

including caterpillars, thrips, sawflies, leafminer, fruit flies, spider mites, beatle larvae, and fire ants, and can be used on vegetables, fruit trees, lawns, and ornamentals. Its environmental attributes include:

- low acute mamalian toxicity (male rat oral LD_{50} = 3738 mg kg^{-1});

- low chronic toxicity (oncogenicity, mutagenicity, teratology, neurotoxicity);

- low toxicity to nontarget organisms;

- synthesis via fermentation (avoids significant wastes, uses no toxic materials, low energy requirements, and employs renewable feedstocks);

- strong binding to soils to prevent leaching into groundwater and surface waters; and

- nonpersistence (readily decomposes in sunlight and is rapidly metabolized by soil microbes).

Spinosad is classified as a *reduced-risk pesticide* by the U.S. EPA.

In order to improve on nature, and discover spinosyns that have all the environmental attributes of spinosad but are more potent and have a wider spectrum of effectiveness on additional insects, extensive computer-aided molecular design research was conducted. Sophisticated calculations revealed that replacement of one of the OCH_3 groups with OCH_2CH_3 improved potency by a factor of 10. In addition, removal of the double bond in the six-membered ring produced enhanced photostability and thus longer effectiveness of insect control. Spinetoram (Figure 13-8b) is a mixture of spinosyns containing these structural features.

The most significant advantage of spinetoram over spinosad is its effectiveness on the major pests—particularly the codling moth—that destroy apple crops. Spinetoram is three times more effective than spinosad in controling codling moth larvae damage to apples and its effectiveness is comparable to the organophosphate pesticide *azinphos-methyl* at only one-tenth the amount of application. Azinphos-methyl has an acute toxicity (oral LD_{50}) more than 1000 times that of spinetoram. The use of spinetoram will reduce the amount of organophophates used by 800,000 kg (1.8 million pounds) over a five-year period.

Green Chemistry Questions 1–10 are based upon the material in the preceding sections.

Herbicides

Herbicides are chemicals that destroy plants. They are usually employed to kill weeds without causing injury to desirable vegetation; e.g., to eliminate broad-leaf weeds from lawns without killing the grass. The agricultural use of herbicides has replaced human and mechanical weeding in developed countries and has thereby sharply reduced the number of people employed in agriculture. Herbicides also are used to eliminate undesirable plants from roadsides, railway and powerline rights-of-way, etc., and sometimes to defoliate entire regions. **Ever since the late 1960s, herbicides have been the most widely used type of pesticide in North America.** Herbicide use is largely concentrated there in corn, soybean, and cotton crops.

In ancient times, armies sometimes used salt or a mixture of brine and ashes to sterilize land that they had conquered, intending to make it unproductive and therefore uninhabitable by future generations of the enemy. In the first half of the twentieth century, several inorganic compounds were used as weed killers—principally *sodium arsenite*, Na_3AsO_3, *sodium chlorate*, $NaClO_3$, and *copper sulfate*, $CuSO_4$. The latter two belong to a large group of salts formerly used as herbicidal sprays that kill plants by the rather primitive action of extracting the water from them, while at the same time leaving the land treated in this way still capable of supporting agriculture.

Organic derivatives of arsenic gradually replaced inorganic compounds as agricultural herbicides since they are less toxic to mammals. However, both inorganic and metal-organic herbicides have been largely phased out because of their persistence in soil. Completely organic herbicides now dominate the market in developed countries; their utility is based partially on the fact that they are much more toxic to certain types of plants than to others, so they can be used to eradicate the former while leaving the latter unharmed.

13.25 Atrazine and Other Triazines

One modern class of herbicides is the **triazines,** which are based upon the symmetric, aromatic structure shown below, which has alternating carbon and nitrogen atoms in a six-membered benzene-like ring:

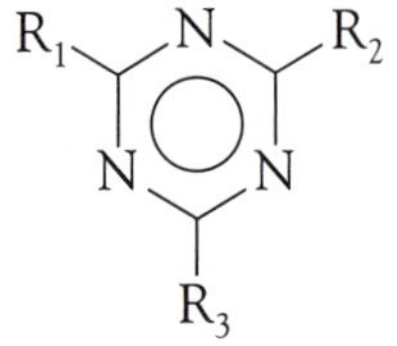

the general formula of the triazines

In triazines that are useful as herbicides, R_1 = Cl and R_2 and R_3 = amino groups, which are nitrogen atoms singly bonded to hydrogens and/or carbon chains.

The best-known member of this group is **atrazine,** which was introduced in 1958 and has been used since that time in huge quantities to destroy weeds in corn fields. Indeed, atrazine is the most heavily used herbicide in the United States—accounting for 40% of all weedkillers applied in the country, including use on 75% of corn crops—and probably the world. In atrazine, R_2 is —NH—CH$_2$CH$_3$ and R_3 is —NH—CH(CH$_3$)$_2$:

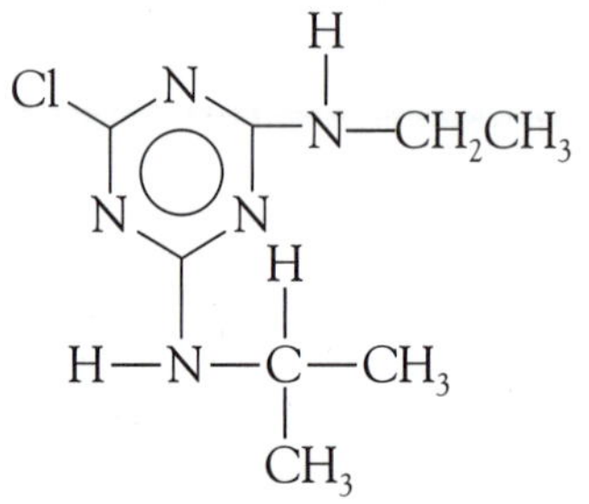

atrazine

It usually is applied to cultivated soils, at the rate of a few kilograms per hectare or one kilogram per acre, in order to kill grassy weeds, mainly in support of corn and soybean cultivation. However, some weeds are becoming atrazine-tolerant.

Biochemically, atrazine acts as a herbicide by blocking photosynthesis in the plant in the photochemical stage that initiates the reduction of atmospheric carbon dioxide to carbohydrate. Higher plants, including corn, tolerate triazines better than do weeds since they rapidly degrade them to nontoxic metabolites. However, if the triazine concentration builds up in a soil—e.g., because of lack of moisture to degrade it—a stage can be reached where no plants will grow. In high concentrations, atrazine has been used to eliminate all plant life, e.g., to create parking lots.

One major ecological risk from the widespread use of atrazine is the death of sensitive plants in water systems close to agricultural fields. Canada has set 2 ppb as its maximum concentration in water for protection of aquatic life. Some controversial recent research regarding the effects of low levels of atrazine on wildlife are discussed in Section 15.10.

While in soil, atrazine is degraded by microbes. One such biochemical reaction results in the replacement of chlorine by a *hydroxyl group*, —OH, yielding a metabolite that is not toxic to plants. The other microbial pathways involve the loss of either the ethyl group or the isopropyl group from an amino unit, with its replacement by hydrogen; these metabolites are toxic to plants.

Atrazine is moderately soluble (30 ppm) in water. During rainstorms, it is readily desorbed from soil particles and dissolved in the water moving through the soil. In waterways that drain agricultural land on which atrazine is used, its concentration typically is found to be a few parts per billion. Usually atrazine is detectable in well water in such regions.

Although it only persists in most soils for a few months, once it or its metabolites enter waterways, atrazine's half-life is several years. For example, in the Great Lakes, its half-life is about 2–5 years, whereas it is less than half a year in Chesapeake Bay, presumably because the water is warmer and so degradation is faster there.

Unfortunately, atrazine is not removed by typical treatments of drinking water unless carbon filtration is used. However, less than 0.25% of the population in the U.S. corn belt states consume atrazine at greater than 3 ppb, its maximum contaminant level, MCL. The European Union has decided not to relist atrazine since it was found to be a persistent groundwater contaminant.

The MCL parameter was defined in Section 11.9.

Since atrazine's measured BCF is less than 10, bioaccumulation does not represent a significant problem. Atrazine is not a very acutely toxic compound (its LOD_{50} is about 2000 mg kg^{-1}). However, some preliminary surveys on the health of farmers and other individuals exposed to it in high concentrations show disturbing links to higher cancer rates and a higher incidence of birth defects. No definitive studies linking atrazine use to human health problems have as yet been reported. Nevertheless, the U.S.

EPA has listed it as a *possible human carcinogen* and has directed states to devise plans to protect groundwater from herbicide contamination. The EPA undertakes ongoing re-evaluations of atrazine with respect to possible human health effects.

In certain American agricultural regions, atrazine usage has been banned outright. For a few years, the triazine called **cyanazine**—which has the same chemical formula as atrazine except that one hydrogen in the isopropyl group is replaced by a cyanide group—became quite popular as an agricultural herbicide, but its manufacturer now has voluntarily phased it out of production because of questions about its effect on human health. Other triazines with similar uses are *simazine* and *metribuzin*.

Atrazine and its metabolite were among the agricultural herbicides most often detected in streams and shallow groundwater in both urban and agricultural areas according to an investigation by the U.S. Geologic Survey in the 1990s. Domestic herbicides found most often were the triazines simazine and prometon. Insecticides found in highest concentrations—including the organophosphates diazinon, malathion, and chlorpyrifos—had higher concentrations in urban than rural regions, presumably because of domestic usage. More than 95% of the streams, and 50% of the groundwater samples, were found to contain at least one pesticide at detectable levels. Research in Switzerland has discovered levels of atrazine and other agricultural pesticides in rainwater that exceed drinking-water standards. Presumably the pesticides evaporated from farm fields.

In some regions where soybeans and corn are grown intensively, atrazine has yielded its status as the herbicide of choice to one of the **chloroacetamides,** which are derivatives of **chloroacetic acid,** $Cl—CH_2COOH$, in which the —OH group is replaced by an amino group $—NR_1R_2$. The most prominent herbicides of this type are *alachlor, metolachlor,* and *acetochlor.* These three compounds differ only in minor variations in the complicated organic groups R_1 and R_2 attached to the amino nitrogen. Some of them have been phased out or banned since they persist in groundwater drained from agricultural fields.

13.26 Glyphosate

Glyphosate is an example of a *phosphonate*, a class of compounds that are structurally similar to organophosphates except that one oxygen of the four that usually surround phosphorus is missing, and is replaced by an organic group, in this case a methylene group, $—CH_2—$, attached to the simple amino acid *glycine.*

glyphosate

Glyphosate is widely used as a herbicide, e.g., as the commercial product *Roundup*.

Glyphosate is rather nontoxic: its LD_{50} values are high for both oral and dermal routes of exposure, although ingestion of or exposure to large quantities of it is fatal. Dermal and oral absorption is small, and it is eliminated essentially unmetabolized.

Glyphosate is nonresidual, and there is no evidence that it bioaccumulates in animal tissue or is carcinogenic. The same is true of its initial breakdown product, the substance corresponding to cleavage of the rightmost NH—CH_2 bond in the structure above.

Glyphosate operates by inhibiting the synthesis of amino acids containing the benzene ring, which in turn prevents protein synthesis from occurring. Although it kills almost all plants, some strains of soybeans, corn, cotton, wheat, canola, alfalfa, and sugar beets have been genetically altered using biotechnology so that they are resistant to glyphosate and consequently it can be used as a weed killer in the growth of these crops (see Box 13-1). Its advantages are that it often replaces several different herbicides and that only one application is required, though the total mass of herbicide-active ingredients is not necessarily reduced substantially. Its greater tendency to stay adsorbed to soil means that it has less tendency to occur in runoff and subsequently in water supplies than do the herbicides atrazine and alachlor, which it often replaces. The evidence gathered so far indicates that glyphosate is a relatively benign herbicide. However, more and more weeds are now developing resistance to it, and some farmers now are also applying other herbicides to counter their growth.

> Glyphosate is in Category III, low toxicity for both acute oral and inhalation toxicity, and very low in dermal toxicity.

BOX 13-1 Genetically Engineered Plants

In 1940 the world population was 2.3 billion people; by 1985 it had more than doubled and it now is over 7 billion. Fortunately, beginning in the 1940s, a "green revolution" in agriculture took place that allowed the world to feed this burgeoning population. Extensive development and use of pesticides (many of which you have seen in this chapter) and fertilizers, along with irrigation and plant breeding programs, led to dramatic increases in the yield per acre of crops. Total worldwide grain production increased from 600 million metric tons in 1950 to more than 1600 million metric tons by 1985. Since 1995, production has leveled off at 1800–2000 metric tons. However, the human population continues to grow and is expected to reach 9 billion by 2050.

Since the 1980s, talk of a second green revolution has centered about genetically engineered plants. Traditional crossbreeding of wheat plants over many years has resulted in plants that yield two to three times more grain than previously existing varieties and that are more resistant to pests and diseases. Genetic engineering of plants offers the possibility of doing these same things and additional feats in much less time and with more selectivity than traditional crossbreeding.

(continued on p. 612)

Genetic engineering involves taking a portion of the DNA from one species and inserting it into the DNA of another, unlike species. Transgenic plants have been produced which have enhanced resistance to herbicides, drought, pests, salinity, and frost, as well as improved taste and nutritional value. The best-known examples of herbicide-resistant plants that have been developed are known as *Roundup Ready.* Roundup, as previously mentioned, is a commonly used broad-spectrum herbicide. Monsanto, its manufacturer, has developed and patented genetically altered seeds for soy, corn, alfalfa, sorghum, canola, and cotton which grow into plants that are resistant to destruction by Roundup. Fields planted with these crops can be sprayed indiscriminately to destroy weeds, with little concern for destruction of the crop.

The use of transgenic plants has been widely adopted in the United States, and to a lesser extent in a number of other countries. In 2009, 93% of all soybean and cotton acreage in the United States was planted with transgenic crops, followed by 86% for corn; corresponding global figures are 77%, 49%, and 26%. The top six countries in growing transgenic crops in 2009 were the United States, Brazil, Argentina, India, Canada, and China.

Although transgenic plants offer the possibility of improving upon what nature has provided us, there are significant concerns about these organisms, especially in Europe. Concerns include

• the use of greater quantities of herbicides since there is less fear of destroying a crop from the indiscriminate application of the herbicide;

• the spread of herbicide resistance to related plants that become "super weeds"; and

• the decrease in genetic diversity of crops as farmers all use the same seeds.

In addition to these concerns, genetically engineered grains have not resulted in substantial increases in crop yields.

13.27 Phenoxy Herbicides

Phenoxy weed killers were introduced at the end of World War II. Environmentally, the by-products contained in some commercial phenoxy herbicide products are of greater concern than the herbicides themselves, as we shall see in Chapter 14. To understand this phenomenon, we begin by discussing the chemistry of **phenol,** the fundamental component of these compounds.

Phenols are mildly acidic; in the presence of concentrated solutions of a strong base like NaOH, the hydrogen of the OH group is lost as H^+ (as occurs with any common acid) and the *phenoxide anion,* $C_6H_5O^-$, is produced in the form of its sodium salt:

phenol   phenoxide

The O^- group is a reactive one, and this property can be exploited in order to prepare molecules containing the C—O—C linkage. Thus if an R—Cl molecule is heated together with a salt containing the phenoxide ion, NaCl is eliminated and the phenoxy oxygen links the benzene ring to the R group:

These reactions are a standard way to produce ethers.

$$C_6H_5O^-Na^+ + Cl—R \longrightarrow C_6H_5—O—R + NaCl$$

Such a reaction is the most direct commercial route to the large-scale preparation of the **phenoxy** group of herbicides, introduced in the mid-1940s. Here (in the reaction immediately above), the R group is **acetic acid,** CH_3COOH, minus one of its methyl-group hydrogens, so that R = —CH_2COOH, and so the Cl—R reactant is Cl—CH_2COOH. Then according to the reaction, C_6H_5—O—CH_2COOH, called **phenoxyacetic acid,** is obtained as an intermediate in the production of the actual herbicides.

In the commercial herbicides, some of the five remaining hydrogen atoms of the benzene ring in phenoxyacetic acid are replaced by chlorine atoms.

phenoxyacetic
acid

2,4-D
2,4-dichlorophenoxyacetic acid

2,4,5-T
2,4,5-trichlorophenoxyacetic acid

Note that the numbering scheme for the benzene ring begins at the carbon attached to the oxygen.

The **2,4-D** compound, **2,4-dichlorophenoxy acetic acid,** is used to kill broad-leaf weeds in lawns, golf course fairways and greens, and agricultural fields. In contrast, **2,4,5-T (2,4,5-trichlorophenoxy acetic acid)** is effective in clearing brush, for instance, on roadsides and powerline corridors.

Like the P—O—C bonds in organophosphates, the O—C bond to the —CH_2— group in 2,4-D and analogous phenoxy herbicides undergoes a hydrolysis reaction in the environment, degrading the compound to a phenol

$$R'—O—CH_2—COOH + H—OH \longrightarrow R'—O—H + HOCH_2—COOH$$

The herbicide MCPA is 2,4-D with the chlorine in the 2 position replaced by a methyl group, CH_3. The herbicides *dichlorprop, silvex, and mecoprop* are identical to 2,4-D, to 2,4,5-T, and to MCPA, respectively, except that their molecules have a methyl group replacing one hydrogen atom of the —CH$_2$— group in the acid chain; thus they are phenoxy herbicides that are based on *propionic acid*, CH$_3$—CH$_2$—COOH, rather than on acetic acid. The herbicide *dicamba* is the same as 2,4-D with a methoxy group at the 5 position of the benzene ring; it is often used as a weed killer in corn fields.

Huge quantities of 2,4-D and its closely related analogs described above are used in developed countries for the control of weeds in both agricultural and domestic settings. In some communities, their continued use on lawns has become a controversial practice—and in some locations, banned—because of their suspected effects on human health. In particular, farmers in the midwestern United States who mix and apply large quantities of 2,4-D to their crops are found to have an increased incidence of the cancer known as non-Hodgkin's lymphoma.

13.28 The Degradation of Pesticides in the Environment

Although some pesticides are very long-lived in the environment, most undergo chemical or biochemical reactions within a few days or months, producing other compounds. Based upon their typical half-lives in the environment, the U.S. EPA classifies pesticides as being

- **non-persistent** if they last less than 30 days;
- **moderately persistent,** for those lasting 30–100 days; and
- **persistent,** for those with lifetimes greater than 100 days.

Like most organic compounds, pesticides in the environment—whether present in air, water, or soil—degrade to other compounds, which in turn decompose further. The complete eventual breakdown of organic compounds to CO$_2$, H$_2$O, and stable inorganic forms of its other elements is called *mineralization*.

In air, the degradation process usually begins either with attack on the organic molecule by the hydroxyl free radical, OH, or by a photochemical reaction if the substance absorbs light with wavelength greater than about 285 nm.

Photochemical decomposition is possible also for pesticides present in water or adsorbed onto soil resident at the Earth's surface. In some instances, adsorption onto soil particles increases the maximum wavelength of light the substance absorbs into the range in sunlight; e.g., the herbicide *paraquat* undergoes photolysis more rapidly when adsorbed on clay than it does in solution. Complexation of organic molecules by metal ions also usually increases their maximum wavelength of absorption, thereby activating them for photochemical decomposition by sunlight in some cases.

As we already have discussed several times in this chapter, pesticides in water and in soil can undergo hydrolysis reactions, especially when the water is somewhat acidic or somewhat basic, since catalysis by H^+ or OH^- can then speed up the processes significantly. Organophosphate insecticides, for example, hydrolyze in alkaline water and soil owing to attack by OH^- on the oxygen atom of the P—O—C link. Even in quite dry soils, hydrated aluminum ions produce hydrogen ions that in the existing moisture can catalyze hydrolysis, e.g., of triazine herbicides, in the conversion of their C—Cl bonds to C—OH ones, thereby eliminating their herbicidal activity.

$$Al(H_2O)_6^{3+} \longrightarrow Al(H_2O)_5OH^{2+} + H^+$$

Organic compounds, including pesticides, can also be transformed in water or soil by oxidation or reduction reactions. Although dissolved O_2 itself can oxidize the compounds, its reactions are often accelerated by the presence of dissolved or adsorbed transition metal ions, which oxidize the pesticide. The reduced form of the metal is subsequently re-oxidized by O_2. For example, Fe^{3+} is a good oxidizing agent for many organic compounds; the Fe^{2+} state to which it is reduced in the process is subsequently oxidized by oxygen back to Fe^{3+}, thereby completing the cycle.

Reducing agents are commonly found in anaerobic waters and soil, and include Fe^{2+} and *sulfide ion*, S^{2-}. For example, pesticides that contain a C—Cl unit are dechlorinated by the iron ion when it abstracts an electron from the C—Cl bond, thereby releasing Cl^- and forming Fe^{3+} and a reactive carbon-based free radical.

Even more important than the chemical processes described above are degradation reactions facilitated by microbial action in water and soil. *Chemoheterotrophs* are microorganisms that derive the energy they require from redox reactions and their carbon from organic compounds. The metabolic reactions proceed in stepwise fashion, the individual steps usually being oxidation, reduction, or hydrolysis. However, the rates of degradation vary over a very wide range, depending on the molecular structure of the pesticide and the properties of the soil. Compounds containing functional groups such as —OH, —NO$_2$, —NH$_2$, and carboxylate degrade most readily in soils since they contain a site for enzymatic attack and are relatively soluble in water, whereas highly chlorinated hydrocarbons are much more resistant since there is no reactive site and their water solubility is very low.

A common example of a microbial oxidation step is enzyme-catalyzed *epoxidation*, a process in which an oxygen atom from an O_2 molecule is added to a C=C bond, even if it is contained within an aromatic benzene ring system:

Following epoxidation, the adduct can undergo further reactions, for example,

- rearrangement to a hydroxylated compound, thereby re-establishing the highly stable aromatic ring, or

- hydrolysis to produce an *ortho*-dihydroxyl compound, or

- the addition of further oxygen and water to other double bonds within the aromatic system.

Review Questions 22–28 are based upon the material in the preceding sections.

Subsequent reactions at an adjacent pair of carbons having —OH groups often leads to ring cleavage at that site, yielding a dicarboxylic acid.

Final Thoughts on Pesticides

In general, there is no pesticide that is completely "safe." All pesticides display *some* acute toxicity to humans and other animals, and many display long-range problems arising from their bioaccumulation in the living environment. However, the elimination of the use of all synthetic pesticides would lead to an increase in the transmission of disease by insects and an increase in the cost and a decrease in the availability of food, all of which would affect human health adversely. Any decision about discontinuing the production and use of a given pesticide must consider whether cheap, safer alternatives are available and, if not, what would be the consequences of both action and inaction. The quandary about whether to ban the use of DDT in tropical developing countries is an excellent illustration.

When a new pesticide, or indeed any other synthetic chemical, is about to be introduced into the market, many environmental groups and some government agencies have proposed that we should err on the side of being too cautious and only allow its introduction if there are no signs that problems could arise. They propose that in such situations, to prevent possible harm to the health of humans and other organisms, we should employ what is known as the **precautionary principle.** One definition of this principle was given at the 1992 Rio conference on Environment and Development: "Where there are threats of serious or irreversible damage, lack of full scientific certainty shall not be used as a reason for postponing cost-effective measures to prevent environmental degradation." Opponents of the use of this principle point out that it is impossible to anticipate *all* possible consequences, positive or negative, of introducing a new substance and that consequently we could become frozen into inaction.

The best technique for predicting where a given pesticide will ultimately end up in the environment is through the types of calculations described in Box 13-2. It should be emphasized that such calculations do *not* predict the rate at which equilibrium will be established.

BOX 13-2 The Environmental Distribution of Pollutants

When a persistent chemical, such as DDT, is released into the environment, we find that later some of it has dissolved in natural bodies of water, some is in the air, some is present in soil and sediments, and some is located in living matter. A constant interchange of the chemical occurs among these various physical phases. **It is possible to predict the amount and concentration of the chemical in each phase once the release of the chemical has stopped and sufficient time has passed that equilibrium among the phases has been achieved.** Even when equilibrium conditions are not yet in place, it is of value to determine the phases where the chemical will ultimately be concentrated.

Recall from your previous background in chemistry that in calculations involving substances participating in *chemical* reactions, we algebraically combine experimental values of equilibrium constants with information concerning initial concentrations in order to determine equilibrium concentrations. A somewhat analogous procedure can be applied to determine the distribution of a substance when by *physical* processes it has achieved equilibrium between several phases. The condition that equilibrium has been achieved in its distribution is that the **fugacity, *f*, of the substance, which is defined as its tendency to escape from a given physical phase, is equal for all phases.** Fugacity has units of pressure, e.g., atmospheres or kilopascals. Thus, for example, when all the DDT in the environment has distributed itself among air, water, sediment, biota, etc., the concentrations in each phase are such that its tendency to escape from any phase (and enter any other) has the same value for all phases.

As you might expect, the fugacity of a substance in a given phase is proportional to its concentration, C, in that phase:

$$f = C/Z$$

where Z is the *fugacity capacity constant* for the substance and the phase. *Generally, the higher the value of Z, the greater the tendency of a chemical to concentrate in that phase.* (These capacity constants are analogous to the equilibrium constants used in chemical reaction calculations.) If we use x to denote the phase of interest, then

$$f_x = C_x/Z_x$$

We can determine the concentration in a phase by rearranging the equation, to give

$$C_x = f_x Z_x$$

At equilibrium, the f_x values for all phases are identical, equal to f, say. Thus, if we know f, we can determine the concentration in each phase from the simplified equation

$$C_x = f Z_x$$

As in chemical equilibrium problems, we usually know the total number of moles, n_{total}, of the material. Also as in other chemical problems, it is useful to state the *mass conservation condition:* **the sum of the equilibrium number of moles, n_x, present in each phase x must add up to n_{total}.** By definition, each n_x is equal to the concentration C_x times the volume V_x for the phase:

$$n_x = C_x V_x$$

Substitution of the next-to-last equation into the last one gives

$$n_x = f Z_x V_x$$

(continued on p. 618)

When we sum the n_x values over all phases x of interest, we must obtain the total number of moles. Thus

$$n_{\text{total}} = f \sum Z_x V_x$$

Rearrangement of this equation allows us to calculate the value of the system fugacity:

$$f = n_{\text{total}} / \sum Z_x V_x$$

An Example of a Fugacity Calculation

As an example of how fugacity calculations are carried out in practice, consider the distribution of 1 mol of DDT among three phases: air, water, and sediment in a model compartment of the Earth (Figure 1). As discussed later, we take the volume of air to be 10^{10} m^3, the water volume to be 7×10^6 m^3, and the volume of accessible sediment to be 2×10^4 m^3. The values of the Z_x constants for DDT, in units of mol atm^{-1} m^{-3}, are determined from experimental data to be

for the air phase, 40.3
for the water phase, 3.92×10^4
for the sediment phase, 2.25×10^9

In the evaluations of Z_x values from experimental data, a temperature of 25°C is usually assumed for simplicity. The Z_x values for sediments (and biota) are assumed to be proportional to the octanol–water partition coefficients K_{ow} discussed earlier in the chapter.

After substitution of the values for Z_x and V_x, the value of the fugacity in this case is

$$f = 1.0/(40.3 \times 10^{10} + 3.92 \times 10^4 \times 7 \times 10^6 + 2.25 \times 10^9 \times 2 \times 10^4)$$

$$= 1.0/(4.03 \times 10^{11} + 2.74 \times 10^{11} + 4.5 \times 10^{13})$$

$$= 2.19 \times 10^{-14} \text{ atm}$$

The concentration of the chemical can now be computed for each phase:

$$C_x = f Z_x$$

so

$$[\text{DDT in air}] = 2.19 \times 10^{-14} \times 40.3$$
$$= 8.8 \times 10^{-13} \text{ mol m}^{-3}$$

$$[\text{DDT in water}] = 2.19 \times 10^{-14} \times 3.92 \times 10^4$$
$$= 8.6 \times 10^{-10} \text{ mol m}^{-3}$$

$$[\text{DDT in sediment}] = 2.19 \times 10^{-14} \times 2.25 \times 10^9$$
$$= 4.9 \times 10^{-5} \text{ mol m}^{-3}$$

Notice the preferential concentration in sediment of DDT, which is hydrophobic due to its carbon content.

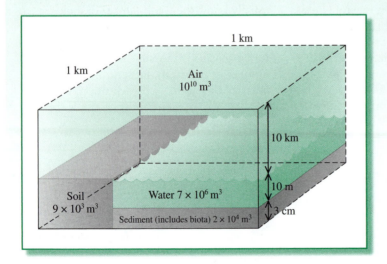

FIGURE 1 Model world parameters used in fugacity calculations.

The *amounts* in each phase are given by the fZV values, i.e., the concentrations multiplied by the respective volumes. Thus the number of moles of DDT

$$\text{in air} = 8.8 \times 10^{-13} \times 1 \times 10^{10}$$
$$= 0.0088 \text{ mol}$$

$$\text{in water} = 8.6 \times 10^{-10} \times 7 \times 10^{6}$$
$$= 0.0060 \text{ mol}$$

$$\text{in sediments} = 4.9 \times 10^{-5} \times 2 \times 10^{4}$$
$$= 0.98 \text{ mol}$$

Thus we see that with air, water, and sediment accessible to it, 98% of the DDT will be found in sediments and about 1% in air and in water. Notice that the concentration of DDT in water is greater than in air, but the total amount of it in air exceeds that in water because the air volume is so much larger. This sort of interchange in ordering between amount and concentration in different phases is common for pollutant chemicals.

The Parameters for the Model World in Fugacity Calculations Are Estimates

The volumes for the various phases used in the above calculation are based upon a model "world" (Figure 1) whose components are able to be in equilibrium with each other. Since only concentrations are obtained in the calculations, it is important only that the *relative* volumes, not their absolute values, be appropriate.

The model world is a square of 1 km by 1 km, whose characteristics are assumed to be average for the real Earth.

- The atmosphere is taken to be 10 km high, which is a reasonable approximation to the troposphere. The air volume then is (1000 m × 1000 m) × (10,000 m) = 10^{10} m^3.

- The 1-km square is assumed to be 70% covered by water and 30% by soil. The average water depth is taken to be 10 m, which is relatively shallow since we are interested only in the part that achieves equilibrium with the air. Thus the water volume is 0.7 × (1000 m × 1000 m) × 10 m = 7 × 10^6 m^3.

- The sediment in equilibrium with this water is assumed to be only 3 cm deep, giving it a volume of 0.7 × 1000 m × 1000 m × 0.03 m = 2.1 × 10^4 m^3. In addition to air, water, and sediment, the model usually also includes soil, whose effective volume is 9 × 10^3 m^3, plus 35 m^3 of solids suspended in the water, and about 3.5 m^3 of biota such as fish. **The Z values for biota are usually of the same order of magnitude as those for sediment, so the concentration of a given chemical in biota is close to that in sediment.**

PROBLEM 1

The Z values for hexachlorobenzene are 4 × 10^{-4} in air, 9.5 × 10^{-5} in water, and 2.3 in sediment (and biota). Using the model world volumes above, calculate the equilibrium concentrations when 1 mol of hexachlorobenzene is distributed between air, water, and sediment.

PROBLEM 2

In fugacity calculations, the Z values for dieldrin are 4 × 10^{-4} in air, 2.0 in water, and 2 × 10^{-5} in sediment (and biota). Using the model world volumes above, calculate the equilibrium concentrations when 1 mol of dieldrin is distributed between air, water, and sediment.

Review Questions

1. What are the three main categories of pesticides? What types of organisms are killed by each category?

2. What is meant by the term *fumigant*?

3. Name three important properties shared by organochlorine pesticides.

4. What does *POP* stand for? Name four pesticides on the United Nations POP list.

5. Draw the structure of 1,4-dichlorobenzene. Which of its properties make it useful as an insecticide?

6. Draw the structure of DDT, and state what the initials stand for.

7. What were the main uses of DDT? Explain why it is no longer used in many developed countries, and why some developing countries wish to continue using it.

8. Explain how DDT functions as an insecticide.

9. Draw the structure of DDE. Is it an insecticide or not? Explain.

10. Explain what is meant by the terms *bioconcentration* and *bioconcentration factor* (BCF).

11. Explain what is meant by the term *biomagnification* and how it differs from *bioconcentration*.

12. Write the defining equation for the partition coefficient K_{ow}. How is it related to a compound's BCF? What is octanol supposed to be a surrogate for in this experiment?

13. Describe one analog of DDT that works in the same fashion but does not bioaccumulate.

14. Define the terms *acute toxicity* and *dose*.

15. Sketch a typical dose–response curve relationship for a toxic chemical using (a) a linear and (b) a logarithmic scale for doses. What is meant by a *threshold* in such plots?

16. Define the terms LD_{50}, LOD_{50}, and LC_{50}.

17. What are the general structures of the three main subclasses of organophosphate insecticides? Give the name of one insecticide in each subclass. Explain how organophosphates function as insecticides and how they decompose in water.

18. In what way are organophosphate insecticides considered superior to organochlorines as pesticides? In what way are they more dangerous?

19. What is the general structure of carbamate insecticides? Name one example.

20. What are *pyrethroids*? What role does *piperonyl butoxide* play in their effectiveness?

21. What are five of the pest control methods that are used in integrated pest management?

22. What is the function of a herbicide? Name a few "old-fashioned" herbicides.

23. What is the general structure of a triazine herbicide? Name the most important commercial example of this class.

24. What is the formula of gyphosate? What are its advantages over other herbicides?

25. What is phenol? Draw its structure and that of 2,4-dichlorophenol.

26. Draw the structures and write out the names of the two most important phenoxy herbicides.

27. Write out three examples of hydrolysis reactions by which pesticides are degraded in the environment.

28. What is meant by the *precautionary principle*?

Green Chemistry Questions

1. The development of the insecticides Confirm, Mach2, and Intrepid won a Presidential Green Chemistry Challenge Award.

(a) Which of the three focus areas (see page xxviii) for these awards does this award best fit into?

(b) List one of the twelve principles of green chemistry (see pages xxiii–xxiv) that are addressed by these new pesticides.

2. What environmental advantage do Confirm, Mach2, and Intrepid offer compared to conventional pesticides?

3. **(a)** What is a U.S. EPA *reduced-risk pesticide*?

(b) Which categories under the reduced-risk criteria do Confirm, Mach2, and Intrepid meet?

4. How do Confirm, Mach2, and Intrepid act to only target specific insects?

5. The development of the Sentricon system won a Presidential Green Chemistry Challenge Award.

(a) Which of the three focus areas (see page xxviii) for these awards does this award best fit into?

(b) List two of the twelve principles of green chemistry (see pages xxiii–xxiv) that are addressed by the hexaflumuron/Sentricon system?

6. What environmental advantages does the hexaflumuron/Sentricon system offer compared to conventional termite control pesticides/methods?

7. Which categories under the reduced-risk criteria does the hexaflumuron/Sentricon system meet?

8. The development of the pesticides spinosad and spinetoram each won a Presidential Green Chemistry Challenge Award.

(a) Into which of the three focus areas (see page xxviii) for these awards do these awards best fit?

(b) List four of the twelve principles of green chemistry (see pages xxiii–xxiv) that are addressed by these discoveries. Justify each of your answers.

9. What are the environmental advantages of using spinodsad and spinetoram as compared to using conventional pesticides?

10. The spinosyns compounds in the pesticides spinosad and spinetoram (Figure 13-8) are referred to as macrocyclic lactones. Identify the *lactone* structure in these compounds.

Additional Problems

1. The fat content of breast milk averages about 4.2 g per 100 mL. Calculate the mass of DDE that would have been ingested by a typical breast-fed infant in 1972 consuming 250 mL of breast milk per feeding if the milk contained about 700 ppb of DDE in its fat component.

2. The BCF for a substance in the body of a particular aquatic species (not just in its fat tissues) can be estimated as the K_{ow} value for the substance times the fraction of body fat in the species of interest. Rainbow trout, which average 5.0% body fat, taken from a particular

lake were found to contain 22 ppb in their tissues. Use the information in Table 13-3 to determine the concentration of parathion in the lake.

3. The threshold/NOEL level found for a particular chemical from animal studies is 0.004 mg per kg body weight per day. The only source for the chemical is fresh-water fish, where it occurs at an average level of 0.2 ppm. What is the maximum average daily consumption of such fish that would keep exposure level below the ADI or RfD for the compound?

4. An approximate mathematical fit to the form of the dose–response curve of Figure 13-4a is

$$R = 1 - e^{-d}$$

where R is the fractional response and d is the dose.

(a) Plot R versus d for values of d ranging from 0 to 5 on both linear and logarithmic scales for d. (Be sure to include some small values of d, from 0.01 to 0.10, in the logarithmic plot to ensure that the form of the curve near zero is displayed.) Do the forms of the curves resemble those in Figures 13-4a and 13-4b, respectively?

(b) Both from your graphs and by solving the equation above analytically, find the dose corresponding to LD$_{50}$.

(c) Does the function R have a nonzero threshold at low doses? Can you confidently predict the answer to this from inspecting your logarithmic dose–response curve?

5. The organophosphate Azinphos-methyl has a 96-hour LC$_{50}$ value of 3 ppb for rainbow trout. In an unfortunate incident, 200 g of this pesticide was sprayed on a field, and a subsequent heavy rainfall washed 35% of it into a nearby lake having a surface area of 30,000 m^2 and an average depth of 0.5 m. Would the pesticide concentration in the lake water have been sufficient to kill a significant fraction of the rainbow trout in the lake?

14

Dioxins, Furans, and PCBs

In this chapter, the following introductory chemistry topics are used:

- Elementary organic chemistry (see online Appendix)
- First- and second-order kinetics rate laws
- Concept of vapor pressure

Background from previous chapters used in this chapter:

- Structure of phenols and phenoxy herbicides
- Carcinogens and toxicology concepts

Introduction

As we have seen in Chapter 13, compounds used as pesticides are somewhat toxic to humans, and can bioaccumulate and cause environmental problems. But **sometimes it is the highly toxic trace impurities in commercial lots of pesticides that are the principal concern regarding human health.** In this chapter, we shall analyze how such hazardous by-products, especially dioxins, arise in the environment, not only from pesticide manufacture but from other anthropogenic processes as well.

In this chapter we also consider PCBs, industrial chemicals of widespread environmental concern with respect to both their own properties and those of their furan contaminants. As we shall see, the toxicity mechanisms by which these contaminants, the PCBs themselves, and dioxins operate all are quite similar. All three groups—PCBs, dioxins, and furans—are listed as *Persistent Organic Pollutants* by the United Nations.

Dioxins

14.1 Dioxin Production in the Preparation of 2,4,5-T

Traditionally, the industrial synthesis of the herbicide called 2,4,5-T (discussed in Section 13.27) started with *2,4,5-trichlorophenol*, which itself was produced by reacting NaOH with the appropriate *tetrachlorobenzene*. The OH group replaces one chlorine atom in the process. Unfortunately, **during this synthesis there occurs an additional reaction that converts a very**

small portion of the trichlorophenol product into "dioxin." In this side reaction, two *trichlorophenoxy anions* react with each other, resulting in the elimination of two chloride ions:

"dioxin"
(tetrachlorodibenzo-*p*-dioxin)

In this process a new six-membered ring is formed which links the two chlorinated benzene rings. This central ring has two oxygen atoms located *para* (i.e., opposite) to each other, as is found in the simple molecule *1,4-dioxin* or *para-dioxin* (*p*-dioxin).

1,4-dioxin

Although the molecule labeled "dioxin" above is correctly known as a *tetrachlorodibenzo-p-dioxin*, it has become popularly known simply as dioxin, with the understanding that it is the most toxic of a class of related compounds.

The side reaction that produces dioxin as a by-product is kinetically second-order in chlorophenoxide. In other words, the rate of the reaction depends on the square (or second power) of the ion's concentration. Consequently, the rate of dioxin production increases dramatically as the initial chlorophenoxide ion concentration increases. Also, the rate of this side reaction increases more rapidly with increasing reaction temperature than does the main reaction. Therefore, the extent to which the trichlorophenol, and consequently the commercial herbicide, become contaminated with the dioxin by-product can be minimized by controlling concentration and temperature in the preparation of the original trichlorophenol. Today, the contamination of commercial 2,4,5-T by this dioxin can be kept to less than 0.1 ppm by keeping both the phenoxide concentration and the temperature low. Nevertheless, its manufacture and use in North America were phased out in the mid-1980s because of concerns about its dioxin content, however small.

A 1:1 mixture of the herbicides 2,4-D and 2,4,5-T called Agent Orange was used extensively as a defoliant during the Vietnam War. Since the mixture contained dioxin levels of about 3 ppm, it is clear that the reaction used to produce the trichlorophenol used for 2,4,5-T preparation was not carefully controlled so as to minimize contamination. As a result, the soil in southern Vietnam is contaminated by dioxins. The consequences of this contamination for the residents, and for the American troops that were exposed while spraying was underway, are still controversial. There is evidence that the rates of certain cancers have increased in Air Force personnel who were involved in the spraying. The defoliation potential of Agent Orange was originally tested in the 1960s by the U.S. armed forces near Gagetown, New Brunswick, Canada, and the soil in the area is polluted by the substance.

The cancers associated with Agent Orange spraying include soft-tissue sarcoma, non-Hodgkin's lymphoma, and chronic lymphocytic leukemia.

Environmental contamination by dioxin also occurred as the result of an explosion in a chemical factory in Meda, a town in northern Italy, in 1976. The factory produced 2,4,5-trichlorophenol from tetrachlorobenzene, as described above. On one occasion the reaction was not brought to a complete halt before the workers left for the weekend. The reaction continued unmonitored, and the heat subsequently released by the reaction eventually resulted in an explosion. Since the trichlorophenol had been heated to a high temperature, considerable dioxin—probably several kilograms—was produced. The explosion distributed the toxin into the environment, and the contamination caused the immediate death of a thousand chickens and rabbits in the area. Much of the dioxin was transported south by winds to the town of Seveso. Although a large number of humans, both adults and children, were also exposed to the chemical as a result of this explosion, no serious health effects to humans were found for many years, although several hundred cases of chloracne occurred over the years.

The top 40 cm of the soil in the zone closest to the plant in Meda was removed, and the area was converted into a park.

More recent studies have established that the rates of several types of cancers are elevated in people who lived in the zones of Seveso most exposed to dioxin from the explosion. Specifically, the risk of contracting breast cancer increased in proportion to the dioxin exposure, as measured by the level of the substance in women's blood samples taken soon after the explosion.

14.2 Dioxin Numbering System

The nomenclature and numbering system used for ring systems like the dioxins is a little unusual. Since the central dioxin ring is connected on either side by benzene rings, the three-ring unit is properly known as **dibenzo-p-dioxin.** The chlorine substitution on the outer rings also should be recognized, so the dioxin shown below is a **tetrachlorodibenzo-p-dioxin,** or **TCDD.**

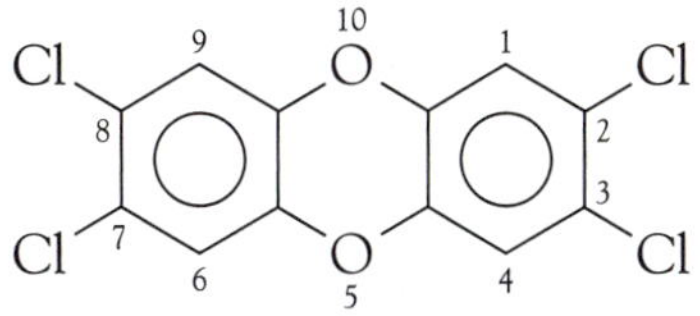

2,3,7,8-tetrachlorodibenzo-p-dioxin
(2,3,7,8-TCDD)

The numbering scheme for the ring carbons in dioxins takes into account the fact that the carbons shared between two rings carry no hydrogen atoms and so need not be numbered. Thus **C-1 is the carbon next to one joining the rings, and the numbering follows a direct path from there.** By convention, the oxygen atoms are also part of the numbered sequence in this scheme, although their locations are not used in naming any of this family of compounds. **The initial (C-1) position for the numbering system is chosen to give the lowest possible value to the first substituent;** if there is a choice after this criterion has been applied, then that which gives the lowest number to the second substituent is used, etc. Applying these rules, the dioxin shown above is named 2,3,7,8-TCDD, or to give it its full title, **2,3,7,8-tetrachlorodibenzo-*p*-dioxin.** No wonder it is simply called "dioxin" in the press!

There are actually 75 different chlorinated dibenzo-*p*-dioxin compounds, when one includes all the possibilities having between one and eight chlorines, given that a number of isomers exist for most of these eight types. **Different members of a chemical family that differ only in the number and position of the same substituent are called congeners.**

All dioxin congeners are planar: all carbon, oxygen, hydrogen, and chlorine atoms lie in the same plane. For convenience, we refer to the carbon atoms closest to the central dioxin ring as alpha carbons, and the outer ones as beta carbons:

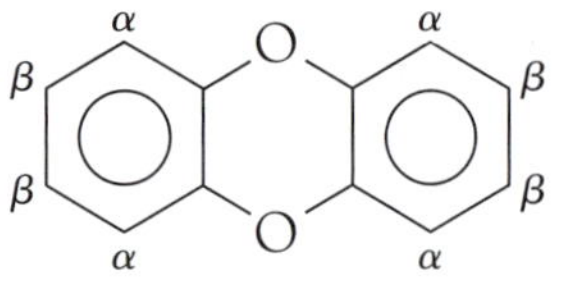

The unsubstituted dibenzo-*p*-dioxin molecule has two types of symmetry that are useful when considering substitution patterns. First there is lateral, or *left–right* symmetry: the carbon atom labeled β at the top of the left-side ring is equivalent to the β carbon at the top of the ring at the right side, and similarly for the two β carbons at the bottom. The dioxin ring also has *up–down* symmetry: the carbon atom labeled β at the top of the left-side ring is equivalent to the β carbon at the bottom of that ring, and similarly for the two β carbons of the right-side ring. Thus all the four β carbons are in fact equivalent in the unsubstituted dioxin. Similarly, the four α carbons all correspond to equivalent positions.

Consequently, for example, there are only two unique monochlorodibenzo-*p*-dioxins; due to the equivalence of the four α positions, those that would otherwise be numbered 4-, 6-, and 9-chlorodibenzo-*p*-dioxin are all equivalent to the 1- molecule. Similarly, 3-, 7-, and 8-chlorodibenzo-*p*-dioxin are all equivalent to the 2- molecule due to the equivalence of the β positions. Some or all of the equivalences can be lost when multiple substitution occurs.

PROBLEM 14-1

By drawing the structures and comparing them, decide whether 1,3-, 2,4-, 6,8-, and 7,9-dichlorodibenzo-*p*-dioxins are unique compounds or whether they are all really the same compound. Are 1,2- and 1,8-dichlorodibenzo-*p*-dioxins unique compounds? Using a systematic procedure, deduce the structures of all unique dichlorodibenzo-*p*-dioxins, keeping in mind that before substitution, the two rings are equivalent and that the molecule has up–down symmetry. ●

14.3 Chlorophenols as Pesticides

In addition to their use as starting materials in the production of herbicides, **chlorophenols find use as wood preservatives (fungicides) and as slimicides.** The most common preservative, in use since 1936, is **pentachlorophenol** (PCP); all the benzene's six hydrogens have been replaced in this compound:

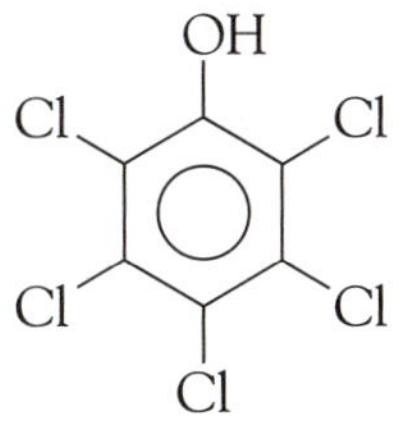

The street drug "angel dust" is *not* the same compound but is sometimes also known as PCP.

pentachlorophenol (PCP)

Commercial PCP is not pure pentachlorophenol but is significantly contaminated with *2,3,4,6-tetrachlorophenol*. This mixture has many pesticidal applications: it is used as an herbicide (e.g., as a pre-harvest defoliant), an insecticide (termite control), a fungicide (wood preservation and seed treatment), and a molluscicide (snail control). Some trichlorophenol isomers and some tetrachlorophenol isomers are also sold as wood preservatives.

Unfortunately, if wood treated with such preservatives is eventually burned, a fraction of the chlorophenols can react to eliminate HCl, thereby producing members of the chlorinated dioxin family. Thus **octachlorodibenzo-*p*-dioxin,** OCDD, is produced as an unwanted by-product in the incomplete combustion of pentachlorophenol products:

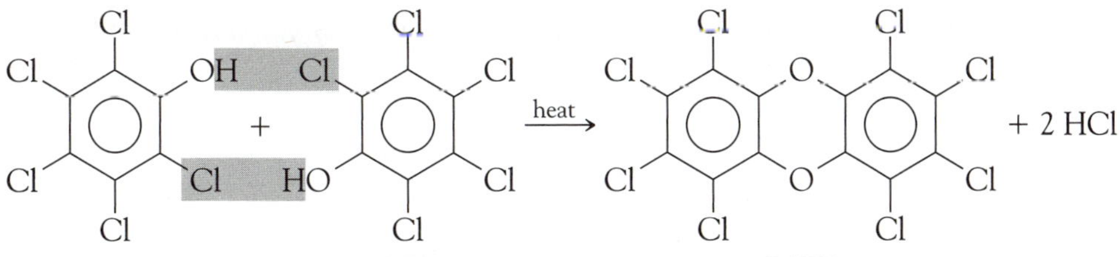

OCDD is the most prevalent dioxin congener found in human fat and in many environmental samples.

Indeed, pentachlorophenols are one of the largest chemical sources of dioxins to the environment; however, the main dioxin they contain, OCDD, is not particularly toxic, as discussed in a later section. Commercial supplies of chlorinated phenols themselves are contaminated with various dioxins.

PROBLEM 14-2

In naming OCDD and pentachlorophenol, no numbers are used to specify the positions of the chlorine substituents. Why is that not necessary here, whereas it is required in, e.g., 2,3,7,8-TCDD?   ●

In general, any two phenol molecules that each have a chlorine on a carbon atom that is next to that having an OH group can combine to produce a dibenzo-_p_-dioxin molecule. The two phenols that combine need not be identical, but simply need to make contact when they have been heated sufficiently to facilitate HCl elimination and dioxin formation. Similarly, coupling of phenoxide anions can occur with Cl⁻ elimination, as discussed previously in the case of 2,4,5-T synthesis. The methodology of problem solving that can be used to deduce the chlorophenolic origin of environmental dioxins is discussed in Box 14-1.

BOX 14-1   Deducing the Probable Chlorophenolic Origins of a Dioxin

The chlorophenolic source of dioxins found in environmental samples can be deduced by reversing the logic used in the text to deduce which dioxin would be produced by the coupling of two specific chlorophenols.

Consider the congener 1,2,7,8-tetrachlorodibenzo-_p_-dioxin; it could have been formed by elimination of two HCl molecules from two chlorophenol molecules in the following two ways (here T stands for trichlorophenol).

If it is assumed that the oxygen atom at the top of the dioxin congener originates with the chlorophenol congener on the left side of the dioxin molecule, then the bottom oxygen must come from the chlorophenol on the right side of the dioxin molecule; with this set of assumptions, the original reactants must have been 2,4,5- and 2,3,4-trichlorophenol. (Notice in the diagram above that the chlorine atoms eliminated must have arisen from positions adjacent to the oxygen atoms.) The alternative possibility, that the oxygen atom at the top of the dioxin structure came from the chlorophenol on the right side of the molecule and the bottom oxygen from the chlorophenol on the left, leads to the possibility that the trichlorophenol molecules that combined were the 2,4,5 and the 2,3,6 congeners. Thus a 1,2,7,8-tetrachlorodibenzo-p-dioxin molecule in the environment could have arisen by combination of a 2,4,5-trichlorophenol molecule with either a 2,3,4- or a 2,3,6-substituted congener.

Unfortunately, some dioxins undergo rearrangement of substituents during their formation, so such a "retrosynthesis" approach is not an infallible guide to the origin of dioxins discovered in the environment.

PROBLEM 1

Deduce the two possible combinations of polychlorophenol molecules that, when coupled through loss of two HCl molecules, would produce a molecule of 1,2,9-trichlorodibenzo-p-dioxin.

PROBLEM 14-3

(a) Deduce the structures and the correct numbering for the two tetrachlorophenol isomers that exist in addition to the 2,3,4,6 isomer mentioned in the text. (b) For each of these two isomers, deduce the structure and names of the dioxin(s) that would result if two molecules of that isomer were to react together.

When both carbon atoms adjacent to the one bonded to —OH or O⁻ bear chlorine atoms on one (or both) of the chlorophenol or chlorophenoxy, several possible dioxins can be formed in some cases. Consider, for example, the possible couplings of *2,3,6-trichlorophenol* with *2-chlorophenol*. The two possible orientations of the trichlorophenol with respect to the chlorophenol are illustrated below. The lower reaction corresponds to the upper one rotated by 180° about the axis (dashed green line) running through the O atom and the carbon atom that is in the para position relative to it. Thus we see that both 1,4- and 1,2-dichlorodibenzo-p-dioxins can be produced, depending upon orientation. In practice, an almost equal mixture of the two isomers would be formed.

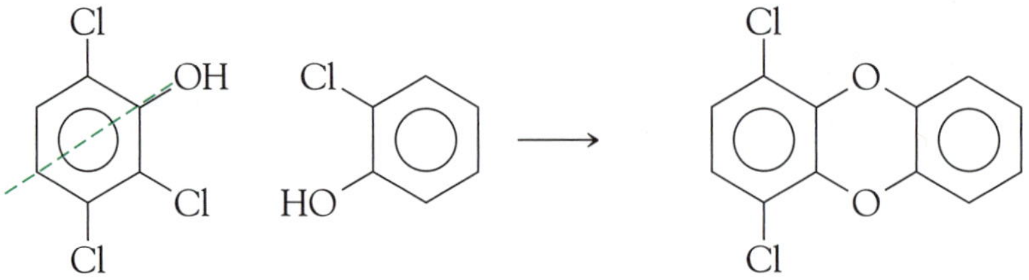

1,4-dichloro isomer

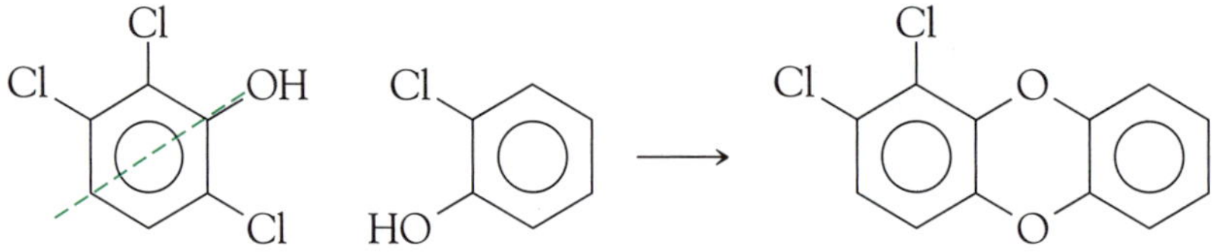

1,2-dichloro isomer

Deduce what dioxin(s) would be produced in side reactions if 2,4-D were to be synthesized from 2,4-dichlorophenol.

14.4 Detecting Dioxins in Food and Water

As a consequence of their widespread occurrence in the environment and their tendency to dissolve in fatty matter, **dioxins bioaccumulate in the food chain. More than 90% of human exposure to dioxins is attributable to the food we eat,** particularly meat, fish, and dairy products. Typically, dioxins and furans (a group of compounds resembling the dioxins in structure, which we'll discuss later) are present in fish and meat at levels of tens or hundreds of picograms (pg, or 10^{-12} g) per gram of the food; in other words, they occur at levels of tens or hundreds of parts per trillion. However, the bulk of dioxins and furans in nature are not present in biological systems: attachment to soil, and to sediments of rivers, lakes, and oceans are their most common sinks.

The ability of chemists to detect TCDD and other organochlorines in environmental samples has improved by orders of magnitude over the past few decades. In the early 1960s, when Carson's *Silent Spring* was published, the lower limit for analysis of DDT and such compounds was the parts-per-million level. Ten years later, detecting such substances at the parts-per-billion level was possible but not common or easy. By 1990, parts-per-trillion detection was possible in soil or biota samples, and parts-per-quadrillion was possible for water samples. Today, a few labs detect some substances at limits up to one thousand times lower than these! At these latter levels, many organochlorines are found in *every* environmental sample, no matter how "clean" an environment it came from. By the late 1990s, researchers at the

Centers for Disease Control in Atlanta were able to detect as little as 10^{-16} g of TCDD in human serum samples.

The potential impact on human health of exposure to dioxins is documented later, following a discussion of the properties of PCBs and furans, two types of chemicals with which dioxins share many properties.

PROBLEM 14-5

Given its formula and Avogadro's constant (6.02×10^{23} molecules mol^{-1}), deduce how many molecules are present in 10^{-16} g of TCDD. ●

PCBs

The well-known acronym **PCB** stands for *polychlorinated biphenyls,* a group of industrial organochlorine chemicals that became a major environmental concern in the 1980s and 1990s. Although *not* pesticides, they found a wide variety of applications in modern society because of certain other properties they possess. Since the late 1950s, over 1 million metric tons of PCBs have been produced, about half in the United States and the rest mainly in France, Japan, and the former Eastern Bloc nations. Like many other organochlorines, **PCBs are very persistent in the environment and they bioaccumulate in living systems.**

As a result of careless disposal practices, PCBs have become a major environmental pollutant in many areas of the world. Indeed, about 10% of the original manufactured one million tonnes remains in the environment. More than 95% of the entire U.S. population has detectable concentrations of PCBs in their bodies. Due both to their own toxicity and to that of their "furan" contaminants, PCBs in the environment have become a cause for concern because of their potential impact on human health, particularly with regard to growth and development.

In the following material, we consider what PCBs are, how they are made, what they are used for, and how they become contaminated and released into the environment.

14.5 The Structure of PCB Molecules

Biphenyl molecules consist of two benzene rings linked by a single bond formed between two carbons that have each lost their hydrogen atom:

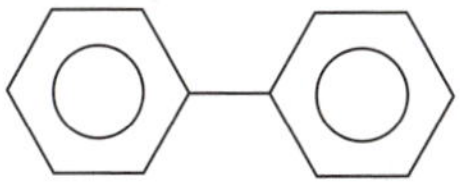

biphenyl

Like benzene, if biphenyl is reacted with Cl_2 in the presence of a *ferric chloride* ($FeCl_3$) catalyst, some of its hydrogen atoms become replaced by chlorine atoms. The more chlorine that is initially present and the longer the reaction is allowed to proceed, the greater the extent (on average) of chlorination of the biphenyl molecule. The products are polychlorinated biphenyls, PCBs. The reaction of biphenyl with chlorine produces a mixture of many of the 209 congeners of the PCB family; the exact proportions depend upon the ratio of chlorine to biphenyl, the reaction time, and the reaction temperature. An example of a PCB molecule is shown below:

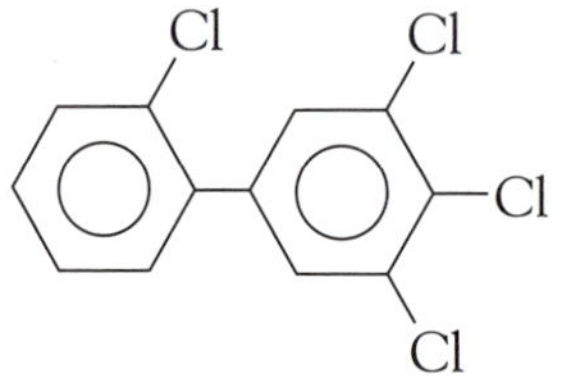

2,3′,4′,5′-tetrachlorobiphenyl

Although many individual PCB compounds are solids, the mixtures are liquids or solids with low melting points. Commercially, individual PCB compounds were not isolated; rather they were sold as partially separated mixtures, with the average chlorine content in different products ranging from 21% to 68%.

PROBLEM 14-6

The general formula for any PCB congener is $C_{12}H_{10-n}Cl_n$, where n ranges from 1 to 10. Calculate the average number of chlorine atoms per PCB molecule in a mixture of congeners that is 60% chlorine by mass, a common value for commercial samples. ●

14.6 The Numbering Systems for PCBs

The numbering scheme used for individual PCB congeners begins with the carbon that is joined to a carbon in the other ring; it is given the number 1, and the other carbons around the ring are numbered sequentially. As illustrated below, the positions in the second ring are also numbered 1 through 6, starting with the ring-joining carbon, but are distinguished by primes. By convention, the 2′ position in the second ring lies on the *same* side of the C—C bond joining the rings as does the 2 position in the first ring, etc.

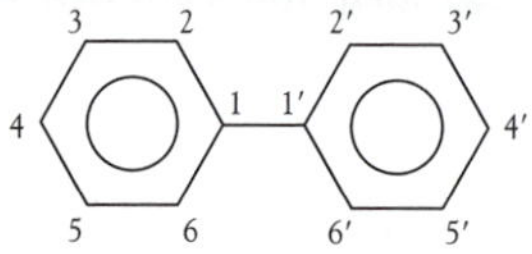

In most instances, the two rings in a chlorinated biphenyl molecule are not equivalent since the patterns of substitution differ. The unprimed ring is chosen to be the one that will give a substituent with the lowest-numbered

carbon. Using all these rules, we can deduce that the name of the PCB molecule shown on page 632 is *2,3',4',5'-tetrachlorobiphenyl*.

Very rapid rotation occurs around carbon–carbon single bonds in most organic molecules, including the C—C link joining the two rings in biphenyl and in most PCBs. Thus **it is not normally possible to isolate compounds corresponding to different relative orientations of the two rings ("rotamers") in a PCB.** For example, *3,3'-* and *3,5'-dichlorobiphenyl* are *not* individually isolatable compounds, since one form is constantly being converted into the other and back again by rapid rotation about the C—C bond linking the rings:

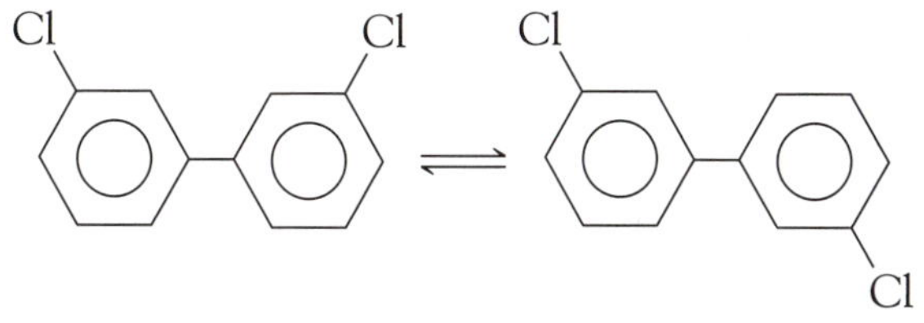

3,3'-dichlorobiphenyl   3,5'-dichlorobiphenyl

The label used for such a compound is that which has the lowest number for the second chlorine, so the system shown above is called the 3,3' isomer. Although the rings rotate rapidly with respect to each other, the energetically optimum orientation is the one having the rings coplanar or close to it, except, as we shall see later, when large atoms or groups occupy the 2 and 6 positions.

PROBLEM 14-7

Using a systematic procedure, draw the structures of all unique dichlorobiphenyls, assuming first that free rotation about the bond joining the rings does *not* occur. Then deduce which pairs of structures become identical if free rotation does occur. ●

14.7 Commercial Uses of PCBs

All PCBs are practically insoluble in water but are soluble in hydrophobic media, such as fatty or oily substances. Commercially, they were attractive because they are

- chemically inert liquids and are difficult to burn,
- have low vapor pressures,
- are inexpensive to produce, and
- are excellent electrical insulators.

As a result of these properties, they were used extensively as the coolant fluids in power transformers and capacitors. Later, they were also employed as *plasticizers*, i.e., agents used to make plastic materials such as PVC products more flexible; in carbonless copy paper; as de-inking solvents for recycling newsprint; as heat transfer fluids in machinery; as waterproofing agents; and even further uses.

PCBs contaminate paint because they are created as by-products when pigments are manufactured.

Because of their stability and extensive usage, together with careless disposal practices, PCBs became widespread, persistent, and common environmental contaminants. For example, PCB-containing transformers and capacitors were often just dumped into landfills in the past, and their PCB contents was allowed to leak into the ground. Indeed, PCBs were inadvertently released into the environment during their production, their use, their storage, and their disposal.

When their accumulation and harmful effects became recognized, **open uses** of PCBs, i.e., uses for which disposal could not be controlled and often was dissipative, were terminated. They are listed by the United Nations as one of the "dirty dozen" persistent organic pollutants and as such are being gradually phased out of all uses worldwide. Although North American production of PCBs was halted in 1977 and was stopped some time ago in other countries as well, the substances often remain in "closed" use in some electrical transformers still in service. As these units are gradually decommissioned, their PCB content usually is stored in order to prevent further contamination of the environment. In some locales, stored PCBs are destroyed by incineration, using techniques discussed in Chapter 16.

14.8 PCBs Cycling Among Air, Water, and Sediments

Only a minority of PCBs manufactured in the past are currently found in the environment or have been destroyed; much of the production lingers in storage or old electrical equipment and may ultimately be released. Electrical transformers and construction sealants represent continuing, slow-release primary sources of PCBs to the atmospheres of many cities in developed countries. PCB releases from older consumer products into indoor air, which is eventually vented outside, is a major source of PCBs in urban air.

If released into the environment, PCBs persist for many years because they are so resistant to breakdown by chemical or biological agents. Although their solubility in water is very slight—indeed, they are more likely to be adsorbed onto suspended particles in the water than dissolved in it—the tiny amounts of PCBs in surface waters are constantly being volatilized and subsequently redeposited on land or in water after traveling in air for a few days. By such mechanisms, PCBs have been transported worldwide. There are measurable background levels of PCBs even in polar regions and at the bottom of oceans. Indeed, the ultimate sinks for PCBs that are mobile are in the deep sediments of oceans and large lakes.

Because deposits of PCBs and other POPs on soil and water can revolatilize and again enter the air, they are sometimes referred to as *environmental capacitors*. Their continued re-entry into the air "buffers" the decline in the atmospheric concentration of such substances, and indeed will eventually become the main source of atmospheric POPs. This environmental load from secondary sources, plus additions of new releases from primary sources, will continue to be recycled among air, land, and water, including the biosphere, for decades to come, as analyzed in greater detail in Chapter 15 when we discuss the *long-range transport of atmospheric pollutants*.

A quantitative measure of the recycling of substances within a water body is provided by the mass balance of current annual inputs and outputs of the compounds. The PCB mass balance for a very large, relatively clean water body—Lake Superior—is illustrated in Figure 14-1. Although it is the least polluted of the Great Lakes, Lake Superior's burden of PCBs in its water and its sediments is substantial. At the time the data were obtained (1986 and 1992), almost all the input of PCBs occurred from the air, with relatively little added from industries or via tributary rivers (Figure 14-1). Overall, Lake Superior was gradually "exhaling" its historical load of PCBs into the air, the output to air being much greater than was the annual input from the atmosphere. Little of Lake Superior's PCBs were being lost to sediments; about as much was redissolved from them as was deposited onto them each year. More recent data indicates that the air concentration now is quite low and is only slowly declining with time.

By contrast, the mass balance of PCBs in Lake Ontario, another of the Great Lakes, was quite different from that of Lake Superior. The PCB concentration in Lake Ontario water substantially exceeds that of Lake Superior,

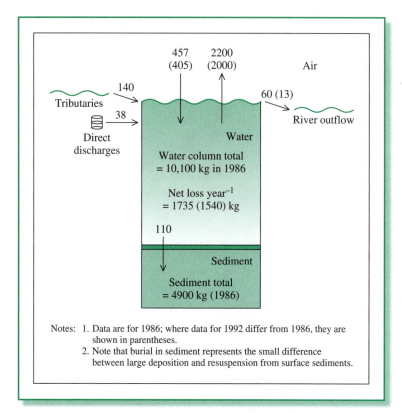

Notes: 1. Data are for 1986; where data for 1992 differ from 1986, they are shown in parentheses.
2. Note that burial in sediment represents the small difference between large deposition and resuspension from surface sediments.

FIGURE 14-1 Mass balance of PCBs in Lake Superior, in kilograms per year. [Source: *The State of Canada's Environment 1996* (Ottawa: Government of Canada, 1996).]

since it is located in a much more industrialized area. In Lake Ontario, the greatest current input comes from land-based sources such as waste dumps that still leach PCBs into the lake and its tributaries. About the same quantity of PCBs were present in the water flowing out of the lake. Approximately equal amounts were lost annually to sediments and to the atmosphere; about one-third of such losses were cancelled by new inputs from the sediments and the air. At the turn of the century, the airborne concentration of PCBs above Lake Ontario was declining with a half-life of about 7 years.

PROBLEM 14-8

The PCB concentration in Lake Michigan is declining according to a first-order rate law having a rate constant of 0.078 year^{-1}. If the PCB concentration in Lake Michigan averaged 0.047 ppt in 1994, what would it have been in 2010? In what year will the concentration fall to 0.010 ppt? What is the half-life period of PCBs in this lake? [*Hint:* Recall that for first-order processes, the fraction f of any sample that still remains after time t has passed is $f = e^{-kt}$.] ●

Because of their persistence and their solubility in fatty tissue, **PCBs in food chains undergo biomagnification;** an example is shown in Figure 14-2. Notice that the ratio of PCBs in the eggs of herring gulls in the Great Lakes was 50,000 times that in the phytoplankton in the water at the time of these

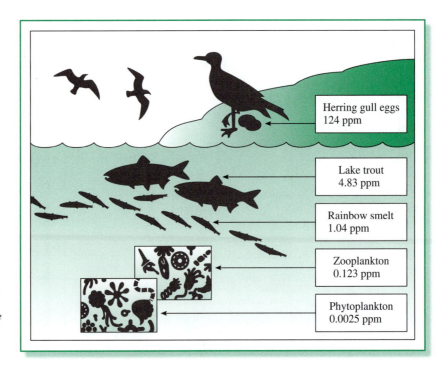

FIGURE 14-2 The biomagnification of PCBs in the Great Lakes aquatic food chain. [Source: *The State of Canada's Environment 1996* (Ottawa: Government of Canada, 1996).]

Herring gull eggs
124 ppm

Lake trout
4.83 ppm

Rainbow smelt
1.04 ppm

Zooplankton
0.123 ppm

Phytoplankton
0.0025 ppm

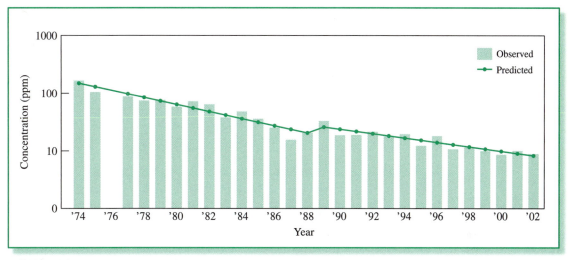

FIGURE 14-3 PCB concentrations in herring gull eggs in Toronto Harbor, 1974–2002. The predicted curves correspond to exponential decay in those time periods. [Source: Dr. Chip Weseloh, Environment Canada.]

measurements. The good news is that the average levels of PCBs in such eggs has fallen with time in many locations as the data in Figure 14-3 illustrates for gull colonies in Lake Ontario around Toronto. The concentrations are plotted in the figure on a logarithmic scale, and the data seem to fit two intersecting straight lines, corresponding to first-order decay sequences with half-lives of first five years and later of seven years, the latter in agreement with the half-life decline of airborne concentrations above the lake for the same period. This complicated double-exponential behavior may arise because of continuing sources of PCBs to the system. PCB levels in fish at the top of the Lake Ontario food chain have also declined since the 1970s, but the current rate of decrease is slow and erratic.

The relative concentrations of the congeners of a PCB mixture begin to change once they enter the environment. Microorganisms in soils and sediments and large organisms such as fish both preferentially metabolize congeners having relatively few chlorine atoms. Thus **the relative concentrations of the more heavily chlorinated congeners increase with time since they are degraded much more slowly.** For example, between 1977 and 1993 the proportion of PCB molecules with 4 or 5 chlorine atoms decreased by 6% each in trout in Lake Ontario, whereas those with 7 or 8 chlorines increased by 7% and 4%, respectively. However, PCBs present in anaerobic soil are eventually dechlorinated microbially at their meta and para positions, leaving congeners that are only chlorinated at ortho positions. Aerobic degradation occurs with congeners having adjacent carbons (ortho + meta, or meta + para) chlorine-free.

14.9 PCB Contamination by Furans

Strong heating of PCBs in the presence of a source of oxygen can result in the production of small amounts of *furans*. These compounds are structurally similar to dioxins; they differ only in that the molecules are missing one oxygen atom in the central ring. **The furan ring contains five atoms, one of which is oxygen and the other four of which are carbon atoms that participate in double bonds:**

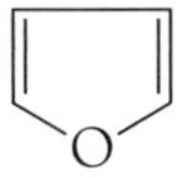

furan

The dibenzofurans (DFs) have a benzene ring fused to opposite sides of the furan ring:

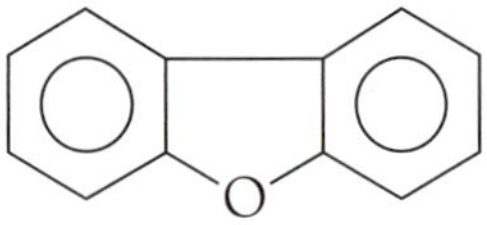

dibenzofuran

As with dioxins, all chlorinated dibenzofuran congeners are planar; i.e., all C, O, H, and Cl atoms lie in the same plane. They are formed from PCBs by the elimination of the atoms X and Y bonded to two carbons that are *ortho* in position to those that link the rings *and* that lie on the same side of the C—C link between the rings:

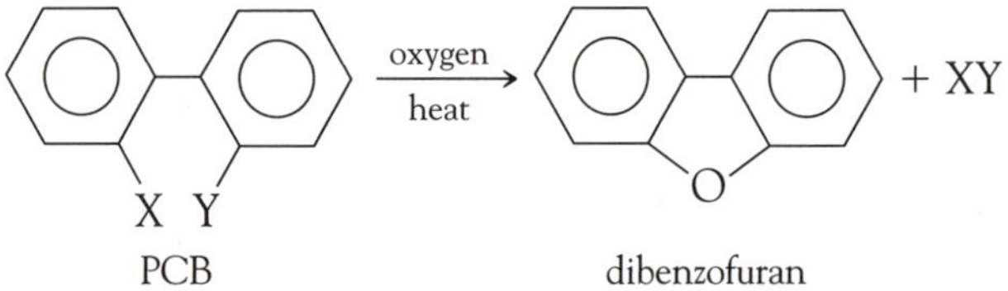

PCB   dibenzofuran

The atoms X and Y can both be chlorine, or one can be hydrogen and the other one chlorine, so the molecule eliminated can be Cl$_2$ or ClH (i.e., HCl), respectively. A more detailed analysis of the nature of the specific furans that result from particular PCB congeners is given in Box 14-2.

BOX 14-2   Predicting the Furans That Will Form from a Given PCB

In deducing the nature of the polychlorinated dibenzofuran (PCDF) that would be formed from a particular PCB, it should be remembered that free rotation occurs about the single bond joining the two rings in the original biphenyl in all PCBs at the elevated temperatures of the reaction. Thus HCl elimination in 2,3'-dichlorobiphenyl gives both 4- and 2-chlorodibenzofuran.

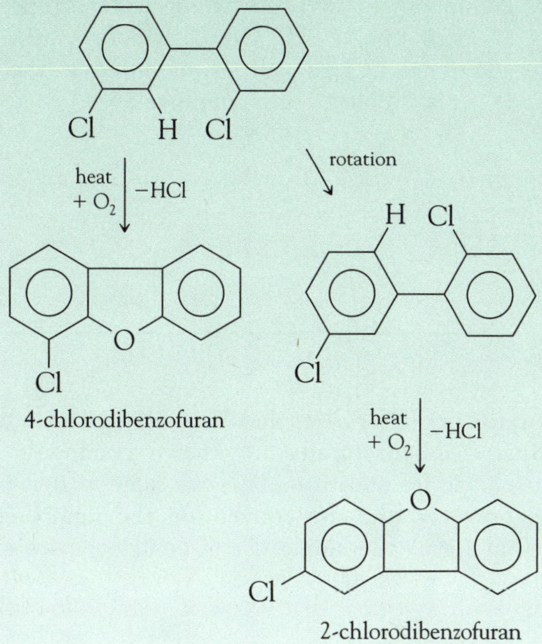

4-chlorodibenzofuran

2-chlorodibenzofuran

At the high temperatures of this reaction, some interchange of the adjacent substituents in the 2 and 3 positions (ortho and meta) of any given ring can occur as a prelude to HCl elimination; in *particular, chlorine can move from an ortho to a meta position, and hydrogen from meta to ortho, preceding HCl elimination.* For example, when 2,6,2',6'-tetrachlorobiphenyl (shown below) is heated in air, some of its molecules lose a pair of ortho chlorines to give a dichlorodibenzofuran, and some first interchange Cl and H in one ring to eliminate HCl and produce a trichlorodibenzofuran. Free rotation about the bond does not occur after the interchange, as presumably the elimination occurs immediately.

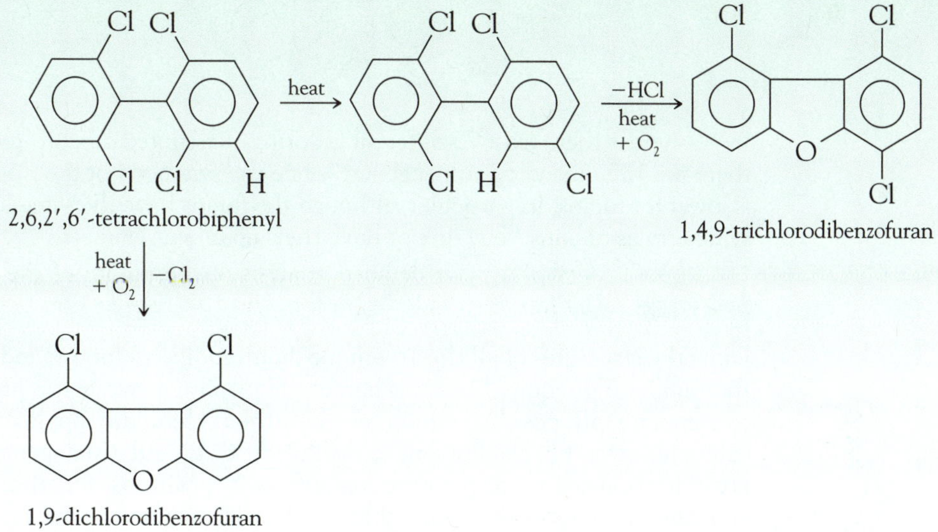

2,6,2',6'-tetrachlorobiphenyl

1,4,9-trichlorodibenzofuran

1,9-dichlorodibenzofuran

(continued on p. 640)

PROBLEM 1

For each PCB shown below, deduce which furans would be expected to be produced by Cl_2 or HCl elimination when the PCB is heated in air. Write the correct name for each polychlorinated dibenzofuran (PCDF).

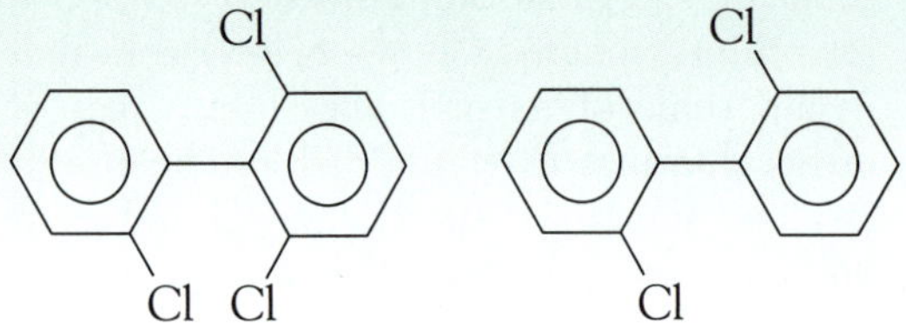

PROBLEM 2

Recently it has been discovered that, upon strong heating in air, PCBs can also react by elimination of two ortho hydrogen atoms (one on each ring) as H_2. Decide which, if any, additional PCDFs will be produced if the PCBs in Problem 1 can eliminate H_2.

Most of the chlorine in the original PCB molecule is still present in the dibenzofuran; **polychlorinated dibenzofurans** are known commonly as **PCDFs. The numbering scheme for substituents is the same as that for dioxins (PCDDs); note, however, that by convention the numbering starts next to a carbon that forms the single C—C bond opposite the oxygen.**

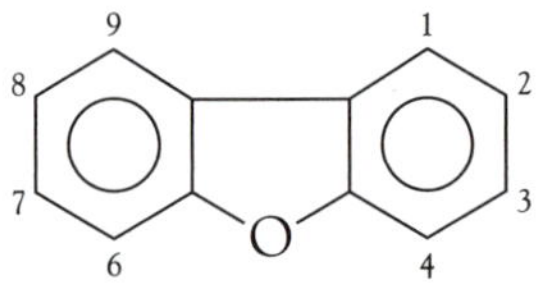

Whereas there exist 75 different chlorine-substituted dibenzo-p-dioxins, there are 135 dibenzofuran congeners, since the symmetry of the ring system is lower for furans. In particular, although the furans have the same left-right symmetry as dioxins, they do not have their up-down symmetry.

PROBLEM 14-9

Draw the structures of all the 16 unique dichlorodibenzofurans, and deduce the numbering required in their names. [*Hint:* Use a systematic procedure to generate all possible structures, but then eliminate duplicates. For example, start by placing one chlorine at C-1, and then generate all possible isomers corresponding to different positions for the second chlorine. Then place the first chlorine at C-2 and repeat the procedure, noting that the 1,2-dichloro isomer is generated both times. Continue the procedure at C-3, etc.]

Almost all commercial PCB samples are contaminated with some PCDFs, but it usually amounts to only a few ppm in the originally manufactured liquids. However, if the PCBs are heated to high temperatures and if some oxygen is present, conversion of PCBs to PCDFs increases the level of contamination by orders of magnitude. The furan concentration in used PCB cooling fluids is found to be greater than in the virgin materials, presumably due to the moderate heating that the fluid undergoes during its normal use. Furan production also occurs if one attempts to burn PCBs with anything but an unusually hot flame.

Other Sources of Dioxins and Furans

In addition to the sources discussed above, polychlorinated dibenzofurans and dibenzo-*p*-dioxins are also produced as by-products in a myriad of processes, including the bleaching of pulp, the incineration of garbage and hospital waste, the recycling of metals and sintering of iron ore, and the production of common solvents such as tri- and perchloroethene.

14.10 Pulp and Paper Mills

Pulp and paper mills that use elemental chlorine, Cl_2, to bleach pulp are dioxin and furan sources. These contaminants, among many other chlorinated compounds, result from the reaction of the chlorine with some of the organic molecules released by the pulp. The tan color of the pulp that has undergone the initial stages of processing is due to the light-absorbing properties of the *lignin* component of the original wood fibers. A generalized structure for lignin is shown in Figure 14-4. In order to make white paper, the residual (~10%) component of the lignin still present after initial processing must be removed, usually by bleaching the pulp with oxidizing agents. If you examine the generalized structure of lignin in Figure 14-4, you can observe several sites of mono-substituted *phenols* and *phenolic ethers*, and ortho-substituted *phenyl diethers*. From these structural components, it is not difficult to imagine how lignin can serve as a precursor to furans and dioxins when it reacts with chlorinating agents such as Cl_2.

More furans than dioxins are formed in bleaching of pulp by elemental chlorine. The furan congeners of highest concentrations in the pulp are *1,2,7,8-TCDF* and the more toxic *2,3,7,8-TCDF*. Unfortunately, the most abundant dioxin produced by the pulp-and-paper bleaching process is the highly toxic 2,3,7,8-TCDD congener. The paper and effluent contain dioxins at parts-per-trillion levels, which resulted in total releases in the past, in North America, of several hundred grams of 2,3,7,8-TCDD annually.

Because of the problems of producing dioxins and furans, the use of elemental chlorine as a bleaching agent for paper was banned in the United States in 2001. Most pulp-and-paper mills there and in other developed countries switched their bleaching agent from elemental chlorine to **chlorine**

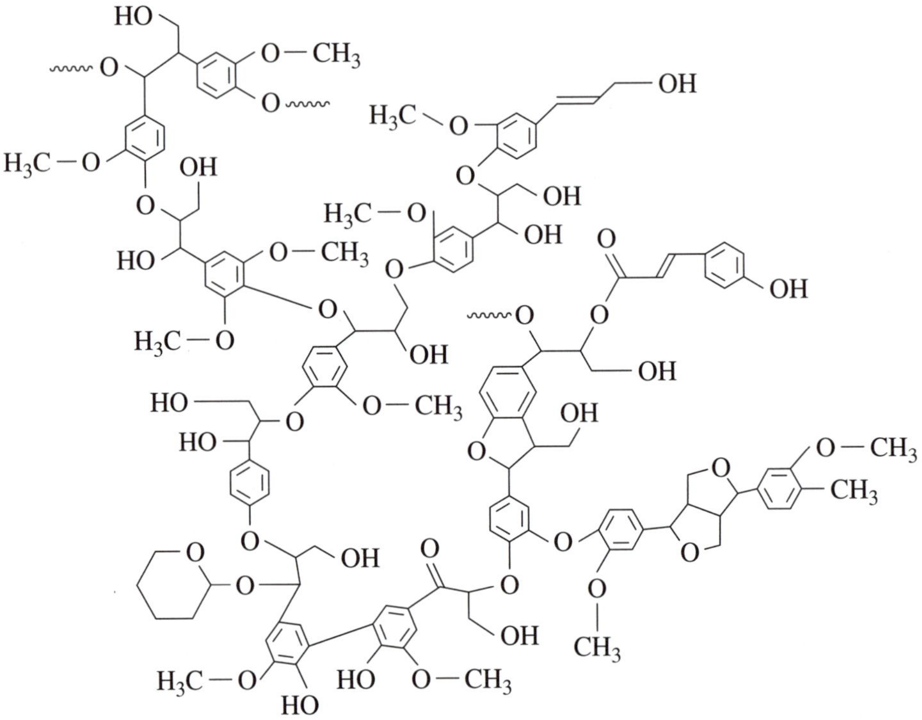

FIGURE 14-4 Generalized structure of lignin. [Source: M. C. Cann and M. E. Connelly, *Real-World Cases in Green Chemistry* (Washington D.C.: American Chemical Society, 2000).]

dioxide, ClO_2, from which the furan and dioxin output is much smaller, even undetectable in many cases. The difference is due to the mechanism by which the compounds attack the pulp's residual lignin. Elemental chlorine reacts to insert chlorine as a substituent on the aromatic rings in lignin, yielding products that are soluble in alkali and that can then be washed away. Experiments suggest that during the oxidation of the lignin, two of the component benzene rings can couple together to form a dibenzofuran or dibenzo-*p*-dioxin system that subsequently is chlorinated and in the process becomes detached from the lignin system. In contrast, chlorine dioxide destroys the aromaticity of the benzene rings by free-radical processes and therefore produces fewer chlorinated products that contain six-membered rings.

Some mills now produce paper pulp without any use of chlorine compounds. **Ozone, hydrogen peroxide, and even high-pressure oxygen are the alternative bleaching agents used in these totally chlorine free (TCF) pulp mills.** Mills that still use chlorine to bleach now remove contaminants from wastewater by treatments such as reverse osmosis (see Section 11.7).

14.11 Green Chemistry: H₂O₂, an Environmentally Benign Bleaching Agent for the Production of Paper

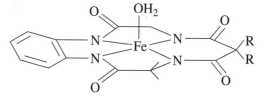

FIGURE 14-5 Tetraamido-macrocyclic ligands (TAML): activators for hydrogen peroxide. [Source: M. C. Cann and M. E. Connelly, *Real-World Cases in Green Chemistry* (Washington D.C.: American Chemical Society, 2000).]

TCF bleaching agents for paper such as **hydrogen peroxide** (H_2O_2), *ozone*, and *diatomic oxygen* have been developed. While TCF agents eliminate the formation of dioxins and furans, these methods in general are problematic in that these oxidizing agents are not as strong as elemental chlorine or chlorine dioxide. Thus they generally require longer reaction times and higher temperatures (more energy input), and they lead to significant breakdown of the cellulose fibers, which weakens the paper, requiring more wood to produce the same amount of paper.

Terry Collins of Carnegie Mellon University earned a Presidential Green Chemistry Challenge Award in 1999 for his development of compounds known as **tetraamido-macrocyclic ligands** (TAML, Figure 14-5), which enhance the oxidizing strength of H_2O_2. Hydrogen peroxide is a particularly enticing oxidizing reagent since its by-products are water and oxygen, which are environmentally benign. The use of TAML in conjunction with hydrogen peroxide reduces the temperature and reaction times normally required for bleaching paper with hydrogen peroxide, thus making hydrogen peroxide a viable alternative for this process.

The TAML ligands can be modified by varying the alkyl groups (R) on the right side of the structure in Figure 14-5. Changing these groups influences the lifetime of these catalysts. In uses such as the bleaching of paper it is important for the TAML catalysts to decompose in a relatively short period of time so they do not become a burden to the environment. However, they must last long enough to fulfill their role as catalysts for hydrogen peroxide. TAML catalysts not only offer significant promise for the bleaching of paper, they are also being considered for use in laundry applications, the disinfection of water, and the decontamination of biological warfare agents such as anthrax.

14.12 Fires and Incineration as Sources of Dioxins and Furans

Fires of many kinds, including forest fires and those in incinerators, release various congeners of the dioxin and furan families into the environment; these chemicals are produced as minor by-products from the chlorine and organic matter in the fuel. **Dioxin and furan production seem unavoidable whenever combustion of organic matter occurs in the presence of chlorine, unless steps are taken to ensure complete combustion by using very high flame temperatures.** Some environmentalists worry particularly about the dioxin emissions when the chlorine-containing plastic PVC is

The chemical composition
of PVC is discussed in
Section 16.11.

incinerated or involved in other fires. Indeed, research on the combustion of newspapers indicates that chlorinated dioxin and furan production rises as the amount of salt or PVC present also increases. In many environmental samples of combustion products, several dozen different dioxin congeners are found, all in comparable amounts. Congeners with relatively high numbers of chlorine substituents usually are the most prevalent.

Incinerators now are the largest anthropogenic source of dioxins in the environment. Dioxins and furans are *formed* in the post-combustion zone of incinerators, where the temperature is much lower (250–500°C) than in the flame itself (see Section 16.32). They are formed during the oxidative degradation of the graphite-like structures in the soot particles that were produced during the incomplete combustion of the waste. Trace metal ions in the original waste probably catalyze the process. The small amounts of chlorine in the waste provide more than enough of this element required to partially chlorinate the furans and dioxins. The dibenzofuran and dibenzo-*p*-dioxin ring systems are formed at high temperatures (>650°C); chlorination progressively occurs when the temperature cools below 650°C and gradually is reduced to 200°C.

Characteristically, incineration produces a greater mass of furans than of dioxins. The yields of specific dioxin congeners increase with the degree of chlorination through to OCDD (the octachloro compound), whereas the peak production of furans occurs with four to six chlorines. In contrast to waste incineration, industrial coal combustion generates little dioxin because it burns much more completely, generating little soot to decompose later into dioxins and furans.

Recent research from Australia indicates that burning wood that has been treated with the preservative *permethrin*, one of the pyrethroids considered to be a "green" insecticide (Chapter 13), results in the formation of a mixture of both chlorinated dioxins and furans, especially if the wood has also been treated with copper-based fungicides. The permethrin structure incorporates both chlorine and diphenyl ether components, which under combustion conditions combine to form various PCDDs and PCDFs. Other Australian research indicates that PCDDs can be formed from the hydroxyl derivatives of chlorinated diphenyl ether contaminants of pentachlorophenol.

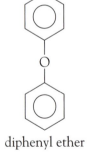

diphenyl ether

14.13 Chlorine Content of Dioxin and Furan Emissions

The profile of estimated annual global PCDF and PCDD emissions for the congeners with 4 to 8 chlorines, i.e., those believed to be toxic, is shown in Figure 14-6a. As discussed previously, furans outnumber dioxins, and furan emissions peak with congeners having four chlorines, whereas the dioxin peak is less pronounced, and occurs with about six chlorines. Furans with these intermediate amounts of chlorine have toxicities similar to that of 2,3,7,8-TCDD, whereas fully chlorinated dioxin molecules have low

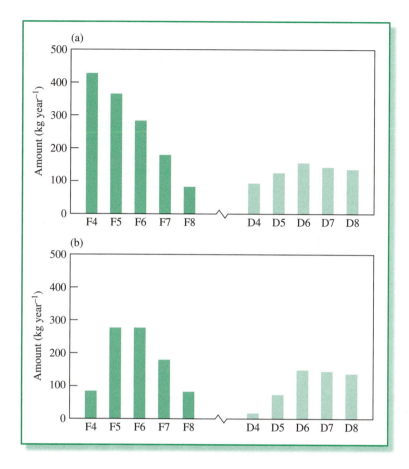

FIGURE 14-6 Annual PCDD and PCDF (a) emissions and (b) deposition rates after reactions with the hydroxyl radical OH. The letters F and D represent furans and dioxins; the numbers indicate the number of chlorine atoms per molecule. [Source: J. I. Baker and R. A. Hites, "Is Combustion a Major Source of Polychlorinated Dibenzo-p-Dioxins and Dibenzofurans to the Environment?" *Environmental Science and Technology* 34 (2000): 2879.]

toxicities. Consequently, the threat to human health from furans in the environment may exceed that from dioxins.

The mass of the dioxin and furan compounds that are eventually deposited from air onto soil and sediments, the primary mechanism by which dioxins eventually enter the food chain, is shown in Figure 14-6b. The loss in mass between emission and deposition is greater the *fewer* the number of chlorines present; hence the significant difference between amounts emitted (Figure 14-6a) and deposited (Figure 14-6b) for the tetrachloro and pentachloro congeners, but not those more heavily chlorinated.

This differentiation occurs because **the principal loss mechanism is attack at an unsubstituted carbon by the hydroxyl free radical, OH,** followed by atmospheric oxidation of the resulting radical (as expected from the principles discussed in Chapters 3 and 17), and the rate of this initial reaction is greater the fewer the chlorines present. The amount of OCDD, the octachloro congener of dioxin, that is found to be deposited greatly exceeds the estimate from this figure, which is based mainly on combustion sources. Some scientists believe that the discrepancy arises because much additional OCDD is created in water droplets in air by the sunlight-initiated photochemical decomposition of PCP, pentachlorophenol, which eventually results in coupling of two PCPs to produce OCDD.

Very small concentrations of dioxins—particularly highly chlorinated ones—were present in the environment in the preindustrial era, presumably as a result of forest fires, volcanoes, etc. Indeed, forest fires are still probably Canada's largest source of dioxins. According to the analysis of soils and the sediments in lakes, the greatest anthropogenic input of dioxins and furans to the environment in developed countries began in the 1930s and 1940s, peaked in about 1970, and are now about half their maximum values. The principal sources were combustion/incineration, the smelting and processing of metals, the chemical industry, and existing environmental reservoirs. Inadvertent production of dioxins continues today, but at a slower rate—about half of the maximum, according to some sediment samples.

The decrease in emissions resulted from deliberate steps taken by industrialized nations to reduce the production and dispersal of these toxic byproducts. In particular, dioxin emissions from large sources in the United States declined by 75% from 1987 to 1995 alone, primarily due to reductions in air emissions from municipal and medical waste incinerators. New regulations should increase the reduction to 95%. However, uncontrolled combustion nonpoint sources such as rural backyard trash burning in barrels—especially when plastics such as PVC are included in the mix—have not yet been brought under control.

Once created, dioxins and furans are transported from place to place mainly via the atmosphere. Eventually they are deposited and can enter the food chain, becoming bioaccumulated in plants and animals. As previously mentioned, our exposure to them arises almost entirely through the foods that we eat. In the next section, we try to answer the question of what effects, if any, this exposure has on our health.

The Health Effects of Dioxins, Furans, and PCBs

Over a billion dollars has been spent on research to determine the extent to which dioxins, furans, and PCBs cause toxic reactions in humans. Nevertheless, conclusions about this issue are still tentative and controversial. Evidence about toxicity is derived from two sources:

- toxicological experiments on animals that have been deliberately exposed to the chemicals, and

- epidemiological studies of humans who have been accidentally exposed.

It is generally agreed that **most PCBs are not very acutely toxic to humans:** the LD_{50} values of most congeners are large. In high doses, PCBs cause cancer in test animals, and consequently they are listed as a "probable human carcinogen" by the U.S. EPA. However, studies of humans exposed to them have produced inconsistent results. Most groups of people who have been exposed to relatively high concentrations of PCBs—e.g., as a result of their employment in electrical capacitor plants—have not experienced a higher overall death rate. (See, however, the comments about human cancer in Section 15.11.) A recent U.S. report connects the concentration of PCBs in household dust with the likelihood of resident children having leukemia. The most common reaction to exposure to PCBs is *chloracne*, a biological response by humans to exposure to many types of organochlorine compounds.

14.14 Inadvertent PCB Poisonings

The most dramatic effects observed to human health from exposure to PCB mixtures occurred when two sets of people, one in Japan in 1968 and the other in Taiwan in 1979, unintentionally consumed PCBs that had accidentally been mixed with cooking oil. In the Japanese incident, and probably in the Taiwanese case as well, the PCBs had been used as a heat exchanger fluid in the deodorization process for the oil. Since the PCB-contaminated oils had been heated, their level of PCDF contamination was much greater than occurs in freshly prepared commercial PCBs. The thousands of Japanese and the Taiwanese people who consumed the contaminated oils suffered health effects far worse than has been found for workers at PCB manufacturing and handling plants, even though the resulting PCB levels in their bodies were about the same.

From this difference, it has been concluded that the main toxic agents in the poisonings were the PCDFs, and that they and dioxins were collectively responsible for about two-thirds of the health effects, with the PCBs themselves responsible for the remainder. Indeed, studies on laboratory animals indicate that the furans involved in these incidents are more than 500 times as toxic on a gram-for-gram basis than are pure PCBs. Cognitive development, as measured by IQ scores, of children born to the most highly exposed Taiwanese mothers—even if births occurred long after the consumption of the contaminated oil—was found to be significantly lower than for their siblings born before the accident occurred and for children of unexposed mothers. Interestingly, children whose fathers but not mothers had consumed the oil showed no detrimental effects. Further effects of this incident are discussed in Section 15.11, in the material concerning environmental estrogens.

An incident comparable to the Asian cooking oil poisonings occurred in early 1999 in Belgium. Several kilograms of a mixture of PCBs that had previously been heated to a high temperature—converting a tiny fraction of the PCBs to furans—was put into a 80,000-kg batch of animal fat, which was then mixed with animal feed and shipped to about 1,000 farmers. Poultry producers subsequently noticed a sudden drop in egg production and in egg hatchability, and high concentrations of dioxins were subsequently found in chicken meat. The contaminated food was withdrawn from the market and destroyed.

14.15 Effects of *in Utero* Exposure to PCBs

Are humans who have been exposed to PCBs through their diets especially susceptible to reproductive problems? To answer this question, Sandra and Joseph Jacobson and their co-workers at Wayne State University in Detroit have spent several decades studying the offspring of people from the Lake Michigan area, including children whose mothers regularly eat fish from the Lake and who, as a result, are expected to have elevated levels of PCBs in their bodies. They have discovered statistically significant differences in children born to women who have high levels of PCBs; these differences were present not only at birth but persisted to the age of at least eleven years.

The *prenatal* (i.e., prebirth) exposure of the infants to PCBs was determined by analyzing the blood of their umbilical cords for these chemicals after birth. Because analytical techniques in the early 1980s were not sufficiently sensitive to detect the amounts of PCBs in all the umbilical cord samples, exposure for many infants was estimated from the PCB levels in their mother's blood and breast milk. The *postnatal* exposure of the children was assessed by analyzing their mother's breast milk and also by analyzing blood samples from the children at the age of four years.

The Jacobsons discovered that, at birth, the children of mothers who had transmitted the highest amounts of PCBs to the children *before* birth had, on average, a slightly lower birth weight and a slightly smaller head circumference; they were also on average slightly more premature than those born to women who passed along lower amounts. The severity of these deficits was larger the greater their prenatal exposure to PCBs. When tested at the age of seven months, many of the affected children displayed small difficulties with visual recognition memory, again with the extent of the problem increasing with prenatal transmission of PCBs. At the age of four years, the lower body weight observed at birth for highly exposed infants still lingered. More serious was the observation that the four-year-olds displayed progressively lower scores on several tests of mental functioning (with respect to verbal and memory abilities) the greater their prenatal PCB exposure (see Figure 14-7a).

At eleven years of age, the effects of the prenatal PCBs were still apparent: the IQ scores of the part of the group that had been the most highly

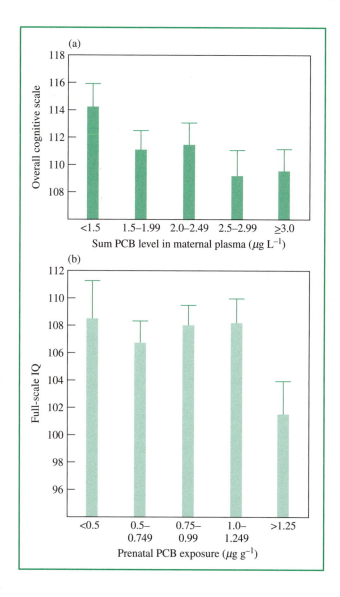

FIGURE 14-7 The effect on the intelligence of children of receiving PCBs prenatally: (a) overall cognitive abilities at 3½ years of age, (b) full-scale IQ at 11 years. [Sources: (a) S. Patandin et al., "Effects of Environmental Exposure to Polychlorinated Biphenyls and Dioxins on Cognitive Abilities in Dutch Children at 42 Months of Age," *Journal of Pediatrics* 134 (1999): 33. (b) J. L. Jacobson et al., "A Benchmark Dose Analysis of Prenatal Exposure to Polychlorinated Biphenyls," *Environmental Health Perspectives* 110 (2002): 393.]

exposed before birth averaged 6 points below the others; the most affected mental functions were memory and attention span. However, prenatal PCB exposure at any but the highest levels did not appear to have affected the IQ of the eleven-year-olds (see Figure 14-7b).

Interestingly, at both four and eleven years of age, the children's total body content of PCBs, which is determined mainly from the breast milk they consumed as infants and to a lesser extent by the fish in their diet rather than from any prenatal transmission of the chemicals, was *not* the relevant factor in determining these physical and mental deficits. Rather, it was the smaller amounts of PCBs that had been transmitted from mother to fetus that were important. **Thus PCBs appear to interfere with the proper prenatal development of the brain and with the mechanisms that ultimately determine physical size.**

The studies by the Jacobsons, and further investigations by other researchers discussed below, are the clearest examples available concerning the influence of toxic environmental chemicals on the health of human beings. It is important to realize that the highest levels of PCBs to which the children in this group were exposed before birth are not much greater than those to which the majority of unborn children in the general population were subjected a few decades ago. And while the chemicals did not produce gross birth defects in the children, they did result in small and consistent deficits of several kinds.

Some doubt has been cast on the Jacobsons' results from an analysis that pointed out the difficulties in establishing the original in utero exposures. However, more recent studies have overcome these analytical difficulties, since the ability to detect very low levels of chemicals such as PCBs in blood serum has improved dramatically over the past few decades. Negative effects on the IQs and/or psychomotor skills of young children by prenatal exposure to background levels of PCBs has been established by researchers in the Netherlands, Germany, Japan, and at several additional sites in the United States.

The various studies differ in the magnitude of the effect, in how long into childhood it lasts, and whether or not it affects only the more highly exposed children. Subtle cognitive deficits have also been found in children in northern Quebec, whose PCB concentrations are high due to the long-range air transport and subsequent deposition and entry into the food chain of the compounds (as discussed in Chapter 15). Overall, it is the *executive functions* of the brain—visual recognition memory, verbal ability, task planning and problem solving, abstract reasoning, information processing speed, etc.—that seem to suffer from prenatal exposure to PCBs.

Recent research has connected prenatal exposure to PCBs to long-lasting reductions in thyroid hormone function among Akwesasne Mohawk native peoples in North America. Dutch researchers have found that PCBs and dioxins transmitted to babies during gestation and via breast milk weakened their immune systems, contributing to more infections in the first few years of life.

The Akwesasne Mohawk Nation is located in an area spanning the St. Lawrence River in New York State and Ontario.

In summary, although PCBs were never intended to be ingested by humans, through their presence in the environment they have found their way into our bodies. Although not acutely toxic, we have seen that PCBs cause numerous health problems, thus illustrating another example of *unin-tended consequences* and the need for *systems thinking*.

The two terms in italics here were defined in Table 0-1.

14.16 The Toxicity Patterns of Dioxins, Furans, and PCBs

Research has shown that single doses of 2,3,7,8-TCDD administered to pregnant laboratory animals cause reproductive effects in their offspring. These results have raised some alarm about the potential effects of dioxins on human reproduction. In this connection, many scientists are worried about the dangers posed by environmental chemicals such as dioxins, furans, PCBs, and other organochlorines that can affect sex hormones, as discussed in Chapter 15.

Test results from studies on animals indicate that **the acute toxicity of dioxins, furans, and PCBs depends to an extraordinary degree on the extent and pattern of chlorine substitution.** The following generalization can be made: the very toxic dioxins are those with four beta chlorine atoms, and few if any alpha chlorines (see diagram on page 626 for definitions of the alpha and beta positions). Thus the most toxic is 2,3,7,8-TCDD, which has the maximum number (four) of beta chlorines and no alpha chlorines.

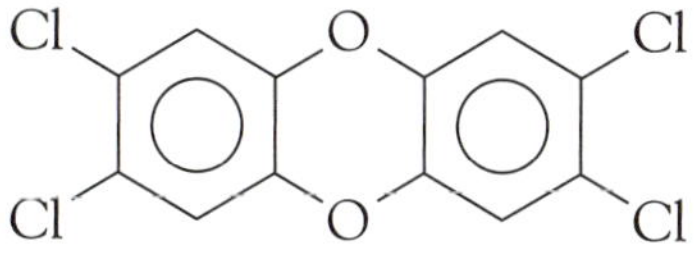

2,3,7,8-TCDD

Dioxin congeners that have three beta chlorines, but no (or only one) alpha chlorine, are appreciably toxic, but less so than the 2,3,7,8 compound. Fully chlorinated dioxin, i.e., octachlorodibenzo-*p*-dioxin (OCDD), has a very low toxicity since all the alpha positions are also occupied by chlorine.

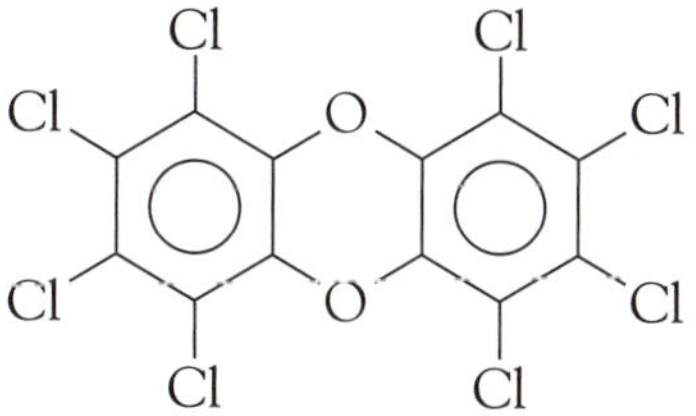

OCDD

Similarly, mono- and dichloro dioxins are usually not considered highly toxic, even if the chlorines are present in beta positions.

Predict the order of relative toxicities of the following three dioxin congeners, given that, for systems not too dissimilar to TCDD, the presence of an
alpha chlorine reduces the toxicity less than does the absence of a beta
chlorine:

2,3,7-trichlorodibenzo-*p*-dioxin

1,2,3-trichlorodibenzo-*p*-dioxin

1,2,3,7,8-pentachlorodibenzo-*p*-dioxin   ●

The toxicity pattern for furans is similar though not identical to that for
dioxins in that the most toxic congeners have chlorines in all the beta positions. However, the most toxic furan, the 2,3,4,7,8 congener, does have one
chlorine atom in an alpha position.

According to animal tests, the most acutely toxic PCBs are those having
no chlorine atoms (or at most one) in the positions that are ortho to the
carbons that join the rings, i.e., on the 2,2',6, and 6' carbons. Without ortho
chlorines, the two benzene rings can easily adopt an almost coplanar configuration, and rotation about the C—C bond joining the rings is rapid. However, because of the large size of chlorine atoms, they get in each other's way
if they are present in both ortho positions on the same side of the two rings;
this *steric* interaction forces the rings to twist away from each other, and
prevents such rings from adopting the *coplanar* geometry:

Consequently, PCB molecules with chlorines at three or four of the ortho
positions cannot adopt a coplanar geometry.

If the rings are not kept from coplanarity by interference between chlorine
atoms, and if certain meta and para carbons have attached chlorine atoms,
then the PCB molecule can readily attain a coplanar geometry that is similar
in size and shape to 2,3,7,8-TCDD. Such PCB molecules are found to be
highly toxic. Apparently 2,3,7,8-TCDD and other molecules of its size and
shape readily fit into the same cavity in a specific biological receptor; the complex of the molecule and the receptor can pass through cell membranes and
thereby initiate toxic action. By comparing molecular models, it is not difficult
to see, e.g., that the most toxic PCB, namely *3,3',4,4',5'-pentachlorobiphenyl*
is almost the same size and shape as 2,3,7,8-TCDD. It is believed that some of
the toxicity of the cooking-oil incidents arose from coplanar PCBs.

Only a very small fraction of commercial PCB mixtures correspond to
coplanar PCBs having no ortho chlorines. Although individually less toxic,

See Additional Problem 4
for a model-building
exercise to illustrate this
point.

PCBs with one ortho chlorine, and with chlorines in both para and at least one meta position, contribute substantially to the overall toxicity of PCB mixtures since they are far more prevalent than those having no ortho chlorines.

In humans, the more highly chlorinated furans, dioxins, and PCBs are stored in fatty tissues and are neither readily metabolized nor excreted. This persistence is a consequence of their structure: few of them contain hydrogen atoms on adjacent pairs of carbons at which hydroxyl groups, OH, can readily be added in the biochemical reactions that are necessary for their elimination. In contrast, those compounds with few chlorines always contain one or more such adjacent pairs of hydrogens and tend to be excreted after hydroxylation, rather than stored for a long time.

14.17 The TEQ Scale

Since most organisms, including humans, have a mixture of many dioxins, furans, and PCBs stored in their body fat, and since all these compounds act qualitatively in the same way, it is useful to have a measure of the *net* toxicity of the mixture. To this end, scientists often report concentrations of these organochlorines in terms of the equivalent amount of 2,3,7,8-TCDD that, if present alone, would produce the same toxic effect. **An international toxicity equivalency factor, or TEQ, has been devised that rates each dioxin, furan, and PCB congener's toxicity relative to that of 2,3,7,8-TCDD, which is arbitrarily assigned a value of 1.0.** Recently, *polybrominated biphenyls* have also been added to this scale.

A summary of the TEQ values for some of the more toxic dioxins, furans, and PCBs is given in Table 14-1. As an example, consider an individual who ingests 30 pg (picograms) of 2,3,7,8-TCDD, 60 pg of 1,2,3,7,8-PCDF, and 200 pg of OCDD. Since the TEQ factors for these three substances are, respectively, 1.0, 0.05, and 0.001, the intake is equivalent to

$$(30 \text{ pg} \times 1.0) + (60 \text{ pg} \times 0.05) + (200 \text{ pg} \times 0.001) = 33.2 \text{ pg}$$

Thus, even though a total of 290 pg of dioxins and furans were ingested by this person, the mixture is equivalent in its toxicity to an intake of only 33.2 pg of 2,3,7,8-TCDD.

PROBLEM 14-11

Using the TEQ values in Table 14-1, calculate the number of equivalent picograms of 2,3,7,8-TCDD that corresponds to an intake of a mixture of 24 pg of 1,2,3,7,8,9-hexachlorodibenzo-*p*-dioxin, 52 pg of 2,3,4,7,8-pentachlorodibenzofuran, and 200 pg of octachlorodibenzofuran.   ●

The TEQ values for environmental samples are sometimes reported in the media as if they represent the concentration of 2,3,7,8-TCDD itself. However, this compound often is not even the dominant contributor to the

TABLE 14-1	Toxicity Equivalence Factors (TEQ) for Some Important Dioxins, Furans, and PCBs	
Dioxin or Furan or PCB		**Toxicity Equivalency Factor**
2,3,7,8-Tetrachlorodibenzo-p-dioxin		1
1,2,3,7,8-Pentachlorodibenzo-p-dioxin		0.5
1,2,3,4,7,8-Hexachlorodibenzo-p-dioxin		
1,2,3,7,8,9-Hexachlorodibenzo-p-dioxin	}	0.1
1,2,3,6,7,8-Hexachlorodibenzo-p-dioxin		
1,2,3,4,6,7,8-Heptachlorodibenzo-p-dioxin		0.01
Octachlorodibenzo-p-dioxin		0.001
2,3,7,8-Tetrachlorodibenzofuran		0.1
2,3,4,7,8-Pentachlorodibenzofuran		0.5
1,2,3,7,8-Pentachlorodibenzofuran		0.05
1,2,3,4,7,8-Hexachlorodibenzofuran		
1,2,3,7,8,9-Hexachlorodibenzofuran	}	0.1
1,2,3,6,7,8-Hexachlorodibenzofuran		
2,3,4,6,7,8-Hexachlorodibenzofuran		
1,2,3,4,6,7,8-Heptachlorodibenzofuran	}	0.01
1,2,3,4,7,8,9-Heptachlorodibenzofuran		
Octachlorodibenzofuran		0.001
3,3',4,4',5-Pentachlorobiphenyl		0.1
3,3',4,4',5,5'-Hexachlorobiphenyl		0.01

TEQ toxicity. As discussed previously, combustion of organic matter produces relatively few toxic dioxins; the TEQ from such sources is often dominated by penta- and hexachlorinated furans. Similarly, the TEQ arising from use of chlorination in bleaching paper is dominated by toxicity from tetrachlorinated furans.

14.18 Dioxins, Furans, and PCBs in Food

Realize that *parts per trillion* and *picograms per gram* are identical scales.

About 95% of human exposure to dioxins and furans arises from the presence of the compounds in food. A bar graph showing the TEQ values for contamination of various types of foods purchased in U.S. supermarkets in the 1990s is shown in Figure 14-8.

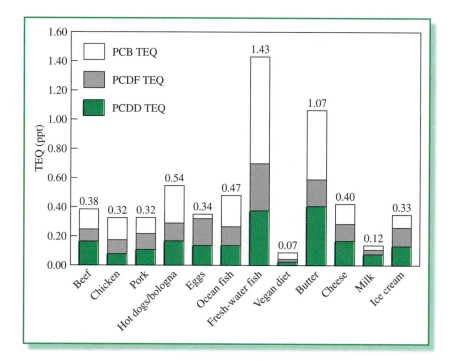

FIGURE 14-8 TEQ values of foods collected from U.S. supermarkets. [Source: A. Schecter et al., "Levels of Dioxins, Dibenzofurans, PCB and DDE Congeners in Pooled Food Samples Collected in 1995 at Supermarkets Across the United States," *Chemosphere* 34 (1997): 1437.]

Notice that fresh-water fish contained the highest levels of both PCB and furan toxicity. Recently, average TEQ levels of several ppt have been found for farmed salmon; these high levels originate with the fishmeal and fish oil the growing salmon had been fed. However, the TEQ levels in young chickens and turkeys and in hogs sold in the United States have declined significantly—by 20–80%—in recent years. The composite vegan diet (i.e., all vegetable, fruit, and grain, with no animal products at all) has a very low TEQ compared to that from animal-based components. A report by the U.S. Institute of Medicine recommended that girls should reduce their consumption of animal products in order to reduce the amount of dioxins that build up in their body fat and that could subsequently affect any children they might have.

PROBLEM 14-12

Given that the average TEQ of animal-based foods was about 0.4 pg of TCDD equivalent per gram when the data in Figure 14-8 were collected, and that the LD_{50} for 2,3,7,8-TCDD is about 0.001 mg per kg of body weight, what mass of animal-based food would you have had to consume to ingest a fatal dose of it?   ●

An incident of dioxin-contaminated food occurred in Germany in early 2011. Contaminated fat had been accidentally added to feed used on farms that raised chicken and turkey, pigs, dairy cattle, and other types of meat. Eggs from the farms were found to contain up to four times the European Union World Health Organization maximum TEQ for dioxins and furans of 3 pg g^{-1} of fat, and chicken meat with more than twice the maximum TEQ of 2 pg g^{-1}. The German government blocked further releases of food from the affected farms. The dioxin-contaminated fat was produced by a biodiesel company, and inadvertently sent to the feed company rather than used for other purposes.

In the mid-1980s, the average total concentration of all dioxins and furans in the fat tissue of adult North Americans was about 1000 ppt. However, because highly chlorinated and therefore less toxic congeners predominated, the TEQ value was much less: about 40 ppt of 2,3,7,8-TCDD equivalent. By the 1990s, the TEQ in human fat had fallen to about 15 ppt. The historical maximum TEQ levels, about 75 ppt, were reached in the 1970s. The variation with time of stored dioxins and furans fits a model in which the daily TEQ dose amounted to about 0.5 pg kg^{-1} body weight in the early decades of the twentieth century, rose to over 6 pg kg^{-1} in the 1940s to 1970s, and has now declined again to 0.5 pg kg^{-1}. The concentration of 2,3,7,8-TCDD itself in human lipids fell by a factor of about seven over the last quarter of the twentieth century in North America.

The average North American adult's body contains about 15 kg of fat, so his/her total body burden of 2,3,7,8-TCDD equivalents now amounts to about 0.2 μg. Given that the average residence time t_{avg} of dioxins and furans in the human body is about 7 years, and using the relationship from Chapter 5 connecting t_{avg} to the total amount C and the input rate R, i.e.,

$$t_{avg} = C/R$$

then the average human rate of intake of 2,3,7,8-TCDD equivalents is calculated to be

$$R = C/t_{avg} = 0.2\ \mu g/7\,y = 0.03\ \mu g\ y^{-1}$$

This value, which corresponds to about 1 pg kg^{-1} body weight per day, is close to the intake estimated over the last few years from the model discussed previously.

14.19 Dioxins as Carcinogens and Acute Toxins

Although there is little argument as to the *relative* acute toxicities of various dioxin and furan congeners, their *absolute* risk to humans is *very* controversial. The amount of 2,3,7,8-TCDD per kilogram body weight required to kill a guinea pig is extraordinarily small—less than 1 μg—making it the most toxic synthetic chemical known for that species. However, the LD$_{50}$ required to kill many other types of animals is hundreds or thousands of times this amount—e.g., the LOD_{50} for hamsters is 1200 $\mu g \, kg^{-1}$; for frogs, 1000 $\mu g \, kg^{-1}$; for rabbits and mice, 115 $\mu g \, kg^{-1}$; for monkeys, 70 $\mu g \, kg^{-1}$; and for dogs it may be as low as 30 $\mu g \, kg^{-1}$. A 2000 EPA report on dioxin suggests that humans fall in the middle range for acute susceptibility to dioxins; humans given doses of 100 $\mu g \, kg^{-1}$ suffered no apparent ill effects beyond chloroacne.

Another widespread exposure of humans to 2,3,7,8-TCDD occurred in the early 1970s in and around Times Beach, Missouri. Waste oil containing PCBs and 2,3,7,8-TCDD from the manufacture of 2,4,5-trichlorophenol was used for dust control on unpaved roads.

The 2,3,7,8-TCDD concentration in the waste oil was later found to be about 300 ppm.

Many horses died due to dioxin exposure in arenas where the dioxin contamination was particularly high since the facilities had been sprayed to keep the dust down. A decade later, widespread contamination of the soil in the town was discovered. In 1997, over 200,000 tonnes of soil from this town and 26 other affected sites in eastern Missouri that had soil 2,3,7,8-TCDD levels of 30–200 ppb were excavated and incinerated in order to remediate the problem. (The town of Times Beach was abandoned in 1982 due in part to a severe flash flood.) Although exposure to the chemical seems to have negatively affected their immune systems, a major study in 1986 did not find evidence of increased disease prevalence in the exposed group of Times Beach residents. Less formal studies and anecdotal evidence, however, indicate problems with seizures and congenital abnormalities, and so the issue remains controversial.

Scientists are more concerned about the long-term effects of exposure to dioxins than about their acute toxicities. The Seveso study discussed in Section 14.1 was the first to show an increased rate of cancer among people exposed accidentally to TCDD. An analysis of the health of American workers who were employed in industries that produced chemicals contaminated with 2,3,7,8-TCDD indicates that exposure to it at relatively high levels may cause cancer. The current theory concerning the action of dioxin predicts that there should be a threshold below which no toxic effects will occur, and recent studies of workers exposed to 2,3,7,8-TCDD supports this hypothesis.

In animal studies, a threshold of about 1000 pg (i.e., 1 ng) of 2,3,7,8-TCDD equivalent per kilogram of body weight per day is observed with respect to the cancer-causing ability of dioxins and furans. In determining the maximum tolerable human exposure to such compounds, many governments apply a safety factor of 100, resulting in a guideline for maximum exposure of 10 pg kg^{-1} day^{-1}, averaged over a lifetime. Currently, the average American ingests less than one-tenth this amount from animal fats in his/her food supply. Exposure levels near the guideline limit are expected for persons consuming large amounts of fish that have elevated dioxin and furan levels.

The U.S. EPA draft report of 2000 concerning the health risks of dioxins concluded that 2,3,7,8-TCDD is a (known) human carcinogen—although this characterization was a point of controversy among members of the expert committee that reviewed the report—and that the mixtures of dioxins to which people are exposed is a "likely human carcinogen." The *International Agency for Research on Cancer* of the *World Health Organization* had previously classified TCDD as a known human carcinogen. Experiments had indicated that 2,3,7,8-TCDD was the most potent multisite carcinogen known in test animals. The EPA estimates that the most sensitive and most highly exposed Americans stand at least a 1-in-1,000 chance of developing cancer from dioxin. The report notes that noncancer effects of dioxin are at least as important as cancer.

However, there is currently no clear indication that the level of cancer has risen in the general American population due to dioxins, though this could well be owing to the inability to relate effects to exposure at current levels. In addition to cancer, the report concludes that dioxins adversely affect the endocrine and immune systems and the development of fetuses, as will be discussed in Sections 15.8 and 15.11.

For furans, direct evidence of human susceptibility is available from the incidents of PCB-contaminated cooking-oil consumption mentioned above. The most common symptoms observed in these groups were chloracne and other skin problems. Unusual pigmentation occurred in the skin of babies born to some of the mothers who had been exposed. The children also often had low birth weight and experienced a rather high infant mortality rate. Children who directly consumed the oil showed retarded growth and abnormal tooth development. Many of the victims also reported aching or numbness in various parts of their bodies, and frequent bronchial problems. Other than chloracne, such symptoms are not observed in workers who are occupationally exposed to PCBs and whose body burdens of PCBs are comparable to those who consumed the contaminated oil, but in which the PCDF concentration is orders of magnitude lower. However, some children of these workers had mild cases of the less serious problems seen in the poisoned group.

Ukrainian president Viktor Yushchenko was apparently the victim of deliberate dioxin poisoning. Since September 2004, he has suffered from ulcers in his stomach and intestines, problems with his liver and spleen, and disfiguring facial cysts that have left him looking far older than he is (Figure 14-9).

14.20 Human Exposure to Dioxins, Furans, and PCBs

There continues to be vigorous debate in scientific, industrial, and medical communities regarding the environmental dangers of dioxins, furans, and PCBs. In one camp are those who feel that the dangers from these chemicals have been wildly overstated in the media and by some special-interest groups. They point to the very low concentrations of these substances that exist in the environment, to the possibility that there is a threshold below

FIGURE 14-9 Former Ukrainian president Viktor Yushchenko before and after he was poisoned with dioxin. [Source: AP Photo/ Efrem Lukatsky.]

which these compounds have no effect on human health, to the lack of known human fatalities resulting from them, and to the enormous economic costs associated with instituting further controls and cleanup measures. At the other extreme are persons who point to the substantial biomagnification and high toxicity per molecule of these substances and to their presence in almost all environments. They consider the detrimental effects such as cancer and birth deformities caused by these chemicals in wildlife to be "warning canaries" that signal potential ill effects in humans. Discovering where the truth lies between these opposing viewpoints presents a challenge even for environmental science students, to say nothing of the public at large!

Review Questions

1. Using structural diagrams, write the reaction by which 2,3,7,8-TCDD is produced from 2,4,5-trichlorophenol.

2. Draw the structure of 1,2,7,8-TCDD. What is the full name for this dioxin?

3. What, chemically speaking, was *Agent Orange*, and how was it used?

4. Draw the structure of *pentachlorophenol*. What is its main use as a compound? What is the main dioxin congener that it could produce?

5. What does *PCB* stand for? Draw the structural diagram of the 3,4′,5′-trichloro PCB molecule.

6. What were the main uses for PCBs? What is meant by an *open use*?

7. Draw the structure of a representative polychlorinated dibenzofuran congener.

8. Other than the chlorophenols and PCBs, what are some of the other sources of dioxins and furans in the environment? What is currently the largest anthropogenic source of dioxins?

9. From what medium—air, food, or water—does most human exposure to dioxin come about? Why is this so?

10. What molecules can be eliminated by a PCB molecule when it is heated to moderately high temperatures?

11. Are PCBs acutely toxic to humans? What is the basis for health concerns about them? Recount recent evidence that shows PCBs can affect human development.

12. Are all dioxin congeners equally toxic? If not, what pattern of chlorine substitution leads to the greatest toxicity? Which is the most toxic dioxin?

13. What is meant by a *coplanar PCB*? What structural features give rise to non-coplanarity?

14. What does *TEQ* stand for? Why is it used?

15. Is dioxin carcinogenic to humans or not? Discuss the evidence for and against.

Green Chemistry Questions

1. The development of TAML catalyst for hydrogen peroxide oxidation by Terry Collins won a Presidential Green Chemistry Challenge Award.

(a) Which of the three focus areas (see page xxviii) for these awards does this award best fit into?

(b) List at least three of the twelve principles of green chemistry (see pages xxiii–xxiv) that are addressed by the green chemistry developed by Collins.

2. What environmental advantages does the TAML/hydrogen peroxide method of bleaching pulp have over the use of elemental chlorine?

Additional Problems

1. Deduce which combination(s) of two different tetrachlorophenol isomers would produce the following hexachlorodibenzo-*p*-dioxins: (a) the 1,2,3,7,8,9 isomer, (b) the 1,2,4,6,8,9 isomer, and (c) the 1,2,3,6,7,9 isomer. [*Hint: See Box 14-1.*]

2. Deduce which dioxins would likely result from the low-temperature combustion of a commercial sample of PCP.

3. In the purification of wastewater contaminated by pentachlorophenol and 2,3,5,6-tetrachlorophenol using ultraviolet light, it was noticed that OCDD and 1,2,3,4,6,7,8-heptachlorodibenzo-*p*-dioxin were formed. Deduce whether the latter was formed by the coupling of a molecule of each of the phenols, or rather must have been formed by photochemical dechlorination of OCDD. What potential flaw exists

in treating water by UV light, given the nature of the products that are formed?

4. Using mechanical ball-and-stick or computer-generated molecular models, construct structures for (a) 2,3,7,8-TCDD, (b) dibenzofuran, and (c) biphenyl. Place chlorines onto the dibenzofuran and biphenyl models so that the space filled by the carbon, oxygen, and chlorine atoms overlaps that of the dioxin as much as possible without occupying much of the space associated with the alpha positions. Do the resulting congeners represent the most toxic furan and biphenyl according to TEQ values?

5. Consider the PCDF shown below. Deduce which PCBs could produce this furan if they are moderately heated in air, given that PCDFs can result from HCl elimination with or without 2,3 interchange, or from Cl_2 elimination. [*Hint:* See Box 14-2.]

6. By comparing the average dioxin levels in humans with those in our food, decide whether or not dioxin is biomagnified in the transition. Given the food typically eaten by domestic animals such as cattle and chickens compared with that of the fresh-water fish we eat, can you explain why dioxin TEQ levels in Figure 14-8 for the fish exceed those for the animals? Why is the TEQ value for butter so much higher than that for milk, and that for hot dogs greater than the general levels for meat? Why is the TEQ value for a vegan diet so very low? Given the vegan diet TEQ and that for the domestic animals, can you predict whether

biomagnification occurs for the latter relative to their diet?

7. Predict the order of toxicity of the three tetrachlorobiphenyls with the following numbering: (a) 2,4,3',4'; (b) 3,4,5,4'; (c) 2,4,2',6'. [*Hint:* Will these PCB molecules be coplanar?]

8. In the potential reactions of chlorinated phenols undergoing incineration in separate facilities, what dioxins could be formed from (a) 2,5-dichlorophenol, and (b) 2,4,6-trichlorophenol? Name the two dioxins, and predict which would be the more toxic.

9. The partitioning of PCBs among air, water, and sediments can be estimated by the fugacity model discussed in Chapter 13. The Z values for a typical environmental PCB are 4×10^{-4} in air, 0.03 in water, and about 10,000 in sediment (and biota). Using the model world volumes in Chapter 13, calculate the equilibrium concentrations when 1 mol of PCBs is distributed among air, water, and sediment.

15

Other Toxic Organic Compounds of Environmental Concern

In this chapter, the following introductory chemistry topics are used:

- ➲ Elementary organic chemistry (see online Appendix)
- ➲ Concept of vapor pressure

Background from previous chapters used in this chapter:

- ➲ Maximum Contaminant Level (MCL)
- ➲ Structures of dioxins, furans, and PCBs
- ➲ Structures of DDT, DDE, and atrazine
- ➲ Concepts of bioaccumulation; K_{ow}
- ➲ Concepts of free-radical and chain reactions
- ➲ LD_{50}

Introduction

In this chapter, we shall look at a series of toxic organic compounds that do not contain chlorine but have become common air and water pollutants. We begin by considering PAHs, pollutants that accompany the combustion of most natural organic materials, and then discuss a wide range of environmental chemicals—most of which have been encountered in other contexts—that may have disruptive effects on our reproductive systems. We then consider the surprising mechanism by which persistent substances become distributed around the world by air currents. Finally, we survey two new classes of commercial products that are causing environmental concern, and are being transported to remote regions by this mechanism. In all these discussions, it should be kept in mind that, given the amazing ability of modern analytical chemistry to quantify very small concentrations of chemicals, the simple detection of a substance in human blood or urine may or may not have implications for the health of the individual.

Polynuclear Aromatic Hydrocarbons (PAHs)

One of the most common and ancient types of environmental pollutant, whether in air, water, or soil, is a large series of hydrocarbons known as PAHs. We first discuss the molecular structure of these compounds, then relate their occurrence in the environment and their health effects on humans.

15.1 The Molecular Structure of PAHs

There is a series of hydrocarbons whose molecules contain several six-membered benzene-like rings connected by the sharing of a pair of adjacent carbon atoms between adjoining **fused** rings. The simplest example is **naphthalene,** $C_{10}H_8$:

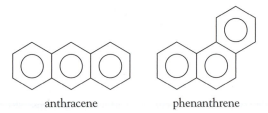

naphthalene

Notice that there are ten, not twelve, carbon atoms in total and that there are only eight hydrogen atoms, since the shared carbons have no attached hydrogen atoms. As a compound, naphthalene is a volatile solid whose vapor is toxic to some insects. It has found use as one form of "moth balls," the other being *1,4-dichlorobenzene.*

Conceptually, there are two ways to fuse a third benzene ring to two carbons in naphthalene; one results in a linear arrangement for the centers (the "nuclei") of the rings while the other is a "branched" arrangement:

anthracene   phenanthrene

The resulting molecules, **anthracene** and **phenanthrene,** are PAH pollutants arising from incomplete combustion, especially of wood and coal. They are also released into the environment from the dumpsites of industrial plants that convert coal into gaseous fuel, and from the refining of petroleum and shale. In rivers and lakes, they are found mainly attached to sediments rather than dissolved in the water; both are subsequently partially incorporated by fresh-water mussels.

PROBLEM 15-1

Draw the full structural diagram for phenanthrene, showing all atoms and bonds explicitly. ●

PROBLEM 15-2

By determining their molecular formulas, show that the molecules below are not additional isomers of $C_{14}H_{10}$:

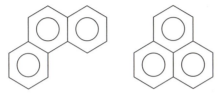

PROBLEM 15-3

Using a systematic procedure, deduce the structural formulas for the five unique isomers of $C_{18}H_{12}$, which contain four fused benzene rings. ●

In general, hydrocarbons that display benzene-like properties are called **aromatic; those that contain fused benzene rings are called *polynuclear* (or polycyclic) *aromatic hydrocarbons*, or PAHs for short.** Like benzene itself, many PAHs possess unusually high stability and a planar geometry. Other than naphthalene, they are not manufactured commercially. However, some PAHs are extracted from coal tar and used in commerce.

15.2 PAHs as Air Pollutants

PAHs are common air pollutants and are strongly implicated in the degradation of human health. Typically, the concentration of PAHs in urban outdoor air amounts to a few nanograms per cubic meter, although it reaches 10 times this amount in very polluted environments. **PAHs are formed when carbon-containing materials are incompletely burned.** Elevated PAH concentrations in indoor air are typically due to the smoking of tobacco and the burning of wood and coal.

The physical state and average airborne lifetime of PAHs depends significantly on their mass and on the ambient temperature, since their lifetime is determined largely by their vapor pressure:

• **PAHs containing four or fewer rings usually exist as gases** if they are released into air, since the vapor pressures of their liquid form is relatively high. After spending on average less than a day in outside air, such PAHs are degraded by a sequence of free-radical reactions that begin, as expected from our previous analysis of air chemistry (Chapters 3 and 17), by the addition of the OH radical to a double bond.

• In contrast to their smaller analogs, PAHs with more than four benzene rings do not exist for long in air as gaseous molecules. Owing to their low vapor pressure, **large PAHs condense and become adsorbed onto the surfaces of suspended soot and ash particles.** In winter, even small PAHs adsorb onto particles, since their vapor pressures decrease sharply at lower temperatures. PAHs adsorb mainly on particles of submicron, i.e., respirable, size; consequently, they can be transported into the lungs by breathing.

By the turn of the millenium, the concentrations of PAHs in rural air in eastern North America were found to have declined by an order of magnitude since reaching their peak in 1985, presumably due to air pollution abatement programs in Canada and the United States. In contrast, the concentrations of PAHs in sediments laid down in urban lakes increased substantially over that period.

PAHs are introduced into the environment from a number of sources: the exhaust of gasoline and especially diesel combustion engines, the "tar" of cigarette smoke, the surface of charred or burned food, the smoke from burning wood or coal, and other combustion processes in which the carbon of the fuel is not completely converted to CO or CO_2. Although PAHs constitute only about 0.1% of airborne particulate matter, their existence as air pollutants is of concern since many of them are carcinogenic, at least in test animals. Vehicle exhaust, especially from diesel engines, older gasoline-powered cars and all vehicles in which the engine has not warmed up are the major contributors to PAH levels in cities. Aluminum smelters are a source of PAHs since their heated graphite anodes deteriorate over time, releasing the hydrocarbons.

PAHs are released into the air along with, and/or adsorbed onto, macroscopic-sized particles of soot (black smoke) from diesel engines and above fires undergoing incomplete combustion. **Soot itself is mainly graphite-like carbon;** it consists of collection of tiny crystals (crystallites), each composed of stacks of planar layers of carbon atoms, all of which occur in fused benzene rings. **Graphite** is the ultimate PAH: its parallel planes of fused benzene rings each contain a vast number of carbon atoms.

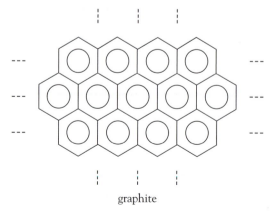

graphite

There are no hydrogen atoms in graphite except at the periphery of the layers. The surfaces of soot particles are excellent adsorbers of gaseous molecules, especially organic ones.

A survey in Taiwan of PAH emissions from exhaust stacks found the following trends for various types of restaurants, the differences presumably arising from the style of cooking employed:

Chinese >> Western > fast-food > Japanese

PAH levels in the air were particularly high near the World Trade Center site in New York for many days after they had been destroyed in September 2001. However, the heat generated by the flames carried most of

Black smoke rising from incomplete combustion in a 2006 field fire in Norfolk, England. (© James Dobson/Alamy)

the smoke aloft. In addition to PAHs, the high temperatures and anaerobic conditions within the piles of debris resulted in the oxidation by chlorine of metals and organic substances, producing a wide variety of contaminants.

15.3 PAHs as Water Pollutants

Polycyclic aromatic hydrocarbons also are serious water pollutants. PAHs enter the aquatic environment as a result of spills of oil from tankers, refineries, and offshore oil drilling sites. In drinking water, the PAH level typically amounts to a few parts per trillion, and usually is an unimportant source of these compounds to humans. The larger PAHs bioaccumulate in the fatty tissues of some marine organisms and have been linked to the production of liver lesions and tumors in some fish. PAHs, PCBs, and the insecticide *mirex* are thought to play a role in the devastation of the populations of beluga whales in the St. Lawrence River; because discharges of these pollutants to the river have decreased substantially in recent years, the health of the belugas may eventually improve.

PAHs are generated in substantial quantity in the production of such coal-tar derivatives as *creosote*, a wood preservative used especially on railway ties. Up to 85% of the 200 compounds in creosote are PAHs, including some carcinogenic ones. In addition to concerns about PAH exposure to workers installing the ties, gardeners who buy used railway ties for landscape design and people burning discarded ties are at risk. The leaching of PAHs from the creosote used to preserve the immersed lumber of fishing docks etc., represents a significant source of seawater pollution to crustaceans such as lobsters.

15.4 Formation of PAHs During Incomplete Combustion

The mechanism of PAH formation during combustion of organic materials is complex, but is due primarily to the repolymerization of hydrocarbon fragments that are formed during the **cracking** (i.e., the splitting into several parts) of larger fuel molecules in the flame. Fragments containing two carbon atoms are particularly prevalent after cracking and partial combustion have occurred. Two C_2 fragments can combine to form a C_4 free-radical chain, which could add another C_2 to form a six-membered ring. Such reactions occur quickly if one of the original C_2 fragments is itself a free radical.

The repolymerization reaction—rather than complete combustion—occurs particularly under oxygen-deficient conditions. Generally, **the PAH formation rate increases as the oxygen-to-fuel ratio decreases.** The fragments often lose some hydrogen, which forms water after combining with oxygen during the reaction steps. The carbon-rich fragments combine to form the **polynuclear aromatic hydrocarbons, which are the most stable molecules that have a high C-to-H ratio.**

Since neither methane, the main component of natural gas, nor methanol molecules contain any C—C bonds with which to form C_2 units, their combustion as fuels produces very little PAH or other soot-based particulates.

15.5 Carcinogenic Properties of PAHs

The most notorious and common carcinogenic PAH is **benzo[*a*]pyrene,** BaP, which contains five fused benzene rings:

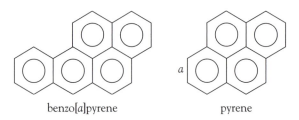

benzo[*a*]pyrene   pyrene

The molecule is named as a derivative of **pyrene,** which has the structure shown above at the right. Conceptually, if an additional benzene ring is added at the bond of pyrene labeled *a*, the benzo[*a*]pyrene molecule is obtained.

Benzo[*a*]pyrene is a common by-product of the incomplete combustion of fossil fuels, of organic matter (including garbage), and of wood. It is a carcinogen in test animals, and a probable human carcinogen. BaP is worrisome since it bioaccumulates in the food chain: its log K_{ow} value is 6.3, comparable to that of many organochlorine insecticides (see Table 13-3). It is considered to be one of the top 15 organic carcinogens in drinking water, as listed in Table 15-1, along with the other compounds in this category, most of which are pesticides.

TABLE 15-1	Top Organic Carcinogens in U.S. Drinking Water*		
Chemical		MCL in ppb	Cancer Risk per 100,000 People
Ethylene dibromide		0.05	12.5
Toxaphene		3	9.6
Vinyl chloride		2	8.4
Heptachlor		0.4	5.2
Heptachlor epoxide		0.2	5.2
Hexachlorobenzene		1	4.6
Benzo[a]pyrene		0.2	4.2
Chlordane		2	2.0
Carbon tetrachloride		5	1.9
1,2-Dichloroethane		5	1.3
PCBs		0.5	0.5
Pentachlorophenol		1	0.3
Di(2-ethylhexyl) phthalate		6	0.2
Benzene		5	0.2–0.8
Dichloromethane (methylene dichloride)		5	0.1

Source: A. H. Smith et al., "Arsenic Epidemiology and Drinking Water Standards," *Science* 296 (2002): 2146.

*Chemicals ordered by their cancer risk, were they to be consumed at their MCL (maximum contaminant level).

A second example of a carcinogenic PAH is the four-ring hydrocarbon **benz[a]anthracene,** which is equivalent conceptually to anthracene with an additional benzene ring fused to the *a* bond:

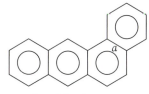

benz[a]anthracene

The relative positions in space of the fused rings in PAHs play a major role in determining their level of carcinogenic behavior in animals. The

PAHs that are the most potent carcinogens each possess a bay region formed by the branching in the benzene ring sequence: the organization of four carbon atoms as an open bay indirectly imparts a high degree of biochemical reactivity to the PAH, as explained in Box 15-1.

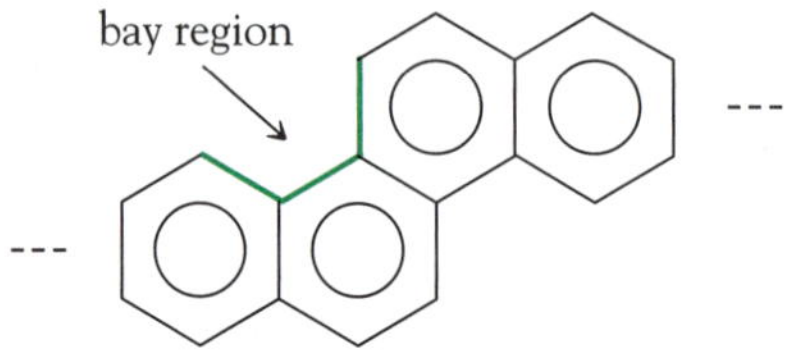

bay region

Thus benzo[*a*]pyrene is expected to be carcinogenic since it has a bay region, whereas pyrene itself is not since it does not possess one.

BOX 15-1   More on the Mechanism of PAH Carcinogenesis

Research has established that the PAH molecules themselves are not carcinogenic agents; rather, they must be transformed by several metabolic reactions in the body before the actual cancer-causing species is produced.

The first chemical transformation that occurs in the body is the formation of an epoxide ring across one C=C bond in the PAH. The specific epoxide of interest to the carcinogenic behavior of benzo[*a*]pyrene is

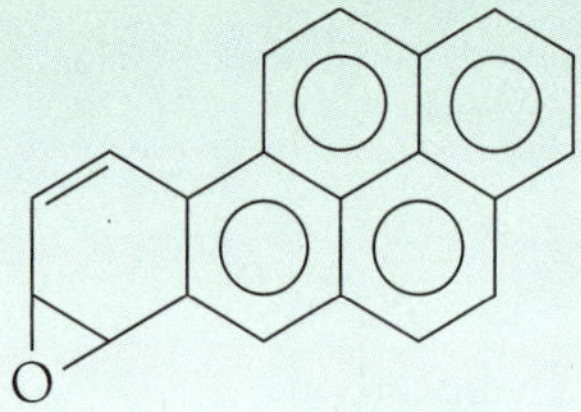

A fraction of these epoxide molecules subsequently add H_2O, to yield two —OH groups on adjacent carbons:

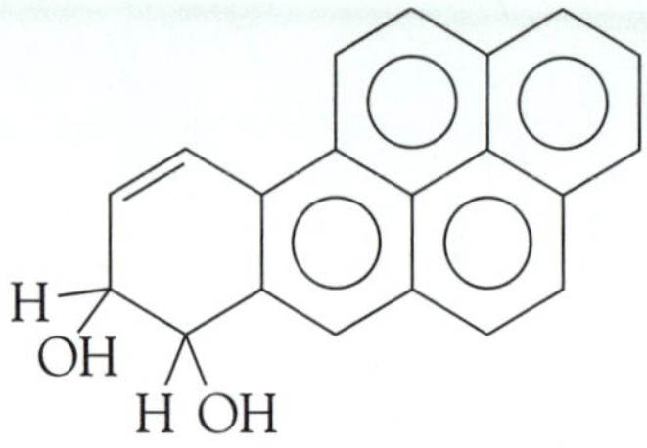

The double bond (shown in the structure) that remains in the same ring as the two groups subsequently undergoes epoxidation, yielding the molecule that is the active carcinogen:

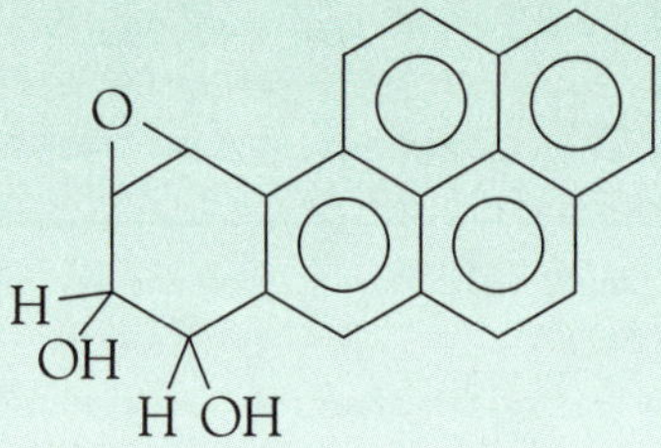

By adding H^+, this molecule can form a particularly stable cation that can bind to molecules such as DNA, thereby inducing mutations and cancer.

The metabolic reactions of epoxide formation and H_2O addition are part of the body's attempt to introduce —OH groups into hydrophobic molecules like PAHs to make them more easily dissolved in water and then eliminated. For BaP and other PAHs that possess a bay region, one of the intermediate products in this multistep process can be diverted instead into the formation of a very stable cation that induces cancer.

PROBLEM 15-4

Based upon the bay region theory, would you expect naphthalene to be a carcinogen? How about anthracene or phenanthrene? What about the PAH called benzo[*ghi*]perylene, whose structure is shown below?

benzo[*ghi*]perylene

Some PAHs with certain of their hydrogen atoms replaced by methyl groups are even more potent carcinogens than are the parent hydrocarbons.

15.6 Environmental Levels of PAHs and Human Cancer

Has exposure to PAHs been demonstrated to produce cancer in humans? The answer is both yes and no. For over 200 years it has been known that prolonged exposure in occupational settings to very high levels of *coal tar*, the principal toxic ingredient of which is benzo[*a*]pyrene, leads to cancer in humans. In 1775 the occurrence of scrotal cancer in chimney sweeps was associated with the soot lodged in the crevices of the skin of their genitalia. Modern workers in coke oven and gas production plants likewise experience increased levels of lung and kidney cancer due to this PAH.

Coal tar is the viscous black liquid produced by the destructive distillation of coal.

The evidence for cancer induction in the general public, whose exposure to PAHs is at levels that are orders of magnitude lower than in these occupational environments, is less clear-cut. The main cause of lung cancer is the inhalation of cigarette smoke, which contains many carcinogenic compounds in addition to PAHs; the deduction from health statistics of the much smaller influences of air pollutants such as PAHs from other sources is difficult to accomplish. Some scientists speculate that the higher death rate from lung cancer in cities as compared to rural areas, found in many countries, is due in part to breathing carcinogenic air pollutants like PAHs, although other factors such as a higher smoking rate also contribute.

Many cities in developing countries have chronic problems with carbon-based particulate air pollution. For example, the serious indoor and outdoor air pollution, which arises primarily from the unvented burning of coal and biomass for cooking and heating and consists primarily of PAHs, sulfur dioxide, and particulate matter, is reputed to be responsible for over 1 million deaths annually in China. The rate of lung cancer in Chinese women is higher than that for men, possibly due to higher exposures to PAHs from coal burning and from cooking-oil fumes.

See also the discussion in Section 4.16.

Diesel engine exhaust (unfiltered) has been labeled a "probable human carcinogen." It contains not only PAHs but also some of their derivatives containing the *nitro group*, —NO_2, as a substituent; these nitrated substances are even more active carcinogens than are the corresponding PAHs. For example, *nitropyrene* and *dinitropyrene* are responsible for much of the mutagenic character of diesel exhaust, i.e., its ability to cause mutations that could ultimately produce cancer. These compounds are formed within the engines by the reaction of pyrene with NO_2 and N_2O_4. There is also evidence that PAHs are nitrated by some of the constituents of photochemical smog.

As discussed in Chapter 3, the emission of particulates from heavy-duty diesel engines can be controlled by means of filter traps in the exhaust stream. These devices temporarily retain the solids, including soot, in the exhaust and eventually oxidize them further. Some scientists have worried that, during their residence in the traps, PAHs could undergo reaction to produce even greater quantities of nitrated PAHs. While this apparently does occur, tests indicate that the total mutagenic activity of a given amount of engine exhaust is actually decreased by the devices, since most of it becomes oxidized.

For most nonsmokers in developed countries, by far the greatest exposure to carcinogenic PAHs arises from their diet, rather than directly from polluted air, water, or soil. As expected from their mode of preparation, charcoal-broiled and smoked meat and fish contain some of the highest levels of PAHs found in food. However, leafy vegetables such as lettuce and spinach can constitute an even greater source of carcinogenic PAHs due to the deposition of these substances from the air onto the leaves of the vegetables while they are growing. Unrefined grains also contribute significantly to the total amount of PAHs ingested from food.

Review Questions 1–4 are based upon the material in the sections above.

Environmental Estrogens

In the last two decades, a new threat to the health of wildlife, and possibly of humans, exposed to synthetic organic chemicals in the environment has been identified. It has been established that certain compounds can affect the reproductive and immune health of higher organisms and may also increase the rate of cancer in reproductive organs. Much of the public interest in this issue was stirred by the 1996 publication of the book *Our Stolen Future* by Theo Colburn and her associates. However, there remains great uncertainty and debate in the scientific community about whether or not there are significant risks to human health from environmental levels of these compounds.

15.7 Mechanism of Action of Environmental Estrogens

The chemicals in question interfere with the system in the organism that operates by transmitting extremely low concentrations—at the parts-per-trillion level—of chemical messenger molecules called **hormones.** The hormones flow through the bloodstream from the point of production and storage to their target organs, including those involved in sexual reproduction in both females and males. The arrival of the hormone at a receptor is a signal for the cell to initiate action of some type. Much of the concern about humans centers upon interference with **estrogens,** the female sex hormones (which also are present in males, but at a smaller concentration). There also exist environmental substances that interfere with **androgens,** the male sex hormones, and with thyroid hormones. Sex hormones, including estrogens and androgens, contain the characteristic four-ring steroid structure (see top of Figure 15-1). They are produced from cholesterol in the ovaries of females and the testes of males in response to signals from the brain and other organs.

Hormone actions are initiated by the binding of the hormone to a specific receptor within a cell. The resulting hormone–receptor complex binds to specific regions of DNA in the cell nucleus, an action that in turn determines the action of the genes. Certain environmental chemicals can also bind to the hormone receptor, and thereby either mimic or block the action of the hormone itself. In particular, the "promiscuous" estrogen receptor will bind to a number of compounds, even ones that bear little structural resemblance to estrogen.

However, most such compounds bind to the receptor with only a small fraction of the strength of estrogen itself. Estrogen itself binds to its main receptor by hydrogen bonding from its two —OH groups and by attractive van der Waals forces from its ring system to amino acid side chains. A second estrogen receptor has recently been discovered. In addition to the interfering mechanisms discussed above, some compounds can accelerate the breakdown of natural hormones, and it has recently been established that some promote the conversion of male hormones into female ones.

As a class, substances that interfere with the **endocrine system** of hormone production and transmission are often referred to as **environmental estrogens** (or *ecoestrogens*). More general names for this class of substances have also been suggested: synthetic *hormonally active agents* (HAAs), *endocrine disrupting chemicals* (EDCs), *environmental hormones*, and *xenoestrogens*.

15.8 The Chemicals That Operate as Environmental Estrogens

Although most ecoestrogens operate by attaching themselves to, or by blocking access to, the hormone receptor, there is not a strong resemblance in overall structure between synthetic substances that have been identified as hormonally active and the natural sex hormones, nor is there much structural similarity of the synthetic ones to each other. It is true that many molecules identified as environmental estrogens contain one or more hydroxyl groups, as does estrogen. Oxygen-containing organochlorine insecticides that are known to act hormonally include *methoxychlor* and *kepone* (Figure 15-1).

However, other environmental estrogens are organochlorine molecules, including some *PCBs, dioxins,* and insecticides (Figure 15-1) that contain no oxygen atoms. In metabolizing them, however, the body itself attaches

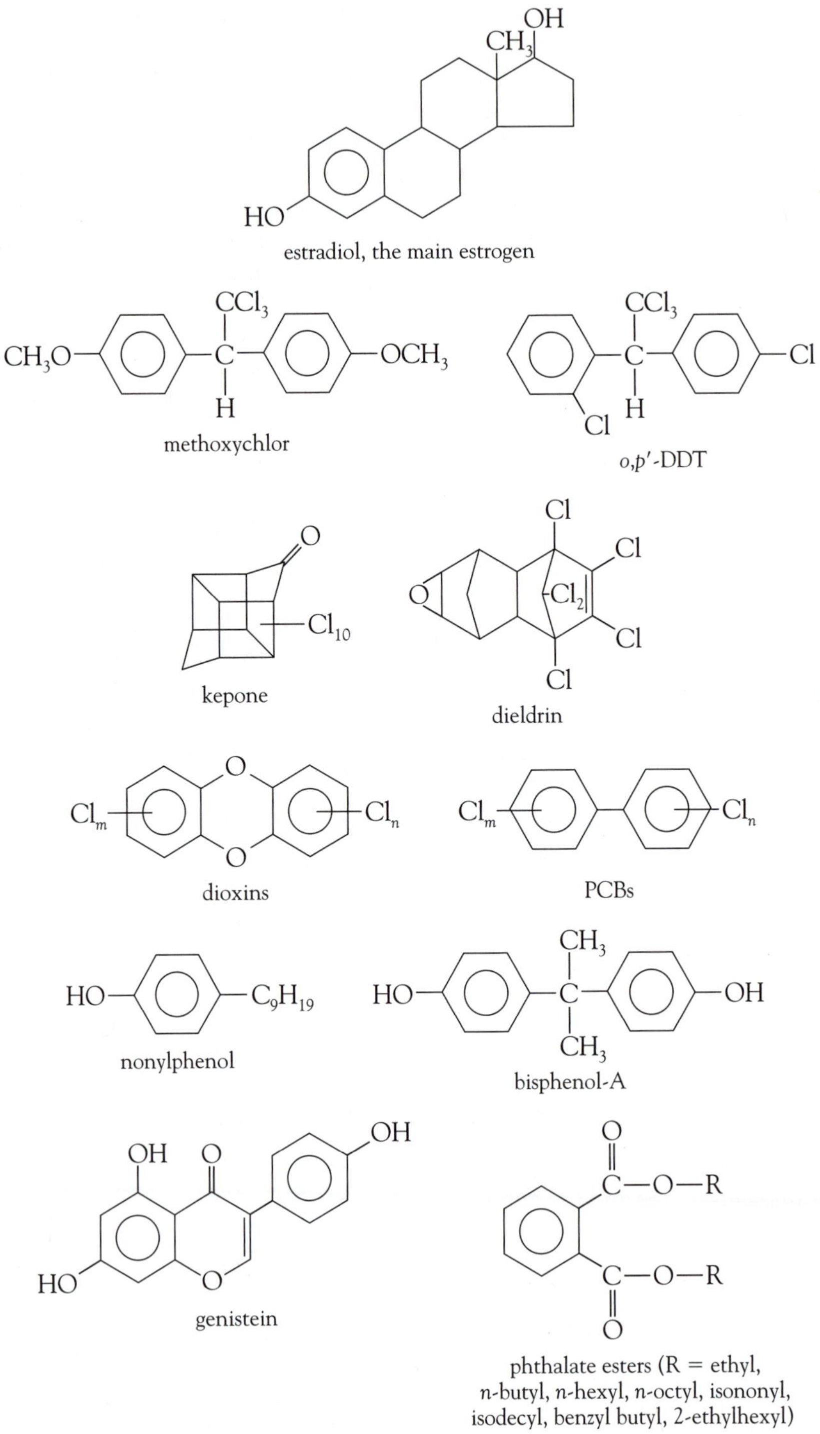

FIGURE 15-1 The structures of estradiol and some environmental estrogens.

FIGURE 15-1 The structures of estradiol and some environmental estrogens.

estradiol, the main estrogen

methoxychlor

o,p'-DDT

kepone

dieldrin

dioxins

PCBs

nonylphenol

bisphenol-A

genistein

phthalate esters (R = ethyl, n-butyl, n-hexyl, n-octyl, isononyl, isodecyl, benzyl butyl, 2-ethylhexyl) hydroxyl groups to some of the nonchlorinated carbon atoms in such molecules. Some of the resulting hydroxylated organochlorines are more hormonally active agents than the original compounds. The same is true with polycyclic aromatic hydrocarbons when they are hydroxylated.

Most nonchlorinated environmental estrogens are alcohols, often with the phenolic ring that is present in the steroidal hormones. **Nonylphenol** (Figure 15-1) is one important example. **Octylphenol** (which has a C_8 rather than a C_9 chain attached to phenol ring) is even more hormonally active. Both these alkylphenols occur in the environment—including drinking water—as a result of the breakdown, for example in sewage treatment plants, of larger *ethoxylate* molecules. The ethoxylates are used in detergents, spermicides, paints, and some plastics. They are also commonly used as emulsifying agents in pesticide formulations, and the nonylphenols produced by their decomposition enter the food chain via sprayed fruits and vegetables. The European Union is moving toward banning the use of ethoxylates, as Norway has done already for applications for which alternatives exist.

A phenolic environmental estrogen of great concern is bisphenol-A; it is discussed at some length in Box 15-2.

BOX 15-2 | **Bisphenol-A**

One of the most controversial industrial chemicals still in production is bisphenol-A **(BPA)**; the prefix bis means "two" and is used here to signify that two phenol rings are connected together:

bisphenol A

BPA is a widely used substance that is polymerized industrially into the *polycarbonate* plastics employed to make lightweight water and baby bottles, as well as other products such as CDs and DVDs that are clear, glass-like, and almost shatterproof. Its presence is identifiable by the number 7 inside the recycling triangle on the plastic products. The concern arises from its behavior as an environmental estrogen.

Bisphenol-A is currently one of the highest volume synthetic chemicals—about 3 billion tonnes are produced annually. Some *epoxy resins* made from derivatives of short bisphenol-A polymer chains are used as linings in food and beverage cans. In forming the polycarbonate polymer, each of the OH groups is reacted with dichlorinated formaldehyde. After HCl eliminations, the bisphenol-A units are joined together via C=O bridges; the O_2C=O unit is formally a carbonate and hence the name of the polymer:

polycarbonate structure

(continued on p. 676)

Bisphenol-A is a controversial chemical, with very different views regarding its potential estrogenic effects in humans being taken by the plastics industry and some of the scientists whose research they support financially compared to those of some academic scientists who have performed tests with the compound on animals. To help resolve the issue, the U.S. National Institute of Environmental Health Sciences has initiated a $30 million research project, mainly consisting of laboratory studies.

The environmental concern arises from the small leakage of bisphenol-A monomer from the mass of the polycarbonate plastic into the liquid contained in the bottle, and from the resins of food and beverage cans lined with epoxy resins into the contents of the can. Additional bisphenol-A is leached if food containers made using the polymerized resin are heated for sterilization purposes. The plastics industry stated in the past that migration of bisphenol-A from plastic food containers does not occur under normal cleaning and use conditions, although some consumer groups claimed evidence that some, albeit small, amount of leaching from polycarbonate plastics such as baby bottles does occur. BPA could potentially leach from dental sealants that are made from resins prepared from it, though the initial evidence that this occurs is now in doubt. Bisphenol-A-based resins are also used to coat carbonless copy paper and cash-register receipts and in some plastic PVC water pipes. Biodegradation over time of discarded polycarbonate products also liberates bisphenol-A.

Urine testing indicates that the great majority of North Americans have been exposed to bisphenol-A, although most of it ingested by humans is metabolized within hours. A 2010 survey found bisphenol-A detectable in the urine of 91% of Canadians aged 6 through 79, similar to the results found previously for Americans. The average level of BPA for Canadians was 1.2 μg L^{-1}, with about one-third of the population having more than double this value. The levels found in younger age groups were higher than for the older cohorts in both the Canadian and American studies. Indeed, teenagers were found to have the highest levels in the Canadian study. Since BPA is known to have a short half-life in the body (5–6 hours) and its excretion by urination is rapid, the high frequency of its detection means that exposure to it by most people is continual and widespread.

An interesting, if small-scale, experiment was performed in 2008 by researchers using as subjects a group of Harvard College students to determine the fraction of BPA that arises from bottles. The 77 students drank cold beverages only from stainless steel (and hence BPA-free) containers and avoided polycarbonate drinking water dispensers for a week, at which point their urine was analyzed. The students then switched to drinking most cold liquids only from (supplied) polycarbonate bottles. The average urinary concentration of BPA had increased by about two-thirds by the end of the second week. This small-scale experiment indicates that polycarbonate drinking bottles were a major source, but not the only significant source, of BPA for the students. In contrast, a 2011 survey of preschool children found that most of their BPA exposure came from solid rather than liquid food.

Test results indicate that BPA is present in soda pop in cans at a level of about 0.5 ppb, with somewhat higher levels in energy drinks. Given that the current safety limits are in the range of 25–50 μg of BPA per kilogram of body weight per day, an adult would have to drink thousands of cans of pop per day to consume an overdose.

Some recent studies on older American adults have linked high levels of blood BPA to increased risk of cardiovascular disease and adult-onset diabetes. Exposure during pregnancy increases the likelihood of giving birth to daughters who display behavioral problems, according to one recent study. Animal studies have found that the chemical can act to modify the expression (though not the sequencing) of genes.

Babies are not able to make the same level of the enzyme that metabolizes BPA in the liver as do adults, and therefore its half-life in infants is greater than in adults. There is some concern from animal studies that fetal and infant exposure to BPA could lead to changes in the brain and to behavior and to an eventual increased risk of prostate cancer. To protect babies, in 2009 Canada became the first country to ban the sale of polycarbonate infant feeding bottles. The European Union has also instituted such a ban. The main American manufacturers of baby bottles have voluntarily moved away from using BPA in baby bottles sold in the United States. Indeed, many plastic consumer products now carry a *BPA FREE* tag.

Another hormonally active diol is **genistein,** a molecule whose structure (Figure 15-1) has similarities to that of estrogen. Genistein is produced naturally by plants, rather than being a synthetic substance. It is a member of a group of natural chemicals called *flavonoids*. Genistein is found in significant concentrations in wood products and in soybeans and soy-based food products. Researchers have discovered that genistein from pulp-mill effluent may cause feminizing and other reproductive effects in fish that swim in the effluent-fed waters. Both nonylphenol and genistein have been found to produce effects at very low doses in test animals. It is not clear yet whether genistein overall has a positive or a negative effect on human health. However, some scientists are worried about the levels of genistein ingested by babies fed with soy milk.

The PCB congeners commonly found in environmental samples and identified recently by screening as showing some, albeit weak, estrogenic activity all have at least one ortho chlorine atom.

15.9 Phthalates

Phthalate esters (Figure 15-1) are widely used as plasticizers in common plastics such as polyvinyl chloride (PVC), from which they can leach into the environment, since they are not chemically bonded to the polymer. The most important example is **di-2-ethylhexyl phthalate** (DEHP), which is present in many plastics found around the home—including some used by children—where phthalate plasticizers can constitute up to 45% of the weight of the object. Examples include shower curtains and raincoats, plastic bags and garden hoses. However, the use of phthalates in baby-bottle nipples and teethers has long been discontinued.

Phthalates are also found in cosmetics and other personal-care products such as shampoos and deodorants. The presence of DEHP in plastic medical devices such as intravenous PVC bags is also of concern since leakage of small amounts the plasticizer into the patient occurs during medical procedures. In the United States, neither DEHP nor other phthalates are used in

The nature of polymers and plastics is discussed in Section 16.11.

The potential health effects of phthalates are discussed in Section 15.11.

food wrap or food packaging. However, diet has been found to be an important source of phthalates for humans.

The European Commission has banned the use of phthalate softeners in PVC toys meant for children under 3 years of age, since such children tend to suck and chew on these toys, particularly rattles and teethers, and would thereby extract and ingest some of the phthalate esters. However, a scientific panel convened by the U.S. Consumer Product Safety Commission concluded that the most common plasticizer used in PVC toys, **diisononyl phthalate,** does not pose a risk to humans. Nevertheless, some companies now are producing phthalate-free toys and other products used by small children. Both the U.S. and Canadian governments have recently begun to regulate maximum levels of phthalates in toys and child-care articles.

The compounds discussed above are thought to be the most significant environmental estrogens uncovered to date. However, the estrogenic compounds in many synthetic substances are not yet known. The U.S. EPA in 1999 began an extensive process of screening potential endocrine disruptors.

In the popular press, phthalates are sometimes known as "rubber duck chemicals."

15.10 Effects of Environmental Estrogens on Wildlife

The most devastating consequences of environmental estrogens commonly are not observed in the mammals that originally ingest them. Rather, these actions result from their transfer from the mother to the fetus or egg. Their presence disrupts the hormone balance in the recipient and causes reproductive abnormalities or produces changes that will result in cancer when the offspring grows to adulthood. During its development, the fetus is particularly sensitive to fluctuations in hormone concentrations. For that reason, exposure to low concentrations of natural or environmental hormones can result in physiological changes that do not occur in adults who are exposed at the same levels.

Abnormalities in the reproduction and/or development of frogs, seals, polar bears, mollusks, as well as several types of birds, have been linked to exposure of the fetuses to endocrine disrupting chemicals.

The most famous example of the environmental effects of hormone-like chemicals upon wildlife involves alligators in Lake Apopka, Florida. In 1980, massive amounts of DDT and its analogs were spilled into the lake. In the mid-1980s, Professor Louis Guilette, Jr., of the University of Florida at Gainesville found that very few alligator eggs were hatching—and few hatchlings survived of those that were born—thereby threatening the future population of the colony. Furthermore, the eggs that did hatch produced alligators with abnormal reproductive systems; therefore they were unlikely themselves to be able to reproduce. The ratio of natural estrogen to the male sex hormone testosterone was greatly elevated in the young alligators. Presumably as a consequence, the penises of male alligators were reduced in size compared to the norm. Apparently these effects were caused by DDE, the metabolite of DDT (see Section 13.6), which has been found by research to inhibit binding of male hormones to their receptor.

Another Florida-based example involves the nearly extinct Florida panthers that live in the Everglades. Their reproductive difficulties may result from the consumption of raccoons whose levels of DDE and other endocrine disruptors are high as a consequence of their consumption of contaminated fish. Some researchers have linked reproductive problems, such as embryo mortality and deformities, of birds in the Great Lakes area to the hormonal activity of pollutants such as PCBs and dioxins.

Tests have revealed that the most estrogenic component of commercial DDT is not the main ($\sim$75%) ingredient, the p,p'-DDT isomer (Chapter 13), but rather the minor (15–20%) o,p' isomer (Figure 15-1), which has one of the ring chlorines in the ortho position and one in the para position. Some scientists have speculated that women who are directly exposed to o,p'-DDT by spraying may be at much higher risk of subsequently developing breast cancer than those in the developed world, whose main exposure instead is through DDE in their diets.

Research has indicated that environmental concentrations, ppb levels, of the herbicide *atrazine* (Chapter 13) could modify the balance of hormones in just-hatched frogs and thereby affect their sexual development. The effect of atrazine was to increase the levels of an enzyme that converts the male hormone testosterone to the female one, estrogen. About 20% of the male tadpoles developed into hermaphrodites, having both testes and ovaries. Because of this, atrazine may be contributing to the worldwide decline in the population of the amphibian. This hormonal action indicates yet another way, in addition to attaching to or blocking a hormone receptor, that environmental chemicals could upset hormonal balances. A report in 2008 found that frogs that had been turned into females by atrazine gave birth exclusively to males in the next generation. However, research reported from other groups failed to duplicate the earlier findings at very low atrazine concentrations, so the issue of whether or not this is a significant environmental problem remains controversial and is as yet unresolved.

Researchers have also found that both natural estrogen, secreted by women as part of their monthly cycles, and the synthetic derivative used in birth control pills are present in wastewater and sewage effluent, and can cause feminization of males in some species of fish. Such feminization was encountered in the 1990s in some British waterways, and was initially blamed on the presence of nonylphenol discharges in the water. A survey of estrogens in coastal waters found much higher levels in shallow bays that receive input from sewage than in the open oceans.

15.11 Effects of Environmental Estrogens on Humans

The areas of human health that are considered to be of potential risk from exposure to environmental hormones are

- reproduction,

- neurobehavior,

- **immune function, and**

- **cancer.**

Much of the human evidence concerning the possible effects of estrogen mimics on developing fetuses was obtained from the experience of women who took the synthetic estrogen *DES* (diethylstilbestrol) in the 1948–1971 period to prevent miscarriage. Many of the daughters of these women are sterile, and a small fraction of them have developed a rare vaginal cancer, as a consequence of their prenatal exposure to DES. The male offspring have an increased incidence of abnormalities in their sexual organs, have decreased average sperm counts, and may have an increased risk of testicular cancer, as a consequence of this exposure, but their fertility is not affected.

Although there is good evidence that high concentrations of environmental estrogens have caused reproductive problems in wildlife and laboratory animals, it is not certain that comparable effects occur in humans at levels to which we are exposed. In the early 1990s, a connection between the rise in environmental contamination by endocrine disruptors and an apparent increase in certain male disorders of the reproductive system was postulated. In particular, the decline in male sperm counts and quality, the increase in the rate of testicular cancer, and the increase in incidence of male reproductive problems in newborns were cited.

However, the changes in the frequency of these human reproductive problems found in some locations may not, in fact, be general phenomena. Both sperm counts and testicular cancer rates vary significantly between geographical regions, and the variations are neither worldwide nor apparently closely linked to differences in pollution levels. Similarly, no clear picture has emerged relating trends in exposure to environmental estrogens to human fertility or the rates of spontaneous abortion, though an association has been found between delayed conception and exposure to high concentrations of environmental contaminants.

Perhaps the most dramatic effects of an environmental hormone on human reproduction is the effect of dioxins and furans in influencing the male:female sex ratio, i.e., the ratio—normally about 0.51—of boys to girls at birth. Epidemiological evidence from the dioxin-exposed group in Seveso, Italy, and from the Taiwanese PCB-poisoned-oil group (both discussed in Chapter 14), indicates that males who were exposed to high concentrations of dioxins, or to furans, dioxins, and PCBs, respectively, during adolescence are in adulthood much less likely to father boys than girls. Most studies of men exposed *as adults* to high levels of dioxins did not show a similar result, and the same is largely true for men who were over age 20 and for all females at the time of the Taiwanese incident. However, research from Austria indicates that workers occupationally exposed to TCDD when under 20 years of age later fathered many more girls than boys, whereas initial dioxin exposure at a later age had no effect on the sex ratio.

Overall, the male:female sex ratio has been dropping in the United States and Japan for decades for reasons that are not clear. The most skewed ratio (almost 1:2) found to date occurs in a native community near Sarnia, Ontario, a region surrounded by petrochemical plants.

Thus it appears that exposure during puberty to environmental hormones from the dioxin-furan-PCB family of compounds can have measurable lifelong effects on subsequent reproduction characteristics of human males. Interestingly, higher-than-normal exposure by males to industrial PCB mixtures through the diet can result in a *higher* male:female sex ratio, perhaps because some PCB congeners exhibit estrogenic and some antiestrogenic or androgenic behavior, the dominant effect depending upon the ratio of one to the other. In addition to PCB congeners, particularly those having one ortho and two para chlorines, other environmental compounds that have been found to interfere with the androgen receptor include the common DDT metabolite *p,p'-DDE*, and the fungicide *vinclozolin*. Recent research indicates that environmental estrogens in humans can affect sexual characteristics: girls who were exposed prenatally to high levels of DDE reach puberty on average almost a full year before those with the lowest exposures. Boys were not affected in this way.

The effects to neurological development of prenatal exposure to PCBs cited in Chapter 14 may arise from disruption to thyroid hormones, but further research is required to establish this connection more definitely.

Concern about human health effects from exposure to phthalates has increased over the past few years. They have been found to adversely affect the development of the male reproductive system in research on laboratory animals since they display antiandrogenic action. Of particular interest—and controversy—has been research linking phthalate metabolite levels in human mothers to the incomplete development of the sex organs in male infants born to them, possibly due to prenatal exposure to phthalates. An expert panel convened by the U.S. government found this research inconclusive, and recommended that the surveys be repeated and expanded. A subsequent report linked phthalate levels in human breast milk to abnormal levels of reproductive hormones in infant boys.

Some researchers have postulated a connection between the increasing exposure of the general population to endocrine disrupting chemicals and the rising incidence of cancers at hormonally sensitive sites, including the breast, the uterus, the testes and the prostate gland. A recent review of the medical literature concluded that the "overall strength of a causal association is weak" between the two, though "there is not enough information to completely reject the hypothesis" that endocrine disruptors could play a role.

Some scientists discount *any* adverse health effects of synthetic chemicals acting as environmental estrogens by pointing out that we all ingest much greater quantities of plant-based estrogen mimics called **phytoestrogens,** including genistein. Common sources of these natural chemicals include all soy products, broccoli, wheat, apples, and cherries. Indeed, there is some evidence that phytoestrogens have a *protective* effect against some types of cancers. It is true that phytoestrogens are quickly metabolized by the body and perhaps do not survive long enough to exert effects on a developing fetus, for example, whereas many synthetic environmental estrogens are stored in body fat rather than being metabolized. However, the current average dioxin level in humans is less than an order of magnitude greater than the lowest dose found by experiments to cause reproductive problems in the offspring of rats.

The greatest difficulty that scientists face in trying to discover whether environmental estrogens affect human health significantly is the lack of exposure data for the suspected substances. Thus, the jury is still out regarding whether environmental estrogens pose a substantial threat to humans or not.

Another puzzle in the environmental estrogen story relates to their dose–response behavior. In contrast to most toxic substances for which the response increases with the dose, eventually leveling off, the response curves for action by estrogen and estrogen mimics have an inverted-U shape (see Figure 15-2). The greatest effects are produced at low dosages; large dosages shut down the responding system to some extent. For example, the effects of atrazine on frogs are found by some researchers to operate only at very low concentrations, not at higher levels.

As an added complication, some estrogenic compounds are found to increase estrogenic activity in some tissues and block it in others! The relevance of the findings discussed above and many other unanswered questions in the environmental estrogen story will only be sorted out once much more research is done in this fascinating new chapter concerning the effects of low environmental levels of toxic organic chemicals upon living organisms.

Review Questions 5–9 are based upon the material in the sections above.

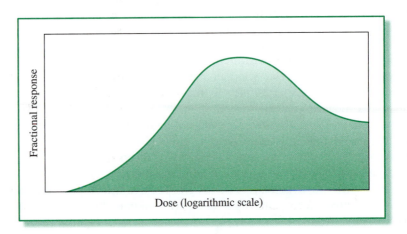

FIGURE 15-2 Dose–response curve (schematic) for estrogen and its mimics.

The Long-Range Transport of Atmospheric Pollutants

At first glance, it seems amazing to discover that relatively nonvolatile organochlorines and PAHs can eventually travel thousands of kilometers by air from their point of release and end up contaminating relatively pristine areas of the world such as the Arctic. Some quantitative understanding of this **long-range transport of atmospheric pollutants** (LRTAP) has been made using principles of physical chemistry.

15.12 The Migration in Air of Organic Pollutants

According to the *global distillation* (or *fractionation*) *hypothesis*, pollutants travel at different rates and are deposited in different geographical regions depending upon their physical properties. **Most persistent organic pollutants have sufficient volatility to evaporate—often rather slowly—at normal environmental temperatures from their temporary locations at the surface of soil or water bodies.** However, because the vapor pressure of any chemical increases exponentially with temperature, **evaporation is favored in tropical and semitropical areas, so these geographic regions are rarely the *final* resting places for pollutants.** In contrast, cold air temperatures favor the condensation and adsorption of gaseous compounds onto suspended atmospheric particles, most of which are subsequently deposited onto the Earth's surface. Thus **the Arctic and Antarctic regions are the final resting places for relatively mobile pollutants that are not deposited at lower latitudes because of their high volatility.** Unfortunately, these compounds degrade even more slowly in these regions because temperatures there are so cold.

Examples of pollutants that migrate to polar regions are the highly *chlorinated benzenes*; PAHs having three rings; and PCBs, dioxins, and furans that have only a few chlorines (see Table 15-2). Equilibrium has been established with these compounds in temperate areas but not yet in polar regions, as reservoirs in warm climates continue to produce new supplies to travel to the north and be condensed. Substances with even greater volatility than these, such as naphthalene and the less chlorinated benzenes, do not deposit even at the cold temperatures of polar regions; consequently they continue their worldwide travels more or less indefinitely until they are chemically destroyed, usually by reaction initiated by the hydroxyl radical.

The mechanisms by which OH reacts with organic compounds are discussed in Chapters 3 and 17.

As implied in Table 15-2, the mobility of a chemical increases as the vapor pressure of its condensed form (as measured by that of the supercooled liquid at 25°C) increases. In addition, mobility increases as the temperature of condensation of the vapor form of the pollutant gas decreases. Thus, substances that do not condense until the temperature drops to $-30°C$ or lower eventually accumulate in polar regions, where such air temperatures are common, according to the distillation hypothesis. Substances having

TABLE 15-2	Predicted Mobilities of Persistent Airborne Pollutants			
Global Transport Behavior	Low Mobility	Relatively Low Mobility	Relatively High Mobility	High Mobility
Property				
Vapor pressure of liquid at 25°C in pascals*	10^{-4}		10^{-2}	1
Condensation temperature	30°C		−10°C	−50°C
Examples				
PAHs	>4 rings	4 rings	3 rings	1–2 rings
Chlorobenzenes	—	—	5–6 Cl	0–4 Cl
PCBs	9–8 Cl	4–8 Cl	1–4 Cl	0–1 Cl
PCDDs/PCDFs	4–8 Cl	2–4 Cl	0–1 Cl	—
Pesticide examples	Mirex	DDT Toxaphene Chlordane	HCB Dieldrin Hexachloro-cyclohexane	Moth balls

Source: Adapted mainly from F. Wania and D. Mackay, "Tracking the Distribution of Persistent Organic Pollutants," *Environmental Science and Technology* 30 (1996): 390A–396A.

*For the supercooled liquid.

condensation temperatures below −50°C remain airborne indefinitely, since not even polar regions sustain such temperatures for long.

DDT is an intermediate case on these transport scales. It does evaporate sufficiently rapidly (supercooled liquid vapor pressure is 0.005 Pascals), but its relatively high condensation temperature of 13°C (55°F) means that much of it becomes permanently deposited at mid-latitudes (especially in the winter) and only a small fraction of it migrates to the Arctic.

Although PCBs are predicted by the model to deposit mainly in temperate areas rather than migrating *en masse* to the Arctic, the migration that does occur is sufficient that animals there are quite contaminated by these chemicals. The world record for PCB contamination, 90 ppm, is held by polar bears in Spitsbergen, Norway. Even breast milk is higher in PCBs for women who live in far northern areas than in more temperate ones, a result partially of their high-fat diet, since organochlorines are known to accumulate in such a medium.

Although the concentration of POPs in air over Arctic regions had been declining for decades owing to the implementation of restrictions on their worldwide use, a partial reversal has begun. Global warming has resulted in the re-evaporation of some POPs from their sinks in Arctic water and ice, and increased their concentrations in air.

PROBLEM 15-5

DDE has a 25°C vapor pressure (for its supercooled liquid) of 0.0032 Pa and a condensation temperature of −2°C. Is DDE more volatile than DDT or less? Predict whether a larger or smaller fraction of the fraction that does vaporize will be deposited at polar latitudes compared to DDT itself. ●

Owing to the variations in air temperature during their transport, most molecules of mobile pollutants experience several successive cycles of evaporation and condensation as they migrate gradually toward colder climates. This "grasshopper effect" is illustrated in Figure 15-3 for a pulse of a relatively mobile pollutant that was emitted near the equator at time t_0. At a later time t_1, the majority of the pollutant mass is still present in tropical regions, but at a subsequent time t_2 it has moved mainly to the subtropics. Whether it eventually ever moves ("hops") from temperate and subpolar regions to polar ones (at a later time t_6) depends upon whether or not its mobility is sufficiently high. The grasshopper effect by which chemicals migrate from the regions where they are produced and used, to far northern and southern regions of the globe is yet another example of the *tragedy of the commons*.

In 2010 an alternative to the *global distillation hypothesis* was proposed by some Swiss and British scientists. Their *differential removal hypothesis* states that fractionation of organic chemicals in the atmosphere results from different rates of loss of the substances as they travel in the air away from their

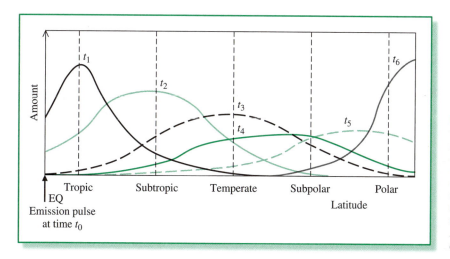

FIGURE 15-3 Calculated variation with time in the geographic distribution of an airborne pollutant released at the Equator (EQ). [Source: F. Wania and D. Mackay, "Tracking the Distribution of an Airborne Pollutant," *Environmental Science and Technology* 30 (1996): 390A.]

point of emission. The important processes that remove a chemical from the air, and thereby prevent its further transport, are

The partition coefficients were defined in Section 13.8.

- chemical reaction, especially with the OH free radical, and

- deposition onto land or water, in accordance with the appropriate partition coefficients such as K_{ow} and K_{oa}.

Consequently, substances that react readily with OH and/or have low volatility will be removed from air before those of lesser reactivity and/or more volatility, and thus travel shorter distances. The authors claim that their removal hypothesis explains the observed fractionation of PCBs in European air better than does the distillation hypothesis with its emphasis on temperature gradients.

Review Questions 10–12 are based upon the material in the sections above.

Fire Retardants

Highly brominated organic compounds are common commercial fire retardants. Most are physically mixed with, rather than chemically bound to, organic polymeric materials; since they are in molecular form, they can and do migrate, albeit slowly, from the products. Large amounts of these fire retardants have been used worldwide, and because of their persistence, they now are accumulating in the environment and have even been detected in the Arctic, to which they presumably migrated by the LRTAP mechanism discussed above. Based upon animal studies, they may have potential for liver toxicity, thyroid-hormone-level disruption, and effects on reproduction and development.

15.13 Fire-Retardant Mechanisms for Brominated Compounds

Many brominated organic compounds function as fire retardants because, when heated to 200–300°C—the approximate temperature range in which many polymers begin to decompose—**they release free bromine atoms, which react with the free radicals of combustion and thereby quench any fire.** Indeed, the brominated compounds chosen for use as fire retardants have decomposition temperatures that lie just below those of the polymers that they protect. At the high temperatures of fire, the compounds will readily vaporize to give denser-than-air gases that also will help extinguish the flames by depriving them of oxygen.

For example, when a molecule of such a retardant has absorbed sufficient energy from the fire and released Br atoms from one or more of its C—Br bonds, the atom may react with one of the free H atoms associated with the free-radical mechanism of combustion:

$$H + Br \longrightarrow HBr$$

The hydrogen bromide molecule so formed may subsequently react with a free hydroxyl radical, which otherwise would also have continued to propagate the combustion:

$$OH + HBr \longrightarrow H_2O + Br$$

The net reaction of these two steps is the indirect formation of a water molecule by the reaction of H and OH, thereby reducing the concentration of highly reactive free radicals in the fire process. Since the bromine atom is reformed, it plays a catalytic role. In this way, energy is withdrawn from the propagation mechanism of the combustion process, and the fire is quenched.

See Section 2.13 for a discussion concerning the bromine-containing halon fire retardants.

Iodine organic compounds would be even more effective in withdrawing energy, but since the C—I bond is so weak, they decompose at too low a temperature to be useful. Fluorinated compounds are generally unsuitable as fire retardants because most C—F bonds are so strong that free fluorine atoms would not be released, and because once formed, HF molecules are so stable that they would not participate in further reaction.

15.14 PBDEs: A New Type of Persistent Pollutant

Brominated diphenyl ethers, especially at levels of 5–30%, are incorporated into polyurethane foam, textiles, ready-made plastic products, and certain electronic equipment to prevent them from ever catching fire. They are found in common domestic products such as carpet padding, mattresses, curtains, and upholstered couches and chairs.

From a conceptual viewpoint, the molecular structure of *diphenyl ether* is analogous to biphenyl, except that an ether oxygen atom joins the two benzene rings. Bromine atoms can occupy any of the 10 other positions on the rings, analogous to the chlorine atoms in PCBs, again giving 209 possible congeners. Molecules with fewer than four bromines are generally not present in commercial mixtures.

diphenyl ether   PBDE-47

As a class, these compounds are known as *polybrominated diphenyl ethers*, PBDEs. The congeners are numbered by a system similar to that used for PCBs, in which the higher the code number, the greater the number of halogen atoms present.

PBDE molecules are of particular concern because some migration of them has occurred from their commercial products into the environment, where they now are widely distributed. Like PCBs, they are persistent and

PBDE-209

lipophilic, they bioaccumulate, and some of them are toxic. PBDEs have been detected in U.S. sewage sludge (much of it destined to be spread on agricultural land), in some fish caught in the wild, and even in sperm whales, which normally feed only in deep ocean waters. Also like PCBs, the commercial products are mixtures of congeners, not pure compounds, although the number of congeners present in each product is relatively small. Unlike most PCBs, PBDEs are solids under ambient conditions rather than liquids.

The *acute toxicity* of PBDEs *decreases* as the number of bromines per molecule increases. Consequently, the *least* acutely toxic PBDE is the fully brominated congener **decabromodiphenyl ether,** known as PBDE-209.

PBDE-209 was the almost exclusive ingredient (> 97%) in *Deca,* the commercial mixture that became the predominant PBDE product on the market, and which was used as the flame retardant in plastic components in computers and TV housings. Some scientists suspect that the PBDEs in Deca, while not themselves highly toxic, may degrade in the environment by loss of some bromine, thereby *increasing* the toxicity of the mixture, since PBDEs having intermediate bromine content are more toxic than is decabromodiphenyl ether. Indeed, there is growing evidence that PBDE-209 can undergo debromination by photochemical decomposition in sunlight, by reduction with elemental iron, and by metabolic processes in fish such as carp and rainbow trout and in rats. It is also debrominated anaerobically in sewage sludge if certain other chemicals are also present.

PROBLEM 15-6

Deduce the structures of (a) the three unique PBDEs formed by loss of one bromine atom, and (b) the fifteen PBDEs formed by loss of two bromine atoms, from a molecule of decabromodiphenyl ether. Assume that the two rings cannot rotate relative to each other around the intermediate oxygen atom. ●

PBDE-47, the tetrabromo compound that was one of the two most abundant in *Penta,* is the dominant congener found in human blood. Its structure was illustrated in the text above.

The product called *Penta* is a mixture mainly of PBDEs having four or five bromine atoms. It was used as a fire retardant in polyurethane foams, such as those used in furniture upholstery and padding in vehicles. The Penta mixture constitutes up to 30% by mass of some polyurethane foams, products that easily deteriorate by outdoor weathering and break into small, easily transportable fragments that can eventually find their way into natural waters. From this source, PBDE molecules can enter the aquatic food chain. Indeed, it is tetra- and pentabromodiphenyl ethers that are most widely distributed in the environment and that are the most bioaccumulative and the most acutely toxic. The commercial PBDE product called *Octa* consists mainly of congeners with 6 or 7 bromine atoms, and is used in thermoplastics.

PBDEs with a large number of bromines (more than 6 Br atoms per molecule) probably do not bioaccumulate because they are not readily incorporated by organisms; instead they bind to particles and accumulate in sediments. However, congeners with 4–6 bromines are taken up by organisms and have the potential to biomagnify in the food chain. They may in the future represent a danger to human health due to our exposure to them in food, especially fish.

Although human exposure to PBDEs occurs through the food we eat, inadvertent digestion of and contact with household dust appears to have been a more important source, especially in North America, for those with high exposure. The much higher blood levels of PBDEs found in Americans and Canadians compared to the concentrations in Europeans may be due in part to the stricter fire regulations in North America and exposure to the consequent higher levels of PBDEs in dust. One small-scale study found a definitive link between PBDE levels in human breast milk and their levels in the dust of the women's homes. Small children are estimated to have a much higher exposure to PBDEs than adults as a consequence of greater dust ingestion resulting from increased contact with floors and "mouthing" activities.

The concentrations of PBDEs in human blood, milk, and tissues had risen exponentially for three decades, until at least the early 2000s, with a doubling time of about 5 years. Though little data is available, the concentration of PBDEs in human breast milk is known to have risen sharply in the 1990s in both U.S. and Canadian women—by more than a factor of 10 from 1982 to 2002 for the latter—and is approaching that of PCBs, though the levels in milk samples from European women are much lower. A study reported in 2011 involving women in California found a 100-fold variation in PBDE levels in their breast milk, with PBDE-47 the dominant congener in most samples but PBDE-209 highest in others.

Cats in North America have been found to have high levels of PBDEs in their blood, probably owing to their continuous grooming habits, by which they could ingest the substances from house dust.

PROBLEM 15-7

The concentration of PBDEs in herring gull eggs from the Great Lakes was about 1100 ppb in 1990, and about 7000 ppm in 2000. What is the doubling time for the PBDEs in this source? If past trends continue, what will be the concentration in 2010? [*Hint:* For exponential growth Ae^{kt}, the doubling time is equal to $0.69/k$.]

The main human health concern about PBDEs is not their acute toxicity but the fact that those having relatively few bromine atoms may affect liver and thyroid hormone systems and interfere with neurodevelopment.

Because of environmental concerns, the European Union banned the Penta and Octa products effective in 2006. The sole North American manufacturer of Penta and Octa voluntarily ceased production of these products at the end of 2004. Deca, the remaining PBDE product, will be banned in the United States after the end of 2013. Its use in electrical and electronic equipment is severely restricted or banned in Canada and European countries. **In 2009, the United Nations Environmental Program added tetra-, penta-, hexa- and heptabromodiphenyl ethers, plus hexabromobiphenyl, to their list of *Persistent Organic Pollutants*.**

15.15 Other Brominated Fire Retardants

Two non-PBDE brominated organic compounds are also widely used as fire retardants. Indeed, the retardant of largest usage volume of all is TBBPA, **tetrabromobisphenol-A,** a compound composed of molecules in which all four carbon atoms that stand ortho to the two hydroxyl groups of bisphenol-A (see Figure 15-1) have been brominated:

TBBPA

Commonly, TBBPA is incorporated *chemically* into the structure of polymers by covalent coupling at the two hydroxyl groups; retardants that are covalently bonded to polymers are called *reactive* ones. Printed circuit boards are a major reactive use for TBBPA. When incorporated into materials in this manner, retardants are much less likely to leach or volatilize into the environment compared to those such as PBDEs, which are only physically dissolved in materials and are said to be *additive* substances. In some uses, however, TBBPA is also used as an additive rather than as a reactive retardant. The compound itself is not very toxic (its LD_{50} is several grams per kilogram), and although it has been found in some biota, it decomposes in air, water, and sediment in weeks or a few months.

Another important brominated fire retardant is the half-brominated cyclic hydrocarbon **hexabromocyclododecane** (HBCD), which is used primarily as an additive in polystyrene foams used in building materials and upholstery, though it is also finding new applications in replacing PBDEs. It is not used as a reactive retardant, since it contains no reactive groups that can bond to the polymer chain. Like TBBPA, it is of low acute toxicity. However, it now is a ubiquitous contaminant in the environment, and undergoes biomagnification in top predators such as birds of prey and marine mammals. To date, it has been detected only in much lower levels in humans. Its environmental levels in Europe are higher than those in North America owing to its greater use there.

PROBLEM 15-8

Draw the structure of HBCD, given that its name is 1, 2, 5, 6, 9, 10-hexabromocyclododecane. Then, knowing that Br_2 molecules will add across C=C bonds, deduce the structure of the cyclic triene hydrocarbon that could be brominated to produce HBCD. [*Hint:* Dodeca means "twelve."]

A number of other polybrominated organic compounds have been introduced into the market now to replace the PBDEs that have been phased out. They all contain heavily brominated benzene rings; indeed, *hexabromobenzene* itself is one of the new ones. Many have log K_{ow} values in the 5–7 range common to organic compounds known to bioaccumulate. One approved replacement for *Penta* in flexible polyurethane foam products—including those meant for babies and children—is a mixture of two derivatives of tetrabromobenzene, one a phthalate and the other a benzoate.

Polybrominated biphenyls (PBBs) were also used as fire retardants, but are now banned in some countries, including the United States. A 1973 industrial accident in Michigan resulted in the widespread contamination of the food supply there with PBBs.

Bioaccumulation and the K_{ow} scale were discussed in Chapter 13.

15.16 Nonbrominated Fire Retardants

Given the environmental persistence, bioavailability, and toxicity of brominated fire retardants, it is not surprising that there is a movement to ban the use of all halogen-containing compounds for this purpose. The argument against such a ban is that they are vital in reducing losses of human lives and of property in fires, although there is not unanimous agreement that retardants are actually necessary.

Several chlorine-containing fire retardants are in widespread use. *Dechlorane Plus* is a chlorinated cyclopentadiene derivative that has been in use as a flame retardant in various plastics for many decades, but was only recently detected in the environment. *TDCPP*, a chlorinated organic phosphate, has been used in upholstery foams for some time as well.

Progress has been made in formulating *nonhalogenated* retardants, some of which are consistent with the principles of green chemistry.

- Phosphorus-based flame retardants are now being used to protect some electronic equipment such as printed wiring boards. The compounds work in a fire by forming a char layer, which is heat-resistant and protects the product from further attack by oxygen and radiant heat.

- *Magnesium hydroxide*, $Mg(OH)_2$, decomposes at high temperatures but in an endothermic manner, thereby removing heat from the burning object and tempering the fire. Unfortunately, the rather large amount of magnesium hydroxide that must be added to some plastics to make them fire resistant can interfere with the properties of the plastic.

- Nanometer-sized particles of layered silicate clay dispersed throughout a polymer material operates as a protective "gauze" against the release of molecular fragments from a polymer to the gas phase, where combustion occurs, and thus retards a fire by slowing it down and reducing the flame temperature. The development of mixtures of the clay with small amounts of either magnesium hydroxide or phosphorus-based compounds may lead to "greener" fire retardants in the near future.

Review Questions 13–15 are based upon the sections above.

Perfluorinated Sulfonates and Related Compounds

All the organic compounds discussed previously in this chapter act either as hydrophobic (water-repelling) or as *oleophobic* (oil-repelling) substances, but not as both. There exists a small class of organic compounds that will dissolve in neither of these classes of compounds. **Fluorinated surfactants are compounds that consist of molecules and ions having a long perfluorinated carbon tail; i.e., a hydrocarbon chain in which each hydrogen atom has been replaced by a fluorine atom.** The best-known example of such a molecule is *perfluorooctane sulfonate* (PFOS):

$$CF_3(CF_2)_7 - \underset{\displaystyle \underset{O}{\|}}{\overset{\displaystyle \overset{O}{\|}}{S}} - OH$$

Notice the similarity in its structure to that of sulfuric acid: one —OH group in the latter has been replaced by an unbranched, perfluorinated 8-carbon octane chain. This substance was used to make the 3M product *Scotchgard*, a fabric protector that, because of the characteristics of the perfluoro chain, repelled both water and oily spills and potential stains. Other compounds based upon PFOS were used in fire-fighting foams, pesticide formulations, cosmetics, lubricants, grease-resistant coatings for paper products, adhesives, and paints and polishes.

The 3M company voluntarily phased out the production of PFOS because it persists long enough in the environment to eventually be detected in human blood samples. Although it is not very toxic, its concentration in some wildlife had reached levels of concern to some scientists. Since 2003, 3M has used the corresponding perfluorosulfonate having a chain of only four, rather than eight, carbon atoms, since such chains do not seem to either bioaccumulate or be toxic. **In 2009, the United Nations Environmental Program added perfluorooctane sulfonic acid, its salts, and perfluorooctane sulfonyl fluoride to their list of *Persistent Organic Pollutants*.**

15.17 Perfluorinated Alkyl Acids

The fully fluorinated 8-carbon carboxylic acid compound *perfluorooctanoic acid* (PFOA), $CF_3(CF_2)_6COOH$, and its associated carboxylate salts have also become of environmental concern since they have no environmental or metabolic degradation pathways. Indeed, the very lack of reactivity that makes perfluorinated compounds so appealing in practical uses also results in their persistence in the environment.

Although its lifetime in rats is only several hours, PFOA is only slowly eliminated in humans, resulting in an average lifetime of about four years. Due to its slow elimination, the acid is now found at the parts-per-billion level in the blood of most humans and wild animals worldwide. It is potentially acutely toxic, potentially carcinogenic, and may cause developmental problems.

Human exposure to PFOA results from its use in producing polymers used to coat surfaces of nonstick cooking-ware, including frying pans, as well as for the membranes of breathable outdoor garments. Fully fluorinated carbon chains are added by covalent bonds to polymer chains in order to make the materials stain-resistant. PFOA has been detected in samples of drinking water in several U.S. states; there is no federal standard for it. In 2006, the U.S. EPA announced a voluntary program, requesting companies using PFOA in consumer products to stop employing the compound.

PFOA is the most prominent member of the family of **perfluoroalkyl acids,** PFAAs. In general, the longer the carbon chain in such molecules, the more persistent is the acid in the human body. 3M is formulating products that use PFAAs with relatively short chains in order to overcome the persistence problems with eight-carbon chain substances. The environmental problems associated with the shortest member of the family, *trifluoroacetic acid*, CF_3COOH, which is the resilient degradation product of some of the hydrofluorocarbons and hydrochlorofluorocarbons used as CFC replacements, was discussed in Section 2.12.

PFOA and similar compounds are now found even in remote regions such as the Arctic, being transported there by LRTAP. Apparently such PFOA results from the atmospheric reaction of **fluorotelomer alcohols,** $CF_3(CF_2)_nCH_2CH_2OH$, industrial compounds that are used to make stain repellants. Unfortunately, a small fraction of the alcohols is released inadvertently into the atmosphere during the manufacturing process. In addition, the tiny concentration of the reactant that was not converted to a repellant but was instead weakly incorporated into the material slowly degasses from it, adding to the atmospheric load of the alcohol.

In air, the fluorotelomer alcohols are converted to the carboxylic acids by a chain reaction that begins when a hydroxyl radical, OH, in air abstracts a hydrogen atom from the —CH_2— group bonded to OH. This process initiates a sequence of free-radical reactions, the net result of which is the oxidation of the terminal —CH_2CH_2OH group to COOH, producing the final perfluorocarboxylic acid $CF_3(CF_2)_nCOOH$. It is also thought that microbial action and animal metabolism play a role in converting the alcohols to acids. Fluorotelomer alcohols have become the main remaining source of PFOA in the environment. In 2006, the Canadian government proposed to ban fluorotelomer polymers that can decompose into long-chain perfluorinated carboxylic acids.

Review Questions 16 and 17 are based upon material in the sections above.

Review Questions

1. What does *PAH* stand for? Draw the structures of two examples.

2. In what processes are PAHs commonly formed?

3. What are the sources of PAHs in air? How is their physical state in air and their lifetime there related to their molecular size?

4. By means of a structural diagram, show what is meant by the *bay region* present in certain PAHs. How is the presence of this region related to the health effects of PAHs?

5. Define the term *environmental estrogen*. How do such compounds operate in the human body? Give two chloroorganic and two nonchloroorganic examples of environmental estrogens.

6. Recount some of the evidence that environmental estrogens affect the health of wildlife and of humans.

7. What are *phthalates* and how are they commonly used?

8. Name the four postulated effects of environmental hormones on human health.

9. What is a *phytoestrogen*?

10. What does *LRTAP* stand for?

11. Which three physical properties are used to predict the ultimate deposition zone of volatile chemicals?

12. How does the *differential removal hypothesis* differ from the *global distillation hypothesis* for pollutants that migrate in the air?

13. What is the mechanism by which brominated organic compounds control fire?

14. What does *PBDE* stand for and what are such compounds used for? What is meant by the commercial terms Penta and Deca? Draw the structure of any PBDE.

15. What are two nonhalogenated fire retardants?

16. Draw the structure of a *perfluorinated sulfonate*. What are such substances used for?

17. What do *PFOA* and *PFAA* stand for? What is the molecular structure of PFOA? What is meant by the term *fluorotelomer alcohol*? What are such alcohols used for?

Additional Problems

1. In an experiment, the level of benzo[*a*]pyrene in hamburgers was found to depend significantly on the cooking method and cooking time. For oven-broiled hamburgers, levels of 0.01 ng g^{-1} were found for both medium and very-well-done burgers. For barbecued burgers, levels of 0.09 and 1.52 ng g^{-1} were found for medium and very-well-done burgers, respectively.

(a) Explain the observed difference in benzo[*a*] pyrene formation by the two cooking methods, and explain the difference in medium versus very-well-done barbecued burgers.

(b) What does 1.52 ng g^{-1} translate to on a *parts-per* scale?

(c) How many micrograms of benzo[*a*]pyrene would be ingested in the consumption of a typical "quarter pounder" hamburger if it was barbecued to the very-well-done stage and there was no loss in mass during its cooking?

2. Which three pairs of octabromo diphenylether isomers identified in Problem 15-6b would interconvert if free rotation were to exist about the C—O bonds?

PART V

ENVIRONMENT AND THE SOLID STATE

Contents of Part V

Introduction

In previous Parts of this book, we have not been much concerned with the ultimate disposal of macroscopic supplies of the various chemicals and other materials that cause pollution when dispersed in the environment. In Part V, we remedy that omission by considering the various techniques used in the disposal of commercial substances ranging from plastic water bottles to PCBs and to chemicals that are highly dangerous. To assist in these deliberations, the nature of soil itself and of freshwater sediments is also described. ●

16

Wastes, Soils, and Sediments

In this chapter, the following introductory chemistry topics are used:

- ⮑ Thermochemistry
- ⮑ Concepts of oxidation and reduction as electron loss or gain; oxidation number; basic electrochemistry
- ⮑ Background organic chemistry (see online Appendix)
- ⮑ Concepts of acids and bases; pH
- ⮑ Phase diagrams

Background from previous chapters used in this chapter:

- ⮑ Adsorption; NO_x; particulates; activated carbon
- ⮑ Aerobic and anaerobic decompositions; methane
- ⮑ Ethanol; BTEX; MTBE; fatty acids
- ⮑ DDT
- ⮑ K_{ow}
- ⮑ PCBs, dioxins, and furans
- ⮑ PAHs; phthalates; chlorinated solvents
- ⮑ BOD, COD, and water carbonate chemistry
- ⮑ Heavy-metal chemistry

Introduction

In this chapter, we turn our attention to the environmental aspects of the solid state—particularly of soil, of the sediments of natural water systems, and of ways that polluted soils and sediments can be remediated. A closely related issue is the nature and disposal of concentrated wastes of all kinds, including both domestic garbage and hazardous waste, and their possible recycling.

The material in this chapter has been ordered in terms of generally increasing toxicity and hazard. Thus we begin with the least toxic—domestic and commercial garbage, and consider its disposal by landfilling or incineration or recycling. We then consider soils and sediments, and their contamination by chemicals. Finally, we look at hazardous wastes, and some of the high-technology methods that are being developed to dispose of them.

Domestic and Commercial Garbage: Its Disposal and Minimization

The great majority of material that we discard and which must be disposed of is not hazardous, but simply corresponds to "garbage" or "refuse." The greatest single generator of this **solid waste**—defined as waste that is collected and transported by a means other than water—is construction and demolition debris, almost all of which is either reused or ends up being buried in the ground. The second largest volume of waste corresponds to that generated by the commercial and industrial sectors, followed by the domestic waste generated by residences. Typically, a North American generates about 2 kg of domestic and commercial waste a day, twice as much as the average European. In these discussions we do not consider the much larger amounts of waste generated by the petroleum industries, by agriculture, as ashes from power plants, or as sewage.

The treatment and disposal of sewage was discussed in Chapter 11.

16.1 The Varying Components of Domestic Garbage

A breakdown by type of the solid waste typically generated in countries at various levels of economic development is shown in Figure 16-1. Notice how the fraction of the waste that corresponds to vegetable matter declines as the level of development increases. Food waste overall, however, is the dominant

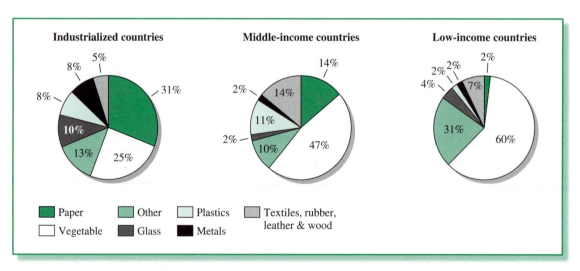

FIGURE 16-1 Typical composition of solid waste for countries at different levels of economic development. [Source: "Waste and the Environment," *The Economist* (29 May 1993): 5 (Environment Survey section).]

component of domestic garbage even in some industrialized countries. The opposite trend is true of **paper, which in industrialized countries is one of the two largest single components of waste, and which dominates waste discarded by the commercial sector.** Historically, the largest component of paper waste was newspapers; now the volume of paper packaging is similar. The amount of packaging has grown, in part, because so many goods now are produced far from their ultimate destination and must be transported safely over long distances.

Plastics, glass, and metals each account for about one-tenth of the volume of solid waste in developed countries, whereas organic matter (food waste) accounts for about twice this amount there and much more in middle-income and low-income countries. These proportions would differ significantly in areas of developed countries that collect materials for recycling or composting: the glass and metals components would be much smaller.

The *Other* category in Figure 16-1 includes ash from domestic fires, which is quite significant in many low-income countries.

16.2 Burying Garbage in Landfills

The main method used for disposal of municipal solid waste, MSW, is to place it in a landfill (also variously called a *garbage dump* or a *rubbish tip*), which is a large hole in the ground that usually is covered with soil and/or clay after it is filled. For example, about 85% of domestic and commercial waste currently is landfilled in the United Kingdom, about 8% is incinerated, and the same fraction is recycled or reused; similar figures apply to many municipalities in North America. Landfilling dominates the disposal methods because its *direct* costs are substantially lower than disposal by any other means.

In the past, landfills were often simply large holes in the ground that had been created by mineral extraction—especially old sand or gravel pits. In many instances, they leaked and contaminated the aquifers that lay beneath them; this was especially true for landfills that used former sand pits, since water easily percolates through sand. These landfills were not designed, controlled, or supervised, and they accepted many types of wastes, including hazardous materials.

FIGURE 16-2
Components of a modern sanitary landfill (in the process of being filled).

Modern municipal landfills in developed countries are much more elaborately designed and engineered, often accept no hazardous waste, and have their sites selected to minimize impact on the environment. The components of a typical modern landfill are illustrated in Figure 16-2.

In a **sanitary landfill,** the MSW is compacted in layers (to reduce its volume) and is covered with about 20 cm (8 in.) of soil at the conclusion of each day's

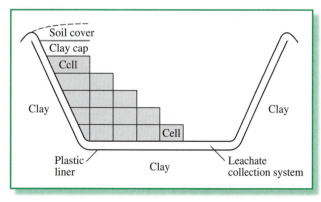

operations. Thus the landfill consists of many adjacent *cells*, each corresponding to a day's waste (Figure 16-2). After one layer of cells is completed, another is begun, and the process continued until the hole is filled. Usually, the landfill is eventually capped by a meter or so of soil, or preferably clay, a material that is fairly impervious to the rain that falls on it. A *geomembrane* made out of plastic may be added as a liner on top instead of the clay, or over it. The system recommended by the U.S. EPA is illustrated in Figure 16-3.

During the time that municipal wastes in a landfill are decomposing—aerobically at first, then anaerobically after a few months or a year—water from precipitation, liquid from the waste itself, and groundwater that seeps into the landfill, all percolate through the garbage, producing a liquid called **leachate.** This liquid contains dissolved, suspended, and microbial contaminants extracted from the solid waste. Leachate volume is relatively high for the first few years after a site is covered.

16.3 Stages in the Decomposition of Garbage in a Landfill

There are three stages of decomposition in a municipal landfill. Operating landfills still receiving garbage undergo all three stages simultaneously in different regions or depths. In practice, only food and yard waste biodegrade.

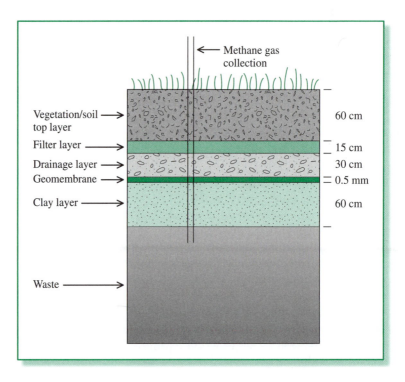

FIGURE 16-3 Landfill design with cover system recommended by the U.S. EPA.

Rubber, plastics, and much of the paper content of garbage are very slow to degrade.

• **In the first, short, aerobic stage, oxygen is available to the waste; it oxidizes organic materials to CO_2 and water, with the release of heat.** The internal temperature can rise to 70–80°C, since the reactions are exothermic. The carbon dioxide released from the organic matter as it decomposes makes the leachate acidic, thereby further facilitating its ability to leach metals encountered in the waste. Since the majority of biodegradable material is *cellulose*, whose empirical formula is approximately CH_2O, we can approximate this phase of the reaction as

$$CH_2O + O_2 \longrightarrow CO_2 + H_2O$$

Some organic matter is partially oxidized to aldehydes, ketones, and alcohols, which give fresh waste its characteristic sweet smell.

• **In the second, anaerobic acid phase, the process of** *acidic fermentation* **occurs, generating** *ammonia, hydrogen,* **and carbon dioxide gases and large quantities of partially degraded organic compounds, especially organic acids.** The pH of the leachate in this phase is 5.5–6.5, so it is chemically aggressive. Other organic and inorganic substances dissolve in this leachate due to its acidity. Again, carbon dioxide is released, but no methane. This phase of the reaction can be approximated by the reaction

$$2\ CH_2O \longrightarrow CH_3COOH$$

although longer-chain fatty acids, which subsequently decompose into **acetic acid,** CH_3COOH, are formed initially, as is hydrogen gas.

In this phase, the leachate has a high oxygen demand (BOD and COD, Chapter 11), as well as relatively high heavy-metal concentrations. Anaerobic decomposition produces volatile carboxylic acids and esters, which dissolve in the water present. The sickly sweet smell that emanates from landfills during this phase is due to these esters and to thioesters.

• **The third, anaerobic—or methanogenic—stage, starts about six months to a year after coverage of the landfill, but can continue for very long periods of time. Anaerobic bacteria work slowly to decompose the organic acids and hydrogen that were produced in the second stage.** Since the organic acids are consumed in the process, the pH rises to about 7 or 8, and the leachate becomes less reactive. The main products of this stage are carbon dioxide and **methane,** CH_4. To a first approximation, the overall reaction here is

$$CH_3COOH \longrightarrow CH_4 + CO_2$$

Methane generation usually continues for a decade or two, and then drops off relatively quickly. Some methane is also formed when hydrogen gas combines with carbon dioxide. Much lower BOD values and a smaller volume are associated with landfills in this phase. Because the leachate is not acidic, the heavy-metal concentrations drop since these elements are not as soluble at higher pH.

The methane gas produced by a landfill can be vented to the atmosphere by being directed into wells or gravel-packed seams in the landfill. However, the gas can be combusted as it emerges through vents (see Figure 16-3) rather than being released into the air. This treatment of the methane is especially desirable given that the greenhouse gas potential of CH_4 is much greater than is that of the CO_2 that is produced by its combustion (Chapter 5). Indeed, regulations in many countries, including the United States, now require that gas from large landfills be either flared or combusted to produce energy so that the methane is not released into the air. However, the collection and control of the gas is impractical in the first few years of operation, and never 100% efficient.

Over the lifespan of the landfill, collecting and burning the emitted gas reduces by slightly more than half its greenhouse gas emissions in terms of the CO_2 equivalent. The heat produced from the combustion of this gas can be used for practical purposes, including the generation of electricity or heating of greenhouses, etc. Indeed, the second and third stages of decomposition in landfills are identical to those used in the deliberate production of **biogas** (biomethane) for energy in reactors using municipal solid waste sludge, food processing waste, livestock waste, and other biodegradable materials (see also Section 5.14).

PROBLEM 16-1

Calculate the volume of methane gas, at 15°C and 1.0 atm pressure, that is released annually by the anaerobic decomposition of 1 kg of garbage, assuming the latter to be 20% biodegradable cellulose, and that decomposition occurs evenly over a 20-year period. [*Hint:* Add together the equations for the two anaerobic stages of decomposition. $R = 0.082$ L atm K^{-1} mol^{-1}.] ●

16.4 Leachate from a Landfill

As a result of the decomposition processes that occur in a landfill, **typical components of the leachate are**

• **volatile organic acids such as acetic acid and various longer chain fatty acids;**

• **bacteria;**

• **heavy metals, usually in low concentration** (those of most concern in leachate are lead and cadmium); **and**

• **salts of common inorganic ions such as Ca^{2+}.**

Micropollutants present in MSW leachate include common volatile organic compounds such as *toluene* and *dichloromethane*.

Engineering is needed to control the leachate from a landfill. Otherwise the liquid can flow out the bottom of the landfill, and percolate through porous soil and contaminate the groundwater below it. Alternatively, if the soil under the landfill is nonporous, the leachate can build up and gradually overflow the site (the "overflowing bathtub effect"), thereby possibly contaminating nearby surface waters.

The typical components used to control the leachate consist of:

- **A leachate collection and removal system,** followed by treatment of the liquid (Figure 16-2). Often the potential effect of leachate on groundwater is monitored by digging and testing several wells in the vicinity.

- **A liner placed around the walls and bottom of the landfill.** The liner material is either synthetic (e.g., a plastic such as 2-mm-thick high-density polyethylene) or natural (e.g., compacted clay). The material chosen is impervious to water and will largely prevent the leakage of the contaminated leachate into the groundwater, especially if and when the collection system fails due to clogging, etc. Since 1991, new landfills in the United States must have at least 6 layers of protection between the garbage and the underlying groundwater! Liners have been developed which consist of *bentonite clay*—an excellent sealant that efficiently binds heavy metals, preventing their migration out of the landfill—sandwiched between two layers of a plastic such as polypropylene.

> The nature of polymers and plastics is described in Section 16.11.

Leachate treatment systems must address all the liquid's major components. The treatment of leachate, usually at a sewage treatment plant, is usually accomplished by aerobic degradation to rapidly decrease the BOD, sometimes using advanced oxidation methods that employ ozone (Chapter 11). In the past, collected leachate was often simply returned to the top of the landfill, since during its second percolation through the waste much of its organic content will be biologically degraded; however, this practice is discouraged in the United States.

16.5 Incineration of Garbage

Besides landfilling, the most common way to dispose of wastes, particularly organic and biological ones, is by **incineration—the oxidation, by controlled burning, of materials to simple, mineralized products such as carbon dioxide and water.** The primary incentive in the incineration of municipal solid waste is to substantially reduce the *volume* of material that must be landfilled. In the case of toxic or hazardous substances, an even more important goal is to eliminate the toxic threat from the material. Incineration of hospital wastes is done to sterilize them as well as to reduce their volume.

Many municipalities throughout the world burn domestic garbage in incinerators. For example, Japan and Denmark burn more than half their domestic waste. However, the practice is banned in some areas. The combustible components of the garbage, such as paper, plastics, and wood provide the fuel for the fire. **The most common domestic MSW incinerators are one-stage mass burn units, and the more modern, two-stage modular type.** In the latter, wastes are placed in the *primary chamber* and burn at a temperature of about 760°C. The gases and airborne particles that result from the first stage are then burned more completely, at temperatures in excess of 870°C, in the *secondary combustion chamber*. The quantity of waste gases that must later be controlled is greatly reduced in the two-stage units compared to those that employ only one stage, although the gases are further heated as they exit the one-stage unit in order to produce more complete combustion.

In some incinerators, the heat of the combustion processes is captured and used to produce steam, hot water, or electricity. Indeed, a recent analysis showed that the combustion of municipal solid waste and the use of energy thereby generated to produce electricity resulted in much less than half the greenhouse gas emissions (CO_2 equivalent) than did the rather inefficient capture and burning of landfill gas for a given mass of garbage. An order of magnitude more electricity is generated by burning the waste directly than by capturing and combusting the landfill gas.

The output from municipal incinerators includes not only the final gases but also solid residues that amount to a small fraction of the initial weight and volume of the garbage. **Bottom ash is the noncombustible airborne material that collects at the bottom of the incinerator. Fly ash is the finely divided solid matter that usually is trapped by environmental pollution controls in the stack to prevent it from being released into the outside air.** Much of the ash corresponds to the inorganic constituents of the waste, which form solids rather than gases even when fully oxidized. Although fly ash accounts for only 10–25% of the total ash mass, it is generally the more toxic component, since heavy metals, dioxins, and furans readily condense onto its small particles.

The low density and small-particle character of the ash make inadvertent dispersal into the environment a significant risk. Of particular concern are heavy metals in the ash, which could potentially be leached from it and pollute nearby surface water and groundwater. For many years, it was common for incinerator ash to be taken to a hazardous waste landfill. Techniques such as the addition of an adhesive, or melting and vitrification, have now been developed whereby ash can be solidified into a leach-resistant material that need not be classified as hazardous waste. In some countries such as Denmark and the Netherlands, the ash is mainly recycled into asphalt.

The main environmental concern about incineration is the air pollution that it generates, consisting of both gases and particulates. The emission controls on MSW incinerators capture a large fraction of, but not all, the toxic substances that would be emitted into the air from the combustion process. About half the capital costs of new incinerators is spent on air pollution control equipment. Typically, the controls include a **baghouse filter,** which is made of woven fabric and is used to filter particulates, especially those of diameters greater than 0.5 μm, from the flow of output gas. Periodically the bags are shaken, or the air flow is reversed, in order to collect the fly ash.

Also typically present in the air pollution control equipment is a **gas scrubber,** which is a stream of liquid or solid that is passed through the emission gas, thereby removing some particles and gases. If the liquid stream consists of *lime*, CaO, mixed with water, or the solid stream consists of lime, then acid gases such as HCl and SO_2 are efficiently removed, since they are neutralized to their calcium salts. Heavy metals are also captured by the alkaline environment formed by the lime, since they form insoluble hydroxides. In some modern installations, *nitrogen oxides* are removed by spraying ammonia or *urea* into the hot exhaust gases. In another new technology used in garbage incinerators, activated charcoal or lignite coke powder is blown into the exhaust gases, which are subsequently filtered by a baghouse; much of the dioxin, furan, and mercury content of the exhaust gases is thereby removed, since these components adsorb onto the charcoal or coke surface.

Although public concern has centered on emissions from *hazardous waste* incinerators (to be discussed in Section 16.32), several U.S. surveys have indicated that more dioxin and furan emissions emanate from medical waste and municipal waste incinerators than from hazardous waste ones, although cement kiln units used for hazardous wastes (discussed in Section 16.31) also make a significant contribution. Emissions to the air from incinerators are likely to happen mainly during start-up and when equipment fails. Because medical waste and backyard barrel garbage incinerators operate in a start-and-stop mode, they tend to produce more airborne pollutants per unit mass of incinerated waste than do larger incinerators. Overall, municipal, backyard, and medical incinerators are believed to be a major anthropogenic source of both mercury and dioxins and furans to the environment, and intermediately important sources of cadmium and lead.

These air pollution control technologies were discussed in greater detail in Chapter 3.

Review Questions 1–6 are based upon the material above.

The Recycling of Household and Commercial Waste

In the past few decades, there has been mounting pressure in developed countries to reduce the amount of materials that are discarded as waste after a single use. The incentives here are to conserve the natural resources, including energy, from which the materials are produced, and to reduce the volume of material that must be buried as garbage or incinerated. In this section, we consider how this ambition can be realized.

16.6 General Features of Recycling

The **four R's** of waste management philosophies designed to reduce the volume of garbage are:

- **Reduce** the amount of materials used (sometimes called "source reduction").

- **Reuse** materials once they are formulated.

- **Recycle** materials to recover components that can be refabricated.

- **Recover** the energy content of the materials if they cannot be used in any other way.

These principles can be and are applied to all types of wastes, including hazardous ones, but in the discussions below we concentrate on their application to domestic materials, particularly in regard to recycling.

A distinction is often made between **pre-consumer recycling, which involves the use of waste generated during a manufacturing process, and post-consumer recycling, which involves the reuse of materials that have been recovered from domestic and commercial consumers. The post-consumer items most often collected for recycling are**

- **paper** (especially newspapers and cardboard),

- **aluminum** (especially beverage cans),

- **steel** (especially food cans), and

- **plastic and glass** containers.

The labor, energy, and pollution costs associated with collecting the materials, sorting them, and transporting them back to facilities where they can be reused must be considered in any analysis of recycling. In addition, historically the demand for recycled materials in these categories—other than aluminium—has been unstable, with prices swinging wildly up and down in response to changes in supply and demand. For these and similar reasons, the recycling of paper, glass, and plastics usually needs to be justified on noneconomic and nonenergy grounds—such as savings in landfill space. The most economically viable forms of recycling of these materials usually involve a minimum of chemical reprocessing, and correspond more closely to **reuse**—e.g., using newspapers to make paperboard or insulation, reusing glass and plastic containers, etc. Debates still rage both in the popular press and in the scientific literature as to whether recycling is a worthwhile activity or not, especially since its net cost is not usually recovered by sale of the collected materials.

16.7 The Recycling of Metals and Glass

From the viewpoints of both economics and energy conservation, the recycling of some metals makes sense. Virgin metal must be obtained by reduction of the oxidized form of the element found in nature. The reduction process requires energy that does not need to be expended again when the metallic form of the element is recycled.

Consider the reduction of aluminum and iron from the oxide ores. By definition, the enthalpies of these processes equal the negative of the enthalpies of formation of the oxides:

$$Al_2O_3 \longrightarrow 2\,Al + \tfrac{3}{2}\,O_2 \qquad \Delta H° = -\Delta H_f°\,(Al_2O_3)$$

$$= +1676 \text{ kJ mol}^{-1} \text{ of oxide}$$

$$= +31 \text{ kJ g}^{-1} \text{ of metal}$$

Recycling aluminum cans saves 95% of the energy that is needed to produce Al metal from bauxite ore. Since the energy required for the aluminum reduction must come in the form of electricity, and since this energy accounts for about 25% of the cost of its production, it makes good economic sense to recycle this metal. However, the recycling rate of aluminum cans had dropped from 65% in 1992 to only 45% in 2005, but increased again to 58% in 2010, in the United States. In contrast, Sweden, which requires deposits on beverage containers that are refunded upon their return, collects 85% of aluminum cans. Indeed, the recycling rates for containers of all types is greater in those parts of the United States where deposits are required.

Considerable savings in aluminum have resulted from the reduction by about one-third of the weight of individual cans over the last few decades. Aluminum can be recycled endlessly without loss in quality. Globally, recycled aluminum provides about one-third of the production of the metal.

Recycling steel cans saves about two-thirds of the energy required to produce them from iron ore.

PROBLEM 16-2

The enthalpy of formation of the principal ore of iron, Fe_2O_3, is -824 kJ mol^{-1}. Calculate the enthalpy of the reaction in which 1.00 g of metallic iron is formed from the ore. Given your result, would you expect that the price that recycling operations are willing to pay per kilogram for scrap iron would be greater or less than that for scrap aluminum? ●

In recent years, recycling of **e-waste**—namely computers, cell phones (called *mobiles* in some countries), printers, photocopiers, TVs, and other widely used electronic items found in offices and households—has become more common. Computers particularly are often reused, sometimes after being refurbished, by those who cannot afford new ones and are often shipped to developing countries. The metals contained in e-waste are the most valuable components to recycle, since on average they comprise about 70% by mass, the remainder being mainly plastics. In developed countries,

Some batteries are recycled
for their metal content.

hazardous components such as mercury-containing light bulbs and batteries are first removed, then the equipment is mechanically shredded. Steel, which often comprises half the metal, can then be isolated by magnets. Aluminum and copper are also common and valuable. Overall, however, e-waste recycling is rarely more than a break-even activity in developed countries. As a consequence, most products are simply discarded as garbage or shipped overseas.

Unfortunately, much of the e-waste recycling is done by hand by individuals in developing countries, exposing the workers and nearby residents to air and water pollution from the operation. For example, wiring is often extracted from computers and burned in the open to remove the plastic casing and recover the copper. The combustion of plastic can produce and release dioxins, furans, PAHs, and fire retardants into the local air. Circuit boards are treated with acids, cyanide, and even mercury to extract metals, with the effluent disposed of in rivers or other water bodies.

In the case of paper, glass, and plastics, there is no significant change in the average oxidation number of the principal materials during their transformation from inexpensive raw-material components—wood, sand and lime, and oil, respectively—to finished products; thus there are no great energy savings when they are recycled.

Modern, low-polluting, and energy-efficient electric furnaces cannot handle as high a proportion of used glass as can their more polluting, more energy-consuming counterparts that use fossil fuels. Consequently, the recycling of too much glass can produce more pollution and use more energy than would otherwise be the case! In any event, the use of glass containers for beverages has fallen sharply, and their recycling rate is rather low unless deposits are mandated.

16.8 The Recycling of Paper

People in developed countries throw away more paper than any of the other components of municipal solid waste (see Figure 16-1), and it seems an obvious material to recycle. However, the production of virgin paper, for example, uses only about one-quarter more energy than does recycling of old paper. The transportation of waste paper to recycling mills, and the deinking process itself, are heavy consumers of energy. Notwithstanding these considerations, tremendous quantities of paper, especially newsprint, currently are recycled. Indeed, corrugated paper products are the most intensely recycled material in North America.

The first step in recycling paper is its mechanical dispersal into its component fibers in water. Then cleaning occurs to remove nonfibrous contaminants, followed by treatment with *sodium hydroxide* or *sodium carbonate* to deink it. A detergent is added to help disperse the pigment, and the ink particles are removed by washing or flotation of ink on air bubbles, which rise to the top. The resulting deinked stock is usually less white than is virgin fiber, so the two types are often blended. If necessary, the whiteness of the recycled stock can be improved by bleaching, usually with *peroxides*. The used ink, which is recovered in a sludge with some pulp fibers, is later pressed to remove water and then can be burned to produce steam for use in mill operations, or can be treated to detoxify it. In general, the use and release into the environment of raw materials such as *chlorine* or other bleaching agents, acids, and organic solvents is significantly smaller in the production of recycled paper than with the creation of the virgin material.

Paper of different types is composed of fibers of very different lengths (long ones in office paper, short ones in newsprint) that cannot be mixed in recycling to produce high-quality paper. In addition, there is a limit to the number of times that paper can be recycled, since with each cycle the pulp fibers become shorter and so lose some of their integrity. Newsprint can be recycled back into newsprint about six to eight times. Food boxes and egg cartons are usually made from recycled pulp fiber. Toilet paper for domestic use is made mainly from virgin fibers, since they are longer than recycled ones and therefore can be "fluffed" to produce a plusher feel.

From 1985 to 2000, the paper industry in the United States spent almost $20 billion on the technology and capital investments required to recycle paper; the ultimate goal was to recycle about half all paper used in the country. Indeed, by the mid-2000s, close to half the paper and paperboard products in North America were being recycled, compared to two-thirds in many European countries. Recycling rates in the United States in 2009 were 89% for newspapers, 81% for corrugated boxes, 74% for office paper, 54% for magazines, and 37% for phone directories; about one-third of recovered paper is exported from the United States.

The issue of whether it is preferable to recycle paper rather than bury it in a landfill is controversial. Since paper is made from wood, which has grown by extracting carbon dioxide from the atmosphere, landfilling paper that will never rot is a form of sequestering CO_2. Research indicates that about 70% of the carbon in paper—and more than 97% of that in other wood products—remains undecomposed in landfills. However, this argument is negated if any significant fraction of the paper decomposes to emit methane, and does not take into account the larger amounts of energy and water require to produce virgin paper compared to recycled paper.

Recall from Chapter 5 that methane is a more potent greenhouse gas than is carbon dioxide.

Perhaps a more clever use of waste paper in the future will be its conversion to fuel *ethanol*, as was discussed in Chapter 7. Paper can be incinerated directly to recover its energy content, a process that reduces the amount of fossil fuels required to be burned in power stations. According to an analysis by Britain's *Centre for Environmental Technology*, recycling paper is environmentally superior to landfilling it but is inferior to burning it for its fuel value, when all factors are included.

16.9 Green Chemistry: Development of Bio-based Toners

You no doubt have pressed the print key on your computer and the copy button on a copy machine. This is done in the United States to the tune of producing about 3 trillion copies per year or approximately 10 thousand copies per person! Worldwide, the total number of copies is estimated to be twice this value. Some people try to lower their ecological footprint by using recycled paper, double-sided copies, and even scrap paper. However, most of us pay little attention to the fact that these trillions of copies require over 180 million kg (400 million pounds) of electrostatic dry toner in the United States alone for printing.

If you have ever refilled the toner cartridge on a copier or a laser printer, you know that that these dry toners are very fine black (and other colors for color printers and copiers) powders. During the printing process these particles are selectively deposited on the paper to produce print. Toners have traditionally been made from petroleum-based resins such as styrene acrylates and styrene butadiene.

They are developed so that their performance properties are maximized to produce high-quality print but with little attention to recycling. One of the major roadblocks to the recycling of printed paper is the removal (deinking) of the print. Deinking processes require several steps including detachment of the ink from the cellulose fibers of the paper, separation, and bleaching.

In 2008 Battelle, in conjunction with partners Advanced Image Resources and the Ohio Soybean Council, won a Presidential Green Chemistry Challenge Award for the development of a bio-based toner. The toner is produced from feedstocks derived from oil and protein from soybeans and carbohydrates from corn. The resins that are made from these renewable materials have all the performance characteristics of their petroleum-derived cousins. In addition, they have built-in sites that allow them to be easily dislodged from the cellulose (paper)–toner interface when exposed to deinking solutions. As a result, the deinking process is simplified, thereby reducing the amount of chemicals used and waste produced and the amount of energy required.

Based upon energy savings from using renewable resources, simpler deinking processes, and increased recycling rates, Battelle estimates that if 25% of users switched to this bio-based toner, it would result in an annual energy savings of 6800 MJ (6.4 million Btu) and an annual reduction in emissions of 0.36 million tons of carbon dioxide.

16.10 The Recycling of Tires

Another consumer commodity that represents a waste-management head-ache is vehicle tires. In North America, about one 10-kg rubber tire per person per year on average is discarded; thus about one-third of a *billion* tires are added to the supply of approximately 3 billion tires presently stored in mountainous piles, awaiting ultimate disposal! Because they are made primarily from oil and consequently are flammable, *tire fires* in these huge piles are not uncommon, and produce tremendous amounts of smoke, *carbon monoxide*, and toxins such as *polycyclic aromatic hydrocarbons* (PAHs) and dioxins. The fires are difficult to extinguish because of air pockets present in and between the tires.

There have been efforts to use the tires either as fuel or as a filler for asphalt, but currently such applications consume only about 10% of the tires that annually are discarded. Some used tires are also used for their rubber content, to produce landscaping products.

A number of attempts have been made to commercially reprocess shredded scrap tires by **pyrolysis—the thermal degradation of a material in the absence of oxygen.** The resulting products are low-grade gaseous and liquid fuels, and a **char** containing minerals and a low-grade version of the material called **carbon black,** which can be further treated and converted into *activated carbon*. It may eventually be possible to convert the liquid component into high-grade char, and thereby make the process economically profitable. The "rubber" in tires consists of about 62% of a hydrocarbon polymer and 31% of carbon black—added to strengthen the tires and reduce wear—so there is a ready market for the latter. Using the liquid component as a fuel is problematic because of its high content of aromatic hydrocarbons.

16.11 Recycling Plastics: Their Constitution

One of the triumphs of industrial chemistry in the twentieth century was the development of a wide variety of useful **plastics. All plastics are composed at the molecular level of polymeric organic molecules, very long units of matter in which a short structural unit is repeated over and over again.** All the raw materials (except chlorine) from which the plastics currently are made are obtained from crude oil.

Conceptually, the simplest organic polymer is **polyethylene** (or **polyethene**), the molecules of which are composed of many thousands of —CH_2— units bonded together:

$$\cdots -\underset{\underset{\displaystyle H}{|}}{\overset{\overset{\displaystyle H}{|}}{C}}-\underset{\underset{\displaystyle H}{|}}{\overset{\overset{\displaystyle H}{|}}{C}}-\underset{\underset{\displaystyle H}{|}}{\overset{\overset{\displaystyle H}{|}}{C}}-\underset{\underset{\displaystyle H}{|}}{\overset{\overset{\displaystyle H}{|}}{C}}-\underset{\underset{\displaystyle H}{|}}{\overset{\overset{\displaystyle H}{|}}{C}}-\underset{\underset{\displaystyle H}{|}}{\overset{\overset{\displaystyle H}{|}}{C}}-\cdots$$

This polymer is prepared by combining together many molecules of *ethylene* (ethene), $CH_2{=}CH_2$ and is an example of an **addition polymer.** Depending upon exactly how the polymerization takes place, either **low-density polyethylene (LDPE,** the plastic given the recycling designation number 4) or **high-density polyethylene (HDPE,** the cloudy white or opaque plastic given the recycling number 2) is formed.

There are several other addition polymers similar in constitution to polyethylene, but in which one (or more) of the four hydrogen atoms in each ethylenic —CH_2—CH_2— unit is replaced by another group, X, giving, for example, the polymer

$$\cdots -\underset{\underset{\displaystyle H}{|}}{\overset{\overset{\displaystyle H}{|}}{C}}-\underset{\underset{\displaystyle X}{|}}{\overset{\overset{\displaystyle H}{|}}{C}}-\underset{\underset{\displaystyle H}{|}}{\overset{\overset{\displaystyle H}{|}}{C}}-\underset{\underset{\displaystyle X}{|}}{\overset{\overset{\displaystyle H}{|}}{C}}- \cdots$$

If the atom X attached to every second carbon atom in each chain is chlorine rather than hydrogen, then the clear (or blue tinted) polymer **polyvinyl chloride** (**PVC,** recycling number 3), is obtained. If the substituent X is a *methyl group*, CH_3, rather than a chlorine, we have **polypropylene** (recycling number 5), and if it is a benzene ring we obtain **polystyrene** (recycling number 6). The plastics formed from all these polymers are used extensively in packaging, as indicated by the original uses listed in Table 16-1.

The other plastic which is commonly recycled (number 1), is the clear plastic **polyethylene terephthalate** (**PET**). Its structure is a short chain of

TABLE 16-1	Commonly Recycled Plastics		
Plastic Recycling Number	Acronym and Name of Plastic	Original Use Examples	Recycle Use Examples
1	PET Poly(ethylene terephthalate)	Beverage bottles; food and cleanser bottles; drugstore product containers	Carpet fibers; fiberfill insulation; nonfood containers.
2	HDPE High-density polyethylene	Milk, juice, and water bottles; margarine tubs; crinkly grocery bags	Oil and soap bottles; trash cans; grocery bags; drain pipes
3	PVC (or V) Polyvinyl chloride	Food, water, and chemical bottles; food wraps; blister packs; construction material	Drainage pipes; flooring tiles; traffic cones
4	LDPE Low-density polyethylene	Flexible bags for trash, milk, and groceries; flexible wraps and containers	Bags for trash and groceries; irrigation pipes; oil bottles
5	PP Polypropylene	Handles, bottle caps, lids, wraps, and bottles; food tubs	Auto parts; fibers; pails; trash containers
6	PS Polystyrene	Foam cups and packaging; disposable cutlery; furniture; appliances	Insulation; toys; trays; packaging "peanuts"
7	Other	Various	Plastic "timber": posts, fencing, and pallets

two CH_2 units alternating with a unit of the organic molecule *terephthalic acid*. PET is used in the form of a film (for magnetic tape as well as photographic film), fiber, and molded resin (e.g., plastic bottles).

16.12 Recycling Plastics: Issues and Rates

In the last quarter of the twentieth century, plastics became the symbol of the throwaway society, since much of the product—especially that used in packaging—was designed to be used once and then quickly discarded. Many environmentalists believed that waste plastic was a major culprit in the "garbage crisis." Indeed, molded plastics take up relatively more space in landfills than their percentage by mass because their densities are low, although they are compressed by the weight of the materials placed upon them, as well as by compacting machinery before placement in the landfill. Plastics are the third most common constituent of municipal garbage, following paper and vegetable matter in developed countries (Figure 16-1). The *per capita* annual use of plastics in North America is approximately 80 kg, compared to 60 kg in the European Union and almost 20 kg worldwide.

For a number of reasons, including the fact that landfills—especially throughout Europe—are reaching their capacity, and that many citizens in developed countries are opposed to their incineration, many plastics now are collected from consumers and recycled. As of the mid-1990s, over 80% of the mass of plastics recycled in the United States consisted of PET and HDPE, in approximately equal amounts, with LDPE the only other one of significance. Some countries, such as Sweden and Germany, have made manufacturers legally responsible for the collection and recycling of the packaging used in their products.

There is little doubt that the public in many developed countries has embraced recycling of plastics. About half the urban communities in the United States have curbside recycling programs that include plastics. However, the U.S. recycling rate for PET soda and bottled water bottles dropped from a high of 40% in 1995 to 20% in 2003, but had slowly risen back to 28% in 2009, again with recovery much higher where deposits are required. As in the case of aluminum, Sweden achieves a much higher recycle rate (almost 90%) by requiring refundable deposits on plastic bottles. The majority of recycled PET bottles are used to make fiber for carpet, strapping, and film, although 14% of it is used in food and beverage containers. Only a fraction of the demand by recyclers for PET bottles is met by the current supply. HDPE containers have about the same recycling rate (26%) as PET ones in the United States, although the total number produced and used now is far fewer. Recycled HDPE products include nonfood bottles, plastic garden products such as pots, and plastic lumber.

Several hundred polyethylene carrier bags are produced annually for every man, woman, and child in developed countries on the planet. Most of them are HDPE, with the ethylene derived mainly from natural gas. About

A tremendous amount of floating plastic trash is collected in a huge whirlpool area—twice the size of Texas—called the North Pacific Ocean Gyre. Four ocean currents meet there, creating a vortex that traps the trash.

80% of them are reused at least once, in kitchens as trash bags or by dog owners. Although they constitute less than 1% of garbage sent to landfills, and their incineration can recover the energy of the oil or gas used to make them, they are highly visible forms of pollution when allowed to blow around and collect on streets and beaches. They have been known to clog drainage systems and lead to flooding. In addition, plastic bags can have a devastating effect on marine animals such as turtles and whales, whose stomachs can be blocked by their inadvertent consumption. Although plastic bags are collected for recycling in some locales, countries such as Denmark, Ireland, and Taiwan, and some municipalities in other countries have imposed a tax on them in order to discourage their use. They are totally banned or highly restricted in some countries, and in certain cities or states of others.

There has been much resistance to plastics recycling in some quarters, including many in the plastics industry. Their argument is that virgin plastics are a low-cost material that is made from a relatively low-cost raw material (oil and natural gas). The energy put into making plastics is very small compared to that used to make aluminum or steel from its raw materials. The cost of cleaning used plastic and converting it back into its monomers so it can be polymerized again is substantial, compared to the current cost of oil or gas. Some executives in the plastics industry believe that the natural disposal method for plastics is simply to burn them and utilize the heat energy so provided, especially given the fact that there is little objection by the public to simply burning the great majority (over three-quarters) of the oil produced—in vehicles, domestic furnaces, and power plants. In addition, experiments indicate that the presence of plastics makes the other materials in domestic garbage burn more cleanly and reduces the need for supplemental fossil fuel to be added. Although plastics account for less than 10% of the mass of garbage, they make up more than one-third of its energy content.

Environmentalists counter these arguments by pointing out that if environmental impacts were to be included in determining the cost of virgin materials, recycled plastic would be the cheaper choice. In addition, the combustion of some plastics, notably PVC, produces dioxins and furans, and releases *hydrogen chloride* gas, HCl.

16.13 Recycling Plastics: Biodegradable and Natural Versions

Attempts have been made to produce plastics which would degrade in the environment and thereby reduce their visual presence and the other negative effects they have. Polyethylene polymer does not biodegrade significantly since very long chains of $-CH_2-$ groups are almost impervious to microorganisms. One technique that has been used is to physically add small amounts of the salt of a metal such as cobalt, iron, or manganese or an organic compound to the plastic, which will slowly promote its fragmentation into small pieces, thereby reducing its visibility, and in some instances initiate the oxidation of carbon chains after exposure to light or heat.

However, true degradation requires that the polymer chains be split into fragments small enough for microorganisms to consume them. This can be accomplished by the incorporation of so-called *weak sites* in the hydrocarbon chain during the polymerization process. The incorporation of some carbonyl groups as part of the chain or in a side chain produces a unit capable of absorbing UV from sunlight and initiating C—C bond cleavage, as occurs with aldehydes in smog reactions (Chapters 3 and 17). The presence of some C=C bonds in the polymer chains also constitute weak sites, since they are capable of undergoing sunlight-promoted oxidation by air with resulting chain cleavage, again in analogy with smog reactions.

> Recall that a carbonyl group contains a C=O bond.

Weak sites capable of reaction with water or soil moisture can be achieved by the incorporation of some ester groups into the polymer chain. The limiting case here is *polylactic acid* (PLA), a polymer consisting of a long repeating chain of ester groups. PLA is made from renewable ingredients, either corn starch or sugarcane, and is an example of *bioplastics*. Microorganisms will decompose PLA in commercial composting operations.

ACTIVITY

Use the internet to research new bioplastics such as *Mirel*. Determine how they are produced and how they decompose or can be recycled, and decide how renewable and recyclable they are.

16.14 Recycling Plastics: Techniques

There are four ways to recycle plastics, one physical and three chemical:

1. Reprocess the plastic (a physical process) by remelting or reshaping. Usually the plastics are washed, shredded, and ground up, so that clean, new products can then be made.

2. Depolymerize the plastic back to its component monomers by a chemical or thermal process so that it can be polymerized again.

3. Transform the plastic chemically into a low-quality substance from which other materials can be made.

4. Burn the plastic to obtain energy ("energy recycling," also known as "quaternary recycling").

Examples of the *reprocessing* option include the production of carpet fibers from recycled PET; of plastic trash cans, grocery bags, etc., from recycled HDPE; and of CD and DVD cases and office accessories such as trays and rulers from recycled polystyrene. Further examples of reprocessing are listed for each category of packaging plastic in the last column of Table 16-1.

The *depolymerization* option can be employed with PET and other polymers of the —A—B—A—B—A—B— type in which two different types of units, A and B, alternate in the structure. These **condensation polymers** are produced by combining small molecules that contain A and B units. During the polymerization process, the reactant molecules each lose a component; these then combine with each other. For example, in the production of PET, the molecule *methanol*, CH_3OH, is formed from the OH unit of one component and the CH_3 of the other. In the chemical *depolymerization* process, a catalysis and heat are applied to a mixture of methanol and the plastics in order to *reverse* the original polymerization process and recover the original components:

$$CH_3-A-CH_3 + HO-B-OH + CH_3-A-CH_3 + HO-B-OH + \cdots$$

$$\underset{\text{depolymerization}}{\overset{\text{polymerization}}{\rightleftharpoons}} CH_3-A-B-A-B-A \cdots -A-B-A-B-OH + \text{many } CH_3OH$$

For PET, B is $-CH_2-CH_2-$ and A is

Physical recycling of PET is currently more economically viable than chemical reprocessing.

One of the difficulties in depolymerization of plastics is the fact that organic and inorganic compounds are often added to the original polymer in forming the plastic to modify its physical properties, such as its flexibility, and these must be removed from the plastic or the monomers before they can be reused.

For many addition polymers, it is difficult to devise a process in which the original monomers can be re-formed. For example, the monomer yield in the thermal depolymerization of polystyrene is about 40%, but it is close to zero for polyethylene because the chain will be broken randomly at almost any position and will not exclusively produce two-carbon units.

Examples of the *transformation* option are:

• **Reductive** processes such as the production of synthetic crude oil by the hydrogenation of the plastics or by heating them to a high temperature to "crack" the polymeric molecules, a process which can be run even with mixed plastics. For example, the pyrolysis of polyethylene plastics to monomers that can be converted to lubricants has been proposed.

• **Oxidative** processes such as the gasification of plastics by adding oxygen and steam in order to produce *synthesis gas* (a mixture of hydrogen and carbon monoxide, discussed in Chapter 7).

16.15 Green Chemistry: Development of Recyclable Carpeting

In the United States there were 2.3 billion square yards of carpeting shipped in 2004. This is enough to cover the entire surface area of Washington, D.C., 11 times, or about ¾ of the state of Rhode Island. The Carpet America Recovery Effort (CARE) reports that in 2005, 89% of used carpeting was disposed of in landfills, 3% was incinerated in cement kilns, and only 7% was recycled. Carpeting is not biodegradable, takes up valuable and rapidly declining landfill space, and is ultimately made from petroleum, a nonrenewable resource. As landfill fees escalate—estimates indicate they will double every five years—and the cost of transportation, installation, and replacement of carpeting rise, the demand for an economical and environmentally responsible alternative to landfilling this product has increased.

The two major components of carpeting are the backing and the face fiber. Since the 1970s, PVC has been the material of choice for carpet backing. Environmental and health concerns about PVC include *vinyl chloride* (the monomer used to produce it) and *phthalate* plasticizers (discussed in Chapter 15). Vinyl chloride, a known carcinogen, is volatile. Some people believe this compound outgases from the polymer, whereas others argue the high temperatures used to process PVC should eliminate virtually all of the volatile monomer. Phthalates, which are added to PVC to make it more flexible, migrate out of the polymer and have become widely dispersed in the environment. The growing concern over the effects of phthalates on the reproductive system of humans has previously been discussed (Chapter 15, Section 15.11). In addition, as indicated previously (Section 16.12), when PVC burns it produces toxic by-products including dioxins, furans, and hydrochloric acid.

Although recycling now is commonplace when it comes to paper, glass, and plastic, the recycling of carpeting to most of us is quite a foreign concept. In the United States, only 7% of carpeting is recycled, in part because the PVC backing interferes with the recycling process. *Shaw Industries* won a Presidential Green Chemistry Challenge Award in 2003 for their development of a new type of carpeting that allows for *closed loop recycling*, i.e., recycling of used carpeting back into carpeting with little loss of material. This new type of carpeting, known as *EcoWorx*, employs polyolefin backing and nylon-6 fibers. In addition to lending itself to recycling, the polyolefin backing has low toxicity and eliminates the significant environmental and health concerns related to PVC.

The process of recycling EcoWorx carpeting begins with grinding the used carpeting and then separating the heavier particles (the polyolefin backing) from the lighter ones (the nylon-6 fibers) with a stream of air. The polyolefin particles can then be reused in the extrusion process to produce new backing. The nylon-6 is depolymerized to its monomer (caprolactam) and repolymerized to virgin nylon-6 fibers. These fibers can then be used to form new carpeting, which completes the cycle. Both the backing and the fibers of EcoWorx carpeting can be used again and again to form new carpeting.

There are several additional environmental advantages of EcoWorx, as well as economic advantages:

• Recycling requires no additional petroleum feedstocks; this benefits not only the environment but also the economic bottom line.

• Recycling reduces the amount of landfill space needed.

• The polyolefin-backed carpeting is 40% lighter in weight than carpeting backed with PVC. More carpeting can be shipped in the same truck without surpassing weight limits, thus lowering fuel consumption, cost, and pollution.

• The use of polyolefins eliminates the need for the energy-intensive heating process that is required for PVC. Again, this results in lowering the fuel consumption, cost, and pollution.

• In carpet backing (including PVC backing), significant amounts of inorganic fillers are used to provide loft and bulk. Traditionally, virgin calcium carbonate is employed for this purpose. EcoWorx contains 60% class C fly ash (a waste by-product from the burning of lignite or subbituminous coal) as a filler. Using fly ash as a filler utilizes an unwanted by-product and precludes the use of a virgin chemical.

Returning goods for recycling is often a significant barrier to recycling. In order to overcome this difficulty, Shaw developed a system for returning the carpeting at the end of its useful life, at no cost to the consumer.

16.16 Life-Cycle Assessments

One technique used in minimizing the production of wastes and in pollution prevention is the **life-cycle assessment (or analysis), LCA—an accounting of all the inputs and outputs in the various steps of a product's life, from raw material extraction to final disposal.** This *cradle-to-grave* analysis for a product (or process) can be used to identify the types and magnitudes of its environmental impacts, including both the natural resources used and the pollution produced.

The results of a life-cycle assessment can be used in two ways:

• **to identify opportunities within the life cycle to minimize the overall environmental burden of a product, and**

• **to compare two or more alternative products, in order to determine which is more environmentally friendly.**

An example of the first use is in the production of motor vehicles; the life cycle and most important inputs and outputs are illustrated in Figure 16-4. In designing new cars, life-cycle assessments are used to help minimize pollution

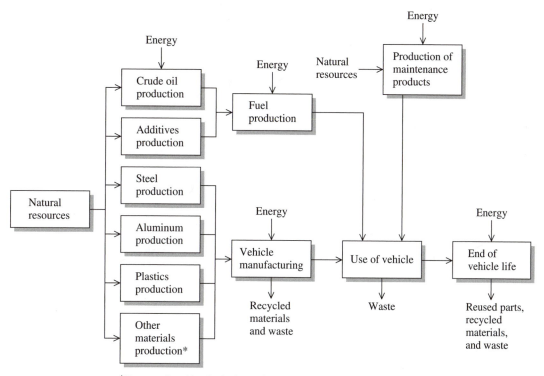

*For example, rubber, lead, glass, paints, and coolants.

FIGURE 16-4 Important input and output components in life-cycle assessments of motor vehicles. [Source: M. Freemantle, "Total Life-Cycle Analysis Harnessed to Generate 'Greener' Automobiles," *Chemical and Engineering News* (27 November 1995): 25.]

while maintaining economic viability. The analysis is particularly useful for identifying new environmental burdens that would arise if other ones are decreased. For example, vehicles could be made much lighter by an increased use of plastics; however, the types of plastics used for this purpose are difficult to recycle, and would increase the eventual burden of solid waste in landfills.

Review Questions 7–12 are based upon the material in the above sections.

Soils and Sediments

The contamination of soil by wastes is not solely a phenomenon of modern times. In Roman times, metal ores were mined and the ores smelted, polluting the surrounding countryside. The production of materials and chemicals in Europe even at the start of the Industrial Revolution produced substantial pollution. However, the extent of contamination and the hazard from the discarded materials expanded greatly in the last century, particularly in the period after World War II.

We begin this section by discussing the nature of soil and of sediments.

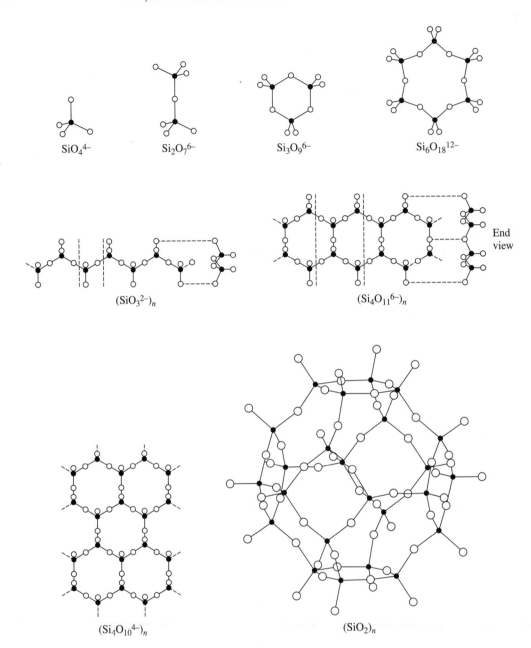

FIGURE 16-5 The common structural units in silicate minerals. Dark circles represent silicon atoms; open circles represent oxygens. [Source: R. W. Raiswell, P. Brimblecombe, D. L. Dent, and P. S. Liss, *Environmental Chemistry* (London: Edward Arnold Publishers, 1980).]

16.17 Basic Soil Chemistry

Soils are composed roughly equally of solid particles, about 90% of which are inorganic in nature and 10% organic matter, and of pore space, about half of which is air and half water. The inorganic particles are residues of weathered rock; chemically they are mainly **silicate minerals.** At the atomic level, these minerals consist of polymeric inorganic structures in which the fundamental unit is a silicon atom surrounded tetrahedrally by four oxygen atoms. Since these oxygen atoms in turn are each bonded to another silicon, etc., the resulting structure is an extended network.

There are many variations on the silicate structural theme. Some networks have exactly twice as many oxygens (formally O^{2-}) as silicons (formally Si^{4+}) and correspond to electrically neutral SiO_2 polymers. In others, some of the tetrahedral sites are occupied by aluminum ions, Al^{3+}, instead of Si^{4+}; the extra negative charge in such networks is neutralized by the presence of other cations such as H^+, Na^+, K^+, Mg^{2+}, Ca^{2+}, and Fe^{2+}. Some common silicon-oxygen structural units are illustrated in Figure 16-5.

The presence of iron oxides can impart a reddish-brown color to soil.

Over time, the weathering of the silicate minerals from rocks can involve chemical reactions with water and acids in which ion substitution occurs. Eventually, these reactions yield substances that are important examples of a class of soil materials known as **clay minerals. A mineral having a particle size less than about 2 μm by definition is a component of the clay fraction of soil.**

In addition to clay, there are several other soil types; the definition of each type depends on particle size, as indicated in Figure 16-6. Notice the factor-of-ten increase in size with each transition of type—the upper boundary for silt is ten times that for clay, and that for fine sand is ten times that of silt, etc. Because the particle size in sand is large, it has a relatively low density and water runs through it easily. In contrast, soils composed of clay are dense and have poor drainage and aeration, since the clay particles form a sticky mass when wet, in contrast to sand and silt particles which do not stick to each other. The best agricultural soils have a combination of soil types present. The range of elemental compositions for major and minor elements in the mineral component of soil is given in Table 16-2.

Clay particles act as colloids in water. Because the clay particles are much smaller than are those of sand or silt, their total surface area per gram

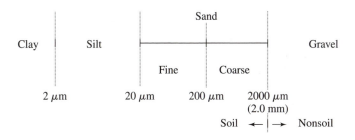

FIGURE 16-6 The soil particle size classification system of the International Society of Soil Science. [Source: G. W. vanLoon and S. J. Duffy, *Environmental Chemistry* (Oxford: Oxford University Press, 2000).]

TABLE 16-2	Element Content of the Mineral Components of Soils			
Major Elements (%)			**Minor Elements (ppm)**	
Si	30–45		Zn	10–250
Al	2.4–7.4		Cu	5–15
Fe	1.2–4.3		Ni	20–30
Ti	0.3–0.7		Mn	~400
Ca	0.01–3.9		Co	1–20
Mg	0.01–1.6		Cr	10–50
K	0.2–2.5		Pb	1–50
Na	Trace–1.5		As	1–20

Source: G. W. vanLoon and S. J. Duffy, *Environmental Chemistry* (Oxford: Oxford University Press, 2000).

is thousands of times larger. Consequently, **most important soil processes in soil occur on the surface of colloidal clay particles.**

Particles of clay possess an outer layer of cations that are bound electrostatically to an electrically charged inner layer, as illustrated in Figure 16-7. The most common cations in soil are H^+, Na^+, K^+, Mg^{2+}, and Ca^{2+}. Depending upon the concentration of cations in the water surrounding the clay particle, the cations on the particle are capable of being exchanged for them. For example, in water rich in potassium ions but poor in other ions, K^+ ions will displace the ions bound to the surface of the clay particle (see Figure 16-7). If, on the other hand, the soil is acidic, the metal ions on the surface will be displaced by H^+ ions and the previously bound metal ions will enter the aqueous phase. Generally, the greater the positive charge on a cation, the more strongly it binds to the particle. Heavy metals dissolved in soil water are often bound to the surface of clay particles.

In addition to minerals, the other important components of soil are organic matter, water, and air. The proportion of each component varies greatly from one soil type to another. The organic matter (1–6% in total, of which about 60% is carbon), which gives soil its dark color, is primarily a material called

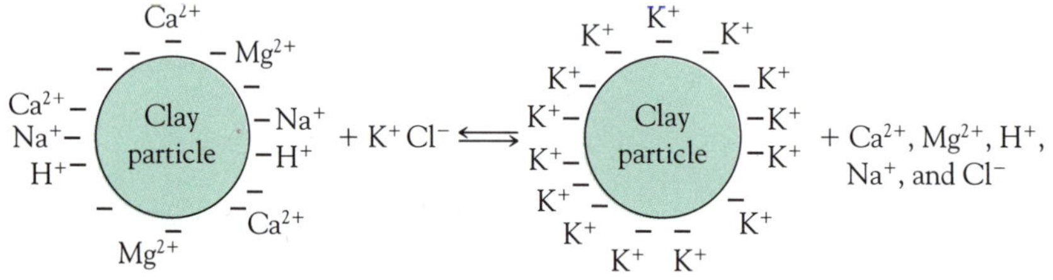

FIGURE 16-7 Ion-exchange equilibria on the surface of a clay particle. The addition of K^+ ions to the soil water displaces the exchange equilibria to the right, whereas removal of K^+ ions from solution displaces it to the left. [Source: R.W. Raiswell, P. Brimblecombe, D. L. Dent, and P. S. Liss, *Environmental Chemistry* (London: Edward Arnold Publishers, 1980).]

humus, which is derived principally from photosynthetic plants, some components of which (such as cellulose and hemicellulose) have already been decomposed by organisms that live in the soil. Much of the organic matter in soil also consists of colloidal particles. The undecomposed plant material in humus is mainly protein and lignin, both of which are polymeric substances that are largely insoluble in water. A significant amount of the carbon in lignin exists in the form of six-membered aromatic benzene rings connected to each other by chains of carbon and oxygen atoms (see Figure 14-4). The nonhumus organic matter in soil consists mainly of carbohydrates of various sizes.

> The concentration of organic matter usually decreases with increasing soil depth.

As a result of the partial oxidation of some of the lignin, many of the resulting polymeric strands contain *carboxylic acid* groups, —COOH. This dark-colored portion of humus consists of **humic** and **fulvic acids** and is soluble in alkaline solutions due to the presence of the acid groups. **Humic acid is *insoluble* in acid solution, whereas fulvic acid is *soluble*.** The humic acid is less soluble in acid than is the fulvic not only because its molecular weight is much greater (range of tens of thousands to millions, compared to hundreds to thousands), but also because its percentage of oxygen content is lower, so there are fewer —OH groups per carbon to hydrogen-bond to water and thereby induce water solubility. The acid groups are often adsorbed onto the surfaces of the clay minerals, to an extent dependent on the distribution of surface charge on the particles. Humic and fulvic acids form colloids that are hydrophilic, whereas those of clay are hydrophobic.

> *Hydrophilic* substances readily dissolve or suspend in water, whereas *hydrophobic* ones do not.

Since some of the carbon in the original plant material is transformed to carbon dioxide and thus lost as a gas to the surroundings, humus is enriched in nitrogen compared to the original plants; its other main components are carbon, oxygen, and hydrogen. Due to decomposition processes occurring in the organic component of soil, the O_2 content of soil air is often only 5–10% rather than 20%, and its CO_2 concentration is often several hundred times that in the atmosphere.

PROBLEM 16-3

The percentage composition of a typical fulvic acid is 50.7% carbon by mass, 45.1% oxygen, and 4.2% hydrogen. Derive the (simplest) empirical formula for the substance.   ●

16.18 The Acidity and Cation Exchange Capacity of Soil

If the soil at the surface contains minerals with elements in a reduced state, their oxidation by atmospheric oxygen can produce acid. An example is the oxidation of sulfur in *pyrite*, discussed previously as acid mine drainage in Chapter 10. Acid rain, of course, provides another source of acidity in certain areas (Chapter 4).

Quantitatively, the ability to exchange cations is expressed as the soil's cation-exchange capacity, CEC, which is defined as the quantity

of cations that are reversibly adsorbed per unit mass of the (dry) material. The quantity of cations is given as the number of moles of positive charge (usually expressed as centimoles or millimoles), and the soil mass is usually taken to be 100 g or 1 kg. Typical values of the CEC for common clay minerals range from 1 to 150 centimoles per kilogram (cmol kg^{-1}). The CEC values are determined in large part by the surface area per gram of the mineral. CEC values for the organic component of soil are high, due to the large number of —COOH groups that can bind to and exchange cations; e.g., that of peat can be as high as 400 cmol kg^{-1}.

PROBLEM 16-4

The CEC for a soil sample is found to be 20 cmol kg^{-1}. What is the CEC value for this sample in units of millimoles per 100 g? ●

Biologically, the exchange of cations by soils is the mechanism by which the roots of plants take up metal ions such as potassium, calcium, and magnesium. Although the roots release hydrogen ions to the soil in exchange for the metal ions, this is not the main reason that soil in which plants grow is often somewhat acidic. Most of the acidity is due to metabolic processes involving the roots and microorganisms in the soil, which result in the production of *carbonic acid*, H_2CO_3, and of weak organic acids.

Rainwater that is acidic releases basic cations from soil particles by exchanging them for H$^+$ ions. The acidity of water flowing through the soil stays low for this reason. However, once the base cations have been exhausted, aluminum ions are released, as discussed in Chapter 4. It is now known that base cations from dust particles, especially those containing calcium and magnesium carbonates, used to neutralize some of the acidity in precipitation, but along with *sulfur dioxide*, their emission from industrial sources has been curtailed in recent decades.

The pH of soil can vary over a significant range for a variety of causes. For example, soils in areas of low rainfall but high concentrations of the soluble salt *sodium carbonate*, Na_2CO_3, become alkaline due to the (hydrolysis) reaction of the *carbonate ion*, CO_3^{2-}, with water, as discussed in Chapter 10. Normal soils containing carbonates usually have pH values from 7.5 to 8.0. Acid precipitation, discussed in Chapter 4, can produce acidic soil, as can the nitrification of ammonia or ammonium ion added as fertilizer (see Section 10.13 for discussion of the acid-producing reaction).

Soils that are too alkaline for agricultural purposes can be remediated by the addition either of elemental sulfur, which releases hydrogen ions as it is oxidized by bacteria to *sulfate ion*, or by the addition of the sulfate salt of a metal such as iron(III) or aluminum, which reacts with the soil's water to extract hydroxide ions and thereby release hydrogen ions:

$$2\,S(s) + 3\,O_2 + 2\,H_2O \longrightarrow 4\,H^+ + 2\,SO_4^{2-}$$
$$Fe^{3+} + 3\,H_2O \longrightarrow Fe(OH)_3(s) + 3\,H^+$$

The pH of water present in the soil is determined by the concentration of hydrogen and hydroxide ions. However, soil has **reserve acidity** owing to the large number of hydrogen atoms present on the —COOH and —OH groups in the organic fraction and on the cation exchange sites on minerals that are occupied by H^+ ions. In other words, **soils act as weak acids, retaining their H^+ ions in a bound condition until acted upon by bases.** Thus the pH of soil tends to be buffered against large increases in pH, since this bound hydrogen ion can be slowly released into the aqueous phase. In the process of **liming, which is the addition of salts such as calcium carbonate to soil,** carbonate ions neutralize acids present in the uppermost soil zones, and thereby produce carbon dioxide and water. Once this process has occurred, calcium ions can replace hydrogen ions in the organic matter or clays. The additional carbonate ions that enter the aqueous phase combine with the newly released H^+ ions, again to produce weakly acidic carbonic acid, which dissociates into carbon dioxide gas and water. Thus liming is a procedure by which the pH of a soil can be raised somewhat, and is the practical method by which acidic soils can be remediated.

16.19 Soil Salinity

In hot, dry climates, salts and alkalinity tend to accumulate in soil since there is little rainfall to leach ions from it and drain them to lower depths. In contrast to other climates, the net movement of water in arid climates is upward, rather than downward in the soil–water evaporation and its loss by transpiration of plants exceeds rainfall. The salts that accompany the upward migration of water remain at or near the surface when the water has escaped. Salt accumulation at the surface also results in semiarid regions due to the use of poor quality irrigation water, whose salt content remains after the water has evaporated.

There is very little organic matter in soils of arid areas.

Ions are also liberated at the surface of soil in the weathering of otherwise insoluble minerals. A simple example is the reaction of *olivine*:

$$Mg_2SiO_4 + 4\ H_2O \longrightarrow Si(OH)_4 + 2\ Mg^{2+} + 4\ OH^-$$

Additional hydroxide ion is produced when the *silicate ion*, SiO_4^{4-}, produced by mineral weathering reacts as a strong base with water.

As a general rule, hydrolysis of silicate minerals at the surface produces cations and hydroxide ions. In nonarid climates, the hydroxide is neutralized by acids that are naturally produced in the soil (see later section), but this does not occur in arid climates. The hydroxide reacts with atmospheric carbon dioxide that dissolves in water, to produce *bicarbonate* and carbonate ions:

$$OH^- + CO_2 \longrightarrow HCO_3^-$$
$$HCO_3^- \longrightarrow H^+ + CO_3^{2-}$$

Consequently, **bicarbonate and carbonate salts accumulate in arid soils.** If the predominant cations in the soil are calcium and magnesium, most of the carbonate ions will be largely locked away as their insoluble carbonate salts. However, if the predominant cations are sodium and potassium, the soil when moist will have a high pH since the carbonate salts of these ions are water soluble and the free CO_3^{2-} will act as a base:

$$CO_3^{2-} + H_2O \rightleftharpoons HCO_3^- + OH^-$$

Increasing soil salinity is a major problem in Australia, especially in regions where wheat and other shallow-rooted crops have replaced natural, long-rooted vegetation. This replacement, plus irrigation of crops such as rice and cotton, has resulted in a rise of the water table and, with it, the salt that was formerly deep in the soil.

16.20 Sediments

Sediments are the layers of mineral and organic particles, often fine-grained, that are found at the bottoms of natural water bodies such as lakes, rivers, and oceans. The ratio of minerals to organic matter in sediments varies substantially, depending upon location. Sediments are of great environmental importance because they are sinks for many chemicals, especially heavy metals and organic compounds such as PAHs and pesticides, from which they can be transferred to organisms that inhabit this region. Thus the protection of sediment quality is a component of overall water management.

The map in Figure 16-8 shows watersheds in the continental United States where sediments are sufficiently contaminated to pose environmental risks. According to a U.S. EPA report, 7% of all watersheds pose a risk to

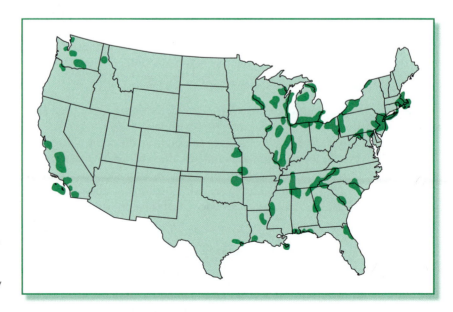

FIGURE 16-8 United States watersheds where contaminated sediments may pose environmental risks. [Source: U.S. EPA, in B. Hileman, "EPA Finds 7% of Watersheds Have Polluted Sediments," *Chemical and Engineering News* (26 January 1998): 27.]

people who eat fish from them, and to the fish and wildlife themselves. The two pollutants found at high levels most frequently at contaminated sites are PCBs and mercury, although *DDT* (and its metabolites) and PAHs were also found at high concentrations at many sampling stations.

The transfer of hydrophobic organic pollutants to organisms may proceed by intermediate transfer to *pore water*, which is the water present in the microscopic pores that exist within the sediment material. Organic chemicals equilibrate between being adsorbed to the solid particles and being dissolved in the pore water. For this reason, pore water is often tested for toxicity in determining sediment contamination levels.

16.21 The Binding of Heavy Metals to Soils and Sediments

The ultimate sink for heavy metals, as well as for many toxic organic compounds, is deposition and burial in soils and sediments. However, heavy metals often accumulate in the top layer of the soil, and are therefore accessible for uptake by the roots of crops. For these reasons it is important to know the nature of these systems and how they function.

Humic materials have a great affinity for heavy-metal cations, and extract them from the water that passes through them by the process of ion exchange. The binding of metal cations occurs largely because of the formation of complexes with the metal ions by —COOH groups in the humic and fulvic acids. For example, for fulvic acids, the most important interactions probably involve a —COOH group and an —OH group on adjacent carbons of a benzene ring in the polymeric structure, where the heavy-metal M^{2+} ion replaces two H^+ ions:

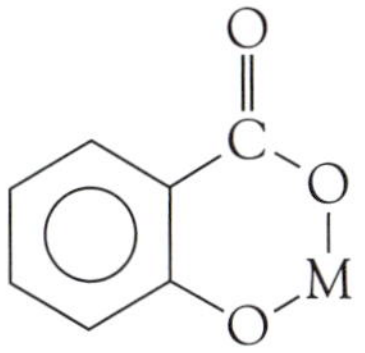

M = heavy metal

Humic acids normally yield water-insoluble complexes, whereas complexes of smaller fulvic acids are water-soluble.

PROBLEM 16-5

Draw the structure that would be expected if a dipositive metal ion M^{2+} were to be bound to two —COO⁻ groups on adjacent carbons on a benzene ring. ●

Heavy metals are retained by soil in three ways:

- **by adsorption onto the surfaces of mineral particles;**
- **by complexation by humic substances in organic particles; and**
- **by precipitation reactions.**

The precipitation processes for mercury ions and cadmium ions involve the formation of the insoluble sulfides HgS and CdS when the free ion in solution encounters *sulfide ion*, S^{2-}. Significant concentrations of aqueous sulfide ion are present near lake bottoms in summer months when the water is usually oxygen-depleted, as discussed in Chapter 10. However, the total concentration of mercury in the water of soils can exceed the limits set by the solubility product of HgS because some of the mercury will take the form of the moderately soluble molecular compound $Hg(OH)_2$ and does not participate in the equilibrium with the sulfide.

In acidic soils, the concentration of Cd^{2+} can be substantial, since this ion adsorbs only weakly onto clays and other particulate materials. Above a pH of 7, however, Cd^{2+} precipitates as the sulfide, carbonate, or phosphate, since the concentration of these ions increases with increasing hydroxide ion levels. Thus the liming of soil to increase its pH is an effective way of tying up cadmium ion and thereby preventing its uptake by plants.

Like many other chemicals, heavy-metal ions often are adsorbed onto the surfaces of particulates, especially organic ones that are suspended in water, rather than simply being dissolved in water as free ions or as complexes with soluble biomolecules such as fulvic acids. The particles eventually settle to the bottom of lakes and become buried when other sediments accumulate on top of them. This burial represents an important sink for many water pollutants, and is a mechanism by which the water is cleansed. Before it is covered by subsequent layers of sediments, however, freshly deposited matter at the bottom of a body of water can recontaminate the water above it by desorption of the chemicals, since adsorption and desorption establish an equilibrium. Furthermore, the adsorbed pollutants can enter the food web if the particles are consumed by bottom-growing and bottom-feeding organisms.

Just as the total concentration of organic material in sediments may not be a good measure of the amounts that are biologically available, the same is true for the levels of heavy-metal ions present. Different sediments having the same total concentration of the ions of a heavy metal can vary by a factor of at least 10 in terms of the toxicities to organisms that arise from the metal. This variability occurs principally because sulfides in the sediments control the availability of the metals. If the concentration of sulfide ions *exceeds* the total of that of the metals, virtually all the metal ions will be tied up as insoluble sulfide salts and will be unavailable biologically at normal pH values. However, if the sulfide concentration is *less* than that of the metals, the difference is biologically available.

The sulfide ion that is available to complex with metals is that which will dissolve in cold aqueous acid, and it is termed acid volatile sulfide, AVS. Industrially polluted sediments may have AVS concentrations of hundreds of micromoles of sulfur per gram, whereas uncontaminated sediments from oxidizing environments can have values as low as $0.01\ \mu\mathrm{mol\ g^{-1}}$.

Recall from Chapter 10 that carbonate predominates over bicarbonate under alkaline conditions.

Recall from Chapter 12 that the toxic heavy metals mercury, cadmium, and lead form very insoluble sulfide salts HgS, CdS, and PbS; analogous insoluble salts are formed by nickel and many other metals.

Although mercury in the form of Hg^{2+} is firmly bound to sediments and does not redissolve much into water, environmental problems have arisen in several bodies of water owing to the conversion of the metal into *methylmercury* and its subsequent release into the aquatic food chain. The overall cycling of mercury species between air, water, and sediments was illustrated in Figure 12-2.

As previously discussed, anaerobic bacteria methylate the *mercuric ion* to form $Hg(CH_3)_2$ and CH_3HgX, which then rapidly desorb from sediment particles and dissolve in water, thereby entering the food web. Although the levels of methylmercury dissolved in water can be extremely low, of the order of hundredths of parts per trillion, a biomagnification factor of 10^8 results in ppm-range concentrations in the flesh of some fish. The devastating consequences of methylmercury poisoning have been already described (Chapter 12).

Excavation and analysis of the sediments at the bottom of a body of water can yield a historical record of contamination by various substances. For example, the curves in Figure 16-9 show the levels of mercury and of lead in the sediments of the harbor in Halifax, Nova Scotia, as a function of depth and therefore also of year. For decades, raw sewage has been dumped into this harbor; consequently, its sediments are a historical record of the levels of pollutants in sewage. The metal pollution peaked about 1970, having begun to increase dramatically about 1900 (Figure 16-9). These trends are typical also of heavy-metal levels in other water bodies, such as the Great Lakes; the characteristic decreases of the past few decades are due to the imposition of pollution controls.

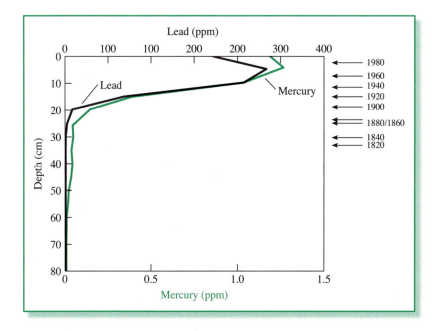

FIGURE 16-9 Lead and mercury concentrations in the sediments of Halifax Harbor versus depth (and therefore year of deposit). [Source: D. E. Buckley, "Environmental Geochemistry in Halifax Harbour," *WAT on Earth* (1992): 5.]

16.22 Mine Tailings

In modern times, many minerals (and in some cases fossil fuels) are extracted from much, much larger quantities of rock (or sand, etc.) than previously, since the remaining supplies of the minerals occur in dilute form. This practice produces huge quantities of unwanted, crushed rock in the form of dry, coarse-grained waste that needs to be disposed of, usually in slag heaps or landfills close to the mine. Eventually these waste piles are covered with soil and vegetation.

A more important environmental problem arises from disposal of the **tailings of mining processing, which are fine-grained slurries that are more mineral-rich and often contain chemicals such as *cyanide* that were used to extract or process the ore.** The toxic components of tailings represent a potential source of pollution to local surface water, groundwater, and soil.

To prevent their dispersal into the environment, tailings are usually deposited as slurries in dams constructed on-site for the purpose. To prevent leakage into the soil, a clay or geomembranic liner is usually incorporated at the landfill's base. Over time, the solid settles to the bottom of the dam and the water evaporates or is drained off, i.e., the tailings become *dewatered*, though this can take a long time to achieve. In some cases, toxic materials such as heavy metals, nitrates, or excess acidity are removed by treatment of the tailings. Eventually, to establish a vegetated cover, organic matter and fertilizer must be added, since the dried tailings themselves contain little or no organic matter and consequently are sterile as well as hostile to plants.

If the crushed rock or tailings contain *iron pyrites*, their exposure to oxygen will produce *sulfuric acid* by the same series of reactions discussed as acid mine drainage in Chapter 10. The acidity can be neutralized by the continuous addition of limestone.

The most important environmental problem associated with a tailings dam is its potential failure, resulting in a catastrophic discharge of the tailings into a waterway and/or onto land. The failure might be due to a flood, earthquake, or simply the loss of stability of the dam to pressure over time. A number of such incidents have occurred within the last decade—e.g., in Spain—with devastating results to wildlife, fish, and, in some cases, agricultural land.

An alternative to the storage of tailings on land is their disposal into the deep ocean by using pipelines reaching down 100 m or more. Lack of oxygen there will slow the process of oxidation, and within a few years the tailings will be covered by other debris. Some biologists are dubious about this plan, given that the environment for bottom-dwelling organisms will suffer. Heavy-metal contamination of fish has occurred in areas where this means of disposal is practiced. Due to the fine-grained nature of the tailings, dispersal over wide areas of the ocean floor occurs.

16.23 The Remediation of Contaminated Soil

Contaminated soil is found near waste disposal sites, especially those accepting toxic wastes, and also near chemical plants, pipelines, and gasoline stations. Even areas thought to be rather pristine can have localized areas of contaminated soil. For example, the large-scale forestry industry in New Zealand has resulted in the contamination of several hundred sites where lumber is treated by the preservative *pentachlorophenol,* PCP (Chapter 14). Furan and dioxin contaminants of the PCP are also found in the soils. All three contaminant types have now leached into the groundwater at some of the sites, and have begun to bioacccumulate in the food chain.

In 1980, the United States federal government created the EPA-administered **Superfund** program to force the cleanup of sites contaminated with hazardous substances. The sites involved are waste dumps that have been abandoned or were illegal and that are potentially endangering public health by their releases, real or potential. By late 2010 there were 1280 sites on their Priority list, 347 having been delisted by that time. The remediation projects have used a wide variety of technologies, many of them innovative, as listed in Table 16-3.

TABLE 16-3	U.S. EPA Source Treatment Projects 1982–2004 and Selected Technologies 2005–2008		
Technology	Projects 1982–2004 Plus Technologies 2005–2008	% of Projects from 1982–2004	% of Treatment Technologies Selected from 2005–2008
In Situ			
Soil vapor extraction	276	26	14
Bioremediation	62	6	4
Solidification and stabilization	56	4	7
Multiphase extraction	54	4	5
Chemical treatment	24	2	4
Thermal treatment	22	1	5
Flushing	19	2	1
Phytoremediation	8	1	1
Ex Situ			
Solidification and stabilization	303	18	14
Incineration (off site)	111	11	3
Physical separation	48	2	13
Chemical treatment	11	1	1
Soil vapor extraction	9	1	1
Neutralization	9	1	1

Source: Adapted from EPA Superfund Remedy Report, 13th Ed., 2010. Office of Solid Waste and Emergency Response.

The general categories of technologies currently available for the remediation of contaminated sites are

- containment or immobilization,

- mobilization, and

- destruction.

In general, the remediation technologies can be applied *in situ*, i.e., in the place of contamination, or *ex situ*, i.e., after removing the contaminated matter to another location. Owing to the costs and risks, such as air pollution arising from excavation, *in situ* processes usually are preferred.

- Among the techniques associated with *in situ* **containment,** i.e., the isolation of wastes from the environment, are **capping of the contaminated site,** especially with clay, and/or the **imposition of cut-off walls of low permeability** that prevent the lateral spread of contaminants. *Ex situ* containment would correspond to the placing of the excavated soil in a special landfill. **Immobilization** techniques, including solidification and stabilization, are especially useful for inorganic wastes, which tend to be difficult to treat by other methods. Stabilization can often be achieved by adding a substance to convert a heavy-metal ion into one of its insoluble salts, such as the sulfide in the case of mercury and lead, or the oxide in the case of chromium. A concentrated waste can be solidified by reaction with *portland cement*, for example, or by entombing the wastes in molten glass in the process of **vitrification.** By these techniques, the solubility and mobility of the contaminants are reduced.

- **Mobilization** techniques are mainly accomplished *in situ*, and include **soil washing and the extraction of contaminant vapor from soil for highly volatile, water-insoluble contaminants** such as gasoline. Heating of the soil in order to increase the rate of evaporation and air injection wells are sometimes used in conjunction with **soil vapor extraction, in which the contaminants are usually removed by drilling wells in the soil and applying vacuum extraction.** As indicated in Table 16-3, this technique has been the most frequently used innovative technology at *Superfund* sites in the United States, although its use as a percentage of all *in situ* technologies has been dropping in recent years. A related technology is **thermal desorption, in which wastes are heated to cause volatile organic compounds to vaporize.** Both soil vapor extraction and thermal desorption are useful to remediate both volatile and semi-volatile organic compounds, the latter including many PAHs.

- *In situ* **soil washing is accomplished by injecting fluids through wells into subsurface soil, and collecting them in other wells.** The fluid can simply be water, which will remove water-soluble constituents, or an aqueous solution that is acidic or basic in order to remove basic and acidic contaminants, respectively. Other options in soil washing include the use of a solution containing chelating agents (defined in Section 11.28) such as *EDTA* in order to remove metals, and of oxidizing agents in order to oxidize and thereby solubilize previously insoluble species. The solvents used to extract the metal–organic complex from the aqueous environment of the soil include hydrocarbons and supercritical carbon dioxide (a substance described previously in Box 3-2). In order for the resulting complex to be electrically neutral and therefore preferentially soluble in the organic phase, a chelating agent with acidic hydrogens that are replaced by the metal is usually employed. If the metal exists as an oxyanion, e.g., chromium as Cr(VI), it may have to first be reduced in oxidation number before it will bind to a chelating agent.

Sometimes the washing solution uses **surfactants,** *surface-active agents.* These are substances such as detergents that possess both hydrophobic and hydrophilic components within the same molecule and which for that reason can increase the mobilization of hydrophobic contaminants into the aqueous phase. **Biosurfactants** produced by microbes have been discovered which can selectively remove certain heavy metals such as cadmium from soil. Currently, soil washing and flushing are the most common innovative technologies used to remove metals at *Superfund* sites.

The containment, mobilization, and immobilization techniques by themselves do not result in the *elimination* of the hazardous contaminants. **Destruction** techniques, principally incineration and bioremediation, do result in permanent elimination because they chemically or biochemically transform the contaminants. Organic contaminants in soil can be oxidized (mineralized) by feeding the excavated soil into the combustion chamber of an incinerator, or by using incineration or one of the specialty oxidation techniques, to be discussed later, to treat the substances that have been extracted from the soil. *Bioremediation* uses metabolic activities of microorganisms to destroy toxic contaminants, and also is discussed in detail later.

• **Electrochemical techniques** are sometimes used to remediate soil. Placing electrodes in the contaminated ground and applying a dc voltage between them results in ion transport within the soil: the ions travel within the groundwater electrolyte. If heavy-metal ions are dissolved in the electrolyte water, they will eventually move to the (negatively charged) cathode and be deposited on it. Indeed, other metal ions tend to be dissolved from their positions on negatively charged clay surfaces in the process, since hydrogen ion is released at the anode (as water is electrolyzed) and subsequently migrates in the groundwater toward the cathode (see Figure 16-10). Recall that heavy-metal ions are much more soluble in an acidic than in a neutral or alkaline environment.

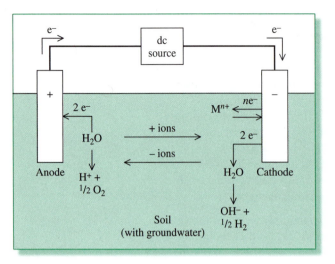

FIGURE 16-10
Electrochemical remediation of metal-contaminated soil.

In some applications of such **electrokinetic** methods, the process is stopped before the metals are deposited on the cathode but after they have migrated most of the distance toward it. The flow of hydroxide ion from the cathode (see Figure 16-10) precipitates many metals in any event. The metal-rich soil surrounding the electrode is then excavated and cleansed by washing it or by other techniques. By repeating this procedure and inserting the electrodes at different locations in the soil, more efficient extraction of the heavy metals is possible. The electrokinetic technique has been used successfully for copper, lead, cadmium, mercury, chromium, and some radioactive metals. The method is high in energy demand, and hence in cost, because so much of the applied electrical potential is lost by the electrolysis of soil water.

- ***In situ* chemical oxidation** can often be used to remediate soils (and groundwater) contaminated with chlorinated solvents and/or with *BTEX* (Chapter 11). The oxidizing agent is injected directly by means of a well into the underground waste and may or may not be extracted at the other side of the contaminated zone. Typically, a salt of *permanganate ion*, MnO_4^-, is used for deposits of MTBE and of the C_2 chlorinated solvents trichloroethene and perchloroethene, whereas *ozone* or *hydrogen peroxide* is used as the oxidizing agent for BTEX and PAHs or, in some cases, for the C_2 chlorinated solvents as well. The MnO_2 product of oxidation by permanganate is a natural constituent of soil. Hydrogen peroxide is often supplemented with ferrous salts (together called *Fenton's Reagent*) in order to create hydroxyl radical, OH, by a reaction described in Chapter 11.

Recall that ferrous ion is Fe^{2+}.

16.24 The Analysis and Remediation of Contaminated Sediments

Many river and lake sediments are highly contaminated by heavy metals and/or toxic organic compounds. Such sediments act as sources to recontaminate the water that flows above them.

One way to determine the extent of contamination of a sediment is to analyze a sample of it for the total amounts of lead, mercury, and other heavy metals that are present. However, this technique fails to distinguish toxic materials that are already present in either their toxic form or in a form potentially able to be resolubilized into the water from those that are firmly bound to sediment particles and that would be unlikely to become resolubilized. Thus a more meaningful test of the sediment involves extracting from a sample the substances that are soluble in water or in a weakly acidic solution, and analyzing the resulting liquid. In this way, the permanently bound and therefore inactive toxic agents can be left out of the account. Finally, the effect of sediments upon organisms that usually dwell in or on them can be determined by adding organisms to a sediment sample and seeing whether they survive and reproduce normally.

Several types of remediation have been used for sediments that are highly contaminated. The simplest solution is often to simply cover the contaminated sediments with clean soil or clay or sediment, thereby placing a barrier between the contaminants and the water system. In other instances, the contaminated sediments are dredged from the bottom of the water body to a depth below which the contaminant concentration is acceptable. If the sediments are high in organic content and inorganic nutrients, they are often used to enhance soil that is used for nonagricultural purposes. In some cases, the sediment can be used for cropland provided that its heavy metals and other contaminants will not enter the growing food. Cadmium is usually the heavy metal of greatest concern in such sediments; if the pH of the resulting soil is 6.5 or greater, most of the cadmium will not be soluble and so a higher total concentration often is tolerated.

Several chemical and biological methods of decontaminating sediments are in use. For example, treatment with calcium carbonate or lime increases the pH of the sediments and thereby immobilizes the heavy metals. In some situations, contaminated sediments are simply covered with a chemically active solid, such as limestone (calcium carbonate), *gypsum* (calcium sulfate), *iron(III) sulfate*, or activated carbon that gradually detoxifies the sediments. In other cases, the sediments are first dredged from the bottom of the water body and then treated. Heavy metals are often removed by acidifying the sediments or treating them with a chelating agent; in both cases, the heavy metals become water-soluble and leach from the solid.

For organic contaminants, extraction of toxic substances using solvents and destruction of them by either heat treatment of the solid or the introduction of microorganisms that consume them are the main options. The cleaned sediments can then be returned to the water body or spread on land. These techniques for removing metals and organics from sediments are also often useful on contaminated soils.

16.25 Bioremediation of Wastes and Soil

Recall from Chapter 11 that bioremediation involves the use of living organisms, especially microorganisms, to degrade environmental wastes. It is a rapidly growing technology, especially in collaboration with genetic engineering, which is used to develop strains of microbes having the ability to deal with specific pollutants. Bioremediation is used particularly for the remediation of waste sites and soils contaminated with semivolatile organic compounds such as PAHs. It is a popular method for use at *Superfund* sites (see Table 16-3).

Bioremediation exploits the ability of microorganisms, especially bacteria and fungi, to degrade many types of wastes, usually to simpler and less toxic substances. Indeed, for many years it was thought that microorganisms could and would eventually biodegrade *all* organic substances, including all pollutants, that entered natural waters or the soil. The discovery that some compounds, chloroorganics especially, were resistant to rapid biodegradation was responsible for correcting that misconception. Substances resistant to biodegradation are termed **recalcitrant** or **biorefractory.** In addition, other substances, including many organic compounds, biodegrade only partially; they are transformed instead into other organic compounds, some of which may be biorefractory and/or even more toxic than the original substances. An example of the latter phenomenon is the potential conversion of the once widely used solvent *1,1,1-trichloroethane* (CH_3CCl_3, methyl chloroform), into carcinogenic vinyl chloride, $CHCl=CH_2$, by a combination of abiotic and microbial steps.

If a bioremediation technique is to operate effectively, several conditions must be fulfilled:

• **The waste must be susceptible to biological degradation and in a physical form that is susceptible to microbes.**

• **The appropriate microbes must be available.**

• **The environmental conditions such as pH, temperature, and oxygen level must be appropriate.**

An example of biodegradation is the degradation of aromatic hydrocarbons by soil microorganisms when land is contaminated by gasoline or oil. The biggest bioremediation project in history before the 2010 Gulf of Mexico oil spill (discussed in Box 6-3) was the treatment of some of the oil spilled by the *Exxon Valdez* tanker in Alaska in 1989. The bioremediation consisted of adding nitrogen-containing fertilizer to more than 100 km of the shoreline that had been contaminated, thereby stimulating the growth of indigenous microorganisms, including those that could degrade hydrocarbons. Both surface and subsurface oil was biodegraded in this operation. Some of the aromatic components in crude oil in marine spills become more susceptible to biodegradation once they are photooxidized by sunlight into more polar species.

In contaminated soils, the biodegradation of PAHs is slow since they are strongly adsorbed onto soil particles and are not readily released into the aqueous phase, where biodegradation could occur. PAH-contaminated soils are especially prevalent at gasworks sites used in the 1850–1950 period for the production of *town gas* (now called synthesis gas) from coal or oil (Chapter 7). The pollution is mainly in the form of deposits of **tars,** which are waste products that are high-molecular-weight organic liquids denser than water that are

Recall that methyl chloroform is now a banned ozone-depleting substance, as discussed in Chapter 2.

mixed with the soil and that contain high levels of both BTEX and PAHs. Unfortunately, groundwater that comes into contact with the tar can become contaminated if some of the tar's more soluble constituents such as *benzene* and *naphthalene* dissolve, although most of the tar is insoluble in water. The other common soil contaminants at gasworks sites are *phenols* and cyanide.

Bioremediation processes take place under either aerobic or anaerobic conditions. In the **aerobic treatment** of wastes, aerobic bacteria and fungi that utilize oxygen are employed; chemically the processes are oxidations, as the microorganisms use the wastes as food sources. In some examples of the aerobic bioremediation of soils, oxygen-saturated water is pumped through the solid to ensure that O_2 availability remains high. For example, about 85,000 tonnes of soil contaminated by gasoline, oil, and grease from a fuel plant in Toronto were decontaminated by first encasing it in plastic and then pumping air, water, and fertilizer into it in order to encourage the population of aerobic bacteria to multiply and devour the hydrocarbon pollutants. The bioremediation process took only three months.

There are many examples of wastes that can be degraded by anaerobic microorganisms, although usually the most rapid and complete biodegradations are obtained with aerobic microorganisms. The **anaerobic treatment** usually works best when there are some oxygen atoms in the organic wastes themselves. In general, the process corresponds to the *fermentation* discussed in Chapter 7, whereby biomass of approximate empirical formula CH_2O decomposes ultimately to methane and carbon dioxide:

$$2\ CH_2O \longrightarrow CH_4 + CO_2$$

An advantage of anaerobic biodegradation is its production of *hydrogen sulfide*, H_2S, which *in situ* precipitates heavy metal ions as the corresponding insoluble sulfides.

Several strategies are used in bioremediation that are based upon the fact that microorganisms evolve quickly—due to their short reproductive cycle—and they develop the ability to use the food source at hand, even if it is chemical wastes.

- One remediation strategy is to isolate the best degradation microorganisms that have flourished at a contaminated site, grow a large population of them in the laboratory, and finally return the enhanced population to the site.

- Another strategy is to introduce microorganisms that were found at other sites to have been useful in degrading a particular type of waste, rather than wait for them to evolve. Unfortunately, microorganisms adapted to one environment may not be capable of surviving at another if additional hostile contaminants are present.

- A third strategy is to encourage an increase in the population of indigenous microorganisms at the site by adding nutrients to the wastes and by ensuring that the acidity and moisture levels are optimum.

Rather than waiting for microorganisms to evolve spontaneously, an alternative strategy is to use genetic engineering to develop microbes that are specifically designed to attack common organic pollutants. However, regulatory authorities are reluctant to allow genetically modified organisms to be released into the environment, since public opposition to such a move would be substantial.

In addition to bacteria, **white-rot fungi** can be used in biodegradation. This species protects itself from pollutants by degrading them *outside* its cell wall by secreting enzymes that catalyze the production of *hydroxyl radicals* and other reactive chemicals. Since OH, particularly, is quite nonspecific about which substances it oxidizes, the fungi are useful in degrading mixtures of wastes including various chlorinated substances such as DDT and 2,4,5-T, as well as PAHs.

16.26 Bioremediation of Organochlorine Contamination

It has been discovered that PCBs (Chapter 14) in sediments undergo some biodegradation. PCB molecules with relatively few chlorine atoms undergo oxidative aerobic biodegradation by a variety of microorganisms. For the reaction to begin, a pair of nonsubstituted carbons—one ortho to the point of connection between the rings and one meta next to it—must be available on one of the benzene rings. After 2,3 hydroxylation at these two sites, the 1,2 carbon–carbon bonds to the ortho carbons split in sequence, thereby destroying the second ring and producing compounds that readily degrade (see Figure 16-11).

PROBLEM 16-6

Deduce the chlorine substitution positions on the benzene ring in the benzoic acid that results from the aerobic degradation of 2, 3′, 5-trichlorobiphenyl. ●

Although PCB molecules that are heavily substituted by chlorines will not undergo this process, since they are unlikely to have adjacent unsubstituted carbons, they will instead undergo **anaerobic degradation,** as will perchlorinated organic compounds. In the absence of oxygen, anaerobic microorganisms facilitate the removal of chlorine atoms and their replacement by

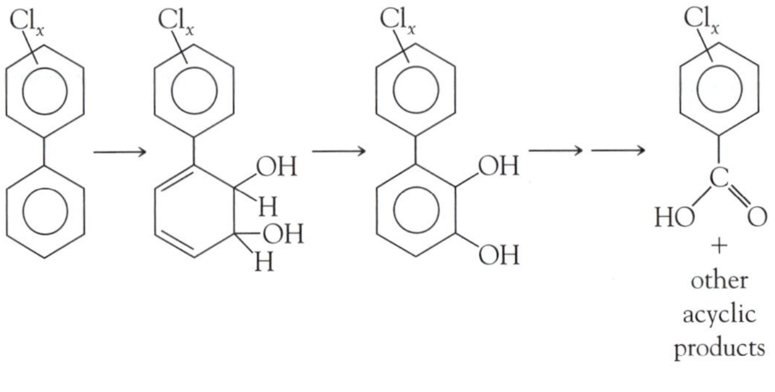

FIGURE 16-11 Example of the aerobic degradation of PCB molecules.

hydrogen atoms, apparently by a *reductive dechlorination* mechanism that initially involves the addition of an electron to the molecule. In the case of PCBs, this reductive dechlorination occurs most readily with meta and para chlorines. Apparently, steric effects block the ortho position from being attacked in most anaerobic mechanisms. Thus the products of anaerobic treatment here are ortho-substituted congeners, ultimately 2-chlorobiphenyl and 2,2′-dichlorobiphenyl especially.

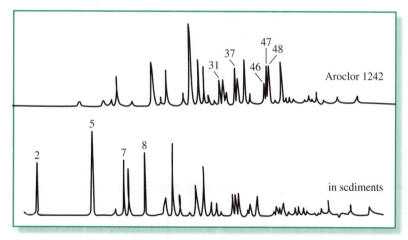

Since dioxin-like toxicity of PCBs requires several meta and para chlorines, **the anaerobic degradation process significantly reduces the health risk from PCB contamination.** Of course, once adjacent ortho and meta sites without chlorine are available, aerobic microorganisms—if available—could degrade the biphenyl structure, as discussed above. Figure 16-12 illustrates the change in composition of a commercial PCB sample after it has resided for some time in sediment that contains anaerobic bacteria.

FIGURE 16-12 Concentration profiles (chromatograms) for various PCB congeners of commercial Aroclor 1242 before and after biodegradation (different scales) in a sediment. The major components of the peaks correspond to the following positions of chlorine substitution: 2 = 2-chlorobiphenyl; 5 = 2,2′ and some 2,6; 7 = 2,3′; 8 = 2,4′ and some 2,3; 31 = 2,2′, 5,5′; 46 = 2,4,4′, 5; 47 = 2,3′, 4′,5; 48 = 2,3′, 4,4′. Notice that biodegradation produces ortho-substituted congeners that are present in low concentrations or absent in the original sample. [Source: D. A. Abramowicz and D. R. Olson, *CHEMTECH* (July 1995): 36–40.]

16.27 Phytoremediation of Soils and Sediments

The technique of **phytoremediation, the use of vegetation for the *in situ* decontamination of soils and sediments of heavy metals and organic pollutants,** is an emerging technology. As illustrated in Figure 16-13, **plants can remediate pollutants via three mechanisms:**

• **the direct uptake of contaminants and their accumulation in the plant tissue ("phytoextraction"),**

• **the release into the soil of oxygen and biochemical substances such as enzymes that stimulate the biodegradation of pollutants, and**

• **the enhancement of biodegradation by fungi and microbes located at the root–soil interface.**

FIGURE 16-13
Mechanisms of phytoremediation by a plant.

Advantages of phytoremediation include its relatively low cost, aesthetic benefits, and nonobtrusive nature. It is used at a few *Superfund* sites (Table 16-3).

Certain plants are **hyperaccumulators of metals—they are able to absorb much higher levels of these contaminants than average** (by a factor of at least 10–100, to yield a contaminant concentration of 0.1% or more) through their roots and to concentrate them much more than do normal plants. This ability probably evolved over long periods of time as the plants grew on natural soils that contained high concentrations of pollutants, especially heavy metals. In bioremediation, these plants are deliberately planted on contaminated sites and then harvested and burned. In some instances, the resulting ash is so concentrated in metal that it can be mined!

Phytoremediation is an attractive technique because metals are often difficult to extract with other technologies because their concentration is usually so small. For example, the shrub called the *Alpine pennycress* has the ability to hyperaccumulate cadmium, zinc, and nickel. Phytoremediation has been successfully used for extraction, for example, of cadmium by both water hyacinths and various grasses; of lead and copper by alfalfa; and of chromium by Indian mustard, sunflowers, and buckwheat. In some cases, a chelating agent is added to the soil to enhance the accumulation of metals by the plant. Scientists are experimenting with various types of plants that can extract lead from soils.

One difficulty with phytoremediation is that hyperaccumulators usually are plants that are slow to grow and therefore slow to accumulate metals. However, fast-growing poplar trees show promise of being efficient phytoremediators. One recently developed type of hybrid poplar efficiently absorbs trichloroethene from hazardous waste sites and from groundwater. In general, there is a need to harvest the plants before they lose their leaves, or begin to decay, in order that contaminants do not become dispersed or return to the soil.

Plants can efficiently take up organic substances that are moderately hydrophobic, with log K_{ow} values (Chapter 13) from about 0.5 to 3, a range that includes BTEX components and some chlorinated solvents. Substances that are more hydrophobic than the upper limit of this range bind so strongly to roots that they are not easily taken up within the plant. Once it absorbs a substance, the plant may store it by transformation within its lignin component, or it may metabolize it and release the products into the air. However, plants usually lack the enzymes required for full metabolism of many organic compounds and so their removal from soil and degradation often are slow processes.

Substances that plants release into the soil include chelating ligands and enzymes; by complexing a metal, chelates can decrease its toxicity, and enzymes in some cases can biodegrade pollutants. For example, it has been found that the plant-derived enzyme *dehalogenase* can degrade trichloroethene. Plants also release oxygen at their roots, thereby facilitating aerobic transformations. As previously discussed, the fungi that exist in symbiotic association with a plant also have enzymes that can assist in the degradation of organic contaminants in soil.

Bioremediation in general, and phytoremediation in particular, are rapidly emerging technologies. The long-term potential for these techniques to be used at many sites that require decontamination is apparent. Experiments at several test sites have shown phytoremediation can be used successfully to degrade petroleum products in soils. The advantages of phytoremediation include its low cost, limited environmental impact, and high degree of public acceptance. However, it is constrained to shallow contamination and to compounds that are not highly hydrophobic. The number of *Superfund* sites applying bioremediation technology is shown in Table 16-3.

Review Questions 13–28 are based upon the material in the above sections.

Hazardous Wastes

In this section, we consider the nature of various types of hazardous wastes and discuss how individual masses of such wastes can be destroyed as an alternative to simply dumping them and thereby deferring the problem to a later date.

Currently there are more than 50,000 hazardous waste sites, and perhaps 300,000 leaking underground storage tanks, in the United States alone. The *Superfund* program of the U.S. EPA was created mainly to remediate such waste sites.

16.28 The Nature of Hazardous Wastes

A substance can be said to be a hazard if it poses a danger to the environment, especially to living things. Thus **hazardous wastes are substances that have been discarded or designated as waste and that pose a danger.** Most of the hazardous wastes we shall discuss are commercial chemical substances or by-products that result from their manufacture; biological materials are not considered herein.

In Chapters 12–15, we paid a great deal of attention to substances that were **toxic,** which is to say that they threaten the health of an organism when they enter its body. Other common types of hazardous materials include those that are

- **ignitable** and burn readily and easily;

- **corrosive** since their acidic or baseic character allows them to easily corrode other materials;

- **reactive** in senses not covered by ignition or corrosion, e.g., by explosion; and

- **radioactive.**

Some waste materials are hazardous in more than a single category.

16.29 The Management of Hazardous Wastes

There are four strategies in the management of hazardous waste. In order of decreasing preference, they are:

- **Source reduction:** The deliberate minimization, through process planning, of generating hazardous waste in the first place. The Green Chemistry cases presented throughout this text provide many examples of this strategy.

- **Recycling and reuse:** The use in a different process as raw materials, whether by the same company or a different one, of hazardous wastes generated in a process.

- **Treatment:** The use of any physical, chemical, biological, or thermal process—including incineration—that reduces or eliminates the hazard from the waste. Examples of such technology are discussed in subsequent sections.

- **Disposal:** Burial of the nonliquid waste in a properly designed landfill. In the past, liquid hazardous wastes were often injected into deep underground wells.

Landfills that are specially designed to accommodate hazardous wastes have several characteristics in addition to those discussed above for sanitary landfills. The locations of such landfills should be

- in an area with clay or silt soil, so as to provide an additional barrier to leachate dispersal, and

- away from groundwater sources.

Often the hazardous wastes in such landfills are grouped according to their physical and chemical characteristics so that incompatible ones are not placed near each other.

16.30 Toxic Substances

As mentioned above, toxic wastes are those which can cause a deterioration in the health of humans or other organisms when they enter a living body. Their characteristics, and many examples, were discussed in Chapters 12 through 15 particularly, and will not be reviewed in detail here. Those of main concern are heavy metals, organochlorine pesticides, organic solvents, and PCBs.

As an example of the magnitude of the problem of waste management of toxic substances, consider the PCBs that are still used in the capacitors which are components of the ballasts used in fluorescent light fixtures. Within the sealed capacitor container is a thick, gel-like liquid of concentrated PCB oil that is absorbed within several layers of paper. A typical capacitor contains about 20 g—about a tablespoon—of liquid PCB.

Although each ballast does not contain much PCB, the number of these lighting fixtures in use in the developed world is huge. Thus the ultimate collection and disposal of PCBs from such sources will be a task that takes many years and dollars. The many methods by which such toxic organic compounds are disposed of are discussed below.

16.31 Incineration of Toxic Waste

Incinerators that deal with hazardous waste are often more elaborate than those that burn municipal waste because it is more important that the initial materials be more completely destroyed and that emissions be more tightly controlled. In some cases, the waste (e.g., PCBs) will not ignite on its own and must consequently be added to an existing fire fuelled by other wastes or by supplemental fuel in the form of natural gas or petroleum liquids. Modern facilities employ very hot flames, ensure that there is sufficient oxygen in the combustion zone, and keep the waste compounds in the combustion region long enough that their **destruction and removal efficiency,** DRE, is essentially totally complete, i.e., >99.9999%, called "six nines." The presence of carbon monoxide at a concentration greater than 100 ppm in the gas emissions is often used as an indicator of incomplete combustion.

About 3 million tonnes of hazardous wastes are burned annually in the United States, though this corresponds to only 2% of that which is generated. Three-quarters of the hazardous waste is dealt with by aqueous treatment, and 12% is disposed of on land or injected into deep wells.

The two most common forms of toxic waste incinerators are the rotary kiln and liquid injection types. The **rotary kiln incinerator** can accept wastes of all types, including inert solids such as soil and sludges. The wastes are fed into a long (>20 m) cylinder that is inclined at a slight angle (about 5° from the horizontal) away from the entrance end. The cylinder slowly rotates so that unburned material is continually exposed to the oxidizing conditions of 650–1100°C; see Figure 16-14. Over a period of about an hour, the waste makes its way down the cylinder and is largely combusted. The hot exit gases from the kiln are sent to a secondary (nonrotating) combustion chamber equipped with a burner, in which the temperature is 950–1200°C. The gas resides in the chamber for at least two seconds so that destruction of organic molecules is essentially complete. In some installations, liquid wastes can be fed directly into this chamber as the fuel. The gases exiting the secondary chamber are rapidly cooled to about 230°C by an evaporating water spray (in some cases with heat recovery), since they would otherwise destroy the air pollution equipment that they next enter.

Rotary kiln and other types of incinerators usually employ the same series of steps to purify the exhaust of particulates and of acid gases before it is released into the air as do garbage incinerators, i.e., a gas scrubber and a baghouse filter (see Section 16.5).

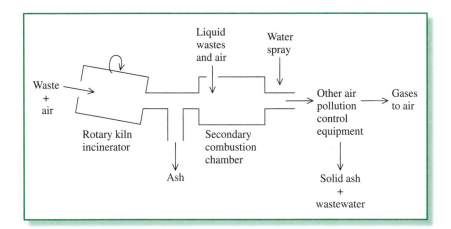

FIGURE 16-14
Schematic diagram of the
components of a rotary kiln
incinerator, including air
pollution equipment.

Cement kilns are a special type of large rotary kiln used to prepare cement from limestone, sand, clay, and shale. Very high temperatures of 1700°C or more are generated in cement kilns in order to drive the carbon dioxide from limestone, $CaCO_3$, in the formation of lime, CaO. In addition to hotter combustion temperatures (compared to incinerators), wastes in cement kilns are burned with the fuel right in the flame; the residence time of the material in the kiln is also longer. Liquid hazardous wastes are sometimes used as part of the fuel for these units, the remainder being a fossil fuel, usually coal. Techniques have been developed that allow kilns to handle sludges and solids. Cement kilns burn more hazardous waste (about a million tonnes a year) than do commercial incinerators in the United States, although even more waste is incinerated on-site by chemical industries.

In the **liquid injection incinerator**—a vertical or horizontal cylinder—pumpable liquid wastes are first dispersed into a fine mist of small droplets. The finer these droplets, the more complete is their subsequent combustion, which occurs at about 1600°C with a waste residence time of a second or two. A fuel or some "rich" (easily combustible) waste is used to produce the high temperatures of the combustion. In contrast to the rotary kiln, only a single combustion chamber is used, although in some modern versions a secondary input of air is introduced to improve oxygen distribution and generate more complete combustion. The exit gases can be passed through a spray dryer to neutralize and remove acid gases, followed by a baghouse filter to remove particulates, before being released into the outside air.

16.32 Air Emissions from Incinerators

The incineration of hazardous waste has garnered much attention from environmentalists and some of the general public because of the potential release of toxic substances—particularly from the stack into the air—resulting from the operation of these units. Of particular concern are the organic **products of incomplete combustion,** PICs, that have been found both in gases and adsorbed on particles emitted from incinerators. The PICs must be formed in the post-flame region because they would not survive the temperatures in the flame itself. Some of the most prevalent PICs are methane and benzene.

In their research to understand the production of these pollutants, scientists and engineers have discovered that reactions can occur downstream of the flame in "quench" zones and in pollution control devices, where temperatures fall below 600°C. Both gas-phase and surface-catalyzed processes occur. For example, trace amounts of dioxins and furans form at 200–400°C on fly ash and soot surfaces where the processes can be catalyzed by transition metal ions. A temperature of about 400°C is optimal for dioxin formation; below this the reaction is slow and above it the dioxins are quickly decomposed. It has not been firmly established whether the dioxins and furans result from the coupling on surfaces of precursor compounds such as chlorobenzenes and chlorophenols, or from the so-called *de novo* synthesis involving chlorine-free furan- or dioxin-like structures reacting with inorganic chlorides.

Concern has also been expressed about increased emissions that could occur when the incinerator is being closed down and during accidents or power failures, when lower temperatures would result for some time, since much greater quantities of dioxins and furans could presumably form under such conditions. **Fugitive emissions,** which include emissions from valves, minor ruptures, incidental spills, etc., are also a concern. The dust emitted from cement kiln incinerators has been found to contain toxic metals and some PICs. In fact, the health risk from toxic metal ion emissions (usually as oxides or chlorides) from the hazardous waste incinerators is found to exceed that from the toxic organics. As in the case of municipal incinerators, the solid residue from hazardous waste units can amount to one-third of the original waste volume and contains traces of toxic materials, as does the wastewater from the scrubber units.

Concerns about incineration have spurred the development of other technologies for disposing of hazardous waste.

• **In molten salt combustion, wastes are heated to about 900°C and destroyed by being mixed with molten sodium carbonate.** The spent carbonate salt contains NaCl, NaOH, and various metals from the combusted waste, which can be recovered so that the Na_2CO_3 can be reused. No acidic gas is evolved, since it reacts within the salt.

• **In fluidized-bed incinerators a solid material such as limestone, sand, or alumina is suspended in air (fluidized) by means of a jet of air, and the wastes are combusted in the fluid at about 900°C. A secondary combustion chamber completes the oxidation of the exhaust gases.**

• **Plasma incinerators can achieve temperatures of 10,000°C by means of passing a strong electrical current through an inert gas such as argon.** The plasma consists of a mixture of electrons and positive ions, including nuclei, and can successfully decompose compounds, producing much lower emissions than traditional incinerators. In such a **thermal** or **hot** plasma, all the particles travel at high speeds and are thermally hot. In a variant that is used to treat municipal solid waste, plasma is first created in air, which is then used to heat a mixture of waste, coke, and limestone to 1500°C or more in a second, oxygen-starved chamber. The inorganic compounds are converted into a slag that is innocuous enough to be used as a construction material. The organic compounds are broken down into *syngas*, the combination of carbon monoxide and hydrogen discussed in Chapter 7, which is then used as a fuel.

16.33 Using Supercritical Fluids to Destroy Waste

The use of **supercritical fluids** is another modern alternative to incineration. The supercritical state of matter is produced when gases or liquids are subjected to very high pressures, and in some cases to elevated temperatures, as previously discussed (Box 5-2). Recall that at pressures and temperatures at or beyond the critical point, separate gaseous and liquid phases of a substance no longer exist. Instead, under these conditions, only the supercritical state, with properties that lie between those of a gas and those of a liquid, exists.

For example, for water, the critical pressure is 218 atm (22.1 megapascals) and the critical temperature is 374°C, as illustrated in the phase diagram in Figure 16-15. Depending upon exactly how much pressure is applied, the physical properties of the supercritical fluid vary between those of a gas (relatively lower pressures) and those of a liquid (higher pressures); the variation of properties with changes in pressure or temperature is particularly acute near the critical point. Thus, the density of supercritical water can vary over a considerable range, depending upon how much pressure (beyond 218 atm) is applied. Other substances that readily form useful supercritical fluids are carbon dioxide, xenon, and argon—see Table 16-4 for their critical temperatures and pressures.

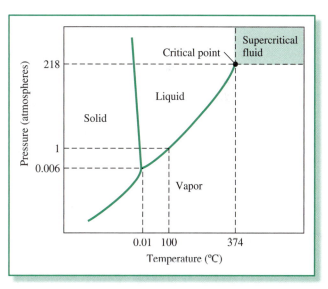

FIGURE 16-15 Phase diagram for water (not to scale). Notice the region for the supercritical state (shaded light green), which exists at temperatures and pressures beyond the critical point.

TABLE 16-4	Supercritical Fluid Characteristics	
Substance	Critical Temperature (°C)	Critical Pressure (atm)
Water	374.1	217.7
Carbon dioxide	31.3	72.9
Argon	150.9	48.0
Xenon	16.6	58.4

One rapidly developing, innovative technology for destruction of organic wastes and hazardous materials such as phenols is **supercritical water oxidation** (SCWO). Initially, the organic wastes to be destroyed are either dissolved in aqueous solution or are suspended in water. The resulting liquid is then subjected to very high pressure and a temperature in the 400–600°C range so that the water lies beyond its critical conditions and so is a super-critical fluid. **The solubility characteristics of supercritical water differ markedly from those of normal liquid water: most organic substances become much *more* soluble and many ionic substances become much *less* soluble.** Similarly, and also because very high pressures are applied, O_2 is much more soluble in supercritical than in liquid water.

At the elevated temperatures associated with supercritical water, the dissolved organic materials are readily oxidized by the ample amounts of O_2 that are pumped into and dissolve in the fluid. Hydrogen peroxide may be added to generate hydroxyl radicals, which initiate even faster oxidation. Because materials diffuse much more rapidly in the supercritical state than in liquids, the reaction is generally complete within seconds or minutes. One practical problem with the SCWO method is that insoluble inorganic salts that are formed in the reactions can corrode the high-pressure equipment and so shorten its lifetime. This corrosion problem can be solved by design-ing the reactor so there are no zones where salts can build up.

The advantages of this technology include the rapidity of the destruction reactions and the lack of gaseous NO_X by-products that are characteristic of gas-phase combustion (Chapter 3). The required pressure and temperature conditions are readily accessible with available high-pressure equipment. However, some intermediate products of oxidation—mainly organic acids and alcohols, and perhaps also some dioxins and furans—are formed with the SCWO process, which raises concerns about the toxicity of the effluent from the process. In the variant of this technology in which a catalyst is used, the percentage conversion to fully oxidized products is increased and the amount of intermediates that persist is decreased.

In the **wet air oxidation** process, temperatures (typically 120–320°C) and pressures lower than those required to achieve supercritical conditions for water are used to efficiently oxidize aqueous wastes (often catalytically). The oxidation is efficient because the greater amount of oxygen that dissolves at the enhanced pressures favors the reaction. The process is generally much slower than in supercritical water, requiring about an hour to occur. The method is cheaper to operate, however, than the relatively expensive SCWO technology.

Supercritical carbon dioxide has been used to extract organic contaminants such as the gasoline additive MTBE from polluted water; once it is extracted, the pressure can be lowered, at which point the carbon dioxide becomes a gas, leaving behind the liquid contaminants to be incinerated or otherwise oxidized. Similarly, supercritical fluids could be used to extract contaminants such as PCBs and DDT from soils and sediments.

Supercritical CO_2 is used for many extractions in the food industry, including the decaffeination of coffee beans.

16.34 Nonoxidative Processes of Waste Destruction

All the processes described above have employed oxidation as the means by which the hazardous wastes are destroyed. However, a closed-loop **chemical reduction process** has been devised with no uncontrolled emissions, using a *reducing* rather than an oxidizing atmosphere to destroy hazardous wastes. One advantage to the absence of oxygen is that there exists no opportunity for the incidental formation of dioxins and furans.

The reducing atmosphere is achieved by using hydrogen gas at about 850°C as the substance with which the preheated mist of wastes reacts. The carbon in the wastes is converted to methane (and some temporarily to other hydrocarbons such as benzene, which are subsequently hydrogenated to produce additional methane). The oxygen, nitrogen, sulfur, and chlorine in the wastes are converted into their hydrides by the process. The reaction is actually enhanced by the presence of water, which under these reaction conditions can act as a reducing agent and form additional hydrogen by the water-gas shift reaction (see Chapter 7). PAH formation, which is characteristic of other processes at high temperatures carried out in the absence of air, is suppressed by maintaining the hydrogen content at more than 50%. The gas output is cooled and scrubbed to remove particulates. The hydrocarbon output from the process is subsequently burned to provide heat for the system; thus there are no direct emissions to the atmosphere.

PROBLEM 16-7

Construct and balance chemical equations for the destruction of the PCB molecule having the formula $C_{12}H_6Cl_4$ (a) by combustion with oxygen, to yield CO_2, H_2O, and HCl, and (b) by hydrogenation, to yield methane and HCl. ●

Chemical dechlorination methods for the treatment of chlorine-containing organic wastes, especially PCBs from transformers, have been developed and used in various parts of the world, although they are rather expensive to operate. The basic idea is to substitute a hydrogen atom or some other nonhalogen group for the covalently bound chlorine atoms on the molecules, thereby detoxifying them. The mostly dechlorinated wastes can then be incinerated or disposed of in landfills. The commonly used reagent for this purpose is MOR, the alkali metal M (sodium or potassium) salt of a polymeric alcohol. In the reaction, an —OR group replaces each of the chlorines, which depart as the salt MCl:

The process is carried out at temperatures above 120°C in the presence of *potassium hydroxide*, KOH, and occurs most efficiently for highly chlorinated PCBs. An alternative dechlorination method is the reaction of the PCBs with dispersions of metallic sodium to give sodium chloride and a polymer containing many biphenyl units joined together.

Review Questions 29–34 are based upon the material in the above sections.

Review Questions

1. Define the term *solid waste* and name the five largest categories in it for developed countries.

2. Describe the components and steps in the creation of a *sanitary landfill*.

3. Describe the three stages of waste decomposition that occur in a sanitary landfill, including the products of each stage. Are all stages equal in production or consumption of acidity?

4. Define the term *leachate*, explain how this substance arises, and list several common components of it. How can leachate be controlled and how is it treated?

5. Explain the difference between the two common types of MSW incinerators.

6. What is the difference between *bottom ash* and *fly ash* in an incinerator? Describe some of the air pollution control devices found on incinerators.

7. What is meant by the *four R's* in waste management?

8. Why can the recycling of metals often be justified by economics alone?

9. Describe the processes by which paper and rubber tires can be recycled.

10. What common types of consumer packaging plastics can be recycled? What four ways are used to recycle plastics?

11. What are some of the arguments for and against the recycling of plastics?

12. What is meant by a *life-cycle assessment* and what are the two main uses for LCAs?

13. Describe the main inorganic constituents of soil. How do clay, sand, and silt particles differ in size?

14. What are the names and origins of the principal organic constituents of soil? Are both types of acids soluble in base? In acid?

15. What is meant by a soil's *cation exchange capacity*? What are the common units for it?

16. What is meant by a soil's *reserve acidity*? How does it arise?

17. Describe several methods, including chemical equations, by which soils that are too acid or too alkaline can be treated.

18. Describe the processes by which soil in arid areas becomes salty and alkaline.

19. What do the terms *sediments* and *pore water* mean?

20. By what three ways are heavy metals bound to sediments?

21. How can mercury stored in sediments become solubilized and enter the food chain?

22. Describe how mine tailings are usually stored and how this represents a potential environmental problem.

23. How can sediments contaminated by heavy metals be remediated so they can be used on agricultural fields?

24. Describe two ways by which contaminated sediments can be treated without removing the sediments themselves.

25. List the three categories of technologies commonly used to remediate contaminated soils. Give examples of each.

26. List three conditions that must be fulfilled if bioremediation of soil is to be successful.

27. Describe the two ways in which PCBs in sediments are bioremediated.

28. Define *phytoremediation* and list the three mechanisms by which it can operate. What are *hyperaccumulators*?

29. Define the term *hazardous waste*. What are the five common types of it?

30. List, in order of decreasing desirability, the four strategies used in the management of hazardous waste.

31. Name and describe the three common types of incinerators used to destroy hazardous waste. What does *DRE* stand for and how is it defined?

32. Define the term *PICs* and describe how they are formed.

33. Explain the advantages and disadvantages of *supercritical water oxidation* for the destruction of hazardous wastes.

34. How is the *chemical reduction process* carried out? What advantages does it have over oxidation methods?

Green Chemistry Questions

1. The development of bio-based toners by Battelle won a Presidential Green Chemistry Challenge Award.

(a) Into which of the three focus areas for these awards (page xxviii) does this award best fit?

(b) List three of the twelve principles of green chemistry (pages xxiii–xxiv) that are addressed by this discovery. Explain each of your answers.

2. What are the environmental advantages of the bio-based toner developed by Battelle?

3. The development of recyclable carpeting by Shaw Industries won a Presidential Green Chemistry Challenge Award.

(a) Which of the three focus areas (see page xxviii) for these awards does this award best fit into?

(b) List three of the twelve principles of green chemistry (see pages xxiii–xxiv) that are addressed by the green chemistry developed by Shaw Industries.

4. What are the environmental advantages of using polyolefin-backed nylon fiber carpeting in place of PVC-backed carpeting?

Additional Problems

1. Consider a small city of 300,000 people in the northern United States or southern Canada or central Europe, and suppose that its residents on average produce 2 kg a day of MSW, about one-quarter of which will decompose anaerobically to release methane and carbon dioxide evenly over a 10-year period. Calculate how many homes could be heated by burning the methane from the landfill, given that residential requirements are about 10^8 kJ per year in that climate zone. Consult page 235 for data on methane combustion energetics.

2. By reference to the general solubility rules for sulfides found in introductory chemistry textbooks, deduce which metals would *not* have their sediment availabilities determined by AVS. Which of these metals would occur in the form of insoluble carbonates instead and thus be biologically unavailable in marine environments?

3. The acid volatile sulfide concentration in a sediment is found to be 10 μmol g^{-1}. The majority of the sulfide is present in the form of insoluble FeS, since the Fe^{2+} concentration in the sediment is 450 μg g^{-1}. What mass of mercury in the form of Hg^{2+} can be tied up by the remaining sulfide in one tonne of such sediment? Assume that a negligible amount of iron is not tied up as FeS.

4. For the PCB congener 2,4,4',5-tetrachlorobiphenyl, deduce

(a) the chlorine substitution pattern on the benzoic acid that results from its aerobic degradation; and

(b) the various PCB congeners with one or two chlorines that could result from its anaerobic degradation.

5. Generate lists of the aspects of production, distribution, and disposal that you would employ in a *life-cycle assessment* to decide which is the more environmentally friendly container for beer: glass bottles or aluminum cans. Do you think the conclusions of such an analysis would depend significantly on the extent of recycling of the containers?

6. Waste having a high cellulose content, such as paper, wood scraps, and corn husks, can be converted to ethanol for use as a fuel. First the cellulose, empirical formula $C_5H_{10}O_5$, is converted by acid hydrolysis to glucose, which then is fermented to ethanol. Write the balanced equation for fermentation of glucose, $C_6H_{12}O_6$, into ethanol, C_2H_5OH and CO_2. What volume of ethanol could be produced from 1.00 tonne of hardwood scraps (46% cellulose), assuming 100% conversion in both steps? Ethanol's density is 0.789 g mL^{-1}.

ADVANCED ATMOSPHERIC CHEMISTRY

Contents of Part VI

Introduction

In Part I, and to some extent in Chapter 16, of this text, chemical reactions involving free radicals such as hydroxyl, OH, have been discussed. For some purposes, including the combating of photochemical air pollution and the understanding of how the stratosphere would be affected by changes to its makeup, it is important to be able to predict detailed mechanisms of free-radical reactions in air. That topic is explored in this final chapter. ●

The Detailed Free-Radical Chemistry of the Atmosphere

In this chapter, the following introductory chemistry topics are used:

- ⮞ Lewis structures (for nonradicals)
- ⮞ Activation energy; reaction mechanisms
- ⮞ Basic organic chemistry (see online Appendix)

Background from previous chapters used in this chapter:

- ⮞ Concept of free radicals
- ⮞ Atmospheric structure
- ⮞ Photochemical reactions; UV categories A, B, C
- ⮞ Stratospheric chemistry; energy versus wavelength
- ⮞ Photochemical air pollution, smog, NO_x

Introduction

In Chapters 1 through 4, we discussed, in broad terms, chemical processes in the stratosphere and troposphere, emphasizing the environmental concerns that have arisen. In this chapter, the reactions that occur in clean tropospheric air and the processes encountered in the polluted air of modern cities are analyzed in more detail. In addition, the processes that deplete ozone in the stratosphere are systematized. It is only by understanding the science underlying such complicated environmental problems that we can hope to solve them.

One of the important characteristics that determines the reactivity of a species is whether

This Mexico City newspaper salesman wears a mask to help protect himself from air pollution during a photochemical smog episode. [Guillermo Gutierrez/AP]

755

or not it has unpaired electrons. To emphasize that an atomic or molecular species is a free radical, **in this chapter we shall place a superscript dot at the end of its molecular formula, signaling the presence of the unpaired electron.** For example, the notation OH˙ is used for the **hydroxyl free radical**. The position of the unpaired electron in the Lewis structure is often important in determining the reaction of a free radical. Box 17-1 discusses the deduction of the Lewis structure for free radicals.

BOX 17-1   Lewis Structures of Simple Free Radicals

Most of the free radicals that are important in atmospheric chemistry have their unpaired electron located on a carbon, oxygen, hydrogen, or halogen atom. In a formula showing the location and position of bonds, the specific atomic location can be denoted by placing a dot above the relevant atom symbol to represent the unpaired electron, e.g., in the notation Ḟ. Characteristically, such an atom forms one fewer bond than usual—its unpaired electron is not in use as a bonding electron. Thus a carbon atom on which an unpaired electron is located forms three rather than four bonds, an oxygen forms one rather than two bonds, and a halogen or hydrogen forms no bonds if it is the radical site. Usually, the unpaired electron exists as a nonbonding electron localized on one atom, not as a bonding electron shared between atoms.

For many polyatomic free radicals, the choice of atom to which the unpaired electron is to be assigned in deducing the Lewis structure is obvious from the atom–atom connections. Thus in the hydroperoxy radical HOO, the hydrogen atom cannot be the radical site, since to be part of the molecule it must form a bond to the adjacent oxygen; neither can the central oxygen be the site since it must form two bonds, one to each neighbor (or one of them would not be part of the molecule). This leaves the terminal oxygen as the radical site, and we can show the bonding network as H—O—Ȯ. If desired, the nonbonded electron pairs can also be shown:

$$\text{H} - \ddot{\text{O}} - \dot{\ddot{\text{O}}}:$$

The procedure is more complicated in molecules that contain multiple bonds. Thus in HOCO it is not initially obvious whether it is the carbon or the terminal oxygen that carries the unpaired electron. After a little manipulation with various bonding schemes, it becomes clear that the unpaired electron could not be located on an oxygen, since to fulfill its valence requirement of four, the carbon would have to form three bonds to the other oxygen. Thus the only reasonable structure is H—O—Ċ=O.

If only a simple formula rather than a partial or complete Lewis structure is drawn for a radical, the superscript dot is placed following the formula and does not indicate which atom carries the unpaired electron. An example is HCO˙, in which the actual location of the unpaired electron is at carbon, not oxygen.

For a few free radicals involving unusual bonding, such as $NO_2^{\cdot}$, these rules generate a Lewis structure that is not the dominant one; further discussion of such systems is beyond the scope of this book.

PROBLEM 1

Draw the Lewis structure, showing the bonds, nonbonded electrons, and unpaired electron, for the following free radicals.

(a) OH˙   (b) $CH_3^{\cdot}$ (c) $CF_2Cl^{\cdot}$
(d) $H_3COO^{\cdot}$ (e) $H_3CO^{\cdot}$ (f) ClOO˙
(g) ClO˙   (h) HCO˙   (i) NO˙

Tropospheric Chemistry

17.1 The Principles of Reactivity in the Troposphere

Most gases in the troposphere are gradually oxidized by a sequence of reactions involving free radicals. For a given gas, the sequence can be predicted from the principles discussed below, which are also systematized in Figure 17-1. These tropospheric reactions are similar in many ways to those encountered in the stratosphere, which are discussed later in the chapter.

As mentioned in Chapter 3, **the usual initial step in the oxidation of an atmospheric gas is its reaction with the hydroxyl free radical rather than with molecular oxygen, O_2. With molecules that contain a multiple bond, the hydroxyl radical usually reacts by** *adding* **itself to the molecule at the position of the multiple bond.** Recall the general principle that radical reactions that are spontaneous are those which produce stable products, i.e., products containing strong bonds. Thus it is understandable that $OH^{\cdot}$ addition does *not* occur to an oxygen atom since the O—O bonds that would result are weak. Similarly, $OH^{\cdot}$ addition does not occur to CO *double* bonds since they are very strong relative to the single O—O or C—O bond that would be produced. For example, the $OH^{\cdot}$ radical adds to the sulfur atom, forming a strong bond, but not to an oxygen atom, in **sulfur dioxide, SO_2:**

When bonds in the products are stronger than those in the reactants, an exothermic reaction results, which for free radicals will have only a small activation energy, as discussed in Box 1-1.

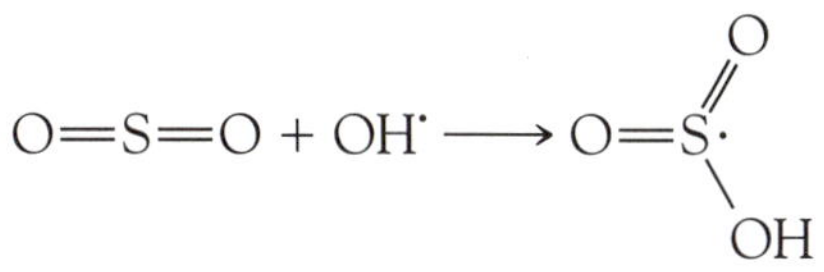

$$O{=}S{=}O + OH^{\cdot} \longrightarrow O{=}\overset{\displaystyle O}{\underset{\displaystyle OH}{\overset{\|}{S}}}\!\!\cdot$$

(Here and elsewhere in this book we write Lewis structures that assume that *d* orbitals in atoms such as sulfur and phosphorus allow these elements to form double bonds.) Hydroxyl radical does not add to **carbon dioxide,** O=C=O, since the molecule contains only very strong C=O bonds. However, $OH^{\cdot}$ addition does occur to the carbon atom in **carbon monoxide, CO,** since the triple bond is thereby converted to the very stable double bond and a new single bond is also formed:

$$^{\ominus}C{\equiv}O^{\oplus} + OH^{\cdot} \longrightarrow HO{-}\dot{C}{=}O$$

This process is exothermic because the third C—O bond in carbon monoxide is weak relative to the other two.

Generally, $OH^{\cdot}$ does not add to multiple bonds in any fully oxidized species such as CO_2, SO_3, and N_2O_5, since such processes are endothermic and therefore are very slow to occur at atmospheric temperatures. Similarly, N_2 does not react with $OH^{\cdot}$ because the component of the nitrogen-to-nitrogen bond that would be destroyed is stronger than the N—O bond that would be formed. It doesn't react with O_2 because a high activation energy is required for this reaction to occur.

(a)

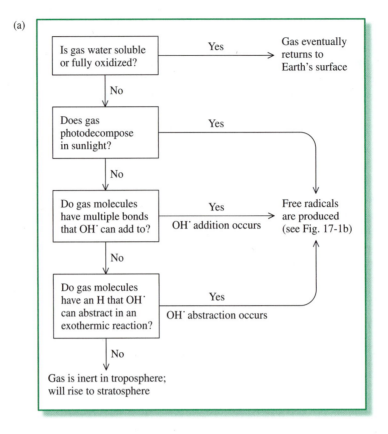

(b)

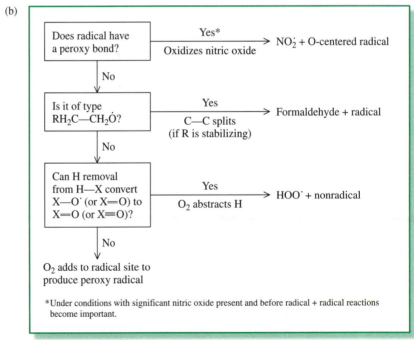

FIGURE 17-1 (a) Decision tree illustrating the fate of gases emitted into the air. (b) Decision tree illustrating the fate of airborne free radicals.

For molecules that do *not* have a reactive multiple bond but do contain hydrogen, OH˙ reacts with them by the **abstraction** of a hydrogen atom to form a water molecule and a new reactive free radical. For CH_4, NH_3, H_2S, and CH_3Cl, for instance, the reactions are

$$CH_4 + OH^{\cdot} \longrightarrow CH_3^{\cdot} + H_2O$$

$$NH_3 + OH^{\cdot} \longrightarrow NH_2^{\cdot} + H_2O$$

$$H_2S + OH^{\cdot} \longrightarrow SH^{\cdot} + H_2O$$

$$CH_3Cl + OH^{\cdot} \longrightarrow CH_2Cl^{\cdot} + H_2O$$

Because the H—OH bond formed in these reactions is very strong, the processes are all exothermic; thus only small activation energy barriers exist to impede these reactions (see Box 1-1).

PROBLEM 17-1

Why aren't gases such as CF_2Cl_2 (a CFC) readily oxidized in the troposphere? Would the same be true for CH_2Cl_2?   ●

PROBLEM 17-2

The abstraction of the hydrogen atom in HF by OH˙ is endothermic. Comment briefly on the expected rate of this reaction: Would it be (at least potentially) fast, or necessarily very slow in the troposphere?   ●

PROBLEM 17-3

The hydroxyl radical does not react with gaseous nitrous oxide, N_2O, even though the molecule contains multiple bonds. What can you deduce about the probable energetics (endothermic or exothermic character) of this reaction from the observed lack of reactivity?   ●

A few gases emitted into air can absorb some of either the UV-A or the visible component of sunlight, and this input of energy is sufficient to break one of the bonds in the molecule, thereby producing two free radicals. For example, most molecules of atmospheric **formaldehyde** gas, H_2CO, react by photochemical decomposition after absorption of UV-A from sunlight:

$$H_2CO \xrightarrow{\text{UV-A } (\lambda < 338 \text{ nm})} H^{\cdot} + HCO^{\cdot}$$

In all the cases discussed, the initial reaction of a gas emitted into air produces free radicals, almost all of which are extremely reactive. **The predominant fate in tropospheric air for most simple radicals is reaction with diatomic oxygen, often by an addition process:** One of the oxygen atoms attaches, or "adds on," to the other reactant, usually at the site of the unpaired electron. For instance, O_2 reacts by addition with the **methyl radical,** $CH_3^{\bullet}$:

$$CH_3^{\bullet} + O_2 \longrightarrow CH_3OO^{\bullet}$$

Notice that $CH_3OO^{\bullet}$ itself is a free radical; the terminal oxygen forms only one bond and carries the unpaired electron:

$$H_3C-\ddot{O}-\dot{\ddot{O}}: \quad \text{or just} \quad H_3C-O-\dot{O}$$

Species such as $HOO^{\bullet}$ and $CH_3OO^{\bullet}$ are called **peroxy** radicals since they contain a peroxide-like $O-O$ bond; recall that $HOO^{\bullet}$ is the **hydroperoxy radical.**

As radicals go, peroxy radicals are less reactive than most. They do *not* readily abstract hydrogen since the resulting peroxides would not be very stable energetically. Since the transfer of H to the peroxy radical would be endothermic and thus would possess a large activation energy, abstraction reactions for peroxy radicals are usually so slow that they are of negligible importance (in contrast to those for $OH^{\bullet}$). Peroxy radicals in the troposphere do not react with atomic oxygen because of the extremely low concentrations of the free atom in this region of the atmosphere. **The most common fate of peroxy radicals in tropospheric air, except for the cleanest type of air, such as that over oceans, is reaction with nitric oxide, $NO^{\bullet}$,** by the transfer of the "loose" oxygen atom (see the later section on stratospheric chemistry), thereby **forming nitrogen dioxide, $NO_2^{\bullet}$, and a radical that has one fewer oxygen atoms:**

$$HOO^{\bullet} + NO^{\bullet} \longrightarrow OH^{\bullet} + NO_2^{\bullet}$$

$$CH_3OO^{\bullet} + NO^{\bullet} \longrightarrow CH_3O^{\bullet} + NO_2^{\bullet}$$

It is by this type of reaction that most atmospheric $NO^{\bullet}$ is oxidized to $NO_2^{\bullet}$, at least in polluted air. Recall that this reaction also is typical of the types encountered in stratospheric chemistry (see Chapters 1 and 2) and that $NO^{\bullet}$ oxidation by ozone in sunlit conditions yields a null reaction.

For free radicals that contain nonperoxy oxygen atoms, the reaction with molecular oxygen frequently involves the abstraction of an H atom by O_2. This process occurs *provided that,* as a result, another new bond within the system is formed: A single bond involving oxygen is converted to a double one, or a double bond involving oxygen is converted to a triple one. As examples, consider the three reactions below in which a $C-O$ single (or double) bond is converted to a double (or triple) one as a consequence of the loss of a hydrogen atom:

$$CH_3-\dot{O} + O_2 \longrightarrow H_2C{=}O + HOO^{\bullet}$$

$$HO-\dot{C}{=}O + O_2 \longrightarrow O{=}C{=}O + HOO^{\bullet}$$

$$H-\dot{C}{=}O + O_2 \longrightarrow {}^{\ominus}C{\equiv}O^{\oplus} + HOO^{\bullet}$$

Such processes do not occur unless a new bond is created in the product free radical, since the strength of the newly created H—OO bond alone is not sufficient to compensate for the breaking of the original bond to hydrogen.

If there is no suitable hydrogen atom for O_2 to abstract, then when it collides with a radical, it instead *adds* to it at the site of the unpaired electron, as was previously discussed for simple radicals. For example, radicals of the type R—$\dot{C}$=O, where R is a chain of carbon atoms, add O_2 to form a peroxy radical:

$$R-\dot{C}=O + O_2 \longrightarrow R-C\overset{\displaystyle O}{\underset{\displaystyle O-\dot{O}}{\big\langle}}$$

The only exception to the generalization that oxygen-containing radicals react with O_2 occurs when the radical can decompose spontaneously in a thermoneutral or exothermic fashion. An example of this rare phenomenon is discussed later in the section on photochemical smog.

These generalizations are summarized in the form of the *decision trees* diagrammed in Figure 17-1. By using these diagrams you can deduce the sequence of reactions by which most atmospheric gases in the troposphere are oxidized.

17.2 The Tropospheric Oxidation of Methane

Gaseous **methane,** CH_4, is released into the atmosphere in large quantities as a result of anaerobic (i.e., O_2-free) biological decay processes and of the use of coal, oil, and, especially, natural gas. It is the predominant hydrocarbon in the atmosphere. Details concerning its production, and the effects on climate of atmospheric methane, were discussed in Chapter 5. Here, however, we shall be concerned with its conversion to carbon dioxide. A similar series of reactions is followed by other alkanes and other gases lacking multiple bonds.

The sequence of reactions by which methane is slowly oxidized in the atmosphere can be deduced by applying the principles outlined above and summarized in Figure 17-1, as discussed below.

Since CH_4 is not very soluble in water, does not absorb sunlight, and contains no multiple bonds, the sequence is initiated by a hydroxyl radical abstracting a hydrogen atom from a methane molecule, giving the methyl radical, $CH_3^{\cdot}$:

$$CH_4 + OH^{\cdot} \longrightarrow CH_3^{\cdot} + H_2O \tag{1}$$

Since the $CH_3^{\cdot}$ radical contains no oxygen, we deduce that it adds O_2, producing a peroxy radical:

$$CH_3^{\cdot} + O_2 \longrightarrow CH_3OO^{\cdot} \tag{2}$$

Further, since $CH_3OO^{\cdot}$ is a peroxy radical, we deduce that except in very clean air, it reacts with $NO^{\cdot}$ molecules in air to oxidize them by transfer of an oxygen atom:

$$CH_3OO^{\cdot} + NO^{\cdot} \longrightarrow CH_3O^{\cdot} + NO_2^{\cdot} \qquad (3)$$

The radical $CH_3O^{\cdot}$ contains a C—O bond that can become C=O upon loss of one hydrogen, so we conclude from our principles that in the next step O_2 abstracts an H atom, producing the nonradical product formaldehyde, H_2CO:

$$CH_3O^{\cdot} + O_2 \longrightarrow H_2CO + HOO^{\cdot} \qquad (4)$$

Thus methane is converted to formaldehyde as the first stable intermediate in its oxidation. Since formaldehyde is reactive as a gas in the atmosphere, the mechanism is not complete at this point. After several hours or days in the sunlight, most formaldehyde molecules decompose photochemically by the absorption of UV-A from sunlight, resulting in the cleavage of a C—H bond and the consequent formation of two radicals:

$$H_2CO \xrightarrow{\text{UV-A } (\lambda < 338 \text{ nm})} H^{\cdot} + HCO^{\cdot} \qquad (5)$$

A minority of formaldehyde molecules react with $OH^{\cdot}$ by H atom abstraction, yielding the same $HCO^{\cdot}$ radical; see Problem 17-7 for the implications of this alternative route.

The hydrogen atom from formaldehyde photolysis is itself a simple radical, and therefore it reacts by addition to O_2 to yield $HOO^{\cdot}$:

$$H^{\cdot} + O_2 \longrightarrow HOO^{\cdot} \qquad (6)$$

Meanwhile, the $H—\overset{\cdot}{C}{=}O$ radical reacts by yielding an $H^{\cdot}$ atom to O_2, to produce carbon monoxide and $HOO^{\cdot}$, since by this route a double bond is converted to a triple one:

$$HCO^{\cdot} + O_2 \longrightarrow CO + HOO^{\cdot} \qquad (7)$$

Thus carbon monoxide also is an intermediate in the oxidation of methane. Indeed, most of the CO in a clean atmosphere is derived from this source. Since CO is not a radical and does not absorb visible or UV-A light, we deduce that it reacts ultimately by hydroxyl radical addition to its triple bond:

$$^{\ominus}C{\equiv}O^{\oplus} + OH^{\cdot} \longrightarrow H—O—\overset{\cdot}{C}{=}O \qquad (8)$$

This radical can convert its O—C bond to O=C by loss of H, so we deduce that O_2 readily abstracts the hydrogen:

$$H—O—\overset{\cdot}{C}{=}O + O_2 \longrightarrow O{=}C{=}O + HOO^{\cdot} \qquad (9)$$

Carbon in its fully oxidized form of carbon dioxide is ultimately produced from methane by this sequence of steps, which are summarized in

Figure 17-2. If we add up the nine steps involved and cancel common terms, the overall reaction is seen to be

$$CH_4 + 5\,O_2 + NO^{\cdot} + 2\,OH^{\cdot} \xrightarrow{\text{UV-A}}$$

$$CO_2 + H_2O + NO_2^{\cdot} + 4\,HOO^{\cdot}$$

If to this result is added the conversion of the four $HOO^{\cdot}$ radicals back to $OH^{\cdot}$ by reaction with four $NO^{\cdot}$ molecules, the revised overall reaction is

$$CH_4 + 5\,O_2 + 5\,NO^{\cdot} \xrightarrow{\text{UV-A}} CO_2 + H_2O + 5\,NO_2^{\cdot} + 2\,OH^{\cdot}$$

We conclude that $NO^{\cdot}$ is oxidized to $NO_2^{\cdot}$ synergistically, i.e., in a mutually cooperative process, when methane is oxidized to carbon dioxide. Note also that the number of $OH^{\cdot}$ free radicals is increased as a result of the process, due to the photochemical decomposition of formaldehyde. Thus hydroxyl radical is not only a catalyst in the overall reaction but also a product of it.

The initial step of the mechanism—the abstraction by $OH^{\cdot}$ of a hydrogen atom from methane—is a slow process, requiring about a decade to occur on average. Once this has happened, however, the subsequent steps leading to formaldehyde occur very rapidly. The slowness of the initial step in methane oxidation, and the increasing amounts of the gas released from the surface of the Earth, have led to an increase in the atmospheric concentration of CH_4 in recent times, as discussed further in Chapter 5.

Under conditions of low nitrogen oxide concentration, such as occur over oceans, the mechanism of methane oxidation differs in some of the steps. In particular, instead of oxidizing $NO^{\cdot}$, the peroxy radicals often react with each other, combining to produce a (non-radical) peroxide:

$$2\,HO_2^{\cdot} \longrightarrow H_2O_2 + O_2$$

Under these conditions, then, the oxidation of methane in clean air decreases, rather than increases, the concentration of free radicals.

In general, during the atmospheric oxidation of any of the hydrides (simple hydrogen-containing molecules such as CH_4, H_2S, and NH_3), one or more stable species are encountered along the reaction sequence before the totally oxidized product is formed. These intermediates are also formed independently by various pollution processes. Figure 17-3 summarizes the sequences for hydrides and partially oxidized materials from the viewpoint of the stable species; close reflection will persuade you that the net result is the $OH^{\cdot}$-induced oxidation of the reduced and partially oxidized gases emitted into the air from both natural and pollution sources. In a few cases, for instance, for methane and methyl chloride, the initiation reaction is sufficiently slow that a few percent of these

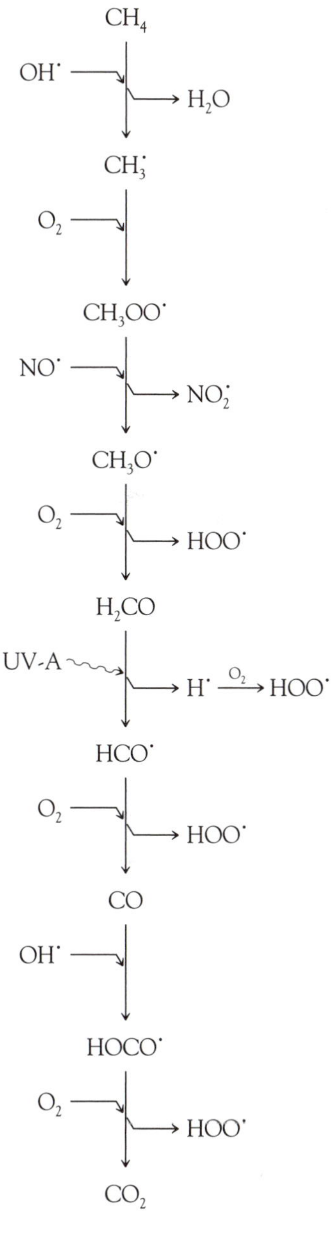

FIGURE 17-2 Steps in the atmospheric oxidation of methane to carbon dioxide.

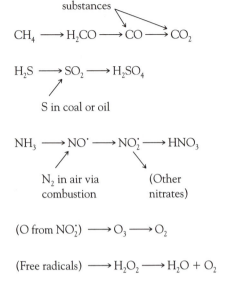

FIGURE 17-3 Stable species (i.e., nontransients) and their additional sources during sequential atmospheric oxidation processes.

gases survive long enough to penetrate to the stratosphere by the upward diffusion of tropospheric air. Most hydrocarbons react much more quickly than methane (since their C—H bonds are weaker, or fast reactions with OH$^\bullet$ other than by hydrogen abstraction are possible) and are classified as **nonmethane hydrocarbons** (NMHC) to emphasize this distinction.

PROBLEM 17-4

Using the reaction principles developed above (the decision trees in Figure 17-1 will help here), predict the sequence of reaction steps by which atmospheric H_2 gas will be oxidized in the troposphere. What is the overall reaction?

PROBLEM 17-5

Deduce two short series of steps by which molecules of methanol, CH_3OH, are converted to formaldehyde, H_2CO, in air. The mechanisms should differ according to which hydrogen atom you decide will react first, that of CH_3 or that of OH.

PROBLEM 17-6

Write equations showing the reactions by which atmospheric carbon monoxide is oxidized to carbon dioxide. Then, by adding the process by which HOO$^\bullet$ is returned to OH$^\bullet$, deduce the overall reaction.

PROBLEM 17-7

Deduce the series of steps, and the overall reaction as well, for the oxidation of a formaldehyde molecule to CO_2, assuming that for the particular H_2CO molecule involved, the initial reaction is abstraction of H by OH$^\bullet$ rather than photochemical decomposition. Overall, is there any increase in the number of free radicals as a result of the oxidation if it proceeds in this manner?

17.3 Photochemical Smog: The Oxidation of Reactive Hydrocarbons

Notwithstanding the great complexity of the process, the most important features of the photochemical smog phenomenon can be understood by considering only its few main categories of reactions; these differ in speed, but not much in type, from those occurring in clean air.

We shall restrict our attention to the most reactive VOCs, namely hydrocarbons that contain a C=C bond. (Readers unfamiliar with the basics of organic chemistry are advised to consult the online Appendix, where the nature of such molecules is explored.) The simplest example is

ethene (ethylene), C_2H_4; its structure can be written in condensed form as $H_2C{=}CH_2$; its full structural formula is

$$\begin{array}{ccc} H & & H \\ \diagdown & & \diagup \\ & C{=}C & \\ \diagup & & \diagdown \\ H & & H \end{array}$$

In similar hydrocarbons, one or more of the four hydrogens are replaced by other atoms or groups, often an alkyl group containing a short chain of carbon atoms such as $CH_3{-}$ or $CH_3CH_2{-}$, which will be designated simply as R, since it is generally not the chain but rather the $C{=}C$ part of the molecule that is the reactive site in atmospheric reactions.

Consider a general hydrocarbon $RHC{=}CHR$. In air it reacts with hydroxyl radical by *addition* to the $C{=}C$ bond:

$$\begin{array}{ccc} R & & R \\ \diagdown & & \diagup \\ & C{=}C & \quad + \; OH^{\bullet} \longrightarrow \\ \diagup & & \diagdown \\ H & & H \end{array} \qquad \begin{array}{ccc} R & & R \\ \diagdown & & \diagup \\ \overset{\bullet\bullet}{C}{-}\overset{}{C}{-}OH & \\ \diagup & & \diagdown \\ H & & H \end{array}$$

This addition reaction is a faster process, due to its lower activation energy, than the alternative of abstraction of hydrogen, so we can neglect the abstraction process in molecules containing a $C{=}C$ link. Because the reaction of addition of $OH^{\bullet}$ to a multiple bond is much faster than H abstraction from methane and other alkane hydrocarbons, $RHC{=}CHR$ molecules in general are much faster to react than are alkanes.

As anticipated from the reaction principles, the carbon-based radical produced from the reaction of hydroxyl radical with the hydrocarbon adds O_2 to yield a peroxy radical, which in turn oxidizes $NO^{\bullet}$ to $NO_2^{\bullet}$:

$$\begin{array}{ccc} R & & R \\ \diagdown & & \diagup \\ \overset{\bullet\bullet}{C}{-}C{-}OH & \xrightarrow{O_2} \\ \diagup & & \diagdown \\ H & & H \end{array} \qquad \begin{array}{c} \overset{\bullet}{O} \\ \diagup \\ O \\ | \\ \begin{array}{ccc} R & & R \\ \diagdown & & \diagup \\ C{-}C{-}OH & \xrightarrow{NO^{\bullet}} NO_2^{\bullet} + \\ \diagup & & \diagdown \\ H & & H \end{array} \end{array} \qquad \begin{array}{c} \overset{\bullet}{O} \\ \diagup \\ O \\ | \\ \begin{array}{ccc} R & & R \\ \diagdown & & \diagup \\ C{-}C{-}OH \\ \diagup & & \diagdown \\ H & & H \end{array} \end{array}$$

Once much of the $NO^{\bullet}$ has been oxidized to $NO_2^{\bullet}$, photochemical decomposition by sunlight of the latter gives $NO^{\bullet}$ plus O, which then quickly combines with molecular oxygen to give ozone, as discussed in Chapter 3. It is a characteristic of air pollution driven by photochemical processes that ozone from $NO_2^{\bullet}$ photodecomposition builds up to much higher levels than are found in clean air. Nitrogen dioxide is the only significant tropospheric source of the atomic oxygen from which ozone can form.

As mentioned previously, the ozone concentration does not build up substantially as a result of this sequence until most of the $NO^{\bullet}$ has been converted to $NO_2^{\bullet}$, since $NO^{\bullet}$ and O_3 mutually self-destruct if both are present in significant concentrations. It is only after most NO˙ has been oxidized to NO$_2^-$ as a result of reactions with peroxy free radicals that the characteristic buildup of **urban ozone** occurs, as can be seen in Figure 3-3 on page 80; the transition that occurred at about 9 A.M. on a particular smoggy day in the 1960s in Los Angeles (when smog levels were higher than in more recent years) is illustrated.

One might anticipate from our reactivity principles (Figure 17-1) that the two-carbon radical mentioned above (RCHOCH(R)OH) would lose H by abstraction by O$_2$, but instead it decomposes spontaneously by cleavage of the C—C bond to give a nonradical molecule containing a C=C bond and another radical, RHCOH:

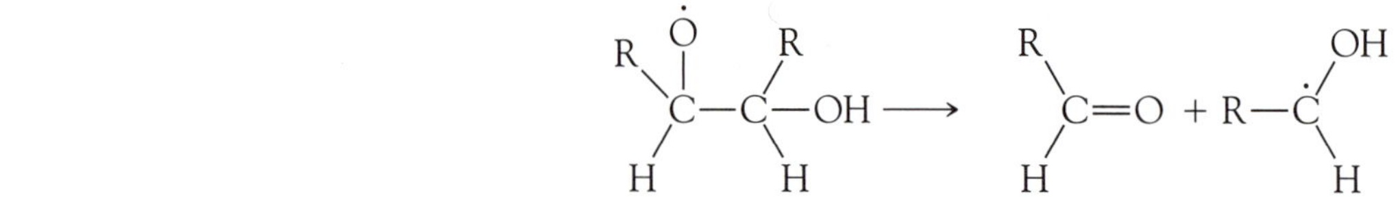

$$
\begin{array}{c}
\text{R} \quad \overset{\cdot\cdot}{\overset{\textstyle O}{|}} \quad \text{R} \\
\diagdown \text{C}-\overset{|}{\underset{|}{\text{C}}}-\text{OH} \longrightarrow \\
\diagup \quad \diagdown \\
\text{H} \qquad \text{H}
\end{array}
\qquad
\begin{array}{c}
\text{R} \qquad\qquad \text{OH} \\
\diagdown \qquad\qquad \overset{\cdot}{\diagup} \\
\text{C}=\text{O} + \text{R}-\text{C} \\
\diagup \qquad\qquad \diagdown \\
\text{H} \qquad\qquad \text{H}
\end{array}
$$

It happens that the reaction requires no energy input, i.e., ΔH is close to zero, because in this case the formation of a C=O bond from C—O compensates energetically for loss of the C—C bond. Since the decomposition of this radical is not endothermic, its activation energy is small and thus the process occurs spontaneously in air.

The carbon-based radical RHĊOH produced in the preceding reaction subsequently reacts with an O$_2$ molecule. Since loss of the hydroxyl hydrogen from this radical allows the C—O bond to become C=O, the oxygen molecule abstracts the H atom and produces an aldehyde.

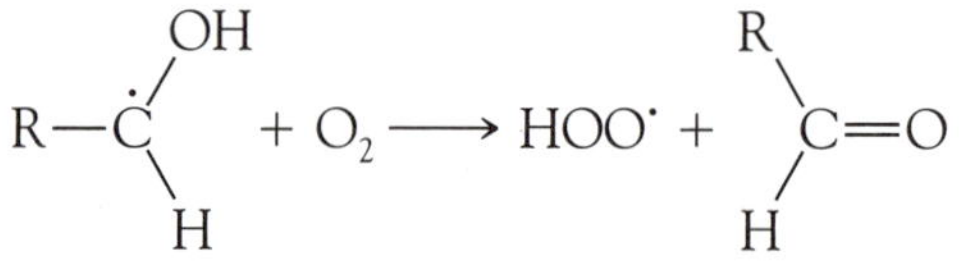

$$
\begin{array}{c}
\text{OH} \\
\overset{\cdot}{\diagup} \\
\text{R}-\text{C} \quad + \text{ O}_2 \longrightarrow \text{HOO˙} + \\
\diagdown \\
\text{H}
\end{array}
\qquad
\begin{array}{c}
\text{R} \\
\diagdown \\
\text{C}=\text{O} \\
\diagup \\
\text{H}
\end{array}
$$

If we add all the above reactions, the net reaction thus far is

$$\text{RHC}{=}\text{CHR} + \text{OH}˙ + 2\,\text{O}_2 + \text{NO}˙ \longrightarrow 2\,\text{RHC}{=}\text{O} + \text{HOO}˙ + \text{NO}_2^-$$

Thus the original RHC=CHR pollutant molecule is converted into two aldehyde molecules, each possessing half the number of carbon atoms. Indeed, as shown in Figure 3-3, by about noon on the very smoggy day in Los Angeles, most of the reactive hydrocarbons emitted into the air by morning rush-hour traffic had been converted to aldehydes. By midafternoon, most of the aldehydes had disappeared, since they had largely been photochemically decomposed into HCO˙ and R˙ (alkyl) free radicals.

$$\text{RHCO} \xrightarrow{\text{sunlight}} \text{R}˙ + \text{HCO}˙$$

The sunlight-induced decomposition of aldehydes and of ozone leads to a huge increase in the number of free radicals in the air of a city undergoing photochemical smog, although in absolute terms the concentration of radicals is still very small.

The steps in the conversion of the original RHC=CHR molecule into aldehydes, and then of the latter to carbon dioxide (see Problem 17-8), are summarized in Figure 17-4. As indicated by the results of Problem 17-8, the net effect of the synergistic oxidation of nitric oxide and RHC=CHR is the production of carbon dioxide, nitrogen dioxide, and more hydroxyl radicals. Thus the reaction is **autocatalytic**—its net speed will increase with time since one of its products, here OH$^{\bullet}$, catalyzes the reaction for other reactant molecules.

PROBLEM 17-8

Rewrite the net reaction shown above for RHC=CHR, assuming that R is H. Deduce the series of steps by which the formaldehyde molecules will subsequently undergo photochemical decomposition and, by a further series of steps, be oxidized to carbon dioxide. Add these steps to the net reaction. Also add the reactions by which HOO$^{\bullet}$ oxidizes NO$^{\bullet}$. What is the final net reaction obtained by adding all these processes together?

PROBLEM 17-9

Repeat Problem 17-8, but this time assume that the alkyl group R in the aldehyde RHCO produced by photochemical smog is a simple methyl group, CH_3, and that, when the aldehyde undergoes photochemical decomposition by sunlight, the radicals $CH_3^{\bullet}$ and HCO$^{\bullet}$ are produced. Using the air reactivity principles, deduce the sequence of reactions by which these radicals are oxidized to carbon dioxide, and determine the overall reaction of conversion of RHCO to CO_2. Assume formaldehyde photolyzes.

17.4 Photochemical Smog: The Fate of the Free Radicals

In later stages of photochemical smog formation, reactions that occur between two radicals are no longer insignificant, since their concentrations have become so high. Because their rates are proportional to the *product* of two radical concentrations, these processes are important when the radical concentrations are high; i.e., they occur quickly under such conditions. Generally, the reaction of two free radicals yields a stable, nonradical product:

$$\text{radical} + \text{radical} \longrightarrow \text{nonradical molecule}$$

One important example of a radical–radical reaction is the combination of hydroxyl and nitrogen dioxide radicals to yield **nitric acid,** HNO_3, a process that, as we saw in Chapter 1, also occurs in the stratosphere:

$$OH^{\bullet} + NO_2^{\bullet} \longrightarrow HNO_3$$

This reaction is the main tropospheric sink for hydroxyl radicals. The average lifetime for an HNO_3 molecule is several days. By then it either has

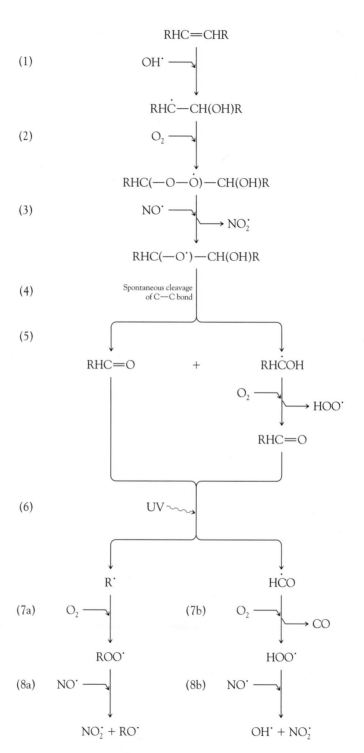

FIGURE 17-4 Mechanism
of the RHC=CHR oxidation
process in photochemical
smog.

dissolved in water and been rained out or has been photochemically decomposed back into its components.

Similarly, combination of OH˙ with NO˙ gives **nitrous acid,** HONO, also written HNO₂. In sunlight the nitrous acid is almost immediately photochemically decomposed back to OH˙ and NO˙, but at night it is stable and therefore its concentration climbs. The observed gigantic increase by dawn in the concentration of OH˙ radicals in the air of smog-ridden cities, which serves to start the oxidation of hydrocarbons, is due largely to the decomposition of the HONO that had been created the previous evening:

The release of HONO from soil is a newly discovered important source of the gas.

$$OH^\bullet + NO^\bullet \longrightarrow HONO \xrightarrow{\text{sunlight}} OH^\bullet + NO^\bullet$$

It is a characteristic of the later stages in the day of a smog episode that oxidizing agents such as nitric acid are formed in substantial quantities. The reaction of two OH˙ radicals, or of two hydroperoxy radicals, HOO˙, produces another atmospheric oxidizing agent, **hydrogen peroxide,** H₂O₂, which, as we have already seen, is also produced in this way in clean atmospheres devoid of nitrogen oxides:

$$2\ OH^\bullet \longrightarrow H_2O_2$$

$$2\ HOO^\bullet \longrightarrow H_2O_2 + O_2$$

The latter reaction occurs also in clean air when the concentration of NO$_X$ is especially low and was encountered in stratospheric chemistry in Chapter 1.

The fate of the R—Ċ=O radicals, produced by H atom abstraction by OH˙ from aldehydes in the ways discussed above, is to combine with O₂ and so produce the free radical

$$R-\overset{\displaystyle O}{\underset{\displaystyle O-\dot{O}}{C}}$$

When NO˙ is plentiful, this complex species, as expected, behaves as a peroxy radical and oxidizes nitric oxide. In the afternoon, when the concentration of NO˙ is very low, the radical reacts instead in a radical–radical process by *adding* to NO˙₂ to yield a nitrate. For the common case for which R is CH₃, the nitrate product formed is **peroxyacetylnitrate,** or PAN, which is a potent eye irritant in humans and is also toxic to plants.

$$CH_3-\overset{\displaystyle O}{\underset{\displaystyle O-\dot{O}}{C}} + NO_2^\bullet \longrightarrow \underset{\text{PAN}}{CH_3-\overset{\displaystyle O}{\underset{\displaystyle O-O-NO_2}{C}}}$$

Overall, then, the afternoon stage of a photochemical smog episode is characterized by a buildup of oxidizing agents such as hydrogen peroxide, nitric acid, and PAN, as well as ozone.

Another important species that is present in the later stages of smog episodes is the **nitrate radical,** $NO_3^{\cdot}$, produced when high concentrations of $NO_2^{\cdot}$ and ozone occur simultaneously:

$$NO_2^{\cdot} + O_3 \longrightarrow NO_3^{\cdot} + O_2$$

Although $NO_3^{\cdot}$ is photochemically dissociated to $NO_2^{\cdot}$ and O rapidly during the daytime, it is stable at night and plays a role similar to $OH^{\cdot}$ in attacking hydrocarbons in the hours following sundown:

$$NO_3^{\cdot} + RH \longrightarrow HNO_3 + R^{\cdot}$$

Thus at night, when the concentration of the short-lived hydroxyl radicals goes almost to zero since no new ones are being produced due to the absence of O^{*}, $NO_3^{\cdot}$ rather than $OH^{\cdot}$ initiates the oxidation of reduced gases in the troposphere. The similarity between $OH^{\cdot}$ and $NO_3^{\cdot}$ is not surprising since both react as $—\dot{O}$ radicals and form very stable O—H bonds when they abstract hydrogens.

In summary, an episode of photochemical smog in a city such as Los Angeles begins at dawn, when sunlight initiates the production of hydroxyl radicals from the nitrous acid and from the ozone left over from the previous day. The initial input of nitric oxide and reactive hydrocarbons from morning rush-hour vehicle traffic reacts first to produce aldehydes (see Figure 3-3), the photolysis of which increases the concentration of free radicals and thereby speeds up the overall reaction. The increase in free radicals in the morning serves to oxidize the nitric oxide to nitrogen dioxide; photolysis of the latter causes the characteristic rise in ozone concentrations about midday. Oxidants such as PAN and hydrogen peroxide are also produced, especially in the afternoons due to the high free-radical concentration present at that time. The late afternoon rush-hour traffic produces more nitric oxide and hydrocarbons, which presumably react quickly under the conditions of high free-radical concentration. The smog reactions largely cease at dusk due to the lack of sunlight, but some oxidation of hydrocarbons continues due to the presence of the nitrate radical. The nitrous acid that forms after dark is stable until dawn, when its decomposition helps initiate the process for another day.

PROBLEM 17-10

Annotate (in pencil) the top of Figure 3-3 to show the *dominant* reaction occurring in the polluted air in the following time segments: (a) 5 A.M.–8 A.M.; (b) 8 A.M.–12 noon.

PROBLEM 17-11

Some formaldehyde molecules photochemically decompose to the molecular products H_2 and CO rather than to free radicals. Deduce the mechanism and overall reaction for the oxidation to CO_2 for formaldehyde molecules that initially produce the molecular products.   ●

PROBLEM 17-12

Deduce the series of steps by which ethylene gas, $H_2C=CH_2$, is oxidized to CO_2 when it is released into an atmosphere undergoing a photochemical smog process. Assume in this case that aldehydes react completely by photochemical decomposition rather than by OH˙ attack.   ●

PROBLEM 17-13

Radical–radical reactions can also occur in clean air, particularly when the nitrogen oxide concentration is very low. Predict the product that will be formed when the $CH_3OO˙$ radical intermediate of methane oxidation combines with the HOO˙ free radical. Note that long oxygen chains are unstable with respect to O_2 expulsion.   ●

17.5 Oxidation of Atmospheric SO_2: The Homogeneous Gas-Phase Mechanism

When the sky is clear or when clouds occupy only a few percent of the tropospheric volume, the predominant mechanism for the conversion of SO_2 to H_2SO_4 is a homogeneous gas-phase reaction that occurs by several sequential steps. As usual for atmospheric trace gases, the hydroxyl radical initiates the process. Since SO_2 molecules contain multiple bonds but no hydrogen, it is expected (see Figure 17-1) that the OH˙ will *add* to the molecule at the sulfur atom:

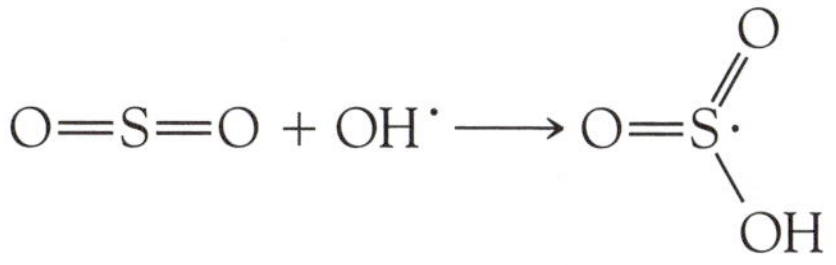

$$O=S=O + OH˙ \longrightarrow O=\overset{\displaystyle O}{\underset{\displaystyle OH}{\overset{\|}{S}}}˙$$

Since a stable molecule, namely **sulfur trioxide,** SO_3, can be produced from this radical by the removal of the hydrogen atom, the reaction principles predict that the next reaction in the sequence is that between the radical and an O_2 molecule to abstract H:

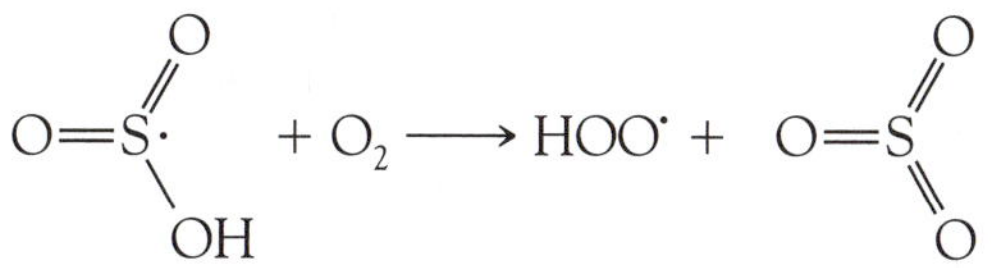

$$O=\overset{\displaystyle O}{\underset{\displaystyle OH}{\overset{\|}{S}}}˙ + O_2 \longrightarrow HOO˙ + O=\overset{\displaystyle O}{\underset{\displaystyle O}{\overset{\|}{S}}}$$

The sulfur trioxide molecule rapidly combines with a gaseous water molecule to form **sulfuric acid.** Finally, the H_2SO_4 molecules react with water, whether in the form of water vapor or as a mist, to form an aerosol of droplets, each of which is an aqueous solution of sulfuric acid. The sequence of steps from gaseous SO_2 to aqueous H_2SO_4 is

$$SO_2 + OH^{\cdot} \longrightarrow HSO_3^{\cdot}$$

$$HSO_3^{\cdot} + O_2 \longrightarrow SO_3 + HOO^{\cdot}$$

$$SO_3 + H_2O \longrightarrow H_2SO_4(g)$$

$$H_2SO_4(g) + \text{many } H_2O \longrightarrow H_2SO_4(aq)$$

The sum of these reaction steps is

$$SO_2 + OH^{\cdot} + O_2 + \text{many } H_2O \longrightarrow HOO^{\cdot} + H_2SO_4(aq)$$

When we include the return of $HOO^{\cdot}$ to $OH^{\cdot}$ via reaction with $NO^{\cdot}$, the overall reaction is seen to be $OH^{\cdot}$-catalyzed co-oxidation of SO_2 and $NO^{\cdot}$:

$$SO_2 + NO^{\cdot} + O_2 + \text{many } H_2O \xrightarrow{OH^{\cdot} \text{ catalysis}} NO_2^{\cdot} + H_2SO_4(aq)$$

For representative concentrations of the $OH^{\cdot}$ radical in relatively clean air, a few percent of the atmospheric SO_2 is oxidized per hour by this mechanism. The rate is much faster for air masses undergoing photochemical smog reactions since the concentration of $OH^{\cdot}$ there is much higher. However, generally only a small amount of sulfur dioxide is oxidized in cloudless air; the rest is removed by dry deposition before the reaction has time to occur.

Dissolved sulfur dioxide, SO_2, is oxidized to **sulfate ion,** SO_4^{2-}, by trace amounts of the well-known oxidizing agents hydrogen peroxide, H_2O_2, and ozone, O_3, that are present in the airborne droplets, as already discussed in Chapter 3. Indeed, such reactions currently are thought to constitute the main oxidation pathways for atmospheric SO_2, except under clear sky conditions when the gas-phase homogeneous mechanism predominates. The ozone and hydrogen peroxide result mainly from sunlight-induced reactions in photochemical smog. Consequently, oxidation of SO_2 occurs most rapidly in air that has also been polluted by reactive hydrocarbons and nitrogen oxides. Since the smog reactions occur predominantly in summer, rapid oxidation of SO_2 to sulfate also is characteristic of the summer season.

Systematics of Stratospheric Chemistry

There are many similarities between the chemical reactions discussed in Chapters 1 and 2 for the stratosphere and those outlined above for the troposphere. For example, a characteristic process in both regions of the atmosphere is hydrogen atom abstraction. The stratosphere and troposphere differ, however, in which reactions are dominant. In the stratosphere $OH^{\cdot}$, O^*, $Cl^{\cdot}$, and $Br^{\cdot}$ are all important in abstracting a hydrogen atom from stable molecules such as methane, whereas only the hydroxyl and nitrate radicals are important in this respect in the troposphere. In the following material we systematize the Chapters 1 and 2 chemistry that is important in the stratosphere, especially in regard to processes of ozone depletion.

17.6 Processes Involving Loosely Bound Oxygen Atoms

Many of the species in the stratosphere have a loosely bound oxygen atom, denoted Y, which is readily detached from the rest of the molecule in several characteristic ways. In Table 17-1 we list the molecules Y—O that contain a loose oxygen. In every case, dissociation of this oxygen atom requires much less energy than is required to break any of the remaining bonds, so the resulting Y units remain intact. Notice that the Y species, except for O_2, are the free radicals that in Chapters 1 and 2 we called X when we discussed ozone destruction catalysts. In terms of electronic structure, all "loose" oxygens are joined by a single bond to another electronegative atom that possesses one or more nonbonding electron pairs. The interaction between the nonbonded electron pairs on this atom and those on the oxygen weakens the single bond.

The characteristic reactions involving loose oxygen are collected below.

- **Reaction with Atomic Oxygen**   Here the oxygen atom detaches the loose oxygen atom by combining with it:

$$Y—O + O \longrightarrow Y + O_2$$

These reactions are all exothermic since the $O{=}O$ bond in O_2 is much stronger than the Y—O bond.

- **Photochemical Decomposition**   The Y—O species absorbs UV-B, and in some cases even longer wavelength light, from sunlight and subsequently releases the loose oxygen atom:

$$Y—O + sunlight \longrightarrow Y + O$$

- **Reaction with NO˙**   Nitric oxide abstracts the loose oxygen atom:

$$Y—O + NO˙ \longrightarrow Y + NO_2˙$$

This reaction is exothermic since the ON—O bond strength (see Table 17-1) is the greatest of those involving a loose oxygen. (Recall the general principle that exothermic free-radical reactions are relatively fast.)

TABLE 17-1	Molecules Containing Loose Oxygen Atoms		
Molecule Y—O	Structure of Y—O	Y—O Bond Energy in kJ mol^{-1}	Comment
O_3	O_2—O	107	The most loose oxygen
BrO˙	Br—O	235	
HOO˙	HO—O	266	
ClO˙	Cl—O	272	
NO$_2$˙	ON—O	305	The least loose oxygen

- **Abstraction of Oxygen from Ozone**   Abstraction of the loose oxygen atom from ozone (only) to form the Y—O species is characteristic of OH$^{\cdot}$, Cl$^{\cdot}$, Br$^{\cdot}$, and NO$^{\cdot}$. Thus all these radicals act as catalytic ozone destroyers, X:

$$O_2{-}O + X \longrightarrow O_2 + XO$$

The reaction involving ozone is exothermic since ozone contains the weakest of the bonds involving a loose oxygen. The other YO species do not undergo this reaction to an important extent either because it is endothermic and therefore negligibly slow or because the atmospheric X species react more quickly with other chemicals.

- **Combination of Two YO Molecules**   If the concentration of YO species becomes high, they may react by the collision of two of them (identical or different species). If at least one is O_3 or HOO$^{\cdot}$, an unstable chain of three or more oxygen atoms is created when they collide and join; in these circumstances, the loose oxygens combine to form one or more molecules of O_2, which are expelled:

$$2\ O_2{-}O \longrightarrow 3\ O_2$$

$$2\ HO{-}O^{\cdot} \longrightarrow HOOH + O_2$$

$$HO{-}O^{\cdot} + O{-}O_2 \longrightarrow OH^{\cdot} + 2\ O_2$$

$$HO{-}O^{\cdot} + {}^{\cdot}O{-}Cl \longrightarrow HOCl + O_2$$

When neither is O_3 or HOO$^{\cdot}$, the two Y—O molecules combine to form a larger molecule, which subsequently often decomposes photochemically:

$$2\ NO_2^{\cdot} \longrightarrow N_2O_4$$

$$2\ ClO^{\cdot} \longrightarrow ClOOCl \xrightarrow{\text{sunlight}} 2\ Cl^{\cdot} + O_2$$

$$ClO^{\cdot} + NO_2^{\cdot} \underset{\text{sunlight}}{\rightleftharpoons} ClONO_2$$

$$ClO^{\cdot} + BrO^{\cdot} \longrightarrow Cl^{\cdot} + Br^{\cdot} + O_2$$

The Y—O—O—Y molecules have little thermal stability, and even at moderate temperatures they may dissociate back to their Y—O components before light absorption and photolysis have time to occur.

PROBLEM 17-14

Which of the following species do(es) *not* contain a loose oxygen?

(a) HOO$^{\cdot}$ (b) OH$^{\cdot}$ (c) NO$^{\cdot}$ (d) O_2 (e) ClO$^{\cdot}$ ●

PROBLEM 17-15

From which Y—O species

(a) does NO$^{\cdot}$ abstract an oxygen atom?
(b) does atomic oxygen abstract an oxygen atom?

(c) does sunlight in the stratosphere detach an oxygen atom?

(d) do the Y—O—O—Y species (with identical Y groups) form in the stratosphere?

(e) is O_2 produced when two identical Y—O species react?   ●

PROBLEM 17-16

Using the principles above, predict what would be the likely fate of $BrO^\cdot$ molecules in a region of the stratosphere that in concentrations was particularly (a) high in atomic oxygen, (b) high in $ClO^\cdot$, (c) high in $BrO^\cdot$ itself, and (d) high in sunlight intensity.   ●

PROBLEM 17-17

Using the principles above, deduce what reaction(s) could be sources of atmospheric (a) $ClONO_2$ (b) $ClOOCl$ (c) $Cl^\cdot$ atoms   ●

PROBLEM 17-18

Draw the Lewis structure for the free radical $FO^\cdot$. On the basis of this structure, could you predict whether it contains a loose oxygen?   ●

PROBLEM 17-19

What is the expected product when $ClO^\cdot$ reacts with $NO^\cdot$? What are the possible fates of the product(s) of this reaction? Devise a mechanism incorporating (a) this reaction, (b) the reaction of $Cl^\cdot$ with ozone, and (c) the photochemical decomposition of $NO_2^\cdot$ to $NO^\cdot$ and atomic oxygen. Is the net result of this cycle, which operates in the lower stratosphere, the destruction of ozone?   ●

Review Questions

1. Explain why $OH^\cdot$ reacts more quickly than $HOO^\cdot$ to abstract hydrogen from other molecules.

2. How does $OH^\cdot$ react with molecules that contain hydrogen but not multiple bonds?

3. What are the two different initial steps by which atmospheric formaldehyde, H_2CO, is decomposed in air?

4. What is the reaction by which most nitric oxide molecules in the troposphere are oxidized to nitrogen dioxide?

5. What are the two common reactions by which diatomic oxygen reacts with free radicals?

6. Explain why photochemical smog is an autocatalytic process.

7. What is the fate of $OH^\cdot$ radicals that react with $NO^\cdot$? With $NO_2^\cdot$? With other $OH^\cdot$?

8. What is the fate of $NO_2^\cdot$ molecules that photodissociate? That react with ozone? That react with $RCOO^\cdot$ radicals?

9. Why does the production of high concentrations of NO_2^- lead to an increase in ozone levels in air? Why does this not occur if much $NO^\bullet$ is present?

10. What is the formula of nitrate radicals? Explain how they are similar in reactivity to hydroxyl radicals.

11. Explain how atmospheric sulfur dioxide is oxidized by gas-phase reactions in the atmosphere.

12. Explain what is meant by the term *loosely bound oxygen*. What are its four characteristic reactions in stratospheric chemistry?

Additional Problems

1. Write the two-step mechanism by which CO is oxidized to CO_2. Also include the sequence of reactions by which the hydroperoxy radical so produced oxidizes $NO^\bullet$ to NO_2^-, the nitrogen dioxide is photolyzed to $NO^\bullet$ and atomic oxygen, and oxygen atoms produce ozone. By adding the steps, show that the atmospheric oxidation of carbon monoxide can increase the ozone concentration by a catalytic process.

2. Using the reactivity principles developed in this chapter, deduce the series of steps and the overall reaction by which ethane, $H_3C{-}CH_3$, is oxidized in the atmosphere. Assume that the aldehydes produced in the mechanism undergo photochemical decomposition to $R^\bullet$ and $HCO^\bullet$.

3. When the concentration of nitrogen oxides in a region of the air is very low, peroxy radicals combine with other species rather than oxidize nitric oxide. Deduce the mechanism, including the overall equation, for the process by which carbon monoxide is oxidized to carbon dioxide under these conditions, assuming that the hydroperoxy radicals react with ozone. From your result, would you predict that ozone levels would be abnormally high or low in air masses having low nitrogen oxide concentration? [*Hint:* See the generalities in the section on the systematics of stratosphere chemistry concerning reactions that produce long oxygen chains.]

4. Predict the most likely reaction (if any) that would occur between a hydroxyl radical and each of the following atmospheric gases:

(a) $CH_3CH_2CH_3$ (b) $H_2C{=}CHCH_3$

(c) HCl   (d) H_2O

5. Draw complete Lewis structures for NO_2, HONO, and HNO_3. (Resonance structures and formal charges are not required.)

6. In Problem 1-2, the longest wavelength of light that could dissociate an O atom from O_3 was calculated and determined to occur in the IR region of the spectrum. Using the information in Table 17-1, calculate the longest wavelength of light that could photolytically cleave the loose hydrogen atom in the case of each of the remaining molecules listed in that table. What region of the electromagnetic spectrum does each correspond to?

Appendix

Oxidation Numbers and Redox Equation Balancing Reviewed

Assigning Oxidation Numbers

A simple way to determine the extent (if any) to which an element is oxidized or reduced in a chemical reaction is to deduce the change in its *oxidation number*, O.N., in the product compared to that in the reactant. The oxidation number of the elements in most compounds and ions can be determined by applying, in sequence, the following set of rules, keeping in mind that **the sum of all the oxidation numbers in a substance must equal the its net charge.** The rules are listed in terms of priority so that, for example, if for a compound, rule (iv) is inconsistent with rule (iii), then rule (iii) takes precedence, since it is higher in the order.

(i) Elements appearing in the free, unbonded form have an O.N. equal to their ionic charge, which is zero if the element is uncharged.

(ii) Fluorine has an O.N. of -1 in compounds. Group 1 and Group 2 metals have O.N. values corresponding to their ionic charges $+1$ and $+2$, respectively, and Al is $+3$.

(iii) Hydrogen has an O.N. of $+1$, except when bonded to a metal, where it is -1.

(iv) Oxygen has an O.N. of -2, except when overridden by a rule higher in the sequence, as an example below illustrates.

(v) Chlorine, bromine, and iodine have O.N.'s of -1, except when overridden by a rule higher in the sequence, as an example below illustrates.

Some examples follow.

HF: F is -1 (rule ii) and H is $+1$ (rule iii); the sum is zero, as required.

H_2O_2: H is $+1$ (rule iii) but O cannot here be -2 (rule iv), since the sum of charges would be $2(+1) + 2(-2) = -2$; the charges must add up to zero for the molecule as a whole. Since rule (iii) takes precedence over rule (iv), each H must be $+1$, so each O here must be -1 in order that the sum be zero.

ClO_2^-: Each O is -2 (rule iv), for a total of -4, so Cl here cannot be -1 (rule v) since the sum of charges would then be $-1 + 2(-2) = -5$, compared to the actual net charge of -1. Since $O.N._{Cl} + 2(-2) = -1$, it follows that $O.N._{Cl} = +3$ here.

As an example of the use of oxidation numbers in reactions, consider the half-reaction in which nitrate ion is converted into nitrous oxide:

$$NO_3^- \longrightarrow N_2O$$

Since in nitrate ion each O is -2, with a sum of -6, and the charge on the ion is only -1, it follows that N in the reactant here is $+5$. In nitrous oxide, the O is -2, and the sum of oxidation numbers is zero, so each N must be $+1$.

Keeping in mind that the reaction requires 2 nitrate ions to supply enough nitrogen for one nitrous oxide molecule, we see that $2(+5)$ N's, total $+10$, here become $2(+1)$ N's, total $+2$. Thus the half-reaction must be a $(10 - 2 =)$ 8-electron reduction:

$$2\ NO_3^- + 8\ e^- \longrightarrow N_2O$$

If it is required to know the numbers of water molecules and H^+ or OH^- ions involved, the detailed balancing scheme discussed below must be employed.

Balancing Redox Equations

There are many equivalent schemes to completely balance redox half-reactions and overall reactions, of which the following method is one example.

• To balance a half-reaction, **first deduce the number of electrons involved in the process,** as in the scheme above, by balancing atoms *other than H and O.*

• Next, **to balance the charge, add sufficient H^+ ions to the side having excess negative charge;** note that only real charges on ions and electrons are considered in the calculations here, *not* oxidation numbers.

• Finally, **balance the number of oxygen atoms by adding H_2O molecules to the side deficient in oxygen.**

Consider, for example, the nitrate to nitrous oxide example discussed above:

$$2\ NO_3^- + 8\ e^- \longrightarrow N_2O$$

The actual charge on the left-hand side is $2 \times (-1) + (-8) = -10$, but that on the right side is zero. Thus we should add 10 positive charges, each in the form of H^+, to the left side, so that its charge also becomes zero:

$$2\ NO_3^- + 8\ e^- + 10\ H^+ \longrightarrow N_2O$$

Finally, since we now have $2 \times 3 = 6$ O atoms on the left side and only one on the right side, we need to add 5 O atoms, each in the form of an H_2O molecule, to the right side:

$$2\ NO_3^- + 8\ e^- + 10\ H^+ \longrightarrow N_2O + 5\ H_2O$$

The half-reaction now is balanced for acidic and neutral solutions.

Answers to Selected Odd-Numbered Problems

Chapter 1

Problems

1-1 (a) 427 kJ mol^{-1}; junction of UV-B with UV-C

(b) 299 kJ mol^{-1}; junction of UV with visible region

(c) 160 kJ mol^{-1}; junction of visible and infrared regions

(d) 29.9 kJ mol^{-1}; beginning of thermal IR region

1-3 390.7 nm; 127.5 nm

1-5 307 nm

1-7

$$OH + O_3 \longrightarrow HOO + O_2$$
$$HOO + O \longrightarrow OH + O_2$$

Overall: $O_3 + O \longrightarrow 2 O_2$

Box 1-2, Problem 1 1.6×10^8 molecules cm^{-3} sec^{-1}

Box 1-3, Problem 1 $[O^*] = k_1 [O_2]/(k_2 [M] + k_3 [H_2O])$

Problem 3 $[NO_2]/[NO] = k_3 [O_3]/k_1$

Additional Problems

1. Net is O_2 + UV photon $\longrightarrow 2 O$
Each O reacts as $O + O_2 \longrightarrow O_3$
Overall $3 O_2$ + UV photon $\longrightarrow 2 O_3$

3. $ClONO_2$ + photon $\longrightarrow Cl + NO_3$
NO_3 + photon $\longrightarrow NO + O_2$
$Cl + O_3 \longrightarrow ClO + O_2$
$NO + O_3 \longrightarrow NO_2 + O_2$
$ClO + NO_2 \longrightarrow ClONO_2$
Overall $2 O_3$ + 2 photons $\longrightarrow 3 O_2$

5. 491 nm; visible

7. 3×10^3 molecules cm^{-3} s^{-1}; 2.5×10^{-12} g cm^{-3} y^{-1}

9. 1.5×10^{-6}

Chapter 2

Problems

2-1

$$Cl + O_3 \longrightarrow ClO + O_2$$
$$Br + O_3 \longrightarrow BrO + O_2$$
$$ClO + BrO \longrightarrow Cl + Br + O_2$$

Overall: $2 O_3 \longrightarrow 3 O_2$

2-5 (a) Mechanism I: $F + O_3 \longrightarrow OF + O_2$
$OF + O \longrightarrow F + O_2$

Mechanism II: $F + O_3 \longrightarrow OF + O_2$
$2 OF \longrightarrow [FOOF]$
$\longrightarrow 2 F + O_2$

(b) $F + O_3 \longrightarrow OF + O_2$
$FO + O_3 \longrightarrow F + 2 O_2$

Net $2 O_3 \longrightarrow 3 O_2$

Green Chemistry Questions

1. (a) 2
(b) 1, 4

3. No, not if the carbon dioxide is a waste by-product from another process.

5. Harpin is applied to the plant, which elicits the plant's own natural defenses.

Additional Problems

1. (b) 4×10^{15} g

3. (a) 140
(b) 10
(c) 141

9. $CHF_3 < CHFCl_2 < CF_3Cl < CFCl_3$

Chapter 3

Problems

3-1 1.4×10^{-14} M; 0.35 ppt

3-3   61.6; 51.3

3-5   80 ppm; yes.

3-7 (a) $2 NO + 2 CO \longrightarrow N_2 + 2 CO_2$
     (b) $38 NO + 2 C_6H_{14} \longrightarrow 19 N_2 + 12 CO_2 + 14 H_2O$

3-9 $8 NH_3 + 6 NO_2 \longrightarrow 7 N_2 + 12 H_2O$
     0.0092 g

3-11 $2 NaOH + SO_2 \longrightarrow Na_2SO_3 + H_2O$
      $Ca(OH)_2$
      $Na_2SO_3 + Ca(OH)_2 \longrightarrow CaSO_3 + 2 NaOH$
      $SO_2 + Ca(OH)_2 \longrightarrow CaSO_3 + H_2O$

3-13 3.88

3-15 (b) 5.53

3-17 $PM_{0.10}$
      $PM_{infinity}$
      Smaller

Box 3-1, Problems 1 (a) 0.032 ppm
                     (b) 7.9×10^{11} molecules cm^{-3}
                     (c) 1.3×10^{-9} M

3 (a) 9.3×10^{11} molecules cm^{-3}
   (b) 74.2 μg O_3 m^{-3} of air

Box 3-3, Problem 1 Two peaks, to the left of each for Mass, but both to the right of that for Number, and closer to the latter distribution than to the former. Peaks would occur at about 0.2 and 2 μm.

Green Chemistry Questions

1. (a) 3
    (b) 1, 7
3. The presence of the 1,4 C=C double bonds results in a diallylic position, at which a hydrogen is easily removed. Yes. No.
5. (a) 2
    (b) 1.4

Additional Problems

1. 4×10^{10} molecules cm^{-3} sec^{-1}; 4×10^5 molecules cm^{-3} sec^{-1}; that with O_3
3. (a) $O + N_2 \longrightarrow NO + N$
         rate $= k [O] [N_2]$
    (b) factor of 2.2×10^3
5. (a) $O_3 + 2 KI + H_2O \longrightarrow I_2 + O_2 + 2 KOH$
    (b) 120 ppb
7. 1.0%
9. 7.8×10^{-9} M; 6.0×10^{-8} M; 5.8×10^{-7} M

Chapter 4

Problems

4-1 $NH_4^+ + 3 H_2O \longrightarrow NO_3^- + 10 H^+ + 8 e^-$
4-3 16% removed

Additional Problems

1. 4.3×10^{-5} M; 0.68; western
3. 3.7×10^9 g
5. 0.18 m^2 y^{-1}
7. 0.024 g

Chapter 5

Problems

5-1 52°C
5-3 CO and NO; their stretching frequencies must lie outside the thermal IR range.
5-5 0.440 tonnes; 0.27 g
5-7 501 Tg
5-9 No; yes; no
5-11 2.5×10^{15} g y^{-1}

Additional Problems

1. (a) Symmetric and antisymmetric stretch and bending vibrations for both.
    (b) Only the SO_2 symmetric stretch will contribute much.
    (c) Short atmospheric lifetimes.
3. 0.48
7. Sharply increased air temperature.
9. $16.7\pi D^3$; $25\pi D^2$

Chapter 6

Problems

6-1 38%; 0.68%
6-3 10 km
6-5 0.00112, 0.00152, and 0.00254 mol CO_2 kJ^{-1}.

Green Chemistry Questions

1. (a) 2
    (b) 1, 5, 7, 9, 10

3. Growing crops requires fertilizers and pesticides. Energy is needed to plant, cultivate, and harvest; to produce, transport, and apply fertilizers and pesticides; to make and run tractors; and to transport seeds, biomass, monomers, and polymers. Use of land to produce crops for chemicals also removes land that could be used to produce food and feed.

Additional Problems

1. 1.7%; 18%
3. Olefins contain C=C bonds, which makes them more reactive in creating smog than aromatics; alkanes are quite unreactive.
5. 1.83 m

Chapter 7

Problems

7-1 −2 in ethanol, 0 in glucose, +4 in CO_2
7-5 $CH_3CH_2CH_2CH_2OH$; a possible biofuel
$CH_3CH_2CH(OH)CH_3$
$(CH_3)_2CHCH_2OH$; a possible biofuel
$(CH_3)_2C(OH)CH_3$
7-7 3:2; 20%
7-9 419 nm; visible
7-11 120 kJ g^{-1}; H_2 is superior by weight but methane is superior by volume.
7-13 23.7 kg; 12.0 kg; Mg
7-15 $O_2 + 2 H_2O + 4 e^- \longrightarrow 4 OH^-$
$2 OH^- + H_2 \longrightarrow 2 H_2O + 2 e^-$
$O_2 + 2 H_2 \longrightarrow 2 H_2O$

Green Chemistry Questions

1. (a) 3
   (b) 1, 3, 6, 7
3. (a) 1
   (b) 1,7

Additional Problems

3. One-third of the CO
   $3 C + 4 H_2O \rightleftharpoons 2 CH_3OH + CO_2$
5. $8 CH + CH_4 + 2 H_2O \rightarrow 8 CH_2 + CO_2$

Chapter 8

Problems

8-1 (a) 4.1
    (b) 8
    (c) The 12 mps speed
8-3 68%
8-5 387°C

Additional Problems

1. 5.6×10^{24} J; 0.009%
3. About 665 and about 2300

Chapter 9

Problems

9-1 (a) $^{218}_{84}Po$
    (b) $^{214}_{84}Po$
    (c) $^{4}_{2}He$
    (d) $^{238}_{92}U$
9-3 49%; 65%; yes, substantially above
9-5 $^{2}_{1}H + ^{3}_{2}He \longrightarrow ^{4}_{2}He + ^{1}_{1}p$ (or $^{1}_{1}H$)
    $2\ ^{3}_{2}He \longrightarrow ^{4}_{2}He + 2\ ^{1}_{1}p$ (or $2\ ^{1}_{1}H$)
9-7 1.69×10^{13} J. One-fifth that from D and T.
Box 5-1, Problem 1   For any member X of the series, $[X]_{SS}/[A]_{SS} = t_C/t_A$

Additional Problems

1. $^{0}_{1}e$; $^{22}_{11}Na \longrightarrow ^{22}_{10}Ne + ^{0}_{1}e$; $^{13}_{7}N \longrightarrow ^{13}_{6}C + ^{0}_{1}e$
3. The effusion rate of $^{235}UF_6$ is 0.43% faster than that of $^{238}UF_6$. Since very little enrichment occurs in a single step, it is necessary to repeat the process hundreds of times to achieve suitable enrichment.

Chapter 10

Problems

1. (a) 4.0×10^{-5} ppm; 0.040 ppb
   (b) 3.0 μg L^{-1}
   (c) 300 ppb
3. 2.2×10^{-3} mol L^{-1} atm^{-1}
5. $NH_3 + 2 O_2 + OH^- \longrightarrow NO_3^- + 2 H_2O$
   Less alkaline
7. 16 mg L^{-1}

9. 7.9×10^{-3} M; 6.3×10^{-20} M; 8.32; 2.51

11. 15.2

13. (a) $NO_3^- + 2\,H^+ + 2\,e^- \longrightarrow NO_2^- + H_2O$
    (b) $+14.9$
    (c) $14.9 - pH - 0.5 \log([NO_2^-]/[NO_3^-])$
    (d) $pE + pH = 15.9$
    (e) 6×10^{-5}

15. 8.3×10^{-5} M; solubility increases with increasing temperature.

17. 8.2×10^{-4} M

19. 6.3; 10.3

21. At pH = 4, the ratio is 2.1×10^6.
    At pH = 8.5, the ratio is 66.

23. 2.6×10^{-4} M

25. 52 mg $CaCO_3$ L^{-1}; less

27. 3.6

Green Chemistry Questions

1. It removes the outermost layer of the raw cotton fiber, known as the cuticle, which allows the fiber to be made wet for bleaching and dyeing.

3. (a) 2
    (b) 1, 3, 4, 6

Additional Problems

3. (a) $SO_4^{2-} + 10\,H^+ + 8\,e^- \longrightarrow H_2S + 4\,H_2O$
    (b) $pE = 5.75 - (5/4)\,pH - (1/8) \log(P_{H_2S}/[SO_4^{2-}])$
    (c) 10^{-136} atm, absolutely negligible

5. 9.17

7. 6.1 mg L^{-1}; polluted

Chapter 11

Problems

11-1 $Ca(HCO_3)_2 + Ca(OH)_2 \longrightarrow 2\,CaCO_3(s) + 2\,H_2O$

  1:1 ratio

11-3 $+1$; OH^-; $NH_3 + HOCl$

11-5 11 ppm nitrogen; slightly less stringent

11-7 $\quad\quad Fe(s) \longrightarrow Fe^{2+} + 2\,e^-$
$2\,H_2O + 2\,e^- \longrightarrow H_2 + 2\,OH^-$
$\overline{\quad\quad\quad\quad\quad\quad\quad\quad\quad\quad\quad\quad\quad}$
$Fe(s) + 2\,H_2O \longrightarrow Fe^{2+} + 2\,OH^- + H_2(g)$

11-11 0.42 g

11-13 Too expensive

11-15 NO, NO_2, NO_2^-, NO_3^-

11-17 $+214.3$ kJ mol^{-1}
  558 nm
  45%

Box 11-3, Problem 1 $t = [\ln(k_1/k_2)]/(k_1 - k_2)$

Green Chemistry Questions

1. It is biodegradable and its synthesis is performed under mild conditions, uses only water as a solvent, and excess ammonia is recycled.

Additional Problems

1. 0.79; 0.54; 0.27; 0.10

3. 3.73, 0.019%; 4.23, 0.059%

5. 8×10^{-12} M

7. 0.441 mL; 11 L

Chapter 12

Problems

12-1 $M^{2+} + H_2S \longrightarrow MS + 2\,H^+$
$M^{2+} + 2\,R\!-\!S\!-\!H \longrightarrow$
$\quad\quad\quad\quad\quad R\!-\!S\!-\!M\!-\!S\!-\!R + 2\,H^+$

12-3 $HgCl_2$

12-5 0.5 mg; 2.0×10^5 g

12-7 $Pb_3(OH)_6$; $Pb_3CO_3(OH)_4$; $Pb_3(CO_3)_2(OH)_2$; $Pb_3(CO_3)_3$, i.e., $PbCO_3$

12-9 $[Hg^{2+}] = 4.8 \times 10^{-17} [H^+]$; yes

Green Chemistry Questions

1. (a) 3
    (b) 4

3. Lower air pollution, better corrosion protection, and reduced waste.

Additional Problems

1. 13 mg m^{-3}, must be greater than the threshold limit.

3. (a) 61%
    (b) There are no lead-free environments.

5. H_3AsO_4 for pH < 2.20; $H_2AsO_4^-$ from 2.20 to 6.89; $HAsO_4^{2-}$ from 6.89 to 11.49; AsO_4^{3-} for pH > 11.49; 17; 0.17

7. H^+ liberates metals from sulfides; liberated metals enter food chain and affect human health.

Chapter 13

Problems

13-1 o, o'; o, m'; o, p'; m, m'; m, p'; p, p'

13-5 About 400 μg; about 30 ppb.

Box 13-2, Problem 1   In air: 9.9×10^{-11} mol m^{-3}
In water 2.3×10^{-11} mol m^{-3}
In sediment 5.7×10^{-7} mol m^{-3}

Green Chemistry Questions

1. (a) 3
    (b) 4
3. (a) A pesticide must meet one or more of
        the following requirements. It
        (1) reduces pesticide risks to human
            health;
        (2) reduces pesticide risks to non-target
            organisms;
        (3) reduces the potential for
            contamination of valued
            environmental resources;
        (4) broadens adoption of IPM or makes
            it more effective.
    (b) 1 and 2
5. (a) 3
    (b) 1, 4
7. 1, 2, and 3
9. Reduction in toxicity to nontarget species,
    prolonged efficacy, renewable feedstocks,
    and reduction in energy.

Additional Problems

1. 7.4 μg
3. About 12 g
5. Yes, since concentration is 4.7 ppb

Chapter 14

Problems

14-1 No, the same
Yes
The unique ones are 1,2; 1,3; 1,4; 1,6;
1,7; 1,8; 1,9; 2,3; 2,7; 2,8.

14-3 (a) 2,3,5,6- and 2,3,4,5-tetrachlorophenols
      (b) Two 2,3,5,6-tetrachlorophenols
          produce the 1,2,4,6,7,9-HexaCDD.
          Two 2,3,4,5-tetrachlorophenols
          produce the 1,2,3,6,7,8-HexaCDD.

14-5 1.87×10^5 molecules

14-7 2,3; 2,4; 2,5; 2,6; 3,4; 3,5; 2,2'; 2,3'; 2,4';
      2,5'; 2,6'; 3,3'; 3,4'; 3,5'; 4,4'.
      With rotation, 2,2' and 2,6' interconvert,
      as do 2,3' and 2,5', as well as 3,3' and 3,5'.

14-9 1,2; 1,3; 1,4; 2,3; 2,4; 3,4; 1,6; 1,7; 1,8;
      1,9; 2,6; 2,7; 2,8; 3,6; 3,7; 4,6;
      dichlorodibenzofurans

14-11 28.6 pg

Box 14-1, Problem 1   2,6-dichlorophenol +
2,3,4-trichlorophenol, or
2,3-dichlorophenol +
2,3,6-trichlorophenol

Box 14-2, Problem 1   1-chlorodibenzofuran,
and 1,4-, 1-6-, and
1-9-dichlorofurans

Green Chemistry Questions

1. (a) 2
    (b) 1, 4, and 6

Additional Problems

1. (a) One 2,3,4,5- and one
        2,3,4,6-tetrachlorophenol
    (b) One 2,3,4,6- and one
        2,3,5,6-tetrachlorophenol
    (c) One 2,3,4,6- and either 2,3,4,5- or
        2,3,5,6-tetrachlorophenol
3. Dechlorination. Dioxins more toxic than
    the originals may be produced.
5. 2,3-, 2,2'-, and 2,3'-dichlorobiphenyls;
    2,3,2'-trichlorobiphenyl
7. 2, 4, 5, 3' > 2, 4, 3', 4' > 2, 4, 2', 6'
9. 2.0×10^{-12} mol m^{-3}; 1.5×10^{-10} mol m^{-3};
    4.9×10^{-5} mol m^{-3}

Chapter 15

Problems

15-5 Less; larger

15-7 3.7 years; 45,000

Additional Problems

1. (b) 1.52 ppb
    (c) 0.172 μg

Chapter 16

Problems

16-1 3.9 L

16-3 $C_3H_3O_2$

16-5

16-7 (a) $2\ C_{12}H_6Cl_4 + 25\ O_2 \longrightarrow$
            $24\ CO_2 + 2\ H_2O + 8\ HCl$
      (b) $C_{12}H_6Cl_4 + 23\ H_2 \longrightarrow$
            $12\ CH_4 + 4\ HCl$

Green Chemistry Questions

1. (a) 1
    (b) 1, 6, 7
3. (a) 1
    (b) 1, 5, 6

Additional Problems

1. About 8000 homes
3. 1.14 kilograms
7. UV-B; need sunlight exposure to decompose;
    399 kJ mol^{-1}

Chapter 17

Problems

17-1 CFCs have no H or multiple bonds, so
      they don't react with OH or light, and
      thus don't oxidize. No, since CH_2Cl_2 has
      H atoms, so it will react with OH.

17-3 Quite endothermic

17-5 If a C-bonded H is abstracted by OH, the
      C-centered radical H_2COH will lose the
      hydroxyl H to O_2 abstraction, producing
      formaldehyde. If instead, OH abstracts

the OH hydrogen initially, the resulting
C-centered H_3CO radical loses H to O_2
abstraction, producing formaldehyde.

17-7 Overall $H_2CO + 2\ NO + 2\ O_2 +$
      sunlight $\longrightarrow CO_2 + H_2O + 2\ NO_2$
      No increase in the number of free radicals

17-9 Overall $CH_3(H)CO + 7\ O_2 +$
      $7\ NO \longrightarrow 2\ CO_2 + 7\ NO_2 + 4\ OH$

17-11 Same as for Problem 17-7

17-13 CH_3OOH and O_2

17-15 (a) O_3, ClO, BrO, HOO
       (b) O_3, ClO, BrO, HOO, NO_2
       (c) O_3, ClO, BrO, HOO, NO_2
       (d) ClO, NO_2, and perhaps BrO
       (e) When $2\ O_3$ or two HOO react

17-17 (a) $ClO + NO_2 \rightarrow ClONO_2$
       (b) $2\ ClO \rightarrow ClOOCl$
       (c) Reaction of ClO with UV or O or
            NO; photolysis of ClOOCl

17-19 Cycle destroys ozone if NO_2 reacts with
       O, but not if NO_2 decomposes in
       sunlight.

Additional Problems

1. $CO + OH \longrightarrow HOCO$
    $HOCO + O_2 \longrightarrow HOO + CO_2$
    $HOO + NO \longrightarrow OH + NO_2$
    $NO_2 + UV \longrightarrow NO + O$
    $O + O_2 \longrightarrow O_3$
    ———————————————————————
    Overall: $CO + 2\ O_2 + UV \longrightarrow CO_2 + O_3$

3. $CO + OH \longrightarrow HOCO$
    $HOCO + O_2 \longrightarrow CO_2 + HOO$
    $HOO + O_3 \longrightarrow OH + 2\ O_2$
    ———————————————————————
    Overall: $CO + O_3 \longrightarrow CO_2 + O_2$

5. NO_2:

    HNO_2:

    HNO_3:

Index

Note: Page numbers followed by f, t, and b indicate figures, tables, and boxes, respectively.